The South China Sea

Developments in Paleoenvironmental Research

VOLUME 13

Aims and Scope:
Paleoenvironmental research continues to enjoy tremendous interest and progress in the scientific community. The overall aims and scope of the *Developments in Paleoenvironmental Research* book series is to capture this excitement and document these developments. Volumes related to any aspect of paleoenvironmental research, encompassing any time period, are within the scope of the series. For example, relevant topics include studies focused on terrestrial, peatland, lacustrine, riverine, estuarine, and marine systems, ice cores, cave deposits, palynology, isotopes, geochemistry, sedimentology, paleontology, etc. Methodological and taxonomic volumes relevant to paleoenvironmental research are also encouraged. The series will include edited volumes on a particular subject, geographic region, or time period, conference and workshop proceedings, as well as monographs. Prospective authors and/or editors should consult the series editors for more details. The series editors also welcome any comments or suggestions for future volumes.

EDITOR AND BOARD OF ADVISORS

For futher volumes:
http://www.springer.com/series/5869

The South China Sea

Paleoceanography
and Sedimentology

Edited by

PinxianWang

State Key Laboratory of Marine Geology, Tongji University,
Shanghai, China

Qianyu Li

State Key Laboratory of Marine Geology, Tongji University,
Shanghai, China
School of Earth and Environmental Sciences, The University of Adelaide,
South Australia 5005, Australia

 Springer

Editors

Pinxian Wang
State Key Laboratory of
Marine Geology
Tongji University
1239 Siping Road
200092 Shanghai
China, People's Republic
pxwang@online.sh.cn
pxwang@tongji.edu.cn

Qianyu Li
State Key Laboratory of
Marine Geology
Tongji University
1239 Siping Road
200092 Shanghai
China, People's Republic
qli01@tongji.edu.cn

School of Earth and Environmental Sciences
The University of Adelaide
South Australia 5005
Australia
qianyu.li@adelaide.edu.au

Cover credit note:

Back ground illustration: Various corals growing in the lagoon of Meiji Reef, Nansha Islands, photo by Kefu Yu

Insert: Pre-Oligocene base topography after younger deposits were removed, image by Huang and Wang

ISBN 978-1-4020-9744-7   e-ISBN 978-1-4020-9745-4

DOI 10.1007/978-1-4020-9745-4

Library of Congress Control Number: 2009920103

Printed on acid-free paper

9 8 7 6 5 4 3 2 1

springer.com

Contents

Contributors

Wei Huang State Key Laboratory of Marine Geology, School of Ocean and Earth Sciences, Tongji University, Shanghai 200092, China, huangwei@tongji.edu.cn

Zhimin Jian State Key Laboratory of Marine Geology, School of Ocean and Earth Sciences, Tongji University, Shanghai 200092, China, zjiank@online.sh.cn; jian@tongji.edu.cn

Jianru Li State Key Laboratory of Marine Geology, School of Ocean and Earth Sciences, Tongji University, Shanghai 200092, China, lijianru@tongji.edu.cn

Qianyu Li State Key Laboratory of Marine Geology, School of Ocean and Earth Sciences, Tongji University, Shanghai 200092, China; School of Earth and Environmental Sciences, The University of Adelaide, South Australia 5005, Australia, qli01@tongji.edu.cn; qianyu.li@adelaide.edu.au

Zhifei Liu State Key Laboratory of Marine Geology, School of Ocean and Earth Sciences, Tongji University, Shanghai 200092, China, lzhifei@tongji.edu.cn

Xiangjun Sun Institute of Botany, Chinese Academy of Sciences, Beijing 100093, China; State Key Laboratory of Marine Geology, School of Ocean and Earth Sciences, Tongji University, Shanghai 200092, China, sunxj@tongji.edu.cn

Jun Tian State Key Laboratory of Marine Geology, School of Ocean and Earth Sciences, Tongji University, Shanghai 200092, China, tianjun@tongji.edu.cn

Pinxian Wang State Key Laboratory of Marine Geology, School of Ocean and Earth Sciences, Tongji University, Shanghai 200092, China, pxwang@tongji.edu.cn; pxwang@online.sh.cn

Rujian Wang State Key Laboratory of Marine Geology, School of Ocean and Earth Sciences, Tongji University, Shanghai 200092, China, rjwangk@online.sh.cn

Kefu Yu South China Sea Institute of Oceanology, Chinese Academy of Sciences, Guangzhou 510301, China, kefuyu@scsio.ac.cn

Jianxin Zhao Centre for Microscopy and Microanalysis, University of Queensland, Queensland 4072, Australia, j.zhao@uq.edu.au

Meixun Zhao State Key Laboratory of Marine Geology, Tongji University, Shanghai 200092, China; Key Laboratory of Marine Chemistry Theory and Technology of the Ministry of Education, College of Chemistry and Chemical Engineering, Ocean University of China, Qingdao 266100, China, maxzhao@tongji.edu.cn; maxzhao@ouc.edu.cn

Quanhong Zhao State Key Laboratory of Marine Geology, School of Ocean and Earth Sciences, Tongji University, Shanghai 200092, China, qhzhaok@online.sh.cn

Guangfa Zhong State Key Laboratory of Marine Geology, School of Ocean and Earth Sciences, Tongji University, Shanghai 200092, China, gfz@tongji.edu.cn

Chapter 1
Introduction

Pinxian Wang and Qianyu Li

The South China Sea (SCS) (Fig. 1.1) offers a special attraction for Earth scientists world-wide because of its location and its well-preserved hemipelagic sediments. As the largest one of the marginal seas separating Asia from the Pacific, the largest continent from the largest ocean, the SCS functions as a focal point in land-sea interactions of the Earth system. Climatically, the SCS is located between the Western Pacific Warm Pool, the centre of global heating at the sea level, and the Tibetan Plateau, the centre of heating at an altitude of 5,000 m. Geomorphologically, the SCS lies to the east of the highest peak on earth, Zhumulangma or Everest in the Himalayas (8,848 m elevation) and to the west of the deepest trench in the ocean, Philippine Trench (10,497 m water depth) (Wang P. 2004). Biogeographically, the SCS belongs to the so-called "East Indies Triangle" where modern marine and terrestrial biodiversity reaches a global maximum (Briggs 1999).

Among the major marginal sea basins from the west Pacific, the SCS presents some of the best conditions for accumulating complete paleoclimatic records in its hemipelagic deposits. These records are favorable for high-resolution paleoceanographic studies because of high sedimentation rates and good carbonate preservation. It may not be merely a coincidence that two cores from the southern SCS were among the first several cores in the world ocean used by AMS ^{14}C dating for high-resolution stratigraphy (Andree et al. 1986; Broecker et al. 1988). Unlike the Atlantic, the western Pacific bottom is bathed with more corrosive waters that weaken carbonate preservation even at relative shallow depths of 2,000–3,000 m. With a large area above CCD (~3,500 m), however, the SCS is unique in the region to yield well-preserved sediment sequences suitable for paleoenvironmental reconstructions.

P. Wang (✉)
State Key Laboratory of Marine Geology, Tongji University, Shanghai 200092, China
e-mail: pxwang@online.sh.cn; pxwang@tongji.edu.cn

Q. Li
State Key Laboratory of Marine Geology, Tongji University, Shanghai 200092, China;
School of Earth and Environmental Sciences, The University of Adelaide,
South Australia 5005, Australia
e-mail: qli01@tongji.edu.cn; qianyu.li@adelaide.edu.au

P. Wang, Q. Li (eds.), *The South China Sea*, Developments in Paleoenvironmental
Research 13, DOI 10.1007/978-1-4020-9745-4_1,
© Springer Science+Business Media B.V. 2009

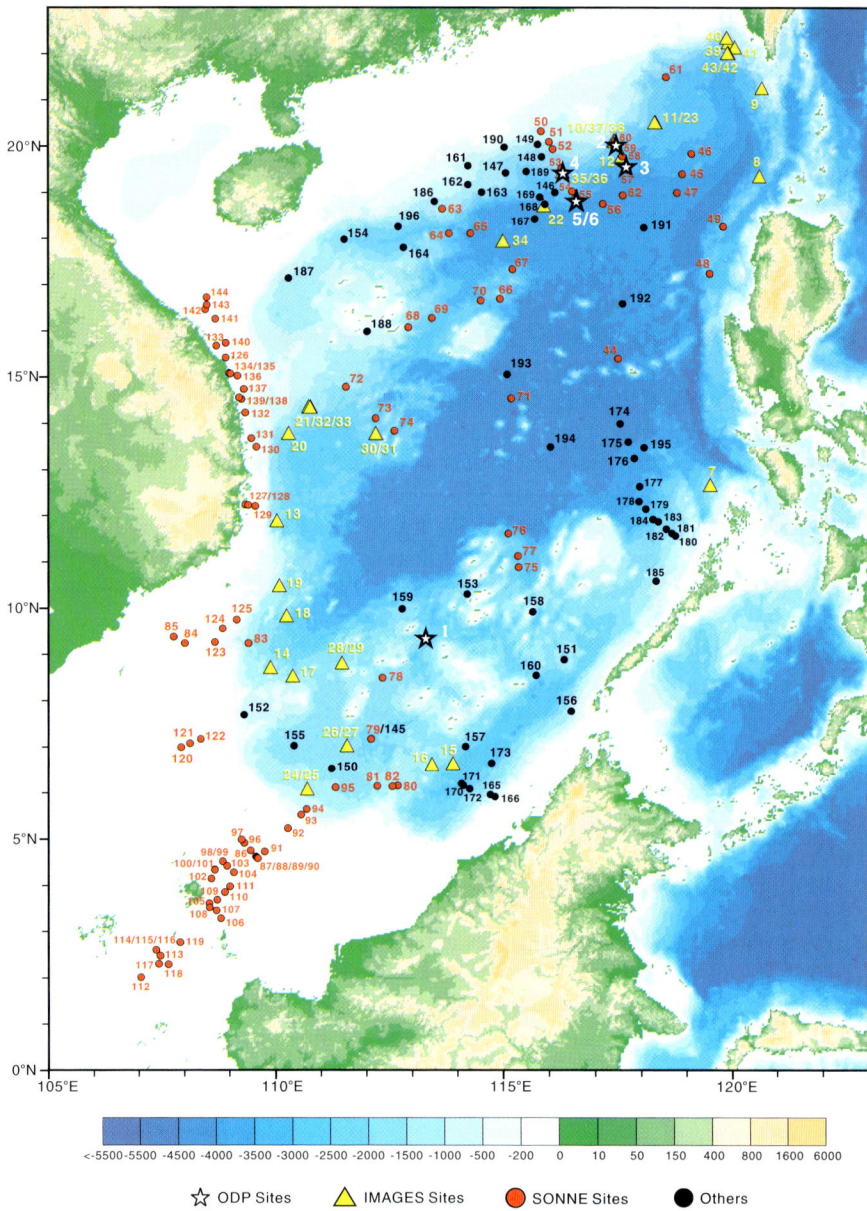

Fig. 1.1 Topography of the South China Sea (SCS) and surrounding areas is superimposed by locations of cores collected for paleoceanographic studies up to 2005. Map was compiled by Enqing Huang based on online topographic data from "http://siovizcenter.ucsd.edu/library.php" using softwares such as Fledermaus, Global Mapper and ArcGIS. Refer to Appendices for site details

Since the 1980s, Chinese geoscientists on both sides of the Taiwan Strait embarked on more active survey and research programs on the SCS mainly driven by petroleum exploration, resulting in a large number of publications mostly not accessible to the global community because of the language barrier and/or limited distribution (see review in Zhou et al. 1994). Several cruises to the SCS were also carried out by scientists from western countries, such as geophysical surveys by French R/V Jean-Charcot in 1984 (Rangin et al. 1988; Pautot and Rangin 1989; Pautot 1992) and by German R/V Sonne in 1987 and 1988 (Kudrass et al. 1992), over 10 years after the earlier expeditions by US marine geophysicists (e.g., Hayes and Ludwig 1967; Ludwig 1970).

In respect of sedimentology and paleoceanography, the SCS did not draw much international attention until the 1990s. Aside from piston coring and shallow profiling by the US R/V Vema and Conrad (e.g., Damuth 1979, 1980), extensive investigations were mainly performed by Chinese expeditions. More than a dozen volumes describing sediment patterns, coral reefs, and late Quaternary history, amongst other topics, were published in China, along with numerous journal papers written in Chinese. Noticeable contributions are serial reports from the Nansha Expeditions of the Chinese Academy of Science to the coral reef area in the southern SCS (Nansha Scientific Expedition 1989–1994), from expeditions of the Ministry of Geology to the Xisha area (He and Zhang 1986), from the South China Sea Institute of Oceanology (1982, 1985, 1987; Su et al. 1989), and from the State Oceanic Administration (1988; Zheng and Chen 1993). At the same time, the isotope-based paleoceanographic studies were first conducted in China (Wang P. et al. 1986; Wang C. et al. 1986), although substantial international publications from SCS studies did not begin until the early 1990s (e.g., Wang and Wang 1990; Wang P. 1990; Thunell et al. 1992; Schönfeld and Kudrass 1993; Miao et al. 1994).

Since the 1990s, the international scientific interest in the SCS has drastically increased. Numerous international expeditions were sent to the region to study topics ranging from climate and sea-level changes in Quaternary glacial cycles, monsoon evolution and variations, to volcanic ash distribution (Table 1.1). Of particular importance is the first paleoceanographic expedition, Sonne-95 cruise under the logo "Monitor Monsoon", which collected 48 piston and gravity cores at 46 sites from the SCS (Sarnthein et al. 1994) and revealed the regional late Quaternary paleoceanographic history for the first time (Sarnthein and Wang 1999). Paleoceanographic expeditions to the SCS were culminated with Ocean Drilling Program (ODP) Leg 184 in the spring of 1999, the first scientific deep-sea drilling expedition off the China coast. A total of 17 holes at 6 sites were cored on the southern and northern continental slopes of the SCS, to explore the late Cenozoic history of the East Asian monsoon (Wang P. et al. 2000). ODP Leg 184 provides the best deep-water stratigraphic sequence in the western Pacific, which archives evidence of the low-latitude oceanic response to orbital forcing (Wang P. et al. 2003c). The growing interest in SCS paleoceanography continues after the ODP cruise, as seen from the joint French-Chinese "Marco Polo" cruise to the SCS in 2005 (Laj et al. 2005) and several other Chinese and international cruises after 2005.

Table 1.1 Major international geological cruises to the SCS since 1990 are listed to show their main themes

Cruise	Time	Theme	References
R/V Sonne, Germany			
SO 72a	Oct. 25–Nov. 18, 1990	Sedimentation	Wong (1993)
SO 95	Apr. 12–Jun. 05, 1994	Monitor Monsoon	Sarnthein et al. (1994)
SO 114	Nov. 20–Dec. 12, 1996	Pinatubo ash	Wiesner et al. (1997)
SO 115	Dec. 13–Jan. 16, 1997	Sunda Shelf	Stattegger et al. (1997)
SO 132	Jun. 17–Jul. 09, 1998	Sedimentation	Wiesner et al. (1998)
SO 140	Apr. 03–May 04, 1999	Sedimentation	Wiesner et al. (1999)
SO 177	Jun. 02–Jul. 02, 2004	Gas Hydrates	Suess (2005)
R/V JOIDES Resolution, USA			
ODP 184	Feb. 11–Apr. 12, 1999	Asian Monsoon	Wang P. et al. (2000)
R/V Marion Dufresne, France			
MD106 (IMAGES III)	Apr. 16–Jun. 30, 1997	"IPHIS"	Chen M. et al. (1998)
MD122 (IMAGES VII)	Apr. 30–Jun. 18, 2001	"WEPAMA"	Bassinot (2002)
MD147 (IMAGES XII)	May 15–Jun. 08, 2005	"Marco Polo 1"	Laj et al. (2005)

Up to 2005, about 200 sediment cores have been retrieved from the SCS primarily for paleoceanographic and sedimentologic studies (Fig. 1.1; Appendices). Except the ODP cores, most of these sediment cores are gravity or piston cores with only a few meters to tens of meters retrieved. However, together with industrial drilling, these sediment cores form a solid base for the picture of the SCS geological and paleoceanographic history to be drawn.

For the global scientific community, the SCS has become one of the hot spots in paleoceanographic and sedimentological studies mainly because of two reasons. First, the SCS now becomes the focus of monsoon studies in East Asia, like the Arabian Sea for South Asia. A series of large scale international field experiments and coring or drilling cruises have been conducted for better understanding the modern and ancient Asian monsoon systems in the two sea regions (see review in Wang P. et al. 2005), and such scientific endeavors continue to develop and expand. Second, the SCS is the major source for off-shore oil and gas in East Asia, a region currently experiencing rapid economic boom also seeking more energy resource to substantiate its growth. The hydrocarbon perspectives of the SCS have enhanced even further with the recent discovery of deep-water gas from the northern slope. Sedimentology of the SCS is a major concern for the petroleum industry not only for exploring source rocks and reservoirs, but also for ground stability of the offshore platforms. All these call for a monograph to summarize available knowledge of the SCS, especially its sedimentology and paleoceanography, two fields with significant progresses over the last two decades.

The present volume is a product of group efforts mainly by the State Key Laboratory of Marine Geology, Tongji University. We tried to pool together all data and research results scattered in various periodicals, in particular those using ODP Leg

Table 1.2 Geographic and basin names frequently referred to in this volume using Chinese "Ping-Yin" are listed to show their synonyms. Figures 2.1 and 2.17 show their localities

Name used in this volume	Synonym	Location
Strait		
Bashi Strait	Luzon Strait	Taiwan—Luzon (Philippines)
Gulf		
Beibuwan or Beibu Gulf	Gulf of Tonkin	Vietnam—China
River		
(Zhujiang)	Pearl River	Mainland China
(Honghe)	Red River	Indochina Peninsula
Reef/Island		
Dongsha	Pratas Islands	Northeastern SCS
Nansha	Spratly Islands and surround	Southern SCS
Yongshu Reef	Fiery Cross or N Investigator	Southern SCS
Meiji Reef	Mischief Reef	Southern SCS
Huanglu Reef	Royal Charlotte Reef	Southern SCS
Sanjiao Reef	Livock Reef	Southern SCS
Xian-e Reef	Alicia Anne Reef	Southern SCS
Xinyi Reef	First Thomas Shoal	Southern SCS
Zhubi Reef	Subi Reef	Southern SCS
Xisha	Paracel Islands	Northern SCS
Yongxin Island	Woody Island	Northern SCS
Shi Island	Rocky Island	Northern SCS
Chenhang Island	Duncan Island	Northern SCS
Dongdao Island	Lincoln Island	Northern SCS
Zhongsha	Macclesfield Bank	Northern SCS
Huangyandao	Scarborough Shoal	Northern SCS
Liyue	Reed Bank	Southern SCS
Nantong Reef	Louisa Reef	Southern SCS
Trench/Trough		
Xisha Trought	Palawan-Borneo Trough	Southern SCS
Basin		
Taixinan (Tainan) Basin	SW Taiwan Basin	Northeastern SCS
Zhujiangkou Basin	Pearl River Mouth Basin	Northern SCS
Qiongdongnan Basin	SE Hainan Basin	Northern SCS
Beibuwan Basin	Beibu Gulf Basin	Northwestern SCS
Yinggehai Basin	Song Hong Basin	Northwestern SCS
Zhongjiannan Basin	Nha Trang Basin	Northwestern SCS
Wan'an Basin	Nam Con Son Basin	W South China Sea
Zengmu Basin	E Natuna–Sarawak basins	Southern SCS

184 data and material and those written originally in Chinese. The volume includes 8 chapters. After this introductory chapter, we will briefly review in Chapter 2 the geomorphologic, oceanographic and tectonic features of the SCS. Chapter 3 provides an updated knowledge of Cenozoic stratigraphy and local sea level changes in the SCS. Distribution and variations of terrigenous, biogenic and volcanic sediments as well as coral reef carbonates are discussed in Chapter 4. Chapter 5 is devoted to evolution and variations of the East Asian monsoon, as recorded in SCS deep-sea sediment archives. Chapter 6 discusses the deep-water story based mainly on benthic faunas and isotopes. Chapter 7 is devoted to changes in productivity

and carbon cycling as a special contribution to biogeochemistry in this volume. Finally, the volume is concluded in Chapter 8 with a synthesis of the history of SCS sedimentology and paleoceanography.

Many geographic names using Chinese Pin-yin are frequently used in the Chinese literature. In this volume, we limit such practice only to a couple of straits, reef islands and basins as shown in Table 1.2.

Acknowledgments This work was supported by the NKBRSF of China (grant 2007CB815900) and grants from the National Natural Science Foundation of China. Recent research results were based on samples and data provided by the Ocean Drilling Program Leg 184 and many other cruises especially SONNE 95 and IMAGES XII (MD-147-MARCO POLO I). The authors express their gratitude to John Chappell, Helvé Chamley, Jilan Su, Kuo-Yen Wei, Wolfgang Kuhnt and Chi-Yu Huang for reviewing the early versions of various chapters and to Michael Sarnthein for proof-reading several chapters. Their critical comments and constructive suggestions significantly improved the final presentation of the contents in this collection. We thank all publishers and copy right holders for granting permissions to use published figures and data as cited in various chapters of this work. Ann Holbourn and Wolfgang Kuhnt provided updated age models for Sites 588, 761 and 1146. Enqing Huang, Jianru Li, Jun Tian, Wei Huang, Wentao Ma, Yang An, Haowen Dang, Xiajing Li, Sui Wan, Qiong Wu and Liming Ye are thanked for technical assistance.

Appendix

Table A1.1 Coring data at ODP Leg 184 Sites are listed together with publications based on studies of relevant site/hole material (compiled by Enqing Huang 2008). Numbers correspond to site localities shown in Fig. 1.1

No.	Site/Hole	Lat. (N)	Long. (E)	W. d. (m)	Drilled (mbsf)	Recov. (m)	References
	ODP Leg 184 (General)				6123	5462	Wang P. et al. (1999, 2000, 2003a,b,d); Tamburini et al. (2003); Prell et al. (2006); Shao et al. (2007); Wang P. et al. (2007); Li Q. et al. (2008)
1	Site1143						Bühring et al. (2000); Li and Jian (2001); Liang et al. (2001); Liu and Cheng (2001); Wang P. et al. (2001, 2003c, 2004); Li J. et al. (2002); Liu et al. (2002, 2008); Tian et al. (2002, 2004a,b,c,d, 2005a,b, 2006); Yang et al. (2002); Huang W. et al. (2003); Liu Z. et al. (2003b, 2006); Nathan and Leckie (2003); Tamburini et al. (2003); Wang and Li (2003); Wehausen et al. (2003); Cheng et al. (2004a); Lee et al. (2004); Hess and Kuhnt (2005); Xu et al. (2005); Li B. et al. (2004, 2005); Jian et al. (2006); Kawagata et al. (2007); Luo and Sun (2007); Wan et al. (2006); Zhang et al. (2007)
	1143A	9°21.720'	113°17.102'	2771.0	400.0	378.28	
	1143B	9°21.717'	113°17.104'	2772.5	258.2	246.37	
	1143C	9°21.713'	113°17.119'	2773.5	500.0	477.54	
2	Site1144						Luo et al. (2001, 2005); Shao et al. (2001); Sun and Luo (2001); Tu et al. (2001); Boulay et al. (2003); Chen et al. (2003); Higginson et al. (2003); Solheid et al. (2003); Sun et al. (2003); Tamburini et al. (2003); Wei G. et al. (2003, 2004); Bühring et al. (2004); Lu et al. (2004); Wang and Lin (2004); Wang R. et al. (2004); Kienast et al. (2005); Zheng F. et al. (2004, 2005)
	1144A	20°3.180'	117°25.133'	2035.7	452.8	468.88	
	1144B	20°3.180'	117°25.143'	2038.5	452.0	445.83	
	1144C	20°3.182'	117°25.152'	2036.9	203.7	198.17	

Table A1.1 (continued)

No.	Site/Hole	Lat. (N)	Long. (E)	W. d. (m)	Drilled (mbsf)	Recov. (m)	References
3	Site 1145						Wehausen and Brumsack (2002); Boulay et al. (2005); Oppo and Sun (2005)
	1145A	19°35.040'	117°37.868'	3175.6	200.0	186.54	
	1145B	19°35.042'	117°37.858'	3174.4	200.0	179.44	
	1145C	19°35.039'	117°37.850'	3176.4	198.1	189.15	
4	Site 1146						Clemens and Prell (2003); Huang B. et al. (2003, 2007); Kissel et al. (2003); Liu Z. et al. (2003a,b); Nathan and Leckie (2003); Trentesaux et al. (2003); Wang P. et al. (2003); Zhu et al. (2003); Arnold (2004); Holbourn et al. (2004, 2005, 2007); Li B. et al. (2004); Su et al. (2004); Wang and Lin (2004); Zheng H. et al. (2004); Jian et al. (2006); Kawagata et al. (2007); Wan et al. (2007b)
	1146A	19°27.402'	116°16.363'	2091.1	607.0		
	1146B	19°27.401'	116°16.376'	2091.7	245.1	241.71	
	1146C	19°27.403'	116°16.385'	2091.7	603.5	606.66	
5	Site 1147						Cheng et al. (2004b); Mercer and Zhao (2004)
	1147A	18°50.108'	116°33.271'	3245.9	81.4	81.7	
	1147B	18°50.108'	116°33.280'	3245.4	85.5	85.5	
	1147C	18°50.109'	116°33.280'	3245.3	78.6	76.6	
6	Site 1148						Jian et al. (2001, 2003); Wang R. et al. (2001); Zhao et al. (2001a,b); Kuhnt et al. (2002); Clift et al. (2002); Huang W. et al. (2003); Jia et al. (2003); Li X. et al. (2003); Wu et al. (2003); Cheng et al. (2004b); Leventhal (2004); Li Q. et al. (2004, 2005, 2006, 2008); Mao et al. (2004, 2007); Mercer and Zhao (2004); Peng et al. (2004); Su et al. (2004); Zhao (2005); Clift (2006); Wei et al. (2006); Tian et al. (2008)
	1148A	18°50.167'	116°33.932'	3297.1	704.0	632.1	
	1148B	18°50.170'	116°33.946'	3291.8	853.2	364.4	

Table A1.2 Details of short piston cores retrieved from the SCS up to 2005 are listed together with publications based on studies of core material (compiled by Enqing Huang 2008). Numbers correspond to site localities shown in Fig. 1.1

No.	Site/Core	Latitude (N)	Longitude (E)	Water depth (m)	Recovery (m)	References
7	MD97-2142	12°41.133′	119°27.9′	1557	35.91	Chen M. et al. (2003); Wei et al. (2003)
8	MD97-2144	19°21.813′	120°32.05′	2680	35.24	Chen M. et al. (1998)
9	MD97-2145	21°15.94′	120°35.27′	1408	24.02	Chen M. et al. (1998)
10	MD97-2146	20°7.019′	117°23.08′	1720	38.69	Lin et al. (2006)
11	MD97-2147	20°31.74′	118°14.7′		49.4	Chen M. et al. (1998)
12	MD97-2148	19°47.804′	117°32.56′	2830	48.72	Chen Y.Y. et al. (1999)
13	MD97-2149	11°54.004′	110°0.56′	1870	27.18	Chen M. et al. (1998)
14	MD97-2151	8°43.73′	109°52.17′	1589	26.72	Lee et al. (1999); Zhao et al. (2006)
15	MD01-2389	6°39.19′	113°52.23′	2534	31.86	Chen M. et al. (1998)
16	MD01-2390	6°38.12′	113°24.56′	1545	43.72	Steinke et al. (2006)
17	MD01-2391	8°32.57′	110°20.94′	1351	42.38	Bassinot (2002)
18	MD01-2392	9°51.13′	110°12.64′	1966	43.2	Zheng et al. (2005)
19	MD01-2393	10°30.15′	110°3.68′	1230	41.793	Liu Z. et al. (2004)
20	MD01-2394	13°47.54′	110°15.56′	2097	39.04	Yu et al. (2006)
21	MD01-2395	14°21.68′	110°44.42′	1418	44.95	Bassinot (2002)
22	MD01-2396	18°43.48′	115°50.73′	3365	39.11	Bassinot (2002)
23	MD01-2397	20°31.64′	118°15.67′	2420	43.4	Bassinot (2002)
24	MD05-2892	6°6.12′	110°40.71′	1183	29.03	Laj et al. (2005)
25	MD05-2893	6°6.07′	110°40.72′	1183	4.52	Laj et al. (2005)
26	MD05-2894	7°2.25′	111°33.11′	1982	10.85	Laj et al. (2005)
27	MD05-2895	7°2.25′	111°33.11′	1982	43.14	Laj et al. (2005)
28	MD05-2896	8°49.5′	111°26.47′	1657	11.03	Laj et al. (2005)
29	MD05-2897	8°49.53′	111°26.51′	1658	30.98	Laj et al. (2005)
30	MD05-2898	13°47.39′	112°11.03′	2395	9.73	Laj et al. (2005)
31	MD05-2899	13°47.66′	112°10.89′	2393	36.68	Laj et al. (2005)
32	MD05-2900	14°22.23′	110°41.74′	1455	9.76	Laj et al. (2005)
33	MD05-2901	14°22.503′	110°44.6′	1454	36.49	Laj et al. (2005)
34	MD05-2902	17°57.7′	114°57.33′	3697	9.42	Laj et al. (2005)
35	MD05-2903	19°27.31′	116°15.06′	2047	11.18	Laj et al. (2005)
36	MD05-2904	19°27.32′	116°15.15′	2066	44.98	Laj et al. (2005)
37	MD05-2905	20°8.17′	117°21.61′	1198	11.98	Laj et al. (2005)
38	MD05-2906	20°8.16′	117°21.59′	1636	36.52	Laj et al. (2005)
39	MD05-2911	22°15.61′	119°51.08′	1085	25	Laj et al. (2005)
40	MD05-2912	22°21.5′	119°48.5′	1090	30.44	Laj et al. (2005)
41	MD05-2913	22°9.23′	119°59.26′	1091	12.68	Laj et al. (2005)
42	MD05-2914	22°1.55′	119°50.97′	1642	35.14	Laj et al. (2005)
43	MD05-2915	22°1.94′	119°50′	1641	3.9	Laj et al. (2005)
44	17922-2	15°25′	117°27.5′	4224	6.63	Sarnthein et al. (1994)
45	17924-2	19°24.7′	118°50.9′	3440	11.50	Sarnthein et al. (1994)
46	17925-3	19°51.2′	119°2.8′	2980	12.42	Sarnthein et al. (1994)
47	17926-3	19°0′	118°44′	3760	10.06	Sarnthein et al. (1994)
48	17927-2	17°15.1′	119°27.2′	2804	5.58	Wang L. et al. (1999a)
49	17928-3	18°16.3′	119°44.7′	2484	3.30	Huang B. et al. (2002)
50	17930-2	20°20′	115°46.9′	629	5.34	Sarnthein et al. (1994)
51	17931-2	20°6′	115°57.8′	1003	3.00	Sarnthein et al. (1994)
	17931-3	20°6′	115°57.8′	1001	4.31	Sarnthein et al. (1994)

Table A1.2 (continued)

No.	Site/Core	Latitude (N)	Longitude (E)	Water depth (m)	Recovery (m)	References
52	17932-2	19°57.1′	116°2.2′	1360	7.56	Sarnthein et al. (1994)
53	17933-3	19°32′	116°13.6′	1970	12.48	Sarnthein et al. (1994)
54	17934-2	19°1.9′	116°27.7′	2665	11.87	Sarnthein et al. (1994)
55	17935-3	18°52.7′	116°31.6′	3148	12.27	Sarnthein et al. (1994)
56	17936-2	18°46′	117°7.2′	3809	13.33	Sarnthein et al. (1994)
57	17937-2	19°30′	117°39.9′	3428	12.92	Wang L. et al. (1999)a
58	17938-2	19°47.2′	117°32.3′	2840	11.78	Wang L. et al. (1999a)
59	17939-2	19°58.2′	117°27.3′	2474	12.74	Wang L. et al. (1999a)
60	17940-2	20°7′	117°23′	1727	13.30	Wang L. et al. (1999a)
61	17941-2	21°31′	118°29′	2200	9.90	Sarnthein et al. (1994)
62	17943-2	18°57′	117°33.2′	919	11.74	Sarnthein et al. (1994)
63	17944-2	18°39.5′	113°38.2′	1217	8.92	Sarnthein et al. (1994)
64	17945-2	18°7.6′	113°46.6′	2403	10.21	Sarnthein et al. (1994)
65	17946-2	18°7.5′	114°15′	3464	11.34	Sarnthein et al. (1994)
66	17948-2	16°42.3′	114°53.8′	2855	13.09	Sarnthein et al. (1994)
67	17949-2	17°20.9′	115°10′	2197	13.34	Sarnthein et al. (1994)
68	17950-2	16°5.6′	112°53.8′	1865	9.91	Lin (2003)
69	17951-2	16°17.3′	113°24.6′	2341	11.97	Sarnthein et al. (1994)
70	17952-3	16°40′	114°28.4′	2883	12.04	Sarnthein et al. (1994)
71	17953-4	14°33′	115°8.6′	4306	12.49	Sarnthein et al. (1994)
72	17954-2	14°47.8′	111°31.5′	1520	11.52	Wang L. et al. (1999a); Huang B. et al. (2002)
	17954-3	14°47.7′	111°31.5′	1515	11.49	Wang L. et al. (1999a)
73	17955-2	14°7.3′	112°10.6′	2393	11.66	Wang L. et al. (1999a)
74	17956-2	13°50.9′	112°35.3′	3388	13.56	Wang L. et al. (1999a)
75	17957-2	10°53.9′	115°18.3′	2195	13.84	Jian et al. (2000)
76	17958-2	11°37.3′	115°4.9′	2581	10.73	Sarnthein et al. (1994)
77	17959-2	11°8.3′	115°17.2′	1959	13.93	Sarnthein et al. (1994)
78	17961-2	8°30.4′	112°19.9′	1968	10.30	Wang L. et al. (1999a); Bühring et al. (2000)
79	17962-2	7°10.9′	112°4.9′	1968	8.29	Sarnthein et al. (1994)
	17962-4	7°10.9′	112°4.9′	1969	8.81	Bühring et al. (2000)
80	17963-3	6°10′	112°40′	1232	8.57	Sarnthein et al. (1994)
81	17964-3	6°9.5′	112°12.8′	1556	9.12	Wang L. et al. (1999a)
82	17965-2	6°9.4′	112°33.1′	890	6.83	Sarnthein et al. (1994)
83	18252-3	9°15.007′	109°23.446′	1273	11.85	Kienast et al. (2001a)
84	18262-3	9°14.999′	107°59.307′	56	9.38	Hanebuth et al. (2000)
85	18265-2	9°23.249′	107°45.022′	47	2.40	Hanebuth et al. (2000)
86	18269-2	4°46.013′	109°26.321′	113	8.815	Hanebuth et al. (2003)
87	18271-2	4°38.33′	109°32.969′	122	5.62	Hanebuth et al. (2004)
88	18273-2	4°37.289′	109°33.931′	126	3.48	Hanebuth et al. (2003)
89	18274-2	4°36.318′	109°34.833′	118	5.61	Hanebuth et al. (2004)
90	18275-2	4°35.727′	109°35.536′	112	5.56	Hanebuth et al. (2004)
91	18276-2	4°44.897′	109°44.837′	116	7.21	Hanebuth et al. (2000, 2003)
92	18282-2	5°14.687′	110°14.605′	151	6.34	Hanebuth et al. (2003)
93	18284-3	5°32.51′	110°32.413′	151	6.34	Steinke et al. (2003)

Table A1.2 (continued)

No.	Site/Core	Latitude (N)	Longitude (E)	Water depth (m)	Recovery (m)	References
94	18287-3	5°39.781′	110°39.689′	598	5.66	Kienast et al. (2001b)
95	18294-4	6°7.809′	111°18.183′	849	6.94	Steinke et al. (2003)
96	18295-2	4°55.587′	109°17.865′	119	8.23	Hanebuth et al. (2004)
97	18296-2	4°59.754′	109°14.446′	118	2.44	Hanebuth et al. (2004)
98	18298-2	4°31.987′	108°49.508′	102	5.87	Hanebuth et al. (2004)
99	18299-1	4°32.004′	108°49.537′	102	5.8	Hanebuth et al. (2000)
100	18300-2	4°21.778′	108°39.215′	91	8.85	Hanebuth et al. (2000, 2003)
101	18301-2	4°21.308′	108°38.811′	93	5.82	Hanebuth et al. (2000)
102	18302-2	4°9.585′	108°34.535′	83	5.98	Hanebuth et al. (2000, 2003)
103	18303-2	4°26.425′	108°55.491′	107	7.36	Hanebuth et al. (2003)
104	18305-2	4°17.318′	109°4.599′	109	5.14	Hanebuth et al. (2000, 2003)
105	18307-2	3°37.626′	108°31.648′	100	9.43	Hanebuth et al. (2000)
106	18308-2	3°17.83′	108°47.143′	80	1.05	Hanebuth et al. (2000)
107	18309-2	3°27.959′	108°41.174′	83	5.97	Hanebuth et al. (2000)
108	18310-2	3°32.131′	108°32.131′	100	5.68	Hanebuth et al. (2000)
109	18312-2	3°42.351′	108°42.38′	101	6.67	Hanebuth et al. (2004)
110	18313-2	3°52.194′	108°52.226′	98	6.2	Hanebuth et al. (2003)
111	18314-2	3°59.469′	108°59.473′	100	3.7	Hanebuth et al. (2003)
112	18315-3	2°1.669′	107°2.041′	69	5.83	Hanebuth et al. (2003)
113	18316-2	2°29.263′	107°27.522′	71	5.97	Hanebuth et al. (2003)
114	18317-3	2°36.596′	107°22.515′	95	1.97	Hanebuth et al. (2003)
115	18318-3	2°36.609′	107°22.508′	87	4.06	Hanebuth et al. (2003)
116	18320-2	2°36.726′	107°22.491′	76	4.92	Hanebuth et al. (2003)
117	18321-2	2°18.453′	107°25.326′	109	5.69	Hanebuth et al. (2003)
118	18322-2	2°18.405′	107°37.881′	70	4.93	Hanebuth et al. (2003)
119	18323-2	2°47.03′	107°53.2′	92	5.4	Hanebuth et al. (2003)
120	18375-2	7°0.2′	107°54.87′	87	3.69	Schimanski and Stattegger (2005)
121	18376-2	7°5.27′	108°6.42′	89	4.82	Schimanski and Stattegger (2005)
122	18377-2	7°10.71′	108°20.28′	98	3.57	Schimanski and Stattegger (2005)
123	18389-3	9°16.45′	108°39.09′	109	2.76	Schimanski and Stattegger (2005)
124	18391-2	9°33.91′	108°49.6′	115	4.78	Schimanski and Stattegger (2005)
125	18393-3	9°45.61′	109°7.95′	155	5.4	Schimanski and Stattegger (2005)
126	18396-3	15°25.51′	108°53.27′	61	4.99	Schimanski and Stattegger (2005)
127	18397-2	12°14.71′	109°19.9′	45	5.32	Schimanski and Stattegger (2005)
128	18398-3	12°14.23′	109°22.8′	59	7.19	Schimanski and Stattegger (2005)
129	18401-3	12°12.9′	109°32.09′	134	7.07	Schimanski and Stattegger (2005)

Table A1.2 (continued)

No.	Site/Core	Latitude (N)	Longitude (E)	Water depth (m)	Recovery (m)	References
130	18404-3	13°30.13′	109°33.67′	169	3.85	Schimanski and Stattegger (2005)
131	18405-3	13°41.1′	109°27.02′	130	5.47	Schimanski and Stattegger (2005)
132	18408-3	14°14.34′	109°19.05′	108	7.54	Schimanski and Stattegger (2005)
133	18409-3	15°41.21′	108°40.79′	40	5.77	Schimanski and Stattegger (2005)
134	18414-3	15°5.83′	108°57.78′	21	4.48	Schimanski and Stattegger (2005)
135	18415-2	15°4.98′	109°0.03′	38	5.59	Schimanski and Stattegger (2005)
136	18416-2	15°2.23′	109°8.97′	66	5.06	Schimanski and Stattegger (2005)
137	18417-3	14°44.81′	109°17.61′	97	7.98	Schimanski and Stattegger (2005)
138	18419-3	14°32.23′	109°14.09′	82	6.41	Schimanski and Stattegger (2005)
139	18420-2	14°34.13′	109°11.36′	62	5.57	Schimanski and Stattegger (2005)
140	18422-3	15°44.92′	108°53.46′	84	5.68	Schimanski and Stattegger (2005)
141	18423-2	16°16.62′	108°39.59′	97	5.48	Schimanski and Stattegger (2005)
142	18424-2	16°28.61′	108°26.15′	90	5.75	Schimanski and Stattegger (2005)
143	18425-2	16°34.53′	108°28.31′	95	5.56	Schimanski and Stattegger (2005)
144	18426-2	16°44.42′	108°27.8′	92	5.13	Schimanski and Stattegger (2005)
145	V35-05	7°11′	112°5′	1953	16.25	Broecker et al. (1988)
146	V36-06-3	19°0.5′	116°5.6′	2809	12.15	Wang and Chen (1990)
147	V36-06-5	19°26′	115°1.1′	2332	10.70	Wang and Chen (1990)
148	V36-07	19°47′	115°48′	1585	11.60	Jian and Wang (1997)
149	V36-08	20°3′	115°43′	1304	12.68	Samodai et al. (1986)
150	RC12-350	6°32.5′	111°13′	1950	11.29	Jian (1992)
151	RC14-79	8°54′	116°18′	706	11.74	Jian and Wang (1997)
152	SCS-12	7°41.99′	109°17.95′	543	1.20	Jian and Wang (1997); Wang L. et al. (1997)
153	SCS-15B	10°19′	114°11′	1500	7.35	Wang and Chen (1990)
154	SCS90-36	17°59.7′	111°29.64′	2050	1.05	Huang et al. (1997)b
155	NS86-43	7°2′	110°23′	1763	3.04	Li et al. (1992)
156	NS87-8	7°47′	116°27′	835	5.14	MOET (1993)
157	NS87-11	7°1′	114°9′	2452	4.73	MOET (1993)
158	NS88-11	9°56′	115°37′	880	4.33	Li et al. (1992); MOET (1993)
159	NS93-5	9°59.94′	112°45.19′	1792	5.34	Chen M. et al. (2005)
160	SO27-91KL	8°33.45′	115°41.48′	2060	14.70	Schönfeld and Kudrass (1993)

Table A1.2 (continued)

No.	Site/Core	Latitude (N)	Longitude (E)	Water depth (m)	Recovery (m)	References
161	SO49-3KL	19°35.23′	114°11.64′	713	3.65	Schönfeld and Kudrass (1993)
162	SO49-8KL	19°11′	114°12′	1040	9.55	Wang and Chen (1990); Schöfeld and kudrass (1993)
163	SO49-12KL	19°1.2′	114°29.91′	1532	9.20	Schönfeld and Kudrass (1993)
164	SO49-37KL	17°49.04′	112°47.09′	2004	13.10	Schönfeld and Kudrass (1993); Mao and Rex (1993)
165	SO49-136KL	5°58.43′	114°41.85′	650	8.05	Schönfeld and Kudrass (1993)
166	SO49-137SL	5°55.77′	114°47.78′	220	1.15	Schönfeld and Kudrass (1993)
167	SO50-29KL	18°26.08′	115°39.22′	3766	9.93	Schönfeld and Kudrass (1993)
168	SO50-31KL	18°45.4′	115°52.4′	3360	3.02	Huang C. et al. (1997a); Chen M. et al. (1998)
169	SO50-37KL	18°54.6′	115°45.78′	2695	8.51	Schönfeld and Kudrass (1993); Winn et al. (1992)
170	SO58-100KL	6°10.73′	114°6.3′	2238	12.08	Schönfeld and Kudrass (1993)
171	SO58-109KL	6°12.73′	114°3.89′	2792	12.23	Schönfeld and Kudrass (1993)
172	SO58-114KL	6°5.88′	114°14.37′	1929	13.11	Schönfeld and Kudrass (1993)
173	SO58-133KL	6°39.41′	114°43.24′	2136	13.00	Schönfeld and Kudrass (1993)
174	GGC1	14°0.3′	117°29.9′	4203	2.75	Miao et al. (1994); Thunell et al. (1992)
175	GGC2	13°36.5′	117°40.5′	4010	1.27	Miao et al. (1994); Thunell et al. (1992)
176	GGC3	13°15.6′	117°48.4′	3725	2.93	Miao et al. (1994); Thunell et al. (1992)
177	GGC4	12°39′	117°55.6′	3530	2.82	Miao et al. (1994); Thunell et al. (1992)
178	GGC5	12°19′	117°55′	3185	2.58	Miao et al. (1994); Thunell et al. (1992)
179	GGC6	12°9.1′	118°3.9′	2975	2.64	Miao et al. (1994); Thunell et al. (1992)
180	GGC8	11°34.6′	118°42.6′	1305	2.47	Miao et al. (1994); Thunell et al. (1992)
181	GGC9	11°37.7′	118°37.9′	1465	2.13	Miao et al. (1994); Thunell et al. (1992)
182	GGC10	11°43.4′	118°30.5′	1605	2.27	Miao et al. (1994); Thunell et al. (1992)
183	GGC11	11°53.2′	118°19.9′	2165	2.03	Miao et al. (1994); Thunell et al. (1992)

Table A1.2 (continued)

No.	Site/Core	Latitude (N)	Longitude (E)	Water depth (m)	Recovery (m)	References
184	GGC12	11°55.9′	118°12.8′	2495	2.51	Miao et al. (1994); Thunell et al. (1992)
185	GGC13	10°36′	118°17.4′	990	2.49	Miao et al. (1994); Thunell et al. (1992)
186	HY4-901	18°49′	113°28′	1120		Li X. et al. (1996)
187	SA12-19	17°9.15′	110°15.42′	1300	6.67	Jiang and Liu (2003)
188	G38	16°0.3′	111°59.7′	1115	4.20	Li C. (1993); Mao and Rex (1993)
189	G76	19°28.05′	115°28.15′	2400		Mao and Rex (1993)
190	G77	19°59.5′	114°59.3′	600	3.15	Li C. (1993)
191	8328	18°15′	118°1′	3860	2.78	Li C. (1993); Mao and Rex (1993)
192	8338	16°36.1′	117°33.1′	3980		Mao and Rex (1993)
193	8345	15°4.083′	115°3.12′	3640		Mao and Rex (1993)
194	8355	13°30.033′	116°0.017′	4095		Mao and Rex (1993)
195	8357	13°29.2′	118°1.3′	3949	3.83	Li C. (1993)
196	KL41	18°16.083′	112°40.117′	2120		Mao and Rex (1993)

References

Andree M., Oeschger H., Broecker W.S., Beavan N., Mix A., Bonani G., Hofmann H.J., Morenzoni E., Nessi M., Suter M. and Wolfli W. 1986. AMS radiocarbon dates on foraminifera from deep sea sediments. Radiocarbon 28(2A): 424–428.

Arnold E. 2004. Data report: Late Miocene–Pleistocene mineralogy, Site 1146. In: Prell W.L., Wang P., Blum P., Rea D.K. and Clemens S.C. (eds.), Proc. ODP, Sci. Results, 184 [Online].

Bassinot F. 2002. IPF les rapports des campagnes à la mer. WEPAMA Cruise MD 122/IMAGES VII on board RV "Marion Dufresne", Leg 1, 301pp.

Boulay S., Colin C., Trentesaux A., Pluquet F., Bertaux J., Blamart D., Buehring C. and Wang P. 2003. Mineralogy and sedimentology of Pleistocene sediments in the South China Sea (ODP Site 1144). In: Prell W.L., Wang P., Blum P., Rea D.K. and Clemens S.C. (eds.), Proc. ODP Sci. Result 184: 1–21 [Online].

Boulay S., Colin C., Trentesaux A., Frank N. and Liu Z. 2005. Sediment sources and East Asian monsoon intensity over the last 450 ky: Mineralogical and geochemical investigations on South China Sea sediments. Palaeogeogr. Palaeoclimatol. Palaeoecol. 228: 260–277.

Briggs J.C. 1999. Coincident biogeographic patterns: Indo-West Pacific Ocean. Evolution 53: 326–335.

Broecker W.S., Andree M., Klas M., Bonani G., Wolfli W. and Oeschger H. 1988. New evidence from the South China Sea for an abrupt termination of the last glacial period. Nature 333: 156–158.

Bühring C., Sarnthein M. and Leg 184 Shipboard Scientific Party 2000. Toba ash layers in the South China Sea: Evidence of contrasting wind directions during eruption ca. 74 ka. Geology 28: 275–278.

Bühring C., Sarnthein M. and Erlenkeuser H. 2004. Toward a high resolution stable isotope stratigraphy of the last 1.1 m.y.: Site 1144, South China Sea. In: Prell W.L., Wang P., Blum P., Rea D.K. and Clemens S.C. (eds.), Proc. ODP, Sci. Results 184: 1–29 [Online].

Chen M.T., Beaufort L. and the Shipboard Scientific Party of the IMAGES III/MD 106-IPHIOS Cruise (Leg II) 1998. Exploring Quaternary variability of the East Asian monsoon, Kuroshio

Current, and Western Pacific Warm Pool systems: High-resolution investigations of paleoceanography from the IMAGES III (MD 106)–IPHIS Cruise. Terr. Atmos. Ocean. Sci. (TAO)
Taipei 9(1): 129–142.

Chen M., Wang R., Yang L., Han J. and Lu J. 2003. Development of east Asian summer monsoon
environments in the late Miocene: radiolarian evidence from Site 1143 of ODP Leg 184. Mar.
Geol. 201: 169–177.

Chen M., Li Q., Zheng F., Tan X., Xiang R. and Jian Z. 2005. Variations of the Last Glacial Warm
Pool: Sea surface temperature contrasts between the open western Pacific and South China Sea.
Paleoceanography 20, PA2005, doi:10.1029/2004PA001057.

Chen Y.Y., Chen M.T. and Fang T.S. 1999. Biogenic sedimentation patterns in the northern South
China Sea: an ultrahigh-resolution record MD972148 of the past 150,000 years from the
IMAGES III-IPHIS Cruise. Terr. Atmos. Ocean. Sci. (TAO) Taipei 10: 215–224.

Cheng X., Tian J. and Wang P. 2004a. Data report: Stable isotopes from Site 1143. In: Prell W.L.,
Wang P., Blum P., Rea D.K. and Clemens S.C. (eds.), Proc. ODP, Sci. Results, 184 [Online].

Cheng X., Zhao Q., Wang J., Jian Z., Xia P., Huang B., Fang D., Xu J., Zhou Z. and Wang P. 2004b.
Data report: Stable Isotopes from Sites 1147 and 1148. In: Prell W.L., Wang P., Blum P., Rea
D.K. and Clemens S.C. (eds.), Proc. ODP, Sci. Results 184: 1–12 [Online].

Clemens S.C. and Prell W. 2003a. A 350,000 year summer-monsoon multi-proxy stack from the
Owen Ridge, Northern Arabian Sea. Mar. Geol. 201: 35–51.

Clemens S.C. and Prell W.L. 2003b. Data report: oxygen and carbon isotopes from Site 1146,
northern South China Sea. In: Prell W.L., Wang P., Blum P., Rea D.K. and Clemens S.C. (eds.),
Proc. ODP, Sci. Results 184: 1–8 [Online].

Clift P. 2006. Controls on the erosion of Cenozoic Asia and the flux of clastic sediment to the
ocean. Earth Planet. Sci. Lett. 241: 571–580.

Clift P., Lee J. I., Clark M.K. and Blusztajn J. 2002. Erosional response of South China to arc rifting
and monsoonal strengthening; a record from the South China Sea. Mar. Geol. 184: 207–226.

Damuth J.E. 1979. Migrating sediment waves created by turbidity currents in the northern South
China Basin. Geology 7: 520–523.

Damuth J.E. 1980. Quaternary sedimentation process in the South China Sea basin as revealed
by echo-character mapping and piston-core studies. In: Hayes D.E. (ed.), The Tectonics and
Geophysical Evolution of Southeast Asian Seas and Islands. AGU Geophys. Monogr. 23:
105–125.

Hanebuth T.J.J. and Stattegger K. 2004. Depositional sequences on a late Pleistocene–Holocene
tropical siliciclastic shelf (Sunda Shelf, Southeast Asia). J Asian Earth Sci. 23: 113–126.

Hanebuth T., Stattegger K. and Groote P.M. 2000. Rapid flooding of the Sunda Shelf: a late-glacial
sea-level record. Science 288: 1033–1035.

Hanebuth T.J.J., Stattegger K., Schimanski A., Lüdmann T. and Wong H.J. 2003. Late Pleistocene
forced-regressive deposits on the Sunda Shelf (Southeast Asia). Mar. Geol. 199: 139–157.

Hayes D.E. and Ludwig W.J. 1967. The Manila Trench and West Luzon Trough, 2, Gravity and
magnetic measurements. Deep-Sea Res. 14: 545–560.

He Q. and Zhang M. 1986. Geology of Xisha Reef Facies, China. China Sci. Press, Bejing (in
Chinese).

Hess S. and Kuhnt W. 2005. Neogene and Quaternary paleoceanographic changes in the southern South China Sea (Site 1143): the benthic foraminiferal record. Mar. Micropaleontol. 54:
63–87.

Higginson M., Maxwell J.R. and Altabet M.A. 2003. Nitrogen isotope and chlorin paleoproductivity records from the Northern South China Sea: remote vs. local forcing of millennial- and
orbital-scale variability. Mar. Geol. 201: 223–250.

Holbourn A., Kuhnt W. and Schulz M. 2004. Orbitally paced climate variability during the middle
Miocene: high resolution benthic stable-isotope records from the tropical western Pacific. In:
Clift P.D., Wang P., Hayes D. and Kuhnt W. (eds.), Continent-Ocean Interactions in the East
Asian Marginal Seas. AGU Geophys. Monogr. 149, pp. 321–337.

Holbourn A., Kuhnt W., Schulz M. and Erlenkeuser H. 2005. Impacts of orbital forcing and
atmospheric carbon dioxide on Miocene ice-sheet expansion. Nature 438: 483–487.

Holbourn A., Kuhnt W., Schulz M., Flores J.A. and Andersen N. 2007. Orbitally-paced climate evolution during the middle Miocene "Monterey" carbon-isotope excursion. Earth Planet. Sci. Lett. 261: 534–550.

Huang B., Cheng X., Jian Z. and Wang P. 2003. Response of upper ocean structure to the initiation of the North Hemisphere glaciation in the South China Sea. Palaeogeogr. Palaeoclimatol. Palaeoecol. 196: 305–318.

Huang B., Jian Z., Cheng X. and Wang P. 2002. Foraminiferal responses to upwelling variations in the South China Sea over the last 220 000 years. Mar. Micropaleontol. 47: 1–15.

Huang B., Jian Z. and Wang P. 2007. Benthic foraminiferal fauna turnover at 2.1 Ma in the northern South China Sea. Chinese Sci. Bull. 52(6): 839–843.

Huang C.Y., Wu S.F., Zhao M., Chen M.T., Wang C.H., Tu X. and Yuan P.B. 1997b. Surface ocean and monsoon climate variability in the South China Sea since the last glaciation. Mar. Micropaleotol. 32: 71–94.

Huang W., Liu Z., Cheng X. and Wang P. 2003. Exploring physical indicators for carbonate contents in deep sea sediments. Earth Sci.-J. China Univ. Geosci. 14(4): 300–305 (in Chinese).

Jia G., Peng P., Zhao Q. and Jian Z. 2003. Changes in terrestrial ecosystem since 30 Ma in East Asia: Stable isotope evidence from black carbon in the South China Sea. Geology 31: 1093–1096.

Jian Z. 1992. Sea surface temperature in the southern continental slope of the South China Sea since last glacial and their comparison with those in the northern slope. In Ye Z. and Wang P. (eds.), Contributions to late Quaternary paleoceanography of the South China Sea. Qingdao Ocean Univ. Press, Qingdao, pp. 78–87 (in Chinese).

Jian Z., Cheng X., Zhao Q., Wang J. and Wang P. 2001. Oxygen isotope stratigraphy and events in the northern South China Sea during the last 6 million years. Sci. China (D) 44(10): 952–960.

Jian Z. and Wang L. 1997. Late Quaternary benthic foraminifera and deep-water paleoceanography in the South China Sea. Mar. Micropaleotol. 32: 127–154.

Jian Z., Wang P., Chen M.P., Li B., Zhao Q., Bühring C., Laj C., Lin H.L., Pflaumann U., Bian Y., Wang R. and Cheng X. 2000. Foraminiferal response to major Pleistocene paleographic changes in the southern South China Sea. Paleoceanography 15: 229–243.

Jian Z., Yu Y., Li B., Wang J., Zhang X. and Zhou Z. 2006. Phased evolution of the south-north hydrographic gradient in the South China Sea since the middle Miocene. Palaeogeogr. Palaeoclimatol. Palaeoecol. 230: 251–263.

Jian Z., Zhao Q., Cheng X., Wang J., Wang P. and Su X. 2003. Pliocene-Pleistocene stable isotope and paleoceanographic changes in the northern South China Sea. Palaeogeogr. Palaeoclimatol. Palaeoecol. 193: 425–442.

Jiang M. and Li X. 2003. Planktonic foraminifera and sea surface temperature (SST) of the Xisha Trough, South China Sea since last glaciation. Sci. China (D) 46: 1–9.

Kawagata S., Hayward B.W. and Kuhnt W. 2007. Extinction of deep-sea foraminifera as a result of Pliocene-Pleistocene deep-sea circulation changes in the South China Sea (ODP Sites 1143 and 1146). Quat. Sci. Rev. 26: 808–827.

Kienast M., Calvert S.E., Pelejero C. and Grimalt J.O. 2001a. A critical review of marine sedimentary $^{13}C_{org}$-pCO_2 estimates: New palaeorecords form the South China Sea and a revisit of other low-latitude $^{13}C_{org}$-pCO_2 records. Global Biogeochem. Cycles 15: 113–127.

Kienast M., Steinke S., Stattegger K. and Calvert S.E. 2001b. Synchronous tropical South China Sea SST change and Greenland warming during deglaciation. Science 291: 2132–2134.

Kienast M., Higginson M.J., Mollenhauer G., Eglinton T.I., Chen M.-T. and Calvert S.E. 2005. On the sedimentological origin of down-core variations of bulk sedimentary nitrogen isotope ratios. Paleoceanography 20: doi:10.1029/2004PA001081.

Kissel C., Laj C., Clemens S. and Solheid P. 2003. Magnetic signature of environmental changes in the last 1.2 Myr at ODP Site 1146, South China Sea. Mar. Geol. 201: 119–132.

Kudrass H.R., Jin X.L., Beiersdorf H. and Cepek P. 1992. Erosion and sedimentation in the Xisha Trough at the continental margin of southern China. In: Jin X., Kudrass H.R. and Pautot G. (eds.), Marine Geology and Geophysics of the South China Sea. China Ocean Press, Beijing, pp. 137–153.

Kuhnt W., Holbourn A. and Zhao Q. 2002. The early history of the South China Sea: evolution of Oligocene-Miocene deep water environments. Rev. Micropaleontol. 45: 99–159.

Laj C., Wang P. and Balut Y. 2005. IPEV les rapports de campagnes à la mer. MD147/MARCO POLO- IMAGES XII à bord du "Marion Dufresne", 59pp.

Lee M.-Y., Chen C.-H., Wei K.-Y., Iizuka Y. and Carey S. 2004. First Toba supereruption revival. Geology 32:61–64.

Lee M.-Y., Wei K.-Y. and Chen Y.-G. 1999. High resolution oxygen isotope stratigraphy for the last 150,000 years in the Southern South China Sea: Core MD972151. Terr. Atmos. Ocean. Sci. (TAO) Taipei 10: 239–254.

Leventhal J.S. 2004. Isotopic chemistry of organic carbon in sediments from Leg 184. In: Prell W.L., Wang P., Blum P., Rea D.K. and Clemens S.C. (eds.), Proc. ODP, Sci. Results, 184 [Online].

Li B. and Jian Z. 2001. Evolution of planktonic foraminifera and the thermocline in the southern South China Sea since 12 Ma (ODP-184, Site 1143). Sci. China (D) 44(10): 889–896.

Li B., Jian Z., Li Q., Tian J. and Wang P. 2005. Paleoceanography of the South China Sea since the middle Miocene: evidence from planktonic foraminifera. Mar. Micropaleontol. 54: 49–62.

Li B., Wang J., Huang B., Li Q., Jian Z. and Wang P. 2004. South China Sea surface water evolution over the last 12 Ma: A south-north comparison from ODP Sites 1143 and 1146. Paleoceanography 19: PA1009, doi:10.1029/2003PA000906.

Li C. 1993. Micropaleontological records, carbonate contents and oxygen-isotopic curves in late Pleistocene deep sea cores from the South China sea. Tropical Oceanol. 12(1): 16–23 (in Chinese).

Li J., Wang R. and Li B. 2002. Variations of opal accumulation rates and paleoproductivity over the past 12 Ma at ODP Site 1143, southern South China Sea. Chinese Sci. Bull. 47: 596–598.

Li L., Tu X., Luo Y. and Chen S. 1992. Planktonic foraminiferal assemblages and paleo-ceanography of South China Sea in late Quaternary. Tropical Oceanol. 11(2): 62–69 (in Chinese).

Li Q., Jian Z. and Li B. 2004. Oligocene-Miocene planktonic foraminifer biostratigraphy, Site 1148, northern South China Sea. In: Prell W.L., Wang P., Blum P., Rea D.K. and Clemens S.C. (eds.), Proc. ODP, Sci. Results, 184 [Online].

Li Q., Jian Z. and Su X. 2005. Late Oligocene rapid transformations in the South China Sea. Mar. Micropaleontol. 54: 5–25.

Li Q., Wang P., Zhao Q., Shao L., Zhong G., Tian J., Cheng X., Jian Z. and Su X. 2006. A 33 Ma lithostratigraphic record of tectonic and paleoceanographic evolution of the South China Sea. Mar. Geol. 230: 217–235.

Li Q., Wang P., Zhao Q., Tian J., Cheng X., Jian Z. Zhong G. and Chen M. 2008. Paleoceanography of the mid-Pleistocene South China Sea. Quat. Sci. Rev. 27: 1217–1233.

Li X., Chen F., Tang R. and Fang X. 1996. Oxygen isotope and paleoclimate in piston core HY4–901 in north part of South China Sea. Chinese Sci. Bull. 41(20): 1722–1725.

Li X., Wei G., Shao L., Liu Y., Liang X., Jian Z., Sun M. and Wang P. 2003. Geochemical and Nd isotopic variations in sediments of the South China Sea: a response to Cenozoic tectonism in SE Asia. Earth Planet. Sci. Lett. 211: 207–220.

Liang X., Wei G., Shao L., Li X. and Wang R. 2001. Records of Toba eruptions in the South China Sea – Chemical charateristics of the glass shards from ODP 1143A. Sci. China (D) 44(10): 871–878.

Lin D.-C., Liu C.-H., Fang T.-H., Tsai C.-H., Murayama M. and Chen M.-T. 2006. Millennial-scale changes in terrestrial sediment input and Holocene surface hydrography in the north-ern South China Sea (IMAGES MD972146). Palaeogeogr. Palaeoclimatol. Palaeoecol. 236: 56–73.

Lin H.-L. 2003. Late Quaternary deep-water circulation in the South China Sea. Terr. Atmos. Ocean. Sci. (TAO) Taipei 14(3): 321–333.

Liu C. and Cheng X. 2001. Variations in upper ocean structure for the last 2 Ma of the Nansha area by means of calcareous nannofossils. Sci. China (D) 44(10): 905–911.

Liu C., Cheng X., Zhu Y., Tian J. and Xia P. 2002. Oxygen and carbon isotopic records of calcareous nannofossils for the past 1 Ma in the southern South China Sea. Chinese Sci. Bull. 47(10): 798–803.

Liu C., Wang P., Tian J. and Cheng X. 2008. Coccolith evidence for Quaternary nutricline variations in the southern South China Sea. Mar. Micropaleontol. 69: 42–51.

Liu Z., Colin C. and Trentesaux A. 2006. Major element geochemistry of glass shards and minerals of the Youngest Toba Tephra in the southwestern South China Sea. J. Asian Earth Sci. 27: 99–107.

Liu Z., Colin C., Trentesaux A., Blamart D., Bassinot F., Siani G. and Sicre M.-A. 2004. Erosional history of the eastern Tibetan Plateau since 190 kyr ago: clay mineralogical and geochemical investigations from the southwestern South China Sea. Mar. Geol. 209: 1–18.

Liu Z., Trentesaux A., Clemens S.C., Colin C., Wang P., Huang B. and Boulay S. 2003a. Clay mineral assemblages in the northern South China Sea: implications for East Asian monsoon evolution over the past 2 million years. Mar. Geol. 201: 133–146.

Liu Z., Trentesaux A., Clemens S.C. and Wang P. 2003b. Quaternary clay mineralogy in the northern South China Sea (ODP Site 1146) -Implications for oceanic current transport and East Asian monsoon evolution. Sci. China (D) 46(12): 1123–1235.

Lu J., Chen M., Wang R. and Pushkar V.S. 2004. Data report: Diatom records of ODP Site 1143 in the southern South China Sea. In: Prell W.L., Wang P., Blum P., Rea D.K. and Clemens S.C. (eds.), Proc. ODP, Sci. Results, 184 [Online].

Ludwig W.J. 1970. The Manila Trench and West Luzon Trough, 3, Seismic refraction measurements. Deep-Sea Res. 17: 553–571.

Luo Y., Cheng H., Wu G. and Sun X. 2001. Records of natural fire and climate history during the last three glacial- interglacial cycles around the South China Sea - Charcoal record from the ODP 1144. Sci. China (D) 44(10): 897–904.

Luo Y. and Sun X. 2007. Deep-sea pollen in the southern South China Sea during 12∼1.6Ma BP and its response to the global climate change. Chinese Sci. Bull. 52(15): 2115–2122.

Luo Y., Sun X. and Jian Z. 2005. Environmental change during the penultimate glacial cycle: a high-resolution pollen record from ODP Site 1144, South China Sea. Mar. Micropaleontol. 54: 107–123.

Mao S., Wu G. and Li J. 2004. Oligocene-early Miocene dinoflagellate stratigraphy, Site 11448, ODP Leg 184, South China Sea. In: Prell W.L., Wang P., Blum P., Rea D.K. and Clemens S.C. (eds.), Proc. ODP, Sci. Results, 184 [Online].

Mao S. and Rex H. 1993. Quaternary organic-walled dinoflagellate cysts from the South China Sea and their paleoclimatic Significance. Palynology 17: 47–65.

Mao S., Li J., Wu G. and Harland R. 2007. Dinoflagellate cycts and environmental evolution of the Oligocene to Lower Miocene at Site 1148, ODP Leg 184, South China Sea. Palynology 31: 37–52.

Mercer J.L. and Zhao M. 2004. Alkenone stratigraphy of the northern South China Sea for the last 35 m.y., Sites 1147 and 1148, ODP Leg 184. In: Prell W.L., Wang P., Blum P., Rea D.K. and Clemens S.C. (eds.), Proc. ODP, Sci. Results, 184 [Online].

Miao Q., Thunell R.C. and Andersen D.M. 1994. Glacial-Holocene carbonate dissolution and sea surface temperatures in the South China and Sulu seas. Paleoceanography 9: 269–290.

MOET (Multidisciplinary Oceanographic Expedition Team of Academia Sinica to the Nansha Islands) 1993. Quaternary biological groups of the Nansha Islands and the neighbouring waters. Zhongshan Univ. Press, Guangzhou, 552pp (in Chinese).

Nathan S.A. and Leckie R.M. 2003. Miocene planktonic foraminiferal biostratigraphy of Sites 1143 and 1146, ODP Leg 184, South China Sea. In: Prell W.L., Wang P., Blum P., Rea D.K. and Clemens S.C. (eds.), Proc. ODP Sci. Results, 184 [Online].

Oppo D.W. and Sun Y. 2005. Amplitude and timing of sea-surface temperature change in the northern South China Sea: Dynamic link to the East Asian monsoon. Geology 33: 785–788.

Pautot G. 1992. Morphostructural analysis of the Central Ridge in South China Sea. In: Jin X., Kudrass H.R. and Pautot G. (eds.), Marine Geoloy and Geophysics of the South China Sea. China Ocean Press, Beijing, pp. 10–20.

Pautot G. and Rangin C. 1989. Subduction of the South China Sea axial ridge below Luzon (Philippines). Earth Planet. Sci. Lett. 92: 57–69.

Peng P., Yu C., Jia G., Hu J., Song J. and Zhang G. 2004. Data report: Marine and terrigenous lipids in the sediments from the South China Sea, Site 1148, Leg 184. In: Prell W.L., Wang P., Blum P., Rea D.K. and Clemens S.C. (eds.), Proc. ODP, Sci. Results, 184 [Online].

Prell W., Wang P., Blum P., Rea D.K. and Clemens S.C. (eds.). 2006. Proceedings of the Ocean Drilling Program, Scientific Results, vol. 184, Texas A&M Univ., College Station, USA.

Rangin C., Stephan J.F., Blanchet R., Baladad D., Bouysee P., Chen M.P., Chotin P., Collot J.Y., Daniel J., Drouhot J.M., Marchadier Y., Marsset B., Pelletier B., Richard M. and Tardy M. 1988. Seabeam survey at the southern end of the Manila trench. Transition between subduction and collision processes, offshore Mindoro Island, Philippines. Tectonophysics 146: 261–278.

Samodai J.P., Thompson P. and Chen C. 1986. Foraminiferal analysis of South China Sea core V36–08 with paleoenviromental implications. Proc. Geol. Soc. China, Taipei 29: 118–137.

Sarnthein M., Pflaumann U., Wang P. and Wong H.K. (eds.). 1994. Preliminary Report on SONNE-95 Cruise "Monitor Monsoon" to the South China Sea. Berichte-Reports, Geol.-Palaont. Inst. Univ. Kiel, 48, pp. 1–225.

Sarnthein M. and Wang P. (eds.). 1999. Response of west Pacific marginal seas to global climate change. Mar. Geol. (Spec. Issue) 156: 1–308.

Schimanski A. and Stattegger K. 2005. Deglacial and Holocene evolution of the Vietnam shelf: stratigraphy, sediments and sea-level change. Mar. Geol. 214: 365–387.

Schönfeld J. and Kudrass H.R. 1993. Hemipelagic sediment accumulation rates in the South China Sea related to late Quaternary sea-level changes. Quat. Res. 40: 368–379.

Shao L., Li X., Wei G., Liu Y. and Fang D. 2001. Provenance of a prominent sediment drift on the northern slope of the South China Sea. Sci. China (D) 44: 919–925.

Shao L., Li X., Geng J., Pang X., Lei Y., Qiao P., Wang L. and Wang H. 2007. Deep water bottom current deposition in the northern South China Sea. Sci. China (D) 50(7): 1060–1066.

Solheid P.A., Laj C. and Banerjee S.K. 2003. Data report: Mineral magnetic properties of sediments from Site 1144, northern South China Sea. In: Prell W.L., Wang P., Blum P., Rea D.K. and Clemens S.C. (eds.), Proc. ODP Sci. Results, 184 [Online].

South China Sea Institute of Oceanology. 1982. Reports of Multidisciplinary Investigations in the South China Sea (I). China Sci. Press, Beijing (in Chinese).

South China Sea Institute of Oceanology. 1985. Reports of Multidisciplinary Investigations in the South China Sea (II). China Sci. Press, Beijing, 432pp (in Chinese).

South China Sea Institute of Oceanology. 1987. Zeng-Mu Reef-Report of Multidisciplinary Investigations. China Sci. Press, Beijing, 245pp (in Chinese).

State Oceanic Administration of China (SOA). 1988. Reports of Multidisciplinary Investigations in Central Part of South China Sea for Resources and Environment. China Ocean Press, Beijing, 419pp (in Chinese).

Stattegger, K., Kuhnt,W., Wong, H.K. and Scientific Party 1997. Cruise Report SONNE 115 SUNDAFLUT. Berichte-Report 86, Institüt für Geowissenschaften, Univ. Kiel, 211pp.

Steinke S., Chiu H.Y, Yu P.S., Shen C.C., Erlenkeuser H., Löwemark L. and Chen M.T. 2006. On the influence of sea level and monsoon climate on the southern South China Sea freshwater budget over the last 22,000 years. Quat. Sci. Rev. 25: 1475 1488.

Steinke S., Kienast M. and Hanebuth T. 2003. On the significance of sea-level variations and shelf paleo-morphology in governing sedimentation in the southern South China Sea during the last deglaciation. Mar. Geol. 201: 179–206.

Su D., White N. and McKenzie D. 1989. Extension and subsidence of the Pearl River mouth basin, northern South China Sea. Basin Res. 2: 205–222.

Su X., Xu Y. and Tu, Q. 2004. Early Oligocene–Pleistocene calcareous nannofossil biostratigraphy of the northern South China Sea (Leg 184, Sites 1146–1148). In: Prell W.L., Wang P., Blum P., Rea D.K. and Clemens S.C. (eds.), Proc. ODP, Sci. Results 184 [Online].

Suess E. 2005. RV SONNE cruise report SO 177, Sino–German cooperative project, South China Sea Continental Margin: geological methane budget and environmental effects of methane emissions and gashydrates. IFM-GEOMAR Reports, 133pp.

Sun X. and Luo Y. 2001. Pollen record of the last 280 ka from deep-sea sediments of the northern South China Sea. Sci. China (D) 44(10): 879–888.

Sun X., Luo Y., Huang F., Tian J. and Wang P. 2003. Deep-sea pollen from the South China Sea: Pleistocene indicators of East Asian monsoon. Mar. Geol. 201: 97–118.

Tamburini F., Adatte T., Föllmi K., Bernasconi S.M. and Steinmann P. 2003. Investigating the history of East Asian monsoon and climate during the last glacial-interglacial period (0–140 000 years): mineralogy and geochemistry of ODP Sites 1143 and 1144, South China Sea. Mar. Geol. 201: 147–168.

Thunell R., Miao Q., Calvert S., Calvert S. and Pedersen T. 1992. Glacial-Holocene biogenic sedimentation patterns in the South China Sea: productivity variations and surface water pCO_2. Paleoceanography 7: 143–162.

Tian J., Pak D.K., Wang P., Lea D., Cheng X. and Zhao Q. 2006. Late Pliocene monsoon linkage in the tropical South China Sea. Earth Planet. Sci. Lett. 252: 72–81.

Tian J., Wang P., Cheng X. and Li Q. 2002. Astronomically tuned Plio-Pleistocene benthic $\delta^{18}O$ records from South China Sea and Atlantic-Pacific comparison. Earth Planet. Sci. Lett. 203: 1015–1029.

Tian J., Wang P. and Cheng X. 2004a. Time-frequency variations of the Plio–Pleistocene foraminiferal isotopes: a case study from the southern South China Sea. Earth Sci.-J. China Univ. Geosci. 15(3): 283–289.

Tian J., Wang P. and Cheng X. 2004b. Responses of foraminiferal isotopic variations at ODP Site 1143 in the southern South China Sea to orbital forcing. Sci. China (D) 47(10): 943–953.

Tian J., Wang P. and Cheng X. 2004c. Pleistocene precession forcing of the upper ocean structure variations in the southern South China Sea. Progr. Nat. Sci. 14(11): 1004–1009.

Tian J., Wang P. and Cheng X. 2004d. Development of the East Asian monsoon and Northern Hemisphere glaciation: Oxygen isotope records from the South China Sea. Quat. Sci. Rev. 23: 2007–2016.

Tian J., Wang P., Chen R. and Cheng X. 2005a. Quaternary upper ocean thermal gradient variations in the South China Sea: Implications for east Asian monsoon climate. Paleoceanography 20: PA4007, doi:10.1029/2004PA001115.

Tian J., Wang P., Cheng X., Wang R. and Sun X. 2005b. Forcing mechanism of the Pleistocene east Asian monsoon variations in a phase perspective. Sci. China (D) 48(10): 1708–1717.

Tian J., Wang P., Zhao Q., Li Q. and Cheng X. 2008. Astronomically modulated Neogene sediment records from the South China Sea. Paleoceanography, doi:10.1029/2007PA001552.

Trentesaux A., Liu Z., Colin C., Boulay S. and Wang P. 2003. Data report: Pleistocene paleoclimatic cyclicity of southern China: clay mineral evidence recorded in the South China Sea (ODP Site 1146). In: Prell W.L., Wang P., Blum P., Rea D.K. and Clemens S.C. (eds.), Proc. ODP, Sci. Results 184 [Online].

Tu X., Zheng F., Wang J., Cai H., Wang P., Büchring C. and Sarnthein M. 2001. A sudden cooling event during the last interglacial in the northern South China Sea. Sci. China (D) 44(10): 865–870.

Wan S., Li A., Clift P.D. and Jiang H. 2006. Development of the East Asian summer monsoon: Evidence from the sediment record in the South China Sea since 8.5 Ma. Palaeogeogr. Palaeoclimatol. Palaeoecol. 241: 139–159.

Wan S., Li A., Stuut J.-B.W. and Xu F. 2007b. Grain-size records at ODP Site 1146 from the northern South China Sea: Implications on the East Asian monsoon evolution since 20 Ma. Sci. China (D) 50(10): 1536–1547.

Wang C.H. and Chen M.P. 1990. Upper Pleistocene oxygen and carbon isotopic changes of Core SCS-15B at the South China Sea. J. SE Asian Earth Sci. 4: 243–246.

Wang C.H., Chen M.-P. and Lo S.-C. 1986. Stable isotope records of late Pleistocene sediments from the South China Sea. Bull. Inst. Earth Sci., Acad. Sinica Taipei 6: 185–195.

Wang L.W. and Lin H.L. 2004. Data report: Carbonate and organic carbon contents of sediments from Sites 1143 and 1146 in the South China Sea. In: Prell W.L., Wang P., Blum P., Rea D.K. and Clemens S.C. (eds.), Proc. ODP, Sci. Results 184 [Online].

Wang L., Jian Z. and Chen J. 1997. Late Quaternary pteropods in the South China Sea: carbonate preservation and paleoenviromental variation. Mar. Micropaleotol. 32: 115–126.

Wang L., Sarenthein M., Erlenkeuser H., Grimalt J., Grootes P., Heilig S., Ivanova E., Kienast M., Pelejero C. and Pflaumann U. 1999a. East Asian monsoon Climate during the late Pleistocene: high- resolution sediment records from the South China Sea. Mar. Geol. 156: 245–284.

Wang L., Sarenthein M., Grootes P. and Erlenkeuser H. 1999b. Millennial reoccurrence of century- scale abrupt events of East Asian monsoon: A possible heat conveyor for the global deglaciation. Paleoceanography 14: 725–731.

Wang P. 1990. Neogen stratigraphy and paleoenvironments of China. Palaeogeogr. Palaeoclimatol. Palaeoecol. 77: 315–334.

Wang P. 2004. Cenozoic deformation and the history of sea-land interactions in Asia. In: Clift P., Wang P., Kuhnt W. and Hayes D. (eds.), Continent-Ocean Interactions in the East Asian Marginal Seas. AGU Geophys. Monogr. 149: 1–22.

Wang P., Clemens S., Beaufort L., Braconnot P., Ganssen G., Jian Z., Kershaw P. and Sarnthein M. 2005. Evolution and variability of the Asian monsoon system: state of the art and outstanding issues. Quat. Sci. Rev. 24: 595–629.

Wang P., Jian Z., Zhao Q., Li Q., Wang R., Liu Z., Wu G., Shao L., Wang J., Huang B., Fang D., Tian J., Li J., Li X., Wei G., Sun X., Luo Y., Su X., Mao S. and Chen M. 2003a. Evolution of the South China Sea and monsoon history revealed in deep-sea records. Chinese Sci. Bull. 48(23): 2549–2561.

Wang P., Min Q., Bian Y. and Feng W. 1986. Planktonic foraminifera in the continental slope of the northern South China Sea during the last 130,000 years and their paleoceanographic implications. Acta Geol. Sinica (Trial English Edition) 60: 1–11.

Wang P., Prell W.L., Blum P. (eds.). 2000. Proc. ODP, Init. Repts, Vol. 184 [CD-ROM]. Ocean Drilling Program, Texas A& M University, College Station TX 77845–9547, USA.

Wang P., Prell W., Blum P. and the Leg 184 Shipboard Scientific Party 1999. Exploring the Asian monsoon through drilling in the South China Sea. JOIDES J. 25(2): 8–13.

Wang P., Tian J. and Cheng X. 2001. Transition of Quaternary glacial cyclicity in deep-sea records at Nansha, the South China Sea. Sci. China (D) 44(10): 926–933.

Wang P., Tian J., Cheng X., Liu C. and Xu J. 2003b. Exploring cyclic changes of the ocean carbon reservoir. Chinese Sci. Bull. 48(23): 2536–2548.

Wang P., Tian J., Cheng X., Liu C. and Xu J. 2003c. Carbon reservoir change preceded major ice-sheets expansion at Mid-Brunhes Event. Geology 31: 239–242.

Wang P., Zhao Q., Jian Z., Cheng X., Huang W., Tian J., Wang J., Li Q., Li B. and Su X. 2003d. Thirty million year deep-sea records in the South China Sea. Chinese Sci. Bull. 48(23): 2524–2535.

Wang P., Tian J., Cheng X., Liu C. and Xu J. 2004. Major Pleistocene stages in a carbon perspective: The South China Sea record and its global comparison. Paleoceanography 19: doi: 10.1029/2003PA000991.

Wang R., Clemens S., Huang B. and Chen M. 2003. Late Quaternary paleoceanographic changes in the northern South China Sea (ODP Site 1146): radiolarian evidence. J. Quat. Sci. 18(8): 745–756.

Wang R., Fang D., Shao L., Chen M., Xia P. and Qi J. 2001. Oligocene biogenetic siliceous deposits on the slope of the northern South China Sea. Sci. China (D) 44(10): 912–918.

Wang R., Jian Z., Xiao W., Tian J., Li J., Chen R., Zheng L. and Chen J. 2007. Quaternary biogenic opal records in the South China Sea: linkages to East Asian monsoon, global ice volume and orbital forcing. Sci. China (D) 50(5): 710–724.

Wang R. and Li J. 2003. Quaternary high resolution opal record and its paleoproductivity implication at ODP Site 1143, southern South China Sea. Chinese Sci. Bull. 48(4): 363–367.

Wang R., Li J. and Li B. 2004. Data report: Late Miocene–Quaternary biogenic opal accumulation at ODP Site 1143, southern South China Sea. In: Prell W.L., Wang P., Blum P., Rea D.K. and Clemens S.C. (eds.), Proc. ODP, Sci. Results 184 [Online].

Wehausen R. and Brumsack H.J. 2002. Astronomical forcing of the East Asian monsoon mirrored by the composition of Pliocene South China Sea sediments. Earth Planet. Sci. Lett. 201: 621–636.

Wehausen R., Tian J., Brumsack H.-J., Cheng X. and Wang P. 2003. Geochemistry of Pliocene sediments from ODP Site 1143 (southern South China Sea). In: Prell W.L., Wang P., Blum P., Rea D.K., and Clemens S.C. (eds.), Proc. ODP, Sci. Results, 184 [Online].

Wei G., Liu Y., Li X., Shao L. and Liang X. 2003. Climatic impact on Al, K, Sc and Ti in marine sediments: Evidence from ODP Site 1144, South China Sea. Geochem. J. 37: 593–602.

Wei G.J., Liu Y., Lia X., Shao L. and Fang D. 2004. Major and trace element variations of the sediments at ODP Site 1144, South China Sea, during the last 230 ka and their paleoclimate implications. Palaeogeogr. Palaeoclimatol. Palaeoecol. 212: 331–342.

Wei G.J., Li X., Liu Y., Shao L. and Liang X. 2006. Geochemical record of chemical weathering and monsoon climate change since the early Miocene in the South China Sea. Paleoceanography 21: PA4214, doi: 10.1029/2006PA001300.

Wiesner M.G., Kuhnt W. and Shipboard Scientific Party 1997. Cruise report, R/V Sonne SO-114, Manila – Kota Kinabalu. Inst. Biogeochem. Mar. Chem., Univ. Hamburg, IBMC Library Ref No IIA 942, 55pp.

Wiesner M.G., Kuhnt W. and Shipboard Scientific Party 1998. Cruise report, R/V Sonne SO-132, Singapore – Manila. Inst. Biogeochem. Mar.e Chem., Univ. Hamburg, IBMC Library Ref No IIA 943, 120pp.

Wiesner M.G., Stattegger K., Kuhnt W. and Shipboard Scientific Party 1999. Cruise report, R/V Sonne SO-140, Singapore–Nha Trang–Manila. Ber. Rep. Inst. Geowiss. Univ. Kiel 7, 157pp.

Winn K., Zheng L., Erlenkeuser H. and Stoffers P. 1992. Oxygen/carbon isotopes and paleoproductivity in the South China Sea during the past 110,000 years. In: Jin X., Kudrass H.R. and Pautot G. (eds.), Marine Geology and Geophysics of the South China Sea. China Ocean Press, Beijing, pp. 154–166.

Wu G., Qin J. and Mao S. 2003. Deep-water Oligocene pollen record from South China Sea. Chinese Sci. Bull. 48(22): 2511–2515.

Xu J., Wang P., Huang B., Li Q. and Jian Z. 2005. Response of planktonic foraminifera to glacial cycles: Mid-Pleistocene change in the southern South China Sea. Mar. Micropaleontol. 54: 89–105.

Yang L., Chen M., Wang R. and Zhen F. 2002. Radiolarian record to paleoecological environment change events over the past 1.2 Ma BP in the southern South China Sea. Chinese Sci. Bull. 47(17): 1478–1483.

Yu P.S., Huang C.-C., Chin Y., Mii H.-S. and Chen M.-T. 2006. Late Quaternary East Asian Monsoon variability in the South China Sea: Evidence from planktonic foraminifera faunal and hydrographic gradient records. Palaeogeogr. Palaeoclimatol. Palaeoecol. 236: 74–90.

Zhang M., He Q., Ye Z., Han C., Li H., Wu J. and Ju L. 1989. Sedimentary Geology of Xisha Reef Carbonates. China Sci. Press, Beijing, 117pp (in Chinese).

Zhang Y., Ji J., Balsam W.L., Liu L. and Chen J. 2007. High resolution hematite and goethite records from ODP 1143, South China Sea: Co-evolution of monsoonal precipitation and El Niño over the past 600,000 years. Earth Planet. Sci. Lett. 264: 136–150.

Zhao M., Huang C.Y., Wang C.C. and Wei G. 2006. A millennial-scale $U_{37}^{K'}$ sea-surface temperature record from the South China Sea (8 °N) over the last 150 kyr: Monsoon and sea-level influence. Paleogeogr. Paleoclimat. Paleoecol. 236: 39–55.

Zhao Q. 2005. Late Cainozoic ostracod faunas and paleoenvironmental changes at ODP Site 1148, South China Sea. Mar. Micropaleontol. 54: 27–47.

Zhao Q., Jian Z., Wang J., Cheng X., Huang B., Xu J., Zhou Z., Fang D. and Wang P. 2001a. Neogene oxygen isotopic stratigraphy, ODP Site 1148, northern South China Sea. Sci. China (D) 44(10): 934–942.

Zhao Q., Wang P., Cheng X., Wang J., Huang B., Xu J., Zhou Z. and Jian Z. 2001b. A record of Miocene carbon excursions in the South China Sea. Sci. China (D) 44: 943–951.

Zheng F., Li Q., Li B., Chen M., Tu X., Tian J. and Jian Z. 2005. A millennial scale planktonic foraminiferal record of mid-Pleistocene climate transition in the northern South China Sea. Palaeogeogr. Palaeoclimatol. Palaeoecol. 223: 349–363.

Zheng F., Li Q., Tu X., Chen T., Li B. and Jian Z. 2004. Abundance variations of planktonic foraminifers during mid-Pleistocene climate transition at ODP Site 1144, northern South China

Sea. In: Prell W.L., Wang P., Blum P., Rea D.K. and Clemens S.C. (eds.), Proc. ODP, Sci. Results 184 [Online].

Zheng H.B., Powell C.M., Rea D.K., Wang J.L. and Wang P.X. 2004. Late Miocene and mid-Pliocene enhancement of the East Asian monsoon as viewed from the land and sea. Global Planet. Change 41: 147–155.

Zheng L. and Chen W. (eds.). 1993. Marine Sedimentation Process and Geochemical Studies in the South China Sea. China Ocean Press, Beijing, 201pp (in Chinese).

Zhou D., Liang Y.B. and Zeng C.K. 1994. Oceanology of China Seas, vol. 2, Kluwer, pp. 345–573.

Zhu Y., Huang Y., Matsumoto R. and Wu B. 2003. Geochemical and stable isotopic compositions of pore fluids and authigenic siderite concretions from Site 1146, ODP Leg 184: implications for gas hydrate. In: Prell W.L., Wang P., Blum P., Rea D.K. and Clemens S.C. (eds.), Proc. ODP, Sci. Results 184 [Online].

.

Chapter 2
Oceanographical and Geological Background

Pinxian Wang and Qianyu Li

Introduction

The South China Sea (SCS) embraces an area of about 3.5×10^6 km^2 and extends from the Tropic of Cancer to the Equator, across over 20 degrees of latitude in the west Pacific. Since the last decade, the SCS has become the focus in studying the East Asian monsoon, like the Arabian Sea for the Indian monsoon (Wang B. et al. 2003). The SCS offers an ideal locality for high-resolution paleoceanographic researches in the low-latitude western Pacific because its hemipelagic sediments often register higher deposition rates and its carbonate compensation depth (CCD) is generally deeper than neighboring sea basins (Wang P. 1999).

Geographically, the SCS is located in between Asia and the Pacific, the largest continent and the largest ocean on earth. To its north and west, the SCS is bordered by South China and Indochina Peninsula; to the east and south, it is surrounded by a chain of islands, ranging from Luzon in the north to Borneo in the south. The continent side to the northwest and west is basically a mountainous land with a narrow coastal plain except for deltas of large rivers such as the Pearl River, the Red River and the Mekong River. Mountains reach 1,000–1,300 m high in South China and generally 1,600–2,000 m in Indochina. A mountain range, Truong Son Ra or Annam Cordillera, extends along the Vietnam coast with a northern peak of 3,142 m. Islands on the southeastern side are also mountainous, peaking at 4,101 m in Borneo and 3,997 m in Taiwan (Liu Y. 1994). The contrast in geodynamics between the NW continental side and the SE island side has given rise to a number of distinguished features in SCS oceanography and sedimentology. This chapter provides an overview of the major oceanographic and geological features of the SCS.

P. Wang (✉)
State Key Laboratory of Marine Geology, Tongji University, Shanghai 200092, China
e-mail: pxwang@online.sh.cn; pxwang@tongji.edu.cn

Q. Li
State Key Laboratory of Marine Geology, Tongji University, Shanghai 200092, China;
School of Earth and Environmental Sciences, The University of Adelaide,
South Australia 5005, Australia
e-mail: qli01@tongji.edu.cn; qianyu.li@adelaide.edu.au

P. Wang, Q. Li (eds.), *The South China Sea*, Developments in Paleoenvironmental Research 13, DOI 10.1007/978-1-4020-9745-4_2,
© Springer Science+Business Media B.V. 2009

2.1 Bathymetry and Geomorphology

The bathymetry of the SCS comprises three parts: the deep basin, the continental slope and the continental shelf, respectively covering about 15%, 38% and 47% of the total area with an average water depth of about 1,140 m. The major feature of the SCS topography is the rhomboid deep basin, which overlies oceanic crust and extends from NE to SW (Fig. 2.1). Water depth in the deep basin averages ~4, 700 m, with a maximum depth of 5,559 m reported from its eastern margin yet to be confirmed by survey. The deep basin is divided by a chain of sea mounts along 15 °N, or "Central Ridge", into two sectors: a relatively shallower northeastern part, and a deeper southwestern part (Wang Y. 1996).

The central deep basin is surrounded by continental and island slopes, topographically dissected and often studded with coral reefs (Fig. 2.2). The northern slope with Dongsha reef island and the western slope with Xisha (Paracel Islands) and Zhongsha reefs (Macclesfield Bank) are separated by the Xisha Trough, while the southern slope is occupied by the Nansha Islands (Spratly Islands), the largest reef area in the SCS. The Nansha Islands are scattered on a carbonate platform known as "Dangerous Grounds", covering a broad area of ~570, 000 km². The eastern slope is narrow and steep, bordered by the deep-water Luzon Trough and Manila Trench (Figs. 2.1 and 2.2).

The continental shelf is well developed on the northern and southern sides of the SCS (Figs. 2.1 and 2.2). Both the northern and southern shelves are narrower to the east and broader to the west. On the northern shelf, there are a number of submarine deltas developed off the Pearl and other river mouths, and fringing reefs are growing along the coasts of Leizhou Peninsula and Hainan Island. Exceeding ~300 km in width, the Sunda Shelf in the southwestern SCS is one of the largest shelves in the world. A number of submerged river valleys observed on the bottom of the Sunda Shelf are relict of a large river network, the Paleo-Sunda Rivers, during the last glacial maximum (LGM).

Although the SCS is open at present on its northeastern and southwestern sides, it has been highly sensitive to sea level changes in the past glacial cycles as a semi-enclosed basin. The southern connection to the Indian Ocean is limited to the uppermost 30–40 m and completely closed at glacial sea level lowstands, as was also the connection to the East China Sea/Okinawa Trough through Taiwan Strait in the northeast. Deeper passages exist only along its eastern side: the Bashi (Luzon) Strait with water depth of ~2, 400 m connecting to the open Pacific or, to be precise, the western Philippine Sea, and both the Mindoro Strait of ~420 m and the Balabac Strait of ~100 m connecting to the Sulu Sea (Fig. 2.3; Table 2.1). During the LGM, therefore, the SCS was largely closed except the two passages along the east boundary which remained open to the Pacific and to the Sulu Sea respectively.

The rhomboid shape of the SCS basin, together with its NE-SW topographic axis and broader northern and southwestern shelves (Fig. 2.3), may exert a significant impact on atmospheric and oceanographic circulation. The coastal ranges block the monsoon wind flows toward the land and give rise to a wind jet, which drives the

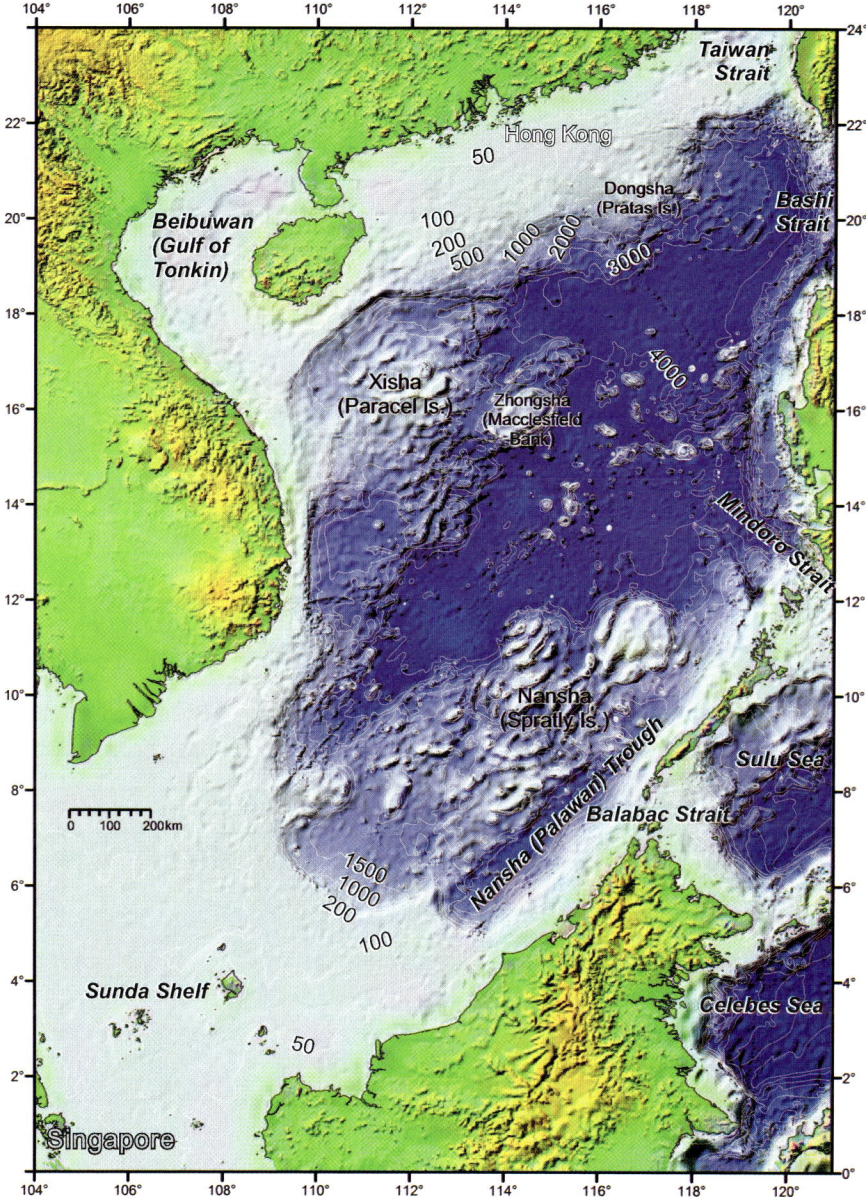

Fig. 2.1 Map shows the major topographic features of the South China Sea (SCS) and neighboring sea basins. Isobaths are in meters (m)

surface currents of the sea roughly along the basin axis with seasonal reversal in direction (Xie et al. 2003; Liu Q. et al. 2004). As shown below, this jet divides the SCS into two halves with different oceanographic features.

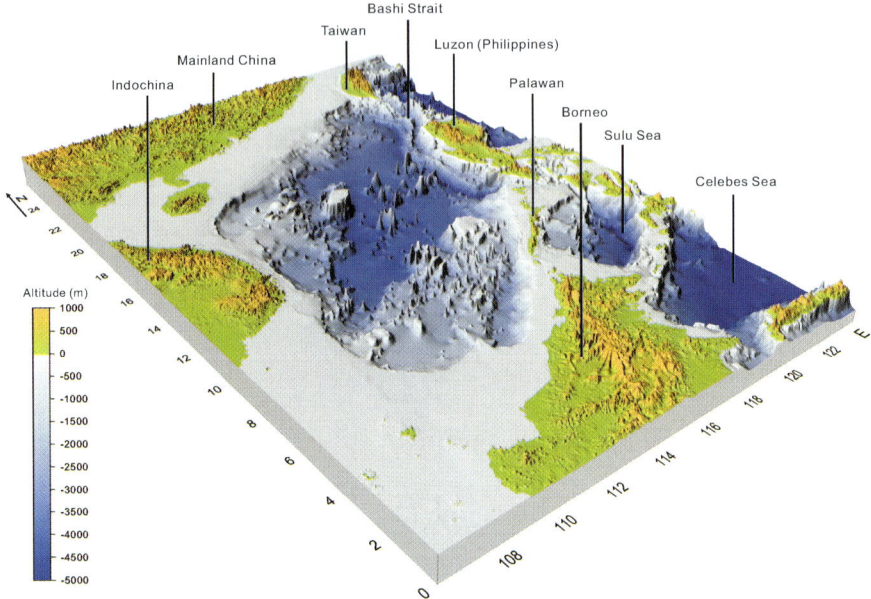

Fig. 2.2 Three-dimensional diagram shows the main geomorphologic features of the SCS (modified from Wang P. 1999)

Table 2.1 Straits of the SCS are mostly shallow except the Bashi Strait (compiled from various sources)

Strait	Connecting to	Sill depth	Minimal width	Fig. 2.3
Taiwan	East China Sea	70 m	130 km	P1
Bashi (Luzon)	Philippine Sea	2,400 m	370 km	P2
Mindoro	Sulu Sea	420 m	125 km	P3
Balabac	Sulu Sea	100 m	~50 km	P4
Karimata	Java Sea	~50 m	150 km	P5
Malacca	Andaman Sea	25 m	37 km	P6

2.2 Oceanography

The oceanography of the upper water of the SCS is largely dictated by two factors: the underlying morphology of the enclosed basin, and the overlying atmospheric circulation especially the East Asian Monsoon. The deep-water oceanography of the SCS is much less studied, but appears to be controlled by its only deep connection to the open ocean, the Bashi Strait. Despite of significant recent progress in the modern SCS oceanography, many paleoceanographic publications are still citing old literature. Here we provide an update of the current knowledge on the SCS oceanography, which is highly relevant to the interpretation of paleo-records, especially those concerning the Asian monsoon, upper water structure, exchanges with the Pacific and other sea basins, surface and deep circulations, productivity and biogeochemistry.

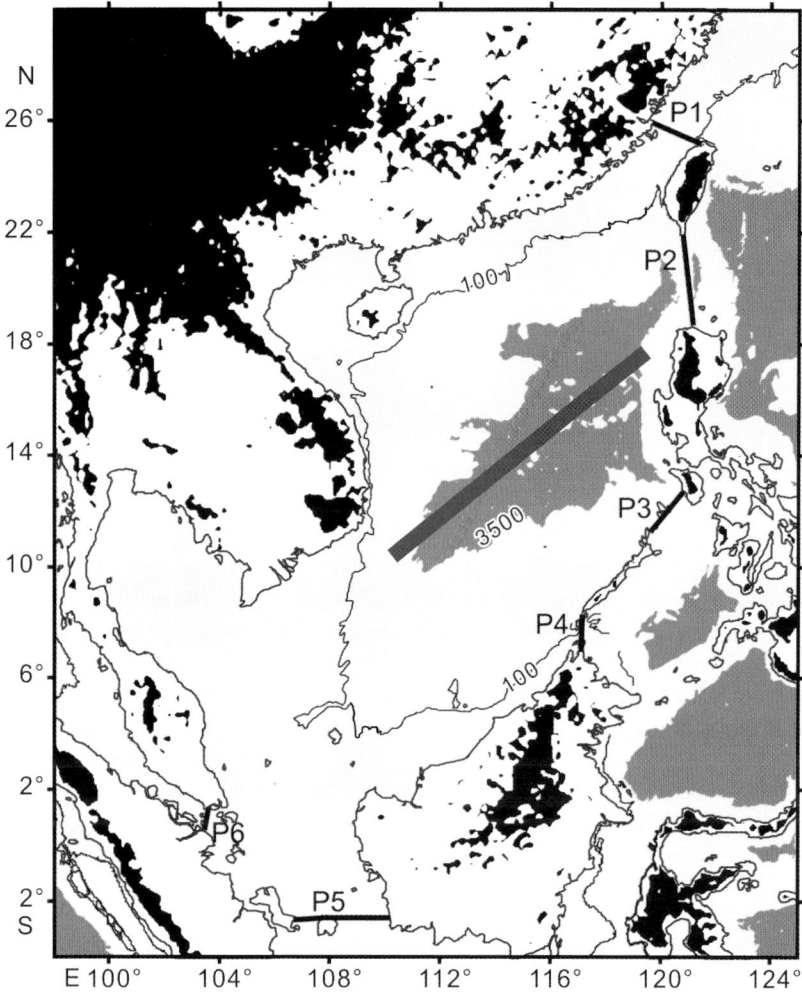

Fig. 2.3 Major topographic features of the SCS basin and environs are shown with black areas indicating +500 m elevation, white areas 500–0 m, light grey areas 0–3,500 m water depth, and dark grey areas deeper than 3,500 m. The NE-SW bar marks the oceanographic divide between the northern and southern parts of the SCS. Short black bars are passages: P1 = Taiwan Strait, P2 = Bashi (Luzon) Strait, P3 = Mindoro Strait, P4 = Balabac Strait, P5 = Karimata Strait, P6 = Malacca Strait

Monsoon

Monsoon is the prevailing climate feature in Asia, including the SCS. The Asian monsoon is the largest monsoon system in the modern world, and its two subsystems, the Indian or South Asian Monsoon and the East Asian Monsoon, are closely related and interact with each other, but they have different structures shaped by diverse sea-land configuration patterns. For East Asia, the ocean is to the east and the Tibetan Plateau to the west; while South Asia has the ocean to the south and the Tibetan Plateau to the north. At least three major distinctions have resulted from these sea-land configuration patterns. (1) For the South Asian subsystem, the winter monsoon is largely blocked by Tibet and hence is much weaker than its summer counterpart, whereas East Asia has the most intensive winter monsoon system in the world. (2) The South Asian summer monsoon is tropical in nature, while the East Asian monsoon has a subtropical in addition to the tropical component (Chen L. 1992). (3) East Asia humidity originates in the western Pacific warm pool (WPWP).

In winter, there are three branches of monsoon flows over the Arabian Sea, the Bay of Bengal (Indian monsoons) and the SCS (East Asian monsoon), forming strong cross-equatorial northerlies there (Fig. 2.4B). Only the northerly over the SCS can be traced backward to its origin over the cold Siberia, from where the world's coldest winter air-mass runs southward along the East Asian coastal zone, generating the strongest winter monsoon, which is in a sharp contrast to that in South Asia.

The East Asian summer monsoon comprises a tropical component connected with cross-equatorial flows from the south and an additional eastern component originated from the subtropical easterlies in the western periphery of the West Pacific subtropical high. Along with the western North Pacific monsoon trough or the ITCZ (the inter-tropical convergence zone) there is an East Asian subtropical front or "Meiyu (plum rain) front", resulting in two major summer monsoon areas in East Asia (Fig. 2.4) (Wang B. et al. 2003). The winter and summer monsoons affect the SCS with different strength and in different time duration. The winter monsoon lasts nearly six months (November to April), while the weaker summer season lasts only nearly 4 months (mid-May to mid-September) (Chu and Wang 2003). The mean surface wind stress over the SCS averages nearly $0.2 \, N/m^2$, reaching $0.3 \, N/m^2$ in the central portion in December but dropping to nearly $0.1 \, N/m^2$ in June. These features underline the necessity to distinguish winter vs summer monsoon in paleo-records which, unsurprisingly, often display a more complicated relationship with the orbital forcing than the Indian and African monsoons. In addition, care is necessary in correlating land-sea monsoon records. For example, the "South China Sea Monsoon Experiment (SCSMEX)" (1996–2001) found that a strong (weak) monsoon over the SCS usually leads to lesser (more) precipitation over the middle and lower reaches of the Yangtze River basin, and more (lesser) precipitation in North China (Ding et al. 2004).

As mentioned above, the East Asian monsoon differs from the Indian monsoon by its intensive winter monsoon flows and an additional subtropical component due to the unique land-ocean configuration (Fig. 2.4). Moreover, the interannual variability of the Asian monsoon associated with El Niño-Southern Oscillations (ENSO) is also much more significant in the East Asian sector than in the Indian sector. Therefore, large changes in the Pacific thermal conditions may greatly alter the intensity of the East Asian monsoon but not the Indian monsoon (Wang B. et al. 2003).

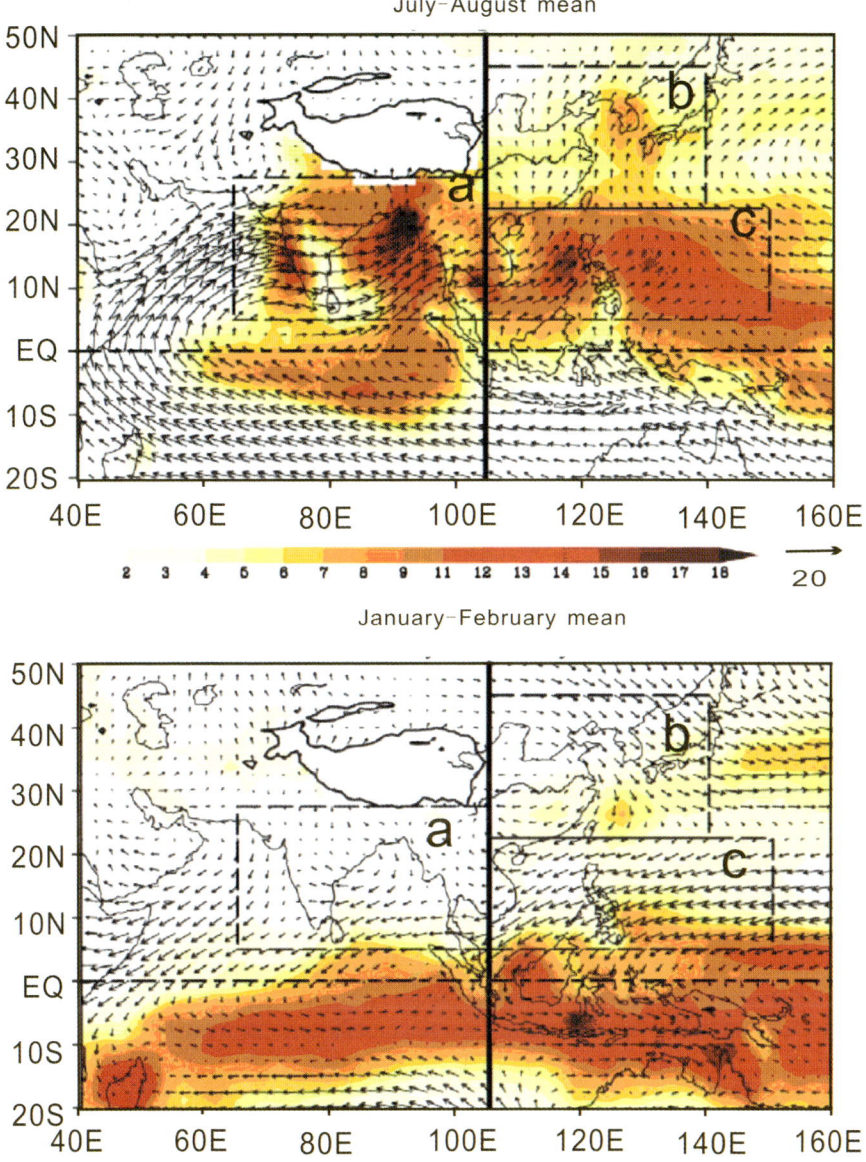

Fig. 2.4 Asian monsoon system is shown in climatological mean precipitation rates (*shaded* in mm/day) and 925 hPa wind vectors (*arrows*) during July–August (*upper panel*) and January–February (*lower panel*). The three major summer precipitation areas include (**a**) Indian tropical monsoon, (**b**) East Asian subtropical monsoon and (**c**) western North Pacific tropical monsoon (modified from Wang B. et al. 2003)

The low-level wind patterns over the SCS are affected by orographic features of the surrounding land. In winter, the northeasterly monsoon prevails the SCS, and two wind speed maxima exceeding $10 \, \mathrm{m \, s^{-1}}$ occur respectively in the Taiwan and

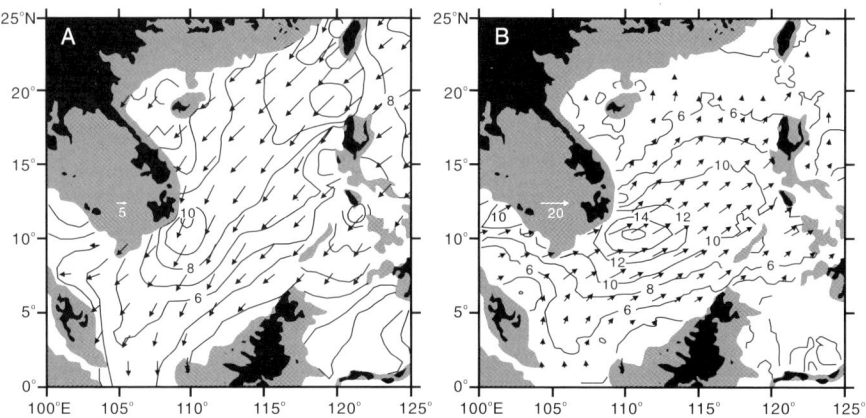

Fig. 2.5 QuikSCAT surface winds over the SCS are shown by (**A**) winter wind velocity (vector) and its magnitude (contours at 2 m/s intervals) averaged for December–February 2000–2002, and (**B**) summer wind stress vectors and their magnitude (contours in $10^{-2} Nm^{-2}$) averaged for June-August 2000–2002. Land topography with elevations >500 m is *shaded* in *black* and 0–500 m in *grey*. Note that the wind jet with maximal velocity is in NE-SW direction (modified from Xie et al. 2003 and Liu Q. et al. 2004)

Bashi straits and around 11 °N off the coast of southern Vietnam near the southern tip of Annam Cordillera (Fig. 2.5A). Both wind maxima result from orographic forcing by mountains. The northeasterly winds are blocked by the Indochina coastal mountain range, giving rise to a wind jet offshore, similar to the southwesterly jet in summer (Fig. 2.5B) (Xie et al. 2003; Liu Q. et al. 2004).

Surface Circulation

Much understanding has been gained about the surface circulation in the SCS since the pioneering work of Wyrtki (1961). Using both hydrographic and ship-drift data, Wyrtki provided the first picture of the surface circulation pattern in the SCS, basically a cyclonic gyre in winter and an anticyclonic gyre in summer, and suggested them to be results of the seasonally reversing monsoon winds. Later, the Wyrtki model of circulation patterns was confirmed by dynamical studies (e.g., Shaw and Chao 1994; Liu Z. et al. 2001) and observations (review by Su 2004). These distinct features are in fact reflections of the upper-layer circulation in response to the seasonal changes of the monsoon wind-stress curl, with additional influence from the Kuroshio in its northern part (Qu 2000). In winter, there is a basin-wide cyclonic gyre (Fig. 2.6A), while in summer the circulation splits into a weakened cyclonic gyre north of ~12 °N and a strong anti-cyclonic gyre in the south (Fig. 2.6B).

The large-scale features of the seasonal surface circulation of the SCS can be demonstrated with the distribution of climatological mean January and July dynamic heights (dyn cm) at 100 m water depth (Fig. 2.7), which shows a large cyclonic gyre in winter (Fig. 2.7B), dominant north of the line roughly from 10 °N, 110 °E to

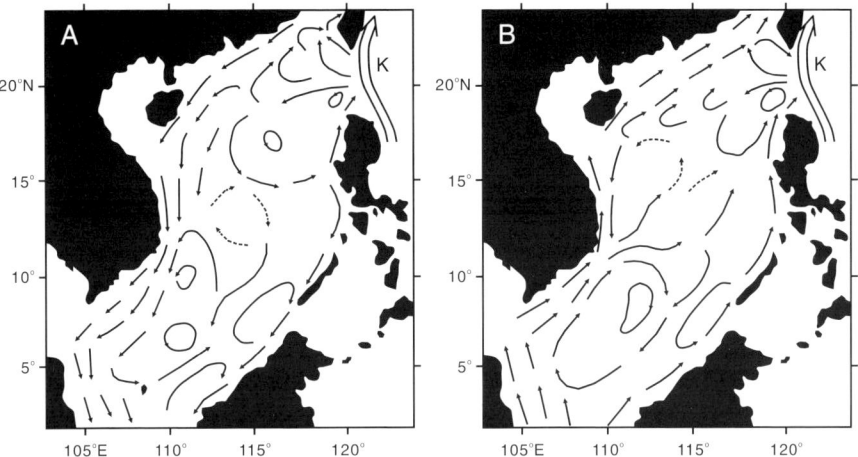

Fig. 2.6 Surface circulation of the modern SCS shows opposite patterns between (A) winter and (B) summer (modified from Fang et al. 1998). The Kuroshio Current is labeled with "K"

15 °N, 120 °E, and two gyres in summer (Fig. 2.7C). A large cyclonic eddy, the West Luzon Eddy, is present west of Luzon in winter and another strong cyclonic eddy, the East Vietnam Eddy, occurring off central Vietnam in both winter and summer (Qu 2000; Su 2004). Because the winter gyre is much stronger, the above-mentioned line separates the SCS into the northern and southern parts with different oceanographic features as discussed below.

Both satellite imageries and dynamic computations reveal a cross-basin wind-induced summer surface jet flowing northeastward from the central Vietnam to Luzon. The maximum speed of the current reaches 0.25 m/s near Vietnam bight at the surface, and declines to 0.20 m/s at 50 m water depth. If the orientation of the maximum velocity is defined as the axis, it extends from 110 °E, 10 °N to 120 °E, 18 °N at 50 m depth (Chu and Wang 2003), which is about the same as the boundary between the southern and northern parts of the SCS (Fig. 2.3; Qu 2000; Su 2004) and close to the axis of maximum monsoonal winds described above (Xie et al. 2003; Liu Q. et al. 2004).

The SCS surface circulation, therefore, can be summarized into a schematic diagram with seasonally alternating basin-scale gyres (Fig. 2.8). Associated with these gyres are strong western boundary currents. In winter, a southward jet flows along the entire western boundary. In summer, a northward jet flows along the western boundary in the southern SCS, apparently veering eastward off central Vietnam near 12 °N″ (Wang G. et al. 2006). This cross-basin eastward jet, or "Summer Southeast Vietnam Offshore Current" of Fang et al. (2002), has a maximum velocity of around 0.8 m/s and plays a significant role in the SCS oceanography by dividing it into the northern and southern parts. The eastward jet is accompanied by a dipole structure with an anticyclonic cell south of the jet and a cyclonic cell north of it. On average the dipole structure begins in June, peaks in strength in August or September, and disappears in October (Wang G. et al. 2006; Liu K. et al. 2002).

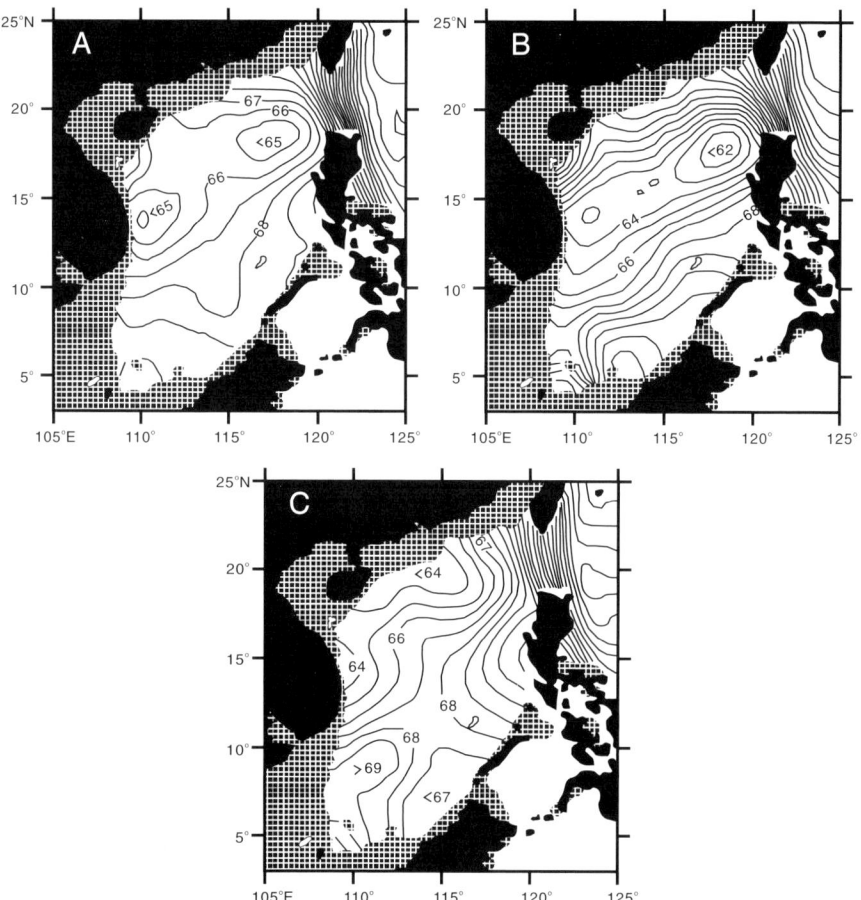

Fig. 2.7 Distribution of the mean dynamic heights (contour in dyn cm) at −100 m in the SCS is shown for three periods: (**A**) annual mean, (**B**) January monthly mean and (**C**) July monthly mean (modified from Qu 2000). Region with water depth shallower than 100 m is stippled

Surface Temperature and Salinity

The seasonal reversal in circulation leads to seasonal contrasts in sea surface temperature (SST) and salinity (SSS) patterns. In summer, surface water temperature in the SCS is uniform, with only insignificant variations between 28.5 and 29.5 °C (Fig. 2.9B). In winter, an intense western boundary current flows southward on the continental slope, separating the Sunda Shelf to the west and the deep basin to the east (dashed line in Fig. 2.8). This current transports cold water from the north and causes a distinct cold tongue south of Vietnam, leading to a steeper temperature gradient in the winter SCS (Fig. 2.9A). Compared to the neighboring Pacific and Indian Oceans, the winter SST in the SCS are considerably lower and form a conspicuous gap in the Indo-Pacific Warm Pool (Liu Q. et al. 2004).

Fig. 2.8 A sketch map illustrates the main surface circulation of the SCS featuring seasonally alternating, basin-scale cyclonic gyres: winter (*dashed line*); summer (*solid line*). The summer pattern involves a cyclonic gyre in the northern and an anticyclonic gyre in the southern parts (based on Wang G. et al. 2006 and Liu K. et al. 2002)

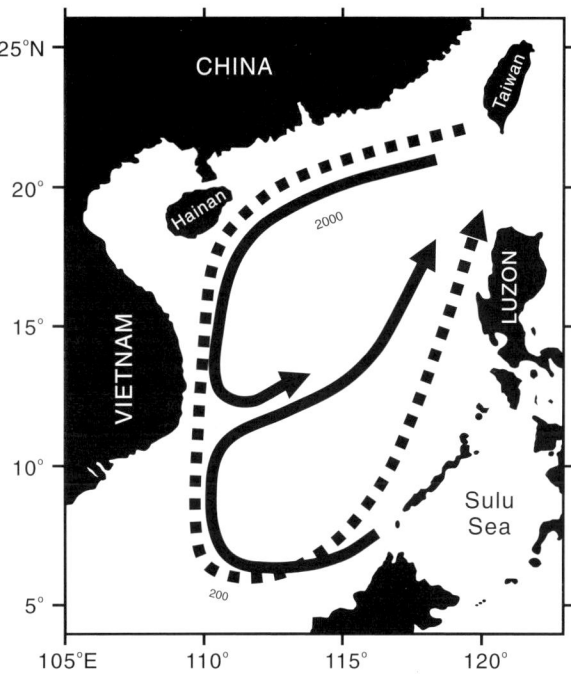

Precipitation exceeds evaporation in the SCS, especially in the southern part. The surface salinity varies between 32.8 and 34.2‰ and thus preconditions the SCS for an "estuarine-basin" circulation. In winter, salinity is higher in the north due to the Kuroshio influence, with a high-salinity tongue in the top 100 m north of 12 °N propagating to the southwest. The winter salinity decreases southward because of increasing precipitation, resulting in isohalines running parallel to latitudes (Fig. 2.9C). In summer, salinity varies less in the northern SCS but very large in the southern part, with higher values off the Vietnam coast due to summer upwelling and a low-salinity tongue outside the Mekong estuary (Fig. 2.9D). Freshwater input is a significant factor for the surface salinity distribution. For example, the Pearl River plume has a distinct monsoonal behavior and influences strongly the hydrographic characteristics nearby the river mouth. Waters near the Pearl River mouth is highly stratified especially during the wet season, with a strong halocline inside the estuary and a plume formed in the surface layer just outside the estuary (Fig. 2.10). Soon after forming, the plume turns toward the west during the dry season, but turns to the east in the wet season due to the southwesterly monsoon (Dong et al. 2004). Once the surface plume reaches beyond the inner shelf, it can easily extend further east across the shelf, possibly driven by the SCS Warm Current and the favorable southwesterly summer monsoon winds (Fig. 2.10E). The bottom current, however, continues to the west outside the estuary along the coast like a density current because of the diminishing effect of the winds below the stratified layer (Su 2004). Therefore, a proper understanding of the plume behavior outside

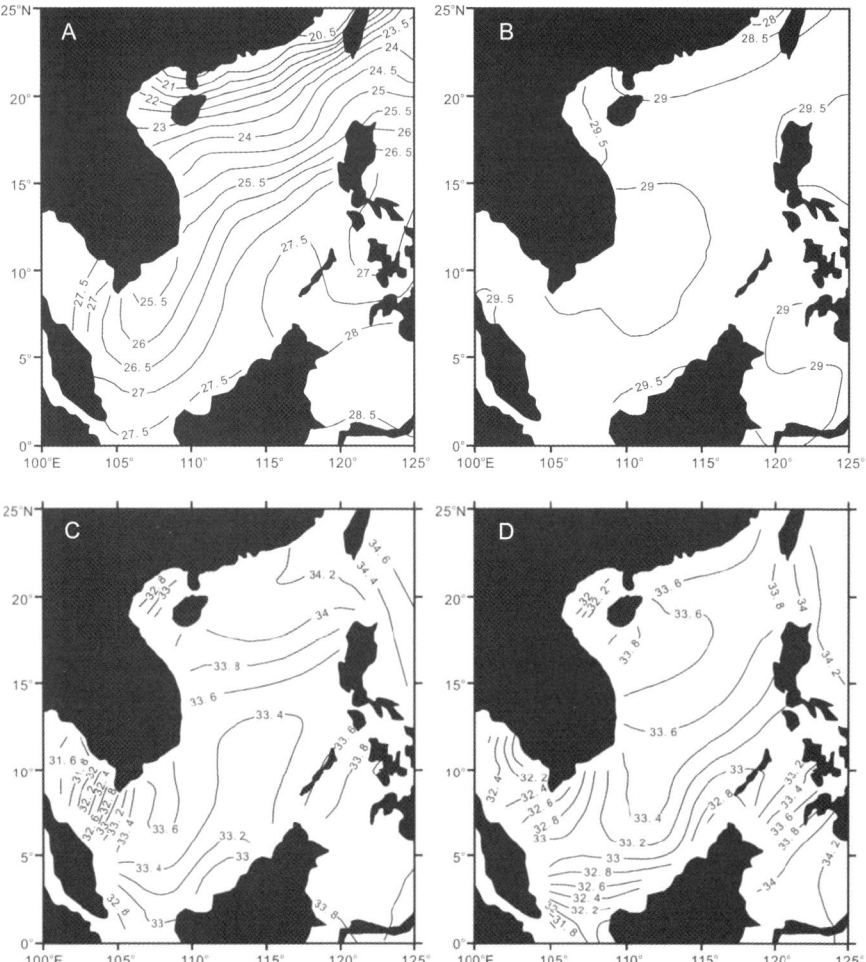

Fig. 2.9 Sea surface temperature (°C) and salinity (‰) in the modern SCS show distinct season-ality patterns: (**A**) SST in winter; (**B**) SST in summer; (**C**) SSS in winter; (**D**) SSS in summer (Wang D. et al. 2002)

the estuary is crucial for interpretation of paleo-data, as many cores off the Pearl River mouth are being used for monsoon reconstruction.

Similar to the wind and current patterns, SST data also clearly show a cross basin, thermal front in winter (roughly along the 25.5 isotherm in Fig. 2.9A), albeit it often disappears in summer (Fig. 2.9B). This upper layer thermohaline front runs across the SCS basin from the South Vietnam coast to Luzon Island. The strength of the front is around 1 °C/100 km at the surface and 1.4 °C/100 km at the subsurface (−50 m) (Chu and Wang 2003). Accordingly, water masses north and south of this front are different, with high temperature, low salinity water in the south and low

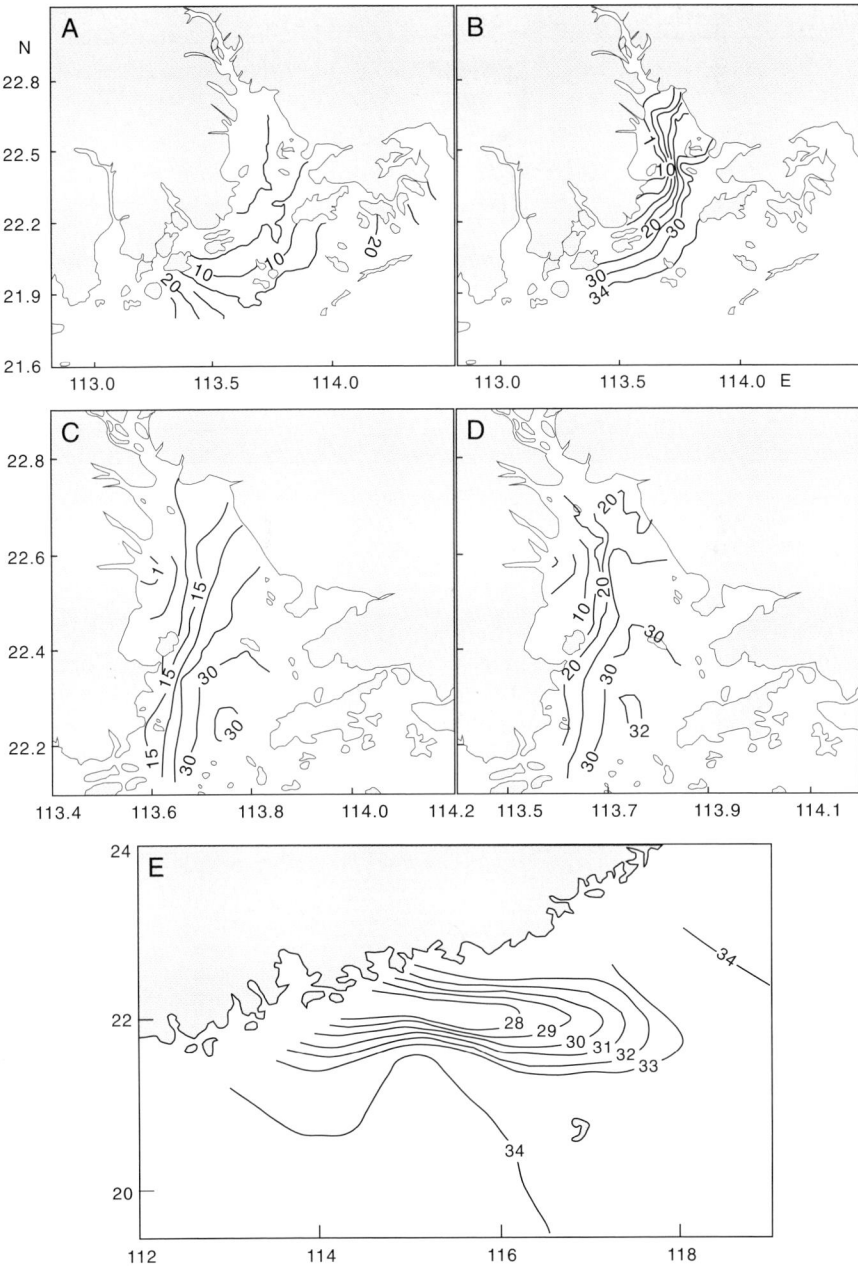

Fig. 2.10 Salinity distribution in the Pearl River estuary is shown by ‰ contours for July 1999 surface (**A**) and bottom (**B**) readings, for January 2000 surface (**C**) and bottom (**D**) readings (Dong et al. 2004), and (E) for June/July 1979 from sea surface outside the estuary (Su 2004)

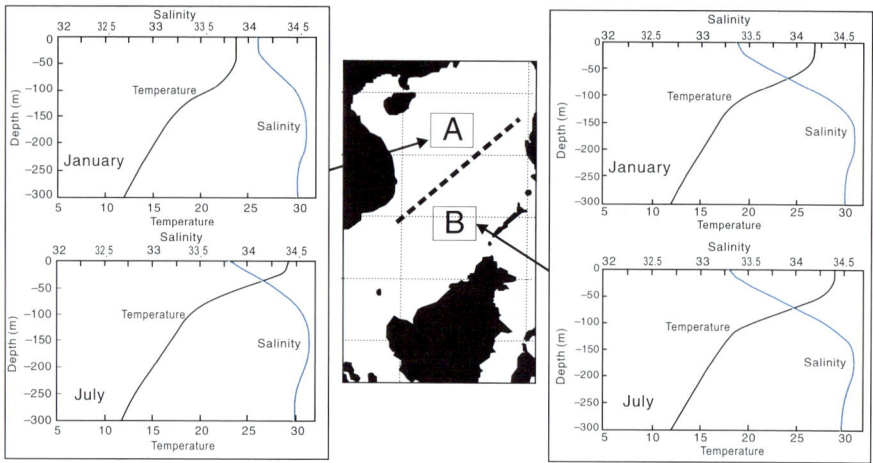

Fig. 2.11 January and July temperature and salinity profiles are compared across the thermohaline front (heavy dash line) between the northern (A) and southern (B) SCS (modified from Chu and Wang 2003)

temperature, high salinity water in the north, as seen clearly from a comparison of the upper layer (0–300 m deep) *T-S* diagrams between the northern and southern SCS shown in Fig. 2.11. Both the temperature and salinity curves show a larger range south of the front than to the north, and the salinity minimum in the north of the front occurs at a shallower depth than in the south. Thus, the thermohaline variability to the south is larger than to the north of the cross-basin current (Fig. 2.11).

Thermocline and Upwelling

In the SCS, a seasonal thermocline exists throughout the year in the deep basin. The thermocline layer is thinnest (< 100 m) in winter in most of the basin, thickest (~ 150 m) in spring, and transitional in summer and autumn (100–120 m). The average thermocline depth is about 100 m in winter, ~ 75 m in spring, 75–85 m in summer and autumn with a west-east difference of over 50 m (Fig. 2.12). The thermocline slope in the deep sea in winter is opposite to that in summer due to monsoon influence. The upper water piles up in the northwestern part of the SCS in winter and in the southeastern part in summer (Shaw et al. 1996; Liu Q. et al. 2000).

Similarly, the layer above the thermocline, or the mixed layer, varies in depth and in temperature under monsoon influence. In winter, it is thickest (~ 80 m) in the NW part, becoming thinner in the south (~ 40 m; Fig. 2.13A). In spring, the mixed layer decreased to about 30 m because of the weakened wind flow

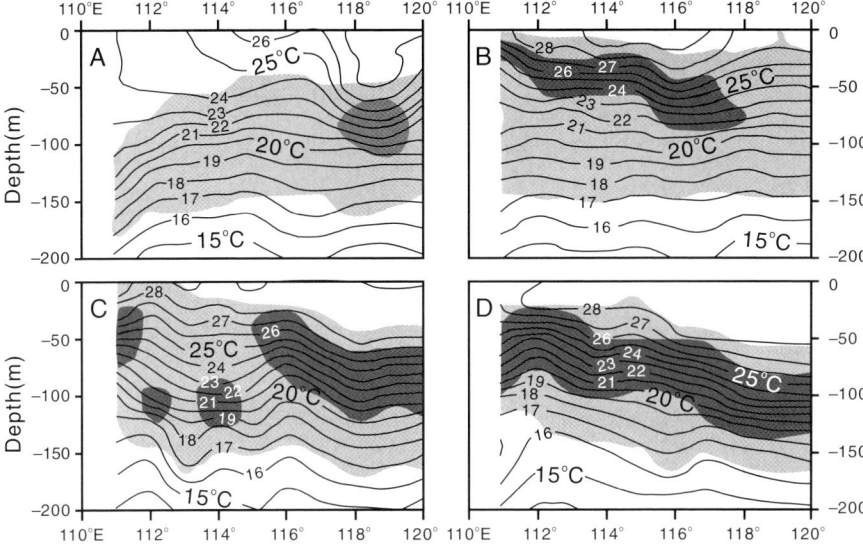

Fig. 2.12 Temperature-depth profiles across the SCS from NW (18 °N, 110 °E) to SE (12 °N, 120 °E) are shown for (**A**) winter, (**B**) spring, (**C**) summer, and (**D**) autumn. Areas with vertical gradient >0.05 °C/m are lightly shaded, and those >0.1 °C/m heavily shaded (Liu Q. et al. 2000)

(Fig. 2.13B). In summer, the mixed layer pattern is opposite to that in winter, becoming thicker in the south (~50 m) and thinner in the northwest (~30 m; Fig. 2.13C). The mixed layer in autumn is relatively thick over the entire SCS basin (~50 m), becoming slightly thinner only off the Vietnam coast (~40 m; Fig. 2.13D) (Shi et al. 2001).

The southern coast of Vietnam is oriented in a SW-NE direction, roughly in parallel with the prevailing southwesterly winds in summer. Due to the Ekman principle, the alongshore wind can easily pump cold waters from beneath the mixed layer to the surface at such a setting, causing the maximum offshore cooling to concur approximately with the summer wind speed maximum along the coast (Xie et al. 2003). A similar phenomenon occurs in winter in the northeast SCS off Luzon, where coastal upwelling is driven by northeasterly monsoon winds. Therefore, two primary regions of deep upwelling exist corresponding to the two eddies in Fig. 2.7A: west of Luzon and east of Vietnam. The winter Luzon cold eddy and the summer Vietnam cold eddy can be easily recognized from the upper layer temperature distribution (Yang and Liu 1998). From October to December, shelf break upwelling along the edge of Sunda Shelf may also occur (Chao et al. 1996).

The development of monsoon-induced upwelling can be better demonstrated using satellite data. It was found, for example, that the summer southwesterly winds impinge on mountain range on the Vietnam eastern coast, and a strong wind jet occurs at the southern tip offshore of Vietnam, resulting in strong wind curls that are

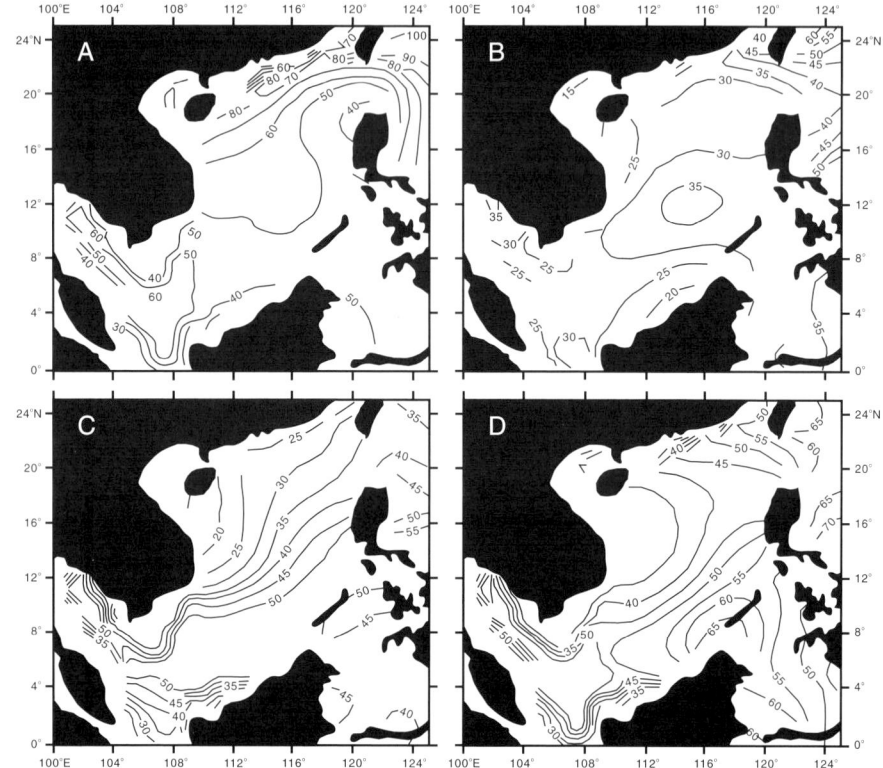

Fig. 2.13 Maps show the mixed layer depth in the SCS in (**A**) January, (**B**) April, (**C**) July, and (**D**) October (Shi et al. 2001)

responsible for upwelling off the coast. In July and August an anticyclonic ocean eddy develops to the southeast, advecting the cold coastal water offshore into the open SCS (Fig. 2.14A). Ocean color observations also revealed a collocated tongue of high chlorophyll concentration in the area (Fig. 2.14B). However, cold water extending to the east in this area is subject to interannual variations. In 1998, when monsoon winds weakened during the El Niño year, this cold water tongue and mid-summer cooling never took place (Xie et al. 2003). Aside from monsoons, tropical cyclones are frequent in the SCS, averaging to 10 typhoons per annum. Tropical cyclones traverse the SCS, also causing upwelling of cold subsurface water into the mixed layer (Wong et al. 2007a).

Water Exchange with Pacific and Kuroshio Intrusion

Only the surface waters in the SCS exchange freely with those in the neighboring seas, while deeper waters flowing into the SCS are primarily from the western

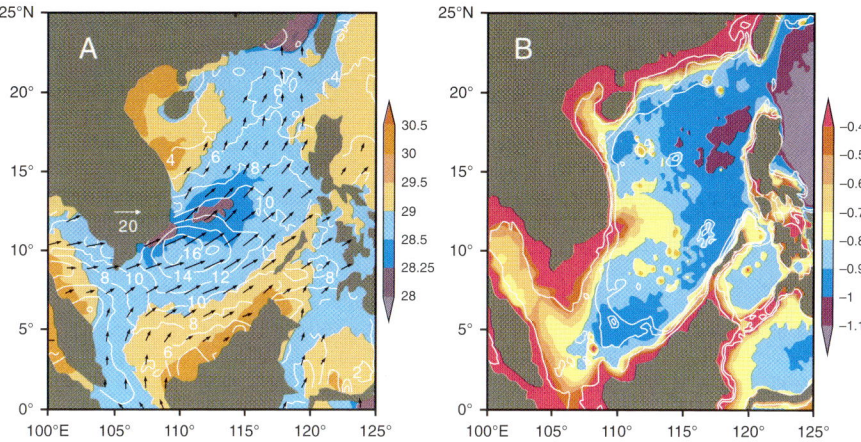

Fig. 2.14 Summer upwelling off Vietnam coast is shown with (**A**) AVHRR August sea surface temperature (color shade in °C) and QuikSCAT surface wind vectors and magnitude (contours in 10^{-2}Nm^{-2}), and (**B**) log^{10} of SeaWiFS chlorophyll concentration averaged for July–August 1999–2002 (color shade in mg/m^3), with bottom isobaths for 50 m, 100 m and 500 m, respectively (Xie et al. 2003)

Philippine Sea through the Bashi Strait. The Tropical Water, originating from the high-salinity pool in the subtropical North Pacific, occurs as a salinity maximum at around 150 m in the SCS. The North Pacific Intermediate Water, flowing from the subpolar North Pacific, occurs as a salinity minimum centered around 500 m. As the Bashi Strait with its sill depth of 2,400 m is the only deep water connection with the Pacific, deep water in the western Philippine Sea overflows the sill and fills the deep SCS (Table 2.2). The water transport through the Bashi Strait influences the circulation and heat budget of the SCS, and determines the nature of its deep water. The annual water budget of the SCS is summarized in Table 2.3.

Wyrtki (1961) was the first to note that waters enter the SCS in winter and flow back to the Pacific in summer. It is now generally accepted that the Bashi Strait transport has a sandwiched vertical structure, with the Philippine Sea water entering the SCS at the surface and in the deeper parts, and with the net Bashi Strait transport out of the SCS at intermediate depths. The net transport of the western Pacific water into the SCS is estimated at the order of 4 Sv by Qu et al. (2000) or in a range of

Table 2.2 Properties of water masses at different depth intervals at the SEATS station (18.3 °N and 115.5 °E), northern SCS (data from Wong et al. 2007b)

	Tropical water	North Pacific Intermediate Water (upper)	Deep water		
Depth (m)	100–150	470–500	1,500–1,600	2,400–2,700	2,800–3,000
Potential T (°C)	16.2–22.1	7.5–10.2	2.60–2.91	2.16–2.20	2.11–2.14
Salinity (‰)	34.56–34.72	34.40–34.42	34.57–34.61	34.59–34.62	34.60–34.63

Table 2.3 Characteristics of annual water budget of the SCS show variations in input and output volumes (modified from Wong et al. 2007a)

Input	Flux (km^3/year)	Output	Flux (km^3/year)
Surface inflow from western Philippine Sea	2.7×10^{-1}	Surface outflow to western Philippine Sea	-2.4×10^{-1}
Deep water inflow from western Philippine Sea	3.7×10^{-2}	Net outflow at upper intermediate depths to W Philippine Sea	-5.8×10^{-2}
Surface inflow from Sulu Sea	1.8×10^{-2}	Net surface outflow to Java Sea	-1.8×10^{-2}
Riverine input	1.7×10^{-3}	Surface outflow to East China Sea	-1.1×10^{-2}
Net precipitation	2.5×10^{-3}		

4.2–5.0 Sv by Su (2004). In any case, the SCS plays the role of a "mixing mill" that mixes surface and deep waters from the western Pacific and returns them through the Bashi Strait at upper intermediate depths (Yuan 2002).

Water transport through the Bashi Strait can be illustrated with results from a field experiment in October, 2005. The distribution of the subinertial flow not only displays different directions between the upper, middle and lower layers, but also a north-south contrast (Fig. 2.15). The results from this experiment also show that the net volume of westward transport is 9 Sv in the upper layer (< 500 m) and 2 Sv in the deep layer ($> 1,500$ m), and the net volume of eastward transport in the intermediate layer (500–1,500 m) is 5 Sv (Tian J.W. et al. 2006).

A more controversial issue is Kuroshio intrusion. As the subtropical west boundary current in the North Pacific, the Kuroshio forms in the east of the Philippines around 13 °N and flows north along the coast of Luzon and, after making a slight excursion into the Bashi Strait, continues northward east of Taiwan. A variety of models have been proposed about how the Kuroshio water invades the SCS. Earlier studies suggested that the Kuroshio could loop around inside the SCS or even intrude as a direct current, but the suggested flow patterns have not been confirmed by any synoptic surveys (Su 2004). However, most numerical models with open boundaries predicted a loop current and a branch current, both from the Kuroshio, into the SCS (e.g., Yuan 2002; Metzger and Hurlburt 1996, 2001). Some authors go even further to believe that impingement of the intruded Kuroshio with the continental margin in the SCS may bring about a SCS Warm Current, which flows northeastward in waters down to 200–400 m along the shelf break from southeast of Hainan Island to the southern end of Taiwan (Guan 1985; Hsueh and Zhong 2004).

The discrepancy between the modeled and observed flow patterns may have been caused by insufficient resolution of topographic features from the Bashi Strait for modeling and by inability of the models to properly represent the physics of Kuroshio intrusion. As Su (2004) argues, a steady branch current from the Kuroshio to the west through the Bashi Strait is physically impossible, and the only possibility for a part of the Kuroshio water to flow through the Strait is via

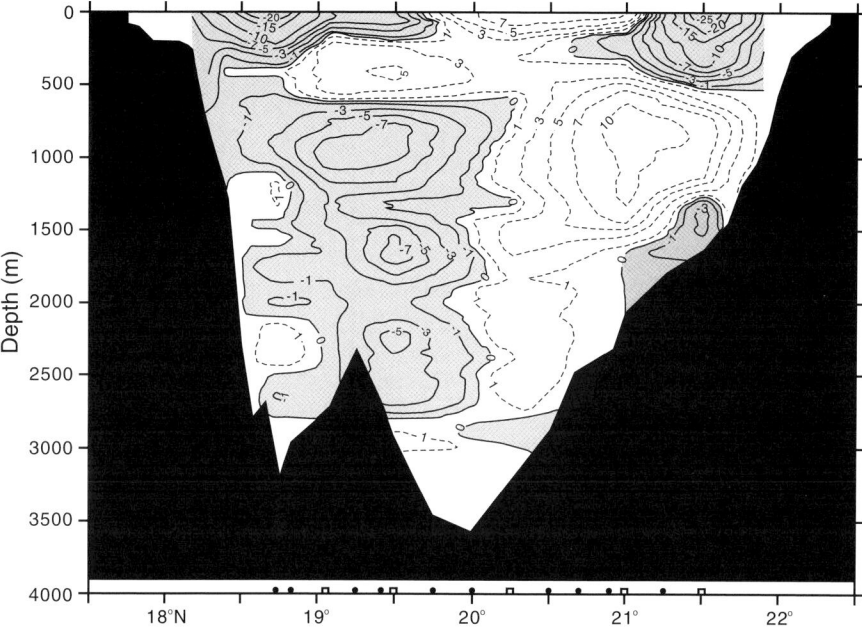

Fig. 2.15 A cross section shows subintertial flow velocity (cm/s) across the Bashi Strait. Dashed lines and positive values denote the eastward flow, solid lines and negative values denote the westward flow (also *shaded gray*). The topography is shaded black (Tian J.W. et al. 2006)

submeso-scale processes such as eddies. Once generated in the Bashi Strait, these submeso-scale eddies will propagate to the west, resulting in a net flow to the SCS. Such an influx is likely from the Kuroshio frontal water rather than its proper water, and the frontal water has hydrographic characteristics close to the SCS water (Su 2004).

Regardless of all the controversy and complexity, the Kuroshio water does enter the SCS one way or another and exchanges with the SCS water through the Bashi Strait. Cumulatively, there is a small (~1 Sv) net outflow of surface water (0–350 m depth) from the SCS in the wet season, but a net inflow (~3 Sv) in the dry season through the Bashi Strait. The differences are mainly made up by inflow and outflow of Sunda Shelf Water in the wet and dry seasons, respectively. Below 350 m, the SCS water flows out, but the West Philippine water again flows into the SCS below 2,500 m, making the deep SCS waters homogenous with the same property as the Philippine Sea (Chen and Huang 1996). In general, a mass balance of inflowing deep waters from the western Philippine Sea is maintained primarily by upwelling and mixing with shallower waters in the SCS and an outflow at intermediate depths to the western Philippine Sea through the Bashi Strait. A rigorous exchange of water between the SCS and the western Philippine Sea has resulted in a residence time of the deep water in the SCS as short as 30–150 years, and the corresponding basin-wide upwelling rate should be as high as about 10–90 m/yr. Such a high upwelling

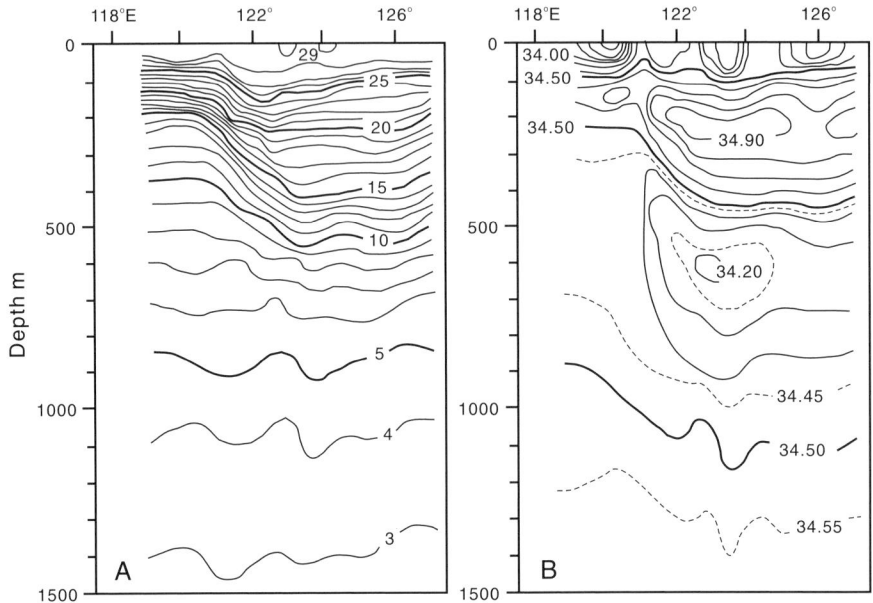

Fig. 2.16 Profiles show variations along 19.5 °N in (**A**) temperature (°C) and (**B**) salinity (‰) across the Bashi Strait (at ~122 °E) in summer 1965 (from Su 2004 after Nitani 1972)

rate gives rise to a thin mixed layer and a shallow top of the nutricline in the SCS (Wong et al. 2007a).

All these result in significant difference in the vertical structure of waters between the two sides of the Bashi Strait. Fig. 2.16 shows the temperature and salinity distribution in the water column across the Strait. In the Philippine Sea, the salinity maximum layer is related to the North Pacific Tropical Water (NPTW), and the salinity minimum to the North Pacific Intermediate Water (NPIW). After crossing the Strait, both salinity maximum and minimum weaken significantly (Fig. 2.16B). Furthermore, the temperature and salinity contours above 600 m shoal up by about 50–200 m to the west, with a continuous shoaling of the salinity contours to about 1,000 m. Thus, at the same depth in the upper 600 m of the ocean, the SCS water is cooler than the western Philippine Sea water. The thickness of the mixed layer in the SCS is also only about half that of the western Philippine Sea. Such apparent "upwelling" hydrographic features in the upper 600 m of water are found over the entire deep basin of the SCS. Below about 600 m, the SCS water becomes slightly warmer than the Philippine Sea water (Su 2004).

The water columns from the two sides of the Bashi Strait differ also in biogeochemistry. The subsurface water of the SCS west of Luzon is more enriched in nutrients and depleted in oxygen relative to the water east of the Strait, but the situation reverses in the intermediate water (Gong et al. 1992). Comparison between dissolved oxygen concentrations at different depth intervals in the Pacific side of the Strait clearly reveals this feature. Specifically, the oxygen content is higher

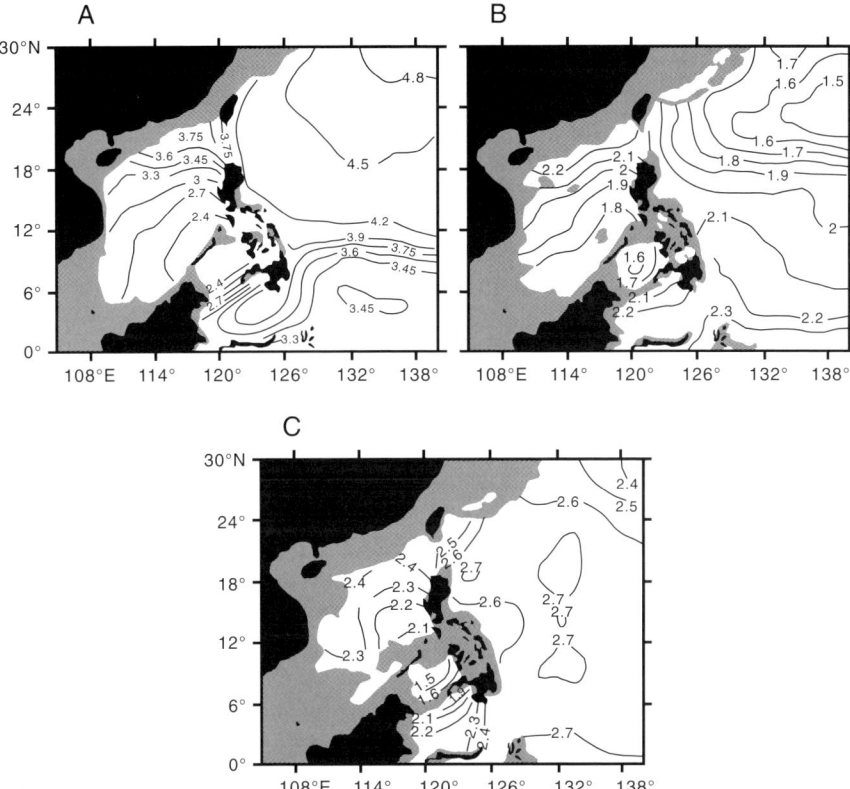

Fig. 2.17 Maps show variations of dissolved oxygen concentration (ml/l) at (**A**) 140 m, (**B**) 1,000 m and (**C**) 2,000 m water depths in the western Pacific on both sides of the Bashi Strait. Note a relative high level in intermediate waters of the SCS compared to other marginal sea basins (Qu 2002)

above 600–700 m, lower between 700 and 1,500 m, then higher again below 1,500 m (Fig. 2.17) (Qu 2002).

Deep Water Circulation

Deep waters in the western Pacific marginal seas, such as the SCS and Sea of Japan, are uniform in properties but different in their origins. In the Sea of Japan, the dense deep water is formed in its northern half under low temperature, whereas in the SCS all waters below 2,000 m are of Pacific origin and have the same hydrographic properties as western Pacific water at about 2,000 m (Su 2004). The incoming Pacific water crosses the Bashi Strait, sinks, spreads out, and fills the SCS deep basin. As a result, the potential temperature in the SCS is above 2.1 °C even at 4,000 m, almost 0.8 °C warmer than in the western Pacific at the same depth (Chen C. et al. 2001).

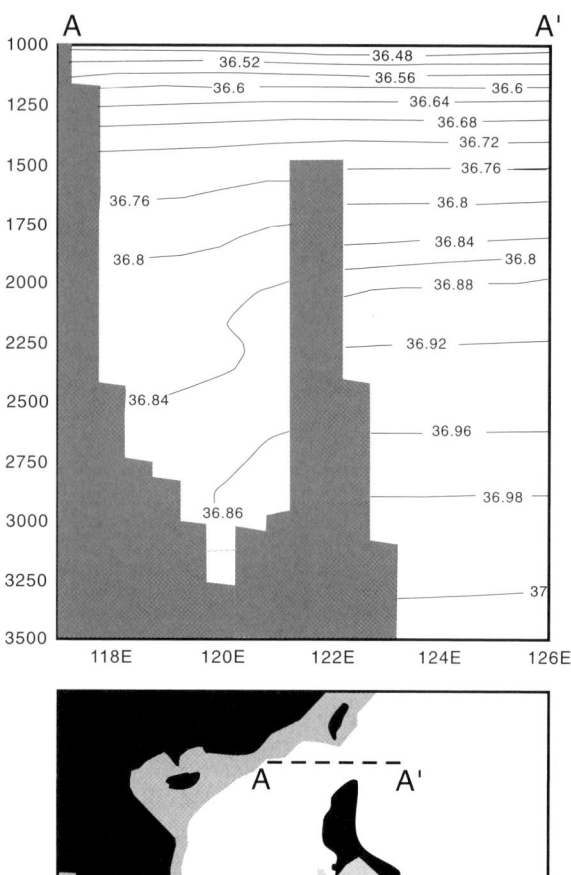

Fig. 2.18 Profile of potential density (kg/m³) is shown along 20 °N, across the Bashi Strait. Bottom topography (*shaded*) is on the 0.5° grid and does not represent the actual water depth range (Qu et al. 2006)

Across the Bashi Strait, a persistent density difference exists between the Pacific and the SCS (Fig. 2.18). Water on the Pacific side is well stratified, with potential density increasing from about 36.48 kg/m³ at 1,000 m to 37.00 kg/m³ at 3,500 m. No deepwater stratification is obvious on the SCS side, where water density is vertically uniform with a density range of only about 0.02 kg/m³ below 2,000 m (Fig. 2.18) (Qu et al. 2006).

Up to now, there is virtually no direct observation indicative of any distribution route of the Pacific water into the deep SCS. However, the density field based on the synthetic salinity data (Fig. 2.19A–C) may throw some light on the deep SCS circulation. Upon entering the SCS through the Bashi Strait, waters of Pacific origin first turn northwestward and then southwestward along the continental margins off southeast China and east Vietnam. This suggests that the deep-layer circulation in the SCS is predominantly cyclonic. The intrusion of deep Pacific water is also evident in oxygen distribution used as a passive tracer (Fig. 2.19D). As it spreads over the SCS, water from the Pacific gradually gets mixed with ambient

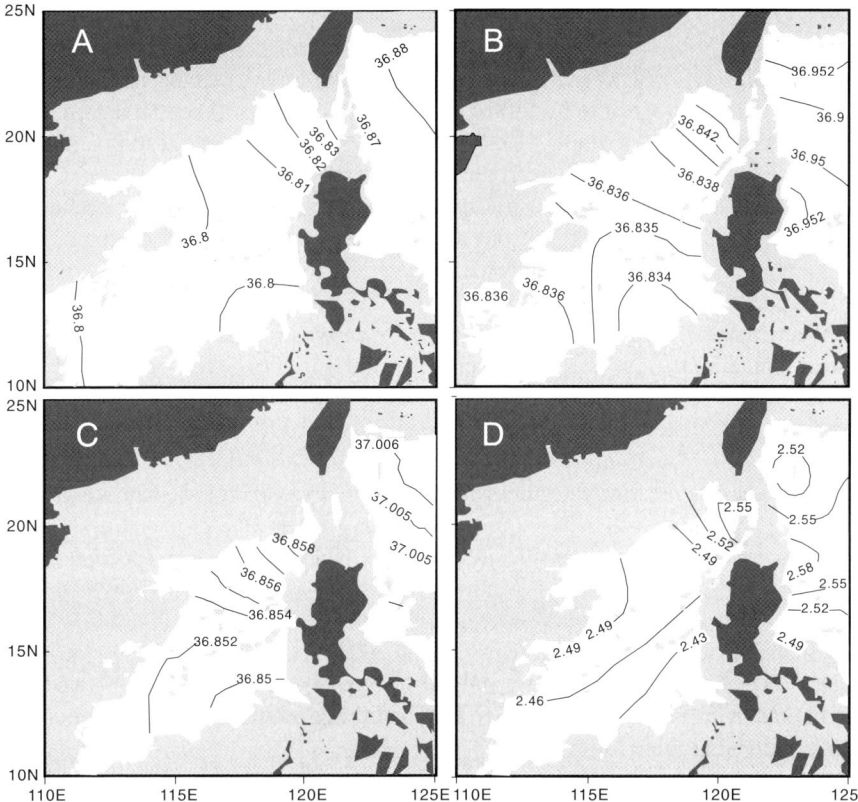

Fig. 2.19 Potential density in kg/m^3 is calculated from the synthetic salinity depths at (**A**) 2,000 m, (**B**) 2,500 m and (**C**) 3,500 m, while (**D**) shows dissolved oxygen concentration (mL/L) along 36.84 density surface, lying at depths roughly between 2,000 and 3,000 m (Qu et al. 2006)

waters, loses its Pacific characteristics, and returns in less dense layers. Therefore, density distributions within the sea basin suggest a cyclonic deep boundary current system, as might be expected for an overflow-driven abyssal circulation (Qu et al. 2006).

Thus, the abyssal basin of the SCS is filled constantly by the Pacific water flowing down the sill of the Bashi Strait, and the deep SCS water is believed to upwell into the intermediate SCS water between 350 and 1,350 m, which is then exported out of the SCS again through the Bashi Strait (Su 2004). The deep SCS water was estimated to have a fast flushing time of 40–50 years (Chen C. et al. 2001) or even less than 30 years (Qu et al. 2006). Therefore, the intermediate, deep and bottom waters are essentially the same age. Because of a short residence time, the amount of particulate matter decomposition in the water column is sufficient to produce only a small maximum of chemical properties in the vertical profiles (Chen C. et al. 2001).

Other Oceanographic Features

Recent studies in the SCS have discovered a number of new oceanographic features which are not yet attended to by geologists but certainly are of potential significance for studies of paleoceanography and sedimentology. Two of these features, namely the inner waves and meso-scale eddies, are very briefly discussed here.

The Bashi Strait is important not only because of its role in water exchanges between the SCS and the Pacific but also as the source region of internal waves observed in the northern SCS. Internal waves can be recognized using synthetic aperture radar (SAR) satellite images. Most SAR images of the northern SCS show nonlinear internal waves (Hsu et al. 2000), from west of Bashi Strait to east of Hainan Island (Liu A. et al. 1998). Generally, the tidal flow over topographic features such as a sill or continental shelf in a stratified ocean can produce nonlinear internal waves. Recent observations in the northeastern SCS show that non-linear internal waves emanate nearly daily from the Bashi Strait during spring tide and rapidly propagate westward to the shallow region west of 118 °E. The amplitude of these waves reach 140 m or more, forming the largest free propagating nonlinear internal waves observed in the internal basin. Their propagation speed at about 2.9 m/s is also faster than similar waves previously observed in the world's oceans. The most likely source region of these giant internal waves is the near 2,000-m-deep ridge in the northern Bashi Strait (Liu C. et al. 2006). These internal waves energize the top 1,500 m of the water column and move large amounts of nutrient up, and their role in enhancing surface productivity and disturbing sedimentation may provide a highly interesting topic for research.

Meso-scale variability has been reported since the 1990s. The upper layer circulation of the SCS is characterized by basin-wide gyres, but embedded in the gyres are many meso-scale eddies, as observed from both hydrographic and altimeter measurements. Altimeter data over the SCS for 5 years from 1992 to 1997 reveal significant meso-scale variability mainly in two narrow strips north of 10 °N: one near the 2,000 m isobath over the NE slope and another as a SW-NE strip of about 450 km wide extending from Luzon to Vietnam (Wang L. et al. 2000). An 8-year time series of satellites data (1993–2000) shows that meso-scale eddies are a constant feature in the SCS, basically in all areas deeper than 2,000 m. Altogether 58 anticyclonic eddies and 28 cyclonic eddies were identified for this period in the SCS, and a typical eddy has a lifetime of about 130 days and a diameter of 300 km. Likely mechanisms causing these eddies include Kuroshio intrusion in SW of Taiwan, vorticity advection from the Kuroshio, Kuroshio-Island interaction, the eastward baroclinic jet off Vietnam and the intense wind-stress curl northwest off Luzon (Wang G. et al. 2003).

Oceanographic Summary

To sum up, the modern oceanography of the SCS is distinguished by the following major features which are highly relevant to interpreting geological records:

1. *Seasonality*. Despite its low-latitude position, the SCS shows strong seasonal variations in water circulation and composition. Driven by East Asian monsoons, the surface circulation is characterized by seasonally alternating basin-scale gyres. Water exchanges with the Pacific and CO_2 exchange with the atmosphere (Chapter 7) are also subject to remarkable seasonal variations. Monsoon winds drive the thermocline to tilt up from NW to SE in winter but tilt to opposite direction in summer. Two coastal upwelling areas are associated with monsoons: west of Luzon in winter and east of Vietnam in summer. Unlike the Indian monsoon, the local winter monsoon exceeds the summer monsoon in strength and in time duration on a yearly basis. Therefore, it is unwise for SCS studies to indiscriminately copy monsoon proxies or climate cyclicity from other regions.

2. *North-South contrast*. The SCS basin is oceanographically dividable into two halves by the monsoon wind jet and the cross-basin current axis in a NE-SW direction, which is associated also with the upper layer thermohaline front. The northern part is controlled by the monsoon-driven cyclonic gyre, while the southern part belongs in the Western Pacific Warm Pool with much less monsoon influence. Accordingly, the northern SCS is distinguished from its southern counterpart by higher salinity and lower temperature. In the past, this N-S contrast became intensified with the growth of the boreal ice sheet and the strengthening of the winter monsoon. Topographical and geological contrasts between the landward side to the northwest and the island arc side to the southeast also enhance many sedimentological and paleogeographic differences between the southern and northern SCS.

3. *Basin-wide upwelling*. The modern SCS exchanges with the open western Pacific, i.e., the western Philippine Sea, through the Bashi Strait with a sill depth of $\sim2,400$ m. Originated from the western Philippine Sea, deep waters in the SCS are relatively uniform below 2,000 m. The deep water inflow results in an estuarine basin-wide upwelling, a shorter residence time, and a shallower thermocline in the SCS than in the open ocean, making the upper water structure and productivity in the SCS more sensitive to monsoon variations.

2.3 Tectonic History and Sedimentary Basins

The tectonic history of the South China Sea (SCS) was genetically related to the deformation of Asia. After India-Asia collision in the Eocene, Asia significantly enlarged its size and increased its altitude. The west-tilting topography of East Asia was reversed with uplift of the Tibetan Plateau and opening of marginal seas, including the SCS (Wang P. 2004). Separated by continental fragments, these marginal sea basins formed as a result of long-lasting extension along the southern margin of mainland Asia (Taylor and Hayes 1980; Holloway 1982) (Fig. 2.20).

Tectonically, the SCS is bounded by the South China (Yangzi) Block in the north, by the Indochina Block in the west, by the Philippine Sea Plate in the east, and in

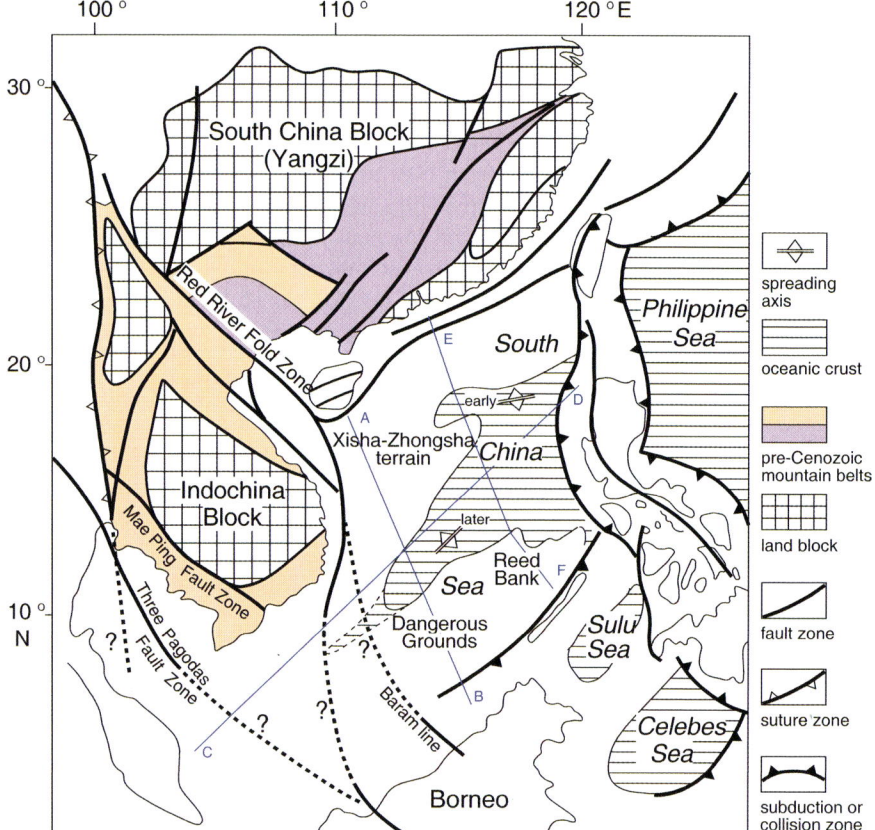

Fig. 2.20 Map shows land blocks and major fault systems surrounding the SCS (modified from Gong and Li 1997; Hall 2002; Pubellier and Chan 2006). Lines A-B, C-D and E-F are locations for sections shown in Fig. 2.22

the south by Borneo Island and the Indonesian Archipelago (Fig. 2.20). This unique setting not only directly controls the general development of the SCS and its basins but also makes it unique in having its own spreading axis among the western Pacific marginal seas.

Up to now, however, no consensus has been reached on the mechanism of the SCS formation (Fig. 2.21). Hypotheses currently gaining relative wider acceptance include (1) India-Eurasia continental collision and the collision-resultant tectonic extrusion process mainly along the Red River Fault Zone (RRFZ), or Ailao Shan-Red River (ASRR) shear zone (Taylor and Hayes 1980, 1983; Tapponnier et al. 1982, 1986; Briais et al. 1993; Lee and Lawver 1994; Zhou et al. 1995), (2) the extensional force from subduction processes of the Pacific Plate along the western Pacific margin (Hawkins et al. 1990; Stern et al. 1990; Hall 2002) (Fig. 2.21), and (3) the extensional force from an upwelling mantle plume (e.g., Zhu and Wang 1989; Fan and Menzies 1992) or movement of the upper asthenosphere

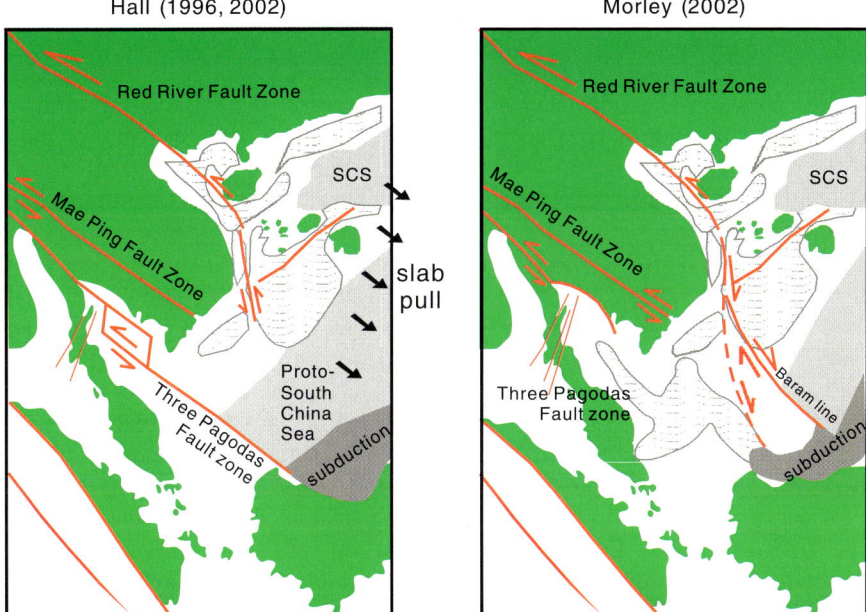

Fig. 2.21 Models show the important role of the Red River Fault Zone and other fault systems for the opening of the SCS (modified from Morley 2002 and Hall 2002). Also shown is the outline of major sedimentary basins and the Proto-South China Sea

(Yao 1996; Yao et al. 2005). It may not be ruled out, however, that a combined process of the above probable driving forces has been in operation especially when the regional tectonic regime changed dramatically after collision between Indo-Australian and Eurasian and Pacific plates became intensified since the Eocene (Hall 2002; Pubellier et al. 2004), as implied also by modeling results (Xia et al. 2006). In this section, we provide a brief account on the tectonic history of the South China Sea and its sedimentary basins.

Prior Terrains and Opening of the SCS

Deep seismic surveys indicate a continental crust thinning from 30 to 35 km along the northern continental margin to ~14 km under the lower slope of the SCS, while the depth of Moho under the deep central basin is generally < 12 km (Nissen et al. 1995; Gong and Li 1997; Yan et al. 2001; Hutchison 2004) (Fig. 2.22). As a failed rift structure, for example, the Xisha Trough between Hainan and Xisha (Paracel) islands has a very thin continental crust of only 8 km with little magmatic underplate in the lower crust (Qiu et al. 2001). A relative thin crust of 8–12 km exists also in the northeast end, between Taiwan and Dongsha Rise (Yeh and Hsu 2004). Hayes and Nissen (2005) modeled the crust thickness and found that crustal extension was much less in the east than in the west where the thermal structure of pre-rift lithosphere was by a factor of two prone to rifting.

Pre-Cenozoic rock types known to occur in the SCS include Jurassic to Cretaceous marine shale and sandstone and Cretaceous or older igneous granite, as well as metamorphic rocks. Overall, igneous rocks may constitute 50% or more of the prior terrains, some of which remain as structural highs in the subsequent basin formation. Around the central deep basin, numerous coral reefs developed on terrain systems including the Nansha Terrain (Reed Bank and Dangerous Grounds) and the Xisha-Zhongsha Terrain (Paracel Island and Macclesfield Bank). Now lying in the south, the Dangerous Grounds, Reed Bank, Nansha (NW Borneo) Trough and Palawan Island were originally terrains of continental origin. They were rifted southward in the late Cretaceous-early Paleogene and later, during the course of seafloor spreading, their edge collided with the northernmost terrain of the proto-Sulu Sea (Schlüter et al. 1996). In the south, apart from granite and dorite, volcanic and shallow marine sediments are also common, indicating a more complex pre-rift lithotopography (ASCOPE 1981; Du Bois 1985).

Wider occurrence of Jurassic and Cretaceous marine shale and sandstone indicates a broad shallow "Proto-South China Sea" during the late Mesozoic to early Cenozoic probably with some deep troughs connecting to the eastern Indian Ocean (e.g., Hall 2002) (Fig. 2.21). This "Proto-SCS" has since been largely subducted into, or uplifted as part of, the southeastern island arcs. The total Mesozoic sequence may reach a maximum thickness of about 2,000–4,000 m (Jin 1989; Xia and Zhou 1993), often with high magnetization on uplifted highs and low magnetization in basins (Li C. et al. 2008).

Initial rifting and graben-forming in the SCS region coincided with the early stage of Indian/Eurasian collision during the Paleocene-early Eocene, which became intensified from about 50 Ma (Harrison et al. 1992; Copeland 1997) (Fig. 2.23). Recent investigations of seismic and provenance data from the Indus Fan region indicate an earlier intensification period from about 55 Ma (Clift et al. 2001). However, it was not until the latest Eocene-earliest Oligocene when the collision-subduction-uplift process became powerful enough to substantiate any significant continental escape (Tapponnier et al. 2001) that ultimately contributed to rifting and spreading of the South China Sea. A total of ~550 km displacement of the Indochina subcontinent due to intensive strike-slip activities along the Red River Fault (Tapponnier et al. 1986; Briais et al. 1993; Leloup et al. 2001) during the Oligocene-Miocene may partly account for seafloor spreading in the SCS. In situ Th-Pb ion microprobe dating of monazite inclusions in garnets from the Red River Fault indicates that synkinematic garnet growth associated with left-lateral shearing between 34 Ma and 21 Ma in the northern region and sinistral deformation along the fault zone between 34 Ma and 17 Ma, confirming the close relationship between Indochina extrusion and SCS spreading (Gilley et al. 2003). Other forces involving in the opening of the SCS may include slab pull due to subduction of the Proto-SCS into the Philippine Sea Plate (Hall 2002; Hall and Morley 2004) (Figs. 2.21 and 2.23), similar to other SE Asian marginal seas due to subduction of the Indian-Australian Plate (Pubellier et al. 2003, 2004). The mechanism causing extensional

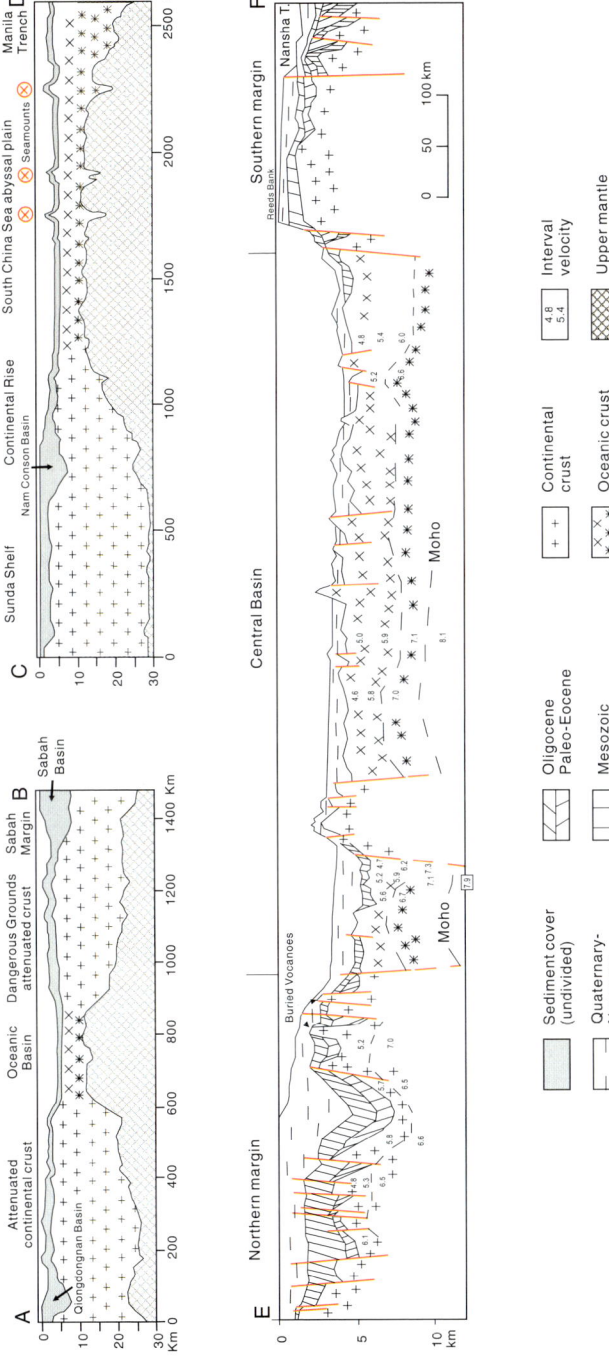

Fig. 2.22 Generalized crustal cross sections along lines A–B, C–D, and E–F (locations in Fig. 2.20) show variations in crust thickness and sediment cover in the SCS. Sections A-B and C-D are modified from Hutchison (2004), and section E-F from Yan et al. (2001). Note that scales are different

Fig. 2.23 Reconstruction
maps for tectonic evolution in
the SE Asian region at 30, 20
and 10 Ma show large scale
movements of plates and land
blocks (simplified from Hall
and Morley 2004)

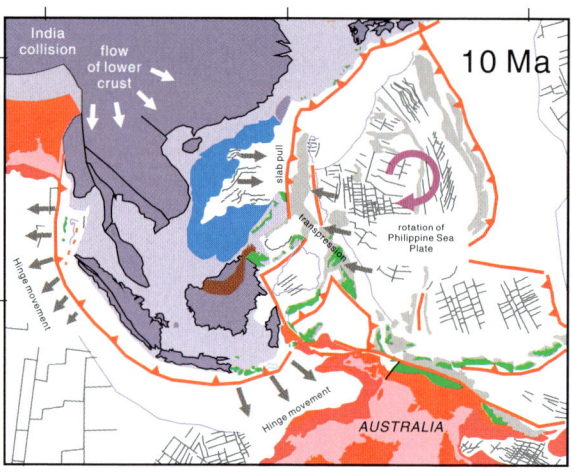

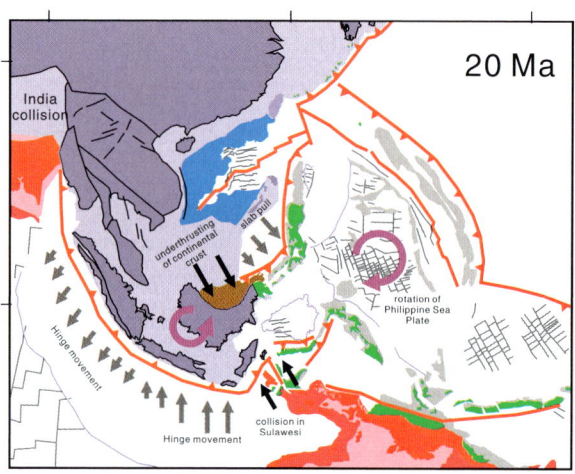

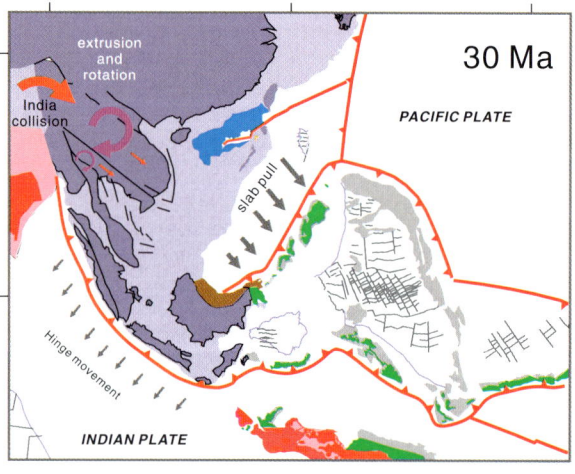

faulting and sea opening is extremely complex, and may have been influenced also by variations of slab dip structure in subduction zones (Lallemand et al. 2005).

Rifting started earlier in the north and northeast and appear to have been periodic during the Paleogene and, in some areas, the early Miocene (Taylor and Hayes 1980, 1983; Ru and Pigott 1986; Jin 1989; Sibuet et al. 2002, 2004). A recent compilation of GIS-based morpho-tectonic data for the northern SCS region has generated many digitized maps showing important fault systems, mountain ranges, and basins in various time periods (Pubellier et al. 2005; Pubellier and Chan 2006). Rifting in the western and southern parts occurred in the Eocene or later, implying two different tectonic and geomorphologic provinces between the then northern and southern + western SCS (Hayes and Nissen 2005). These two provinces show different lithofacies marking different early developing histories largely relating to different strike-slip faults in the SE along the so-called "Baram line", a proposed extension of the Red River Fault, as well as to local fault systems since the Eocene time (Morley 2002; Hayes and Nissen 2005) (Figs. 2.20 and 2.21).

Early studies of magnetic anomalies clearly reveal three important stages in the development of the SCS (Fig. 2.24). Seafloor spreading began at least from Chron 11 and ended at Subchron 5C (e.g., Briais et al. 1993), corresponding to the time

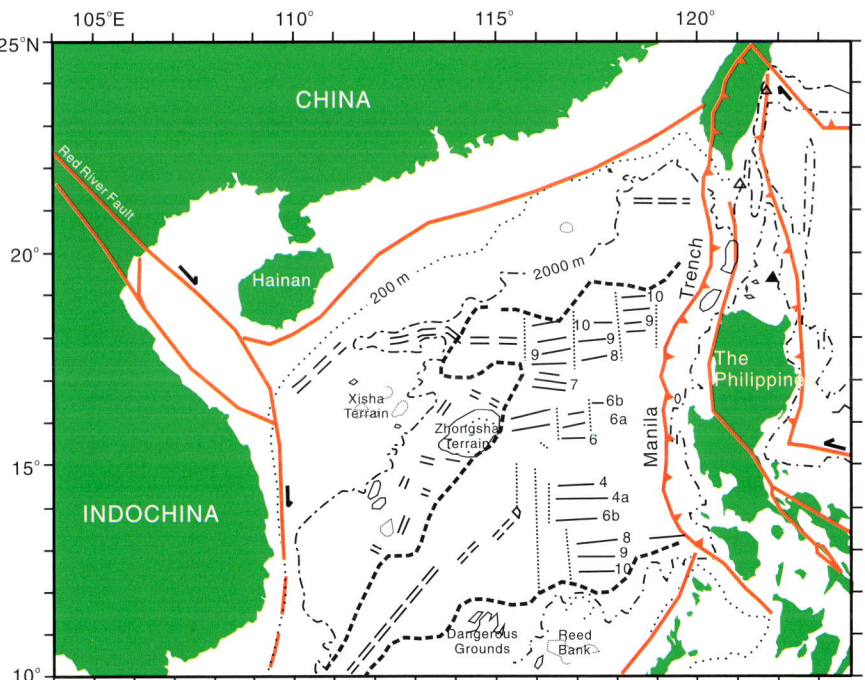

Fig. 2.24 Major tectonic elements of the northern and central parts of the SCS are shown with early interpretation of magnetic anomaly lineaments in the central deep basin as outlined by the thick dash line (modified from Hayes et al. 1995)

interval of 30–16.7 Ma using the new paleomagnetic time scale of Gradstein et al.
(2004), or 32–16 Ma on older time scales. A southward ridge jump during the course
of spreading occurred close to anomaly 7/6b boundary with an age of 23–25 Ma (old
age 24–26 Ma), resulting in the separation of the Nansha from the Xisha-Zhongsha
terrains (Fig. 2.24). However, because the igneous oceanic crust has not been
drilled, there is no unanimity about the exact age of SCS spreading. For example,
Barckhausen and Roeser (2004) argues that seafloor spreading in the SCS started at
~31 Ma (anomaly 12) and ended at ~20.5 Ma (anomaly 6A1), with a ridge jump at
25 Ma when a second spreading center also developed in the southwest.

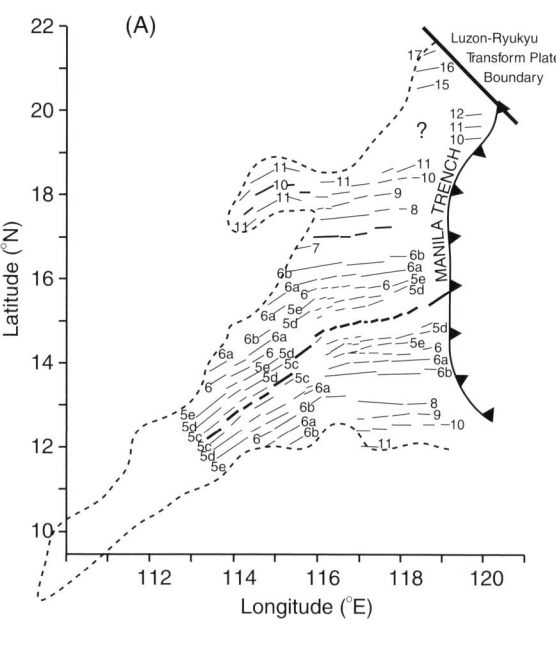

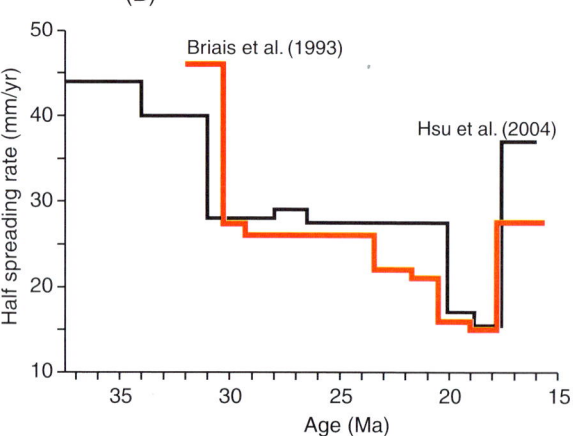

Fig. 2.25 (**A**) New
interpretation of major
magnetic anomaly lineaments
in the central deep basin
reveals an oceanic crust as
old as 37 Ma at anomaly C17
in the NE corner of the deep
basin. (**B**) Variations of
half-spreading rates show a
relative stable rate of
~28 mm/yr during the
interval from 31 to 20 Ma.
Both **A** and **B** are modified
from Hsu et al. (2004)

A recent study of magnetic lineaments discovers that seafloor spreading in the northeastern corner of the SCS occurred much earlier at ~37 Ma, during anomaly C17 (Hsu et al. 2004) (Fig. 2.25). The newly identified continent-ocean boundary (COB) is marked by the presence of a relatively low magnetization zone at the base of the continental slope, corresponding to the thinned portion of the continental crust. Hsu et al. (2004) suggested that the northern extension of the SCS oceanic crust is terminated by an inactive NW-SE trending trench-trench, left-lateral transform fault, called the Luzon-Ryukyu Transform Plate Boundary (LRTPB) (Fig. 2.25). The LRTPB was connected to the former southeast- dipping Manila Trench in the south and the northwest-dipping Ryukyu Trench in the north, and probably became inactive at ca. 20 Ma. These results indicate a half-spreading rate of 40–44 mm/yr from 37 to 31 Ma before stabilizing at ~28 mm/yr in the early spreading history of the SCS (Fig. 2.25).

In the south, rifting lasted from Eocene (~46 Ma) to early Miocene (~21 − 19 Ma). The subsequent seafloor spreading caused a break-up hiatus of 3–5 myr, or the Mid-Miocene Unconformity widely recorded in many southern areas (ASCOPE 1981; Hutchison 2004) (Fig. 2.26). However, Mesozoic thrust anticlines recently found in the hinterland of the Nansha Block (Yan and Liu 2004) seem to support the existence of a NW-SE subduction zone already in the Mesozoic, as proposed originally by Holloway (1982). This subduction zone may further support the hypothesis that at least part of the present SCS, like many marginal seas in the western Pacific, was once a back-arc basin (Honza and Fujioka 2004). Nevertheless, the southeasterly extrusion of the Indochina subcontinent along the Red River Fault since the late Paleogene and the subsequent collision and subduction along the eastern SCS margin may have overridden some pre-existing geotectonic features (Hall 2002;

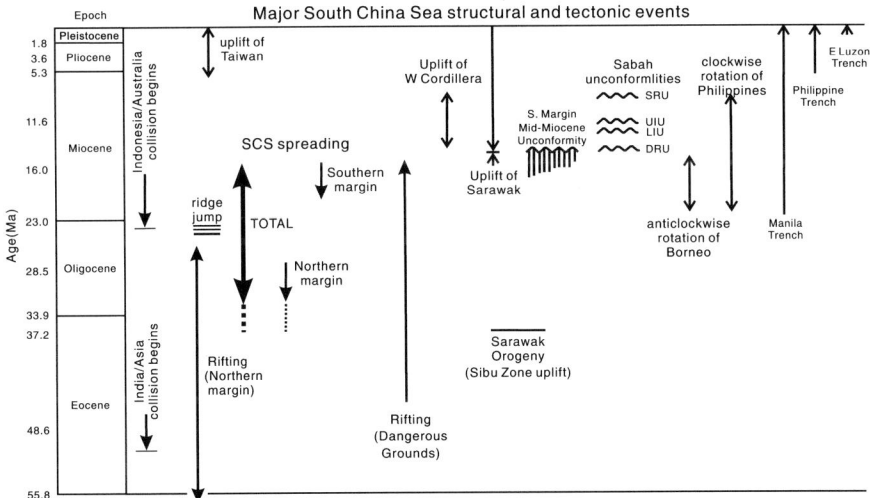

Fig. 2.26 Major structural and tectonic events in the SCS region since the Eocene are compiled mainly from Hall (2002), Hutchison (2004), and Yumul et al. (2003). Earliest seafloor spreading is now dated at ~37 Ma, magnetic anomaly C17 (refer to Fig. 2.25)

Morley 2002) (Fig. 2.23). The southward extension of the Red River Fault to central Salawak along the Baram Line (Morley 2002) (Fig. 2.21) subdivided the relatively stable southwestern margin from the subduction zone in the southeastern and eastern SCS since the middle Miocene (Hutchison 2004; Hall and Morley 2004). Therefore, while post-rift sediment fill characterized the northern and southwestern margin, collision and subduction were the main features along the eastern margin especially in the last 5 myr (Fig. 2.26).

Therefore, available data indicate a stepwise opening of the SCS: ~37 Ma in the northeast, ~33−~31 Ma in the north and ~21 Ma or younger in central and southern parts (Fig. 2.26). The main driving force of seafloor spreading in the SCS may include the SE extrusion of the Indochina subcontinent due to Tibet uplift and slab pull from subduction of the eastern Asian margin including the Proto-SCS into the Philippine Sea Plate (Fig. 2.23).

Step-Wise Closure of the Sea Basin

However, the present shape and semi-enclosed nature of the SCS has not resulted from seafloor spreading, but from the post-spreading tectonic development. In the south, Australia-Asia collision began in Sulawesi at about 25 Ma, leading to the anti-clockwise rotation of Borneo and closure of the SCS in the south (Hall and Morley 2004) (Fig. 2.23). The rotation of the Philippine Sea Plate and the collision of the Luzon Arc with the Asian margin of the Eurasian Plate gave rise to the uplift of Taiwan, closing the SCS basin on the eastern side (Figs. 2.23 and 2.25). Sibuet et al. (2002, 2004) argued that the rise of Taiwan and the closure of the SCS in general may not require a rotating Philippine Sea Plate but a stronger push from the joint force of the Pacific and Australian plates. Subduction along the eastern margin leading to oblique arc-continent collision in northern Taiwan began at about 6.5 Ma, in the late Miocene (Huang et al. 1997).

The closure of the SCS basin is closely related to a change of the tectonic regime in SE Asia region, which also contributed to the closure of the water way between the Pacific and Indian oceans, the Indonesian seaway. Five barriers may have acted more significantly than others in blocking the Indonesian seaway as well as the closure of the SCS: (1) Banda Arc colliding with New Guinea at 11–9 Ma, (2) south Banda colliding with NW Australian margin at ~6 Ma, (3) Sula and (4) Seram blocks both migrating and repositioning during 5–3 Ma, and (5) the closing of the Molucca Sea at ~3 Ma (Linthout et al. 1997; Hall 2002; Zhou et al. 2004). A significant shoaling at ~10 Ma around Sulawesi was caused by collision of the continental fragments of Buton (part of the Indo-Australian Plate) and the Sundaland (Eurasian Plate). In Sulawesi, contemporary geochemical signatures shifted from indicating arc-volcanics to a subduction-modified mantle with a significant contribution from the subcontinental lithospheric mantle (Elburg and Foden 1999).

The emergence of Taiwan and the present Bashi Strait resulted from Luzon Arc/Eurasian Plate collision beginning ~6.5 Ma ago (Huang et al. 1997) (Fig. 2.26). A series of events were involving in the westward migration of collision (Sibuet et al. 2004). In a review of collision, subduction and accretion events in the Philippines, Yumul et al. (2003) noted that Luzon onramped the SCS crust in the course of rotation and effectively converted the shear zone and bounded them into a subduction zone. Rangin et al. (1985) found the SCS oceanic crust of Oligocene age jammed into the Mindoro collision zone. The Taiwan accretionary prism includes a collision prism in the west and a subduction wedge in the east, due to a change of tectonic modes from intra-oceanic subduction to arc-continent collision and arc accretion (Huang et al. 1997). By ~5 Ma, the northeastern end of the SCS oceanic crust at the present position off southern Taiwan and probably also some area in the Bashi Strait and further south along the Manila Trench had been completely subducted (Fig. 2.26).

Accompanying the rise of Taiwan was a series of neotectonic activities relating to collision and subduction in the SCS region (e.g., Liu Y. 1994). Lüdmann and Wong (1999) identified two main collision phases in the northern region: 5–3 Ma and 3–0 Ma. Pliocene faults are mainly ENE-WSW to NE-SW trending and strike-slip in nature, while recent faults are generally oriented NE-SW subparallel to the syn-rift faults. Most likely, the underwater part of the Bashi Strait, now at ~2, 400 m water depth, has been acting as a barrier blocking Pacific bottom waters flowing into the SCS since ~5 Ma. Therefore, the isolated SCS bottom water became colder and more corrosive, subsequently causing more severe carbonate dissolution at deep water localities (Chapter 6). Continued subduction along the east margin over the last 5 myr has turned the once open SCS to a semi-enclosed basin with the Bashi Strait as the only deep passage remaining to connect the sea basin with the West Pacific.

Arc-continent collision in the region since the middle Miocene has contributed to frequent volcanic activities along the eastern margin, although volcanism at the time of rifting was also common in the western Pacific arc region (Taylor and Natland 1995). Among a string of volcanoes in a long, mountainous Bicol Peninsula in southern Philippines, the peak of Mayon Volcano rises to a height of 2,525 m. Volcanoes in the north and west are relative rare and mostly confined to the Pliocene and Quaternary. As revealed by ODP Leg 184 drilling, volcanic ash layers appear to concentrate in three periods: ~10 Ma, ~6 Ma, and since ~2 Ma (Wang P. et al. 2000). These periods also saw significant collision and uplift of island arcs and/or expansion of foreland basins along the east. For example, an unconformity of up to 10 myr in the Western Taiwan Foreland Basin separates the Quaternary sediments from the underlying eroded Miocene strata, indicating the development of a flexural forbulge associated with the transformation of tectonic settings from shelf-slope to foreland (Yu and Chou 2001).

Formation of Shelf-Slope Sedimentary Basins

About 40 Cenozoic sedimentary basins or basin-groups have been recognized in the SCS (Fig. 2.27). Some large basins are listed in Table 2.4 to reflect nomenclature complexity due to unsettled jurisdiction.

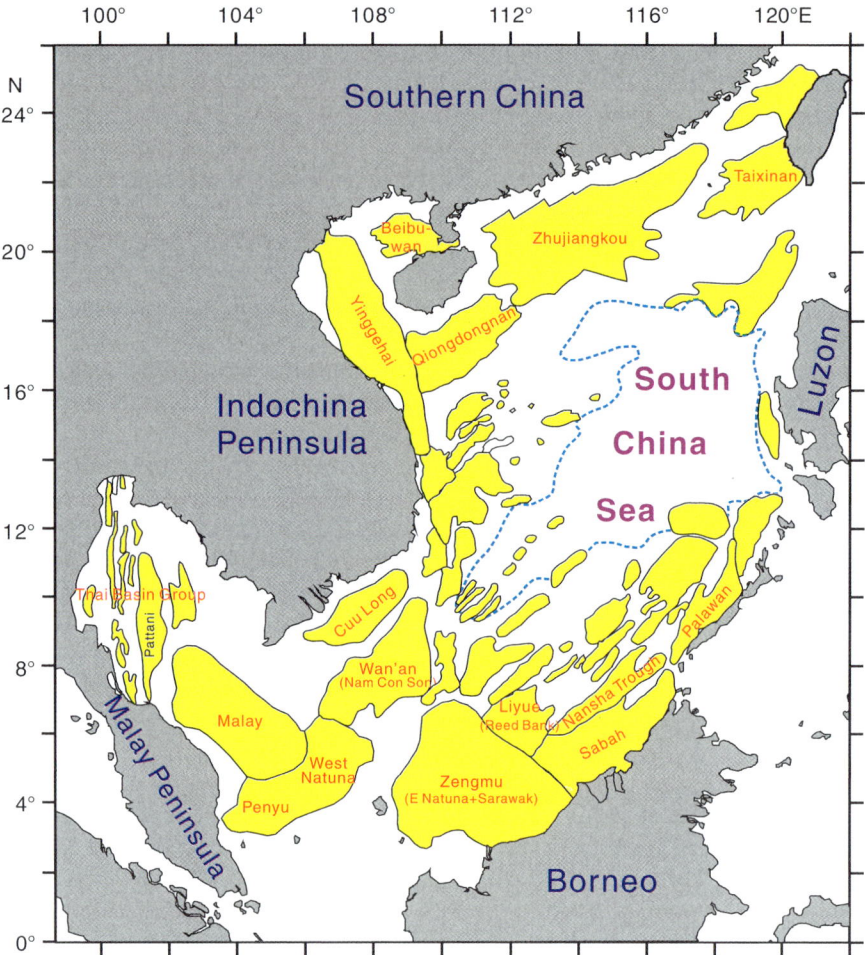

Fig. 2.27 Map shows distribution of Cenozoic sedimentary basins in the SCS as compiled from various sources. Refer to Table 2.4 for basin names and their synonyms

The development of shelf-slope basins closely accompanied Cretaceous-Paleogene rifting and subsequent seafloor spreading of the SCS (Fig. 2.26). A series of normal faults developed mostly in parallel to the shelf break, leading to the formation of most basins in the northern and southern SCS (Table 2.4). Faults in the west and northwest are often sheared or even strike-slip mainly in responding to the impact of the Red River Fault, and they control the general shape and development of the Yinggehai Basin and basins along the coast of Indochina Peninsula. Both sheared-faulted, the Yinggehai Basin in the northwest and Zengmu Basin in the south have a thickest sediment cover of over 14 km. Compressional or fore-arc basins are limited to the eastern margin, including the Sabah and Palawan basins (Jin 1989; Gong and Li 1997; Balaguru and Nichols 2004) (Table 2.4). Neotectonic activities along the northern margin include the development of ENE-WSW to

Table 2.4 Characteristics of major SCS basins are compiled from various sources

Center (°E/°N)	Basin	Area (km²)	Formation character	Basement	[1]Syn-rift sediment	Post-rift sediment	Other phases
	Northern South China Sea basins:						
119/22	Taixinan (SW Taiwan)	66,000	normal to foreland	Mz clastics	Paleogene	Oligo-Mio	Latest Mio (foreland)
114/20	Zhujiangkou (Pearl R Mouth)	147,000	normal faulting	Mesozoic granite	L Cret-E Oligo	L Oligo-Recent	
109/20	Beibuwan (Beibu Gulf)	38,000	normal faulting	Pz-Mz granite and clastics	Paleogene	Neogene	
108/18	Yinggehai (Song Hong)	113,000	transverse to sheared	Mz granite and clastics	Paleogene	Neogene	
110/17	Qiongdongnan (SE Hainan)	45,000	normal faulting	Mz granite and clastics	Paleogene	Neogene	
	Southern South China Sea basins:						
107/10	Cuu Long	25,000	normal faulting	Late Mz granite	Eocene-Oligocene	L Oligo-Recent	
109/09	Wan'an (Nam Con Son)	90,000	normal to sheared	Mz granite, volc & clastics	Eocene-Miocene	Pliocene-Recent	
110/05	Zengmu (E Nat una + Sarawak)	170,000	foreland to compressed	Cret-Eocene turbidite			Oligo-Rec (foreland)
117/11	Liyue (Reed Bank)	20,000	foreland to compressed	Cretaceous-Eo turbidite			Oligo-Rec (foreland)
107/05	West Natuna	70,000	normal to inversional	Pre-Cenozoic meta volcanics	L Eo-E Oligo	L Oligo-Recent	L Oligo-Mio (inversion)
104/04	Penyu	10,000	normal faulting	Mesozoic volcanics	Oligocene	Miocene to Rec	M - L Mio (inversion)
103/07	Malay	80,000	normal to transverse	Pz-Mz igneous, meta clastics	Late Eo-E Oligo	L Oligo-E Mio	M - L Mio (inversion)
101/11	Thai Basin Gp (incl. Pattani)	75,000	normal to transverse	Pz-Mz igneous, clastics	Late Eo-E Mio	M Mio-Rec	
115/05	Sabah-Borneo	50,000	compressed to foreland	Mz-Pg marine sediment	Oligocene		M Mio-Rec (foreland)
118/10	Palawan	40,000	compressed to foreland	Mz-Pg marine sediment	Oligocene		M Mio-Rec (foreland)
115/07	Nansha Trough (NW Borneo T)	25,000	compressed to foreland	Mz-Pg marine sediment	Oligocene		M Mio-Rec (foreland)

[1]Pz = Paleozoic; Mz = Mesozoic; Pg = Paleogene; volc = volcanics; meta = metamorphosed.

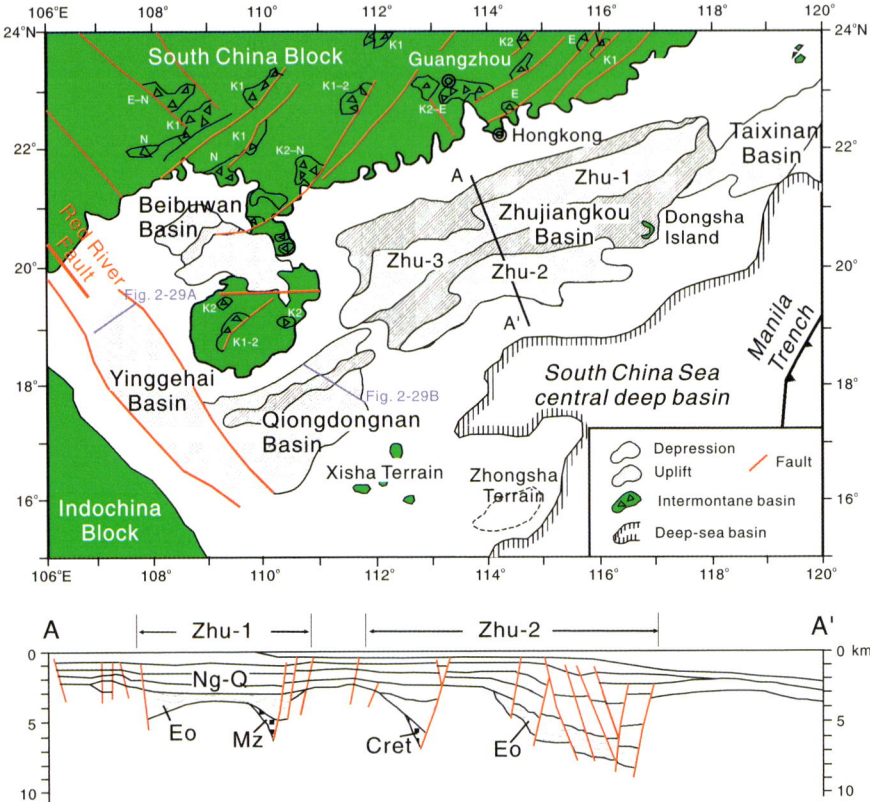

Fig. 2.28 Sketch map (*upper panel*) shows sedimentary basins in the northern margin. A cross section along line A-A' (*lower panel*) illustrates the general characteristics of structural units and sediment packages in Zhu-1 and Zhu-2 depressions of the Zhujiangkou Basin (Ru et al. 1994)

NE-SW faults in the Pliocene and NE-SW faults in the Quaternary (Lüdmann and Wong 1999).

Major characteristics of several northern basins are illustrated in Figs. 2.28 and 2.29. Clearly, various stages of basin formation can be recognized for a specific basin, from rifting to thermal subsidence or even subsequent inversion (Pigott and Ru 1994; Madon and Watts 1998; Xie et al. 2006). Some stages are likely related also to preferential mantle lithospheric extension under the South China margin (Clift et al. 2001).

Although basins in a structural zone show similarities in both age and phases of formation, their subsidence rate and amplitude are often different (Fig. 2.30). Total subsidence over 14 km in the Yinggehai and Zengmu basins is highest, about half of which was attributed to structural subsidence. In other basins, total subsidence varies mostly between 3 and 8 km. Total subsidence < 5 km exists not only in more

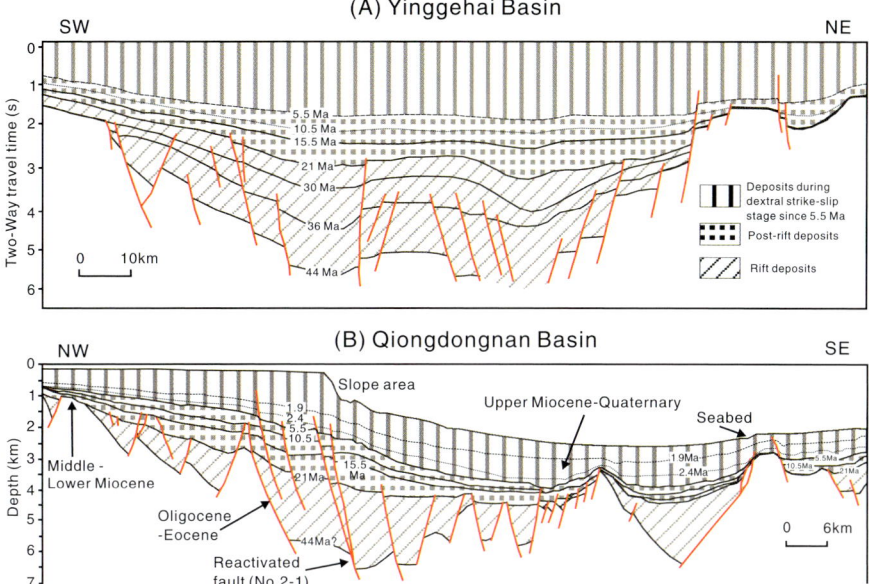

Fig. 2.29 Interpreted seismic sections for (**A**) the Yinggehai Basin and (**B**) the Qiongdongnan Basin show major structural features and reactivated faults along the slope break area (Xie et al. 2006). Refer to Fig. 2.28 for line locations

offshore basins in the north (such as those neighboring the Taixinan Basin) but also in the south including some sub-basins of the Malay Basin and Thai Basin Group (Fig. 2.30). Xie et al. (2006) recently found anomalous subsidence of 300–1,200 m in several northern basins probably induced by a late thermal cooling and renewed faulting since the late Miocene. However, late Miocene basin inversion was a main feature in the Malay Basin (Madon and Watts 1998) and other basins in the south and southeast, as well as in the Taixinan Basin (Lee et al. 1993). Inversion also caused dolomitization in the Malay Basin (Fig. 2.30).

Increases in subsidence rates concurred with major steps in the evolution of the SCS, especially at ∼25 Ma and ∼5 Ma. The 25 Ma event was related to the spreading ridge jump to southwest (Fig. 2.26), leading to a large scale regional subsidence and marine transgression along the northern margin (Gong and Li 1997). The 5 Ma event was largely responding to intensive collision and subduction along the eastern margin that also contributed to the rise of Taiwan. Smaller-scaled increases in subsidence rates appear to represent variations in individual basins due to local factors (Madon and Watts 1998; Wheeler and White 2002) (Fig. 2.30; see also Chapter 3). Increased subsidence often leads to increase in sedimentation rates if erosion in source areas was also strong (Clift et al. 2004).

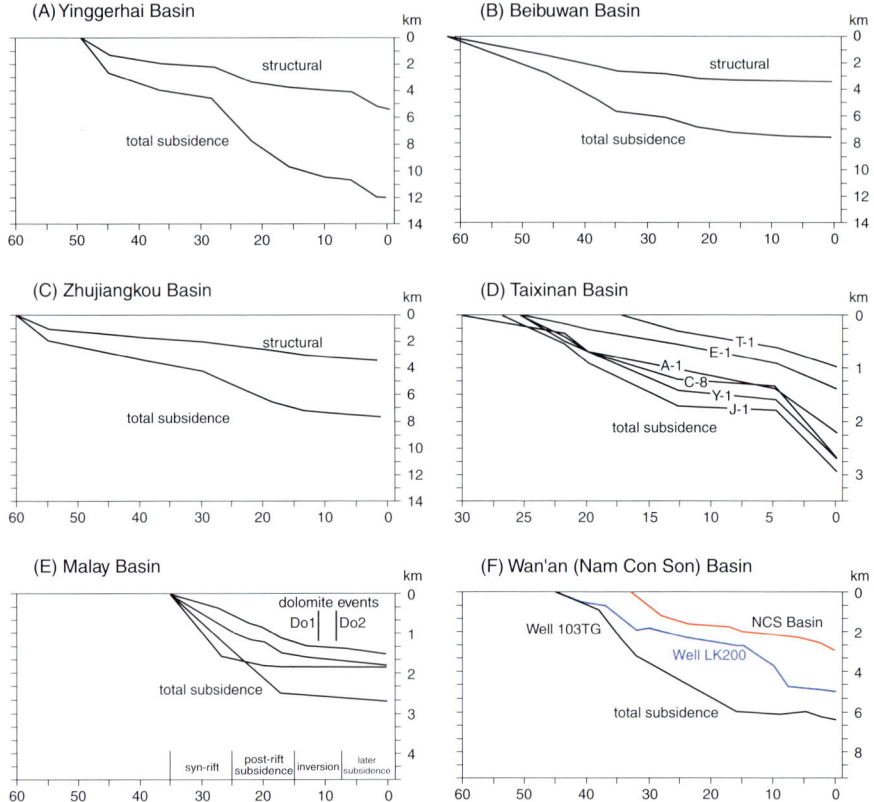

Fig. 2.30 Various subsidence trends are observed in SCS basins: (**A**) Yinggehai Basin (Gong and Li 1997), (**B**) Beibuwan Basin (Gong and Li 1997), (**C**) Zhujiangkou Basin (Gong and Li 1997), (**D**) Taixinan Basin (Lee et al. 1993), (**E**) Malay Basin (Madon and Watts 1998; dolomite events after Ali 1995), (**F**) Wan'an Basin and offshore Vietnam (based on Wheeler and White 2002; Nielsen et al. 1999)

Sediments of the SCS Shelf-Slope Basins: An Overview

The basins of the northern SCS shelf and slope often show a two-layer structure: the lower section characterized by syn-rift half-grabens filled with non-marine Paleogene sequences, and the upper section by a wide range of terrigenous and marine sediments deposited during Neogene subsidence (Figs. 2.28, 2.29, and 2.31). In the south and southwest, similar sedimentary structure can also be observed although the boundary between the rift and post-rift sediments mostly lies in the middle Miocene due to a delay in seafloor spreading there. As a prelude to Chapter 3 and Chapter 4, several important points on the major sedimentation characteristics of shelf to slope basins in the SCS can be generalized:

(1) The development of the marginal basins was associated with rifting and spreading, but the basal ages of basin sediments indicate that many basins were

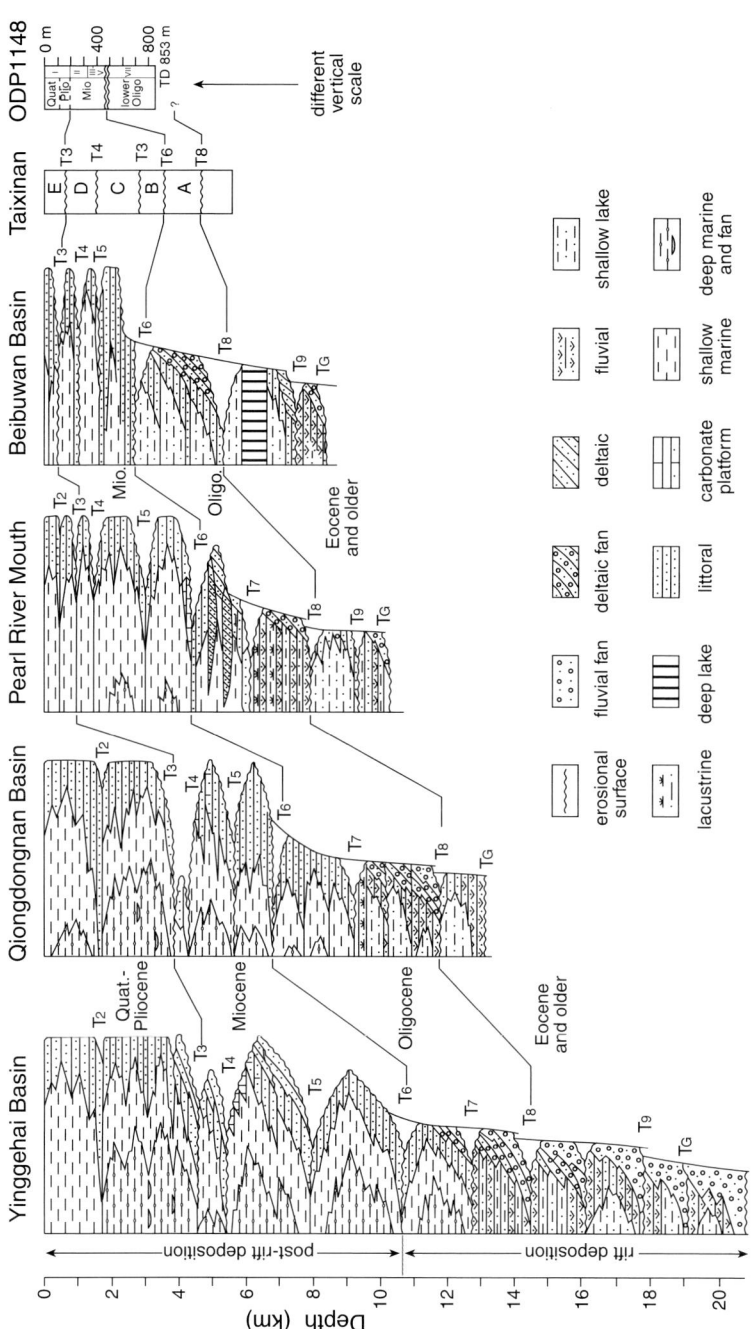

Fig. 2.31 Generalized lithostratigraphic columns show variations of lithofacies and sediment thickness in northern SCS basins (modified from Gong and Li 1997). T1 to TG are seismic reflectors representing unconformities. Taixinan Basin supersequences are after Lee et al. (1993). ODP Site 1148 sediment ages and lithostratigraphic units I to VII are after Wang P. et al. (2000) and shown with a different vertical scale

initiated before spreading and that the ages of individual basins vary (Table 2.4). For example, sedimentation began in the Paleocene-lower Eocene in the northwest and northeast, during the lower Oligocene in the north and in the south (Sunda shelf), and during the Paleocene in the southeastern Reeds Bank basin. Sediment accumulation was determined by basin locations relative to major sediment sources. The Zhujiangkou Basin received sediments mainly from the Pearl, while the Yinggehai Basin/Song Hong Basin and basins in the southwest were mainly fed by clastic sediments from the Red and Mekong rivers, respectively. Weathering of mountain ranges along the eastern-southeastern arc system also contributed a significant amount of sediment material to the nearby basins, especially during the Neogene (Clift et al. 2004) (Chapter 4).

(2) Non-marine fluvial and lacustrine sedimentation dominates the shelf basins in pre- and syn-rifting periods. Transition to marine sedimentation varies from late Oligocene-earliest Miocene in the north (Fig. 2.31) to middle Miocene in southern basins (Chapter 3). While global sea level has been generally declining throughout the late Eocene-Holocene period (Haq et al. 1987), these data indicate asynchronous shelf subsidence during the later phase of rifting in various parts of the SCS.

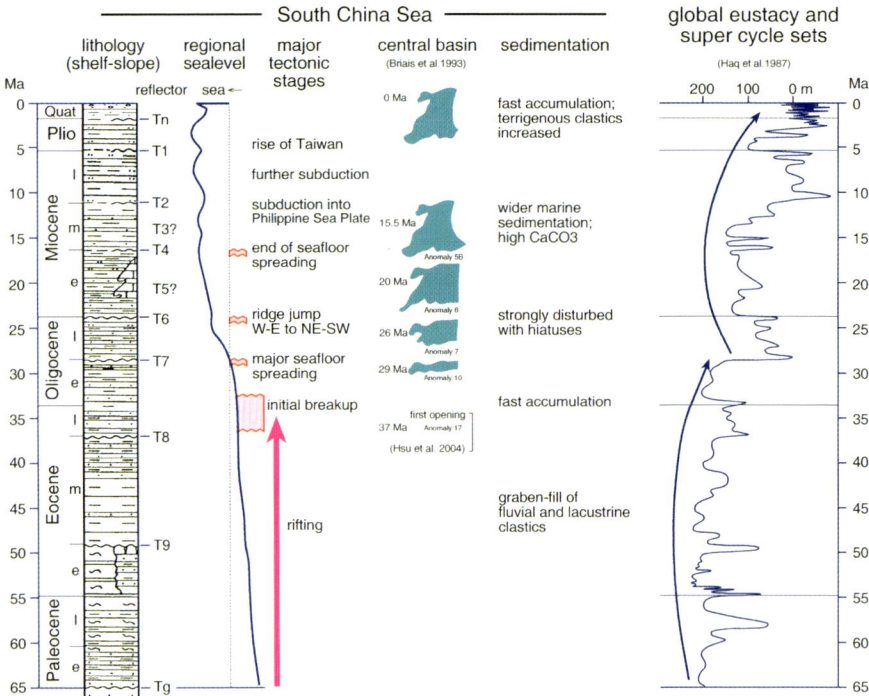

Fig. 2.32 Major features of sedimentary and tectonic events and deep basin formation in the SCS (synthesized from various sources) are compared with global eustatic curves (of Haq et al. 1987), showing tectonic imprints on regional sea level change

(3) Buried faults within basin sediments indicate that shelf-basin tectonics declined toward the end of rifting (Figs. 2.28 and 2.29). Meanwhile, marine inundation toward the end of rifting marked by widespread development of carbonate banks on the S-SE shelf (Chapter 3) was often accompanied by widespread unconformities (Fig. 2.26). Marine sedimentation interrupted by unconformities/seismic reflectors (Fig. 2.31) apparently reflect global sea level variations superimposed on differential local subsidence (Fig. 2.32).

Summary of Tectonics and Basin Formation

The semi-enclosed South China Sea evolved from late Cretaceous to Paleogene rifting, late Eocene to middle Miocene seafloor spreading, and post-spreading subduction and closing since the late Miocene. The opening of the SCS was mainly influenced by the extrusion of Indochina subcontinent due to Tibet uplift and slab pull due to subduction of the Asian continental crust into the Philippine Sea Plate. A Proto-SCS existed before rifting and spreading mainly along the present southeastern margin although its size and connection with other paleo-oceans are not clear. Based on magnetic anomaly lineament data from the northeast, the SCS first opened at ∼37 Ma (anomaly C17), in the late Eocene (Fig. 2.25). A spreading ridge jump from nearly W-E to SW at 25–23 Ma caused widespread subsidence especially in the north. The central deep basin lying on an oceanic crust of ∼8 km appears to have subsided to a maximum depth during the late Miocene-Pliocene (Chapter 6) when collision and subsidence along the eastern margin intensified.

Rifting in the south delayed until the Eocene-Miocene, and a breakup unconformity of 3–5 myr often underlies strata of 16 Ma and older ages. Since the middle Miocene, the southern margin west of the Baram line, a proposed extension of the Red River Fault, was tectonically more stable while the eastern margin has been subducting, subsequently causing the closure of the sea basin and rise of Taiwan in the last 5 myr.

Rifting opened sedimentary basins of various depths and sizes that are often characterized by distinctive syn-rift and post-rift sediment packages. Three kinds of basins can be identified: those normal faulted often parallel to the coast, those sheared often controlled by major lateral-strike or transverse fault zones, and those compressional often relating to subduction. The Yinggehai and Zengmu basins are deepest, with sediment fill of over 14 km, about half of which was induced by structural subsidence. Basin inversion is common in tectonically active regions since the late Miocene. In the north, faulting direction changed from NE-SW in the Miocene to ENE-WSW or NE-SW in the Pliocene to NE-SW in the Quaternary. Terrigenous clastic sediments are predominant in shelf-slope basins, while hemipelagic components increase down-slope to over 50% in the central deep basin.

Acknowledgments This work was supported by the Ministry of Science and Technology of China (NKBRSF Grant 2007CB815900) and the National Natural Science Foundation of China (Grants 40576031, 40621063 and 40631007).

References

ASCOPE (Asean Council on Petroleum) 1981. Tertiary Sedimentary Basins of the Gulf of Thailand and South China Sea: Stratigraphy, Structure and Hydrocarbon Occurrences. ASCOPE, Jakarta, Indonesia, 72pp.

Ali M.Y. 1995. Carbonate cement stratigraphy and timing of diagenesis in a Miocene mixed carbonate-clastic sequence, offshore Sabah, Malaysia: constraints from cathodoluminescence, geochemistry and isotope studies. Sedimentary Geol. 22: 191–214.

Balaguru A. and Nichols G. 2004. Tertiary stratigraphy and basin evolution, southern Sabah (Malaysian Borneo). J. Asian Earth Sci. 23: 537–554.

Barckhausen U. and Roeser H.A. 2004. Seafloor spreading anomalies in the South China Sea revisited. In: Clift P., Wang P., Kuhnt W. and Hayes D. (eds.), Continent-Ocean Interactions within East Asian Marginal Seas. AGU Geophys. Monogr. 149: 121–125.

Briais A., Patriat P. and Tapponnier P. 1993. Update interpretation of magnetic anomalies and seafloor spreading stages in the South China Sea: Implications for the Tertiary tectonics of southeast Asia. J. Geophys. Res. 98(B4): 6299–6328.

Chao S.Y., Shaw P.T. and Wu S.Y. 1996. El Nino modulation of the South China Sea circulation. Progr. Oceanogr. 38: 51–93.

Chen L. 1992. Features of the East Asian monsoon. In: Murakami M. and Ding Y. (eds.), Studies of Asian Monsoon in Japan and China. Meteorol. Res. Inst., Ibaraki, Japan, pp. 220–235.

Chen C.T.A., Wang S.L., Wang B.J. and Pai S.C. 2001. Nutrient budgets for the South China Sea basin. Mar. Chem. 75: 281–300.

Chen C.T.A. and Huang M.H. 1996. A mid-depth front separating the South China Sea water and the Philippine Sea water. J. Oceanogr. 52: 17–52.

Chu P.C. and Wang G. 2003. Seasonal variability of thermohaline front in the central South China Sea. J. Oceanogr. 59: 65–78.

Clift P.D. and Lin J. 2001. Preferential mantle lithospheric extension under the South China margin. Mar. Petroleum Geol. 18: 929–945.

Clift P.D., Shimizu N., Layne G.D. and Blusztajn J. 2001. Tracing patterns of unroofing in the Early Himalaya through microprobe Pb isotope analysis of detrital K-feldspars in Indus Molasse, India. Earth Planet. Sci. Lett. 188: 475–491.

Copeland P. 1997. The when and where of the growth of the Himalaya and the Tibetan Plateau. In: Ruddiman W.F. (ed.), Tectonic Uplift and Climate Change. Plenum, New York, pp. 19–40.

Clift P.D., Layne G.D. and Blusztajn J. 2004. Marine sedimentary evidence for monsoon strengthening, Tibetan uplift and drainage evolution in East Asia. In: Clift P., Wang P., Kuhnt W. and Hayes D. (eds.), Continent-Ocean Interactions within East Asian Marginal Seas. AGU Geophys. Monogr. 149: 255–282.

Ding Y., Li C. and Liu Y. 2004. Overview of the South China Sea Monsoon Experiment. Adv. Atmospheric Sci. 21: 343–360.

Dong L., Su J., Wong L.A., Cao Z. and Chen J.C. 2004. Seasonal variations and dynamics of the Pearl River plume. Continental Shelf Res. 24: 1761–1777.

Du Bois E.P. 1985. Review of principal hydrocarbon-bearing basins around the South China Sea. Bull. Geol. Soc. Malaysia 18: 167–209.

Elburg M.A. and Foden J. 1999. Geochemical response to varying tectonic settings: An example from southern Sulawesi (Indonesia). Geochim. Cosmochim. Acta 63: 1155–1172.

Fan W. and Menzies M.A. 1992. The lithospheric mantle composition of volcanism in extension settings: the geochemical evidence of Cenozoic basalts in Leiqiong area. In: Liu R. (ed.), The Ages and Geochemistry of Cenozoic Volcanic Rocks in China. Seismological Press, Beijing, pp. 320–329.

Fang G.H., Fang W.D., Fang Y. and Wang K. 1998. A survey of studies on the South China Sea upper ocean circulation. Acta Oceanogr. Taiwanica 37: 1–16.

Fang W., Fang G., Shi P., Huang Q. and Xie Q. 2002. Seasonal structures of upper layer circulation in the southern South China Sea from in situ observations. J. Geophys. Res. 107: C11 3202, doi:10.1029/2002 JC001343.

Gilley L.D., Harrison T.M., Leloup P.H., Ryerson F.J., Lovera O.M. and Wang J.H. 2003. Direct dating of left-lateral deformation along the Red River shear zone, China and Vietnam. J. Geophys. Res. 108(B2): 2127, doi:10.1029/2001JB001726.

Gong G.C., Liu K.K., Liu C.T. and Pai S.C. 1992. The chemical hydrography of the South China Sea west of Luzon and a comparison with the West Philippine Sea. Terr. Atmos. Ocean. Sci. (TAO) Taipei 13: 587–602.

Gong Z. and Li S. (eds.). 1997. Continental Margin Basin Analysis and Hydrocarbon Accumulation of the Northern South China Sea. China Sci. Press, Beijing, 510pp (in Chinese).

Gradstein F., Ogg J. and Smith A. (eds.). 2004. A Geologic Time Scale 2004. Cambridge Univ. Press, Cambridge, 589pp.

Hall R. 2002. Cenozoic geological and plate tectonic evolution of SE Asia and the SW Pacific: computer-based reconstructions, model and animations. J. Asian Earth Sci. 20: 353–431.

Hall R. and Morley C.K. 2004. Sundaland basins. In: Clift P., Wang P., Kuhnt W. and Hayes D. (eds.), Continent-Ocean Interactions within East Asian Marginal Seas. AGU Geophys. Monogr. 149: 55–85.

Harrison T.M., Copeland P., Kidd W.S.F. and Yin A. 1992. Raising Tibet. Science 255: 1663–1670.

Haq B.U., Hardenbol J. and Vail P.R. 1987. Chronology of fluctuating sea levels since the Triassic (250 million years ago to present). Science 235: 1156–1167.

Hawkins J.W., Lonsdale P.F., Macdougall J.D. and Volpe A.M. 1990. Petrology of the axial ridge of Mariana Trough backarc spreading center. Earth Planet. Sci. Lett. 100: 226–250.

Hayes D.S. and Nissen S.S. 2005. The South China Sea margins: Implications for rifting contrasts. Earth Planet. Sci. Lett. 237: 601–616.

Hayes D.E., Nissen S.S., Buhl P., Diebold J., Yao B., Zeng W. and Chen Y. 1995. Throughgoing crustal faults along the northern margin of the South China Sea and their role in crustal extension. J. Geophys. Res. 100: 22435–22446.

Holloway N.H. 1982. North Palawan Block, Philippines, its relation to Asian mainland and role in evolution of South China Sea. AAPG Bull. 66: 1355–1383.

Honza E. and Fujioka K. 2004. Formation of arcs and backarc basins inferred from the tectonic evolution of Southeast Asia since the Late Cretaceous. Tectonophysics 384: 23–53.

Hsu M.K., Liu A.K. and Liu C. 2000. A study of internal waves in the China Seas and Yellow Sea using SAR. Continental Shelf Res. 20: 389–410.

Hsu S.K., Yeh Y.C., Doo W.B. and Tsai C.H. 2004. New bathymetry and magnetic lineations identifications in the northernmost South China Sea and their tectonic implications. Mar. Geophys. Res. 25: 29–44.

Hsueh Y. and Zhong L. 2004. A pressure-driven South China Sea warm current. J. Geophys. Res. 109: C09014, doi:10.1029/2004JC002374.

Huang C.Y., Wu W.Y., Chang C.P., Tsao S., Yuan P.B., Lin C.W. and Xia K.Y. 1997. Tectonic evolution of accretionary prism in the arc-continent collision terrane of Taiwan. Tectonophysics 281: 31–51.

Hutchison C.S. 2004. Marginal basin evolution: the southern South China Sea. Mar. Petroleum Geol. 21: 1129–1148.

Jin Q. (ed.). 1989. The Geology and Petroleum Resources in the South China Sea. Geol. Publ. House, Beijing, 417pp (in Chinese).

Lallemand S., Heuret A. and Boutelier D. 2005. On the relationships between slab dip, back-arc stress, upper plate absolute motion, and crustal nature in subduction zones. Geochem. Geophys. Geosyst. 6: Q09006, doi:10.1029/2005GC000917.

Lee T.-Y. and Lawver L.A. 1994. Cenozoic plate reconstruction of the South China Sea region. Tectonophysics 235: 149–180.

Lee T.-Y., Tang C.-H., Ting J.-S. and Hsu Y.-Y. 1993. Sequence stratigraphy of the Tainan Basin, offshore southwestern Taiwan. Petroleum Geol. Taiwan 28: 119–158.

Leloup P.H., Arnaud N., Lacassin R., Kienast J.R., Harrison T.M., Phan Trong T.T., Replumaz A. and Tapponnier T. 2001. New constraints on the structure, thermochronology, and timing of the Ailao Shan-Red River shear zone, SE Asia. J. Geophys. Res. 106: 6683–6732.

Li C., Zhou Z., Hao H., Chen H., Wang J., Chen B. and Wu J. 2008. Late Mesozoic tectonic structure and evolution along the present-day northeastern South China Sea continental margin. J. Asian Earth Sci. 31: 546–561.

Linthout K., Helmers H. and Sopaheluwakan J. 1997. Late Miocene obduction and microplate migration around the southern Banda Sea and the closure of the Indonesian Seaway. Tectonophysics 281: 17–3.

Liu Z., Wang P., Wang C., Shao L. and Huang W. 2001. Paleotopography of China during the Cenozoic: a preliminary study. Geol. Rev. 47(5): 467–475 (in Chinese).

Liu A.K., Chang Y.S., Hsu M.-K. and Liang N.K. 1998. Evolution of nonlinear internal waves in the East and South China Seas. J. Geophys. Res. 103(C4): 7995–8008.

Liu Q., Yang H. and Wang Q. 2000. Dynamic characteristics of seasonal thermocline in the deep sea region of the South China Sea. Chinese J. Oceanol. Limnol. 18(2): 104–109.

Liu K.-K., Chao S.-Y., Shaw P.-T., Gong G.-C., Chen C.-C. and Tang T.Y. 2002. Monsoon-forced chlorophyll distribution and primary production in the South China Sea: observations and a numerical study. Deep-Sea Res. I 49: 1387–1412.

Liu Q., Jiang X., Xie S.-P. and Liu W.T. 2004. A gap in the Indo-Pacific warm pool over the South China Sea in boreal winter: Seasonal development and interannual variability. J. Geophys. Res. 109: C07012, doi: 10.1029/2003JC002179.

Liu C.-T., Pinkel R., Hsu M.-K., Klymak J.M., Chen H.-W. and Villanov C. 2006. Nonlinear internal waves from the Luzon Strait. EOS, Trans. AGU 87(42): 449–451.

Lüdmann T. and Wong H.K. 1999. Neotectonic regime on the passive continental margin of the northern South China Sea. Tectonophysics 311: 113–138.

Madon M.B. and Watts A.B. 1998. Gravity anomalies, subsidence history and the tectonic evolution of the Malay and Penyu Basins (offshore Peninsular Malaysia). Basin Res. 10: 375–392.

Metzger E.J. and Hurlburt H.E. 1996. Coupled dynamics of the South China Sea, the Sulu Sea, and the Pacific Ocean. J. Geophys. Res. 101: 12331–12352.

Metzger E.J. and Hurlburt H.E. 2001. The nondeterministic nature of Kuroshio penetration and eddy shedding in the South China Sea. J. Phys. Oceanogr. 31: 1712–1732.

Morley C.K. 2002. A tectonic model for the Tertiary evolution of strike-slip faults and rift basins in SE Asia. Tectonophysics 347: 189–215.

Nielsen L.H., Mathiesen A., Bidstrup T., Vejbñk O.V., Dien P.T. and Tiem P.V. 1999. Modeling of hydrocarbon generation in the Cenozoic Song Hong Basin, Vietnam: a highly prospective basin. J. Asian Earth Sci. 17: 269–294.

Nissen S.S., Hayes D.E., Buhl P., Diebold J., Bochu Y., Weijun Z. and Chen Y. 1995. Deep penetration seismic soundings across the northern margin of the South China Sea, J. Geophys. Res. 100(B11): 22407–22434.

Nitani, H. 1972. Beginning of the Kuroshio. In: Stommel H., Yoshida K.(eds.), Kuroshio: Its Physical Aspects. University of Washington Press, Seattle, pp. 129–163.

Pigott J.D. and Ru K. 1994. Basin superposition on the northern margin of the South China Sea. Tectonophysics 235: 27–50.

Pubellier M. and Chan L.S. (eds.). 2006. Morphotectonic Map of Cenozoic Structures of the South China-Northern Vietnam Coastal Region. Output-express Print Office. Hong Kong, 16pp (CD-ROM).

Pubellier M., Ego F., Chamot-Rooke N. and Rangin C. 2003. The building of pericratonic mountain ranges: structural and kinematic constraints applied to GIS-based reconstructions of SE Asia. Bull. Soc. Geol. France 174: 561–584.

Pubellier M., Monnier C., Maury R. and Tamayo R. 2004. Plate kinematics, origin and tectonic emplacement of supra-subduction ophiolites in SE Asia. Tectonophysics 92: 9–36.

Pubellier M., Rangin C., Ego F., et al. 2005. Altas of the Margin of SE Asia. Soc. Geol. France/AAPG Spec. Publ. 176 (CD-ROM).

Qiu X., Ye S., Wu S., Shi X., Zhou D., Xia K. and Flueh R. 2001. Crustal structure across the Xisha Trough, northwestern South China Sea. Tectonophysics 341: 179–193.

Qu T. 2000. Upper-layer circulation in the South China Sea. J. Phys. Oceanogr. 30: 1450–1460.

Qu T. 2002. Evidence for water exchange between the South China Sea and the Pacific Ocean through the Luzon Strait. Acta Oceanol. Sinica 21(2): 175–185.

Qu T., Girton J.B. and Whitehead J.A. 2006. Deepwater overflow through Luzon Strait. J. Geophys. Res. 111: C01002, doi: 10.1029/2005JC003139.

Qu T., Mitsudera H. and Yamagata T. 2000. Intrusion of the North Pacific waters into the South China Sea. J. Geophys. Res. 105(C3): 6415–6424.

Rangin C., Stephan J.F. and Mueller C. 1985. Middle Oligocene oceanic crust of South China Sea jammed into Mindoro collision zone (Philippines). Geology 13: 425–428.

Ru K. and Pigott J.D. 1986 Episodic rifting and subsidence in the South China Sea. AAPG Bull. 70: 1136–1155.

Ru K., Zhou D. and Chen H. 1994. Basin evolution and hydrocarbon potential of the northern South China Sea. In: Zhou D., Liang Y. and Zeng C. (eds.), Oceanology of China Seas. Kluwer Press, New York, vol. 2, pp. 361–372.

Schlüter H.U., Hinz K. and Block M. 1996. Tectono-stratigraphic terranes and detachment faulting of the South China Sea and Sulu Sea. Mar. Geol. 130: 39–78.

Shaw P.T. 1996. Winter upwelling off Luzon in the Northeastern South China Sea. J. Geophys. Res. 101: 16435–16448.

Shaw P.T. and Chao S.Y. 1994. Surface circulation in the South China Sea. Deep-Sea Res. I 41: 1663–1683.

Shi P., Du Y.,Wang D.X. and Gan Z.J. 2001. Annual cycle of mixed layer in South China Sea. Tropical Oceanol. 20: 10–17 (in Chinese).

Sibuet J.C., Hsu S.K. and Debayle E. 2004. Geodynamic context of the Taiwan orogen. In: Clift P., Wang P., Kuhnt W. and Hayes D. (eds.), Continent-Ocean Interactions within East Asian Marginal Seas. AGU Geophys. Monogr. 149: 127–158.

Sibuet J.C., Hsu S.K., Le Pichon X., Le Formal J.P. and Reed D. 2002. East Asia plate tectonics since 15 Ma: constraints from the Taiwan region. Tectonophysics 344: 103–134.

Su J. 2004. Overview of the South China Sea circulation and its influence on the coastal physical oceanography outside the Pearl River Estuary. Continental Shelf Res. 24: 1745–1760.

Tapponnier P., Peltzer G. and Armijo R. 1986. On the mechanics of the collision between India and Asia. In: Coward M.P. and Ries A.C. (eds.), Collision Tectonics. Blackwell, Oxford, pp. 115–157.

Tapponnier P., Peltzer G., Le Dain A.Y., Armijo R. and Cobbold P. 1982. Propagating extrusion tectonics in Asia: new insights from simple experiments with plasticine. Geology 10: 611–616.

Tapponnier P., Xu Z., Roger F., Meyer B., Arnaud N., Wittlinger G. and Yang J. 2001. Oblique stepwise rise and growth of the Tibet Plateau. Science 294: 1671–1677.

Taylor B. and Hayes D.E. 1980. The tectonic evolution of the South China Sea Basin. In: Hayes D.E. (ed.), The Tectonic and Geologic Evolution of Southeast Asian Seas and Islands. AGU Geophys. Monogr., Washington, D.C., pp. 89–104.

Taylor B. and Hayes D.E. 1983. Origin and history of the South China Sea basin. In: Hayes D.E. (ed.), The Tectonics and Geologic Evolution of Southeast Asian Seas and Islands: Part 2. AGU Geophys. Monogr., Washington, D.C., pp. 23–56.

Taylor B. and Natland J. (eds.). 1995. Active Margins and Marginal Basins of the Western Pacific. AGU Geophys. Monogr., Washington, D.C., 88, p. 417.

Tian J.W., Yang Q., Liang X., Xie L., Hu D., Wang F. and Qu T. 2006. Observation of Luzon Strait transport. Geophys. Res. Lett. 33: L19607, doi: 10.1029/2006GL026272.

Wang C. 1996. Sequence stratigraphic analysis of marine Miocene formations in the Pearl River Mouth Basin and its significance. China Offshore Oil Gas (Geol.) 10(5): 279–288 (in Chinese).

Wang P. 1999. Response of Western Pacific marginal seas to glacial cycles: Paleoceanographic and sedimentological features. Mar. Geol. 156: 5–39.

Wang L., Koblinsky C.J. and Howden S. 2000. Mesoscale variability in the South China Sea from the TOPEX/Poseidon altimetry data. Deep-Sea Res. I 47: 681–708.

Wang P., Prell W.L., Blum P. (eds.). 2000. Proc. ODP, Init. Repts, Vol. 184 [CD-ROM]. Ocean Drilling Program, Texas A&M University, College Station TX 77845–9547, USA.

Wang D., Du Y. and Shi P. (eds.). 2002. Climatological Atlas of Physical Oceanography in the Upper Layer of the South China Sea. Meteorol. Press, Beijing, 168pp (in Chinese).

Wang G., Su J. and Chu P.C. 2003. Mesoscale eddies in the South China Sea observed with altimeter data. Geophys. Res. Lett. 30(21): 2121, doi: 10.1029/2003GL018532.

Wang R., Clemens S., Huang B. and Chen M. 2003. Late Quaternary paleoceanographic changes in the northern South China Sea (ODP Site 1146): radiolarian evidence. J. Quat. Sci. 18(8): 745–756.

Wang B., Clemens S.C. and Liu P. 2003. Contrasting the Indian and East Asian monsoons: implications on geological timescales. Mar. Geol. 201: 5–21.

Wang P. 2004. Cenozoic deformation and the history of sea-land interactions in Asia. In: Clift P., Wang P., Kuhnt W. and Hayes D. (eds.), Continent-Ocean Interactions in the East Asian Marginal Seas. AGU Geophys. Monogr. 149: 1–22.

Wang G., Chen D. and Su J. 2006. Generation and life cycle of the dipole in the South China Sea summer circulation. J. Geophys. Res. 111: C06002, doi: 10.1029/2005JC003314.

Wheeler P. and White N. 2002. Measuring dynamic topography: An analysis of Southeast Asia. Tectonics 21: 1040, doi:10.1029/2001TC900023.

Wong G.T.F., Ku T.L., Mulholland M., Tseng C.M. and Wang D.P. 2007a. The South East Asian Time-series Study (SEATS) and the biogeochemistry of the South China Sea-An overview. Deep-Sea Res. II 54: 1434–1447.

Wong G.T.F., Tseng C.M., Wen L.S. and Chung S.W. 2007b. Nutrient dynamics and nitrate anomaly at the SEATS station. Deep-Sea Res. II 54: 1528–1545.

Wyrtki K. 1961. Physical oceanography of the Southeast Asian waters. NAGA, La Jolla, Calif., Rept. 2: 1–195.

Xia K.Y. and Zhou D. 1993. The geophysical characteristics and evolution of northern and southern margins of the South China Sea. Geol. Soc. Malaysia Bull. 33: 223–240.

Xia B., Zhang Y., Cui X.J., Liu B.M., Xie J.H., Zhang S.L. and Lin G. 2006. Understanding of the geological and geodynamic controls on the formation of the South China Sea: A numerical modelling approach. J. Geodynamics 42: 63–84.

Xie S.P., Xie Q., Wang D. and Liu W.T. 2003. Summer upwelling in the South China Sea and its role in regional climate variations. J. Geophys. Res. 108(C8): 3261, doi: 10.1029/2003JC00 1867.

Xie X., Müller R.D, Li S., Gong Z. and Steinberger B. 2006. Origin of anomalous subsidence along the Northern South China Sea margin and its relationship to dynamic topography. Mar. Petroleum Geol. 23: 745–765.

Yang H.J. and Liu Q.Y. 1998. The seasonal features of temperature distributions in the upper layer of the South China Sea. Oceanol. Limn. Sinica 29: 501–507 (in Chinese).

Yan P. and Liu H. 2004. Tectonic-stratigraphic division and blind fold structures in Nansha waters, South China Sea. J. Asian Earth Sci. 24: 337–348.

Yan P., Zhou D. and Liu Z. 2001. A crustal structure profile across the northern continental margin of the South China Sea. Tectonophysics 338: 1–21.

Yao B. 1996. Tectonic evolution of the South China Sea in the Cenozoic. Mar. Geol. Quat. Geol. 16(2): 1–13 (in Chinese).

Yao B., Wan L. and Wu N. 2005. Cenozoic tectonic evolution and the 3D structure of the lithosphere of the South China Sea. Geol. Bull. China 24: 1–8 (in Chinese).

Yeh Y.C. and Hsu S.K. 2004. Crustal structures of the northernmost South China Sea: Seismic reflection and gravity modeling. Mar. Geophys. Res. 25: 45–61.

Yuan D.L. 2002. A numerical study of the South China Sea deep circulation and its relation to the Luzon Strait transport. Acta Oceanol. Sinica 21: 187–202.

Yumul G.P. Jr., Dimalanta C.B., Tamayo R.A. Jr. and Maury R.C. 2003. Collision, subduction and accretion events in the Philippines: A synthesis. Island Arc 12: 77–91.

Yu H.S. and Chou Y.W. 2001. Characteristics and development of the flexural forebulge and basal unconformity of Western Taiwan Foreland Basin. Tectonophysics 333: 277–291.

Zhou D., Ru K. and Chen H.Z. 1995. Kinematics of Cenozoic extension on the South China Sea continental margin and its implications for the tectonic evolution of the region. Tectonophysics 251: 161–177.

Zhou Z., Jin X., Wang L., Jian Z. and Xu C. 2004. Two closures of the Indonesian seaway and its relationship to the formation and evolution of the western Pacific warm Pool. Mar. Geol. Quat. Geol. 24: 7–14 (in Chinese).

Zhu B. and Wang H. 1989. Nd-Sr-Pb isotope and chemical evidence for the volcanism with MORB-OIB source characteristics in the Leiqiong area, China. Geochimica 18: 193–201.

.

Chapter 3
Stratigraphy and Sea Level Changes

Qianyu Li, Guangfa Zhong and Jun Tian

Introduction

Over the last three decades, intensive industrial drilling and marine geological research climaxing with ODP Leg 184 have generated voluminous data on the variations and distribution of sediment sequences in the South China Sea (SCS). Lithostratigraphy, biostratigraphy and sequence stratigraphy provide evidence of regional tectonic activities and basin evolution. In addition to these classical approaches toward stratigraphic correlation, isotopic stratigraphy based on tuned astronomical cycles is playing an increasingly important role in refining the regional stratigraphy, especially at deep-sea settings. ODP Site 1148 is unique worldwide in providing a continuous isotope sequence for the last 23 myr, and ODP Site 1143 represents one of the best ODP sites worldwide with the highest time-resolution isotope record for the last 5 myr. This chapter summarizes the characteristics of the regional biofacies and lithofacies including sequence stratigraphy in major northern and southern basins, as well as deep-sea oxygen isotope records and regional sea level changes.

Q. Li (✉)
State Key Laboratory of Marine Geology, Tongji University, Shanghai 200092,
China; School of Earth and Environmental Sciences, The University of Adelaide,
South Australia 5005, Australia
e-mail: qli01@tongji.edu.cn; qianyu.li@adelaide.edu.au

G. Zhong
State Key Laboratory of Marine Geology, Tongji University, Shanghai 200092, China
e-mail: gfz@tongji.edu.cn

J. Tian
State Key Laboratory of Marine Geology, Tongji University, Shanghai 200092, China
e-mail: tianjun@tongji.edu.cn

P. Wang, Q. Li (eds.), *The South China Sea*, Developments in Paleoenvironmental
Research 13, DOI 10.1007/978-1-4020-9745-4_3,
© Springer Science+Business Media B.V. 2009

3.1 Lithostratigraphic Overview (Li Q. and Zhong G.)

Pre-Cenozoic Basement

Pre-Cenozoic rock types encountered in exploration wells in the SCS include Jurassic to Cretaceous marine shales and sandstones, Cretaceous or older igneous granites, and metamorphic rocks. Igneous, sedimentary and metamorphic rock types may respectively comprise 55%, 25% and <20% of all known basement rocks (Gong and Li 1997). Their burial depths differ greatly between basins due to differential subsidence and uplift in various rifting and postrifting stages of basin evolution. For example, the NE–SW stretching Zhujiangkou Basin sits on a basement at 3,000–4,000 m depths, which comprises mainly granite in the east and metamorphic quartzous sandstones and siltstones in the west. Further east, however, the Taixinan Basin has a basement formed by Jurassic to lower Cretaceous marine shale sometimes even bearing deep-sea radiolarians (Jin 1989; Gong and Li 1997; Shao et al. 2007; Wu et al. 2007). In the south, granite and dorite are dominant, but volcanic and shallow marine sediments are also common, indicating a more complex pre-rift lithotopography (ASCOPE 1981; Du Bois 1985).

The oldest sedimentary rock so far recovered in northern shelf-slope basins is limestone of Devonian–Carboniferous age (Gong and Li 1997). Early Mesozoic marine sandstone and mudstone mainly occur in areas near Taiwan in the northeast and to the east of the Dangerous Grounds in the south, although whether they belonged to contemporary deposition in a single pre-rift shelf basin is not clear. Wider occurrence of Jurassic and Cretaceous marine shale and sandstone indicates a broad shallow "Proto-South China Sea" during the late Mesozoic and early Cenozoic. Mesozoic sequences can be defined by seismic intervals at 4.0–5.5 km/s (TWTT) from the Zhujiangkou Basin (Huang et al. 1993; Yao and Zeng 1994), at 2.0–3.5 km/s from the eastern Zhujiangkou–Taixinan basins (Su et al. 1995), and at >5.0 km/s in various southern basins. The thickness of Mesozoic sequences may reach 2,000–4,000 m and more (Jin 1989; Xia and Zhou 1993).

Lithostratigraphy of Syn-Rift Sediments

Rifting started earlier in the Paleocene in the northern and northeastern SCS and in the Eocene or later in southern areas (Chapter 2). Many half-grabens developed in areas now occupied by the present shelf and slope, subsequently leading to the formation of shelf-slope basins (Fig. 3.1).

Sediments deposited during the early phase of rifting in the Paleocene and Eocene are characterized by alluvial clastic, lacustrine to littoral marine facies, occasionally interrupted by volcanoclastic rocks. The terrigenous-dominated sediment sequences suggest a wide range of lacustrine, estuary to shallow marine environments over many parts of the SCS. The South China and Indochina landmasses provided the primary source to sediment deposition in half-grabens. Often unconformably overlain by post-rift sequences of mainly marine origin, these syn-rift

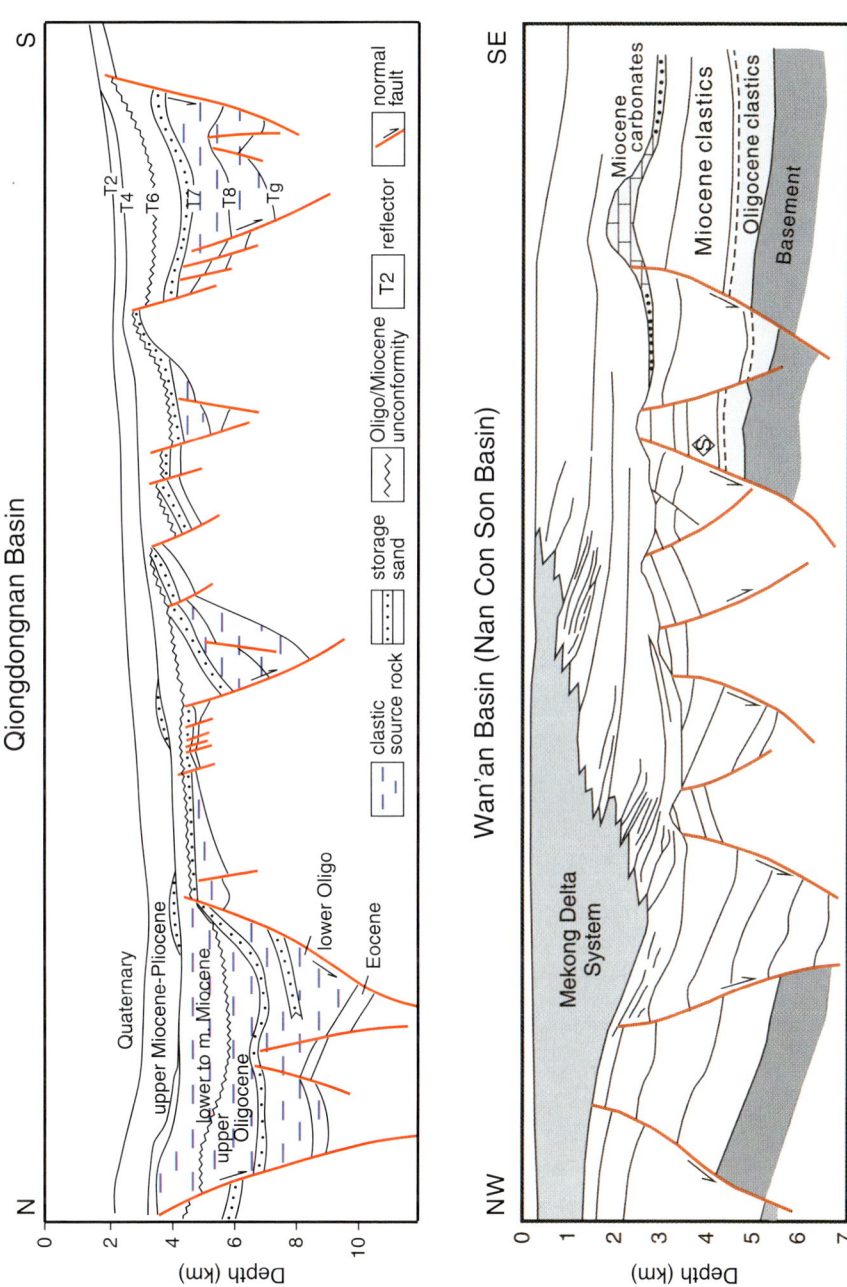

Fig. 3.1 Schematic cross sections show the distribution of the syn-rift/post-rift boundary and major sediment systems in the southern part of the Qiongdongnan Basin (upper panel, modified from Gong and Li 1997) and in the near shore part of the Wan'an (Nam Con Son) Basin (lower panel, modified from Matthews et al. 1997). See Fig. 3.27 for locations

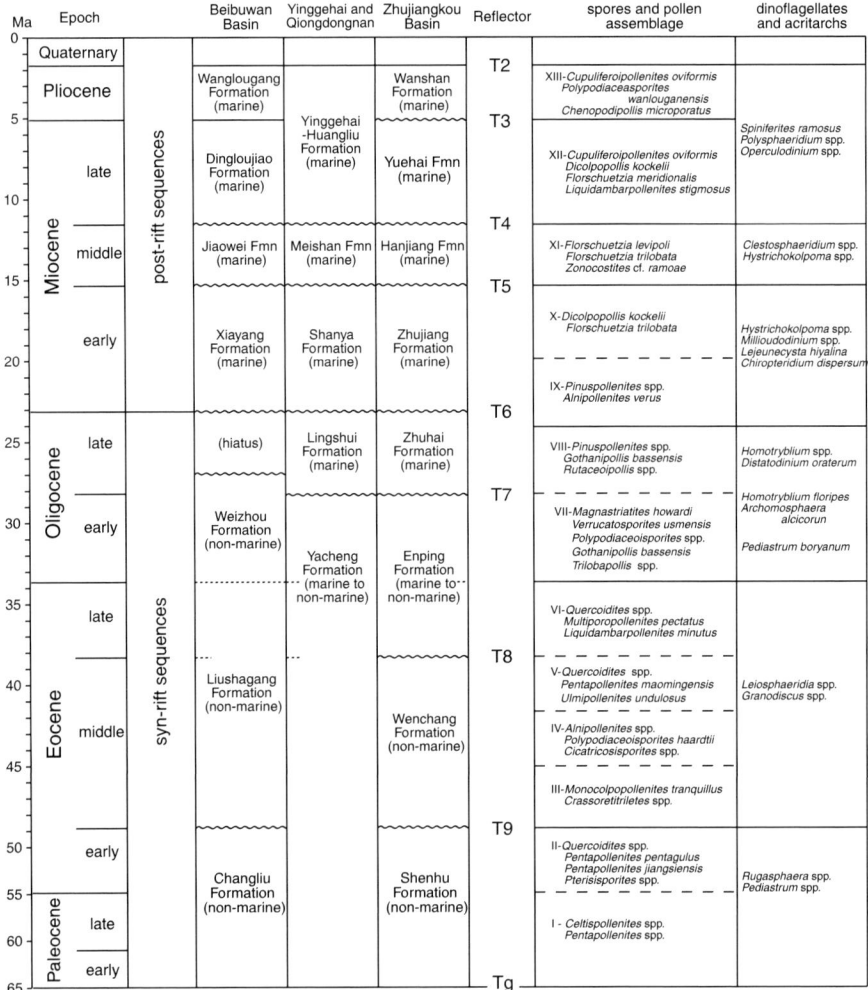

Fig. 3.2 Major lithostratigraphic units from northern SCS basins are bounded by unconformities (reflectors T2 to Tg), shown together with important palynologic assemblages (modified from Jiang Z. et al. 1994). Note that vertical scale changes at 50 Ma

sequences show similar features across basins in both lithological features and sediment packaging (Figs. 3.1, 3.2, 3.3). Containing rich hydrocarbon resource, some intervals of these sequences have been subject to industrial exploration for many decades.

Representing the syn-rift sequence in the northern SCS are Changliu Formation, Liushagang Formation, and Weizhou Formation in the Beibuwan Basin, Yacheng Formation and Lingshui Formation in the Qiongdongnan and Yinggehai basins, and Shenhu Formation, Wenchang Formation and Enping Formation in the Zhujiangkou Basin (Fig. 3.2). They are all fluvial to lacustrine deposits with various thicknesses (Fig. 3.3). The upper Oligocene Yacheng and Lingshui formations in the Yinggehai

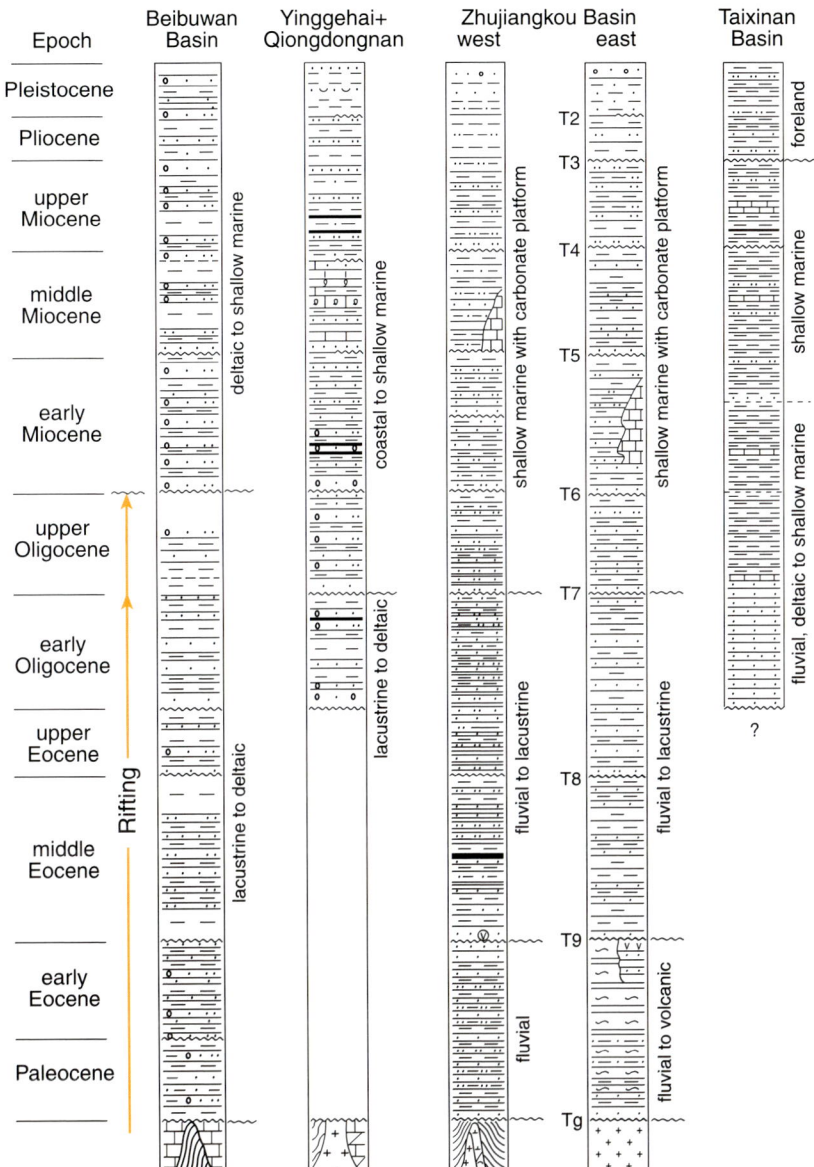

Fig. 3.3 Simplified lithostratigraphic columns show variations of syn-rift and post-rift lithofacies in northern SCS basins (modified from Jiang Z. et al. 1994). The Taixinan Basin data are from Lee et al. (1993). Refer to Fig. 3.5 for lithofacies explanations

Basin often disconformably overlie pre-Cenozoic granites or dolomites. Although stratigraphically relatively young, these two formations from the Yinggehai and Qiongdongnan basins may attain a thickness of over 3,000 m, similar to all syn-rift formations combined from other northern basins.

Many of these lithological units can be differentiated not only by sediment/rock types but also by physical property variations. A number of seismic surfaces are consistent over long distance and therefore have become standard reference reflectors for subdividing sequence stratigraphic units and lithological units in northern SCS basins (Fig. 3.2).

However, seismic reflectors of the same age from different basins are often labeled with different names, and vice versa for those with the same names but from different basins. For example, reflector T2 is recognized as marking the Pliocene/Pleistocene boundary in the Yinggehai and Qiongdongnan Basins (Fig. 3.1), but T2 is used for the Oligocene/Miocene boundary reflector in the Beibuwan Basin (Kang et al. 1994; Ying 1998). Therefore, care should be exercised when applying these nomenclatures to specific sequences. In this chapter, unless otherwise explicitly mentioned, we use the reflector sequence listed in Figs. 3.2 and 3.3 for northern basins.

In the south, near the Dangerous Grounds, the main rift sequence accumulated also during the late Eocene–Oligocene (Xia and Zhou 1993; Mohd Idrus et al. 1995). Among others, ASCOPE (1981) and Du Bois (1985) overviewed the general lithostratigraphy, structure and hydrocarbon occurrences in various southern basins. Their compilations still remain good sources for understanding the regional stratigraphic framework in the southern SCS. Madon et al. (1999a,b) summarized stratigraphy of the Malay Basin and neighboring areas, providing correlation data that were previously available only within the industry circle. Hutchison (2004) provided an updated general account of southern SCS geology using core and seismic data and emphasized sedimentary features from the southwest surrounding the Sundaland and the southeast collision zone region separated by the Baram line. The Oligocene and early Miocene syn-rift sequences from these southern basins show a rifting-spreading process clearly distinct from northern basins (Figs. 3.4 and 3.5). The main features include: (1) syn-rift sequences are younger in age, Eocene to early Miocene, (2) lacustrine and deltaic sediments are dominant, (3) more clastic deposition occurred in basins from the southwest than the southeast, and (4) carbonate platforms started to form offshore Palawan and Reed Bank.

Post-Rift Sediments in Shelf-Slope Basins

General Characteristics

A prominent regional unconformity separates the post-rift sediment sequences from the underlying syn-rift sequences, approximating the Oligocene/Miocene boundary in northern basins (Figs. 3.1, 3.2, 3.3) and the early/middle Miocene boundary in southern basins (Figs. 3.1, 3.4, 3.5). In the south, it is often referred to "mid-Miocene unconformity (MMU)" or "deep regional unconformity (DRU)" (Fig. 3.6) (ASCOPE 1981; Yumul et al. 2003; Hutchison 2004). Up to ~5 myr sediment record may have been erased by this break-up unconformity (Chapter 2).

Shallow marine deposition was the main feature of post-rift sedimentation in most shelf basins, but coastal and deep water depositions were also taken place

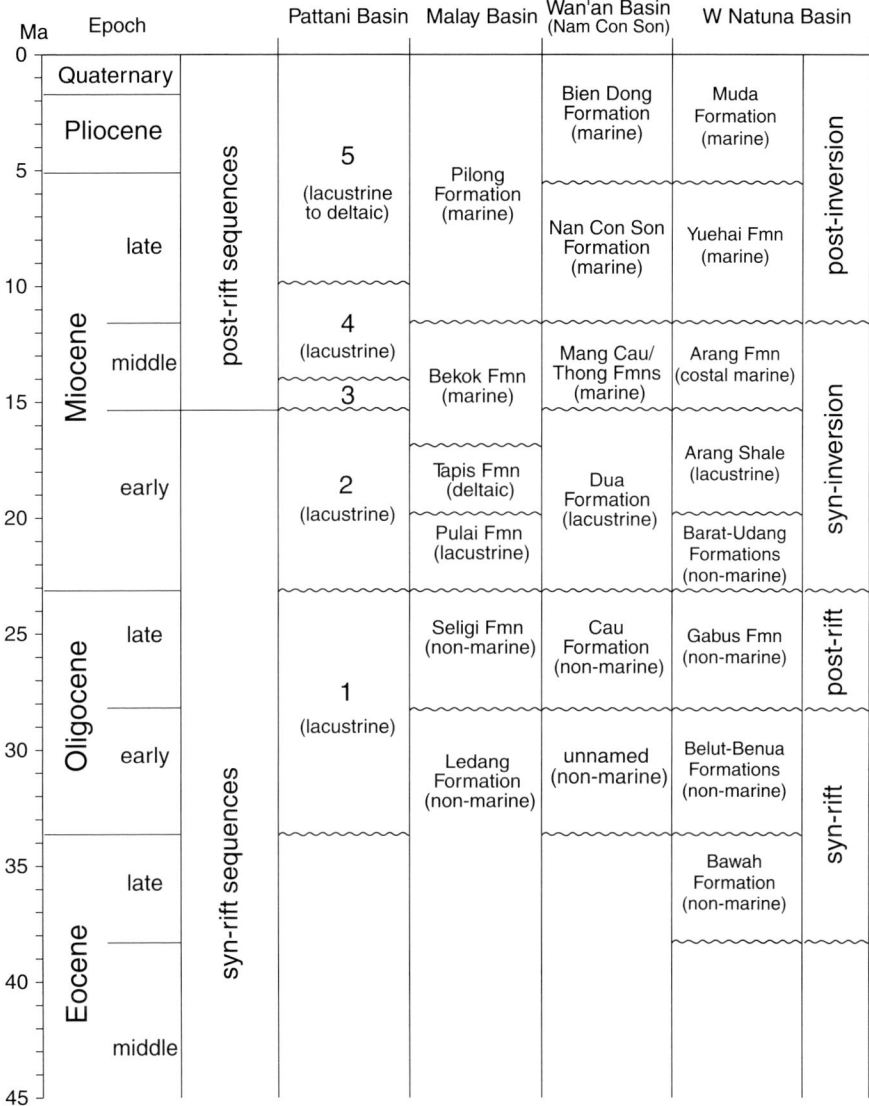

Fig. 3.4 Major lithostratigraphic units from southern SCS basins are separated by unconformities (compiled from Jardine 1997; Matthews et al. 1997; Madon et al. 1999a,b; Darmadi 2005). Younger sequences in the West Natuna Basin and basins offshore from Borneo-Parawan-Luzon-Taiwan are often associated with an inversion process due to subduction along the eastern margin

widely in more near-shore and more off-shore basins respectively (Figs. 3.2–3.5). Lithofacies in individual basins remain highly variable in space and time, although marine components now became more common compared to the underlying syn-rift sequences. As in many other marginal seas (Clift et al. 2004; Wang 2004), several factors have been acting together on the abundance, packaging and distribution of

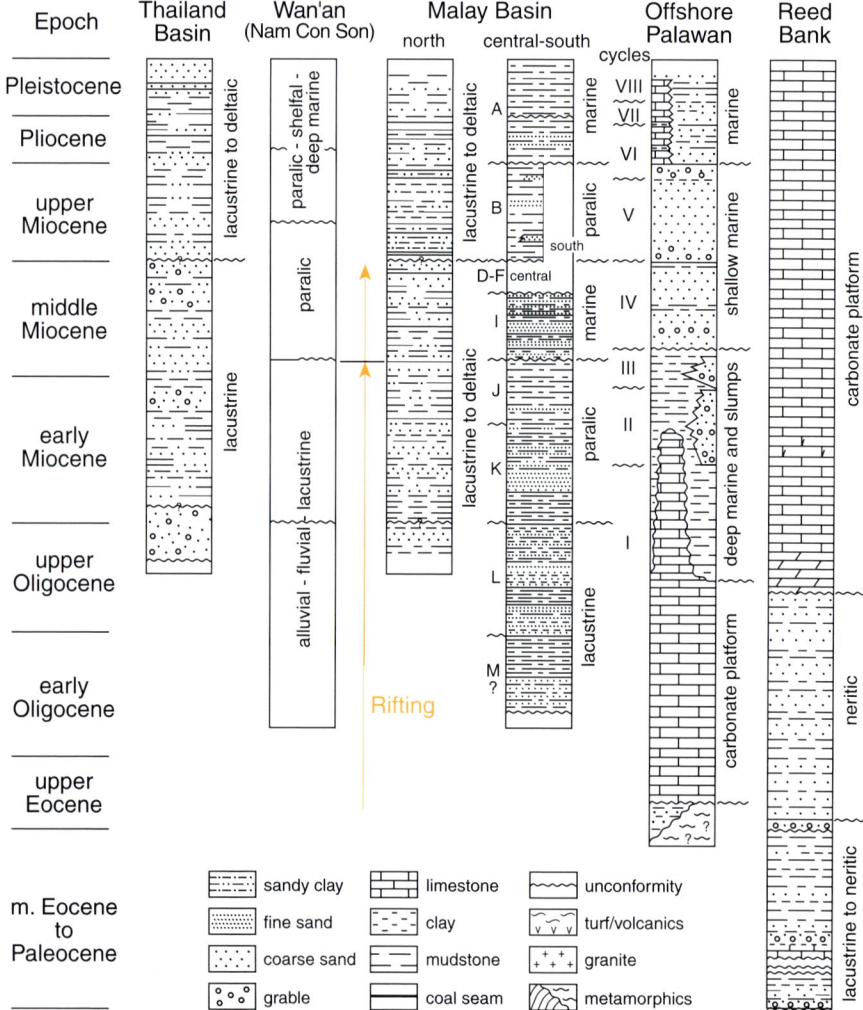

Fig. 3.5 Simplified lithostratigraphic columns show variations of syn-rift and post-rift lithofacies in southern SCS basins (modified from ASCOPE 1981; Jiang Z. et al. 1994). Carbonate platforms were well developed in basins near the Dangerous Grounds (Reed Bank) and offshore Sarawak, representing tectono-environmental regimes different from those in the west and in the north

sands, silts, clays and biological components in this marginal sea basin. The most important factors include geographic setting (proximity to source and depositional position), climate (sea level, monsoons and weathering), and tectonics (Jin 1989; Gong and Li 1997).

In the northern SCS, marine facies first established during the late Eocene–early Oligocene in both the Qiongdongnan and (outer) Zhujiangkou basins (Jiang Z. et al. 1994). However, full marine influence in all northern basins did not start until the early Miocene (Figs. 3.2, 3.3). Representing the first post-rift sediment package are Xiayang Formation in the Beibuwan Basin, Shanya Formation in the Yinggehai and

Qiongdongnan Basins, and Zhujiang Formation in the Zhujiangkou Basin (Fig. 3.2). All these sediment packages contain abundant terrigenous sands and clays as well as marine components at various proportions. Marine influence reached a maximum during the middle Miocene–Pliocene, resulting in the accumulation of deeper water lithofacies and biofacies (discussed below). Since the late Miocene, sedimentation started to become more or less continuous even across basins, although variations in physical property of sediments between basins are huge due to local factors such as accumulation rate, heat, compaction and reservoir distribution. The maximum thickness of Oligocene to Quaternary sequences may reach 10 km or more in the Yinggehai Basin and Zengmu Basin, and often 3–8 km in other basins. There is no doubt, therefore, that variability in the thickness of these sequences and their burial depths has been strongly affecting on local heat and fluid flow and the evolution of the thermal regime in the SCS at large (He et al. 2001).

Postrift lithostratigraphy in southern basins is characterized by cycles of paralic sediments presumably resulting from frequent tectonic activities and sea level fluctuations (Figs. 3.4, 3.5). The I to VIII cycle concept, first proposed by the Shell Company for Oligocene to Pleistocene sequences from offshore of Sarawak in the 1970s (Fig. 3.6), has been widely applied to neighboring regions (Figs. 3.5, 3.6). Due to structural differences and complex depositional environments between basins strongly affected by syn-depositional tectonic deformation, however, these "cycles" rarely show a clear cyclic relationship between them, making intra- and inter-basin correlation difficult. The complexity of tectono-stratigraphy is well illustrated in several sections from offshore Palawan by Schlüter et al. (1996), in addition to those generalized in Figs. 3.5 and 3.6. To avoid confusion, supersequences or stages bounded by finite unconformities (seismic reflectors) have been introduced to aid correlation between the clastic-dominated lithofacies in southern basins (e.g., Lovell 1987; Mat-Zin and Tucker 1999). These stages, supersequences or seismic groups (Fig. 3.6) are so named that seismic sequences can be differentiated or otherwise be called cycles earlier. For example, seismic supersequences T1S to T7S and seismic groups M to A were proposed for the Oligocene to Pleistocene succession in offshore Borneo and in the Malay Basin respectively (Fig. 3.6). However, to unify the complex stratigraphic nomenclatures currently used in the southern region for a more easily identified, well-architected correlation scheme still remains a challenge (Madon et al. 1999b).

The Miocene–Pliocene sediment sequences in several basins from the northeast, southeast and southwest have been strongly deformed due to an inversion process and other neotectonic activities associating especially with the collision and subduction along the eastern SCS margin. These deformed sequences not only have their sediment properties altered but also pose a great challenge to exploration for oil and gas as well as other resources (Ingram et al. 2004).

Carbonate Platforms

Oldest shallow-water carbonate platforms of latest Eocene to earliest Oligocene age are known to occur offshore Palawan and neighboring islands (Fig. 3.5) (ASCOPE 1981; Jiang Z. et al. 1994; Qiu 1996; Qiu and Wang 2001). They continued to

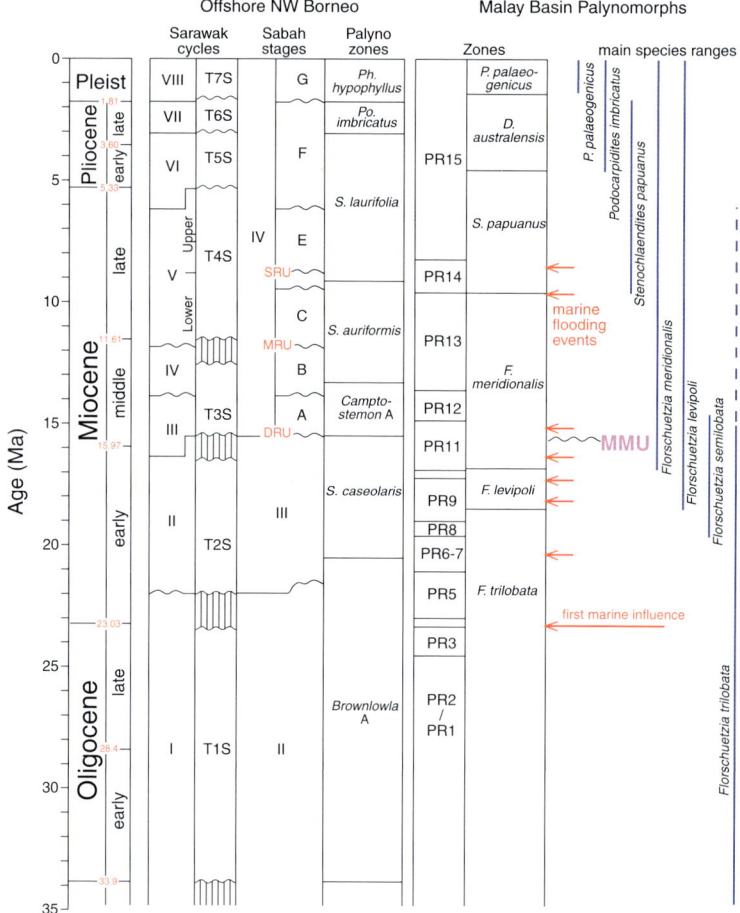

Fig. 3.6 Lithostratigraphic cycles from offshore NW Borneo are correlated to palynomorph zones from the neighboring Malay Basin (modified from Madon et al. 1999a,b; Mat-Zin and Tucker 1999). SRU, MRU and DRU are abbreviations for shallow, middle and deep regional unconformities, respectively. The latter is also referred to "mid-Miocene unconformity" or MMU

grow and expand until the late Oligocene or early Miocene before these areas inter-mittently subsided due to seafloor spreading and subduction. Among the Miocene platforms from the south are the Loconia–NW Sabah platform and Kudat platform from offshore Sabah (Madon and Redzuan 1999b; Madon et al. 1999c). Drowned platform forms a prominent feature under the Dangerous Grounds, Nansha (NW Borneo) Trough and Palawan slope (Fig. 3.7) (Hutchison 2004). By late Oligocene time, the center of carbonate growth shifted to area now called the Reed Bank (Liyue) and further north (Taylor and Hayes 1980). In the Zengmu Basin (East Natuna), a distinct middle to late Miocene carbonate complex called the Terumbu Formation L-structure is mainly composed of boundstones, grainstones and pack-stones that have been homogenized by organic activity (May and Eyles 1985).

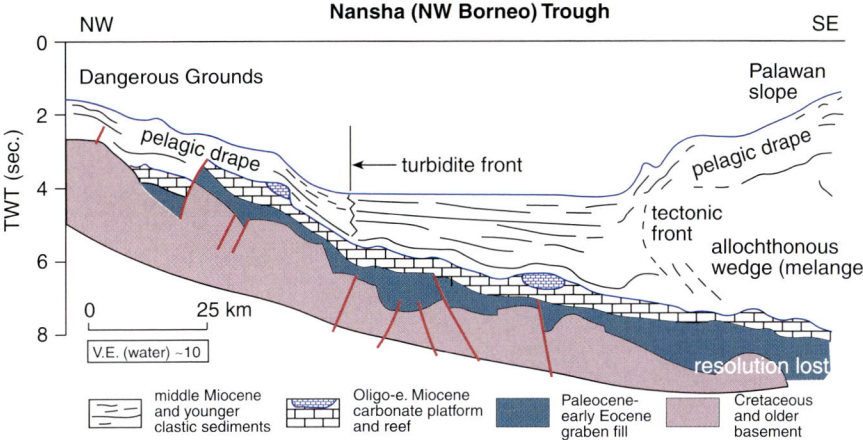

Fig. 3.7 Cross section shows drowned Oligocene–early Miocene carbonate platform and younger turbidite deposition along the Nansha (Borneo) Trough, offshore from Sabah (modified from Hutchison 2004)

During the late Miocene, carbonate platforms were widely developed also in the southeastern Wan'an (Nam Con Son) Basin not affected by the Mekong Delta (Fig. 3.1; Fig. 3.5) (Matthews et al. 1997), as well as offshore Sabah (Madon and Redzuan 1999b; Madon et al. 1999c). Apart from the Reed Bank, reef platforms now scattering in the southern region all appear to have been building on a Miocene or younger base.

In the north, carbonate platforms developed along the shelf edge since the late Oligocene and achieved a maximum distribution during the early Miocene (Fig. 3.8). Like its southern counterparts, however, most carbonate platforms in the north were subsequently drowned during the latest early Miocene and early middle Miocene, responding to further local subsidence along the northern margin and a general rise in sea level. Stratigraphic subdivision of these limestones has been mainly based on variations in colors and biogenic contents especially larger benthic foraminifera.

Sediment Accumulation Rates in Shelf-Slope Basins

Sediment accumulation rates vary considerably between basins and even between different structural units within a single basin (Fig. 3.9). Sedimentation rate changes are largely controlled by sediment input (erosion), basin type, and topographic and structural positioning of the site locality (Clift et al. 2004). In the northern SCS, high accumulation rates of ~500 m/myr occurred in the following intervals: early middle Eocene (Yinggehai Basin), latest Eocene (Beibuwan Basin), late Oligocene (Zhujiangkou Basin), and since the early Pliocene (Qiongdongnan Basin) (Fig. 3.9A) (Gong and Li 1997). About half of these are attributed to structural accumulation rates. In a recent study, Clift et al. (2002) found high accumulation rates near the middle/late Miocene boundary in the Zhujiangkou Basin, when sea

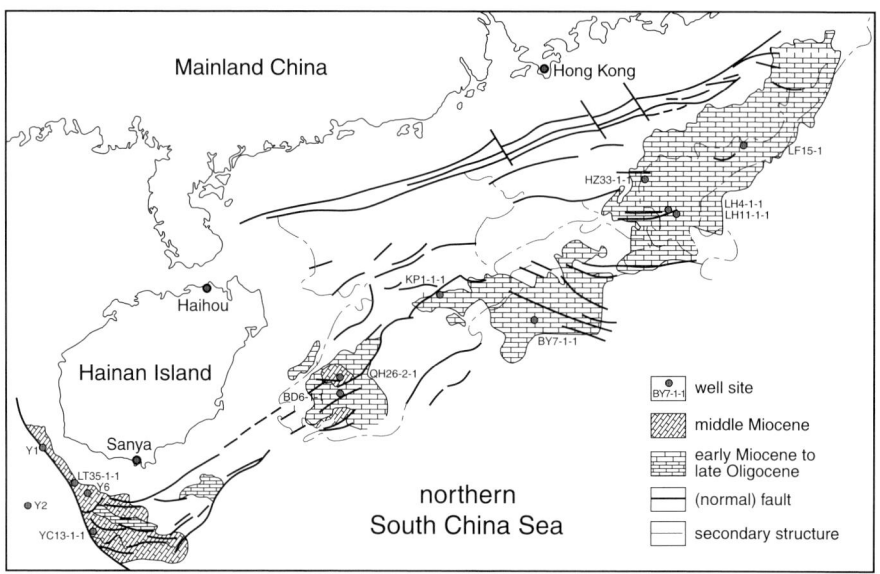

Fig. 3.8 Sketch map shows areal distribution of Oligocene–Miocene carbonate platforms along the northern SCS margin (modified from Gong and Li 1997). Major structural units and selected drilling wells are also shown

level was lower (Fig. 3.9B). According to these authors, average accumulation rates changed from 20–40 m/myr at ∼25 Ma to ∼10 m/myr or less during 24–19 Ma before increasing to ∼40 m/myr or more near 11 Ma. On a regional scale, the late Pliocene–Quaternary is another period with high sediment accumulation, contributing to the deposition of a large part of the Sunda Shelf and contemporary sediment packages along the northern margin.

Deep Water Lithostratigraphy

Seismic stratigraphy indicates a sediment cover of ∼1 km in the SCS deep basin except for areas close to seamounts (Fig. 2.22). The deepest section so far drilled is at ODP Site 1148 in the lower slope of the northern SCS, at a water depth of 3,292 m. Hole 1148A penetrated to 853 m and recovered a 632 m sediment sequence (∼74% core recovery), over the last ∼33 Ma (Wang et al. 2000; Li et al. 2006).

As in other marginal seas, terrigenous sediments attain very high percentages even at ODP sites: >50% for the Oligocene, 40–60% for the Miocene–early Pliocene, and up to 70–85% for the late Pliocene–Quaternary (Wang et al. 2000). Increase in biogenous components in the Miocene deep sea matches well with a wider marine deposition and carbonate platform development on shelf. Together, these lithostratigraphic features imply significant subsidence in the SCS during the Miocene and early Pliocene. Compared to sediments in shelf-slope basins, however, deepwater

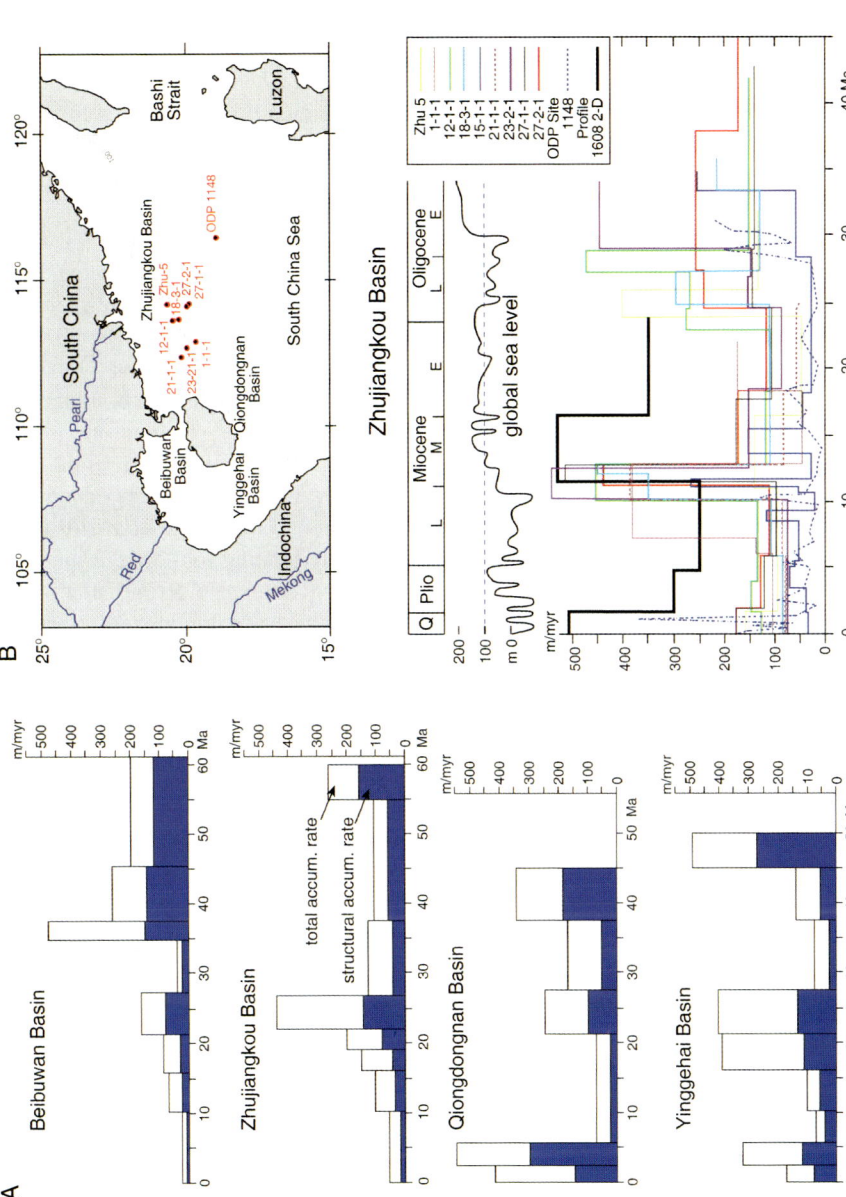

Fig. 3.9 (**A**) Bar diagrams show average total (*unshaded*) and structural (*shaded*) accumulation rates in various northern SCS basins (modified from Gong and Li 1997). (**B**) Sedimentation rate variations in different wells from the Zhujiangkou Basin and ODP Site 1148 are shown against the global sea level (modified from Clift et al. 2002)

sediments not only contain more biogenous elements but also form a much thinner Oligocene–Quaternary sequence. At ODP Site 1148, the 853 m sequence recovered includes about 350 m sediments from early Oligocene rifting (Fig. 3.10). From this sequence, seven lithostratigraphic units were identified based on sediment composition (especially clay vs. nannofossils), depositional facies, and color variations (Wang et al. 2000). Trace fossils including *Zoophycos* and *Chondrites* are common. As summarized in Table 3.1, the Pliocene–Holocene section is represented by Unit I, the Miocene by Units II–V, and the Oligocene by Units VI and VII.

The Oligocene section exceeds 390 m, about 90% of which, ~350 m, are monotonous grayish to olivergreen, quartz-rich clay that accumulated during the early Oligocene (Unit VII) at a sedimentation rate of over 60 m/myr. This section is mixed between deep water and shallow water biofacies of ostracods, dinoflagellates, spores and pollen, benthic and planktonic foraminifers, nannoplankton, etc. Nannofossil chalk was mainly confined to the upper Oligocene in Core 1148A-50X, ~473–480 mcd (meter composite depth), although pieces of chalk mixed with other sediments can be found up to about 420 mcd. The Oligocene/Miocene boundary coincides with the unconformity at the top of the slumped unit, and sediment-mixing due to slumps characterizes the upper Oligocene (Unit VI) sediments that bracket the double seismic reflectors (Fig. 3.11) (Wang et al. 2000). This slumped unit and associated unconformities signal a large-scale tectonic transition from rifting to spreading (Clift et al. 2001; Li et al. 2005a) probably relating to the ridge jump centered at ~25 Ma (Briais et al. 1993). The missing section spans a time interval of about 3 myr although all the slumped sediment could have accumulated in a short time period (Li et al. 2005a). The lower Miocene (Units V and IV) consists mainly of greenish to grayish brown nannofossil clay mixed sediments, with common iron sulfide particles. Due to a greater water depth at the site, biogenous skeletons are strongly affected by dissolution (Li et al. 2006).

A better preserved Miocene to Pliocene succession was recovered at the shallower northern Site 1146 (2,092 m) and southern Site 1143 (2,772 m) with oldest sediments of 18.6 Ma and ~11 Ma, respectively (Wang et al. 2000). Unlike those from the north, the late Miocene section at Site 1143 contains many thin turbidite layers (Fig. 3.12), similar to those found in the Nansha (NW Borneo) Trough offshore from Sabah and Palawan (Fig. 3.7) (Mohd Idrus et al. 1995; Hutchison 2004).

Up to 2600 m of turbidite deposits overlying a thin pelagic-semipelagic layer and oceanic basement occurs in the northern part of the Manila Trench. These trubidites were transported from the uplifted collision zone of Taiwan, southward along the Manila Trench by gravity-controlled processes (Lewis and Hayes 1984). These authors also suggested that thinner turbidite deposits in the southern end of the trench were transported northward from the collision zone involving the North Palawan block and the central Philippines.

As in other basins, turbidites in the Taixinan Basin are better developed at slope settings (Lee et al. 1993; Huang C.-Y. 2007, personal communication). Two distinct turbidite systems, namely the late Miocene system and the Pliocene–Quaternary system without any obvious kinetic relationship, occur in Taiwan. Late Miocene turbidites found in the Hengchun peninsula were deposited in the northeastern SCS

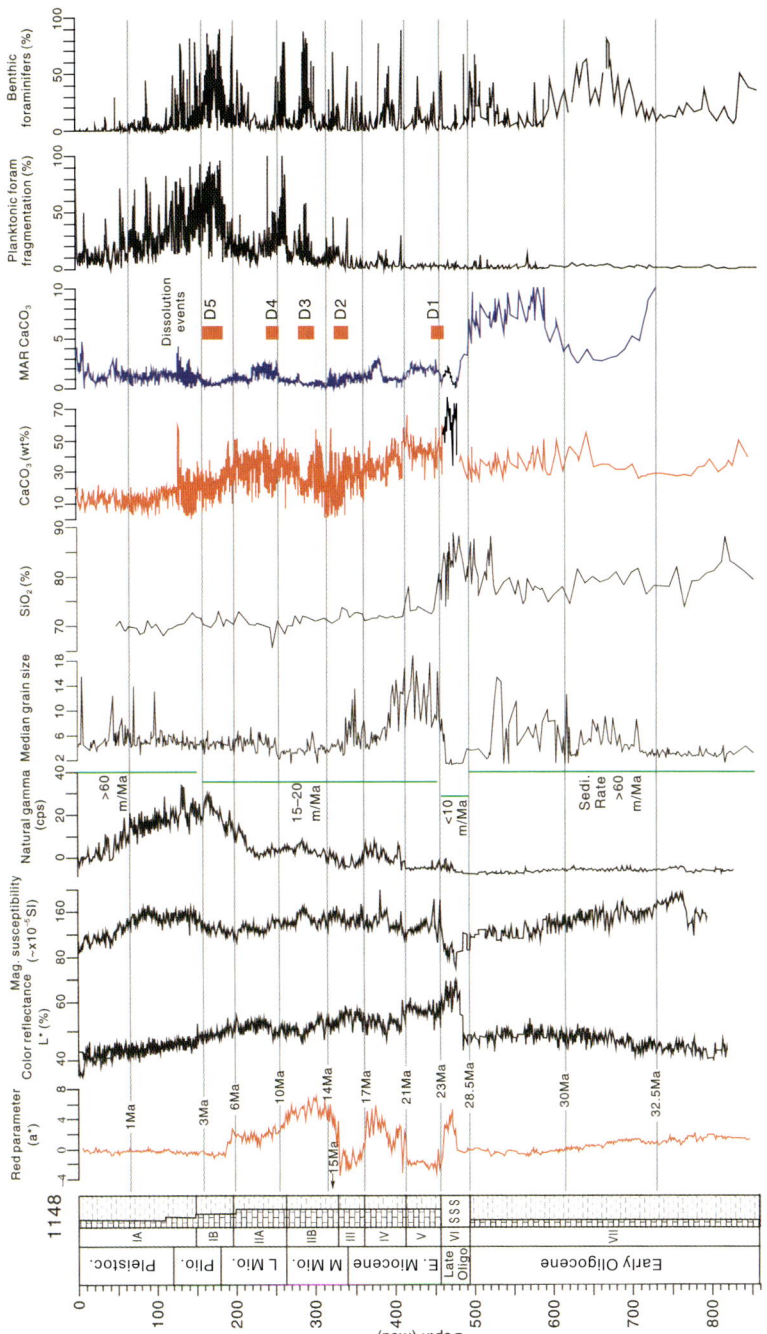

Fig. 3.10 Lithostratigraphic units, geophysical logs, major sediment components and foraminiferal features from ODP Site 1148 show variations in sediment properties including 5 dissolution events D1 to D5 that mark major evolutionary stages of the South China Sea deep water over the last 33 Ma. Note that high sedimentation rates (>60 m/myr) occurred in two intervals: the early Oligocene and late Pliocene–Pleistocene. Data are mainly from Wang et al. (2000), Chen et al. (2002), Shao et al. (2004), and Li et al. (2006)

Table 3.1 Major features of Site 1148 lithostratigraphic units and calculated sedimentation rates are based on data from Wang et al. (2000)

Litho unit	Depth at base (mcd)[1]	Lithological characteristics	Sedi Rate (m/myr)	Age
I	194.02	well-bioturbated nannofossils clay with quartz and abundant pyrite concretions. Subunit IA (0–146.02 mcd) is more clay-rich, with more green clay layers and irregular green clay mottles, and common "iron sulfide" in the middle part; Subunit IB contains more frequent light intervals marking the increase of nannofossils.	~60 (IA) ~15 (IB)	Pleistocene to Pliocene
II	328.82	dominated by clay with frequent light-colored, carbonate-rich layers of nannofossils, few quartz grains or siliceous microfossils. Subunit IIA (194–260 mcd) is olive-gray colored; Subunit IIB is reddish brown colored.	~22 (IIA) ~15 (IIB)	late to middle Miocene
III	360.22	grayish green clayey nannofossil ooze with 10–50 cm thick intercalations of dark reddish brown clayey nannofossil ooze or clay with nannofossils, with a sudden jump in the a* value at the base.	~30	middle Miocene
IV	412.22	brownish nannofossil clay mixed sediment with a minor amount of greenish gray nannofossil clay intercalations; a significant increase in the a* value at the base.	~20	early middle to early Miocene
V	457.22	greenish gray nannofossil clay mixed sediment interbedded with nannofossil clay and very minor amounts of clay with nannofossils.	~30	early Miocene
VI	494.92	slumped clay–nannofossil mixed sediment and nannofossil clay especially between 457 and 472 mcd; a thin layer of chalk at 473–480 mcd.	<5	late Oligocene
VII	859.45	intensely bioturbated sequence of monotonous grayish olive-green nannofossil clay.	>60	early Oligocene

[1]mcd = meter composite depth.

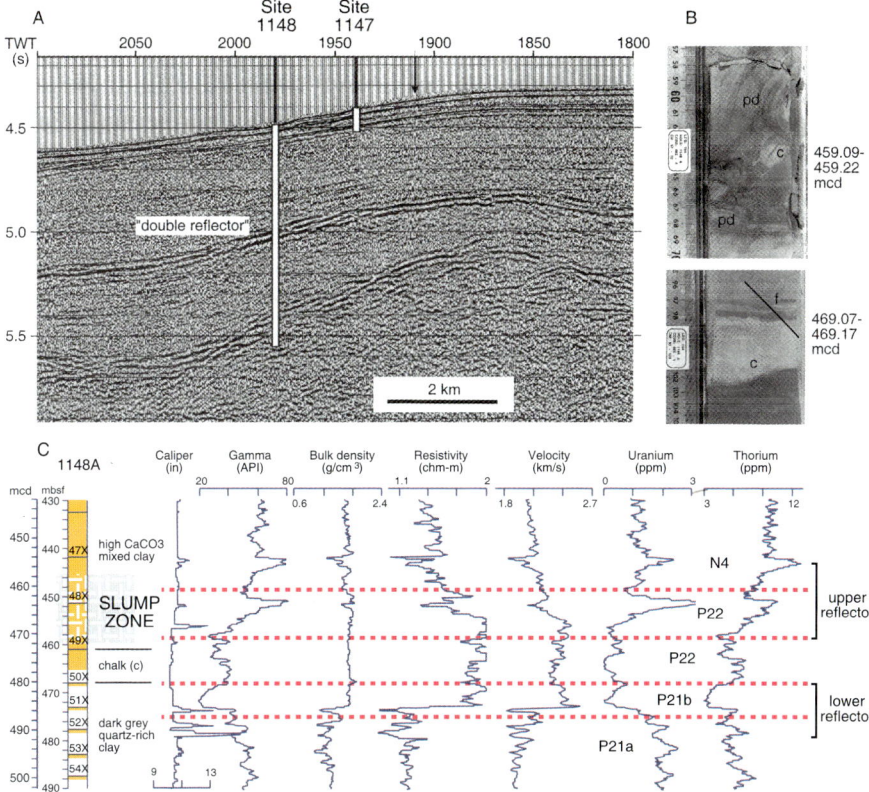

Fig. 3.11 Geophysical and lithological characteristics marking late Oligocene slumped deposits and unconformities at ODP Site 1148 are documented in (**A**) a double reflector (DR) on seismic profile, (**B**) core sediments showing displaced chalk (c), microfaults (f), and plastic deformation (pd), and (**C**) strong swings in all physical log readings (Wang et al. 2000). Dashed horizontal lines in C indicate unconformities defined by biostratigraphy, with planktonic foraminifer zones P21, P22 and N4 indicated (Li et al. 2005a)

passive margin and later accreted into the pre-collision accretionary prism when subduction was active in the late Miocene. In the Coastal Range of Taiwan, huge Plio–Pleistocene turbidite layers were deposited in the former and modern North Luzon Trough forearc basin before being uplifted (Huang et al. 1997c). Although they were developed in different tectonic settings, at least the uplifts of these two turbidite systems may have respectively related to the first phase of tectonic subduction in the late Miocene and then to the uplift of Taiwan and formation of island arcs since the Pliocene.

Therefore, the concentration of turbidite deposition along the eastern margin was responding to intensive collision and subduction since the late Miocene (Sibuet et al. 2002, 2004; Lin et al. 2003). Tectonic activities alter sediment load and sediment transfer dynamics in individual basins, contributing to most turbidite deposition in the SCS. However, whether the thin turbidite layers at ODP Site 1143 are generated

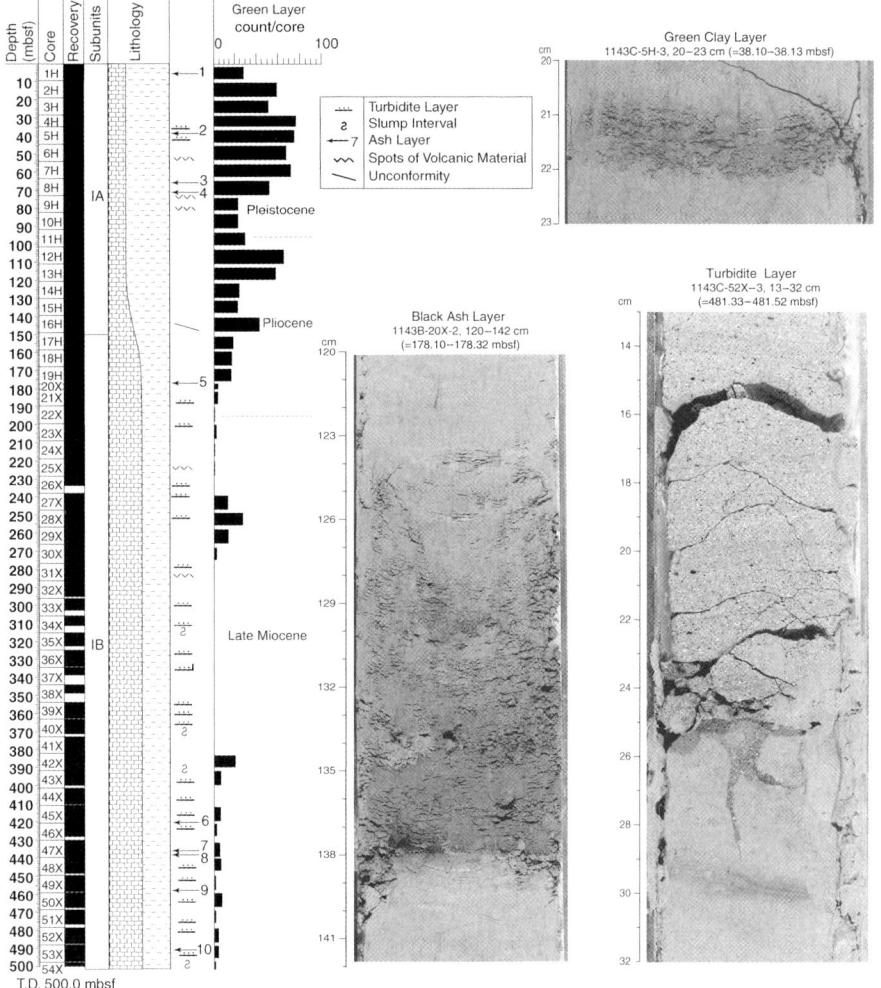

Fig. 3.12 Core recovery and lithofacies variations at ODP Site 1143 from the southern SCS are shown together with the distribution of green layers, ash layers, slumped intervals and turbidite layers, and some core photos (modified from Wang et al. 2000)

by periodic intrusions of turbidity fronts further afield from the east or by slope failures from the Sunda Shelf due to rapid discharge of the Mekong River especially during sea level lowstands is not clear.

The hemipelagic sediments in ODP 184 cores also contain some thin volcanic ash layers (Fig. 3.12). They are mostly confined to last 2 myr sediment sections at southern and since ~1 Ma at northern sites. The light colored ashes reflect the dominant dacitic-rhyolitic composition of the nearby arc's explosive fraction, and their increase uphole may imply either more volcanic eruptions during the Pleistocene or the diagenetic alteration (loss) of chemically unstable volcanic glass in the older, deeper parts of the section (Wang et al. 2000) (Fig. 3.12).

Thin green clay layers form another lithostratigraphic feature of ODP Leg 184 cores, and they appear to be of non-glauconite origin (Fig. 3.12). Most green layers are confined to the Pliocene–Pleistocene interval except for a lower Miocene set recovered at Site 1148. Their common association with burrows and patches caused by burrowing suggests that they may have a link with the former presence of organic matter, or some reducing conditions as suggested by their green color. As the green clay layers are interbedded with clear tephra-bearing unaltered volcanic glass and because no appreciable change in the background sediment is noted over these intervals, the diagenetic environment seems uniform between beds (Wang et al. 2000; Tamburini et al. 2006). Further studies on their true origin are clearly needed.

3.2 Biostratigraphic Framework (Li Q.)

Biostratigraphic work in the 1960s and 1970s by Chinese researchers led to the publication in 1981 of two monographs respectively on the Tertiary paleotology and depositional systems of the northern continental shelf of the SCS (Hou et al. 1981; Zeng et al. 1981). Various fossil assemblages mainly from the Beibuwan, Yinggehai and Zhujiangkou basins were described and illustrated, including spores and pollen, dinoflagellates, foraminifera, nannofossils, and ostracods. In the 1980s, many more papers and monographs on biostratigraphy and general micropaleontology were published, mostly also in Chinese except the monograph of "Marine Micropaleontology of China" edited by Wang (1985). These early studies laid a solid foundation for subsequent detailed biostratigraphic work, and provided information for reconstruction of paleoenvironmental evolution in the region (Wang 1990). Decades of stratigraphic research indicate that microfossils remain as one of the most powerful tools in correlating the regional stratigraphy. Specifically, floral assemblages are important for correlation and environmental interpretation of neritic and paralic deposits, while foraminifera and nannofossils are more useful for biostratigraphy and correlation of deeper water depositional sequences. In contrast, magnetostratigraphic endeavors have been less successful due to signal overprinting, often with only the topmost few magnetic (sub)chrons identified (e.g., Wang et al. 2000). We follow the timescale of Gradstein et al. (2004) in the following discussion, and we will refer to original age models only if correlation to this new timescale is too difficult.

Floral and Shallow-Water Faunal Assemblages

Associated with terrigenous clastics in many shallow water sedimentary units of the SCS are various spores and pollen assemblages (Figs. 3.2 and 3.6) and ostracod-dominated faunas (Table 3.2). Together with dinoflagellates, these floral and faunal

Table 3.2 Ostracod assemblages from the northern SCS are compiled from Jiang Z. et al. (1994) and Zhao (2005)

Epoch	Beibuwan Basin	Ying-Qiong basins	Age (Ma)	ODP Site 1148
Quaternary			0–14	*Krithe* (40%)–*Abyssocythereis* (10%)–*Henryhowella* (9%)
Pliocene	*Hemikrithe foveata–Neomonoceratina delicata*	*Neomonoceratina delicate–Ambocythere elliptic*		
Late Miocene	*Spinileberis inflexicostata–Cytheropteron striatituberculata*	*Argilloecia–Krithe–Xestoleberis*		
Middle Miocene	*Spinileberis longicaudata*		14–23	*Krithe* (44%)–*Pelecocythere* (10%)–*Bradleya* (7%)
Early Miocene	*Puriana? Nanhaiensis–Psammocythere? luminosa*			
Late Oligocene				
Early Oligocene	*Chinocythere*		>28	*Krithe* (29%)–*Argilloecia* (28%)
Middle to Late Eocene	*Chinocythere* and *Eucypris stagnalis*			
Early Eocene and older	*Sinocypris*			

assemblages are extremely useful for differentiate syn-rift sequences and post-rift sequences of shallow-water origin. A total of up to 15 palynologic zones and subzones have been identified from the northern and southern margins (Figs. 3.2 and 3.6), providing a fairly reliable biostratigraphic framework for those paralic sequences that are otherwise lithologically too similar to be divided. Especially, sediments in shelf-slope basins from the south are heavily relied on palynology for correlation (Morley 1991). Unlike those subtropical to warm temperate associations in the north (Fig. 3.2), palynological assemblages along the southern region show distinctive tropical characteristics. From the Malay Basin, West Natuna Basin, Nan Con Son Basin, to basins offshore of Borneo and Sabah, *Retitriporites curvimurati* assemblage characterizes the late Eocene, *Florschuetzia trilobata* assemblage the Oligocene, and *Florschuetzia levipoli* the early and early middle Miocene (Fig. 3.6) (ASCOPE 1981; Jiang Z. et al. 1994; Madon et al. 1999b). Interestingly, *F. trilobata* and its allied forms achieved a wider distribution in the Miocene not only in the south but also in the northern SCS (Fig. 3.2) (Hao et al. 1996), indicating a more stabile warm climate than in earlier periods. Woodruff (2003) highly appraised the paleoclimatic significance of this palynomorph assemblage when reviewing sea level and biogeographic changes since the Miocene in Thai–Malay Peninsula.

Shallow marine influence during the Oligocene in northern shelf basins can be detected by dinoflagellate *Homotryblium* and benthic foraminifers, as well as a coastal *Verrucatosporites usmensis* pollen assemblage (Fig. 3.2). An earlier marine ingression event in the late Eocene has been reported in wells drilled in deeper waters. For example, in Well BY7-1-1 from Zhu II Depression of the Zhujiangkou Basin, Mao et al. (in Hao et al. 1996) distinguished three dinoflagellate assemblages. *Homotryblium tenuispinosum– Hystrichosphaeridium tubiferum* assemblage for the Enping Formation (3,441–3,117 m) bears an age of early Oligocene–late Eocene, *Homotriblium plectilum–Corfodosphaeridium gracile* assemblage for the Zhuhai Formation (3,099–2,817 m) is late Oligocene, and *Polysphaeridium zoharryi–Lingulodinium machaerophorum* assemblage for the middle and upper parts of the Zhujiang Formation (2,685–2,142 m) lies in the early Miocene. The interval from 2,800 m to 2,685 m between Zhuhai and Zhujiang Formations lacks dinoflagellates or other marine fossils. Clearly, marine environments expanded to wider areas of the Zhujiangkou and Qiongdongnan basins during the late Oligocene, and started to invade the Yinggehai Basin, which was then limited to the northern Beibu Gulf.

Mainly based on industrial well data from various northern basins, Jiang Z. et al. (1994) summaried the local biostratigraphy in a monograph devoted to Cenozoic lithostratigraphy. For the Zhujiangkou Basin, biostratigraphy of several key deposit units was updated in a group of papers edited by Hao et al. (1996). However, because of poor preservation of diagnostic species and strong variations of lithofacies between localities, age for some older lithostratigraphic units is still in dispute. For example, the placement of the Zhuhai Formation in the lower Miocene (Qin in Hao et al. 1996; Qin 2002) or in the middle to upper Oligocene (Wan et al. in Hao et al. 1996) is yet to be resolved.

Sediments from shelf-slope basins also contain numerous benthic foraminifera, which are good indicators of paleoenvironmental changes including sea level fluctuations (Thompson and Abbott 2003). Among them, some larger foraminfiera are useful for subdividing shallow-water carbonate sequences into "Letter Stages" as defined by their range zones (Fig. 3.13). Compared to planktonic foraminiferal zonation, however, these "Letter Stages" provide a relatively coarse biostratigraphic resolution.

Planktonic Foraminiferal and Nannofossil Biostratigraphy

Similar to other normal marine settings, the basic biostratigraphic framework in the SCS region is established using planktonic foraminiferal and nannoplankton skeletons. Planktonic foraminifera and nannofossils provide a standard base for regional and global biostratigraphic correlation of marine sequences (Fig. 3.14). These pelagic organisms provide crucial evidence of waxing and waning of marine environments as well as the intensity of marine influence. Biostratigraphic correlation is achieved by using some easily identified and widely accepted biotic events

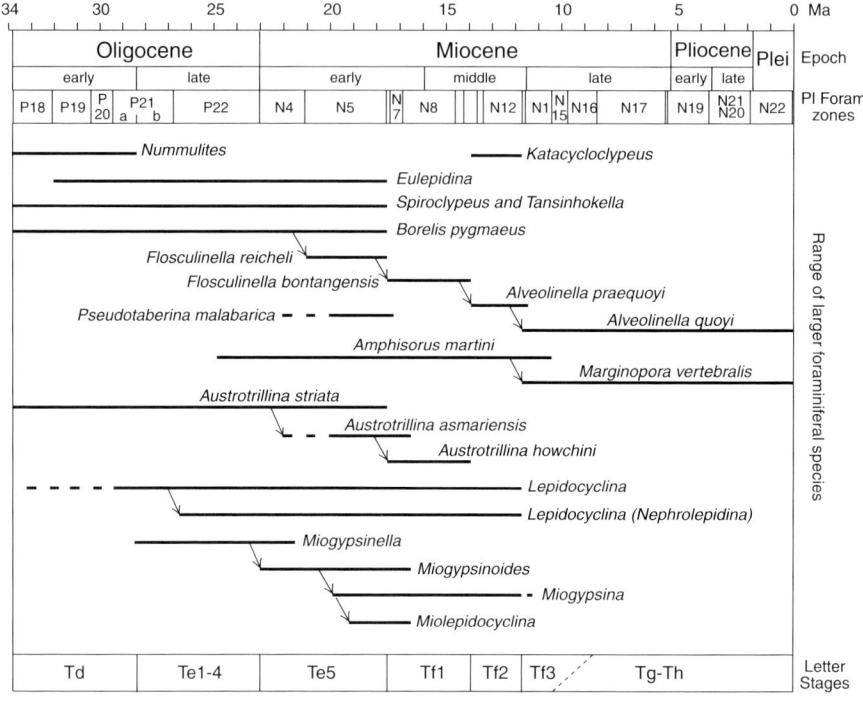

Fig. 3.13 "Letter Stages" defined by ranges of key larger benthic foraminifera are used for subdividing shallow-water carbonate sequences (modified from Adams 1970; Boudagher-Fadel and Banner 1999). Arrows indicate phylogenetic relationship between taxa. Planktonic foraminiferal N/P zones are shown for correlation

or datums, mostly the first or last occurrences of age-diagnostic species (Fig. 3.14). However, to apply these global standards to local sediment deposits sometimes is difficult because of poor preservation or lacking index species due to mixing by drilling or inadequate sampling. Therefore, work over the last three decades has generated a regional scheme for shelf-slope sequences by using mainly the last occurrence bio-events that can be readily recognized (Fig. 3.15). Together with seismic and physical well data, these planktonic foraminiferal and nannofossil events and assemblages not only improve stratigraphic correlation but also help resolving the age of unconformities (and reflectors) induced by tectonic disruption.

Huang (1997b) reported a middle Eocene nannofossil assemblage characterized by *Sphenolithus furcatolithoides* and *Helicosphaera seminulum* in core sediments from outer Zhujiangkou Basin, which represents the oldest marine microfossils found in the northern region. It is not clear, however, whether these species belonged to a middle Eocene nannoplankton association of the Proto-South China Sea or of a short-term marine ingression in the earliest stage of seafloor spreading. At most localities, continuous, well-preserved nannofossils and planktonic foraminifera so far recovered are mainly those of Oligocene and younger ages.

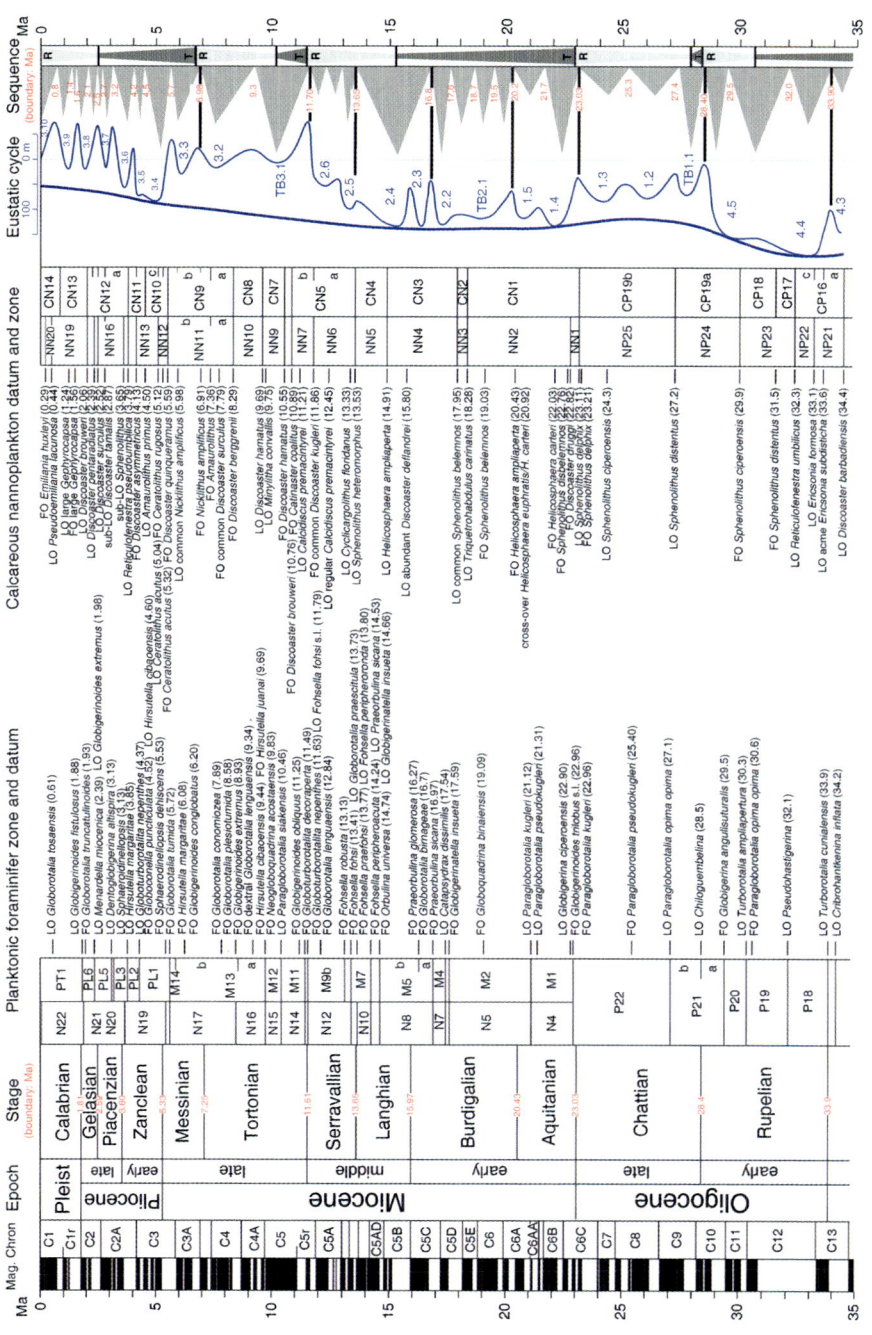

Fig. 3.14 Oligocene to Pleistocene planktonic foraminifer and nannoplankton biostratigraphy on the timescale of Gradstein et al. (2004) is correlated with the sea level curves of Haq et al. (1987) and sequences of Hardenbol et al. (1998). FO = first occurrence; LO = last occurrence

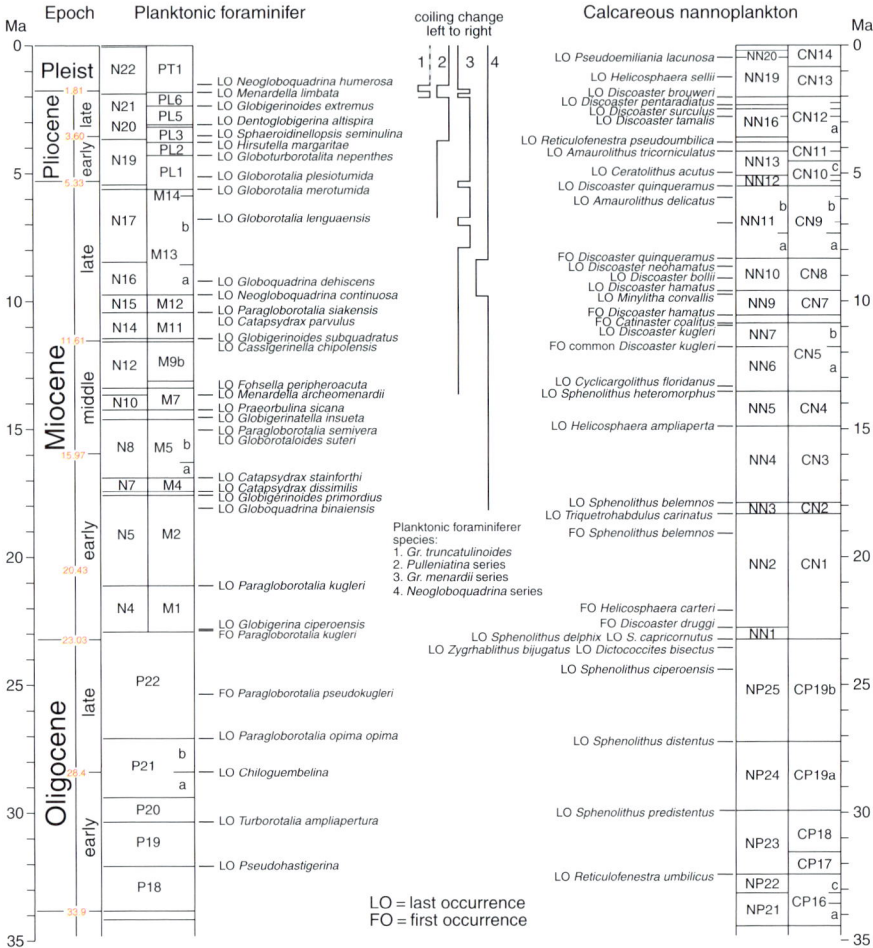

Fig. 3.15 Occurrence datums and coiling direction changes of some important planktonic foraminifera and nannofossils in northern SCS basins (updated from Jiang Z. et al. 1994) are correlated with zonations shown in Fig. 3.14

Quaternary Lithobiostratigraphic Events

Table 3.3 lists some Quaternary lithobiostratigraphic events plus geomagnetic chrons and subchrons that are widely recorded in the SCS region.

The most distinct Quaternary lithostratigraphic events in and around the region are microtektite accumulation in the mid-Pleistocene and volcanic ash deposition in the late Pleistocene (Table 3.3). Abundant microtektites occur close to the base of Brunhess at many localities, representing an extraordinary astronomical event at ∼0.8 Ma in Australasian region (Glass and Wu 1993; Zhao and Wang 1999; Zhao et al. 2004). At deeper water settings such as ODP Leg 184 sites, the microtektite layer is often ∼10–15 cm thick (Table 3.4).

Table 3.3 Important Quaternary lithobiostratigraphic events from the SCS are listed from data given in Zhao and Wang (1999) and Li et al. (2007)

Type	Datum/Event[1,2]	Age (Ma)
Volcanic ash	Toba ash (MIS4/5 transition)	0.074
Coccolith	FO *Emiliania huxleyi* (acme)	0.090
Planktonic foram	LO *Globigerinoides ruber* (pink)	0.12
Radiolaria	FO *Buccinosphaera invaginata*	0.21
Radiolaria	FO *Collosphaera tuberosa*	0.42
Planktonic foram	FO *Globigerinoides ruber* (pink)	0.42
Coccolith	LO *Pseudoemiliania lacunosa*	0.44
Benthic foram	LO *Stilostomella*	0.75
Geomagnetism	Brunhes/Matuyama transition	0.78
Microtektite	Microtektite (MIS20)	0.80
Coccolith	LO/FO *Recticulofenstra asanoi*	0.91/1.14
Geomagnetism	Jaramillo Event (top/base)	0.99/1.07
Geomagnetism	Cobb Mountain Event (base)	1.24
Coccolith	FO *Gephyrocapsa* (>0.004 mm)	1.69
Geomagnetism	Olduvai Event (top/base)	1.77/1.95
Coccolith	LO *Discoaster brouweri*	1.93
Planktonic foram	FO *Globorotalia truncatulinoides*	1.93
Planktonic foram	LO *Globigerinoides extremus*	1.98
Geomagnetism	Reunion Event (top/base)	2.14/2.23
Coccolith	LO *Discoaster pentaradiatus*	2.39

[1]FO = first occurrence; LO = last occurrence.
[2]Other radiolarian events were discussed in Wang and Abelmann (1999, 2002).

Many layers of volcanic ash found at SCS localities record past eruptions of volcanoes such as Toba Caldera on Sumatra and Mt. Pinatubo on west Philippines (Wang et al. 2000; Bühring et al. 2000; Lee et al. 2004). The earliest Toba eruption has been recently dated at 788 ka ± 2.2 ka across the MIS 20/19 boundary

Table 3.4 Distribution of the younger Toba volcanic ash and microtektites at selected sites of the SCS

Site	Latitude (N), Longitude (E)	Site water depth (m)	Core depth (m)	References
Toba ash (74 ka)				
17961-2	08°30.4′, 112°19.9′	1,968	7.82	Bühring et al. (2000)
17962-4	07°10.9′, 112°04.9′	1,969	10.78	Bühring et al. (2000)
MD972151	08°43.73′, 109°52.17′	1,550	15.58	Song et al. (2000)
MD01-2392	09°51.13′, 110°12.64′	1,966	12.45	Zheng et al. (2005)
MD01-2393	10°30.15′, 110°03.68′	1,230	19.48	Liu et al. (2006)
ODP 1143	09°21.72′, 113°17.11′	2,772	5.55	Liang et al. (2001)
Microtektites (800 ka)				
17957-2	10°53.9′, 115°18.3′	2,195	7.8–8.15	Zhao et al. (1999)
17959	11°08.3′, 115°17.2′	1,959	13.6	Wang J. et al. (2000)
ODP 772	16°39′, 119°42′	1,530	66.3–66.7	Wang J. et al. (2000)
ODP 1144	20°3.18′, 117°25.14′	2,037	386.30	Zhao et al. (2004)
ODP 1146	19°27.4′, 117°16.4′	2,092	115.85	Li B. et al. (2004)
ODP 1143	09°21.72′, 113°17.11′	2,772	42.82	Li B. et al. (2004)

(Lee et al. 2004), compared to 840 ka±30 ka commonly cited earlier. The prominent youngest Toba tephra layer occurred across the MIS 4/5 transition at ages commonly assigned to ∼74 ka. Like its older counterparts, the younger Toba volcanic ash layer is limited to northeastern Indian Ocean and southern SCS localities, and Song et al. (2000) reported its most eastern disposal in core MD972151 (Table 3.4). In numerous piston cores taken from the southern SCS, the 74 ka ash layer is so distinct that correlation of the upper sediment sections between cores can be relatively easy (Zhao and Wang 1999; Bühring et al. 2000). Geochemical analyses indicate that the Toba eruption contributed at least partly to cooling ∼74 ka ago in the region (Huang et al. 2001). The ashes are mostly light colored, with high total alkali content and high $^{87}Sr/^{86}Sr$ ratio, indicating the dominant dacitic-rhyolitic composition of the arc's explosive fraction. A contemporary sulfuric acid spike recorded at GISP2 core appears to have resulted also from aerosols flow of Toba ash (Schulz et al. 2002).

3.3 Isotopic and Astronomical Stratigraphy (Tian J. and Li Q.)

Oxygen isotopic stratigraphy relies on a solid chronobiostratigraphic framework and correlation of the results with published standards such as Zachos et al. (2001) who compiled a composite isotopic record from some 40 DSDP/ODP profiles. Continuous $\delta^{18}O$ and $\delta^{13}C$ sequences have been obtained from ODP Site 1148 for the entire Neogene with 5–10 kyr resolution (Zhao et al. 2001a,b), from Site 1143 for the last 5 Ma with 2 kyr resolution (Tian et al. 2002) (Fig. 3.16), and from Site 1146 for the middle Miocene interval between 17 and 12 Ma (Holbourn et al. 2004, 2005). In addition to these long sequences there is a 200 yr resolution record for the last 0.7 Ma from Site 1144 (Bühring et al. 2004). These continuous records are more advantageous over the composite curves of Zachos et al. (2001) especially in providing detailed information from a single site to discuss paleoceanographic trends, although they may have been affected by local factors such as the opening and closing of the South China Sea and East Asian monsoons (Tian et al. 2004d). This section provides a brief account on isotopic stratigraphies of ODP Sites 1148 and 1143, which are becoming the standard references for deep sea research in the region.

Neogene Isotopic Records at Site 1148

The Site 1148 isotopic series involve 975 planktonic foraminifer measurements from the upper 409.58 mcd and 1755 benthic foraminifer measurements from the upper 476.68 mcd, with average sampling resolution of ∼16 kyr for the Miocene and ∼8 kyr for the Pliocene–Holocene (Zhao et al. 2001a,b; Jian et al. 2003; Tian et al. 2008). In most cases, planktonic species *Globigerinoides ruber* (above 165 mcd) and *Globigerinoides sacculifer* (below 165 mcd) and benthic species *Cibicidoides wuellerstorfi* and *Uvigerina* sp. were used, and their results were adjusted to represent *G. ruber* and *C. wuellerstorfi*, respectively.

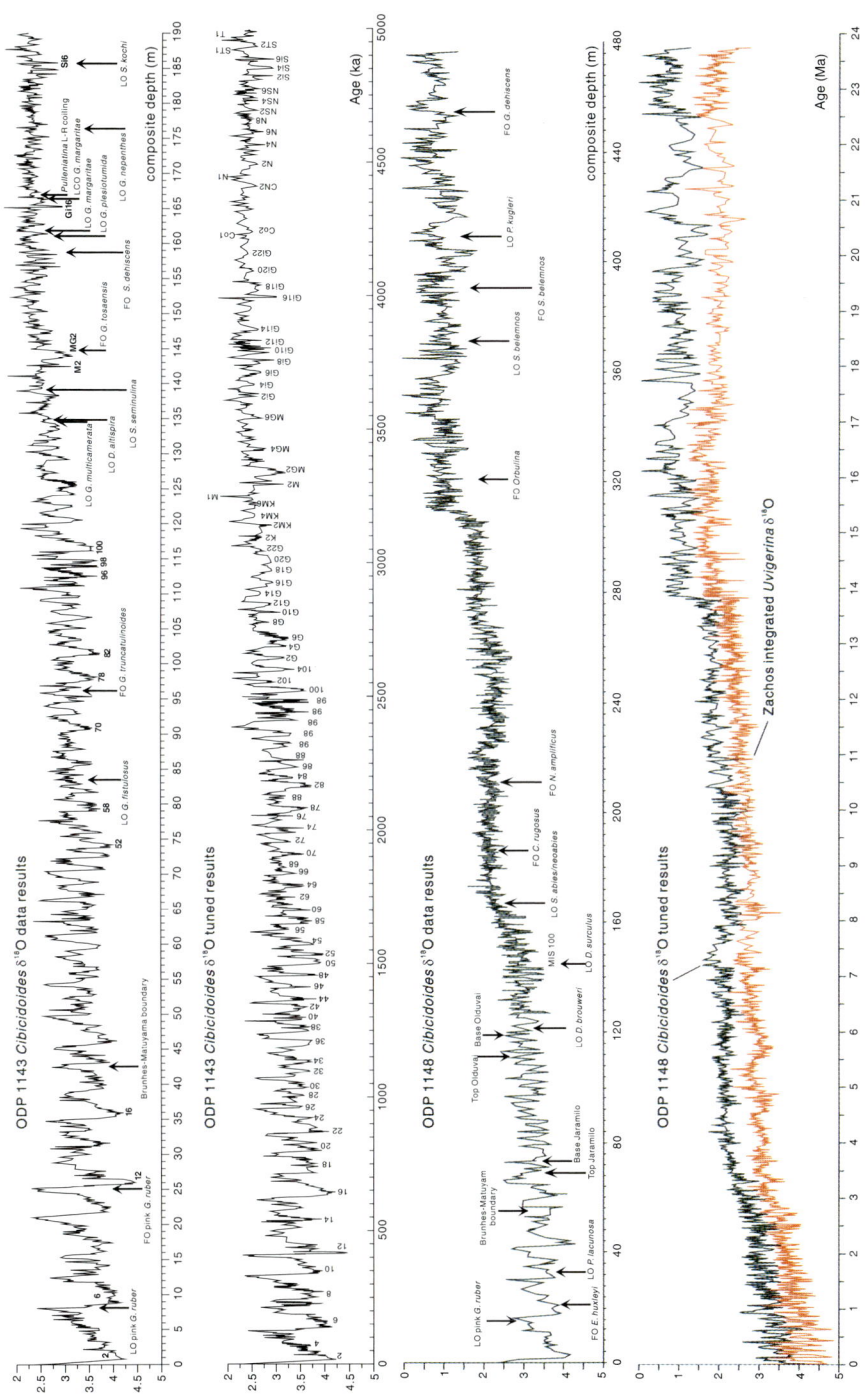

Fig. 3.16 Benthic foraminifer oxygen isotopic sequences from ODP Sites 1143 and 1148 are shown against depth and age respectively (modified from Tian et al. 2002, 2008). Also shown is the composite record of Zachos et al. (2001) (*red curve*), bio-events and marine isotopic stages (MIS)

The chronobiostratigraphic timescale of Gradstein et al. (2004) was used as an initial age model, which was then slightly modified by comparing between the benthic $\delta^{18}O$ curves of Site 1148 and that of Zachos et al. (2001). Increase by >1 per mil of the benthic $\delta^{18}O$ from the interval of 305–316 mcd at Site 1148 (Fig. 3.16) evidences the relative abrupt global cooling at ~14 Ma, when East Antarctic ice sheet expanded. Similarly, gradual $\delta^{18}O$ increase from the interval of 143.36–161.98 mcd (Fig. 3.16) record the mid-Pliocene global cooling and the enlargement of the Northern Hemisphere ice sheet. Other prominent features of the Site 1148 benthic $\delta^{18}O$ curve include strong fluctuating cycles from the interval of 0–44.51 mcd and much stronger fluctuations below 320 mcd (Fig. 3.16), respectively indicating a shift to 100 kyr climate cycle dominance since the mid-Pleistocene and more variable climate and bottom conditions in the early Miocene. Because of many well-studied records now available for comparison (Tiedemann et al. 1994; Shackleton et al. 1995; Tian et al. 2002), post-8 Ma benthic $\delta^{18}O$ stratigraphy is more reliable than for older time intervals. Choosing well-dated, consistent litho-, bio- and chemostratigraphic datums is essential for establishing an orbital timescale for the rest of the Neogene.

Postcruise studies of Site 1148 material identified 53 nannofossil datums (Su et al. 2004) and 34 Oligocene–Miocene planktonic foraminifer datums (Li et al. 2004), including three local foraminifer events affected mainly by dissolution. Table 3.5A lists the most important bio- and magneto-stratigraphic events according to the timescale of Gradstein et al. (2004) for orbital tuning.

Several $\delta^{18}O$ and $\delta^{13}C$ events correlate well with their counterparts identified originally from other localities, including Mi (Miocene oxygen isotopic maxima) events and CM (carbon maxima) events (Miller et al. 1987, 1991; Woodruff and Savin 1991; Zhao et al. 2001a,b), providing extra points in tightening the ODP Site 1148 chronostratigraphy (Table 3.5A; Fig. 3.16). Compared to Zachos et al. (2001), the Site 1148 benthic $\delta^{18}O$ results show a shift at 16–15 Ma, characterized by overall lighter $\delta^{18}O$ values before 15 Ma and lighter $\delta^{13}C$ values after 15 Ma (see also Chapter 6). This isotopic shift at ~15 Ma may reflect the impact from the termination of seafloor spreading in the South China Sea. Lighter $\delta^{18}O$ values from ~23 to 15 Ma appear to support the paleobotanic evidence for an early Miocene development of the East Asian monsoon (Sun and Wang 2005) and relative humid conditions surrounding ODP Site 1148 (Clift 2006). In contrast, lighter $\delta^{13}C$ values after ~15 Ma could have been caused by bottom water ventilation described above, or they may indicate the development of a relatively isolated local bottom water since the end of spreading (Chapter 6).

Following the work of Shackleton and Crowhurst (1997), a tuned orbital age model for Site 1148 can be established by tuning the magnetic susceptibility (MS) record from the interval of 8–18 Ma (280.74–370.54 mcd) to the Northern Hemisphere insolation (Fig. 3.16) (Tian et al. 2008). As shown in Table 3.5A, the MS-tuned ages and initial ages for the biotic datums have discrepancies averaging to ~200 ka, affected likely by sampling resolution, or distortion of datums due to local dissolution, or simply an improper selection of MS maxima for insolation minima. For example, the base of Olduvai at 118.50 mcd corresponds to an age of 1.95 Ma

Table 3.5A Neogene nannofossil datums found at ODP Site 1148 are used for biostratigraphy (Age 1, Gradstein et al. 2004) and for orbital tuning (Age 2, orbitally adjusted by Tian et al. 2008)

Datum[1]	Nannofossil datum	mcd[2]	Age 1 (Ma)	Age 2 (Ma)
FO	*Emiliania huxleyi* Acme	13.05	0.09	0.09
LO	*Pseudoemiliania lacunosa*	34.93	0.44	0.49
LO	*Reticulofenestra asanoi*	59.75	0.91	0.85
FO	*Reticulofenestra asanoi*	78.20	1.14	1.18
LO	*Calcidiscus macintyrei*	102.52	1.61	1.60
LO	*Discoaster brouweri*	122.33	1.93	2.00
LO	*Discoaster pentaradiatus*	136.60	2.39	2.37
LO	*Reticulofenestra pseudoumbilicus*	169.67	3.70	3.95
LO	*Amaurolithus* spp.	175.11	4.80	4.30
LO	*Ceratolithus acutus*	177.77	5.04	4.50
FO	*Ceratolithus rugosus*	187.37	5.23	5.23
LO	*Triquetrorhabdulus rugosus*	190.37	5.28	5.41
FO	*Ceratolithus acutus*	190.37	5.35	5.41
LO	*Discoaster quinqueramus*	193.31	5.58	5.57
LO	*Nicklithus amplificus*	198.57	5.98	5.88
FO	*Nicklithus amplificus*	211.27	6.91	6.59
FO	*Amaurolithus primus*	219.37	7.36	7.16
FO	*Discoaster berggrenii*	242.61	8.29	9.18
FO	*Discoaster quinqueramus*	242.61	8.29	9.18
FO	*Discoaster pentaradiatus*	253.37	8.55	9.81
LO	*Discoaster hamatus*	256.37	9.40	10.00
LO	*Catinaster calyculus*	261.81	9.67	10.34
LO	*Catinaster coalithus*	267.47	9.69	10.93
FO	*Discoaster hamatus*	275.57	10.38	11.51
FO	*Catinaster calyculus*	279.21	10.41	11.87
FO	*Catinaster coalithus*	281.01	10.89	12.04
LO	*Discoaster kugleri*	286.67	11.58	12.39
FO	*Discoaster kugleri*	293.27	11.86	12.80
FO	*Triquetrorhabdulus rugosus*	302.87	13.20	13.45
LO	*Cyclicargolithus floridanus*	302.87	13.33	13.46
LO	*Sphenolithus heteromorphus*	308.81	13.53	13.83
LO	*Helicopontosphaera ampliaperta*	331.87	14.91	15.42
FO	*Sphenolithus heteromorphus*	370.17	17.71	18.24
LO	*Sphenolithus belemnos*	371.57	17.95	18.30
FO	*Sphenolithus belemnos*	390.97	19.03	19.30

[1]LO = last occurrence; FO = first occurrence. [2]mcd = meter composite depth (datums from Su et al. 2004).

in the initial and tuned age models, but the base of planktonic foraminifer *Globoro-talia truncatulinoides* at 130.63 mcd cannot have an age of 1.93 Ma according to Gradstein et al. (2004) but should be tuned to ~2.23 Ma, implying a local datum for this species (Table 3.5AB). An abrupt increase in benthic $\delta^{18}O$ at ~13.9 Ma, which marks the final formation of the east Antarctic ice sheet (Zachos et al. 2001; Holbourn et al. 2005), falls at 309.28 mcd of Site 1148. Accordingly, the top of *Prae-orbulina glomerosa* found at 312.38 mcd with an initial assigned age of 13.80 Ma should be changed to 14.07 Ma (Table 3.5AB).

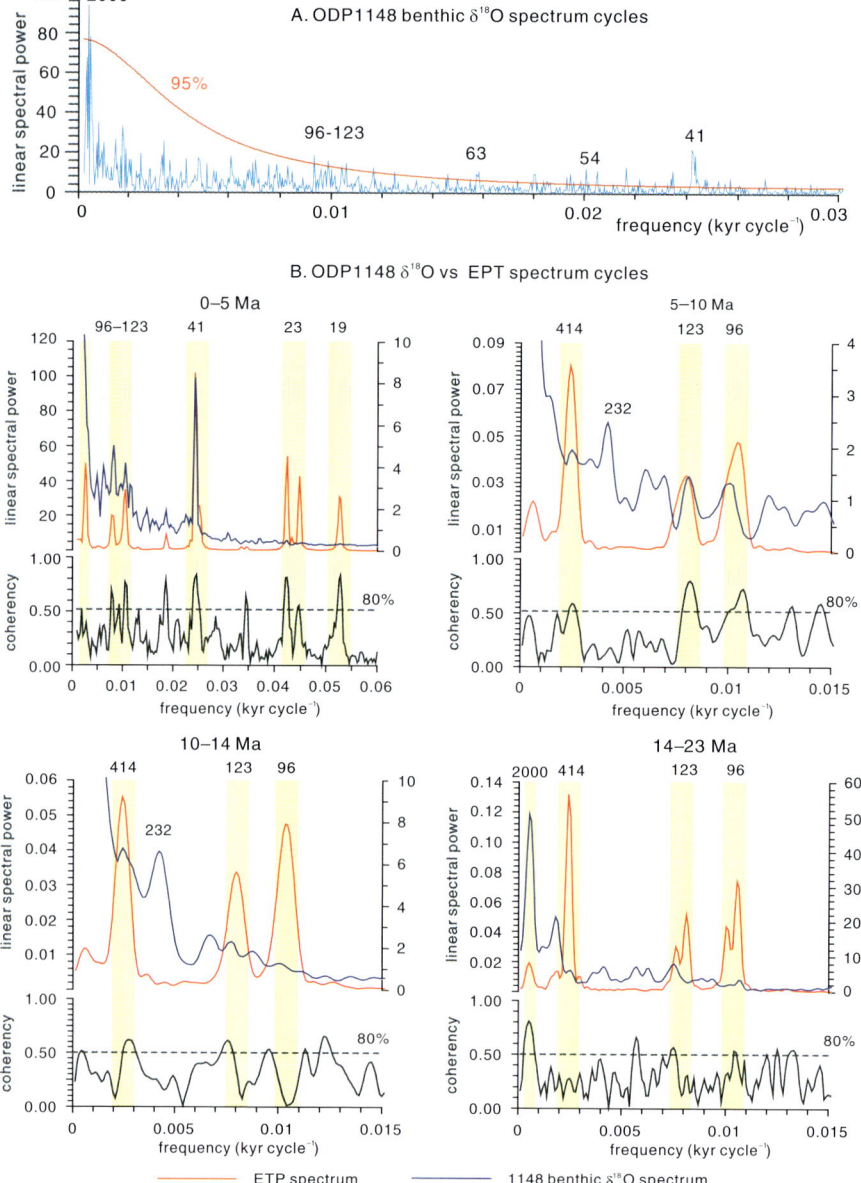

Fig. 3.17 The Site 1148 δ^{18}O record was further analyzed by (**A**) bias-corrected spectrum with red line representing 95% statistical significance level, and (**B**) cross spectral analyses with ETP (*red line*) for 4 intervals using the Laskar (1990) solution

Table 3.5B Neogene planktonic foraminifer datums and paleomagnetic events found at ODP Site 1148 are used for biostratigraphy (Age 1, Gradstein et al. 2004) and for orbital tuning (Age 2, orbitally adjusted by Tian et al. 2008)

Datum[1]	Planktonic foraminifer datum	mcd[2]	Age 1 (Ma)	Age 2 (Ma)
FO	*Globigerinoides ruber* Pink	29.54	0.40	0.39
FO	*Globorotalia truncatulinoides*	130.63	1.93	2.23
LO	*Globoquadrina altispira*	160.38	3.13	3.18
LO	*Globoturborotalita nepenthes*	176.91	4.37	4.43
FO	*Sphaeroidinellopsis dehiscens*	188.16	5.53	5.28
FO	*Globorotalia tumida*	196.08	5.72	5.75
FO	*Globigerinoides conglobatus*	206.79	6.20	6.34
FO	*Globigerinoides extremus*	244.26	8.93	9.31
LO	*Globoquadrina dehiscens* (local datum)	257.16	9.80	10.05
FO	*Neogloboquadrina acostaensis*	259.70	9.83	10.21
LO	*Paragloborotalia mayeri*	275.22	10.46	11.47
FO	*Globoturborotalita nepenthes*	283.78	11.63	12.22
LO	*Globorotalia fohsi* (local datum)	301.02	13.00	13.33
FO	*Globorotalia fohsi*	303.28	13.41	13.49
FO	*Globorotalia praefohsi*	308.68	13.77	13.83
LO	*Praeorbulina glomerosa*	312.38	13.80	14.07
FO	*Globorotalia praemenardii*	317.98	14.38	14.46
LO	*Orbulina*	320.37	14.74	14.63
FO	*Globigerinatella insueta*	320.97	14.66	14.66
FO	*Praeorbulina glomerosa*	344.18	16.27	16.28
FO	*Praeorbulina curva*	352.98	16.40	16.92
FO	*Praeorbulina sicana*	355.39	16.97	17.11
LO	*Catapsydrax dissimilis*	364.88	17.54	17.81
FO	*Globigerinatella insueta* (local datum)	367.37	18.00	18.02
LO	*Globoquadrina binaiensis*	377.18	19.09	18.64
FO	*Globorotalia praescitula*	379.53	19.10	18.77
FO	*Globigerinoides altiapertura*	406.38	20.40	20.69
LO	*Paragloborotalia kugleri*	408.83	21.12	21.30
FO	*Globoquadrina dehiscens*	454.17	22.70	23.21
	Paleomagnetic events			
PM	*B/M boundary*	55.20	0.78	0.78
PM	*Top Jaramilo*	69.10	0.99	0.99
PM	*Base Jaramilo*	73.00	1.07	1.07
PM	*Top Olduvai*	111.40	1.77	1.77
PM	*Base Olduvai*	118.50	1.95	1.95

[1]LO = last occurrence; FO = first occurrence; PM = paleomagnetic datums (from Wang et al. 2000). [2]mcd = meter composite depth (datums from Li et al. 2006).

Spectral analysis using REDFIT (Schulz and Mudelsee 2002) show three peaks with 95% statistical significance: 41 ka, 100 ka and 2,000 ka (Fig. 3.17). The 41 kyr cycle is the most prominent component of the benthic $\delta^{18}O$, indicating a continuous obliquity-related influence on climate change. Due to the low time resolution before the Pliocene (>10 ka), this peak should mainly reflect the spectral features after the early Pliocene. The eccentricity is represented by 96 kyr, 100 kyr and 123 kyr cyclicities with a relative high level of the combined statistical significance. Thus, both the 41 kyr and 100 kyr cycles are important cycle not only for the Plio–Pleistocene

(Tiedemann et al. 1994; Shackleton et al. 1995; Tian et al. 2002, 2004b) but also for the Miocene (Holbourn et al. 2005). Also shown are a prominent 2,000 kyr mega-cycle and cycles ranging from 402 to 550 kyr with lower than the 80% statistical significance level. Analyses of different time intervals show good 100 kyr and 414 kyr eccentricity cycles, as well as 41 kyr obliquity and 19–23 kyr precession cycles, conforming the accuracy and reliability of this tuned isotopic record (Fig. 3.17).

Continuous Wavelet Transform (CWT) analysis shows very strong and quite stable 100 kyr and 414 kyr cycles for the ETP (eccentricity, obliquity or tilt, and the negative precession) but very variable and discontinuous for the Site 1148 benthic $\delta^{18}O$ record (Fig. 3.18). Whether these are erratic variations of the long eccentricity

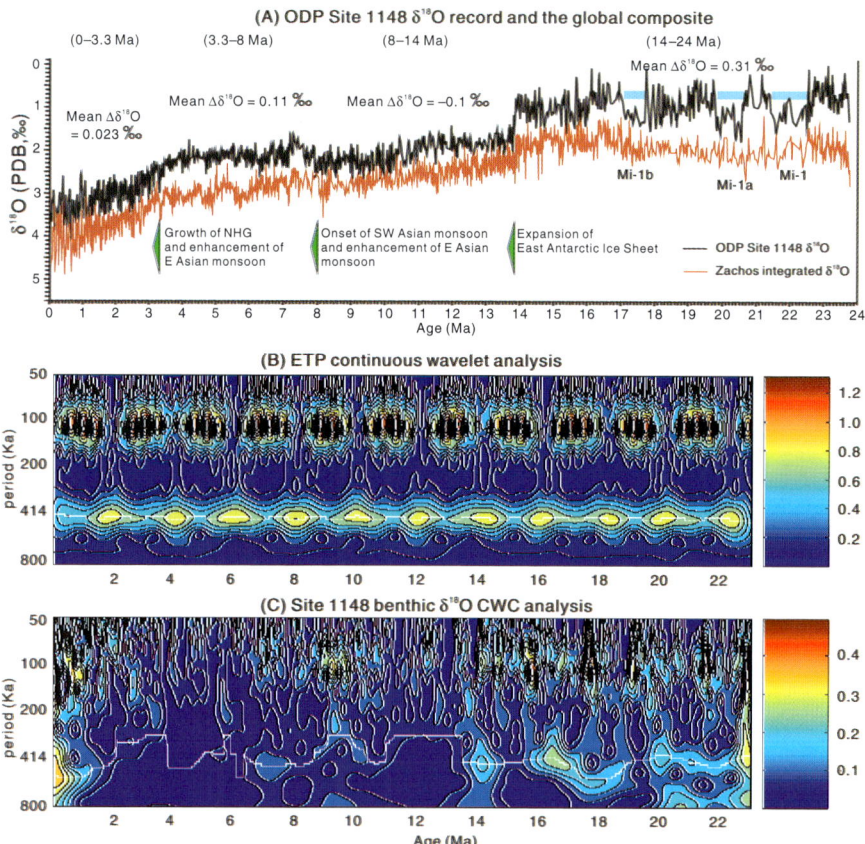

Fig. 3.18 (**A**) The Site 1148 benthic $\delta^{18}O$ curve is shown alongside with the composite $\delta^{18}O$ record of Zachos et al. (2001) and major paleoclimatic events. (**B**) Results of Continuous Wavelet Transform analyses of ETP are compared with (**C**) those of Site 1148 $\delta^{18}O$. The white lines are the ridges of the low-frequency variability centered around 1/400–1/500 kyr. The multicolor bar on the right indicates the modulus density, with higher values (*red*) indicating higher density

related climate cycle or artificial noise due to an incomplete sediment succession from Site 1148 is not clear. Further cross-spectral analysis on different time slices reveals a high coherency ($>$80%) at the 414 kyr cycle only for the 5–14 Ma interval but very low coherencies for other periods (Tian et al. 2008). Distinct 40 kyr, 100 kyr and 400 kyr cycles are also found between 12.3 and 14.7 Ma at Site 1146 (Holbourn et al. 2005, 2007), marked similarly by a major $\delta^{18}O$ change corresponding to the mid Miocene global cooling and significant buildup of ice sheet on east Antarctica at ∼13.9 Ma. The "Monterey" carbon isotope excursion (16.9–13.5 Ma) consists overall of nine 400 kyr cycles superimposed by 100 kyr-frequency variations, indicating the control of mid-Miocene climate change by eccentricity cycles through the modulation of long-term carbon budget (Holbourn et al. 2007).

Pliocene–Pleistocene Isotopic Records at Site 1143

ODP Site 1143 lies within the northwestern part of the Dangerous Grounds or Nansha Islands area, in the southern SCS. The Pleistocene to upper Pliocene sediments recovered at the site consist mostly of olive, greenish, and light gray green and greenish gray clayey nannofossil mixed sediment, clay with nannofossils, and clay, with average linear sedimentation rates of 30–70 m/myr (Wang et al. 2000). Planktonic and benthic foraminifers in 1992 samples at 2 cm spacing from 0 to 190.77 mcd were measured for stable isotopes (Tian et al. 2002, 2004b).

An initial age model was based on biostratigraphy of 14 planktonic foraminifer datums, as well as the Brunhes/Matuyama paleomagnetic polarity reversal at 43.2 mcd of Hole 1143C that represents an age of 780 ka. By comparing the Site 1143 $\delta^{18}O$ curve to the composite curve of Shackleton et al. (1995) for the last 5 myr, the corresponding marine isotopic stages (MIS 1 to MIS T1) were identified (Fig. 3.16). Apart from these distinct glacial cycles, an increasing trend in isotopic values from MIS MG2 to MIS 96 marks the growth of Northern Hemisphere ice sheet between 3.3 and 2.5 Ma.

Orbital tuning of the Site 1143 benthic $\delta^{18}O$ record was performed using the astronomical solution of Laskar (1990) for the obliquity and precession as the tuning target (Tian et al. 2002). Because previous studies have found $\delta^{18}O$ lagging the Northern Hemisphere 65° summer insolation maxima by ∼69° at the obliquity band and ∼78° at the precession band, equivalent to 7.8 ka and 5 ka respectively, an 8-kyr lag for the obliquity curve and 5-kyr lag for the precession curve from Laskar (1990) were applied for tuning using the Dynamic Optimization technique (Yu and Ding 1998). A final age model (Fig. 3.16) was obtained by repeatedly tuning the initial age model to obliquity until coherencies between the filtered $\delta^{18}O$ signals and orbital signals reached a maximum fit.

Cross-spectral analyses of the tuned $\delta^{18}O$ record with the ETP for intervals of 0–1 Ma, 1–2 Ma, 2–3 Ma, 3–4 Ma and 4–5 Ma all show good coherency (most $>$95%) at the 41 kyr obliquity band and 23 kyr and 19 kyr precession bands (Fig. 3.19). The spectral densities of 100 kyr, 23 kyr and 19 kyr periodicities in $\delta^{18}O$ increase significantly through time, as also their corresponding coherencies. A

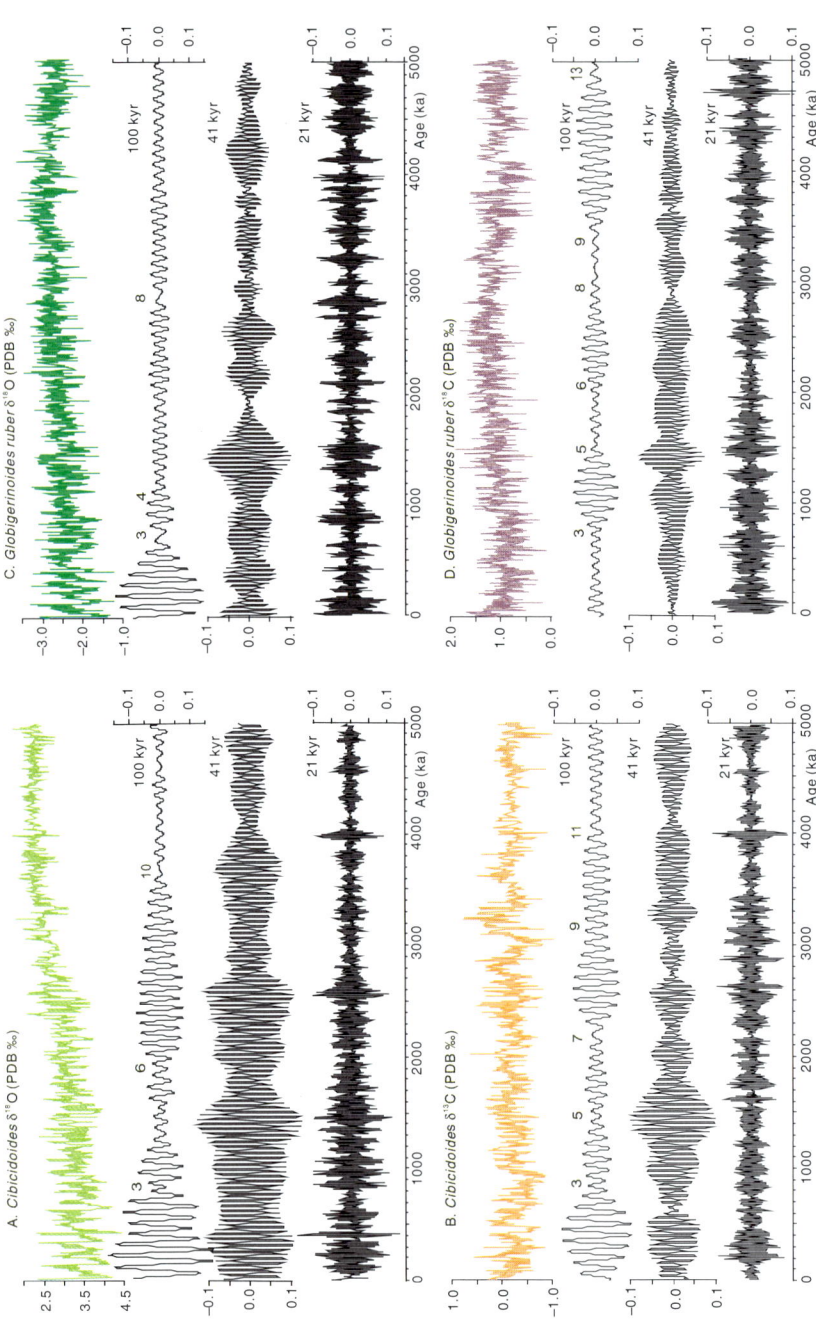

Fig. 3.19 Filtered orbital signal components are revealed in (**A, B**) benthic and (**C, D**) planktonic foraminifer $\delta^{18}O$ and δ^{13} from ODP Site 1143 over the eccentricity (100 ka), obliquity (41 ka) and precession (21 ka) bands (modified from Tian et al. 2004b). Note the discontinuous distribution of the ~400 ka long eccentricity cycle numbered 3 to 13

strong 41 kyr cycle highly coherent with ETP in all the intervals clearly demonstrate its importance not only in the Pleistocene but also in the Pliocene. The dominance of the 41 kyr cycle was replaced by the 100 kyr eccentricity cycle at about 0.9 Ma, the so-called mid-Pleistocene revolution (Prell 1982; Berger et al. 1992; Imbrie et al. 1993). The ~100 kyr short- and ~400 kyr long-eccentricity cycles, however, appear to be discontinuous (Fig. 3.19), although the ~400 kyr cycle has been proposed as a basic astronomical unit for correlating stratigraphic sequences (Wade and Pälike 2004; Li et al. 2005b).

The tuned age model can be assessed by comparing the filtered 41-kyr component of $\delta^{18}O$ with the lagged obliquity and comparing the filtered 23-kyr component of $\delta^{18}O$ with the lagged precession. As shown in Fig. 3.17, good matches between the filtered signals and the orbital parameters confirm the accuracy of the tuned timescale. Some minor discrepancies exist between the filtered 23-kyr component and the lagged precession especially at ages of ~400 ka, 2.8–2.9 Ma, 3.95–4.3 Ma, and 4.65–5.0 Ma. Although further studies are needed to clarify the cause of these discrepancies, the tuning strategy by fixing on to a good 41-kyr obliquity cycle may have forced some mismatch on the 23-kyr precession cycle.

To sum up, the orbitally tuned timescales from ODP Site 1143 and Site 1148 provide some unprecedented means for the first time for high-resolution paleoceanographic studies in the SCS region. These timescales lay a solid foundation not only for more accurate stratigraphic correlation of deep sea sediment sequences but also for exploring the relationship between regional paleoceanographic events and global climate change and their probable driving mechanisms.

3.4 Stratigraphy of Major Shelf and Slope Basins (Zhong G. and Li Q.)

Over the past half century, petroleum exploration has been carried out in most sedimentary basins of the SCS. As a result, litho-, bio- and sequence stratigraphies in many shelf and slope basins have been intensively investigated. However, only a few syntheses have been attempted on stratigraphy of individual basins and much lesser on the overall stratigraphy of the SCS. Important regional stratigraphic references include ASCOPE (1981) on stratigraphy of the Gulf of Thailand and southern SCS basins, Wang (1985) on the Cenozoic biostratigraphy of China seas, Jin (1989) on various aspects of SCS geology, Zeng et al. (1981) and Jiang Z. et al. (1994) on Tertiary stratigraphy of the northern SCS basins, Gong and Li (1997) on sedimentology, sequence stratigraphy and petroleum resources of the Zhujiangkou, Yinggehai and Qiongdongnan basins, Fraser et al. (1997) on petroleum geology of Southeast Asia, and PETRONAS (Leong 1999a,b) on straitgraphy and petroleum resources of basins offshore of Malaysia.

The accumulation of vast data over decades provides the base for an overview on the stratigraphy of SCS basins. In this section, we summarize the major stratigraphic characteristics of marjor shelf-slope basins mainly based on references scattered in various Chinese and English publications. The selected basins can be broadly grouped into two categories: northern SCS basins and southern SCS basins. From northeast to southwest, they are Taixinan (=SW Taiwan or Tainan), Zhujiangkou (=Pearl River Mouth), Beibuwan (=Beibu Gulf), Yinggehai (=Song Hong) and Qiongdongnan (=SE Hainan), Cuu Long and Wan'an (=Nam Con Son), Zengmu (including Sarawak and East Natuna), West Natuna, Malay basins, and the Thai Basin Group (including the Pattani Basin) (Fig. 3.20).

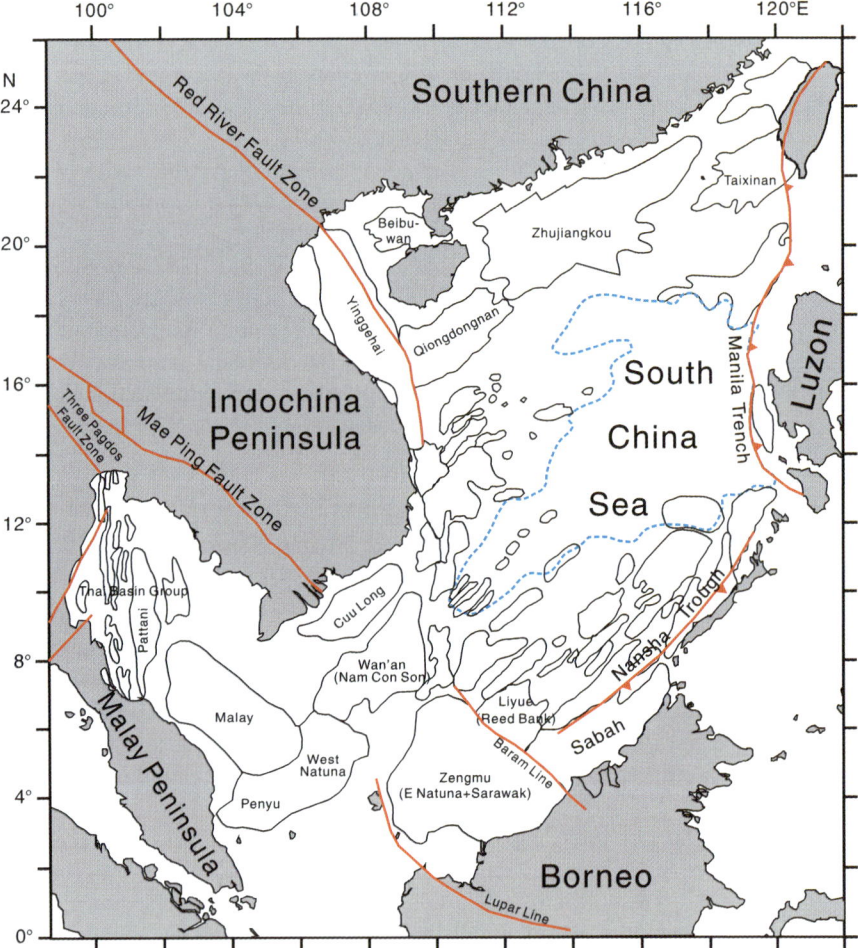

Fig. 3.20 Sketch map shows Cenozoic sedimentary basins in the SCS and major tectonic lines surrounding the SCS, as compiled from various data sources: ASCOPE (1981), Gong et al. (1989), Du (1994), Wu and Yang (1995), Jardine (1997), Phillips et al. (1997), Ying (1998), Madon (1999a,b), Watcharanantakul and Morley (2000), Lee et al. (2001), Qiu and Wang (2001), Bishop (2002), Lin et al. (2003), Qiu et al. (2005) and Binh et al. (2007). The oceanic crust is outlined by the *dashed blue line*

Northern South China Sea Basins

Taixinan Basin

Four sedimentary basins exist to the west of the Taiwan Island, in the northeastern end of the SCS. From north to south, they are the Nanjihtao, Taihsi, Penghu and Taixinan (=SW Taiwan or Tainan) basins, all NE-SW trending (Fig. 3.21). As the largest in the group, the Taixinan Basin covers an area of about $70,000\,km^2$, among which 90% lies offshore, in present water depths from $\sim200\,m$ to $>1,500\,m$ (Du 1991, 1994; Huang et al. 1998).

Tectonically, the Taixinan Basin is located on the southeast tip of the South China Block. It is separated to the north from the Penghu and Taihsi basins by the Penhu- Peikang High (a pre-Cenozoic basement uplift), and to the west from the Zhujiangkou Basin by the Dongsha Swell Area. It extends northeastward from offshore to the onshore Western Foothills before terminating by the N-S trending

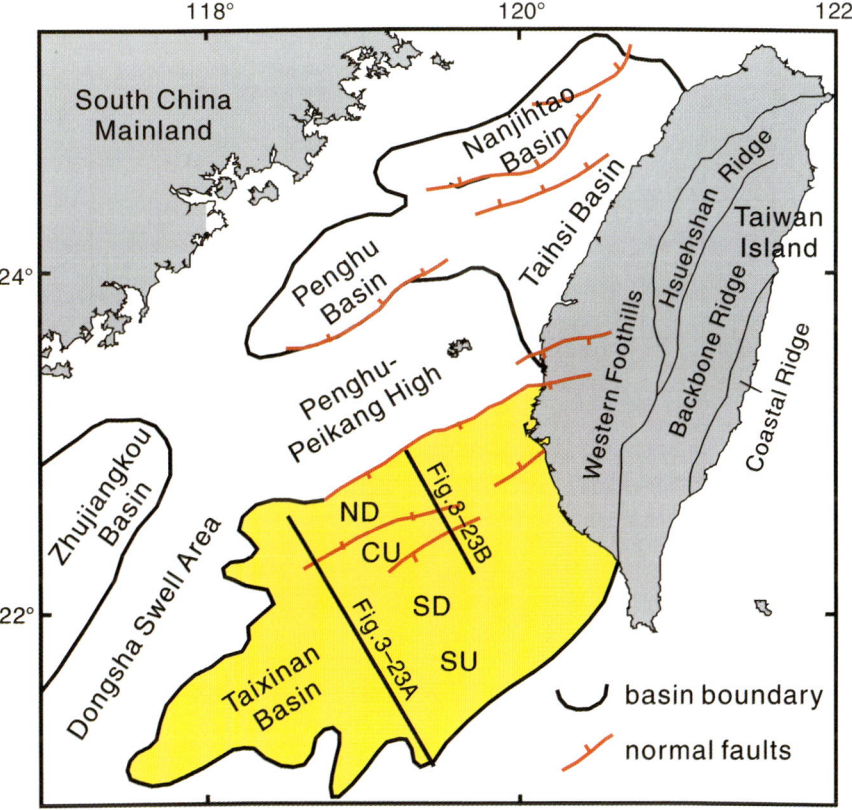

Fig. 3.21 Sketch map shows major structural elements in the Taixinan and neighboring basins (compiled from Du 1994; Lin et al. 2003). Sub-structural units in the Taixinan Basin include Northern Depression (ND), Central Uplift (CU), Southern Depression (SD) and Southern Uplift (SU)

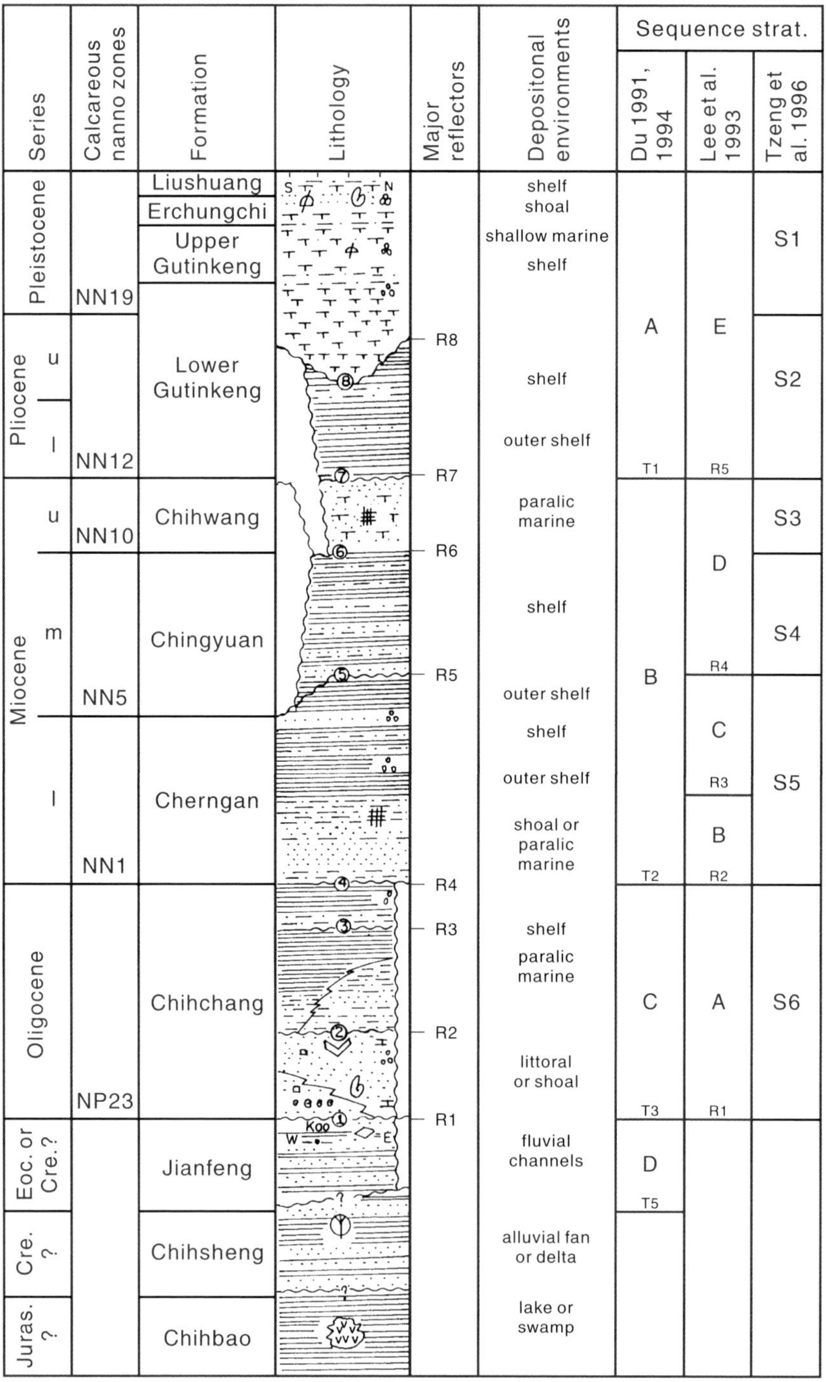

Series	Calcareous nanno zones	Formation	Lithology	Major reflectors	Depositional environments	Sequence strat.		
						Du 1991, 1994	Lee et al. 1993	Tzeng et al. 1996
Pleistocene		Liushuang			shelf			S1
		Erchungchi			shoal			
		Upper Gutinkeng			shallow marine			
	NN19				shelf	A	E	
Pliocene u		Lower Gutinkeng		R8	shelf			S2
Pliocene l	NN12			R7	outer shelf	T1	R5	
Miocene u	NN10	Chihwang		R6	paralic marine		D	S3
Miocene m	NN5	Chingyuan		R5	shelf	B		S4
					outer shelf		R4	
					shelf		C	
Miocene l	NN1	Cherngan			outer shelf		R3	S5
					shoal or paralic marine	T2	B	
				R4	shelf		R2	
Oligocene		Chihchang		R3	paralic marine	C	A	S6
				R2	littoral or shoal			
	NP23			R1		T3	R1	
Eoc. or Cre.?		Jianfeng			fluvial channels	D		
						T5		
Cre. ?		Chihsheng			alluvial fan or delta			
Juras. ?		Chihbao			lake or swamp			

Fig. 3.22 (continued)

Taiwan Orogeny in Taiwan Island. The latter is resulting from the oblique collision between the Luzon volcanic arc and the southeastern continental margin of China since the late Miocene (Chapter 2) (Yang et al. 1990; Du 1991, 1994; Tzeng et al. 1996; Lin et al. 2003). The basement of the Taixinan Basin mainly consists of late Mesozoic continental to paralic clastic rocks as revealed by drilling on local basement highs (Jin 1989; Lee et al. 1993; Tzeng et al. 1996; Lin et al. 2003). Four main structural elememts, trending NE-SW, have been recognized in the basin: the Northern Depression, Central Uplift, Southern Depression, and Southern Uplift (Fig. 3.21) (Du 1991, 1994; Chang 1997; Tang et al. 1999).

The Taixinan Basin is a Cenozoic rift basin superimposed by a latest Miocene to Recent foreland basin, respectively relating to the opening of the SCS and the later arc-continental collision (Du 1991, 1994; Lee et al. 1993; Chen et al. 2001; Lin et al. 2003; Yu and Hong 2006). The maximum thickness of Cenozoic sediments in the basin is estimated from 6,000 to 10,000 m by different authors using geophysical survey and drilling data (Du 1991, 1994; Lee et al. 1993; Huang et al. 1998; Lin et al. 2003). Oligocene sediments are the oldest Cenozoic strata recovered by drilling on the Central Uplift (e.g., Lee et al. 1993; Tzeng et al. 1996). However, the existence of probable Paleocene to Eocene strata in the northern and southern deep faulted sags has been suggested, because interpretation of seismic data reveals a set of reflectors below the Oligocene section with reflection characteristics very similar to the drilled contemporary sequence in the nearby Zhujiangkou Basin (Jin 1989; Du 1991, 1994; Lee et al. 1993; Chang 1997). In addition, in the Penghu and Taihsi basins and in Taiwan Island, Paleocene to Eocene sediments have been found in wells or at outcrops (Du 1994; Lin et al. 2003).

Oligocene and younger deposits in the Taixinan Basin mainly consist of marine sandstones, siltstones, shales, and clays (Tzeng et al. 1996; Oung 2000) (Fig. 3.22). They have been subdivided into lithostratigraphic units, including Chihchang, Cherngan, Chingyuan, Chinwang, Gutinkeng, Erchungchi, and Liushuang formations upwards (Fig. 3.22). The Oligocene Chihchang Formation is composed of sandstones and siltstones in the lower part, with minor shales and limestones, and mainly shales in the upper part. The lower Miocene Cherngan Formation consists of shales with minor siltstones, sandstones and limestones. The middle Miocene Chingyuan Formation consists of shales interbedded with siltstones and some thin limestone beds. The upper Miocene Chihwang Formation is composed of shales intercalated with coarse sands, limestone, and thin coal beds. The Pliocene lower Gutinkeng Formation consists of clays with minor sandstone and siltstone interbeds, while the Pleistocene, including (upper) Gutinkeng, Erchungchi, and Liushuang formations, mainly consists of clay intercalated with thin-bedded sandstones.

← _____

Fig. 3.22 Correlation between different stratigraphic classification schemes in the Taixinan Basin shows major stages in basin development (modified from Oung 2000 and Chang 1997). Capital letters A to E and S are sequence names of various authors, and R and T represent unconformity reflectors, as also in the following figures

The stratigraphic framework for the Taixinan Basin was established on seismic and sequence stratigraphy, supplemented by limited lithostratigraphic and paleotological calibrations (Du 1991, 1994; Lee et al. 1993; Tzeng et al. 1996; Lin et al. 2003). About eight regional unconcormities have been defined in the Taixinan Basin by interpreting seismic data tied to well control (Chang 1997; Oung 2000) (Fig. 3.22). These unconformities enable a sequence stratigraphic framework for the basin, though sequence classification schemes by different authors vary (Figs. 3.22 and 3.23) (Du 1991, 1994; Lee et al. 1993; Tzeng et al. 1996). Three megasequences respectively characterize the three stages in basin development: the Paleocene(?)–early Miocene syn-rift, middle to late Miocene post-breakup and latest Miocene-Recent foreland stages. Some authors, however, suggest that the breakup unconformity sepatating the syn-rift from the post-rift megasequences lies either at the Eocene/Oligocene boundary (Lin et al. 2003) or Oligocene/Miocene boundary.

Exploration in the offshore Taixinan Basin dated back to the early 1970s (Jin 1989; Oung and Tsao 1991; Lee et al. 1993; Oung 2000). Well F-1 drilled on the "F" Structure of the Central Uplift in 1974 marks the first oil discovery in the basin. Subsequent drilling proved that the Oligocene and Miocene sandstones, as well as the feactured Cretaceous basement sandstones, are the main reservoirs in the Taixinan Basin (Jin 1989; Lee et al. 1993; Tzeng et al. 1996; Tang et al. 1999).

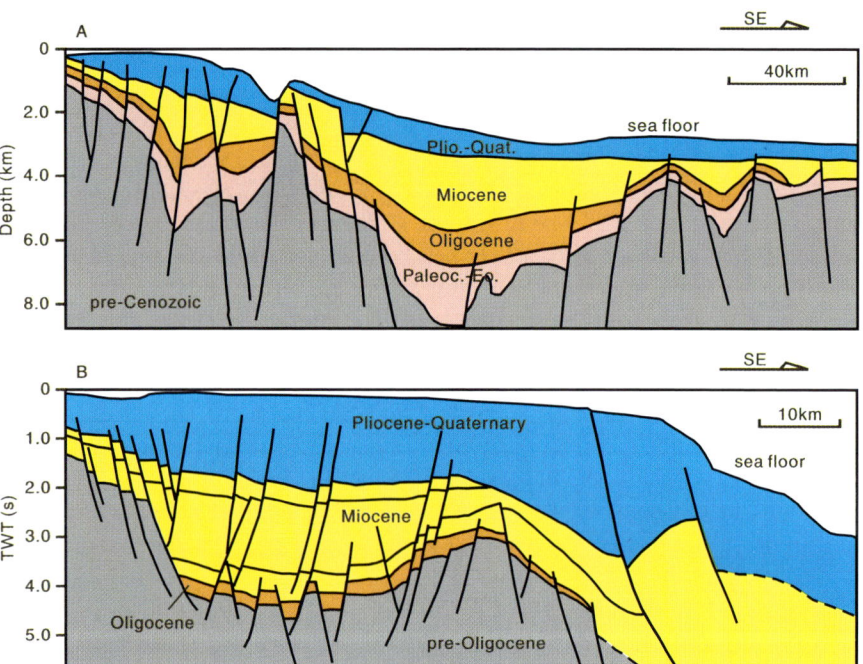

Fig. 3.23 Interpreted cross sections show major faults, structural units and seismic sequences in the Taixinan Basin: (**A**) modified from Du (1991, 1994) and (**B**) modified from Lee et al. (1993). See Fig. 3.21 for line locations. TWT = two-way travel time, as also in the following figures

Zhujiangkou Basin

The Zhujiangkou (=Pearl River Mouth) Basin is located in the shelf and slope area offshore from the Pearl River mouth. The NE-SW trending basin is 800 km long and 100–300 km wide, and covers an area of approximately 14.7×10^4 km^2 as estimated by the 1,000 m sediment thickness contour (Gong et al. 1989). Structurally, Zhujiangkou Basin is bordered to the north by the Wanshan Swell Area, to the west by the Hainan Uplift Zone and the Qiongdongnan Basin, and to the east by the Dongsha Swell Area and the Taixinan Basin. To its south lies the central deep basin of the SCS (Fig. 3.24).

The Zhujiangkou Basin is a Cenozoic rift basin overlain by a post-rift thermal subsidence sag, developed on a down-faulted basement consisting of Mesozoic granites with minor volcanic rocks and Paleozoic metasedimentary rocks (Gong et al. 1989; Jin 1989; Chen 2000). Very thick marine Jurassic-Cretaceous sedimentary rocks have been encountered in wells recently drilled in the eastern area of the basin (Shao et al. 2007; Wu et al. 2007). Breakup unconformity separating syn-rift sediments from the overlying thermal sag fill is dated at 28–30 Ma, close to the early/late Oligocene boundary. Six major structural units exist in the basin, including

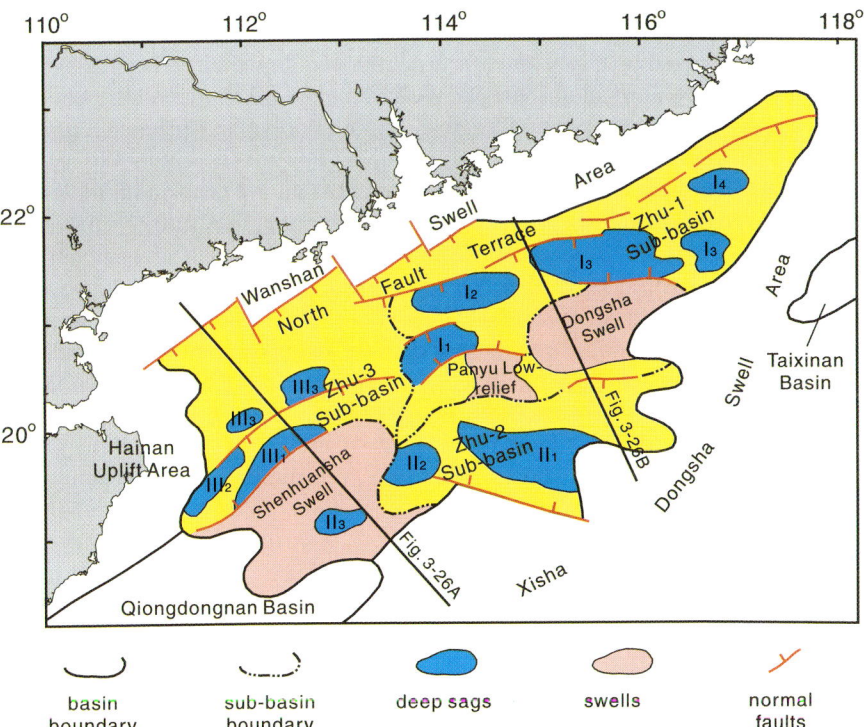

Fig. 3.24 Sketch map shows the location of major structural units in the Zhujiangkou Basin (Gong et al. 1989). I$_1$, Enping sag; I$_2$, Xijiang sag; I$_3$, Huizhou sag; I$_4$, Lufeng sag; II$_1$, Baiyun sag; II$_2$, Kaiping sag; II$_3$, Shunde sag; III$_1$, Wenchang sag; III$_2$, Qionghai sag; III$_3$, Yangjiang sag

Zhu I, Zhu II, and Zhu III sub-basins, and Dongsha, Shenhuansha and Panyu uplifts (Fig. 3.24) (Gong et al. 1989; Jin 1989; Chen 2000), and the Cenozoic sequence may reach a thickness of more than 10,000 m (Gong et al. 1989; Chen 2000).

The development of the Zhujiangkou Basin involved two major phases: a rifting phase and a post-rift passive continental marginal basin development phase (e.g., Chen et al. 1994). At least three stages of rifting have been recognized: late Cretaceous to Paleocene, late Eocene to early Oligocene, and during the middle Miocene (Ru and Pigott 1986; Li and Rao 1994; Pigott and Ru 1994; Wang and Sun 1994; Chen et al. 1994). During the rifting phase, individual half-grabens and faulted centers were filled by fluvial, lacustrine and other clastic rocks. After the mid-Oligocene unconformity, rapid subsidence caused a rise in local sea level and marine transgressions from south to north. As a result, the previous exposed structural highs (Shenhuansha, Dongsha, and Panyu) were gradually submerged and became the sites for carbonate platform growth (e.g., Zhu 1987; Moldovanyi et al. 1995; Zampetti et al. 2005). Sediments transported by the paleo-Pearl River were largely trapped in depressions as a huge, mainly destructive delta (paleo-Pearl River delta) affected strongly by waves and tides (Chen et al. 1994).

The sediment sequence in the Zhujiangkou Basin is divided into the Paleogene Shenhu, Wenchang, Enping and Zhuhai formations, and the Neogene Zhujiang, Hanjiang, Yuehai, and Wanshan formations (Fig. 3.25) (Zeng et al. 1981; Hou et al. 1981; Wang 1985;Wang et al. 1985a,b; Duan and Huang 1991; Jiang Z. et al. 1994). Characterized by a lacustrine facies developed during an early rift stage, the Eocene Wenchang Formation shales are believed to be oil source rocks, while the coal-bearing strata in the more widespread Enping Formation (Eocene-Oligocene) are often considered as gas sources. A coarsening upward succession observed in the Enping Formation suggests a lake filling process during deposition. Intensified faulting in this syn-rift stage created an unconformity before further subsidence, which increased accommodation space and unified the lake half-grabens into one shallow marine basin. The Zhuhai Formation is an up to 1,500 m thick, wave- or fluvial-dominated delta-shelf system (Chen et al. 1994). Sheetlike sandstones are common, mainly glauconitic quartzose sandstone and arkosic quartzose sandstone with very little mudstone, presumably deposited in a wave-dominated enviroment. The overlying Zhujiang Formation also accumulated at delta to shelf settings in slightly deeper waters, prograding to carbonate platform reef facies on the Dongsha Swell Zone. The middle Miocene Hanjiang Formation contains mudstones and sandstones similar to those of the Zhujiang Formation, probably deposited also in an overall transgressive environment (Chen et al. 1994; Jiang Z. et al. 1994).

Numerous seismic and sequence stratigraphic studies have been done in the Zhujiangkou Basin (e.g., Zhou 1987; Jin 1989; Xu et al. 1995; Wang C. 1996; Gong and Li 1997; Xu 1999; Huang 1999; Wu et al. 2000), with about 10 seismic

Fig. 3.25 Correlation between different stratigraphic classification schemes in the Zhujiangkou Basin show major stages in basin development. The lithologic columns and major reflectors are modified from Jiang Z. et al. (1994). MS, megasequences, and see Fig. 3.22 for other abbreviations

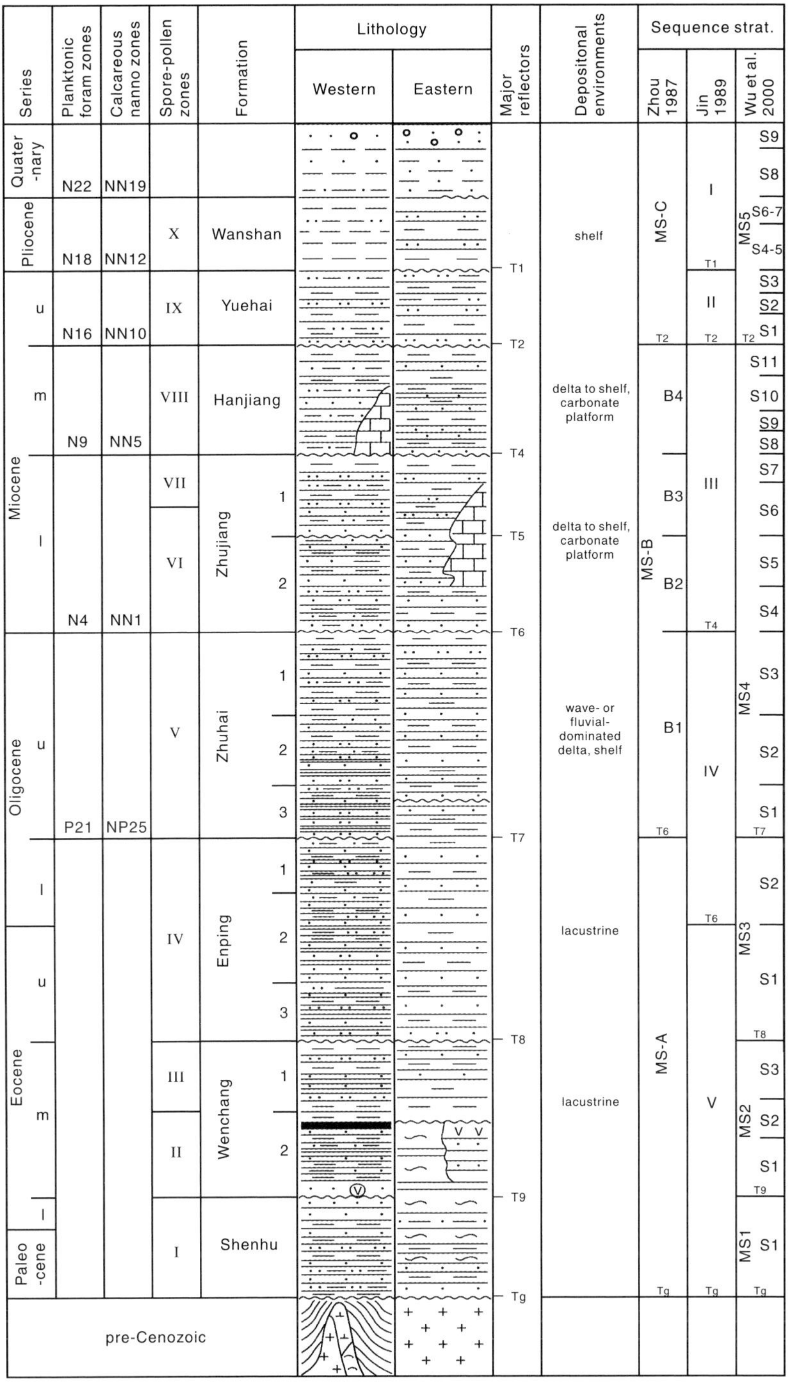

Series	Planktonic foram zones	Calcareous nanno zones	Spore-pollen zones	Formation	Lithology Western	Lithology Eastern	Major reflectors	Depositional environments	Sequence strat. Zhou 1987	Jin 1989	Wu et al. 2000
Quaternary	N22	NN19						shelf	MS-C	I	S9
											S8
Pliocene			X	Wanshan							MS5 S6-7
	N18	NN12					T1			T1	S4-5
u			IX	Yuehai						II	S3
	N16	NN10					T2		T2	T2	S2 / S1
m			VIII	Hanjiang				delta to shelf, carbonate platform	B4		S11
											S10
	N9	NN5									S9
							T4				S8
Miocene			VII	Zhujiang 1				delta to shelf, carbonate platform	B3	III	S7
							T5				S6
l			VI	Zhujiang 2					B2		S5
	N4	NN1					T6			T4	S4
				Zhuhai 1				wave- or fluvial-dominated delta, shelf	B1		S3
u			V	Zhuhai 2						IV	S2
Oligocene	P21	NP25		Zhuhai 3			T7		T6		S1
				Enping 1						T7	S2
l			IV	Enping 2				lacustrine		T6	MS3 S1
u				Enping 3							
							T8			T8	
m			III	Wenchang 1				lacustrine	MS-A	V	MS2 S3
			II	Wenchang 2							S2 / S1
Eocene						v v	T9				T9
l			I	Shenhu							MS1 S1
Paleocene							Tg		Tg	Tg	Tg
pre-Cenozoic						+ + +					

Fig. 3.25 (continued)

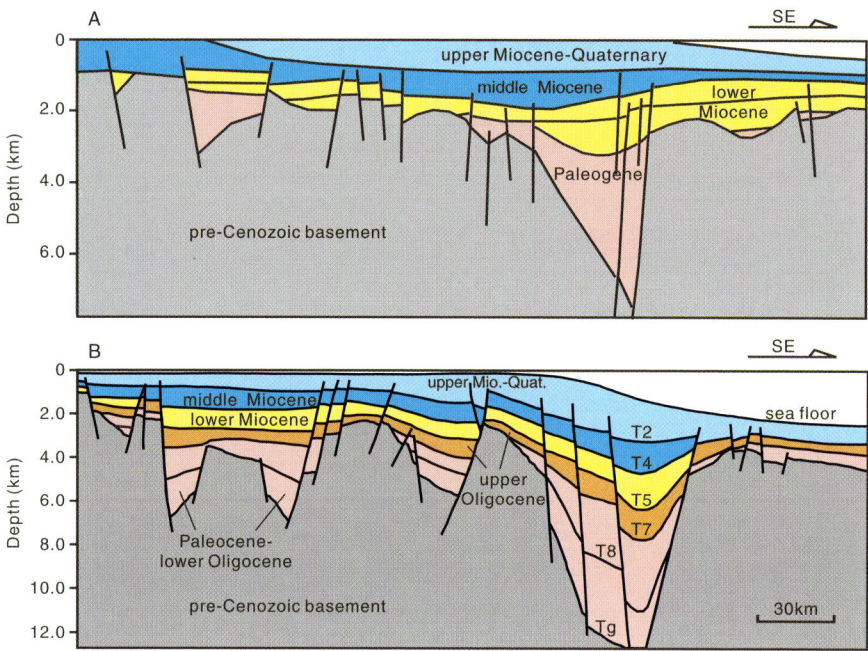

Fig. 3.26 Interpreted cross sections show distributions of major faults and Cenozoic seismic sequences in the Zhujiangkou Basin, as modified after (**A**) Gong and Li (1997) and (**B**) Chen (2000). See Fig. 3.24 for line locations

reflectors, T0 to T9 and Tg, identified as marking regional unconformities (Figs. 3.25 and 3.26). Reflector T7 has often been considered as representing the breakup unconformity separating the lower syn-rift from the upper post-rift megasequences in the basin. Among the designated megasequences and sequences, megasequences 1 to 3 with six sequences characterize the syn-rift nonmarine phase (Xu 1999; Huang 1999), and megasequences 4 and 5 with about twenty sequences the post-rift marine phase (Fig. 3.25) (Gong and Li 1997; Wu et al. 2000). Most sequence boundaries can be correlated with deep-sea hiatuses and positive excursions in deep sea oxygen isotope curves, suggesting an origin from climatic cooling and glacio-eustatic fall (Gong and Li 1997; Wu et al. 2000).

Petroleum exploration since the mid-1970s (Jin 1989) in the Zhujiangkou Basin has discovered abundant oil and gas resources mainly in the deltaic to littoral sandstones and reefal carbonates of the Zhuhai and Zhujiang Formations (e.g., Zhu et al. 1999). Industrial drilling over decades has confirmed its status as the largest and the most productive hydrocarbon-bearing basin in the northern South China Sea margin.

Beibuwan Basin

The Beibuwan Basin (=Beibu Gulf Basin) is situated in the northeastern Gulf of Beibu (Tokin) and covers an area of about 38,000 km^2 mainly in modern water depths <50 m. The shallow water basin is bounded by 108°–110.5 °E and

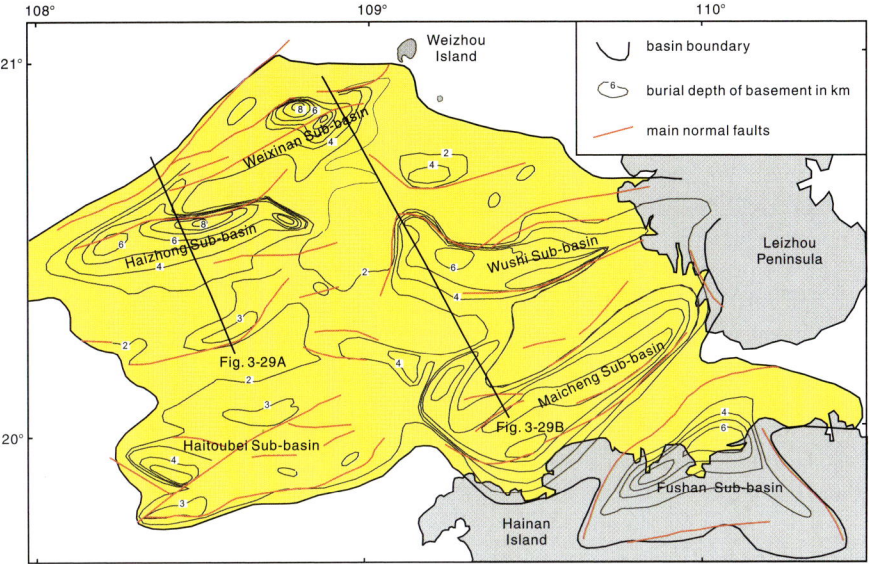

Fig. 3.27 Sketch map shows major structural units and thickness (in km) of Cenozoic sediments in the Beibuwan Basin (after Ying 1998)

19.5°–21 °N, with part of the eastern and southern margin respectively joint with Leizhou Peninsula and Hainan Island (Fig. 3.27) (Zhang and Su 1989; Jiang Z. et al. 1994).

The Beibuwan Basin is a Cenozoic rift basin containing several thousand meters of Paleogene syn-rift sediments and relatively thin Neogene post-rift sediments, with a combined maximum thickness of up to 7,000 m in depocenters (Kang et al. 1994; Ying 1998). The pre-Cenozoic basement consists of slightly metamorphosed Paleozoic carbonate and clastic rocks, Mesozoic granitic plutonic rocks and andesitic volcanic rocks and Mesozoic sedimentary rocks (Zhang and Kou 1989; Zhang and Su 1989; Ying 1998). Six W-E stretching sub-basins are recognized: Weixinan, Haizhong, Wushi, Haitoubei, Maicheng, and Fushan sub-basins, bounded by NE to ENE trending fault systems with dip-slip offsets of 2 km or more and separated by uplifted fault blocks (Fig. 3.27) (Zhang and Su 1989; Ying 1998).

At least two distinct phases of extension exist in the basin: the Paleocene to late Eocene (first) and the Oligocene (second) phases (Ying 1998). In the first phase, major basin-bounding fault systems evolved along previously existing NE trending structural lineations and resulted in large simple half grabens. In the second phase, the previously existing Eocene graben-bounding faults were reactivated. Newly generated secondary faults crosscut and rotated the syn-rift sequence deposited in the first extension phase, causing structural reconfiguration of the basin and shifting of subsidence centers. Numerous secondary faults developed with increasing extension rates during this later phase of extension. The relative quiet post-rift structural period since the Miocene was punctuated by two significant events: a short pulse of renewed subsidence at about 15 Ma and basin inversion since the late Pliocene.

Basin inversion was concentrated in the western part of the basin and was most likely related to a change of motion along the Red River strike-slip fault system (Rangin et al. 1995; Ying 1998).

Stratigraphy of the Beibuwan Basin is classified as, form bottom to top, Changliu (upper Paleocene to lower Eocene), Liushagang (middle Eocene to lower Oligocene), Weizhou (middle and upper Oligocene), Xiayang (lower Miocene), Jiaowei (middle Miocene), Dengloujiao (upper Miocene), and Wanglougang (Pliocene-Quaternary) formations (Fig. 3.28). The Changliu, Liushagang, and Weizhou formations, constituting the Paleogene syn-rift strata, are mainly nonmarine and restricted to the separated grabens and half-grabens, with ages coarsely dated by spore-pollen assemblages (Kou and Ye 1981; Sun et al. 1981; Zhang and Kou 1989). The overlying Neogene post-rift strata, including Xiayang, Jiaowei, Dengloujiao, and Wanglougang formations, are all marine facies, and thus their chronostratigraphy has been well defined by marine microfossils, primarily planktonic foraminfera and calcareous nannofossils (Fig. 3.28) (e.g., Wang 1985; Jiang Z. et al. 1994).

As described by Hu and Su (1981), Wang (1985), Zhang and Kou (1989) and Ying (1998), the Changliu Formation consists of reddish sandstones, conglomerates, and shales deposited at alluvial and alluvial fan settings, and contains *Celtispollenites triporatus-Pentapollenites* palynomorphs with an age of Paleocene to early Eocene. The Changliu may reach a maximum thickness of 840 m or more, unconformably overlying pre-Cenozoic basement rocks (Zhang and Kou 1989). The Liushagang Formation features very thick lacustrine dark gray shales subdividable into three members by variations in intercalated components: an upper Liu-1 member having interbeded sandstones, an intermediate Liu-2 member, and a lower Liu-3 member with more sandstones. The Liushagang may reach a total thickness of over 1,800 m, and contain middle to late Eocene *Quercoidites* palynomorphs. The Weizhou Formation is composed of interbedded purple, green, gray variegated mudstone and gray sandstone and conglomerate, with locally distributed coal and marine sandstone interbeds and containing Oligocene *Magnastriatites howardi-Gothanipollis bassensis- Hydrocotaepites pachydermus* palynomorphs. The >700 m thick Xiayang Formation is dominated by gray sandy conglomerate, pebbly sandstone with minor mudstone deposited in coastal to shallow marine environments during a major marine transgression, as indicated by early Miocene planktonic foraminifera *Globigerina* spp., *Cassigerinella chipolensis*, and *Globigerinatella insueta*. The overlying Jiaowei Formation is composed of greenish gray fine sandstone, muddy sandstone and (mainly in the upper part) mudstone deposited in neritic environments. The formation attains ~100 m to >570 m in thickness, and contains middle Miocene foraminifera including *Paragloborotalia siakensis*, *Orbulina* and *Cassigerinella chipolensis*. The Dengloujiao Formation comprises nearly 600 m thick interbedded gray sandstone, sandy conglomerate, and gray mudstone, and contains a late Miocene foraminiferal association featuring *Globoquadrina dehiscens, Globorotalia lenguaensis, Dentoglobigerina altispira* and *Neogloboquadrina acostaensis*. The ~300 m thick Wanglougang Formation is dominated by greenish gray to gray mudstone intercalated with sandstones, and bears *Globigerinoides fistulosus, Globigerinoides extremus, Dentoglobigerina*

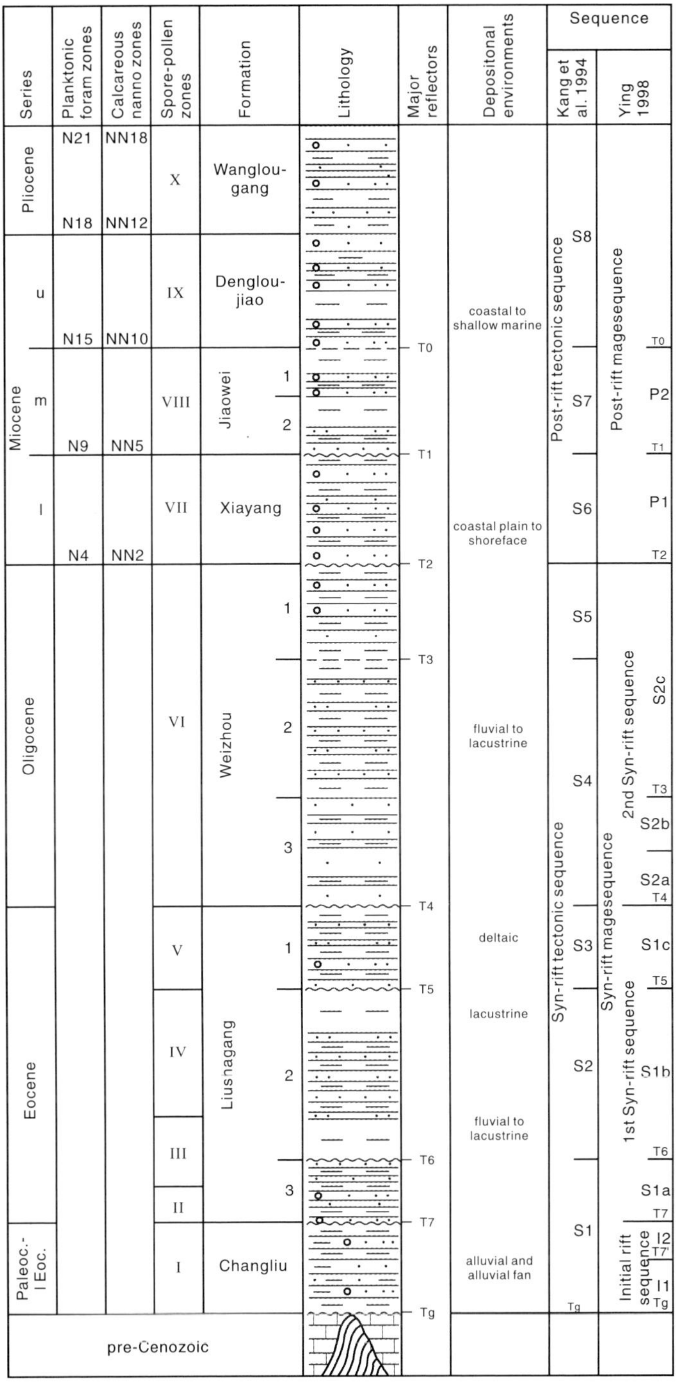

Fig. 3.28 Composite lithostratigraphy of the Beibuwan Basin was modified from Jiang Z. et al. (1994). See Fig. 3.22 for sequence abbreviations

altispira, Globorotalia margaritae, Globorotlia tumida, and other Pliocene planktonic foraminifera and nannofossils. The Quaternary sediment in the Beibuwan
Basin seldom excesses 35 m and is dominated by littoral sand and clay.

Studies by Kang et al. (1994), Ying (1998) and Du and Wei (2001) identified several major seismic horizons as marking sequence boundaries, including
Tg (basement unconformity), T7 (top Changliu Formation), T4 (top of Liushagang Formation), and T2 (top of Weizhou Formation, also Oligocene/Miocene
boundary) (Fig. 3.28). The breakup unconformity (T2) separates the Paleogene
syn-rift megasequence (below T2) and the Neogene–Quaternary post-rift megasequence (above T2) (Figs. 3.28, 3.29). The syn-rift megasequence, consisting of the
Changliu, Liushagang, and Weizhou Formations with a combined thickness of several thousand meters, is confined to isolated grabens and half grabens, as revealed
by drilling and seismic survey in various sub-basins. Ying (1998) further divided
the Paleogene syn-rift magasequence into two initial rifting sequences I1 and I2
(Tg-T7), and two syn-rifting sequences S1 (T7-T4) and S2 (T4-T2) (Fig. 3.28). The
Neogene post-rift megasequence includes the Xiayang, Jiaoweo, Dengloujiao, and
Wanglougang Formations and is characterized by a series of progradational clastic sequences deposited in coastal to shallow marine environments and by locally
developed lower Miocene carbonate reef buildups.

Petroleum exploration started in the mid-1960s in the Beibuwan Basin, one of
the earliest explored offshore basins in the northern SCS margin. Reservoirs have
been found in Liushagang, Weizhou, and Jiaowei sandstones (Jin 1989, 2005).

Yinggehai and Qiongdongnan Basins

The Yinggehai Basin (=Song Hong Basin) (Fig. 3.30) is a NW-SE trending elongated basin in the southern Gulf of Beibu between Hainan Island and Indochina
Peninsula in modern water depths mostly <100 m, and extends toward the northwest
into the Hanoi Basin. It is about 500 km long and 50–60 km wide and covers an area
of approximately $11.3 \times 10^4 \, \text{km}^2$ (Zhang and Zhang 1991; Jiang Z. et al. 1994).
The Qiongdongnan Basin (=Southeast Hainan Basin) is a northeast trending basin
situated between Hainan Island in the north and the Xisha Islands in the south, and
bordered by the Zhujiangkou Basin to the east and Yinggehai basin to the west.
With an area of about $4.5 \times 10^4 \, \text{km}^2$ in water depths mostly <200 m (Jiang Z.
et al. 1994), the Qiongdongnan Basin is separated from the Yinggehai Basin by
the so-called No. 1 fault, a southern extension of the NW-SE trending Red River
Fault System (Fig. 3.30). Seismic data indicate that the No. 1 fault is a normal fault
controlling the northeastern boundary of the Yinggehai Basin, and it fades gradually
southeastward (Zhang and Zhang 1991), although it could have been of right lateral
strike-slip in nature over various periods.

Both the Yinggehai and Qiongdongnan basins consist of a similar rift succession
in the lower part, overlain by a similar post-rift succession. The breakup unconformity separating these two successions has been assigned to various ages by different
authors: base of the Miocene (T6) by Gong and Li (1997), base of upper Oligocene
by Anderson et al. (2005), and in the lower Miocene by Xie et al. (2006) (Fig. 3.31).

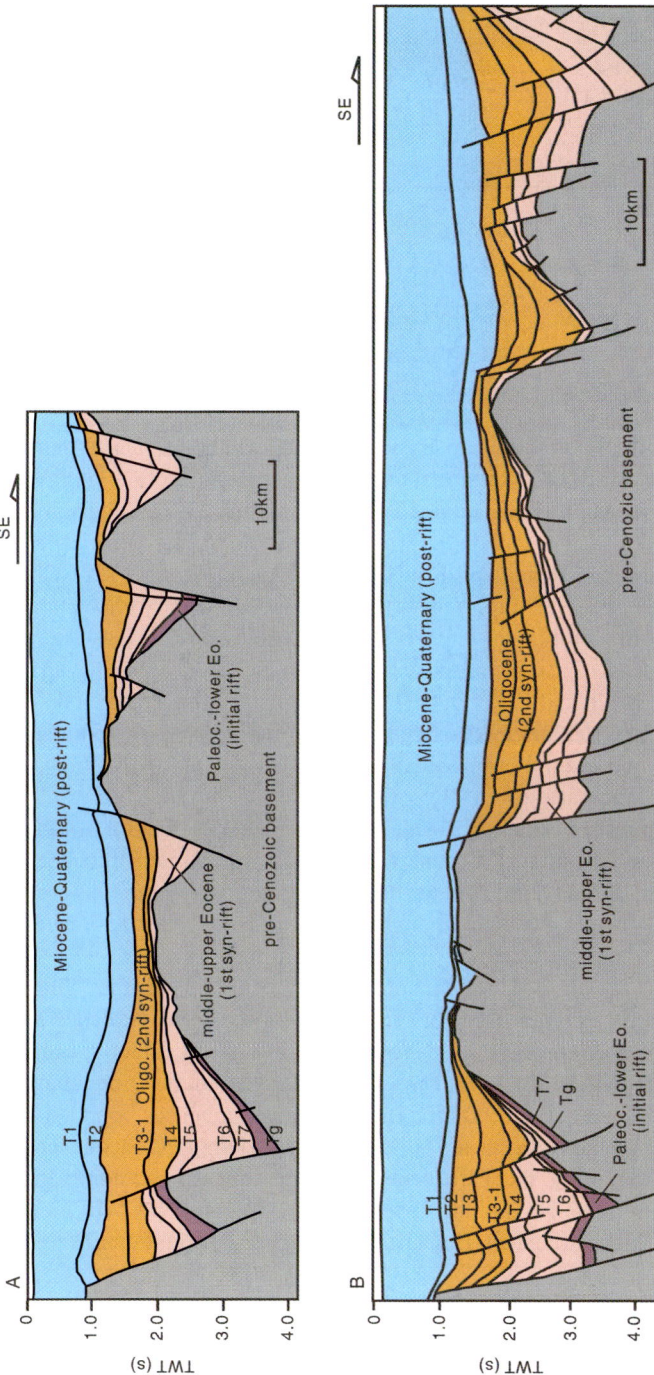

Fig. 3.29 Interpreted cross sections show major Cenozoic seismic sequences in the Beibuwan Basin (modified after Ying 1998). See Fig. 3.27 for line locations

Fig. 3.30 Sketch map show
the location of fault systems
in the Yinggehai and
Qiongdongnan basins
(modified after Qiu and Wang
2001; Ying 1998; Gong et al.
1989)

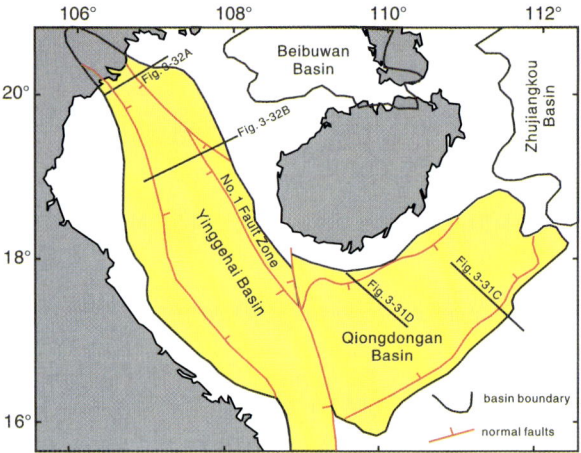

Compared to the Qiongdongnan Basin, the Yinggehai Basin has been influenced strongly by transformed extensional tectonic processes closely relating to activities of the Red River Fault (Chen et al. 1993; Gong and Li 1997; Andersen et al. 2005; Xie et al. 2006). Some authors suggest that the Yinggehai Basin is a strike-slip tensional basin (e.g., Zhang and Kou 1989; Zhang and Zhang 1991). Clift and Sun (2006) recently reviewed important structural and sedimentary characteristics of the Yinggehai Basin.

Very thick Cenozoic deposits have been found in these two basins: up to 17 km in the Yinggehai Basin and 12 km in the Qiongdongnan Basin (Zhang and Zhang 1991; Gong and Li 1997; Zhang 1999). Because of similar lithologies, sediment packages of similar age in these two basins are often given with a single formation name. From bottom to top, they are lower Oligocene to older aged Yacheng Formation, upper Oligocene Lingshui Formation, lower Miocene Sanya Formation, middle Miocene Meishan Formation, upper Miocene Huangliu Formation and Pliocene Yinggehai Formation (Fig. 3.31) (Jiang Z. et al. 1994). The later two formations are sometimes lumped together as Yinggehai-Huangliu (or Ying-Huang) formations due to their similar neritic characteristics.

A number of basin-wide unconformities identified in the Yinggehai and Qiongdongnan basins are used for reconstructing the local sequence stratigraphy (Chen et al. 1993; Wang et al. 1998; Hao et al. 2000; Wu et al. 2000; Lin et al. 2001; Wei et al. 2001; Xie et al. 2008) (Figs. 3.31 and 3.32). In the Qiongdongnan Basin, for example, Chen et al. (1993) identified five megasequences bounded by regional unconformities, and Wang et al. (1998) recognized fourteen sequences, with the lower six representing the Paleogene rift megasequence including the Yacheng and Lingshui formations, and the upper eight the Neogene post-rift megasequence including the Sanya, Meishan, Huangliu and Yinggehai formations. Wei et al. (2001)

Fig. 3.31 Composite lithobiostratigraphy of the Yinggehai and Qiongdongnan basins show important stages in the development of these two basins. The lithologic column and major reflectors are modified from Jiang Z. et al. (1994). See Fig. 3.22 for sequence abbreviations

Series	Planktonic foram zones	Calcareous nanno zones	Spore-pollen zones	Formation	Lithology	Major reflectors	Depositional environments	Sequence stratigraphy
								Chen et al. 1993 / Gong and Li 1997 / Wang et al. 1998 / Wu et al. 2000 / Hao et al. 2000 / Wei et al. 2001
Quaternary	N22	NN19				T2	littoral to bathyal	
Pliocene (u)	N18	NN12	VI	Yinggehai /Huangliu		T4	littoral to bathyal	
Miocene (m)	N16	NN10	V	Meishan			littoral to shallow marine	
	N9	NN5	IV	Sanya		T5-1	littoral to shallow marine	
Miocene (l)	N4	NN1	III			T5-2 / T6		
Oligocene (u)		NP25	II	Lingshui		T6-1 / T6-2 / T7	coastal to neritic	
Oligocene (l)			I	Yacheng		T8 / Tg	coastal to neritic / lacustrine	
				pre-Cenozoic				

Sequence stratigraphy (reference schemes):

- **Chen et al. 1993:** MS6, MS5, MS4 (T4), MS3 (T6), MS4 (T8), MS2, MS1 (T7, T8, T10)
- **Gong and Li 1997:** S5-12, S1-4 (T2, T4); Post-rift tectonic sequence; Syn-rift tectonic sequence; S1–S18, T6, T8
- **Wang et al. 1998:** MS1 (S1–S7, T2), MS2 (S8–S13, T4, T6), MS3 (S15, S16), MS4 (S17, S18, T8), MS2 (S1, S2), Ms1 (S1, T9, T10); Tg
- **Wu et al. 2000:** S8-10, S7, S6, S3-5, S2, S1 (T2); MS5; S6, S5, S4, S3, S2, S1 (T4, T6); MS4; S5, S4, S3, S2, S1 (T6); MS3; S3, S2, S1 (T7, T8); MS2; S2, S1 (T9)
- **Hao et al. 2000:** T2, S4, S3, S2, S1 (III, T4); S4, S3, S2, S1 (II, T6); S3, S2, S1 (I, T7); TA markers
- **Wei et al. 2001:** S5-8, TB2; S14; S12-13, TB1; S9-11; TA3 S5-8; TA2 S3-4; TA1 S1-2

Fig. 3.31 (continued)

also identified five super-sequences, super-sequences TA1 to TA5, to mark 5 major stages of basin evolution: (1) TA1 from initial rifting in the Eocene, (2) TA2 from middle rifting in the early Oligocene, (3) TA3 from later rifting during the late Oligocene and earliest Miocene, (4) TA4 indicating an early post-rift ramp setting from the late early Miocene to middle Miocene, and (5) TA5 indicating a late post-rift passive margin setting in the late Miocene and Pliocene (Fig. 3.31).

Based on micropaleontologic, lithologic, well log and seismic data, Hao et al. (2000) assigned three second-order sequences SS1, SS2 and SS3 and 19 third-order sequences for the upper Oligocene-Pliocene succession in these two basins (Fig. 3.31). SS1 is equivalent to the Lingshui Formation and dividable into three third-order sequences with upper P21 to lower N5 planktonic foraminifera. SS2 is equivalent to the Sanya and Meishan formations with N5 to N15 planktonic foraminifera. SS3 is equivalent to the Yinggehai and Huangliu formations with N16 to N21 planktonic foraminifera, subdividable into six third-order sequences (Fig. 3.31).

All the above-mentioned studies did not offer any clear clue to whether any sediment sequences have been deformed. However, Anderson et al. (2005) found strongly deformed sequences from the northwestern Yinggehai Basin likely affected by lateral strike-slip movement of the Red River Fault during the late Miocene-Pliocene (Fig. 3.32A). Further studies are needed to clarify whether this is a localized or basinal phenomenon.

The extremely thick upper Miocene to Quaternary sediment succession in the Yinggehai and Qiongdongnan basins offers opportunities for high-resolution seismic sequence stratigraphic studies. Chen et al. (1993) described 12 sequences (SQ1–SQ12 in Fig. 3.32D) bounded by type-1 sequence boundaries in the Qiongdongnan Basin, with 8 sequences (SQ5 to SQ12) for the late Pliocene to Quaternary interval, which are significantly out number the four third-order sequences identified by Haq et al. (1988). Similarly, Lin et al. (2001) recognized 9 unconformity-bounded sequences in the Ying-Huang formations to represent late Miocene to Pliocene lowstand and highstand systems tracts, as they found that transgressive systems tracts were generally too thin to be recognized on seismic profiles. Lin et al. (2001) discovered incised valleys and accompanying slope and basin-floor gravity flow deposits in lowstand systems tracts, suggesting multiple sea level fall events since the late Miocene. In another study, Xie et al. (2008) identified 7 third-order sequences for the same late Miocene to Pliocene interval based on regional 2D seismic profiles tied to well control. Therefore, many different schemes of sequence stratigraphy have been proposed using different data sets with different resolution, and to acknowledge which scheme is better or more practical is very difficult.

Fig. 3.32 Interpreted cross sections show major faults and Cenozoic seismic sequences in (**A**) Yinggehai Basin (modified after Andersen et al. 2005), (**B**) Yinggehai Basin (modified after Gong and Li 1997), (**C**) Qiongdongnan Basin (modified after Xie et al. 2006), and (**D**) Qiongdongnan Basin (modified after Chen et al. 1993). See Fig. 3.30 for line locations

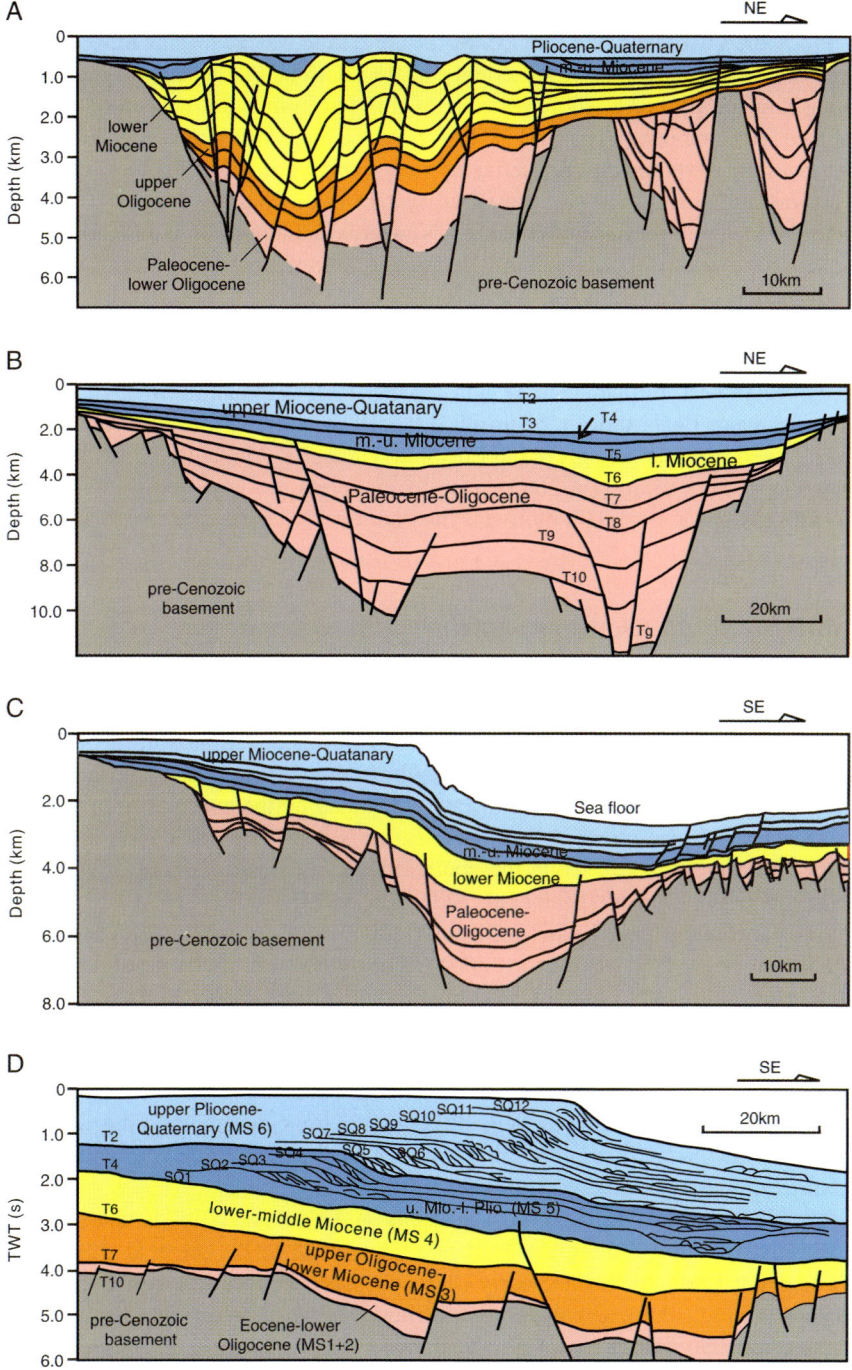

Fig. 3.32 (continued)

Since the Oligocene, the depocenter in the Yinggehai Basin has moved southward due to stepwise greater subsidence in the south and/or sequential availability of basin space further offshore because of rapid accumulation nearshore, close to the Red River mouth (Gong and Li 1997). Associated with depocenter movement there was the development of a series of mud diapiric structures in close relation to overpressured shales. The overpressured system in this basin was caused by rapid sediment load and greater burial depths. Now at burial depths of 3–5 km, for example, even the relatively young Ying-Huang formations (lower part, 11–5 Ma) may produce high pressure of 60–80 MPa (Hao et al. 1995).

Industrial exploration in the Yinggehai Basin dated back to 1957, when a survey for oil and gas seepages was undertaken along the coast of Yinggehai village in the southwest Hainan (Zhang and Zhang 1991). Exploration in the deeper Qiongdongnan Basin did not begin until the 1980s. Oil and gas have been found and pumped from sandstones of the Lingshui, Huangliu, Yinggehai formations, and even Pleistocene strata (Jin 2005). Due to overpressure, however, the Yinggehai sandstones appear to contain more gas than oil.

Southern South China Sea Basins

Cuu Long and Wan'an (Nam Con Son) Basins

The Cuu Long and Wan'an (=Nam Con Son) Basins are located in the shelf and slope area between the Mekong River mouth to the northwest and Nansha Islands (the Dangerous Ground) to the southeast. Separated by the northeast-trending Con Son Swell Zone, the northern Cuu Long Basin is nearshore and covers an area of about 250 km × 100 km, while the southern Wan'an Basin is more offshore and covers about 90,000 km^2 (Binh et al. 2007) (Fig. 3.33).

Both are extensional basins initiated from Eocene to early Oligocene rifting (Matthews et al. 1997; Lee et al. 2001). The maximum thickness of Eocene to Quaternary sediments is about 8 km in the Cuu Long Basin (Binh et al. 2007) and 12.5 km in the Wan'an Basin (Chen and Peng 1995). The basement in the Cuu Long Basin consists of late Jurassic-early Cretaceous granites and granodiorites as revealed by drilling in the Bach Ho field (Areshev et al. 1992), while the basement in the Wan'an Basin is formed by highly variable pre-Cenozoic granitic, volcanic, and low-grade metasedimentary units, which have been penetrated by wells drilled on structural highs (Matthews et al. 1997).

The structural evolution of the Cuu Long and Wan'an basins is characterized by rifting and regional thermal subsidence (Dien et al. 1998; Lee et al. 2001). One rifting phase during the Eocene-early Oligocene has been identified in the Cuu Long Basin (Lee et al. 2001), and at least two rifting phases occurred in the Wan'an Basin (Canh et al. 1994; Fraser et al. 1996; Matthews et al. 1997; Lee et al. 2001). Timings for the two rifting phases in the Wan'an Basin have been assigned to Eocene-Oligocene and early to late Miocene (24-8 Ma) by Lee et al. (2001), and to Eocene-early Oligocene and middle to late Miocene by Matthews et al. (1997), respectively.

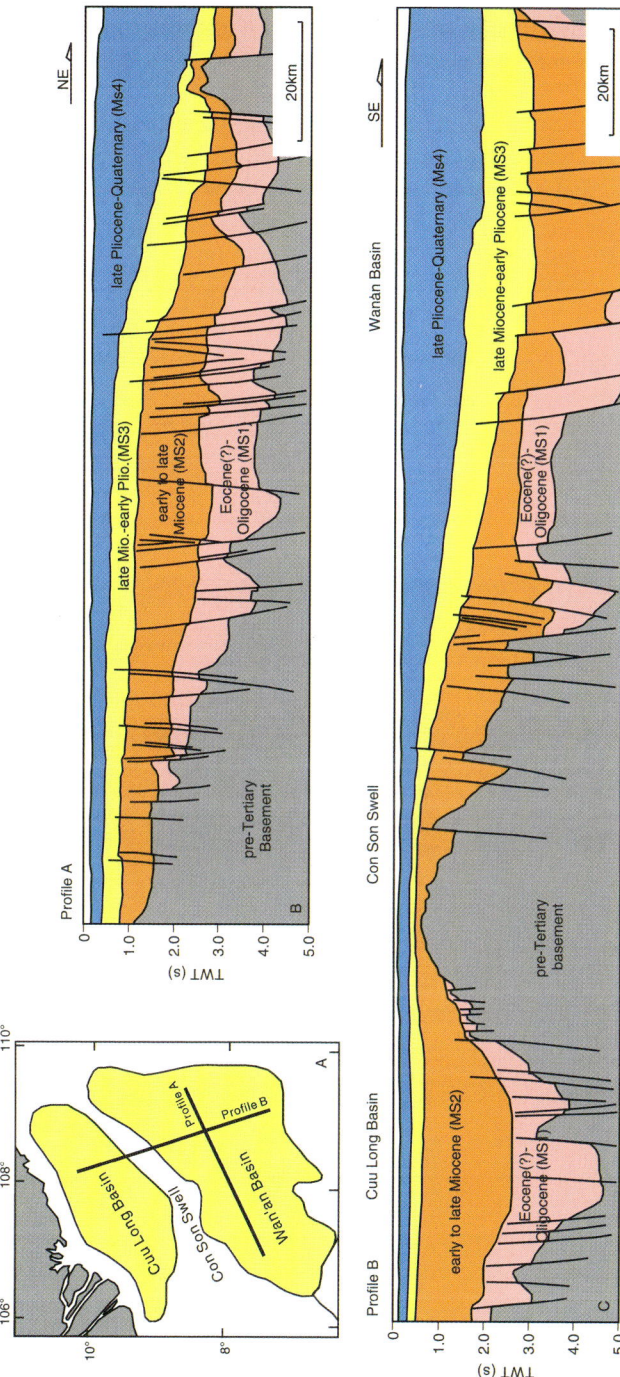

Fig. 3.33 (**A**) Sketch map shows that the NE-SW trending Cuu Long and Wan'an (Nam Con Son) Basins are separated by a narrow Con Son Swell off the Mekong River mouth (compiled from Lee et al. 2001; Binh et al. 2007). (**B**) and (**C**) Interpreted cross sections show distributions of major faults and Cenozoic seismic sequences in the Wan'an and Cuu Long basins (modified after Lee et al. 2001)

The discrepancy in the age of the two rifting phases is probably due to poor dating, but the existence of a third rifting phase is also likely if all areas of the Wan'an Basin were fully surveyed (Olson and Dorobek 2002).

According to Matthews et al. (1997) and Binh et al. (2007), Wan'an Basin strata include Eocene-lower Oligocene deposits, Oligocene Cau Formation, lower Miocene Dua Formation, middle Miocene Thong and Mang Cau Formations, upper Miocene Nam Con Son Formation and Pliocene to Quaternary Bien Dong Formation (Fig. 3.34).

These formations are mainly composed of clastic rocks, although the middle Miocene to lower Pliocene formations of Thong, Mang Cau, Nam Con Son and

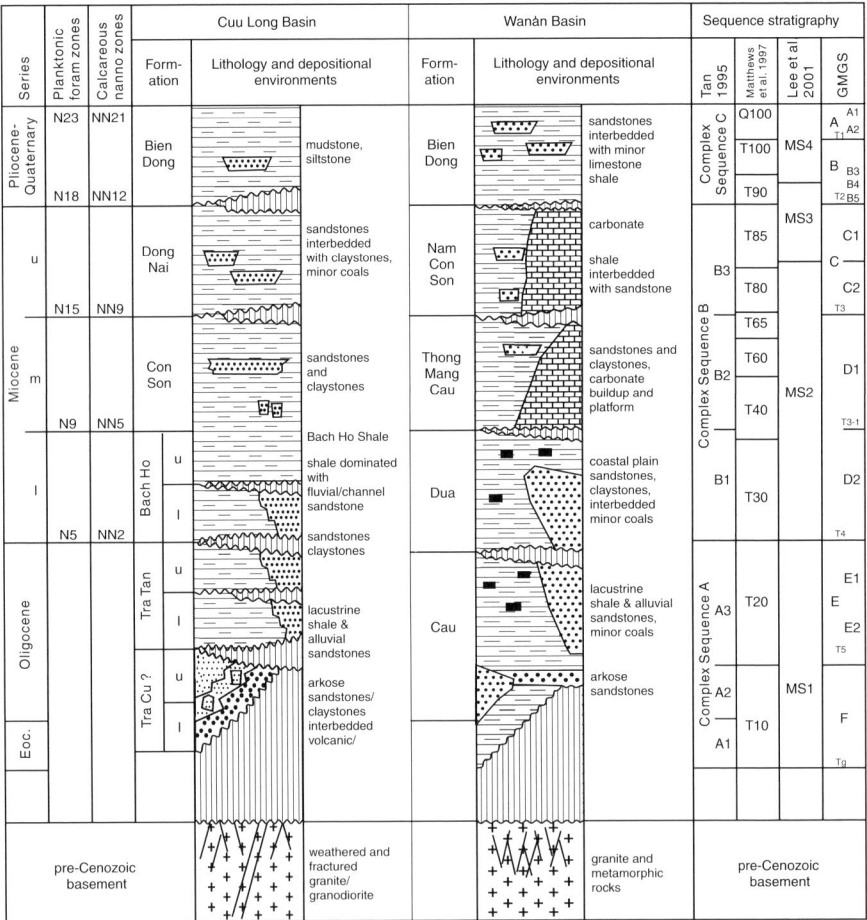

Fig. 3.34 Generalized lithobiostratigraphic columns (Binh et al. 2007) are correlated with different sequence stratigraphic schemes for the Cuu Long and Wan'an (Nam Con Son) Basins. The GMGS (Guangzhou Marine Geological Survey) scheme is integrated from Wu and Yang (1994), Chen and Peng (1995), Yang et al. (1996), Wang et al. (1996, 1997) and Wu et al. (1997)

lower Bien Dong from eastern structural highs may comprise carbonate platform and shallow marine facies. In contrast, all sediment deposits in the Cuu Long Basin are of clastic origin, mainly from Mekong discharge. From bottom to top, they are the Tra Cu Formation (late Eocene to early Oligocene), Tra Tan Formation (later early to late Oligocene), Bach Ho Formation (early Miocene), Con Son Formation (middle Miocene), Dong Nai Formation (late Miocene) and Bien Dong Formation (Pliocene to Quaternary).

Many sequence stratigraphic studies have been published, including Tan (1995), Wu and Yang (1994), Chen and Peng (1995), Yang et al. (1996), Wang et al. (1996, 1997), Wu et al. (1997), Matthews et al. (1997), and Lee et al. (2001) (Fig. 3.34). With various sequence numbers and names by different authors, the Cenozoic deposits in the Cuu Long and Wan'an basins can be grouped into a syn-rift and a post-rift megasequences, although the boundary for the two is placed at the Oligocene/Miocene contact in the Cuu Long and at the Miocene/Pliocene contact in the Wan'an (Figs. 3.33 and 3.34). The syn-rift succession in the Wan'an Basin can be further divided into an older, Eocene-Oligocene syn-rift megasequence, which is dominated by alluvial and fluvial sandstones and lacustrine shales intercalated occasionally with coal beds, and a younger, Miocene syn-rift megasequence, which mainly comprises coastal to shallow shelf sandstones, shales, and platform carbonate rocks (Matthews et al. 1997; Lee et al. 2001). The post-rift megasequence in the Wan'an Basin is characterized by shelfal to deep marine shales intercalated with siltstones and sandstones, but in the Cuu Long Basin it mainly comprises littoral to shallow marine clastics and mudstones.

The Cuu Long Basin is oil-prone, especially in fractured basement highs, whereas the Wan'an Basin is now known containing more gas trapped in Miocene sands and in middle Miocene to lower Pliocene carbonates (Tan 1995; Lee et al. 2001).

Zengmu Basin

Zengmu Basin, including the Sarawak and East Natuna basins, is located in the southern margin of the SCS between the Nansha Islands (the Dangerous Ground) to the north, the Natuna Islands to the west, and Borneo (onshore Sarawak) to the south (Fig. 3.35). Covering an area of about $17 \times 10^4 \, km^2$ (Qiu et al. 2005), the Zengmu Basin is the largest shelf-slope basin in the SCS region.

Tectonically, the Zengmu Basin is a late Eocene to Recent sedimentary basin on the Zengmu (Locunia) Block, which is a micro-continental block rifted off and drifted southward from the South China continent and later collided with the northern Borneo margin similar to other neighbouring continental blocks, such as the Nansha (Dangerous Ground), Liyue (Reed Bank) and North Palawan, apparently responding to the Oligocene to early Miocene opening of the SCS (e.g., Huang 1997; Madon 1999b). Other authors, however, considered the Zengmu Block as an eastward extension of the Indochina continental block (e.g., Wu 1991; Jin and Li 2000).

The Zengmu Basin is structurally separated from the West Natuna Basin to the west by the Natuna Arch, from the Wan'an Basin to the northwest by Xiya Uplift,

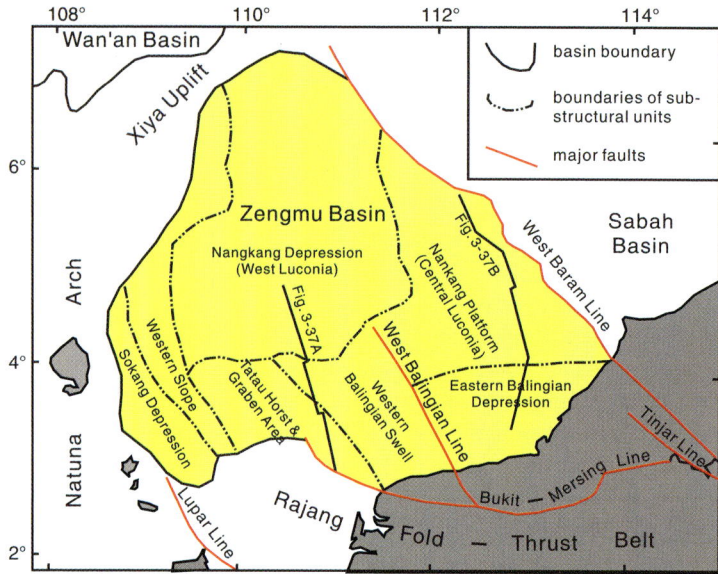

Fig. 3.35 Sketch map shows main structural-stratigraphic provinces in the Zengmu Basin (modified after Qiu et al. 2005; Madon 1999a,b)

and from the Sabah Basin to the east by the West Baram Line, a NW-SE extending tectonic discontinuity interpreted as a major transform fault by some authors (Fig. 3.35) (Madon 1999b). On the south of the Zengmu Basin lies the Rajang Fold-Thrust Belt (Sibu Zone) onshore the northern Borneo, which consists of highly deformed, low-grade metamorphosed late Cretaceous to late Eocene deep marine shales and turbidites, with some radiolarian chert, spilite and dolerite (Madon 1999b). Therefore, the early Zengmu Basin probably represents part of the remnants of the late Cretaceous to early Miocene "Proto-South China Sea" oceanic basin, the majority of which has been since subducted southward beneath the Borneo and along the island arc in the east (Madon 1999d).

The Zengmu Basin is widely believed to have originated as a peripheral foreland basin that formed in the north front of Rajang collisional orogen after the collision of the Zengmu Block with the West Borneo, and the closure of the Proto-SCS during the late Eocene (Madon 1999b). Deformation and uplift of the Rajang Fold-Thrust Belt (Fig. 3.35) provide sediment supply to the Zengmu Basin. The basin's main formation stage since the Oligocene is mainly extensional, and active extensional and strike-slip tectonics played an important role in its formation and evolution. Some authors (e.g., Mat-Zin and Swarbrick 1997) proposed that the Zengmu Basin was formed by strike-slip movement of major NW-SE faults during the late Oligocene to early Miocene, while others suggested at least two basin formation phases occurring in the Zengmu Basin (Zhong et al. 1995; Jin and Li 2000; Chen 2002). The first phase is marked by a phripheral foreland basin during the late Eocene-early Miocene, while the second phase by a strike-slip or shear extensional pull-apart basin during the middle and late Miocene, which was apparently related to movement of the strike-slip fault zones around the basin. Since the middle Miocene or late Miocene (Zhong et al. 1995; Jin and Li 2000; Chen 2002), the Zengmu Basin entered a new stage of regional subsidence and outbuilding of a passive continental margin.

The Zengmu Basin has been subdivided by different authors (e.g., ASCOPE 1981; Jiang Z. et al. 1994; Madon 1999b; Chen 2002; Qiu et al. 2005) into different structural- stratigraphic units or provinces. For example, PETRONAS (Madon 1999b) identifies seven provinces (excluding the East Natuna Basin), respectively the SW Sarawak, West Luconia, Tatau, Balingian, Tinjar, Central Luconia, and the North Luconia provinces. These provinces are mainly separated by basement-involved extensional, strike-slip, or gravity-driven detached tectonics, and experienced a variety of structural and sedimentological evolution phases (Madon 1999a). Other schemes recognize Sokang Depression, Tatau Horst and Graben Area, Western Balingian Swell, Eastern Balingian Depression (=Balingian+Tinjar Provinces), Nankang Platform (=Central Luconia Province), Nangkang Depression, and Western Slope (Fig. 3.35) (Qiu et al. 2005; ASCOPE 1981).

Most of the structural-stratigraphic provinces consist of clastic rocks, except for the Central Luconia Province that contains a huge thickness of carbonate rocks (Ho 1978; Ali and Abolins 1999; Ismail and Hassan 1999; Madon and Redzuan 1999a,b; Madon 1999c; Madon and Abolins 1999). The maximum thickness of the late Eocene to Recent sediments in the Zengmu Basin is estimated by different authors as varying from 12 km to more than 15 km (Madon 1999b; Jin and Li 2000; Chen 2002).

The stratigraphic subdivision in the Malaysian side of the Zengmu Basin was established by Shell Oil Company in the 1970s. The Eocene and younger sedimentary succession in offshore Sarawak were divided into eight regional regressive cycles separated by brief transgressive phases based on a sedimentary cycle concept (Ho 1978) (Fig. 3.36). The cycles are determined by marine transgressive units using biostratigraphic assemblages or zones (foraminiferal zonation for marine sediments or palynological zonation for coastal and nonmarine deposits; see "Biostratigraphic framework" section above). Each cycle starts with a transgressive marine shale, followed by a regressive unit in which sand percentage increases upwards and terminates with deposits indicative of coastal conditions.

The main characteristics of the cycles are as follows (ASCOPE 1981; Doust 1981; Madon et al. 1999b): Cycles I and II, upper Eocene to lower Miocene, are mainly fluvial and esturine channel sands, overbank clays and coals, and minor carbonate sediments. Cycle III, lower to middle Miocene, consists of shale with thin limestone and sandstone beds, with patchy but thicker limestones in the upper part. Cycle IV, middle Miocene, represents open shallow marine conditions throughout most parts of the Sarawak shelf, coeval to a period of extensive carbonate buildups in the Central Luconia Province. Cycle V, middle to upper Miocene, was developed with extensive reefoid cabonates in Central Luconia and with clastics in the east and west of Central Luconia, as a result of progradation of the Baram and Rajang paleo-deltas. Cycles VI to VIII, upper Miocene to Recent, comprise open marine

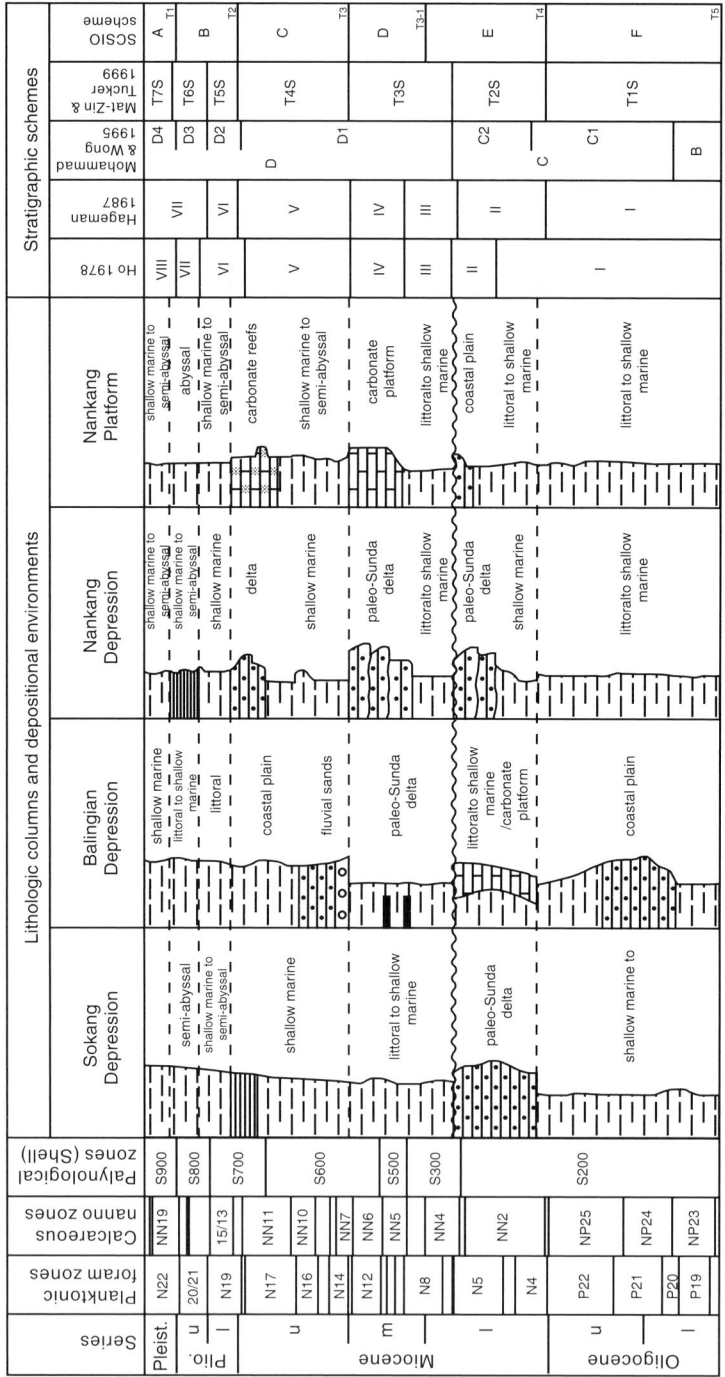

Fig. 3.36 Correlation between various litho- bio- and sequence stratigraphic schemes and interpreted depositional environments in the Zengmu Basin (modified Qiu et al. 2005 and various sources). SCSIO scheme is integrated from NISIT (1989), Jiang et al. (1990), Yao and Bai (1991), Jiang S. et al. (1994), and Wu and Yang (1994). Palynomorph zones of Shell were adopted from Mat-Zin and Tucker (1999). I to VIII represent sedimentary cycles

to coastal clays and sands during delta progradation, sometimes completely covering carbonate buildups in the Central Luconia Provience. Shell's concept of cycles is equivalent to the "T-R sequence" of Embry (1988), representing probably the earliest attempt of sequence stratigraphy in the southern SCS.

However, limitations of Shell's cycle concept have been noted (Hageman 1987; Mat-Zin and Tucker 1999) because cycles are difficult to define in coastal and nonmarine sediments where foraminifera are often absent and in open marine successions where cycles are less developed. In addition, since a cycle is bounded by marine flooding surfaces, the possible presence of a sequence boundary within it can be easily overlooked. Practically, cycle boundaries associating with major erosional events along basin margins are more easily identified on seismic sections and well logs than those relating to marine flooding surfaces. Attempting to resolve these, Hageman (1987) modified Shell's scheme by defining new cycle boundaries at the known unconformities for the bases of Cycles II, III and V (Fig. 3.36). However, this modified scheme incidently forms a hybrid of the T-R cycle concept and the Exxon sequence concept (Mat-Zin and Tucker 1999).

Whatever cycles or sequences are used, sediment packages in the Zengmu Basin can be divided and correlated using seimic data tied to well control, although the number of sequences may vary between authors. For example, Mohammad and Wong (1995) identified 8 seismic horizons in the offshore Sarawak deepwater area for grouping the local sequences into 4 supersequences (Fig. 3.36): A for the upper Cretaceous-middle Eocene, B for the upper late Eocene-upper Oligocene, C for the upper Oligocene-lower Miocene and D for the lower Miocene-Recent. These supersequences are definable also by biofacies and lithofacies data as well as GR logs. Mat-Zin and Swarbrick (1997) and Mat-Zin and Tucker (1999) identified 7 seismic sequences for the Sarawak Basin: T1S (mainly upper Oligocene), T2S (mainly lower Miocene), T3S (mainly middle Miocene), T4S (mainly upper Miocene), T5S (mainly Pliocene), T6S (Pliocene-Pleistocene), and T7S (Pleistocene-Recent) (Figs. 3.36 and 3.37). They suggest that all the Oligocene to Miocene sequence boundaries are probably tectonically induced but Pliocene-Pleistocene sequence boundaries likely derived from eustatic sea-level changes or a combination of eustasy and tectonics forcing. Based on a large number of seismic profiles, researchers in the South China Sea Institute of Oceanology (SCSIO) and Guangzhou Marine Geological Survey (GMGS) recognized 6 regional unconformities in the Zengmu Basin and neighbouring areas (Fig. 3.36): T1 (base of the Pleistocene), T2 (base of the Pliocene), T3 (base of the upper Miocene), T3-1 (base of the middle Miocene), T4 (base of lower Miocene), and T5 or Tg (top of pre-Cenozoic basement). Therefore, 6 seismic sequences A to F bounded by these horizons were defined to respresent various sedimentary stages in the Zengmu Basin (NISIT 1989; Jiang et al. 1990; Yao and Bai 1991; Jiang Z. et al. 1994; Wu and Yang 1994).

Since the mid-1950s, drilling for oil and gas has been carried out in the Zengmu Basin, especially in the Balingian and Nankang Platform provinces. Main producing reservoirs have been found in Cycle II and Cycle III sandstones in the Balingian

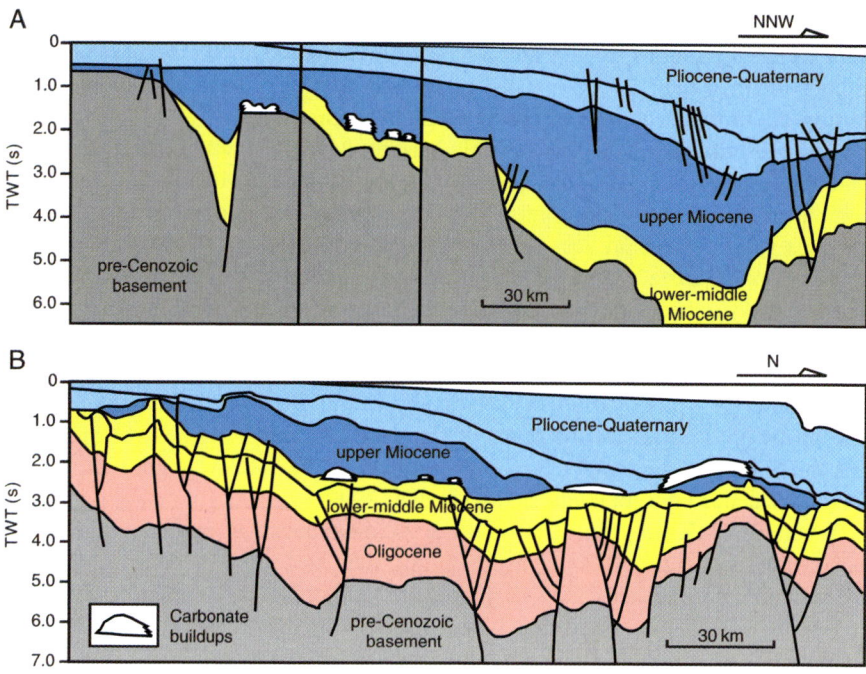

Fig. 3.37 Interpreted cross sections show the distributions of major faults and Cenozoic seismic sequences in the Zengmu Basin (modified after Mat-Zin and Swarbrick 1997). See Fig. 3.35 for line locations

Province and in Miocene carbonate rocks of Cycles IV and V from the Nankang Platform (Mahmud and Saleh 1999).

West Natuna Basin

To the west of Natuna Island and north of the Anambas Islands lie two intensely faulted basins (Fig. 3.38): the NE-SW-trending West Natuna Basin and the W-E-trending Penyu Basin, mostly in present-day water depths of 50–60 m. Structurally, the West Natuna Basin belongs to the Sundaland, the cratonic core of SE Asia, and is enclosed by a series of fault-bounded basement highs, including the Khorat Swell to the north, the Natuna Arch to the east, shallow parts of the Sunda "Craton" to the south, and a gradual transition into the Malay Basin to the west (ASCOPE 1981; Pollock et al. 1984).

The West Natuna Basin is generally considered as originating from a rift or a pull-apart basin. It is characterized by a series of small, almostly W-E oriented depocenters with intervening basement ridges that formed during Paleogene rifting (Wongsosantiko and Wirojudo 1984; Daines 1985; Ginger et al. 1993). Many half-graben depocenters in the basin experienced significant contraction during the late Oligocene and Miocene, leading to the formation of "Sunda Folds", some classical examples of inverted basins (Ginger et al. 1993; Darmadi et al. 2007). The

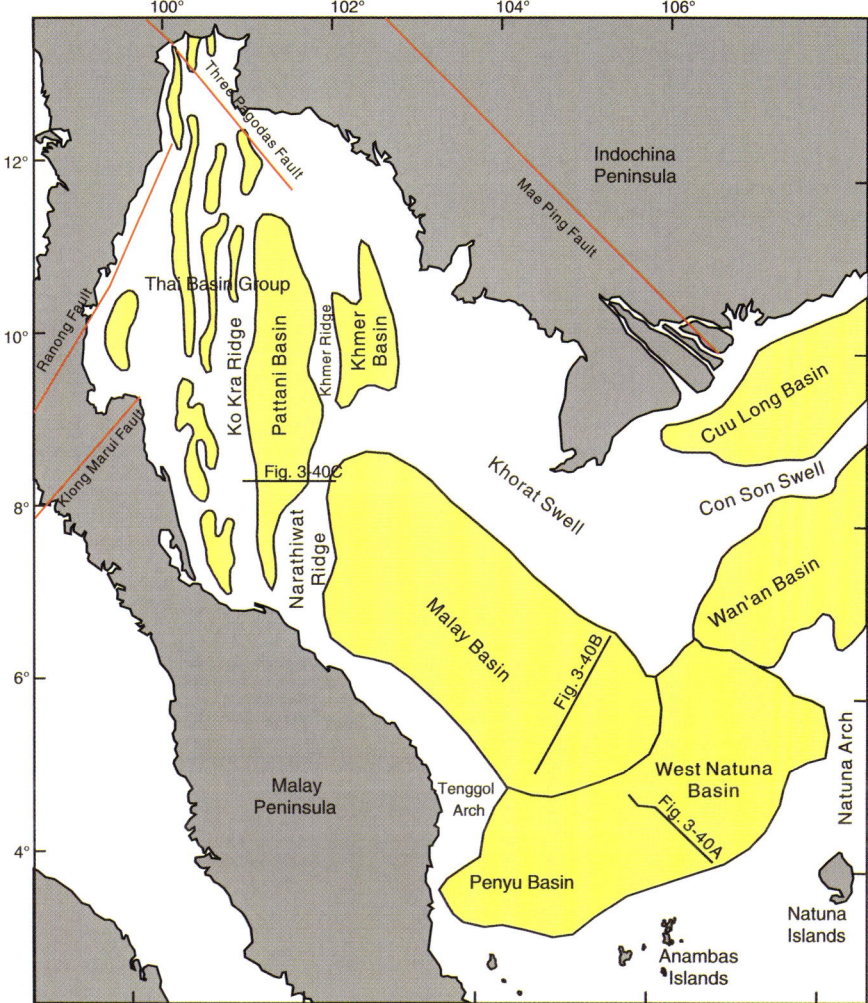

Fig. 3.38 Sketch map shows the location of the West Natuna, Malay and Thai basins in the Gulf of Thailand (compiled from ASCOPE 1981; Phillips et al. 1997; Madon et al. 1999a; Bishop 2002; Jardine 1997)

basin is filled by a thick sediment pod of 5 km or more, unconformably overlying pre-Cenozoic basement of mainly metamorphosed lavas, amphibolites and "gray effusive rocks" (ASCOPE 1981).

The Cenozoic strata include the lower Oligocene Belut, uppermost lower Oligocene to upper Oligocene Gabus, upper Oligocene to lower Miocene Barat, lower to middle Miocene Arang and upper Miocene to Recent Muda formations (Fig. 3.39) (ASCOPE 1981; Pollock et al. 1984; Phillips et al. 1997; Darmadi et al. 2007).

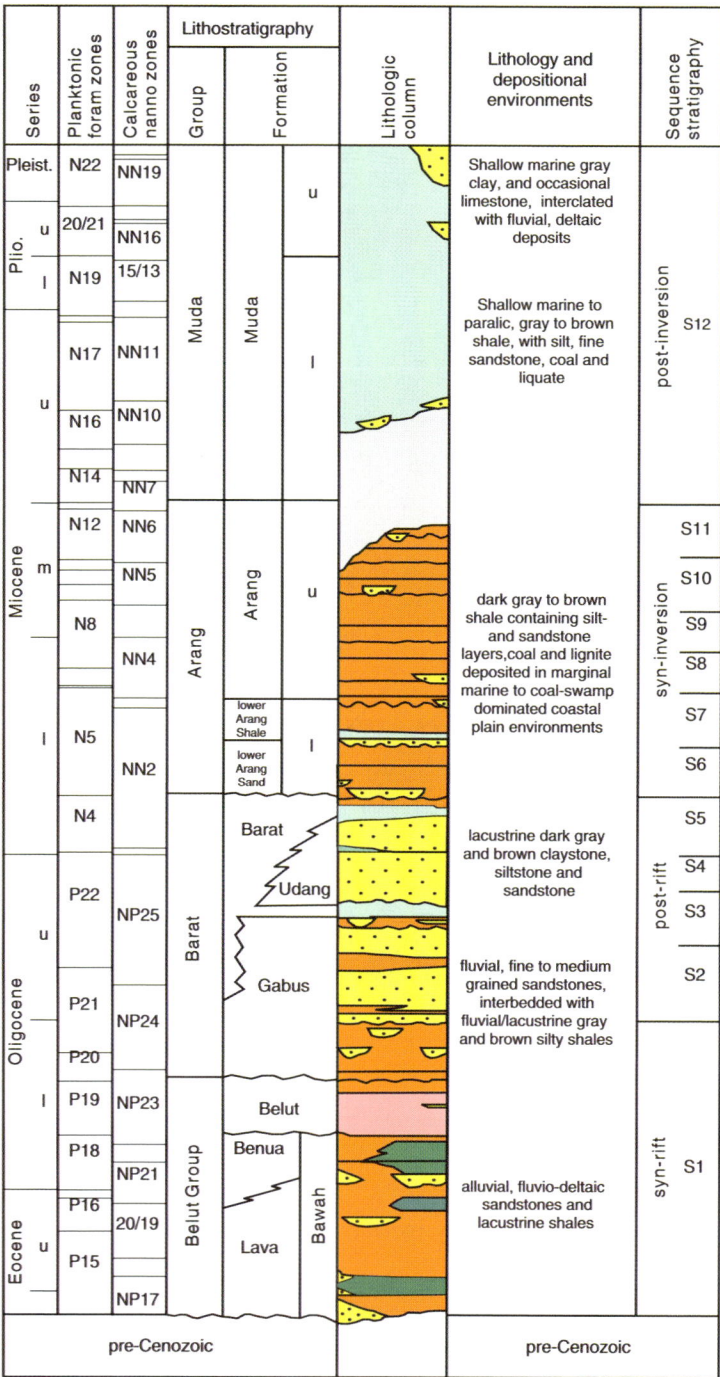

Fig. 3.39 (continued)

The widely distributed Gabus Formation consists of fluvial, fine to medium grained sandstones interbedded with gray and brown silty shales, forming a primary exploration target in the basin. The overlying Barat Formation consists of dark gray and brown claystone, as well as clean sandstones (Udang Formation). The upper limit of the Barat is truncated locally by an unconformity (reflector "e") but, in general, it is transgressively overlain by the Arang Formation with several locally distinct members. The disconformably overlying Muda Formation is characterized by a late Miocene to lower Pliocene gray to brown shale, with silt, fine sandstone, coal and liquate, and a late Pliocene and younger aged soft gray clay with intercalations of silt and occasional coal and limestone. Recent studies indicate that the upper Muda Formation is dominantly a fluvial succession, which changes upward from offshore shelf deposits, through a thin interval of deltaic deposits and a succession of fluvial deposits, and finally to a thin transgressive succession directly beneath the modern sea floor (Darmadi et al. 2007).

As in other southern SCS basins, to correlate the highly variable strata in the West Natuna Basin is difficult. Micropaleontology is only of occasional use, palynological markers are long ranging and imprecise, and lithology is subject to rapid vertical and horizontal variations. Therefore, correlation depends heavily on seismic interpretation, supplemented by limited paleontology, palynology, well log and lithology data (Fig. 3.39) (ASCOPE 1981).

In the West Natuna Basin, four megasequences have been recognized as marking the four tectonic stages: syn-rift, post-rift, syn-inversion and post-inversion (Fig. 3.39). The Oligocene syn-rift megasequence is equivalent to the Belut Formation, the late Oligocene to early Miocene post-rift megasequence to the Gabus, Udang and Barat formations, the early to middle Miocene syn-inversion megasequence to the Arang Formation, and the late Miocene to Recent post-inversion megasequence to the Muda Formation (Phillips et al. 1997; Ginger et al. 1993) (Figs. 3.39, 3.40). The Miocene inversion re-activated pre-existing faults and caused large scale folding/faulting and uplift in the West Natuna and neighboring basins (Fig. 3.40A).

Malay Basin

The Malay Basin is situated in the southern part of the Gulf of Thailand between Indochina Peninsula to the north and Malay Peninsula to the south. The NW-SE elongate basin extends to approximately 500 km long and 200 km wide, and covers an area of approximately 80,000 km^2 (Madon et al. 1999a; Bishop 2002). The basin runs almost parallel to the Malay Peninsula but perpendicular to the NE-SW trending West Natuna and Penyu basins to its southeast. To the north, the Malay basin

Fig. 3.39 Correlation between different bio- litho- and sequence stratigraphic schemes for the West Natuna Basin (lithostratigraphy from Darmadi 2005; lithology and depositional environments from ASCOPE 1981, Meirita 2003 and Darmadi et al. 2007; sequence stratigraphy from Phillips et al. 1997)

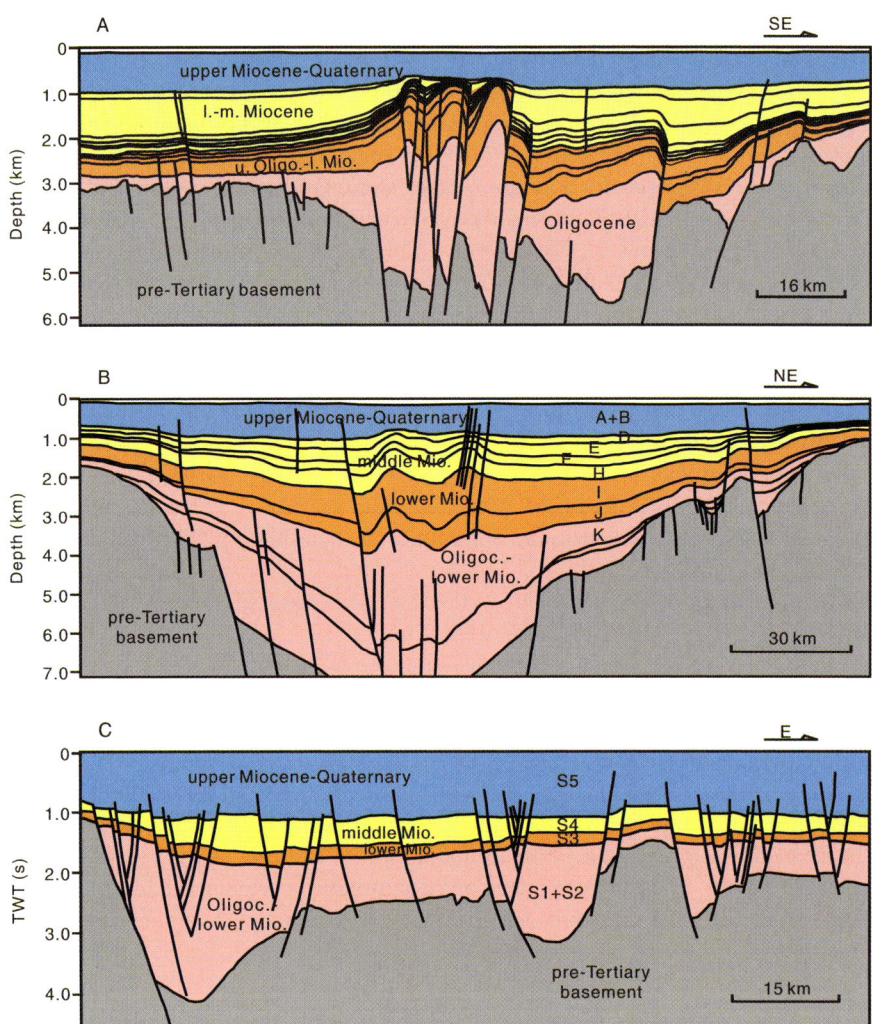

Fig. 3.40 Interpreted cross sections show major seismic sequences in (**A**) West Natuna Basin (redrawn after Phillips et al. 1997), (**B**) Malay Basin (redrawn after Madon 1999a), and (**C**) Pattani Basin (redrawn after Watcharanantakul and Morley 2000). See Fig. 3.38 for line locations

holds a strikingly different structural strike from the N–S trending Pattani Basin in the Gulf of Thailand (Fig. 3.38).

Structurally, the Malay Basin is located at the centre of the Sundaland, separated from the Penyu Basin to the southeast by the Tenggol Arch, from the Pattani Basin to the northwast by the Narathiwat Ridge, and bounded by the Terengganu Platform to the southwest and the Con Son Swell Zone to the northeast (Fig. 3.38) (Madon et al. 1999a; Bishop 2002). It sits on a pre-Cenozoic basement featuring Carbonifierous to Mesozoic igneous, metamorphic and sedimentary rocks (Liew 1994; Leo 1997; Madon et al. 1999a).

The Malay Basin consists of numerous grabens and half-grabens, most of which have been rarely drilled because of their great depths (Madon et al. 1999a). Magnetic, gravity, and seismic data indicate that the basement normal faults largely occur in the southern and central parts of the basin. These faults extend east-westly, oblique to the overall basin trend, and align *en echelon* into two large basement-involved fault zones, i.e., the Axial Malay Fault Zone (AMFZ) in the central basin and the Western Hinge Fault (WHF) in the southwestern margin (Madon 1997; Madon et al. 1999a).

The Malay Basin is an extensional basin, but its development may have also involved a significant strike-slip component of extension (Madon and Watts 1998). Several models have been proposed for the basin formation, including back-arc basin (e.g., Kingston et al. 1983), extrusive pull-apart basin (e.g., Tapponnier et al. 1982), regional thinning of the continental crust (White and Wing 1978), crustal extension above a hot spot (Hutchison 1989), and failed rift basin (Madon et al. 1999a). The extrusion model emphasizes a direct link between the formaion of the Malay Basin and the Three Pagodas Fault, a major strike-slip fault zone in the Gulf of Thailand (Fig. 3.38) (Madon et al. 1999a).

Rifting of the Malay Basin started during the late Eocene-early Oligocene, and later experienced a thermal subsidence since the Miocene (Madon 1999a). Similar to the West Natuna and Penyu basins, the Malay Basin had undergone a period of basin inversion during the middle and late Miocene, producing major anticlinal structures that trap large amounts of oil and gas in the basin (Tjia 1994; Madon 1997). Basin inversion was likely caused by a dextral shear regime following a change in the regional stress field (Madon 1997).

The maximum thickness of Cenozoic sediments in the Malay Basin is estimated as varying from 12 km to 15 km by different authors (Madon 1997; Madon and Watts 1998; Madon et al. 1999a; Morley and Westawayw 2006). In general, Oligocene deposition was in a lacustrine environment, followed by an extensive fluvial to deltic system in the Miocene. After the middle to late Miocene inversion, the basin has been prevailed by fully marine conditions.

Stratigraphy of the Malay Basin was initially established by Esso Production Malaysia Inc. in the late 1960s (Figs. 3.40B, 3.41) (Madon et al. 1999b). Sediment succession was subdivided into units called "groups" bounded by unconformities, which are seismically defined packages of strata similar to "depositional sequences" defined by Mitchum et al. (1977) in modern sequence stratigraphy. The seismic groups are labeled A to M from top to bottom, in which units M and older represent the syn-rift succession, consisting of alternating sand- and shale-dominated fluvial and lacustrine sequences. Units L and younger are interpreted as post-rift deposits of nonmarine and blackish origin (Madon and Watts 1998). Although the "groups" were subsequently given formation names by Armitage and Viotti (1977) (Fig. 3.41), their formational nomenclature has been very rarely used.

Some petroleum companies developed their own stratigraphic nomenclatures for their respective exploration acreages. During the late 1960s Conoco operated in the southern part of the Malay Basin and established a different stratigraphic scheme using formations, roughly correlatable with Esso's seismic groups

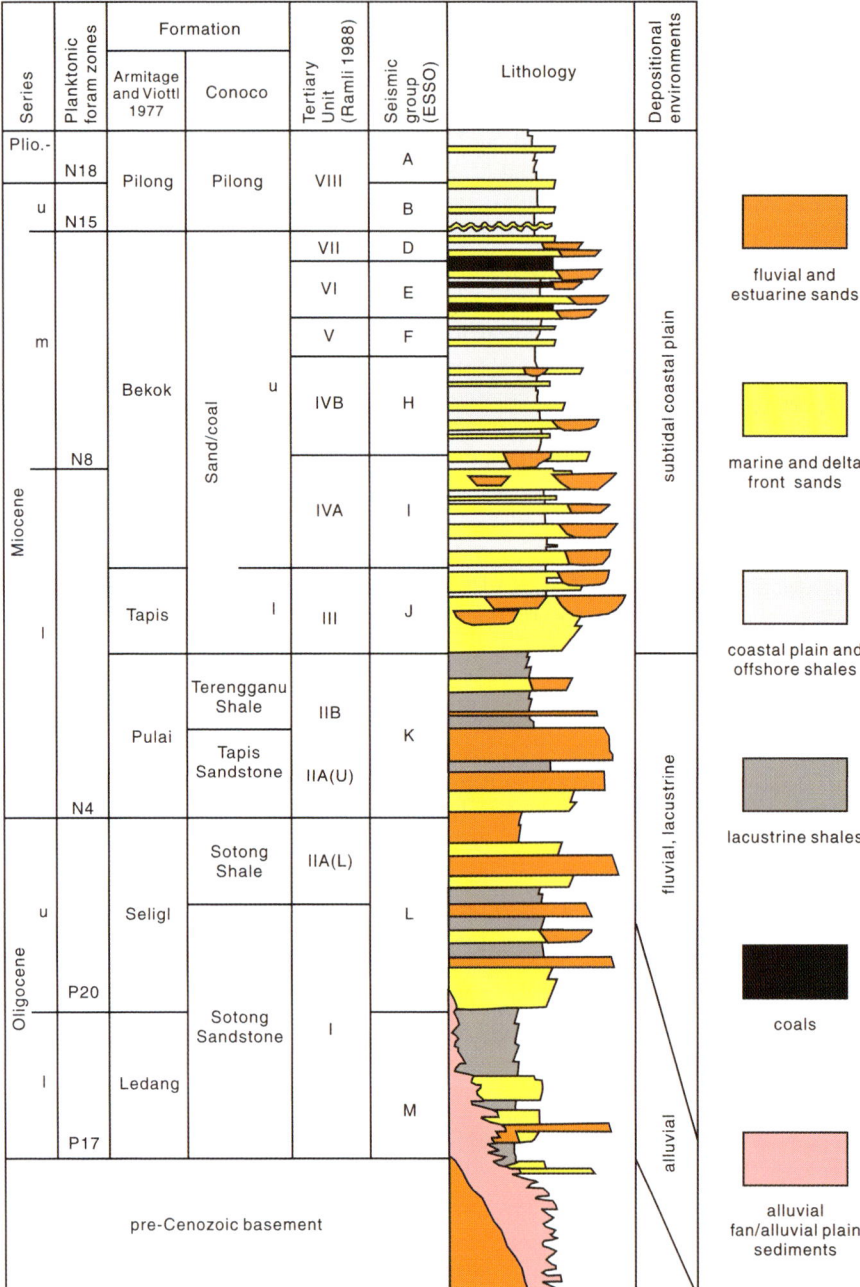

Fig. 3.41 Correlation between various stratigraphic schemes in the Malay Basin is based on Madon et al. (1999a,b)

(Fig. 3.41). Conoco's nomenclature was later superseded by a scheme introduced by PETRONAS Carigali when they assumed the operatorship of Conoco's exploration blocks in 1979 (Ramli 1988). In the Cargali scheme described by Ramli (1988), the stratigraphic column is subdivided into eight "Tertiary units", I to VIII up-section (Fig. 3.41), defined either by the top of regional shale markers (as in Units I and II) or by micropaleontologically defined marine transgressive pulses (as in Units III and VIII), or a combination of both, in the same way as the "genetic stratigraphic sequence" of Galloway (1989). This scheme has been used only in PETRONAS Carigali's exploration blocks and in some neighbouring areas (Madon et al. 1999b).

Biostratigraphy for all sediment sections of the Malay Basin relies heavily on palynomorphs (Fig. 3.6). Since paleo-water depths rarely reached 100 m, reliable index species of planktonic foraminifera and nannofossils are rare. Thus the occurrence of a relatively higher number of these planktonic skeletons can be used to indicate water deepening or even a marine flooding event.

The first industry well in the Malay Basin, Tapis-1, was drilled in 1968 and gas was found in Groups J and K sandstones. Since then, hydrocarbons have been found in sandstone reservoirs of Group D down to Group K, in depositonal environments varying from nonmarine fluvial channels (Groups K, J, and M), shoreface and subtidal shelf sands (Group J), to fluvial-deltaic and estuarine channel complexes (Group I and younger) (Madon et al. 1999a).

Thai Basin Group (TBG)

The Thai Basin Group (TBG) is located in the northern part of the Gulf of Thailand (maximum water depth 86 m), between the Indochina Peninsula to the east and Peninsular Thailand to the west (Fig. 3.38). The Group comprises about 12 basins of various sizes, mostly narrowly elongated and neatly in parallel N-S trending, among which the Pattani Basin is the largest.

Structurally, the Thai Basin Group includes a series of Cenozoic fault-bounded troughs (grabens) and ridges (horsts), and all structures are longitudinally arranged, which is in sharp contrast to the adjacent NW-SE extending Malay Basin. From the central Ko Kra Ridge, the Thai Basin Group can be divided into the Western and the Eastern Graben areas (Polachan and Sattayarak 1991; Pigott and Sattayarak 1993; Watcharanantakul and Morley 2000). The Western Graben Area consists of about ten grabens, namely Sakhon, Paknam, Hua Hin, Prachuap, N. Western, Western, Kra, Chumphon, Nakhon and Songkhla basins, and the East Graben Area include two larger basins: Pattani and Khmer Basins. The Pattani Basin (Fig. 3.38), which is separated to the east from the Khmer Basin by the Khmer Ridge and to the southeast from the Malay Basin by the Narathiwat Ridge, is not only the largest but also the most productive hydrocarbon-bearing basin in the Group.

The Thai Basin Group lies near the intersection of three major strike-slip fault systems (Fig. 3.38). The NW trending Three Pagodas Fault runs from the Myanmar border and appears to extend into the northern part of the Gulf. The NE-SW trending Ranong and Klong Marui Faults cut across Malay Peninsula and probably extend into the northwestern part of the Gulf (Pigott and Sattayarak 1993; Watcharanantakul and Morley 2000). These major fault systems bound a series of N-S-trending grabens and half grabens that constitute the Thai Basin Group. The pre-Cenozoic basement mainly comprises Mesozoic clastics and granites, as well as Paleozoic clastics and carbonate rocks (Lian and Bradley 1986; Trevena and Clark 1986; Jardine 1997).

The structural origin of the Thai Basin Group has been attributed to be extensional by some authors (e.g., McCabe et al. 1988; Jardine 1997). However, others (e.g., Polachan and Sattayarak 1991) assign them to transtensional pull-apart basins resulting from progressive northward collision between the Indian Plate and the Eurasian Plate during the Eocene-Oligocene. Right lateral strike slip movement along the Three Pagodas Fault zone and left lateral movement along the conjugate NE Ranong fault system were the main driving force in the formation of the N-trending northern grabens and the NW-trending Malay Basin. Though the responsible mechanism remains in dispute, basins in the Thai Basin Group are often considered as dominantly extensional, either as pure shear or rifted basins (Wollands and Haw 1976; Achalabhuti and Udom-Ugsorn 1978; Hellinger and Sclater 1983) or as simple shear pull-apart basins along the master strike-slip faults (Polachan and Sattayarak 1991).

Several stages of structural evolution in the Pattani Basin have been recognized by Jardine (1997), including (1) pre-rift folding and uplift of pre-Cenozoic accreted basement terranes, (2) initial rifting and creation of localized sub-basins (half-grabens) from late Eocene to Oligocene, (3) structural inversion and erosion at the end of the Oligocene, (4) rifting and basin formation in the early Miocene, (5) post-rift collapse and basin subsidence in the middle Miocene, (6) widespread erosion from middle to early late Miocene, and (7) continued basin subsidence from the late Miocene to Recent.

The thickness of the Cenozoic sequence varies greatly between these basins, more than 8 km in the Pattani Basin, 4 km in the Western Graben Area, and less than 300 m over the shallow subcropping pre-Cenozoic highs (Lian and Bradley 1986; Pigott and Sattayarak 1993; Watcharanantakul and Morley 2000). The sediment consists almost exclusively of nonmarine, fluvial to deltaic sediments which become shallow marine only in the youngest intervals (Pigott and Sattayarak 1993). Because of a high terrestrial content and lacking diagnostic microfossil species, stratigraphic correlation is often by means of palynological zonation and recognition of sediment cycles, as in the neighboring Malay, Penyu, and West Natuna basins described above. Critical palynomorphs include *Florschuetzia* from the Oligocene-Miocene, and *Dacrydium* and *Podocarpus* from the Pliocene and Quaternary (Fig. 3.6) (ASCOPE 1981).

Similar to those found in the contiguous northwestern Malay Basin, three sedimentary cycles are recognized particularly in the Pattani Basin (Woollands and Haw 1976). Cycles I and II are offlap regressive packages, respectively representing syn-rift sediment fill during the Oligocene-early Miocene and during the early and middle Miocene, while the younger Cycle III is generally a marine transgressive package since the later middle Miocene (ASCOPE 1981).

Lian and Bradley (1986) divided the strata into four seismically and/or petro-physically mappable sequences based on region-wide unconformities and variations in well logs and acoustic characters (Fig. 3.42). However, the sequence stratigraphic scheme established by Unocal for the Pattani Basin (Jardine 1997) appears to have gained wider acceptance, as the major depositional sequences 1 to 5 recognized are considered better reflecting Neogene low-order relative sea-level changes (Fig. 3.40C). These sequences are truncated by two periods of non-deposition: one in the late Oligocene (the mid Tertiary unconformity, or MTU), and the other at around 10 Ma (the mid-Miocene unconformity, or MMU). Sequences 1 and 2 are dominated by lacustrine, fluvial and delta plain deposits during Oligocene-early Miocene rifting, and sequences 3–5 are post-rift fluvial and marginal marine deposits that show increasing marine influence from the east. According to Morley and Westawayw (2006), the main syn-rift succession represented by sequences 1 and 2 in the northern Pattani Basin probably span the Eocene and early Oligocene.

Active petroleum exploration since the late 1960s has found the main reservoir rocks in the lower to middle Miocene fluvial and fluvial-deltaic sandstone in the Pattani Basin (Trevena and Clark 1986; Jardine 1997).

Other Southern Basins

Several other basins along the southeastern margin of the SCS are also important for industrial exploration and marine geological research alike, including the Sabah Basin, Palawan Shelf, and Reed Bank area (Fig. 3.20; Table 3.4).

The Sabah Basin is located offshore from Sabah-Borneo coast and forms part of the northwest Sabah shelf. Together with the neighboring Zengmu (Sarawak) Basin (Fig. 3.35), it represents a major segment of the greater Sarawak-Brunei-Sabah continental margin that evolved since at least the late Cretaceous time. The location now occupied by the Sabah Basin was part of the Proto-South China Sea or "Rajang Sea" until the Eocene (Tongkul 1991; Leong 1999b) when the Proto-SCS started subducting under the NW margin of Borneo. Subduction caused deformation and uplift in the late Eocene, subsequently giving birth to new depocenters in the so-called "Oligocene basins" filled by Oligocene-early Miocene turbidites and deepwater sediments often with a chaotic character. The younger Sabah Basin started to develop since the middle Miocene probably as a foreland basin following the collision of a SCS microcontinental fragment (now forming the NW Sabah Platform) with western Sabah (Madon et al. 1999c). Mainly middle Miocene and younger sediments accumulated in the Neogene Sabah Basin, truncated by several regional unconformities, principally the Shallow, Middle and Deep Regional Unconformities (or SRU, MRU and DRU) at about 9 Ma, 11–12 Ma and 15–16 Ma, respectively (Fig. 3.6). Subsidence was rapid (>500 m/myr) between 15 and 9 Ma, followed by a generally slower rate accompanied by pronounced uplift along the eastern margin of the basin (Madon et al. 1999c). As a result, the middle Miocene and younger sediments (as Stage IV in Fig. 3.6) are dominantly fluvial to shallow marine sandstone, siltstone and mudstone/shale, with limestone developed mainly on the Kudat Platform in the

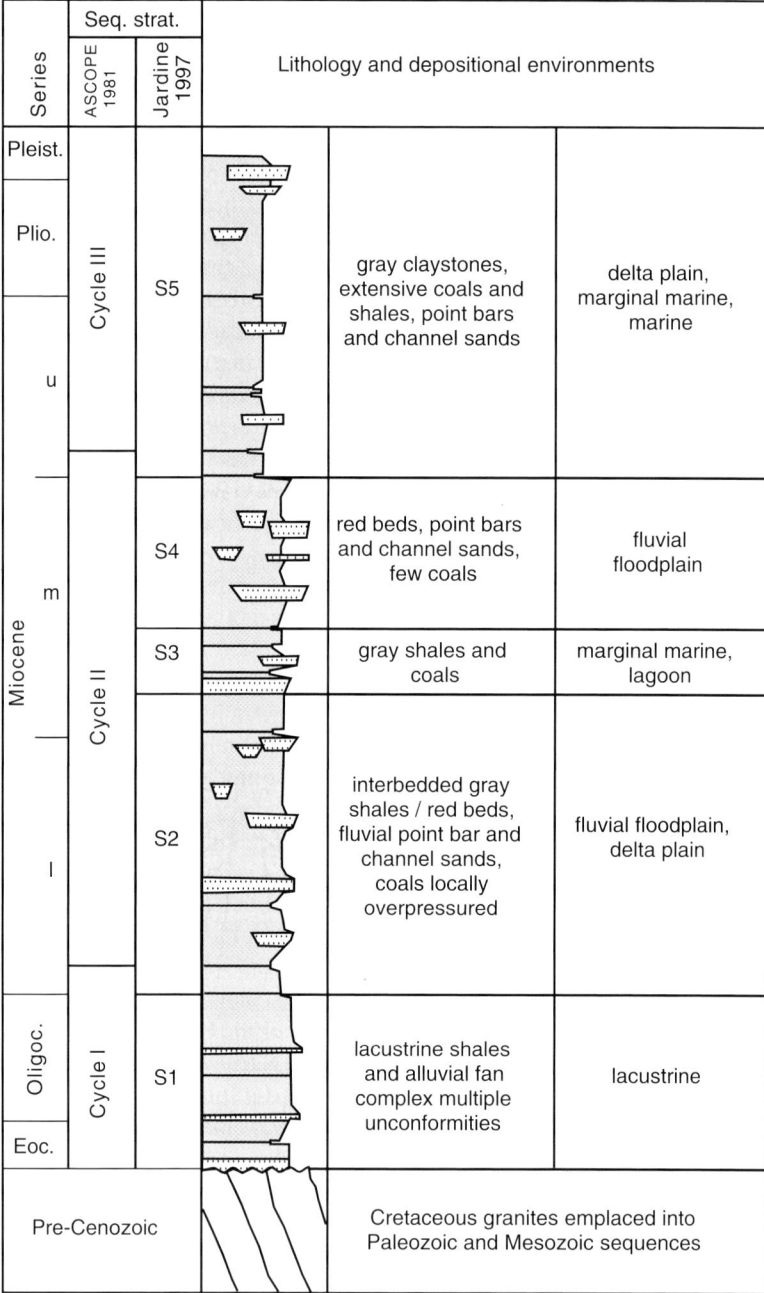

Fig. 3.42 Generalized lithostratigraphy in the Thai Basin Group is shown with divisions of sequences and cycles, as well as the corresponding depositional environments (compiled from Jardine 1997; Pigott and Sattayarak 1993; Lian and Bradley 1986)

north since the late Miocene. The lithology in the southern Sabah Basin progradationally integrates into the Baram Delta, offshore Brunei and Sarawak (Du Bois 1985). The post-middle Miocene succession has been subdivided into 7 substages, A to G up-section, bounded by unconformities (Fig. 3.6). Similar to the neighboring basins along the southern margin, the Sabah Basin and basins of older age onshore Sabah are rich in hydrocarbon resource. Madon et al. (1999c) provides a detailed discussion on oil and gas reservoirs, as well as tectonics and stratigraphy in the area.

The western Palawan Shelf parallels the NE-SW trending Palawan Island and borders to the west with the Palawan Trough (part of the Nansha Trough) (Fig. 3.20). Across the Trough and further northwest is the Reed Bank, an area of active carbonate platform growth since the late Oligocene time (Fig. 3.5). Several NE-trending fault systems control the development of the shelf basin as part of the NW Borneo Geosyncline. The pre-Cenozoic basement in the area mainly consists of Mesozoic marine facies of shallow to bathyal environments. Shallow marine deposition continued into the Paleogene, leading to carbonate platform buildups first on NW Palawan shelf (late Eocene to early Miocene) and then in the Reed Bank area (ASCOPE 1981; Du Bois 1985). The shallow Paleogene sea likely formed part of the Proto-SCS before the area was rifted and subsided, drowning the platform on the NW Palawan Shelf in the early Miocene. The break-up unconformity lies close to the early/middle Miocene boundary in Palawan area, while the Reed Bank area remained relatively stable as a microcontinent not affected by collision and compressional extention like other coastal regions along the southeastern SCS margin. The post-middle Miocene sediment sequences in the NW Palawan Shelf bear a close similarity with those in the southwest, in Sabah and Sarawak basins (Figs. 3.5 and 3.6). Therefore, the Sarawak cycle concept can also be applied to sediment sequences in the NW Palawan Shelf if cycle boundaries can be clearly identified.

3.5 Regional Sea Level Changes (Zhong G. and Li Q.)

Far from the polar ice sheets, the SCS provides a favorite site for reconstructing the history of late Cenozoic sea level changes. Many efforts have been made to document local sea level changes in the northern, western and southern shelves by using geomorphological, geological, geochemical, paleobiological and geophysical methods, with a primary focus on the late Quaternary sea level fluctuations (e.g., Pelejero et al. 1999; Hanebuth et al. 2000; Nguyen et al. 2000; Voris 2000; Lüdmann et al. 2001; Steinke et al. 2003; Sun et al. 2003; Tanabe et al. 2003; Wong et al. 2003; Hori et al. 2004; Schimanski and Stattegger 2005; Yim et al. 2006; Wang et al. 2007). Largely due to lacking information, however, long-term sea level changes in the region have been discussed only in several studies (Xu et al. 1997b; Li et al. 1999; Hao et al. 2000; Woodruff 2003; Zhong et al. 2004). This section summarizes the main characteristics of sea level changes in the South China Sea in the late Quaternary and its long-term changes since the Oligocene.

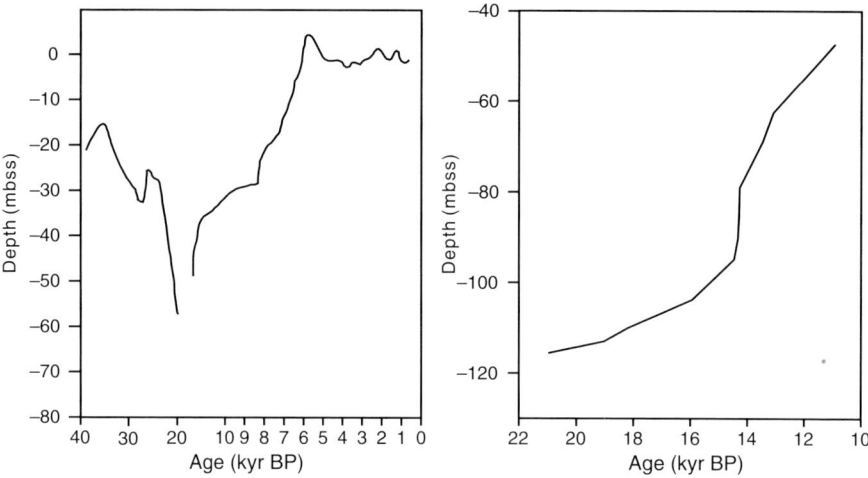

Fig. 3.43 Late Quaternary sea level curves are shown for (*left panel*) the South China margin over the last 40 kyr (based on Huang et al. 1986) and (*right panel*) the Sunda Shelf between 21 and 11 ka BP (based on Hanebuth et al. 2000)

Late Quaternary Sea Level Changes

The SCS is characterized by broad northern and southern shelves, and the Sunda Shelf in the south, known as "the Great Asian Bank", is one of the largest shelves in the world (Chapter 2). During late Quaternary glacial stages, a large part of the shelves were subaerially exposed to become land bridges (e.g., Voris 2000). The glacial SCS was strongly reduced in size, and several large river systems, including the North Sunda or Molengraaff and Thai rivers, drained the shelf area of the modern Sunda Shelf during the last glacial maximum (LGM), when sea level dropped to about −120 m (Pelejero et al. 1999; Hanebuth et al. 2000; Eong 2001) (Fig. 3.43), or more precisely, −123 ± 2 m (Hanebuth et al. 2008). Chen et al. (1990) found last glacial paleo-coastline on the continental shelf off the Pearl River mouth, featuring terraces, channels, sand dunes and marshes at ∼130 m below the present sea level. During the LGM, although land area increased significantly and land bridges widely developed, rain forest remained dominant along the southern SCS coast (Morley 2000; Kershaw et al. 2001), as confirmed by a recent study of spore and pollen assemblages in core GC 18287-3 (5°39′N, 110°39′E) collected during R/V SONNE cruise 115 (Wang et al. 2007).

Post-glacial sea-level rise has been intensively studied in the western Pacific margin region including northern Australia (e.g., Biwas 1973; Chappell and Veeh 1978; Chappell 1982; Emmel and Curray 1982; Somboon and Thiramonkol 1992; Woodroffe et al. 1985, 1989; Woodroffe 1993; Lambeck and Chappell 2001). These studies demonstrate that relative sea level rapidly rose during the early Holocene and reached a maximum height, which varies from place to place due to differential hydro-isostatic effects. The maximum Holocene sea level occurred at around 6,000 − 5,000 yr BP in Australia (Woodroffe et al. 1985, 1989; Woodroffe

1993) and at 4,500 − 4,000 yr BP in Malaysia and Indonesia (Haile 1971, 1975; Tjia 1977; Thommeret and Thommeret 1978) although the accuracy of dating needs to be evaluated. Sea level rise to +5 m has been also reported at around 5,000 − 4,000 yr BP in Kulala Kurau of Peninsula Malaysia (Geyh et al. 1979; Kamaludin 1993), +3.5 m at ∼6,000 yr BP in northern West Java (Rimbaman 1992), +4.5 m at ∼6,000 − 5,000 yr BP in the Mekong River Delta (Nguyen et al. 2000), and +2 − +3 m at ∼6,000 − 4,000 yr BP in the Red River delta (Tanabe et al. 2003). Studies by Chinese scientists (e.g., Huang et al. 1986) indicate that sea level along the South China coast was −40 to −30 m during 10,000 − 8,000 yr BP before rapidly increasing to +4 m at ∼6,000 yr BP (*left panel* in Fig. 3.43).

Hanebuth et al. (2000) presented a detailed analysis of post-glacial transgression along a 600 km, NE-SW transect on the northern Sunda Shelf (*right panel* in Fig. 3.43). They found sea level dramatically rose by as much as 16 m and the whole shelf was flooded within 300 years from 14.6 to 14.3 ka, equivalent to the first melt-water pulse (MWP 1A). At the time, sea surface temperature (SST) in the southern sea region rose by about 1.5 °C, accompanied by a significant decline in clay content and n-nonacosane concentrations (Pelejero et al. 1999). A second flooding at MWP 1B from about 11.5 to 10 ka involved the establishment of modern hydrographic conditions and further SST increase in the region. A detailed review of sea level changes world-wide in the last glacial cycle was given by Lambeck and Chappell (2001).

In the northern SCS, *Pinus* pollen was dominant in interglacials at ODP Site 1144, and herbs in glacials when the local climate on the exposed shelf was dry due to stronger winter monsoons (Sun et al. 2003). As herbs are often transported within a short distance, their dominance in glacials may denote an origin from the exposed shelf. A high Herb/*Pinus* ratio occurring briefly in glacial MIS 12 and consistently since MIS 6 may further indicate a narrower shelf before but a broad shelf after MIS 6 (Sun et al. 2003) (Chapter 5). However, the growth of a broad northern shelf only since MIS 6 needs further studies because the earlier shelf may have largely submerged and evidence of its existence may have been wiped out by erosion during younger glacial cycles.

Long-Term Sea Level Changes Since the Oligocene

Although few studies have been dedicated to long-term sea level changes in the South China Sea, results from several northern basins (Fig. 3.44) show a continuous sea level rising trend from ∼40 to 5 Ma although fluctuations were closely in pace with global patterns depicted by Haq et al. (1987, 1988) and Miller et al. (2005). Sea level fluctuations often differ between basins in magnitude, largely responding to local tectonics and basin dynamics especially sediment load and subsidence rate.

Qin (1996a) and Xu et al. (1997b) deduced a relative sea level curve for the Zhujiangkou Basin from analyzing foraminiferal sequence biostratigraphy in 13 wells, using maximum abundance and diversity values for maximum flooding surfaces (MFS) and minimum values for sequence boundaries. As shown in Fig. 3.44, the relative sea level curve generated by these authors displays a similar

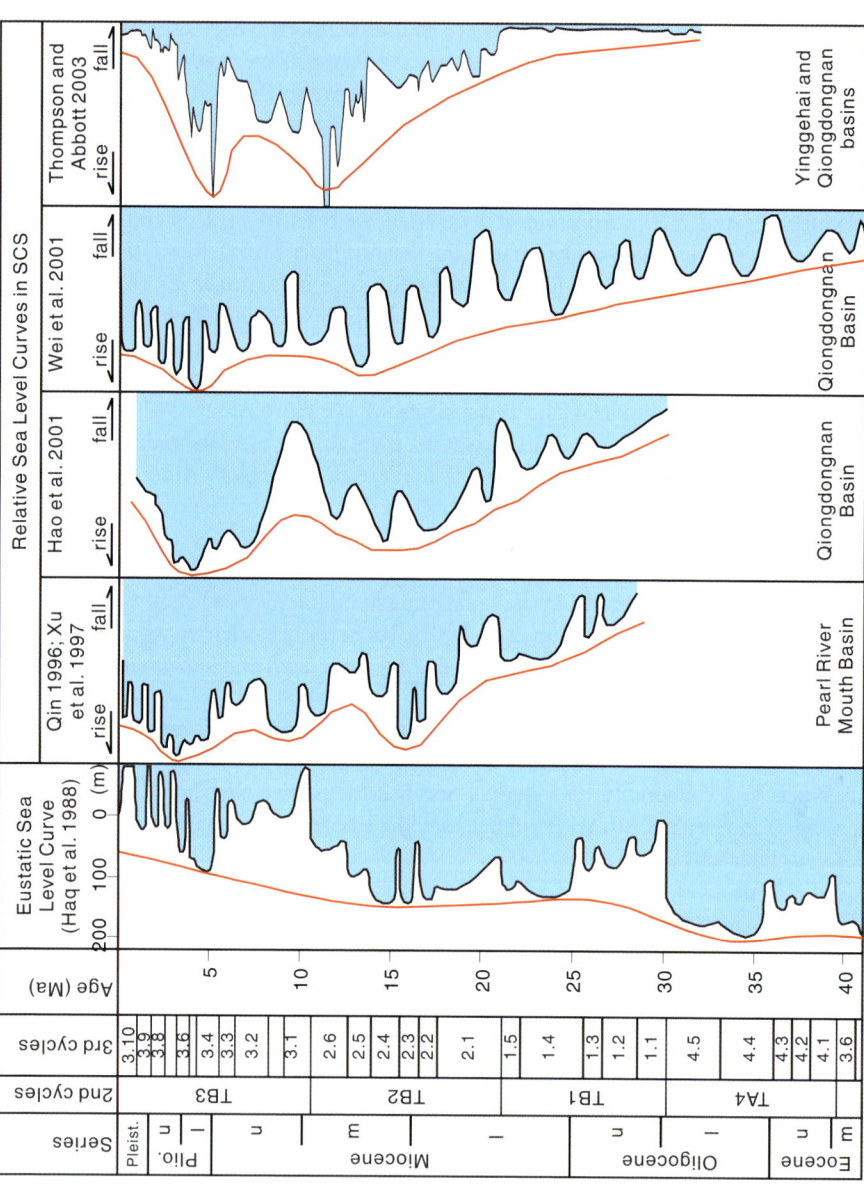

Fig. 3.44 Comparison between regional relative sea level curves from the northern SCS and the eustatic sea level curve of Haq et al. (1988) shows similar third-order frequencies but a continuous rising trend in the local sea level from Oligocene to Pliocene

overall trend with the global curve of Haq et al. (1988) except an obvious discrepancy across the middle/late Miocene boundary near TB3.1 when tectonic subsidence was prevalent in the Zhujiangkou Basin soon after the end of seafloor spreading in the SCS (Xu et al. 1997b).

Based on foraminiferal biofacies and stable isotope data, Li et al. (1999) compiled a sea level curve from seven transgressive-regressive cycles recognized for the late Miocene to Pliocene Yinggehai-Huangliu Formations in the Qiongdongnan Basin. Similarly, Hao et al. (2000) presented a low-resolution sea level curve for the Yinggehai-Qiongdongnan basins to show an overall rising sea level punctuated by large fluctuations since the Oligocene (30 Ma) (Fig. 3.44). The study by Wei et al. (2001) extended a third-order relative sea level curve to ~40 Ma in the Qiongdongnan Basin (Fig. 3.44), in which three major marine flooding events respectively at the earliest early Miocene, middle Miocene and early Pliocene match well with the corresponding global changes shown in Haq et al. (1988). Similar findings are reported by Thompson and Abbott (2003) who analyzed microfossil-derived sea level history in 31 wells from the Qiongdongnan and Yinggehai basins although several intervals of extraordinary inferred sea level rise were probably caused by turbidites (Fig. 3.44).

In the south, Woodruff (2003) reviewed marine transgression history in and around Malay Peninsula and found no evidence of high sea level in the middle Miocene and early Pliocene as shown in Haq et al. (1988), further confirming a greater role of local tectonics on regional sea level changes.

New Approach Toward Fine-Scale Sea Level Magnitude

Vail and his Exxon colleagues (Vail et al. 1977) developed a method to estimate the magnitude of sea level changes by reflection seismic data. In their method, "coastal aggradation" (the difference in elevation between adjacent onlap/offlap point pairs) is used as a measure of the magnitude of relative sea level rise and fall. In any given area, measurements of coastal aggradation are sequentially summed to produce a coastal onlap curve, which was then called a relative sea level curve. Vail et al. (1977) constructed a eustatic sea level curve on the basis of a "modal average" of the curves of relative change in sea level from individual regions around the world. Haq et al. (1987, 1988) published a revised version of the eustatic sea level curve mainly by augmenting it with data from world-wide outcrop studies. Publication of Exxon's two generations of eustatic sea level curves and the methodology was a milestone in the development of seismic stratigraphy. But it also provoked a wide range of response: while many consider their curves as a global standard, some are skeptical (De Graciansky et al. 1998; Hallam 1992; Miall 1997).

Sea level is defined by the position of the sea surface with reference to a datum. Eustatic or global sea level refers to the position of the sea surface on a worldwide basis relative to a fixed datum, such as the center of the earth, and is therefore a function of time and is independent of local factors (Burton et al. 1987; Posamentier et al. 1988). Relative sea level refers the position of the sea surface to the position of a moving datum (e.g. , basement of the basin) at or near the sea floor. It is influenced by local subsidence, including tectonic subsidence, flexural response of the lithosphere to sediment and water loading, and subsidence caused by compaction of the pre-existing sediment column. Therefore, relative sea level is a function of eustasy and movement of seafloor or basement. Because of spatial variations in tectonic and loading subsidence, relative sea level can change from location to location (Posamentier and James 1993). In addition, coastal aggradation is subjected to other factors such as paleo-bathymetry, post-depositional structural deformation and post-depositional erosion of onlap and offlap points. All these factors apparently have not been taken into account quantitatively in the study by Vail et al. (1977, 1984). In practice, the incremental relative sea level change of Vail et al. (1977) is actually a measure of the increment in relative sea level at the onlap points in the time interval of deposition of the layer bounded by these points. Therefore, their coastal onlap curve or relative sea level curve represents neither the individual relative sea level change at a certain location, nor the average change in the area where it was documented.

Considering the relationship between eustasy, basement tectonic and loading subsidence, compaction, sedimentation, and paleo-bathymetry during the deposition of a stratigraphic unit i between times $t1$ and $t2$ at a given position p, Zhong and co-workers (Zhong 2002; Zhong et al. 2003, 2004) deduced the general expression for the incremental change in eustatic sea level during the deposition of unit i from the definitions of eustatic and relative sea level changes:

$$\Delta Ei = (\Delta Wd_p + \Delta S_p) - (\Delta Y_{tec,p} + \Delta Y_{load,p} + \Delta H_p) \qquad (3.1)$$

where ΔE_i is the incremental change in eustatic sea level during the deposition of unit i, ΔWd_p is the increment of water depth between $t1$ and $t2$ at position p, ΔS_p is the initial or depositional thickness of unit i at position p. $\Delta Y_{tec,p}$ and $\Delta Y_{load,p}$ are increments in tectonic subsidence and basement subsidence due to isostatic loading, respectively, at position p during the period of deposition of unit i. ΔH_p is incremental thickness of the underlying sediment column of the newly deposited unit i at position p during the period of deposition of unit i.

Theoretically, according to equation (3.1), any point in a basin can be used as a reference point for the calculation of the incremental change in eustatic sea level (ΔE_i) as long as all quantities on the right hand side of equation (3.1) can be determined. These quantities, however, are very difficult to estimate for an arbitrary point. Therefore, for the sake of convenience, the onlap point (A) on the bottom surface of an onlap layer or the offlap point (B) on the top surface of an offlap layer is selected as reference point. An onlap layer is defined as a layer for which the onlap point shifts landward during the deposition of the layer, and an offlap layer is defined as a layer for which the onlap point shifts basinward. In the case of a coastal onlap/offlap (not marine onlap/offlap), the vertical difference between an onlap/offlap point and its coeval sea surface is assumed constant, although it may actually vary within a typical range of several meters or more (Vail et al. 1977). Then we would have:

$$\Delta E_i = \begin{cases} Wd_{t2,A} + \Delta S_A - (\Delta H_A + \Delta Y_{load,A} + \Delta Y_{tec,A}) & (for\ case\ of\ onlap) \\ -Wd_{t1,B} - (\Delta Y_{load,B} + \Delta Y_{tec,B}) & (for\ case\ of\ offlap) \end{cases}$$

$$(3.2)$$

where subscripts A and B represent the onlap/offlap points at the bottom and top surfaces, respectively, of the layer considered. Reconstructing the original or depositional position of an eroded onlap or offlap point is the first and critical step in estimating incremental changes in eustatic sea level using equation (3.2), which can be recovered by fitting the thickness data in an area of non-erosion, the so-called "effective stratal thickness", and then extrapolating to the point with zero thickness landwards. Depositional thickness ΔS_A and compaction subsidence ΔH_A in equation (3.2) can be calculated by using backstripping and decompaction algorithms. In addition, paleo-water depths $Wd_{t2,A}$ and $Wd_{t1,B}$ in equation (3.2) can be estimated by using relevant biostratigraphic, sedimentological, and seismo-stratigraphic techniques. The incremental changes in basement tectonic subsidence, $\Delta Y_{tec,A}$ and $\Delta Y_{tec,B}$ of equation (3.2), are critical parameters and are the most difficult to estimate. For details of solving the quantities in equation (3.2), see Zhong et al. (2004).

The Sunda Shelf in the southern South China Sea is an excellent site for the documentation of late Cenozoic sea level fluctuations (Sea Level Working Group 1992). Relatively complete records of late Cenozoic coastal onlap/offlap points from offshore Natuna Islands (Fig. 3.45) provide excellent material for testing the method. In the region, a total of about 2,200 km of high-resolution seismic lines were available, of which about 1,550 km were obtained during Cruise 115 of the German R/V SONNE in 1997 (Stattegger et al. 1997; Wong et al. 2003). Other 650 km were from Cruise 93-1 of the R/V FENDOU IV carried out by the Guangzhou Marine Geological Survey.

Based on a regional seismic sequence interpretation of the data set, two up-dip composite cross profiles for the Pleistocene and the Pliocene respectively were selected as reference profiles for estimating the increments in eustatic sea level (Fig. 3.45). These cross profiles show relatively strong reflections and little tectonic activity. A total of 122 reflectors were traced on the interpreted regional seismic data set. They comprise 23 seismic sequences, including 17 from the Pliocene and 6 in the Pleistocene (Fig. 3.45).

The resultant eustatic sea level curve (Fig. 3.46) shows 36 cycles of sea level fluctuations with periods ranging from 0.08 to 0.29 myr. They are 4th order eustatic sea level cycles as defined by Vail et al. (1991). Although the general trends are similar between the SCS eustatic sea level curve and that of Haq et al. (1988) (Fig. 3.46), major differences exist in the magnitude and number of eustatic cycles. The SCS curve shows apparently more sea level cycles than that of Haq et al., suggesting that the time resolution of the curve is higher. Morever, the SCS curve matches well with the high-resolution deepsea stable oxygen isotope records measured on benthic foraminifera from ODP Site 1148 (northern SCS, 18° 50′N, 116° 34′E) (Jian et al. 2001a; Zhao et al. 2001a) and ODP Site 846 (eastern Pacific, 3° 06′S, 90° 49′W)

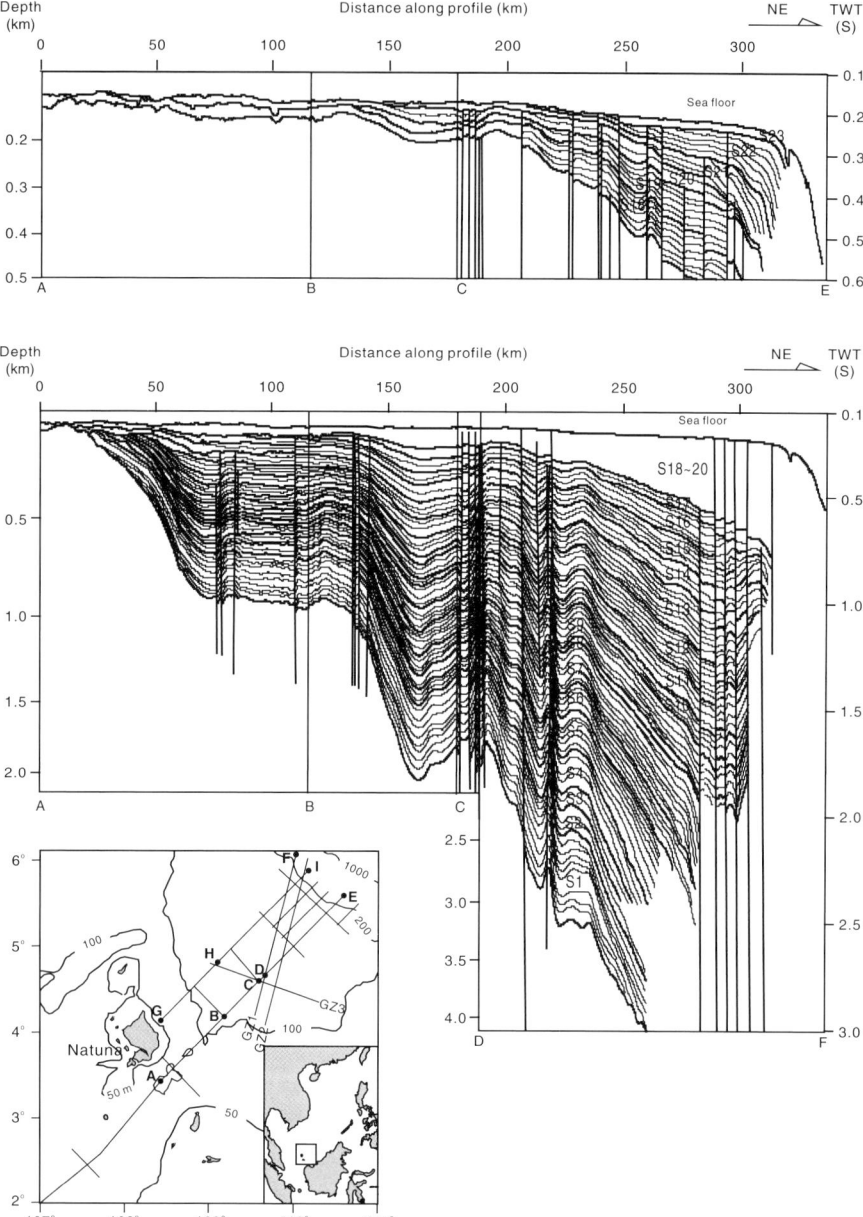

Fig. 3.45 A high-resolution study of the Pliocene-Pleistocene succession in the southern South China Sea margin reveals 23 fourth-order seismic sequences (Zhong et al. 2004). TWT = two − way travel time

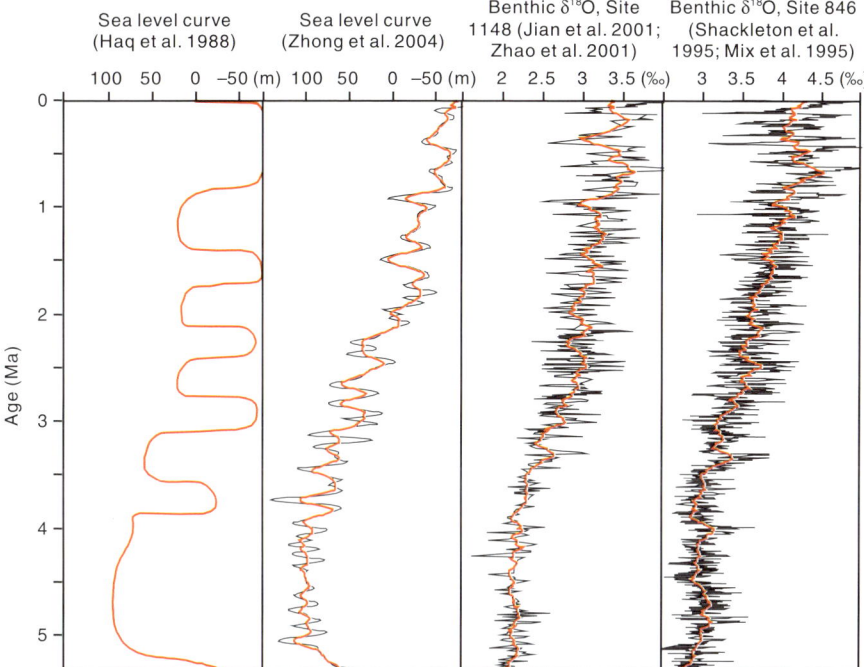

Fig. 3.46 The reconstructed eustatic sea level curve for the northern Sunda Shelf (Zhong et al. 2004) is compared with the eustatic curve of Haq et al. (1988) and ODP Sites 846 and 1148 benthic isotope curves (Shackleton et al. 1995; Mix et al. 1995; Jian et al. 2001a; Zhao et al. 2001a)

(Shackleton et al. 1995; Mix et al. 1995). The Site 846 record is often considered as a proxy for global continental ice volume and hence represents an indirect record of the history of sea-level change (Chappell and Shackelton 1986; Zachos et al. 2001).

Summary of South China Sea stratigraphy

The development of about 40 Cenozoic sedimentary basins was directly related to three major stages in the evolution of the SCS: rifting, seafloor spreading and post-spreading deformation. Litho-, bio- and sequence stratigraphic analyses reveal considerable variations in the timing and mode of basin formation, sediment packaging and sequence distribution between the northern and southern SCS as well as between individual basins. In the north, rifting occurred earlier in the Paleocene and Eocene before seafloor spreading taking place in the Oligocene. In the south, rifting continued into the early Miocene. Syn-rift sequences in many northern and southern SCS basins are similarly composed of fluvial, lacustine and deltaic sediments, although shallow marine facies were common along the present SE margin, in areas belonging to the Proto-SCS. The contact between syn-rift and post-rift sequences often coincides with a basin-wide unconformity, marking significant changes in depositional environments during the rifting/spreading transition. Seafloor spreading opened up the deep central basin first in the northeast and caused subsidence in pre-existing basins and in areas of the present outer shelf and slope. However, rapid subsidence did not occur until the latest Oligocene when spreading accelerated after the spreading axis shifted to southeast. At ODP Site 1148 from the lower northern slope, this event is recorded by a double reflector on the seismic profile and by a slumped unit recovered in cores. Several unconformities were associated with the slump, indicating the missing of sediment record of >3 myr.

With a rise in local sea level since the early Miocene, terrigenous components reduced while carbonate content increased in deposits in many parts of the SCS. Carbonate reefs, which were mainly confined to the Dangerous Grounds and offshore Palawan, now expended to wider regions also in the northern continental margin. Further subsidence and accumulation of full marine deposits continued in many northern basins even after spreading ceased near the early/middle Miocene boundary. Starting from this time, basins in the south entered a period of post-rift deposition although clastic and shallow-marine sediments remained common except at more offshore settings along the SE margin where subsidence was prevailing due to subduction. Further subduction and collision during the late Miocene and Pliocene caused basin inversion, deformation and frequent turbidite deposition in the south and in the east. In general, therefore, stratigraphic correlation between sediment packages from the south is difficult because of their shallow-water origin and subsequent deformation.

Planktonic foraminifera and nannofossils play a key role in age dating and correlation of sediment sequences deposited in open marine environments. Stratigraphic correlation of paralic sequences, however, depends primarily on spore and pollen assemblages, dinoflagellates, ostracods and other microfossils. Together with lithostratigraphy, faunal and floral assemblages indicate continuous increases in local sea level from the Oligocene to Pliocene largely as a result of tectonic subsidence due to seafloor spreading. Variations in spatiotemporal distribution of sediment sequences are better revealed by sequence stratigraphy, and high resolution sequence stratigraphic analyses open a new window for better understanding the evolution of sediment packaging on a fourth-order or finer scale. More importantly, stratigraphic correlation has been advanced by astronomically tuned timescales based on isotopic records from ODP Site 1143 and isotopic and magnetic records from ODP Site 1148, which not only provide a more accurate standard for refining deep sea stratigraphy but also offer information useful for reconstructing the history of the SCS.

Acknowledgments This work was supported by the Ministry of Science and Technology of China (NKBRSF Grants 2007CB815900 and 2007CB411702) and the National Natural Science Foundation of China (Grants 40476030, 40576031, 40621063 and 40631007).

References

Achalabhuti C. and Udom-Agsorn P. 1978. Petroleum potential in the Gulf of Thailand. J. Geol. Soc. Thailand 3: 1–12.

Adams C.G. 1970. A reconsideration of the East Indian Letter Classification of the Tertiary. Bull. British Mus. (Nat. Hist.) 19(3): 87–137.

Ali M.Y.B. and Abolins P. 1999. Central Luconia Province. In: Leong K.M. (ed.), The Petroleum Geology and Resources of Malaysia, PETRONAS, Kuala Lumpur, pp. 371–392.

Andersen C., Mathiesen A., Nielsen L.H., Tiem P.V., Petersen H.I. and Dien P.T. 2005. Distribution of source rocks and maturity modelling in the northern Cenozoic Song Hong Basin (Gulf of Tonkin), Vietnam. J. Petroleum Geol. 28: 167–184.

Anderson D.M., Overpeck J.T. and Gupta A.K. 2005. Increase in the Asian Southwest monsoon during the past four centuries. Science 297: 596–599.

Areshev e.g., Dong T.L., San N.T. and Shnip O.A. 1992. Reservoirs in fractured basement on the continental shelf of southern Vietnam. J. Petroleum Geol. 15: 451–464.

Armitage J.H. and Viotti C. 1977. Stratigraphic nomenclature – southern end Malay Basin. Proc. 6th Annual Conv. Indonesian Petroleum Ass., Jakarta, pp. 69–94.

ASCOPE (Asean Council on Petroleum) 1981. Tertiary Sedimentary Basins of the Gulf of Thailand and South China Sea: Stratigraphy, Structure and Hydrocarbon Occurrences. ASCOPE, Jakarta, Indonesia, 72pp.

Berger A., Loutre M.F. and Laskar J. 1992. Stability of the astronomical frequencies over the Earth's history for paleoclimate studies. Science 255: 560–566.

Binh N.T.T., Tokunaga T., Son H.P. and Binh M.V. 2007. Present-day stress and pore pressure fields in the Cuu Long and Nam Con Son Basins, offshore Vietnam. Mar. Petroleum Geol. 24: 607–615.

Bishop M.G. 2002. Petroleum systems of the Malay Basin Province, Malaysia. Open-File Report 99–50T. U.S. Geol. Surv.

Biwas B. 1973. Quaternary changes in sea level in the South China Sea. Bull. Geol. Soc. Malaysia 6: 229–255.

Boudagher-Fadel M.K. and Banner F.T. 1999. Revision of the stratigraphic significance of the Oligocene-Miocene "Letter-Stages". Rev. Micropaléontologie 42: 93–97.

Briais A., Patriat P. and Tapponnier P. 1993. Update interpretation of magnetic anomalies and seafloor spreading stages in the South China Sea: Implications for the Tertiary tectonics of southeast Asia. J. Geophys. Res. 98(B4): 6299–6328.

Bühring C., Sarnthein M. and Erlenkeuser H. 2004. Toward a high resolution stable isotope stratigraphy of the last 1.1 m.y.: Site 1144, South China Sea. In: Prell W.L., Wang P., Blum P., Rea D.K. and Clemens S.C. (eds.), Proc. ODP, Sci. Results 184: 1–29 [Online].

Bühring C., Sarnthein M. and Leg 184 Shipboard Scientific Party 2000. Toba ash layers in the South China Sea: Evidence of contrasting wind directions during eruption ca. 74 ka. Geology 28: 275–278.

Burton R., Kendall C.G. St.C. and Lerche I. 1987. Out of our depth: on the impossibility of fathoming eustasy from the stratigraphic record. Earth-Sci. Rev. 24: 237–277.

Canh T., Ha D.V., Carstens H. and Berstad S. 1994. Vietnam - attractive plays in a new geological province. Oil Gas J. 92(11): 78–83.

Chang M. 1997. Studies on the unconformities of the Tainan Basin, offshore southwestern Taiwan. Petroleum Geol. Taiwan 31: 155–168.

Chappell J. 1982. Evidence for smoothly falling sea levels relative to north Queensland, Australia, during the past 6,000 years. Nature 302: 406–408.

Chappell J. and Shackelton N.J. 1986. Oxygen isotopes and sea level. Nature 324: 137–140.

Chappell, J. and Veeh H.H. 1978. Late Quaternary tectonic movements and sea level changes at Timo and Atauro Island. GSA Bull. 89: 356–368.

Chen C. 2000. Petroleum geology and conditions for hydrocarbon accumulation in the eastern Pearl River Mouth Basin. China Offshore Oil Gas (Geol.) 14(2): 73–83 (in Chinese).

Chen J., Xu S. and Sang J. 1994. The depositional characteristics and oil potential of paleo Pearl River delta systems in the Pearl River Mouth Basin, South China Sea. Tectonophysics 235: 1–11.

Chen L. 2002. Geologic structural feature in west of Zengmu Basin, Nansha Sea Area. Geophys. Prospect. Petroleum 37: 354–362 (in Chinese).

Chen L. and Peng X. 1995. Preliminary study of seismic stratigraphy of Wan'an Basin in Nansha waters. Geophys. Prospect. Petroleum 34(2): 57–70 (in Chinese).

Chen P.P.H., Chen Z.Y. and Zhang Q.M. 1993. Sequence stratigraphy and continental margin development of the northwestern shelf of the South China Sea. AAPG Bull. 77: 842–862.

Chen W.S., Ridgway K.D., Hong C.S., Chen Y.K., Shea K.S. and Yeh M.G. 2001. Stratigraphic architecture, magnetostratigraphy, and incised-valley systems of the Pliocene-Pleistocene collisional marine foreland basin of Taiwan. GSA Bull. 113: 1249–1271.

Chen X., Zhao Q. and Jian Z. 2002. Carbonate content changes since the Miocene and paleoenvironmental implications, ODP Site 1148, northern South China Sea. Mar. Geol. Quat. Geol. 22: 69–74 (in Chinese).

Clift P. 2006. Controls on the erosion of Cenozoic Asia and the flux of clastic sediment to the ocean. Earth Planet. Sci. Lett. 241: 571–580.

Clift P., Lee J. I., Clark M.K. and Blusztajn J. 2002. Erosional response of South China to arc rifting and monsoonal strengthening; a record from the South China Sea. Mar. Geol. 184: 207–226.

Clift P.D. and Sun Z. 2006. The sedimentary and tectonic evolution of the Yinggehai-Song Hong basin and the southern Hainan margin, South China Sea: Implications for Tibetan uplift and monsoon intensification. J. Geophys. Res. 111: B06405, doi:10.1029/2005JB004048.

Clift P.D., Layne G.D. and Blusztajn J. 2004. Marine sedimentary evidence for monsoon strengthening, Tibetan uplift and drainage evolution in East Asia. In: Clift P., Wang P., Kuhnt W. and Hayes D. (eds.), Continent-Ocean Interactions within East Asian Marginal Seas. AGU Geophys. Monogr. 149: 255–282.

Clift P.D., Lin J. and ODP Leg 184 Scientific Party 2001. Patterns of extension and magmatism along the continent-ocean boundary, South China Margin. In: Wilson R.C.L., Whitmarsh R.B., Taylor B. and Froitzheim N. (eds.), Non-volcanic Rifting of Continental Margins: a Comparison of Evidence from Land and Sea. Geol. Soc. London, Spec. Pub. 187, pp. 489–510.

Daines S.R. 1985. Structural history of the West Natuna Basin and the tectonic evolution of the Sunda region. Proc. 14th Ann. Conv. Indonesian Petroleum Ass., Jakarta, pp. 39–61.

Darmadi Y. 2005. Three-dimensional fluvial-deltaic sequence stratigraphy Pliocene-Recent Muda Formation, Belida field, West Natuna Basin, Indonesia. MSc Thesis, Texas A&M University.

Darmadi Y., Willis B.J. and Dorobek S.L. 2007. Three-dimensional seismic architecture of fluvial sequences on the low-gradient Sunda Shelf, offshore Indonesia. J. Sedimentary Res. 77: 225–238.

De Graciansky P.C., Hardenbol J., Jacquin T. and Vail P.R. 1998. Mesozoic and Cenozoic Sequence Stratigraphy of European Basins. SEPM Spec. Publ. 60, 786pp.

Dien P.T., Phung S.T., Nguyen V.D. and Do V.N. 1998. Late Mesozoic to Tertiary basin evolution along the southeast continental shelf of Vietnam. AAPG Ann. Conv. Abst. CD-ROM.

Doust H. 1981. Geology and exploration history of offshore central Sarawak. In: Halbouty M.T. (ed.), Energy resources of the Pacific Region. AAPG Studies Geol. 12: 117–132.

Du Bois E.P. 1985. Review of principal hydrocarbon-bearing basins around the South China Sea. Bull. Geol. Soc. Malaysia 18: 167–209.

Du D. 1991. Characteristics of geologic structure and hydrocarbon potential of the Southwest Taiwan Basin. Mar. Geol. Quat. Geol. 11(3): 21–33 (in Chinese).

Du D. 1994. Tectonic evolution and analysis of oil-gas accumulation in Southwest Taiwan Basin. Mar. Geol. Quat. Geol. 14(3): 5–18 (in Chinese).

Du Z. and Wei K. 2001. Sequence stratigraphic framework and its characteristics of the Weizhou Formation in North Sag of Beibuwan Basin. Acta Sedimentol. Sinica 19: 563–568 (in Chinese).

Duan W. and Huang Y. 1991. Tertiary calcareous nannofossil biostratigraphy in the north part of the South China Sea. Acta Geol. Sinica 65: 86–101 (in Chinese).

Embry A.F. 1988. Triassic sea-level changes: evidence from the Canadian Arctic archipelago. In: Wilgus C.K., Hastings B.S., Kendall C.G.St.C., Posamentier H.W., Ross C.A. and Van Wagoner

J.C. (eds.), Sea Level Changes – An integrated Approach. SEPM Spec. Publ., vol. 42, pp. 249–259.

Emmel F.J. and Curray J.R. 1982. A submerged Late Pleistocene delta and other features related to sea level changes in the Malacca Strait. Mar. Geol. 47: 197–216.

Eong O.J. 2001. Geology of the seas. In: Eong O.J. (ed.), The Seas. The encyclopedia of Malaysia, 31–33.

Fraser A.J., Matthews S.J. and Murphy R.W. (eds.). 1997. Petroleum Geology of Southeast Asia. Geol. Soc. London Spec. Publ., vol. 126, 442pp.

Fraser A.J., Matthews S.J., Lowe S., Todd S.P. and Peel F.J. 1996. Structure, stratigraphy and petroleum geology of the South East Nam Con Son Basin, offshore Vietnam. AAPG Ann. Conv. Program Abst., p. A49.

Galloway W.E. 1989. Genetic stratigraphic sequence in basin analysis I: architecture and genesis of flooding-surface bounded depesitional units. AAPG Bull. 73: 125–142.

Geyh M.A., Kudrass H.R. and Streif H. 1979. Sea-level changes during the late Pleistocene and Holocene in the Strait of Malacca. Nature 278: 441–443.

Ginger D.C., Ardjakusumah W.O., Hedley R.J. and Pothecary J. 1993. Inversion history of the West Natuna Basin: examples from the Cumu-Cumi PSC. Proc. 22nd Ann. Conv. Indonesian Petroleum Ass., Jakarta, pp. 635–658.

Glass B.P. and Wu J. 1993. Coesite and shocked quartz discovered in the Australasian and North American microtektite layers. Geology 21: 435–438.

Gong Z. and Li S. (eds.). 1997. Continental Margin Basin Analysis and Hydrocarbon Accumulation of the Northern South China Sea. China Sci. Press, Beijing, 510pp (in Chinese).

Gong Z., Jin Q., Qiu Z., Wang S. and Meng J. 1989. Geology tectonics and evolution of the Pearl River Mouth Basin. In: Zhu X. (ed.), Chinese Sedimentary Basins. Elsevier, Amsterdam, pp. 181–196.

Gradstein F., Ogg J. and Smith A. (eds.). 2004. A Geologic Time Scale 2004. Cambridge Univ. Press, Cambridge, 589pp.

Hageman H. 1987. Palaeobathymetrical changes in NW Sarawak during Oligocene to Pliocene. Bull. Geol. Soc. Malaysia 21: 91–102.

Haile N.S. 1971. Quaternary shorelines in West Malaysia and adjacent parts of the Sunda Sheft. Quaternaria 15: 333–343.

Haile N.S. 1975. Postulated late Cenozoic high sea levels in the Malay Peninsula. J. Malays Branch R. Asiatic Soc. 48: 78–88.

Hallam A. 1992. Phanerozoic Sea-Level Changes. Columbia Univ. Press, New York, pp. 11–46.

Hanebuth T., Stattegger K. and Groote P.M. 2000. Rapid flooding of the Sunda Shelf: a late-glacial sea-level record. Science 288: 1033–1035.

Hanebuth T.J.J., Stattegger K. and Bojanowski A. 2008. Termination of the Last Glacial Maximum sea-level lowstand: The Sunda-Shelf data revisited. Global Planet. Change, doi:10.1016/j.gloplacha.2008.03.011.

Hao F., Sun Y.C., Li S.T. and Zhang Q.M. 1995. Overpressure retardation of organic-matter maturation and hydrocarbon generation: a case study from the Yinggehai and Qiongdongnan basins, offshore South China Sea. AAPG Bull. 79: 551–562.

Hao Y., Chen P., Wan X. and Dong J. 2000. Late Tertiary sequence stratigraphy and sea level changes in Yinggehai-Qiongdongnan Basins. Earth Sci. – J. China Univ. Geosci. 25: 237–245 (in Chinese).

Hao Y., Xu Y. and Xu S. 1996. Research on Micropalaeontology and Paleoceanography in Pearl River Mouth Basin, South China Sea. China Univ. Geosci. Press, Beijing, 136pp (in Chinese).

Haq B.U., Hardenbol J. and Vail P. R. 1988. Mesozoic and Cenozoic chronostratigraphy and eustatic cycles. In: Wilgus C.K., Hastings B.S., Kendall C.G. St.C., Posamentier H.W., Ross C.A. and Wagoner J.C. (eds.), Sea-Level Changes: an Integrated Approach. SEPM Spec. Publ. vol. 42, pp. 71–108.

Haq B.U., Hardenbol J. and Vail P.R. 1987. Chronology of fluctuating sea levels since the Triassic (250 million years ago to present). Science 235: 1156–1167.

Hardenbol J., Thierry J., Farley M.B., Jacquin T., De Graciansky P.C. and Vail P.R. 1998. Cenozoic sequence chronostratigraphy. In: De Graciansky P.C., Hardenbol J., Jacquin T. and Vail P.R. (eds.), Mesozoic and Cenozoic Sequence Chronostratigraphic Framework of European Basins. SEPM Spec. Publ. vol. 60, pp. 3–13, Charts 1–8.

He L., Wang K., Xiong L. and Wang J. 2001. Heat flow and thermal history of the South China Sea. Physics Earth Planet. Interiors 126: 211–220.

Hellinger S.J. and Sclater J.G. 1983. Some comments on two-layer extensional models for the evolution of sedimentary basins. J. Geophys. Res. 88: 8251–8269.

Ho K.F. 1978. Stratigraphic framework for oil exploration in Sarawak. Bull. Geol. Soc. Malaysia 10: 1–14.

Holbourn A., Kuhnt W. and Schulz M. 2004. Orbitally paced climate variability during the middle Miocene: high resolution benthic stable-isotope records from the tropical western Pacific. In: Clift P.D., Wang P., Hayes D. and Kuhnt W. (eds.), Continent-Ocean Interactions in the East Asian Marginal Seas. AGU Geophys. Monogr. 149, pp. 321–337.

Holbourn A., Kuhnt W., Schulz M. and Erlenkeuser H. 2005. Impacts of orbital forcing and atmospheric carbon dioxide on Miocene ice-sheet expansion. Nature 438: 483–487.

Holbourn A., Kuhnt W., Schulz M., Flores J.A. and Andersen N. 2007. Orbitally-paced climate evolution during the middle Miocene "Monterey" carbon-isotope excursion. Earth Planet. Sci. Lett. 261: 534–550.

Hori K., Tanabe S., Saito Y., Haruyama S., Nguyen V. and Kitamura A. 2004. Delta initiation and Holocene sea-level change: example from the Song Hong (Red River) delta, Vietnam. Sedimentary Geol. 164: 237–249.

Hou Y., Li Y., Jin Q. and Wang P. (eds.). 1981. Tertiary Paleontology of the Northern Continental Shelf of South China Sea. Guangdong Sci. Tech. Press, Guangzhou, 274pp.

Hu Z. and Su H. 1981. Tertiary of the Beibu Gulf and Yinggehai depressions. In: Zeng D., Guo B., Huo C., Zhong S., Huang X., Hu P. and Su H. (eds.), Tertiary System of the Northern Continental Shelf of South China Sea. Guangdong Sci. Tech. Press, Guangzhou, pp. 35–143 (in Chinese).

Huang C. 1997. Tectonic elements and analysis of Cenozoic sedimentary basins in the South China Sea and its surrounding areas. In: Gong Z. and Li S. (eds.), Continental Margin Basin Analysis and Hydrocarbon Accumulation of the Northern South China Sea. China Sci. Press, Beijing, pp. 44–52 (in Chinese).

Huang C.Y., Wu W.Y., Chang C.P., Tsao S., Yuan P.B., Lin C.W. and Xia K.Y. 1997c. Tectonic evolution of accretionary prism in the arc-continent collision terrane of Taiwan. Tectonophysics 281: 31–51.

Huang C.Y., Zhao M., Wei G. and Wang C.C. 2001. Cooling of the South China Sea by the Toba eruption and correlation with other climate proxies ~71,000 years ago. Geophys. Res. Lett. 28: 3915–3918.

Huang F.F.W., Liang S.C. and Juang H.J. 1998. Depositional model for the Oligocene sandstone in the southwest Taiwan offshore. Petroleum Geol. Taiwan 32: 69–86.

Huang L. 1997b. Calcareous nannofossil biostratigraphy in the Pearl River Mouth basin, South China Sea, and Neogene reticulofenestra coccolith size distribution pattern. Mar. Micropaleontol. 32: 3–29.

Huang L. 1999. The application of sequence stratigraphy for analysis in nonmarine depression with example of Enping Sag, Pearl River Mouth Basin. China Offshore Oil and Gas (Geol.) 13(3): 159–168 (in Chinese).

Huang M., Li F. and Li P. 1993. Characteristics of pre-Tertiary and early Tertiary depressions in the eastern Pear River Mouth Basin and their hydrocarbon perspective. China Offshore Oil Gas (Geol.) 7(3): 1–9 (in Chinese).

Huang Z., Li P., Zhang Z. and Zong Y. 1986. Sea level changes along the coastal area of South China since the late Pleistocene. In: Qin Y. and Zhao S. (eds.), Late Quaternary Sea-Level Change. China Ocean Press, Beijing, pp. 142–154 (in Chinese).

Hutchison C.S. 1989. Geological Evolution of South-East Asia. Oxford Monogr. Geol. Geophys. No. 13, Clarendon Press, Oxford, 368pp.

Hutchison C.S. 2004. Marginal basin evolution: the southern South China Sea. Mar. Petroleum Geol. 21: 1129–1148.

Imbrie J., Berger A., Boyle E., Clemens S.C., Duffy A., Howard W.R., Kukla G., Kutzbach J., Martinson D.G., McIntyre A., Mix A.C., Molfino B., Morley J.J., Peterson L.C., Pisias N.G., Prell W.L., Raymo M.E., Shackleton N.J. and Toggweiler J.R. 1993. On the structure and origin of major glaciation cycles 2: The 100000-year cycle. Paleoceanography 8: 699–735.

Ingram G.M., Chisholm T.J., Grant C.J., Hedlund C.A., Stuart-Smith P. and Teasdale J. 2004. Deepwater North West Borneo: hydrocarbon accumulation in an active fold and thrust belt. Mar. Petroleum Geol. 21: 879–887.

Ismail M.I.B. and Hassan R.B.Abu. 1999. Tinjar Province. In: Leong K.M. (ed.), The Petroleum Geology and Resources of Malaysia, PETRONAS, Kuala Lumpur, pp. 395–410.

Jardine E. 1997. Dual petroleum systems governing the prolific Pattani Basin, offshore Thailand. In: Howes J.V.C. and Noble R.A. (eds.), Proc. Int. Conf. Petroleum Syst. SE Asia and Australia, Indonesian Petroleum Ass., Jakarta, pp. 351–363.

Jian Z., Cheng X., Zhao Q., Wang J. and Wang P. 2001a. Oxygen isotope stratigraphy and events in the northern South China Sea during the last 6 million years. Sci. China (D) 44(10): 952–960.

Jian Z., Zhao Q., Cheng X., Wang J., Wang P. and Su X. 2003. Pliocene-Pleistocene stable isotope and paleoceanographic changes in the northern South China Sea. Palaeogeogr. Palaeoclimatol. Palaeoecol. 193: 425–442.

Jiang S., Zhou X., Huang C. and Xia K. 1994. The stratigraphy, structure and evolution of Zengmu Basin. Tectonophysics 235: 51–62.

Jiang S., Zhou X., Huang C., Xia K. and Liu D. 1990. Characters of the tectonic sequence in Zengmu Basin. Tropical Oceanol. 9(2): 23–30 (in Chinese).

Jiang Z., Zeng L., Li M. and Zhong Q. (eds.). 1994. The North Continental Shelf Region of South China Sea. Tertiary in Petroliferous Regions of China, vol. 8. Petroleum Industry Press, Beijing, 145pp (in Chinese).

Jin Q. (ed.). 1989. The Geology and Petroleum Resources in the South China Sea. Geol. Publ. House, Beijing, 417pp (in Chinese).

Jin Q. 2005. Overview on offshore peteoleum exploration of the northern South China Sea. Fisheries Sci. Tech. 3: 1–5 (in Chinese).

Jin Q. and Li T. 2000. Regional geologic tectonics of the Nansha Sea Area. Mar. Geol. Quat. Geol. 20: 1–8 (in Chinese).

Kamaludin B.H. 1993. The change of mangrove shorelines in Kuala Kurau, Peninsular Malaysia. Sedimentary Geol. 83: 187–197.

Kang X., Zhao W., Pan Z., Zhang Q. and Chen Z. 1994. Study on architecture of sequence stratigraphic framework of Beibuwan Basin. Earth Science-J. China Univ. Geosci. 19: 493–502 (in Chinese).

Kershaw A.P., Penny D., Van der Kaars S., Anshari G. and Thamotherampilai A. 2001. Vegetation and climate in lowland Southeast Asia at the Last Glacial Maximum. In: Metcalfe I., Smith J.M.B., Morwood M. and Davidson I. (eds.), Faunal and Floral Migrations and Evolution in SE Asia-Australasia. Balkema, Lisse, pp. 227–236.

Kingston D.R., Dishroom C.P. and Williams P.A. 1983. Global basin classification system. AAPG Bull. 67: 2175–2193.

Kou C. and Ye G. 1981. Tertiary of the Leizhou Peninsula. In: Zeng D., Guo B., Huo C., Zhong S., Huang X., Hu P. and Su H. (eds.), Tertiary System of the Northern Continental Shelf of South China Sea. Guangdong Sci. Tech. Press, Guangzhou, pp. 149–177.

Lambeck K. and Chappell J. 2001. Sea level change through the last glacial cycle. Science 292: 679–686.

Laskar J. 1990. The chaotic motion of the solar system: A numerical estimate of the size of the chaotic zones. Icarus 88: 266–291.

Lee G.H., Lee K. and Watkins J.S. 2001. Geologic evolution of the Cuu Long and Nam Con Son basins, offshore southern Vietnam, South China Sea. AAPG Bull. 85: 1055–1082.

Lee M.-Y., Chen C.-H., Wei K.-Y., Iizuka Y. and Carey S. 2004. First Toba supereruption revival. Geology 32:61–64.

Lee T.-Y., Tang C.-H., Ting J.-S. and Hsu Y.-Y. 1993. Sequence stratigraphy of the Tainan Basin, offshore southwestern Taiwan. Petroleum Geol. Taiwan 28: 119–158.

Leo C.T.A. 1997. Exploration in the Gulf of Thailand in deltaic reservoirs, related to the Bongkot Field. In: Fraser A.J., Matthews S.J. and Murphy R.W. (eds.), Petroleum Geology of Southeast Asia. Geol. Soc. London Spec. Pub., vol. 126, pp. 77–87.

Leong K.M. (ed.). 1999a. The Petroleum Geology and Resources of Malaysia. PETRONAS, Kuala Lumpur, 665pp.

Leong K.M. 1999b. Geological setting of Sabah. In: Leong K.M. (ed.), The Petroleum Geology and Resources of Malaysia. PETRONAS, Kuala Lumpur, pp. 475–497.

Lewis S.D. and Hayes D.E 1984. A geophysical study of the Manila Trench, Luzon, Phillippines 2. Force Arc basis structural and stratigraphic evolution. J. Geophys. Res. 89(B11): 9196–9214.

Li B., Wang J., Huang B., Li Q., Jian Z. and Wang P. 2004. South China Sea surface water evolution over the last 12 Ma: A south-north comparison from ODP Sites 1143 and 1146. Paleoceanography 19: PA1009, doi:10.1029/2003PA000906.

Li J., Lin C. and Chen P. 1999. Sea level change and sequence chronostratigraphy of the Yinggehai-Huangliu Formations in the Qiongdongnan Basin. Geol. Rev. Beijing 45: 514–520 (in Chinese).

Li P. and Rao C. 1994. Tectonic characteristics and evolution history of the Pearl River Mouth Basin. Tectonophysics 235: 13–25.

Li Q., Jian Z. and Li B. 2004. Oligocene-Miocene planktonic foraminifer biostratigraphy, Site 1148, northern South China Sea. In: Prell W.L., Wang P., Blum P., Rea D.K. and Clemens S.C. (eds.), Proc. ODP, Sci. Results, 184 [Online].

Li Q., Jian Z. and Su X. 2005a. Late Oligocene rapid transformations in the South China Sea. Mar. Micropaleontol. 54: 5–25.

Li Q., Lourens L. and Wang P. 2007. New ages for Neogene marine biostratigraphic events. J. Stratigr. 31: 197–208 (in Chinese).

Li Q., Tian J. and Wang P. 2005b. Recognizing the stratigraphic and paleoclimatic significance of the eccentricity cycles. Earth Sci. – J. China Univ. Geosci. 30: 519–528 (in Chinese).

Li Q., Wang P., Zhao Q., Shao L., Zhong G., Tian J., Cheng X., Jian Z. and Su X. 2006. A 33 Ma lithostratigraphic record of tectonic and paleoceanographic evolution of the South China Sea. Mar. Geol. 230: 217–235.

Lian H.M. and Bradley K. 1986. Exploration and development of natural gas, Pattani Basin, Gulf of Thailand. Trans. 4th Circum-Pacific Energy and Mineral Res. Conf., Singapore. pp. 171–181.

Liang X., Wei G., Shao L., Li X. and Wang R. 2001. Records of Toba eruptions in the South China Sea – Chemical charateristics of the glass shards from ODP 1143A. Sci. China (D) 44(10): 871–878.

Liew K.K. 1994. Structural development at the west-central margin of the Malay Basin. Geol. Soc. Malaysia Bull. 36: 67–80.

Lin A.T., Watts A.B. and Hesselbo S.P. 2003. Cenozoic stratigraphy and subsidence history of the South China Sea margin in the Chinese Taipei region. Basin Res. 15: 453–479.

Lin C., Liu J., Cai S., Zhang Y., Lu M. and Li J. 2001. Depositional architecture and developing settings of large-scale incised valley and sub-marine gravity flow systems in the Yinggehai and Qiongdongnan basins, South China Sea. Chinese Sci. Bull. 46: 690–693.

Liu Z., Colin C. and Trentesaux A. 2006. Major element geochemistry of glass shards and minerals of the Youngest Toba Tephra in the southwestern South China Sea. J. Asian Earth Sci. 27: 99–107.

Lovell B.K. 1987. The nature and significance of regional unconformity in the hydrocarbon-bearing Neogene sequence offshore West Sabah. Bull. Geol. Soc. Malaysia 21: 55–90.

Lüdmann T., Wong H.K. and Wang P. 2001. Plio-Quaternary sedimentation processes and neotectonics of the northern continental margin of the South China Sea. Mar. Geol. 172: 331–358.

Madon M.B. and Watts A.B. 1998. Gravity anomalies, subsidence history and the tectonic evolution of the Malay and Penyu Basins (offshore Peninsular Malaysia). Basin Res. 10: 375–392.

Madon M.B.Hj. 1997. The kinematics of extension and inversion in the Malay Basin, offshore Peninsular Malaysia. Bull. Geol. Soc. Malaysia 41: 127–138.

Madon M.B.Hj. 1999a. Basin types, tectono-stratigraphic provinces, and structural styles. In: Leong K.M. (ed.), The Petroleum Geology and Resources of Malaysia. PETRONAS, Kuala Lumpur, pp. 77–111.

Madon M.B.Hj. 1999b. Geological setting of Sarawak. In: Leong K.M. (ed.), The Petroleum Geology and Resources of Malaysia. PETRONAS, Kuala Lumpur, pp. 275–290.

Madon M.B.Hj. 1999c. North Luconia Province. In: Leong K.M. (ed.), The Petroleum Geology and Resources of Malaysia. PETRONAS, Kuala Lumpur, pp. 443–454.

Madon M.B.Hj. 1999d. Plate tectonic elements and the evolution of Southeast Asia. In: Leong K.M. (ed.), The Petroleum Geology and Resources of Malaysia. PETRONAS, Kuala Lumpur, pp. 59–76.

Madon M.B.Hj. and Abolins P. 1999. Balingian province. In: Leong K.M. (ed.), The Petroleum Geology and Resource of Malaysia. PETRONAS, Kuala Lumpur, pp. 443–454.

Madon M.B.Hj. and B.A.H. 1999a. Tatau province. In: Leong K.M. (ed.), The Petroleum Geology and Resources of Malaysia. PETRONAS, Kuala Lumpur, pp. 413–426.

Madon M.B.Hj. and Redzuan B.A.H. 1999b. West luconia province. In: Leong K.M. (ed.), The Petroleum Geology and Resources of Malaysia. PETRONAS, Kuala Lumpur, pp. 429–439.

Madon M.B.Hj., Abolins P., Hoesni M.J.B. and Ahmad M.B. 1999a. Malay Basin. In: Leong K.M. (ed.), The Petroleum Geology and Resources of Malaysia. PETRONAS, Kuala Lumpur, pp. 173–217.

Madon M.B.Hj., Karim R.Bt.Abd. and Fatt R.W.H. 1999b. Tertiary stratigraphy and correlation schemes. In: Leong K.M. (ed.), The Petroleum Geology and Resources of Malaysia. PETRONAS, Kuala Lumpur, pp. 113–137.

Madon M.B.Hj., Leong K.M. and Anuar A. 1999c. Sabah Basin. In: Leong K.M. (ed.), The Petroleum Geology and Resources of Malaysia, PETRONAS, Kuala Lumpur, pp. 501–542.

Mahmud O.A.B. and Saleh S.B. 1999. Petroleum resources, Sarawak. In: Leong K.M. (ed.), The Petroleum Geology and Resources of Malaysia. PETRONAS, Kuala Lumpur, pp. 457–472.

Matthews S.J., Fraser A.J., Lowe S., Todd S.P. and Peel F.J. 1997. Structure, stratigraphy, and petroleum geology of the SE Nam Con Son Basin, offshore Vietnam. In: Fraser A.J., Matthews S.J. and Murphy R.W. (eds.), Petroleum Geology of Southeast Asia. Geol. Soc. London Spec. Publ., vol. 126, pp. 89–106.

Mat-Zin I.C. and Swarbrick R.E. 1997. The tectonic evolution and associated sedimentation history of Sarawak Basin, eastern Malaysia: a guide for future hydrocarbon exploration. In: Fraser A.J., Matthews S.J. and Murphy R.W. (eds.), Petroleum Geology of Southeast Asia. Geol. Soc. London Spec. Publ., vol. 126, pp. 237–245.

Mat-Zin I.C. and Tucker M.E. 1999. An alternative stratigraphic scheme for the Sarawak Basin. J. Asian Earth Sci. 17: 215–232.

May J.A. and Eyles D.R. 1985. Well log and seismic character of Tertiary Terumbu carbonate, South China Sea, Indonesia. AAPG Bull. 69: 1339–1358.

McCabe R., Celeya M., Cole J., Han H.C., Ohnstadt T., Paijitprapapon V. and Thitipawan V. 1988. Extension tectonics: the Neogene opening of the N-S trending basins of Central Thailand. J. Geophys. Res. 93: 11899–11910.

Meirita M.F. 2003. Structural and depositional evolution, KH Field, West Natuna Basin, offshore Indonesia. MSc thesis, Texas A&M Univ., 56pp.

Miall A.D. 1997. The Geology of Stratigraphic Sequences. Springer-Verlag, Berlin, 433pp.

Miller K.G., Fairbanks R.G. and Mountain G.S. 1987. Tertiary oxygen isotope synthesis, sea level history, and continental margin erosion. Paleoceanography 2: 1–19.

Miller K.G., Kominz M.A., Browning J.V., Wright J.D., Mountain G.S., Katz M.E., Sugarman P.J., Cramer B.S., Christie-Blick N. and Pekar S.F. 2005. The Phanerozoic record of global sea-level change. Science 310: 1293–1298.

Miller K.G., Wright J.D. and Fairbanks R.G. 1991. Unlocking the ice house: Oligocene-Miocene oxygen isotopes, eustasy, and margin erosion. J. Geophys. Res. 96(B4): 6829–6848.

Mitchum R.M. Jr., Vail P.R. and Thompson S. III 1977. Seismic stratigraphy and global changes of sea level, Part 2: Depositional sequence as a basic unit for stratigraphic analysis. In: Payton C.E (ed.), Seismic Stratigraphy – Applications to Hydrocarbon Exploration. AAPG Mem., vol. 26, pp. 53–62.

Mix A.C., Le J. and Shackleton N.J. 1995. Benthic foraminiferal stable isotope stratigraphy of Site 846: 0–1.8 Ma. Proc. ODP Sci. Results 138: 839–854.

Mohammad A.M. and Wong R.H.F. 1995. Seismic sequence stratigraphy of the Tertiary sediments offshore Sarawak deepwater area, Malaysia. Bull. Geol. Soc. Malaysia 37: 345–361.

Mohd Idrus B.I., Abdul R.E., Abdul Manaf M., Sahalan A.A. and Mahendran B. 1995. The geology of Sarawak deepwater and surrounding areas. Bull. Geol. Soc. Malaysia 37: 165–178.

Moldovanyi E.P., Waal F.M. and Yan Z.J. 1995. Regional exposure events and platform evolution of Zhuijang Formation carbonates, Pearl River Mouth Basin: evidence for primary and diagenetic seismic facies. In: Budd D.A., Saller A.H. and Harris P.M. (eds.), Unconformities and Porosity in Carbonate Strata. AAPG Memoir, vol. 63, pp. 133–45.

Morley C.K. and Westawayw R. 2006. Subsidence in the super-deep Pattani and Malay basins of Southeast Asia: a coupled model incorporating lower-crustal flow in response to post-rift sediment loading. Basin Res. 18: 51–84.

Morley R.J. 1991. Tertiary stratigraphic palynology in Southeast Asia: current status and new directions. Bull. Geol. Soc. Malaysia 28: 1–36.

Morley R.J. 2000. Origin and Evolution of Tropical Rain Forests. J. Wiley, Chichester, 378pp.

Nguyen V.L., Ta T.K.O. and Tateishi M. 2000. Late Holocene depositional environments and coastal evolution of the Mekong River Delta, southern Vietnam. J. Asian Earth Sci. 18: 427–439.

NISIT (Nansha Integrated Scientific Investigation Team of Chinese Academy of Sciences). 1989. Reports on the Integrated Scientific Investigations in the Nansha Islands and its adjoining offshore region. China Sci. Press, Beijing, vol. 1, pp. 180–198 (in Chinese).

Olson C. C. and Dorobek S. L. 2002. Comparison of structural styles and regional subsidence patterns across the Nam Con Son and Cuu Long Basins, offshore Southeast Vietnam. Chapman Conf. Continent-Ocean Interactions within the East Asian Marginal Seas (Abst.), San Diego, CA.

Oung J.-N. 2000. Two-dimensional basin modeling - a regional study on hydrocarbon generation, offshore Taiwan. Petroleum Geol. Taiwan 34: 33–54.

Oung J.-N. and Tsao C.-Q. 1991. Biomarker studies on oil and rock samples in the Tainan Basin, offshore Taiwan. Petroleum Geol. Taiwan 26: 231–238.

Pelejero C., Kienast M., Wang L. and Grimalt J.O. 1999. The flooding of Sundaland during the last deglaciation: imprints in hemipelagic sediments from the southern South China Sea. Earth Planet. Sci. Lett. 171: 661–671.

Phillips S., Little L., Michael E. and Odell V. 1997. Sequence stratigraphy of Tertiary petroleum systems in the West Natuna Basin, Indonesia. In: Howes J.V.C. and Noble R.A. (eds.), Proc. Int. Conf. Petroleum Syst. SE Asia and Australia, Indonesian Petroleum Ass., Jakarta, pp. 381–401.

Pigott J.D. and Ru K. 1994. Basin superposition on the northern margin of the South China Sea. Tectonophysics 235: 27–50.

Pigott J.D. and Sattayarak N. 1993. Aspects of sedimentary basin evolution assessed through tectonic subsidence analysis. Example: northern Gulf of Thailand. J. SE Asian Earth Sci. 8: 407–420.

Polachan S. and Sattayarak N. 1991. Development of Cenozoic Basins in Thailand. Mar. Petroleum Geol. 8: 84–87.

Pollock R.E., Hayes J.B., Williams K.P. and Young R.A. 1984. The petroleum geology of the KH Field, Kakap, Indonesia. Proc. 13th Ann. Conv. Indonesian Petroleum Ass., Jakarta, vol. 1, pp. 407–423.

Posamentier H.W. and James D.P. 1993. An overview of sequence stratigraphic concepts: uses and abuses. In: Posamentier H.W., Summerhayes C.P., Haq B.U. and Allen G.P. (eds.), Sequence Stratigraphy and Facies Associations. Int. Ass. Sedimentol. Spec. Publ., Blackwell Sci. Publ., Oxford, vol. 18, pp. 3–18.

Posamentier H.W., Jervey M.T. and Vail P.R. 1988. Eustatic controls on clastic deposition I-conceptual framework. In: Wilgus C.K., Hastings B.S., Kendall C.G.St.C., Posamentier H.W., Ross C.A. and Van Wagoner J.C. (eds.), Sea-Level Changes: an Integrated Approach. SEPM Spec. Publ., vol. 42, pp. 109–124.

Prell W. L. 1982. Oxygen and carbon isotope stratigraphy of the Quaternary of Hole 502B: evidence for two modes of isotopic variability. Proc. DSDP, Init. Repts. 68: 455–464.

Qin G. 1996a. Application of micropaleontology to the sequence stratigraphic studies of Late Cenozoic in the Zhujiang River Mouth Basin. Mar. Geol. Quat. Geol. 16(4): 1–18 (in Chinese).

Qin G. 1996b. Biostratigraphic zonation and correlation of the late Cenozoic planktonic foraminifera in Pearl River Basin. In: Hao Y. (ed.), Research on Micropalaeontology and Paleo-ceanography in Pearl River Mouth Basin, South China Sea. China Univ. Geosci. Press, Beijing, pp. 19–31 (in Chinese).

Qin G. 2002. Late Cenozoic sequence stratigraphy and sea-level changes in the Pearl River Mouth Basin, South China Sea. China Offshore Oil and Gas 16: 1–10 (in Chinese).

Qiu X., Ye S., Wu S., Shi X., Zhou D., Xia K. and Flueh R. 2001. Crustal structure across the Xisha Trough, northwestern South China Sea. Tectonophysics 341: 179–193.

Qiu Y. 1996. Sequence stratigraphic interpretation of carbonate sequences in major basins of southwest South China Sea. Geol. Res. South China Sea 8: 62–73 (in Chinses).

Qiu Y. and Wang Y.-M. 2001. Reefs and paleostructure and paleoenvironment in the South China Sea. Mar. Geol. Quat. Geol. 21: 65–73 (in Chinese).

Qiu Y., Chen G., Xie X., Wu L., Liu X. and Jiang T. 2005. Sedimentary filling evolution of Ceno-zoic strata in Zengmu Basin, Southwestern South China Sea. Tropical Oceanol. 24(5): 43–52 (in Chinese).

Ramli Md.N. 1988. Stratigraphy and palaeofacies development of Carigali's operating areas in the Malay Basin, South China Sea. Bull. Geol. Soc. Malaysia 22: 153–188.

Rangin C., Klein M., Roques D., Le Pichon X. and Trong L.V. 1995. The Red River fault system in the Tonkin Gulf, Vietnam. Tectonophysics 243: 209–222.

Rimbaman I. 1992. The role of sea-level changes on the coastal environment of northern West Java (case study of Eretan, Losarang and Indramayu). J. SE Asian Earth Sci. 7: 71–77.

Ru K. and Pigott J.D. 1986. Episodic rifting and subsidence in the South China Sea. AAPG Bull. 70: 1136–1155.

Schimanski A. and Stattegger K. 2005. Deglacial and Holocene evolution of the Vietnam shelf: stratigraphy, sediments and sea-level change. Mar. Geol. 214: 365–387.

Schlüter H.U., Hinz K. and Block M. 1996. Tectono-stratigraphic terranes and detachment faulting of the South China Sea and Sulu Sea. Mar. Geol. 130: 39–78.

Schulz H., Emeis K.-C., Erlenkeuser H., von Rad U. and Rolf C. 2002. The Toba volcanic event and interstadial/stadial climates at the Marine Isotopic Stage 5 to 4 transition in the northern Indian Ocean. Quat. Res. 57: 22–31.

Schulz M. and Mudelsee. M. 2002. REDFIT: estimating red-noise spectra directly from unevenly spaced paleoclimatic time series. Computers Geosci. 28: 421–426.

Sea Level Working Group 1992. Working Group Report. JOIDES J. 18: 28–36.

Shackleton N.J. and Crowhurst S. 1997. Sediment fluxes based on an orbitally tuned time scale 5 Ma to 14 Ma, Site 926. In: Shackleton N.J., Curry W.B., Richter C., et al. (eds.), Proc. ODP Sci. Results 154: 69–82.

Shackleton N.J., Hall M.A. and Pate D. 1995. Pliocene stable isotope stratigraphy of Site 846. Proc. ODP Sci. Results 138: 337–355.

Shao L., Li X., Wang P., Jian Z., Wei G., Pang X. and Liu Y. 2004. Sedimentary record of the tectonic evolution of the South China Sea since the Oligocene. Advances Earth Sci. 19: 539–544 (in Chinese).

Shao L., You H., Hao H., Wu G., Qiao P. and Lei Y. 2007. Petrology and depositional environments of Mesozoic strata in the northeastern South China Sea. Geol. Rev. 53: 164–169 (in Chinese).

Sibuet J.C., Hsu S.K. and Debayle E. 2004. Geodynamic context of the Taiwan orogen. In: Clift P., Wang P., Kuhnt W. and Hayes D. (eds.), Continent-Ocean Interactions within East Asian Marginal Seas. AGU Geophys. Monogr. 149: 127–158.

Sibuet J.C., Hsu S.K., Le Pichon X., Le Formal J.P. and Reed D. 2002. East Asia plate tectonics since 15 Ma: constraints from the Taiwan region. Tectonophysics 344: 103–134.

Somboon J.R.P. and Thiramonkol N. 1992. Holocene highstand shoreline of the Chao Phraya Delta, Thailand. J. SE Asian Earth Sci. 7: 53–60.

Song S.R., Chen C.H., Lee M.Y., Yang T.F., Iizuka Y. and Wei K.Y. 2000. Newly discovered eastern dispersal of the youngest Toba Tuff. Mar. Geol. 167: 303–312.

Stattegger K., Kuhnt W., Wong H.K., Bühring C., Haft C., Hanebuth T., Kawamura H., Kienast M., Lorenc S., Lotz B., Lüdmann T., Lurati M., Mühlhan N., Paulsen A.-M., Paulsen J., Pracht J., Putar-Roberts A., Hung N.Q., Richter A., Salomon B., Schimanski A., Steinke S., Szarek R., Nhan N.V., Weinelt M. and Winguth C. 1997. Sequence stratigraphy, late Pleistocene-Holocene sea level fluctuations and high-resolution record of the post-Pleistocene transgression on the Sunda Shelf. Cruise Rept. Sonne 115, pp. 24–36.

Steinke S., Kienast M. and Hanebuth T. 2003. On the significance of sea-level variations and shelf paleo-morphology in governing sedimentation in the southern South China Sea during the last deglaciation. Mar. Geol. 201: 179–206.

Su N., Zeng L. and Li P. 1995. Geological features of Mesozoic sags in the eastern part of Pearl River Mouth Basin. China Offshore Oil and Gas (Geol.). 9(4): 228–236 (in Chinese).

Su X., Xu Y. and Tu, Q. 2004. Early Oligocene–Pleistocene calcareous nannofossil biostratigraphy of the northern South China Sea (Leg 184, Sites 1146–1148). In: Prell W.L., Wang P., Blum P., Rea D.K. and Clemens S.C. (eds.), Proc. ODP, Sci. Results 184 [Online].

Sun X. and Wang P. 2005. How old is the Asian monsoon system? – Palaeobotanical records from China. Palaeogeogr., Palaeoclimatol., Palaeoecol. 222: 181–222.

Sun X., Li M., Zhang Y., Lei Z., Kong Z., Li P., Ou Q. and Liu Q. 1981. Palynology and plant fossils. In: Hou Y., Li Y., Jin Q. and Wang P. (eds.), Tertiary Paleontology of the Northern Continental Shelf of South China Sea. Guangdong Sci. Tech. Press, Guangzhou, pp. 1–59 (in Chinese).

Sun X., Luo Y., Huang F., Tian J. and Wang P. 2003. Deep-sea pollen from the South China Sea: Pleistocene indicators of East Asian monsoon. Mar. Geol. 201: 97–118.

Tamburini F., Adatte T. and Föllmi K.B. 2006. Origin and nature of green clay layers, ODP Leg 184, South China Sea. In: Prell W.L., Wang P., Blum P., Rea D.K. and Clemens S.C. (eds.), Proc. ODP, Sci. Results 184: 1–23 [Online].

Tan M. 1995. Seismic-stratigraphic studies of the continental of Southern Vietnam. J. Petroleum Geol. 18: 345–354.

Tanabe S., Hori K., Saito Y., Haruyama S., Vu V. P. and Kitamura A. 2003. Song Hong (Red River) delta evolution related to millennium-scale Holocene sea-level changes. Quat. Sci. Rev. 22: 2345–2361.

Tang F.S.L., Oung J.N., Hsu J.Y.Y. and Yang C.N. 1999. Elementary study of structural evolution in Tainan Basin in southwest Taiwan Strait. Petroleum Geol. Taiwan 33: 125–149.

Tapponnier P., Peltzer G., Le Dain A.Y., Armijo R. and Cobbold P. 1982. Propagating extrusion tectonics in Asia: new insights from simple experiments with plasticine. Geology 10: 611–616.

Taylor B. and Hayes D.E. 1980. The tectonic evolution of the South China Sea Basin. In: Hayes D.E. (ed.), The Tectonic and Geologic Evolution of Southeast Asian Seas and Islands. AGU Geophys. Monogr., Washington, D.C., pp. 89–104.

Thommeret J. and Thommeret Y. 1978. [14]C dating of some Holocene sea level on the north of the islands of Java (Indonesia). Quat. Res. SE Asia. 4: 51–56.

Thompson P.R. and Abbott W.H. 2003. Chronostratigraphy and microfossil-derived sea-level history of the Qiongdongnan and Yinggehai basins, South China Sea. In: Olson H.C. and Leckie R.M. (eds.), Microfossils as Proxies of Sea-Level Change and Stratigraphic Discontinuities. SEPM Spec. Publ., Tulsa, vol. 75, pp. 97–117.

Tian J., Wang P. and Cheng X. 2004a. Time-frequency variations of the Plio–Pleistocene foraminiferal isotopes: a case study from the southern South China Sea. Earth Sci.-J. China Univ. Geosci. 15(3): 283–289.

Tian J., Wang P. and Cheng X. 2004b. Responses of foraminiferal isotopic variations at ODP Site 1143 in the southern South China Sea to orbital forcing. Sci. China (D) 47(10): 943–953.

Tian J., Wang P. and Cheng X. 2004c. Pleistocene precession forcing of the upper ocean structure variations in the southern South China Sea. Progr. Nat. Sci. 14(11): 1004–1009.

Tian J., Wang P. and Cheng X. 2004d. Development of the East Asian monsoon and Northern Hemisphere glaciation: Oxygen isotope records from the South China Sea. Quat. Sci. Rev. 23: 2007–2016.

Tian J., Wang P., Cheng X. and Li Q. 2002. Astronomically tuned Plio-Pleistocene benthic $\delta^{18}O$ records from South China Sea and Atlantic-Pacific comparison. Earth Planet. Sci. Lett. 203: 1015–1029.

Tian J., Wang P., Zhao Q., Li Q. and Cheng X. 2008. Astronomically modulated Neogene sediment records from the South China Sea. Paleoceanography, doi:10.1029/2007PA001552.

Tiedemann R., Sarnthein M. and Shackleton N.J. 1994. Astronomic timescale for the Pliocene Atlantic $\delta^{18}O$ and dust flux records from Ocean Drilling Program Site 659. Paleoceanography 9: 619–638.

Tjia H.D. 1977. Sea level variations during the last six thousand years in Peninsular Malaysia. Sains Malaysia 6: 171–183.

Tjia H.D. 1994. Inversion tectonics in the Malay Basin: evidence and timing of events. Bull. Geol. Soc.Malaysia 36: 119–126.

Tongkul F. 1991. Tectonic evolution of Sabah, Malaysia. J. SE Asian Earth Sci. 6: 395–405.

Trevena A. S. and Clark R. A. 1986. Diagenesis of sandstone reservoirs of Pattani Basin, Gulf of Thailand. AAPG Bull. 70: 299–308.

Tzeng J., Uang Y.C, Hsu Y.Y. and Teng L.S. 1996. Seismic stratigraphy of the Tainan Basin. Petroleum Geol. Taiwan 30: 281–307.

Vail P.R., Audemard F., Browman S.A., Eisner P.N. and Perez-Crus C. 1991. The stratigraphic signatures of tectonics, eustasy and sedimentology – an overview. In: Einsele G., Ricken W. and Seilacher A (eds.), Cycles and Events in Stratigraphy. Springer-Verlag, Berlin, pp. 617–659.

Vail P.R., Hardenbol J. and Todd R.G. 1984. Jurassic unconformities, chronostratigraphy and sea-level changes from seismic stratigraphy and biostratigraphy. In: Schlee J.S (ed.), Interregional Unconformities and Hydrocarbon Accumulation. AAPG Mem. 36: 129–144.

Vail P.R., Mitchum R.M. Jr., Todd R.G., Widmier J.M., Thompson S. III, Dangree J.B., Bubb J.N. and Hatlelid W.G. 1977. Seismic stratigraphy and global changes of sea level. In: Payton C.E (ed.), Seismic Stratigraphy–Applications to Hydrocarbon Exploration. AAPG Mem. 26: 49–212.

Voris H.K. 2000. Maps of Pleistocene sea levels in Southeast Asia: shorelines, river systems and time durations. J. Biogeogr. 27: 1153–1167.

Wade B.S and Pälike H. 2004. Oligocene climate dynamics. Paleoceanography 19: PA4019, doi:10.1029/2004PA001042.

Wang C. 1996. Sequence stratigraphic analysis of marine Miocene formations in the Pearl River Mouth Basin and its significance. China Offshore Oil Gas (Geol.) 10(5): 279–288 (in Chinese).

Wang C. and Sun Y. 1994. Development of Paleogene depressions and deposition of lacustrine source rocks in the Pearl River Mouth Basin, northern margin of the South China Sea. AAPG Bull. 78: 1711–1728.

Wang G., Wu C., Zhou J. and Li S. 1998. Sequence stratigraphic analysis of the tertiary in the Qiongdongnan Basin. Experimental Petroleum Geol. 20: 124–128 (in Chinese).

Wang J., Zhao Q., Cheng X., Wang R. and Wang P. 2000. Age estimation of the mid-Pleistocene microtektite event in the South China Sea: a case showing the complexity of the sea-land correlation. Chinese Sci. Bull. 45: 2277–2280.

Wang L., Jian Z. and Chen J. 1997. Late Quaternary pteropods in the South China Sea: carbonate preservation and paleoenviromental variation. Mar. Micropaleotol. 32: 115–126.

Wang L., Liu Z., Wu J. and Zhong G. 1996. Depositional history and its relationship to hydrocarbon in Wan'an Basin. China Offshore Oil and Gas (Geol.) 10(3): 144–152 (in Chinese).

Wang L., Zhong G. and Wu J. 1997. Sequence stratigraphic analysis of Wan'an Basin. Geol. Res. South China Sea 9: 67–77 (in Chinese).

Wang P. (ed.). 1985. Marine Micropaleontology of China. China Ocean Press, Beijing and Springer-Verlag, Berlin, 370pp.

Wang P. 1990. Neogen stratigraphy and paleoenvironments of China. Palaeogeogr. Palaeoclimatol. Palaeoecol. 77: 315–334.

Wang P. 2004. Cenozoic deformation and the history of sea-land interactions in Asia. In: Clift P., Wang P., Kuhnt W. and Hayes D. (eds.), Continent-Ocean Interactions in the East Asian Marginal Seas. AGU Geophys. Monogr. 149: 1–22.

Wang P., Min Q. and Bian Y. 1985a. Foraminiferal biofacies in the northern continental shelf of the South China Sea. In: Wang P. (ed.), Marine Micropaleontology of China. China Ocean Press, Beijing, pp. 151–175.

Wang P., Prell W.L., Blum P. (eds.). 2000. Proc. ODP, Init. Repts, Vol. 184 [CD-ROM]. Ocean Drilling Program, Texas A& M University, College Station TX 77845–9547, USA.

Wang P., Xia L. and Chen X. 1985b. Neogene biostratigraphy in the northern shelf of the South China Sea. In: Wang P. (ed.), Marine Micropaleontology of China. China Ocean Press, Beijing, pp. 291–303.

Wang R. and Abelmann A. 1999. Pleistocene radiolarian biostratigraphy in the South China Sci. China (D) 42: 536–543.

Wang R. and Abelmann A. 2002. Radiolarian responses to paleoceanographic events of the southern South China Sea during the Pleistocene. Mar. Micropaleontol. 46: 25–44.

Wang X., Sun X., Wang P. and Stattegger K. 2007. A high-resolution history of vegetation and climate history on Sunda Shelf since the last glaciation. Sci. China (D) 50: 75–80.

Watcharanantakul R. and Morley C.K. 2000. Syn-rift and post-rift modelling of the Pattani Basin, Thailand: evidence for a ramp-flat detachment. Mar. Petroleum Geol. 17: 937–958.

Wei K., Cui H., Ye S., Li D., Liu T., Liang J., Yang G., Wu L., Zhou X. and Hao Y. 2001. High-precision sequence stratigraphy in Qiongdongnan Basin. Earth Science – J. China Univ. Geosci. 26: 59–66 (in Chinese).

White J.M.Jr. and Wing R.S. 1978. Structural development of the South China Sea with particular reference to Indonesia. Proc. 7th Ann Conv. Indonesian Petroleum Ass., Jakarta, pp. 159–178.

Wollands M.A. and Haw D. 1976. Tertiary stratigraphy and sedimentation in the Gulf of Thailand. Seappex Offshore SE Asia Conf., Singapore, pp. 1–22.

Wong H.K., Lüdmann T., Haft C., Paulsen A.M., Hübscher C. and Geng J. 2003. Quaternary sedimentation in the Molengraaff paleo-delta, northern Sunda Shelf (southern South China Sea). In: Sidi F.H., Nummedal D., Imbert P., Darman H. and Posamentier H.W. (eds.), Tropical Deltas of Southeast Asia – Sedimentology, Stratigraphy, and Petroleum Geology. SEPM Spec. Publ., vol. 76, pp. 201–216.

Wongsosantiko A. and Wirojudo G.K. 1984. Tertiary tectonic evolution and related hydrocarbon potential in the Natuna Area. Proc. 13th Ann. Conv. Indonesian Petroleum Ass., Jakarta, vol. 1, pp. 161–183.

Woodroffe C.D. 1993. Late Quaternary evolution of coastal and lowland riverine plains of Southeast Asia and northern Australia: an overview. Sedimentary Geol. 83: 163–175.

Woodroffe C.D., Chappell J., Thom B.G. and Wallensky E. 1989. Depositional model of a macrotidal estuary and floodplain, South Alligator River, northern Australia. Sedimentology 36: 737–756.

Woodroffe, C.D., Thom B.G. and Chappell J. 1985. Development of widespread mangrove swamp in mid-Holocene times in northern Australia. Nature 317: 711–713.

Woodruff D.S. 2003. Neogene marine transgressions, palaeogeography and biogeographic transitions on the Thai-Malay Peninsula. J. Biogeogr. 30: 551–567.

Woodruff F. and Savin S.M. 1991. Mid-Miocene isotope stratigraphy in the deep sea: high-resolution correlations, paleoclimatic cycles and sediment preservation. Paleoceanography 6: 755–806.

Wu F., Lu Y. and Li S. 2000. Cenozoic sequence stratigraphy and sea-level changes in the eastern and southern sea regions of China. In: Wang H., Shi X., Yin H., Qiao X., Liu B., Li S. and

Chen J (eds.), Research on the Sequence Stratigraphy of China. Guangdong Sci. Tech. Press, Guangzhou, pp. 330–351 (in Chinese).

Wu G.X., Wang R.J., Hao H.J. and Shao L. 2007. Microfossil evidence for development of marine Mesozoic in the north of South China Sea. Mar. Geol. Quat. Geol. 27: 79–85 (in Chinese).

Wu J. 1991. Structural characteristics and perspective on petroleum resources in the Nansha Islands. Geol. Res. South China Sea, Memoir 3, pp. 24–38 (in Chinese).

Wu J. and Yang M. 1994. Seismic sequence analysis in the southern South China Sea. Geol. Res. South China Sea 6: 16–29 (in Chinese).

Wu J. and Yang M. 1995. Stratigraphic zonation and characteristics in the south of South China Sea. Mar. Geol. Quat. Geol. 15(Suppl.): 95–106 (in Chinese).

Wu J., Liu B., Zhou W. and Liu B. 1997. Sequence stratigraphic characteristics and prediction of non-structural traps in the Wan'an Basin. In: Xu H. (ed.), From Seismic Stratigraphy to Sequence Stratigraphy. Petroleum Industrial Press, Beijing, pp. 65–72 (in Chinese).

Xia K.Y. and Zhou D. 1993. The geophysical characteristics and evolution of northern and southern margins of the South China Sea. Geol. Soc. Malaysia Bull. 33: 223–240.

Xie X., Müller R. D., Ren J., Jiang T. and Zhang C. 2008. Stratigraphic architecture and evolution of the continental slope system in offshore Hainan, northern South China Sea. Mar. Geol. 247: 129–144.

Xie X., Müller R.D, Li S., Gong Z. and Steinberger B. 2006. Origin of anomalous subsidence along the Northern South China Sea margin and its relationship to dynamic topography. Mar. Petroleum Geol. 23: 745–765.

Xu D., Liu X. and Zhang X. (eds.). 1997a. China Offshore Geology. Geol. Publ. House, Beijing, 310pp (in Chinese).

Xu S. 1999. Sequence stratigraphic theory and practice in exploration prospect prediction: example from the Pearl River Mouth Basin. China Offshore Oil Gas (Geol.) 13(3): 152–158 (in Chinese).

Xu S., Qing G. and Yang S. 1997b. High-resolution bio-stratigraphy and curve of relative sea level changes in northern South China Sea. In: Gong Z. and Li S. (eds.), Basin analysis and petroleum accumulation on the northern margin of the South China Sea. China Sci. Press, Beijing, pp. 128–136 (in Chinese).

Xu S., Yang S. and Huang L. 1995. The application of sequence stratigraphy to stratigraphic correlation. Earth Sci. Frontiers 2: 115–123 (in Chinese).

Yang K.-M., Ting H.-H., Yuan J. 1990. Structural styles and tectonic modes of Neogene extensional tectonics in southwestern Taiwan: implications for hydrocarbon exploration. Petroleum Geol. Taiwan 26: 1–31.

Yang M., Wu J., Yang R. and Duan W. 1996. Stratigraphic division and nomenclature of the southwestern Nansha sea area. Geol. Res. South China Sea 8: 37–46 (in Chinese).

Yao B. and Bai Z. 1991. The seismic reflection characteristics in Zengmu Basin, Nansha sea region. Geol. Res. South China Sea 4: 97–109 (in Chinese).

Yao B. and Zeng W. 1994. The Geological Memoir of South China Sea Surveyed Jointly by China and USA. China Univ. Geosci. Press, Wuhan, 204pp (in Chinese).

Yim W.W.S., Huang G., Fontugne M.R., Haled R.E., Paterne M., Pirazzoli P.A. and Thomas W.N.R. 2006. Postglacial sea-level changes in the northern South China Sea continental shelf: Evidence for a post-8200 calendar yr BP meltwater pulse. Quat. Internat. 145/146: 55–67.

Ying D. 1998. Syn-rift structural style, basin fill and sequence stratigraphy and their control on development of organic facies in the Beibu Gulf Basin, South China Sea. PhD dissertation, Stanford Univ., 336pp.

Yu H.S. and Hong E. 2006. Shifting submarine canyons and development of a foreland basin in SW Taiwan: controls of foreland sedimentation and longitudinal sediment transport. J. Asian Earth Sci. 27: 922–932.

Yu Z. and Ding Z. 1998. An automatic orbital tuning method for paleoclimate records. Geophys. Res. Lett. 25: 4525–4528.

Yumul G.P. Jr., Dimalanta C.B., Tamayo R.A. Jr. and Maury R.C. 2003. Collision, subduction and accretion events in the Philippines: A synthesis. Island Arc 12: 77–91.

Zachos J., Pagani M., Sloan L., Thomas E. and Billups K. 2001. Trends, rhythms, and aberrations in global climate 65 Ma to present. Science 292: 686–693.

Zampetti V., Sattler U. and Braaksma H. 2005. Well log and seismic character of Liuhua 11–1 Field, South China Sea: relationship between diagenesis and seismic reflections. Sedimentary Geol. 175: 217–236.

Zeng D., Guo B., Huo C., Zhong S., Huang X., Hu P. and Su H. (eds.). 1981. Tertiary System of the Northern Continental Shelf of South China Sea. Guangdong Sci. Tech. Press, Guangzhou (in Chinese).

Zhang Q. 1999. Evolution of the Ying-Qiong basin and its tectonic-thermal system. Natural Gas Industry 19: 12–18 (in Chinese).

Zhang Q. and Kou C. 1989. Petroleum geolgy of Cenozoic basins in the northwestern continental shelf, South China Sea. In: Zhu X. (ed.), Chinese Sedimentary Basins: Sedimentary Basins of the World. Elsevier, Amsterdam, vol. 1, pp. 197–206.

Zhang Q. and Su H. 1989. Petroleum geology of the Beibu Gulf Basin. Mar. Geol. Quat. Geol. 9(3): 73–81 (in Chinese).

Zhang Q. and Zhang Q. 1991. A distinctive hydrocarbon basin – Yinggehai Basin, South China Sea. J. SE Asian Earth Sci. 6: 69–74.

Zhao Q. 2005. Late Cainozoic ostracod faunas and paleoenvironmental changes at ODP Site 1148, South China Sea. Mar. Micropaleontol. 54: 27–47.

Zhao Q. and Wang P. 1999. Pregress in Quaternary paleoceanography of the South China Sea: a review. Quat. Sci. 6: 481–501 (in Chinese).

Zhao Q., Jian Z., Cheng X., Liu Z., Xu J. and Xia P. 2004. Mid-Pleistocene impact and marine environmental changes – a high-resolution record from ODP Site 1144, South China Sea. Acta Micropalaeontol. Sinica 21: 130–135 (in Chinese).

Zhao Q., Jian Z., Wang J., Cheng X., Huang B., Xu J., Zhou Z., Fang D. and Wang P. 2001a. Neogene oxygen isotopic stratigraphy, ODP Site 1148, northern South China Sea. Sci. China (D) 44(10): 934–942.

Zhao Q., Wang P., Cheng X., Wang J., Huang B., Xu J., Zhou Z. and Jian Z. 2001b. A record of Miocene carbon excursions in the South China Sea. Sci. China (D) 44: 943–951.

Zheng F., Li Q., Li B., Chen M., Tu X., Tian J. and Jian Z. 2005. A millennial scale planktonic foraminiferal record of mid-Pleistocene climate transition in the northern South China Sea. Palaeogeogr. Palaeoclimatol. Palaeoecol. 223: 349–363.

Zhong G. 2002. Late Cenozoic seismic onlap-offlap sequences and sea level changes on the northern Sunda Shelf, South China Sea. PhD dissertation, Tongji Univ., Shanghai (in Chinese).

Zhong G., Geng J., Wong H.K., Ma Z. and Wu N. 2004. A semi-quantitative method for the reconstruction of eustatic sea level history from seismic profiles and its application to the southern South China Sea. Earth Planet. Sci. Lett. 223: 443–459.

Zhong G., Xu H. and Wang L. 1995. Structure and evolution of Cenozoic basins in the southwest area of South China Sea. Mar. Geol. Quat. Geol. 15(Suppl.): 87–94 (in Chinese).

Zhong G., Zhou Z. and Geng J. 2003. Extrapolating the original positions of removed onlap-offlap points using a piecewise-constructed model of effective stratal thickness. Acta Sedimentol. Sinica 21: 614–618 (in Chinese).

Zhou C. 1987. A preliminary analysis on seismic sequences of the Zhujiangkou Basin. Geophys. Prospect. Petroleum 26(4): 30–36 (in Chinese).

Zhu W. 1987. Study of Miocene carbonate and reef facies in the Pearl River Mouth Basin. Mar. Geol. Quat. Geol. 7: 11–19 (in Chinese).

Zhu W., Li M. and Wu P. 1999. Petroleum systems of the Zhu III Subbasin, Pearl River Mouth Basin, South China Sea. AAPG Bull. 83: 990–1003.

Chapter 4
Sedimentology

Zhifei Liu, Wei Huang, Jianru Li, Pinxian Wang, Rujian Wang, Kefu Yu and Jianxin Zhao

Introduction

The South China Sea (SCS) receives approximately 700 million tons of deposits annually in modern times, including about 80% of terrigenous matters provided by surrounding rivers and 20% of biogenic carbonate and silicates and volcanic ash. A similar scenario has been indentified also in the geological past. Since the early Oligocene, the sea has accumulated about 14.4 thousand trillion tons of deposits,

Z. Liu (✉)
State Key Laboratory of Marine Geology, Tongji University, Shanghai 200092, China
e-mail: lzhifei@tongji.edu.cn

W. Huang
State Key Laboratory of Marine Geology, Tongji University, Shanghai 200092, China
e-mail: huangwei@tongji.edu.cn

J. Li
State Key Laboratory of Marine Geology, Tongji University, Shanghai 200092, China
e-mail: lijianru@tongji.edu.cn

P. Wang
State Key Laboratory of Marine Geology, Tongji University, Shanghai 200092, China
e-mail: pxwang@tongji.edu.cn; pxwang@online.sh.cn

R. Wang
State Key Laboratory of Marine Geology, Tongji University, Shanghai 200092, China
e-mail: rjwangk@online.sh.cn

K. Yu
South China Sea Institute of Oceanology, Chinese Academy of Sciences,
Guangzhou 510301, China
e-mail: kefuyu@scsio.ac.cn

J. Zhao
Centre for Microscopy and Microanalysis, The University of Queensland,
Queensland 4072, Australia
e-mail: j.zhao@uq.edu.au

P. Wang, Q. Li (eds.), *The South China Sea*, Developments in Paleoenvironmental
Research 13, DOI 10.1007/978-1-4020-9745-4_4,
© Springer Science+Business Media B.V. 2009

171

which contain 63% terrigenous matters and 37% biogenic carbonate with negligible biogenic silicates and volcanic materials (Huang 2004). Most of these deposits accumulated on the SCS shelf (43% of total sediment mass) and slope (52%). Such a huge deposition cover makes the SCS an ideal place to study terrigenous input, paleoceanograhy, and regional and global climate evolution as well as sedimentary evolution of the SCS.

This chapter provides an overview of SCS sedimentology by synthesizing surface and geological deposition patterns in terms of terrigenous, biogenic, and volcanic deposits, and presents a systematic outline on sedimentation evolution since the SCS began to form in the early Oligocene.

4.1 Surface Deposition Patterns (Liu Z.)

Deposit Distribution Patterns

The SCS surface deposits have been intensively investigated over the past decades for both region-wide patterns (Chen P. 1978; Su and Wang 1994; Wang P. et al. 1995; Xu et al. 1997; Liu Z. et al. 2003a) and for deposit distribution in individual basin (Su et al. 1989; The Multidisciplinary Oceanographic Expedition Team of Academia Sinica to Nansha Islands 1993; Li X. 2005). In general, the surface deposit distribution is closely tied to water depth and topography, such as shallow shelf terrigenous clastic deposits, hemi-pelagic slope ooze, and abyssal basin clay (Su and Wang 1994) (Figs. 4.1 and 4.2).

(1) Shallow shelf terrigenous clastic sediments: mainly modern terrigenous clayey silt, silty clay, and bioclasts. The shelf sediments are mainly fluvial clasts with clay content higher than 80% and sand less than 15%. Bioclasts are mostly fragments of shell, gastropods, foraminifera, ostracods, and sponge spicules.

(2) Hemi-pelagic slope ooze: biogenic and terrigenous sediments above the carbonate compensation depth (CCD), including clayey silt, silty clay, and calcareous ooze. The ooze is grey or yellowish grey with planktonic foraminifera and other calcareous skeletons. The carbonate content is generally higher than 30% or even up to 62%. The carbonate content decreases with increasing water depth due to dissolution.

(3) Coral sediments: mainly biogenic sand and gravel, which are mainly restricted to shallow settings (0–400 m) of the Dongsha, Xisha, and Nansha islands. Fragments of coral and associating organisms such as bivalves, algae and foraminifera constitute the major components of coral sediments.

(4) Abyssal basin clay: mixed biogenic and terrigenous ooze below the CCD. Biogenic tests are much less than those in shelf and slope sediments, here mainly containing radiolarians and lesser amounts of diatoms and agglutinated foraminifera.

Among the terrigenous sediments in the surface SCS, clay minerals are most important components. The clay mineral distribution can be divided into six provinces

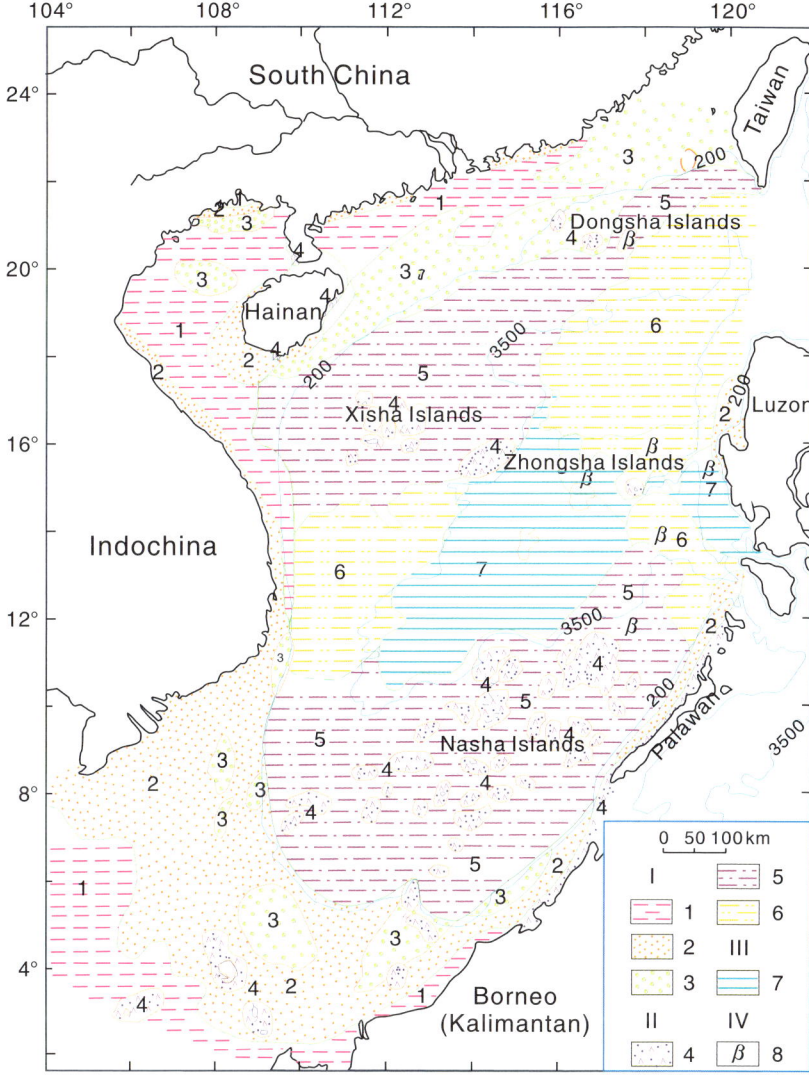

Fig. 4.1 Map shows genetic classification of surface sediments in the SCS (Su and Wang 1994). I Terrigenous type includes (1) nearshore modern terrigenous mud, (2) nearshore modern terrigenous sand and silt, and (3) neritic (paleo-littoral) relict sand. II Biogenic type includes (4) neritic coral sand and gravel, (5) hemi-pelagic and abyssal calcareous ooze, and (6) abyssal siliceous ooze. III Mixed type is represented by (7) abyssal clay. IV Volcanic-biogenic-terrigenous type includes (8) volcanic material (about 5% of the sediment)

(Fig. 4.3) (Chen P. 1978; Liu Z. et al. 2003a). Province A is dominated by illite and chlorite averaging 65% and 25%, respectively, and covers the northern shelf of the SCS, the Taiwan Strait, and the East China Sea, but not in the inner estuary of the Pearl River. Province B includes the central part of the SCS and extends

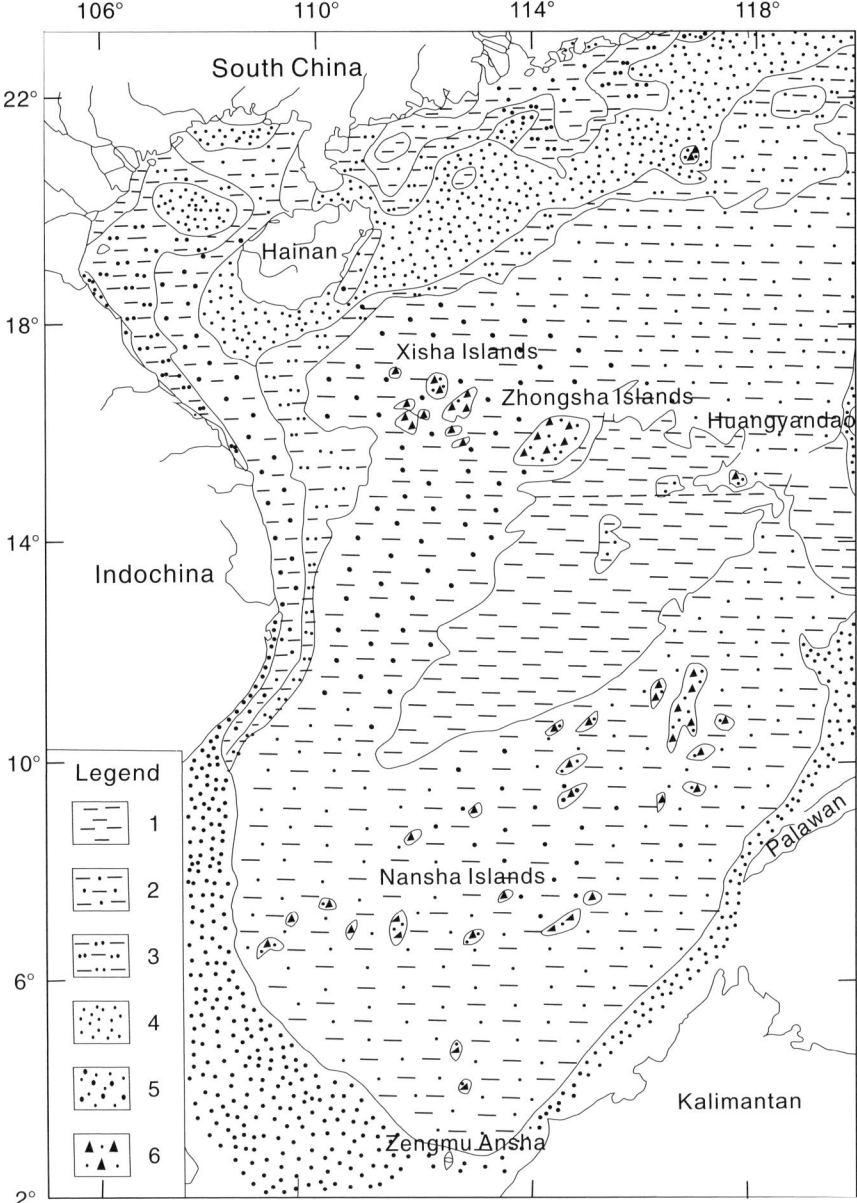

Fig. 4.2 Map shows grain-size distribution in surface sediments of the SCS (Su and Wang 1994). Sediment types are: (1) clay, (2) silty clay, (3) clayey silt, (4) sand, (5) gravel-containing sand, and (6) coral sand and gravel

northeastward through the Bashi (Luzon) Strait. Relative to province A, illite and chlorite in province B are reduced by 5–10%, whereas smectite and kaolinite increase by 6–8%. In province C, surrounding the Luzon islands, smectite increase to 23%, twice the amount in province B. Province D covers most of the Sunda

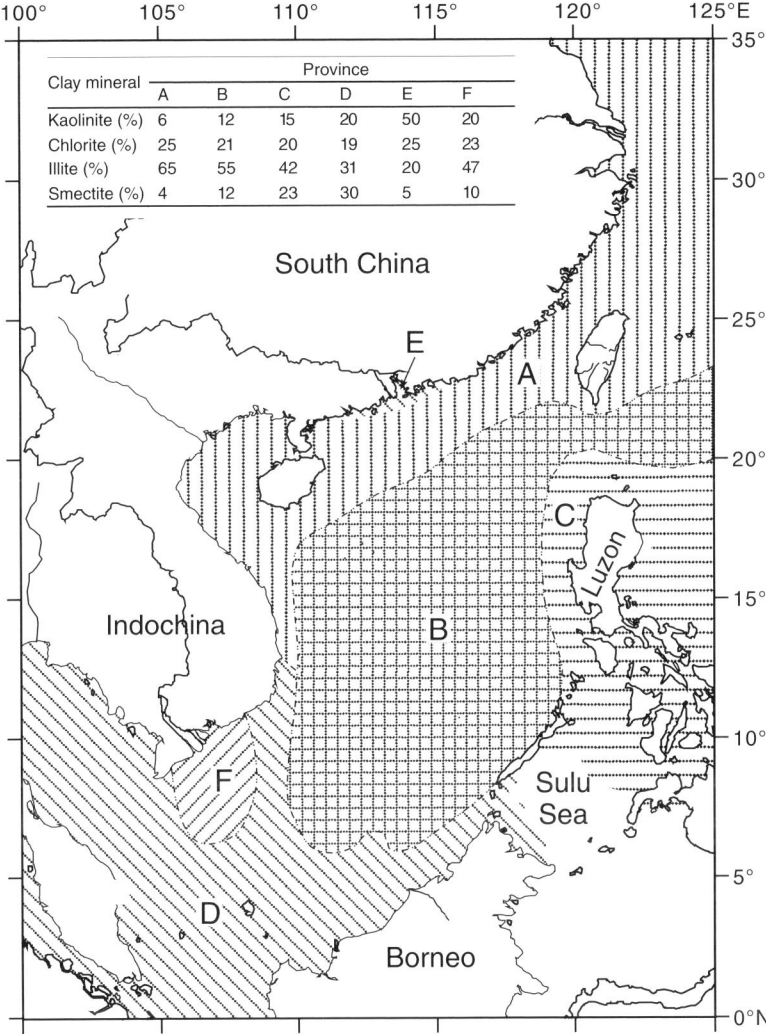

Fig. 4.3 Map shows clay mineral distribution in surface sediments of the SCS (modified after Chen P. 1978 and Liu Z. et al. 2003a)

shelf and extends to the Gulf of Thailand and the Malacca Strait. Smectite becomes very prominent in this province, ranging from 20% to 50%, with an average of 30%. Coastal province E, including the bay and inner estuary of the Pearl River, is characterized by high kaolinite (50%) which rapidly decreases slope-ward. Province F, the Mekong River estuary, is relatively rich in illite and chlorite, 47% and 23% respectively. The kaolinite content in provinces C, D, and F averages about 20%, but reaches 50% in the Malacca Strait.

Distributions of chemical elements in surface sediments of the SCS are closely correlated to the grain-size distribution. Here, we take two major chemical components of SiO_2 and $CaCO_3$ as examples (Figs. 4.4 and 4.5). SiO_2 is the major

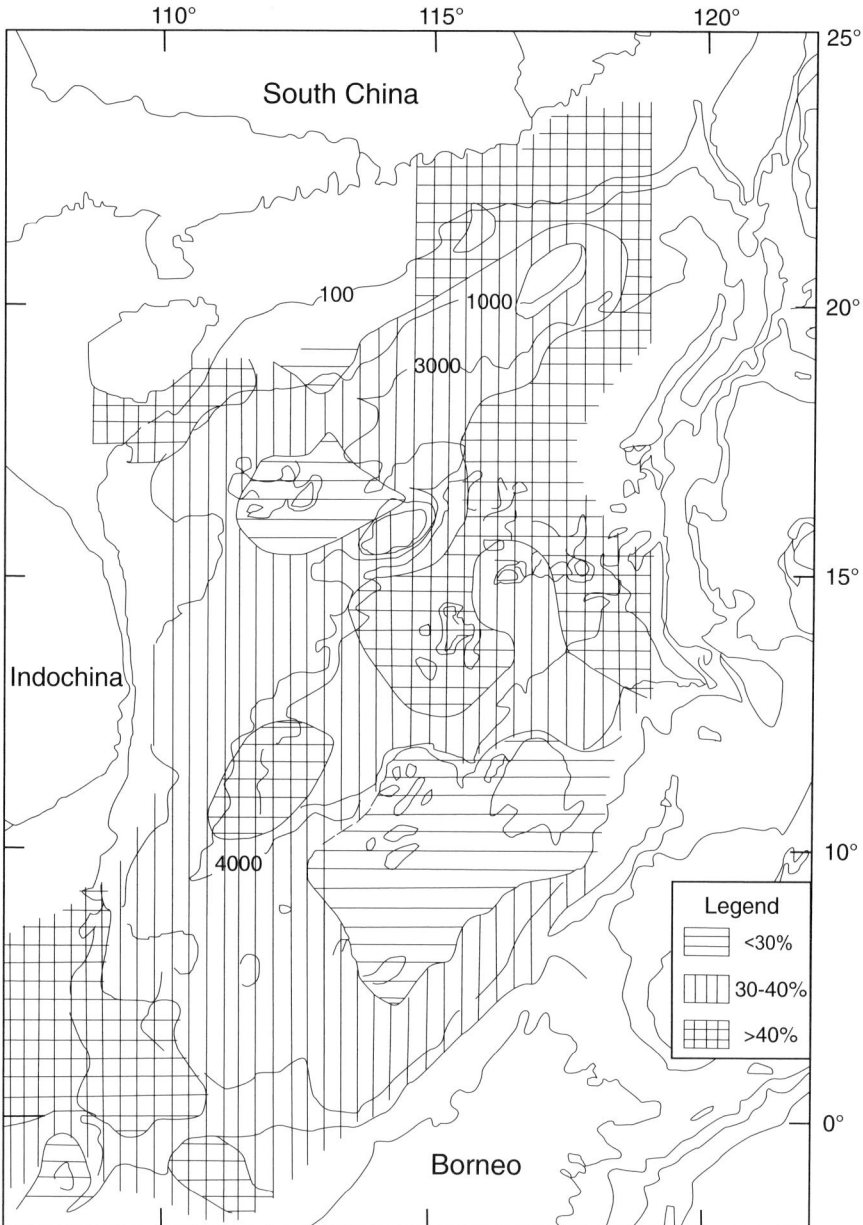

Fig. 4.4 Map shows distribution of SiO₂ abundance in surface sediments of the SCS, based on data from Su et al. (1989) and The Multidisciplinary Oceanographic Expedition Team of Academia Sinica to Nansha Islands (1993)

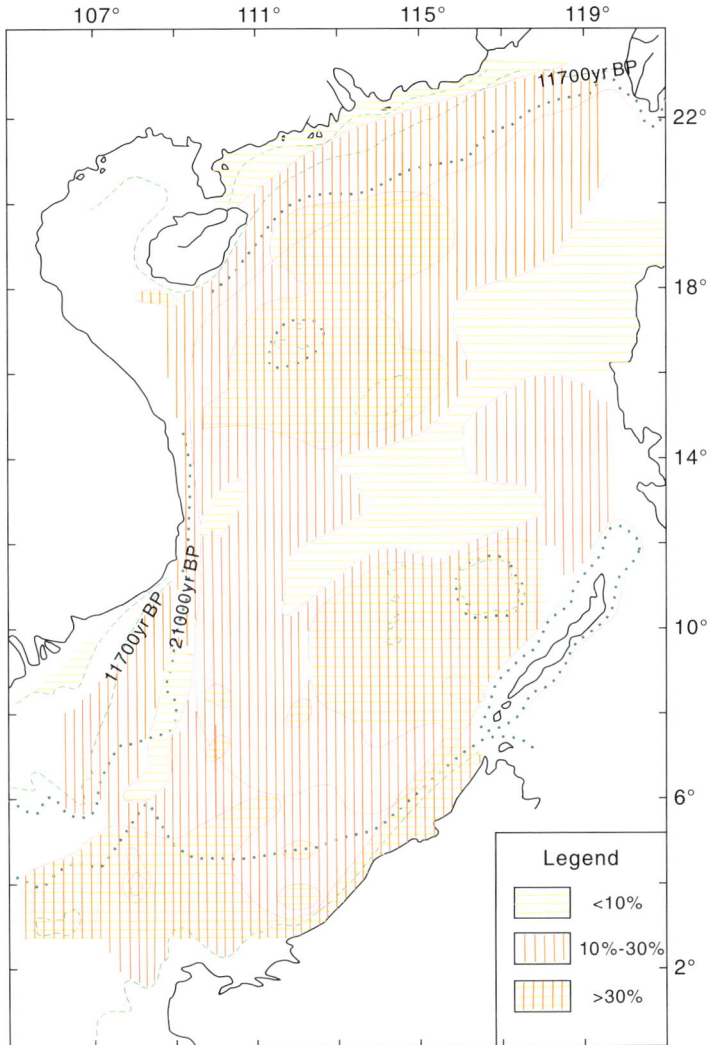

Fig. 4.5 Map shows distribution of carbonate content in surface sediments of the SCS (Su et al. 1989; The Multidisciplinary Oceanographic Expedition Team of Academia Sinica to Nansha Islands 1993; Li and Yang 1997)

component of terrigenous sediments and therefore is distributed throughout the surface SCS. The shallow shelf contains most abundant SiO_2, such as the abyssal basin, with major terrigenous sediments for the former and clay and radiolarian and other silicate organisms for the latter. On the hemi-pelagic slope, calcareous ooze is abundant and therefore SiO_2 is relatively low. Variations of the $CaCO_3$ content are generally opposite to the SiO_2 content. When $CaCO_3$ content is high, SiO_2 content becomes low, and vice versa. The highest $CaCO_3$ content is in the coral reef regions, e.g., the regions of the Dongsha, Xisha, and Nansha islands. On the slope of the northern and southern SCS, CaCO$_3$ contents generally exceed 30%. However, abyssal basin sediments contain less than 10% CaCO$_3$.

Sediment Transport

Sediment transport in the SCS is largely dependant on its circulation system (Chapter 2). It has been suggested that the surface circulation transported clay minerals throughout the SCS, by carrying more smectite to the north during interglacials and more illite and chlorite to the south during glacials (Liu Z. et al. 2003a). In addition, the surface circulation in the northern SCS is significantly affected by the flow of the Kuroshio Current through the Bashi Strait (Caruso et al. 2006). The current is one of most important carriers accounting for smectite contribution to the northern SCS from the Luzon Arc and the western Philippine Sea (Wan et al. 2007; Liu Z. et al. 2008).

The pioneering work on bottom sedimentation processes in the SCS was based on "sonar mapping and piston-core studies". Down-slope transporting processes, e.g., turbidity currents and mass wasting, were found to be predominant in the SCS basin (Damuth 1979, 1980). With technical progress in the recent years, sediment transport caused by bottom currents are widely reported in the SCS, and one examples is the high sedimentation-rate drifts on the slope southeast of the Dongsha Islands in the northern SCS (Fig. 4.6) (Lüdmann et al. 2001; Shao et al. 2001).

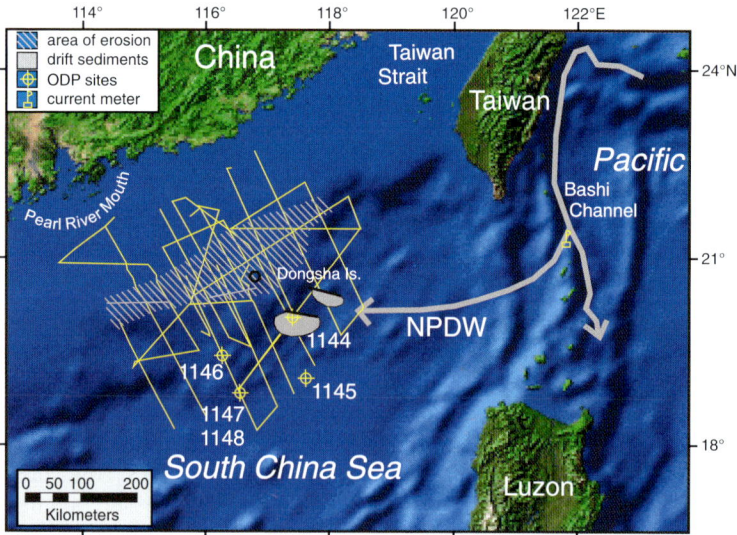

Fig. 4.6 Sediment drift near the Dongsha Islands in the northern SCS was probably caused by the intrusion of the North Pacific Deep Water (NPDW) through the Bashi Channel (arrows), according to Lüdmann et al. (2005)

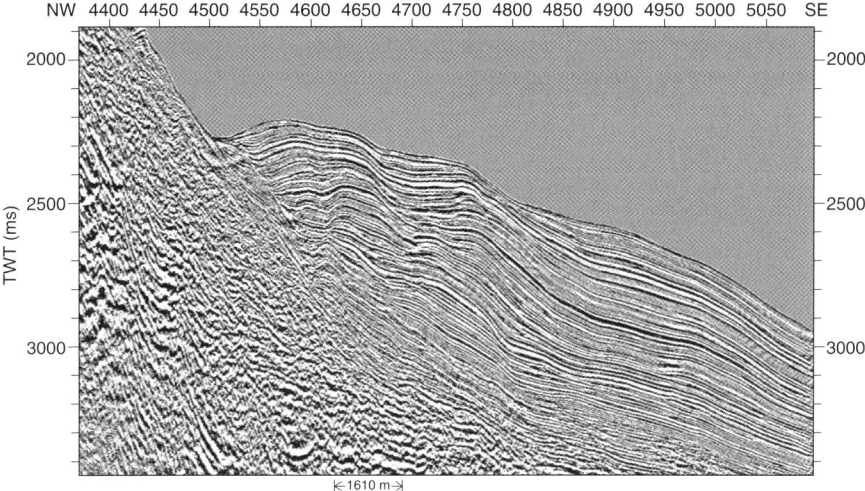

Fig. 4.7 Seismic profile of the southeastern slope of the Dongsha Islands shows the stacking pattern of layers in a sediment drift (Shao et al. 2007)

Multi-channel reflection seismic data indicate a lenticular sediment drift with very thick deposition existing on the southeast slope of the Dongsha Islands (Fig. 4.7). The sediment decreases in thickness SE-wards and finally vanishes into normal deep ocean depositional environment (Shao et al. 2007). No slumping structure, turbidity deposition, and other gravity-flow transports have been found in sediment cores of SO17940, ODP Site 1144, and MD05-2905 (Wang L. et al. 1999; Wang P. et al. 2000; Laj et al. 2005). Sedimentation rates in the sediment drift are very high, from 33 cm/ka at core SO17940 (Wang L. et al. 1999), to 49 cm/ka at Site 1144 (Bühring et al. 2004), to 97 cm/ka at core MD05-2905 (Laj et al. 2005). Lüdmann et al. (2005) postulated that these drift deposits were generated by upward directed eddies near the Dongsha Islands of the North Pacific Deep Water (NPDW), which enters the SCS via the Bashi Channel (sill depth >2500 m) (Fig. 4.6). This flow results in deposition of slope sediments resuspended off East and South Taiwan on the slope southeast of the Dongsha Islands. However, newly-acquired high-resolution seismic data revealed that the sediment drift is actually composed of a series of sediment waves migrating upslope (Fig. 4.8) (Zhong et al. 2007).

A recent study discovered a Pearl River deep-water fan system, located in the deep water slope area of Baiyun Sag in the Zhujiangkou (Pearl River Mouth) Basin (Fig. 4.9). With the rich sediments supplied by the Pearl River, bottom currents are suggested to have accounted for the sediment transport and formation of the paleo-fan system (Pang et al. 2006). Bottom currents in the SCS can also drive deep-water turbidity deposition. In the southern SCS, a highly graded foraminifera-rich turbidite in core MD05-2894 (07°02′N, 111°33′E, water depth 1982 m) originates from the Sunda slope (Fig. 4.10). The turbidite layer has an abrupt contact with the underlying clay and calcareous ooze, with grain sizes fining upwards.

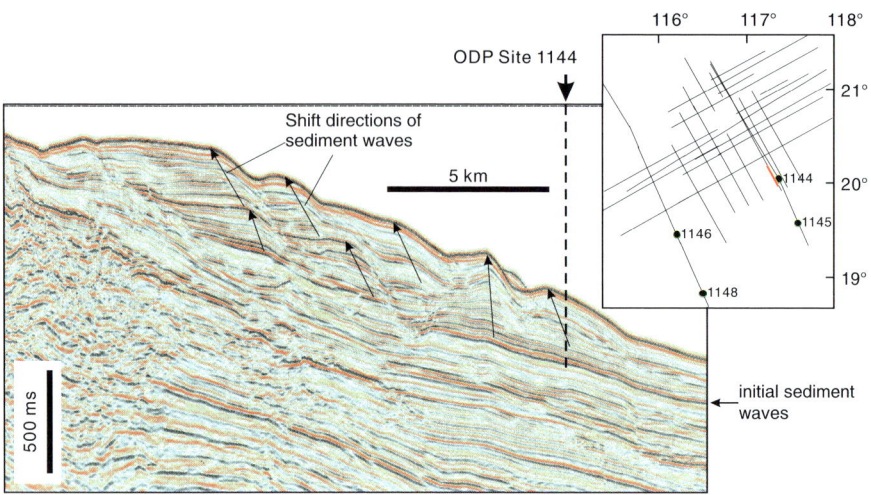

Fig. 4.8 The sediment drift on the slope near ODP Site 1144 was generated likely by sediment waves with sediments coming from Taiwan (Zhong et al. 2007)

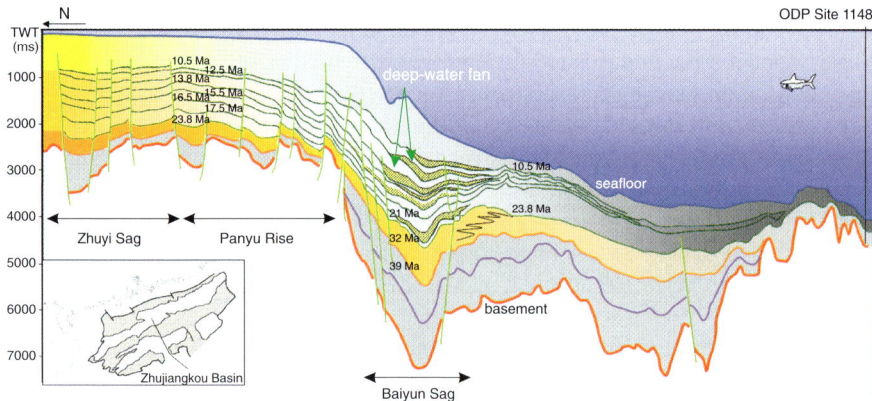

Fig. 4.9 Interpretation profile of sequence stratigraphy of the Zhujiangkou (Pear River Mouth) Basin reveals a deep-water fan system (Pang et al. 2006)

4.2 Terrigenous Deposition (Liu Z.)

Clay Mineralogy and Geochemistry of Source Areas

The primary control on mineralogical and geochemical variations of sediments in the modern and past SCS is exerted by different sediment provenances (Chen P. 1978; Liu Z. et al. 2004). Two main source areas with markedly different geological characteristics contribute terrigenous sediments to the SCS. The northern and western source is mainly the Asian continent and Taiwan, while the southern and eastern source consists of islands or volcanic arcs lying along the eastern margin of the SCS

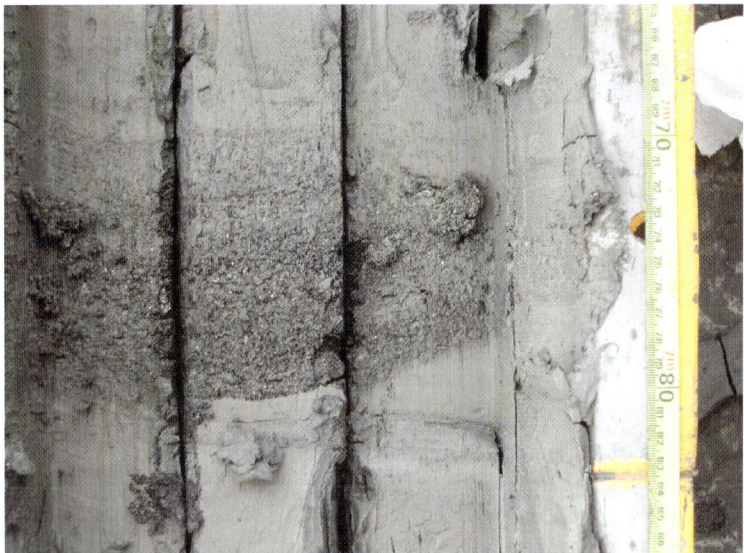

Fig. 4.10 A graded foraminifera-rich turbidite layer in core MD05-2894 shows a sharp lower boundary and fining upward grain sizes (Laj et al. 2005)

(Fig. 4.11). Weathering products from these land-source areas are transported to the SCS chiefly by larger rivers, mainly the Mekong River, Pearl River, and Red River, and small mountainous rivers especially those in southwestern Taiwan (Table 4.1). Here we characterize clay mineralogy and geochemistry of these source areas.

Clay Minerals

Detrital sediments in the northern SCS are mainly derived from the Pearl River and the Red River, as well as rivers in southern Taiwan and in Luzon (Philippines). In the Pearl River drainage basin, kaolinite (30–67%) is the most dominant clay mineral with an average of 46%, while chlorite (15–37%) and illite (6–40%) have lesser abundance with a similar average content of about 25%. Smectite (0–11%) is very scarce with an average content of about 3% (Table 4.2, Fig. 4.12) (Boulay et al. 2005; Liu Z. et al. 2007a,b). For the Red River sediments, illite (31–57%) is the dominant clay mineral, with an average content of 43%; kaolinite (17–38%) and chlorite (6–29%) are less abundant, with a similar average content of about 25%; smectite (1–14%) remains as a minor component, but its average 7% content is higher than in the Pearl River basin (Liu Z. et al. 2007a). In southwestern Taiwan, river and lake sediments contain high percentages of illite (44–66%) and chlorite (33–48%), with very rare kaolinite (0–4%) and smectite (0–8%). The average percentage is 55% for illite, 42% for chlorite, 2% for kaolinite, and 1% for smectite (Fig. 4.12) (Liu Z. et al. 2008). For Luzon rivers, the clay is predominantly smectite (83–92%, average 88%), with minor kaolinite (average 9%) and scarce chlorite and illite (Fig. 4.12).

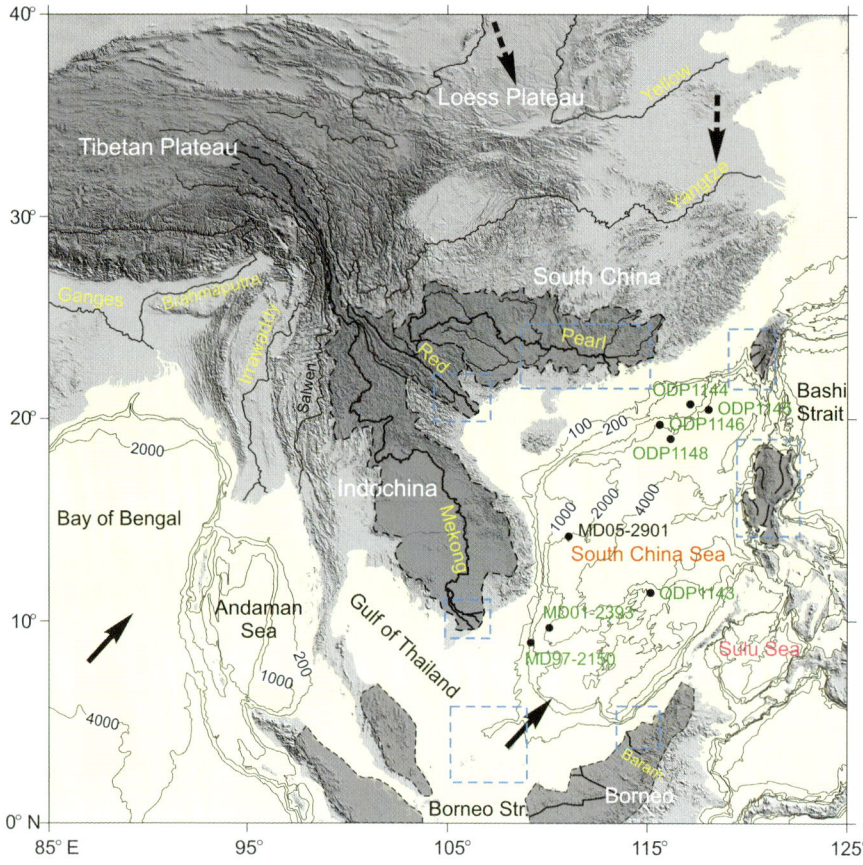

Fig. 4.11 Map shows major river drainage basins surrounding the SCS, with *dashed squares* outlining sediment source areas and dots indicating locations of IMAGES cores and ODP Leg 184 sites mentioned in this section. Also shown are prevailing wind flows during summer (*solid arrows*) and winter (*dashed arrows*)

In the southern SCS, the Mekong River, northern Borneo, and Indonesian volcanic arcs are the major suppliers of terrigenous sediments. Samples from the Mekong River delta have similar clay mineral assemblages between various locations: kaolinite (24–41%), illite (21–38%), and chlorite (21–30%) making up the dominant clay minerals with an average content of 28%, 35%, and 26%, respectively (Fig. 4.12) (Liu Z. et al. 2007a). Smectite (6–18%) is a minor clay mineral with an average percentage of 11%, a value considerably higher than the Pearl and Red records. In northern Borneo, Baram and Trusan rivers provide high contents of illite (63%), lesser abundant chlorite (24%) and kaolinite (12%), but no smectite (Liu Z. et al. 2007c). Surface sediments from the northern Sunda Shelf show a high content of smectite (average 40%) and moderate contents of kaolinite, illite, and chlorite (average ~20%, respectively) (recalculated from Jagodziński 2005). The Indonesian islands are characteristic in having abundant smectite (41–76%, average

Table 4.1 Drainage area, runoff, and suspended sediment discharge are listed for major rivers flowing directly into the SCS

River	Drainage area (km^2)	Runoff (m/yr)	Suspended sediment discharge (Mt/yr)	Data source
Pearl (South China)	440, 000	6.9	69.0	Milliman and Syvitski (1992)
Red (Vietnam)	120, 000	10.0	130.0	idem
Mekong (Vietnam)	790, 000	5.9	160.0	idem
Chao Phraya (Thailand)	160, 000	1.9	11.0	idem
Ta-An (Taiwan)	633	1.6	7.1	Dadson et al. (2003)
Wu (Taiwan)	1, 981	1.9	9.8	idem
Cho-Shui (Taiwan)	2, 989	1.2	54.1	idem
Pei-Kang (Taiwan)	597	1.3	2.2	idem
Pa-Chang (Taiwan)	441	1.5	6.3	idem
Tseng-Wen (Taiwan)	1, 157	1.1	25.1	idem
Erh-Jen (Taiwan)	175	1.8	30.2	idem
Kao-Ping (Taiwan)	3, 067	2.5	49.0	idem
Tung-Kang (Taiwan)	175	2.9	0.4	idem
Lin-Pien (Taiwan)	310	2.5	3.3	idem
Baram (Malaysia)	22, 800	2.5	12	Hiscott (2001), Lambiase et al. (2002)

54%) and kaolinite (17–41%, average 28%), frequent chlorite (6–24%, average 17%) and scarce illite (0–2%, average 1%) (Fig. 4.12) (recalculated from Gingele et al. 2001).

Significantly different clay mineral assemblages exist in surface sediments from these fluvial drainage basins, as shown by the illite+chlorite–kaolinite–smectite ternary diagrams (Fig. 4.13). Apparently, the differences in clay mineral distribution are related to the intensity of weathering, geology and weathering regime of the drainage basins (Chen P. 1978; Gingele et al. 2001; Liu Z. et al. 2007a).

In general, weathering intensity is mainly controlled by lithology, climate, and morphology and, in turn, determines the formation of clays (Chamley 1989). A simple scenario is that parent rocks with different lithological compositions in the upper reach of a river attribute to various clay mineral distributions in its lower drainage basin. This scenario, however, does not apply to most of the drainage basins surrounding the SCS. Although a large dissimilarity in their parent lithologies exists in individual river basin, the clay mineral assemblage is similar basin-wide (Liu Z. et al. 2007a, 2008), implying a primary influence of hydrolytic conditions and topography on clay formation. The reason for this basin-wide clay similarity appears to be upon the different hydrolytic weathering degrees occurring in each river basin.

Kaolinite is readily found in monosialitic soils, and therefore displays a strong climatic dependence controlled mainly by the intensity of continental hydrolysis (Chamley 1989). Kaolinite is also common on steep slopes within a drainage basin with relative good drainage conditions. There is much more kaolinite in the Pearl River sediments than in the sediments of other rivers. Warm and humid climate conditions combined with a low-relief, stable (cratonic) morphology in the Pearl River

Table 4.2 Average clay mineral assemblages are listed for various regions surrounding the SCS

Region	No. of samples	Chlorite (%)	Illite (%)	Smectite (%)	Kaolinite (%)	Illite chemistry	Illite crystal-linity (°$\Delta 2\theta$)	Data source
Pearl (S China)	37	25	26	3	46	0.62	0.30	Liu Z. et al. (2007a)
Red (Vietnam)	39	24	43	7	26	0.40	0.19	idem
Mekong (Vietnam)	17	26	35	11	28	0.47	0.21	idem
Taiwan	24	42	55	1	2	0.34	0.16	Liu Z. et al. (2008)
Luzon (Philippines)	7	2	1	88	9			idem
Borneo	5	24	63	0	12	0.61	0.33	Liu Z. et al. (2007c)
Sunda Shelf	22	19	20	40	21			Recalculated from Jagodziński (2005)
Indonesian islands	4	17	1	54	28			Recalculated from Gingele et al. (2001)

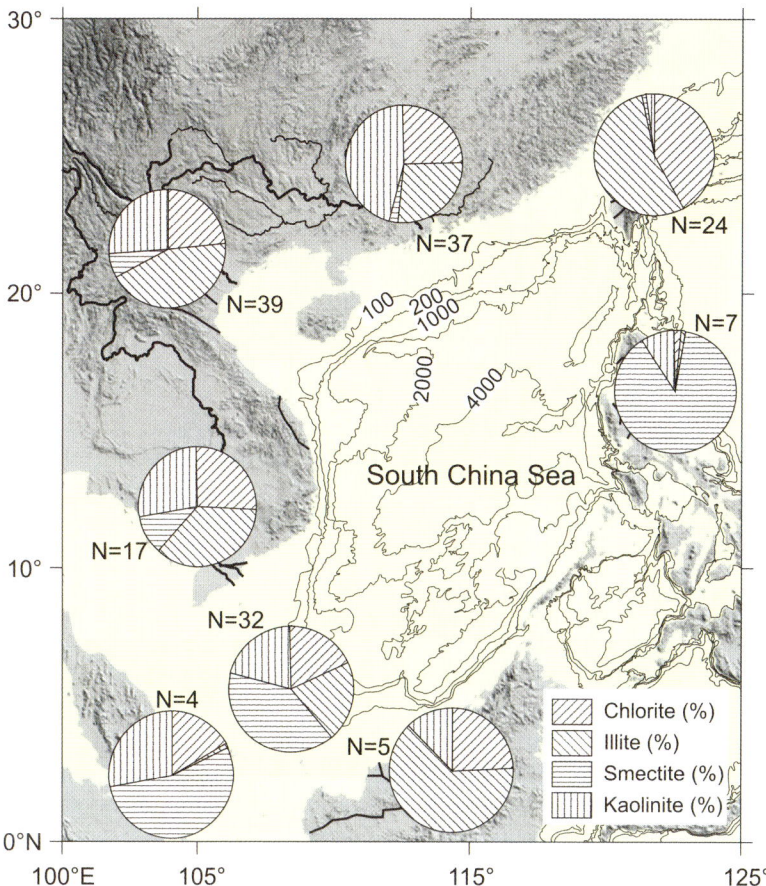

Fig. 4.12 The four major clay mineral assemblages with average percentage in various regions surrounding the SCS are shown using data from Table 4.2. N denotes number of samples

basin favour significant chemical weathering processes of monosialitization and alitization. Both granites in the east and sedimentary rocks in the entire basin produce a significant amount of kaolinite in modern clay mineral assemblages. Parent aluminosilicates associated with calcite and dolomite in the carbonate rocks in the west have undergone weathering processes similar to granitic and other sedimentary rocks in the eastern part of the Pearl River basin.

Illite is the most abundant clay mineral in the southwestern Taiwan and northern Borneo drainage basins, and in the Red and Mekong drainage basins as well, compared to the Pearl River sediments. Illite is considered as a primary mineral, which reflects decreased hydrolytic processes in continental weathering and increased direct rock erosion under cold and arid climatic conditions. In our various river basins, illite could have been derived from physical erosion of metamorphic and granitic parent rocks, mainly located at mountain belts of Taiwan and Borneo and high elevation in eastern Tibet. Or it may have been formed by weathering of

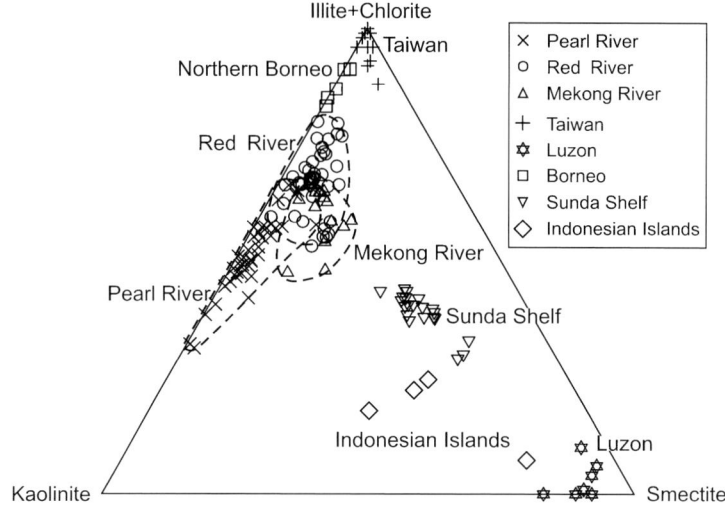

Fig. 4.13 Ternary diagram displays the relative proportions of kaolinite, smectite and illite+chlorite components in surface sediments from various regions surrounding the SCS

nonlayer silicate, such as feldspar from granites under moderate hydrolysis conditions, and by degradation of micas (mainly in Taiwan). Although warm and humid climate conditions in the middle-lower reaches of Taiwan, Borneo and Indochina are similar to the Pearl River basin, these river basins have a different history of tectonic activities and river incision, as well as humid and cold climate in their headwaters, which may have significantly increased physical erosion and decreased hydrolytic weathering processes as bisialitization and monosialitization. The major and trace element geochemistry indicates only a moderate chemical weathering in Taiwan due mainly to extremely high erosion rates triggered by earthquake and storm activities (Selvaraj and Chen 2006). These weathering processes produce the high content of illite (primary mineral) and minor or reasonable amount of kaolinite and smectite (secondary minerals) as found in samples analyzed.

Smectite is a secondary mineral, which is derived from chemical weathering of parent aluminosilicate and ferromagnesian silicate under warm and humid conditions. Smectite could be also produced by chemical weathering of extrusive basaltic rocks in various climatic conditions (Griffin et al. 1968). Smectite is formed in confined environments by recombination of released cations. In Luzon (Philippines) and the southern Sunda Shelf, volcanic activities and basaltic rocks are widely distributed. Their weathering must have produced very high contents of smectite, as indicated from the results of clay mineral assemblages (Fig. 4.12) (Liu Z. et al. 2008). In the Mekong drainage basin, widely distributed bisiallitic soil in the middle and lower reaches is a potential source for smectite in the delta (Liu Z. et al. 2004). Distribution of bisiallitic soil in the Pearl and Red river basins, however, is very limited (Ségalen 1995). But the minor extrusive igneous rocks exposed along the Red River fault zone and several large Neogene basalt bodies outcropped in the lower Mekong River basin do not produce a significant amount of smectite.

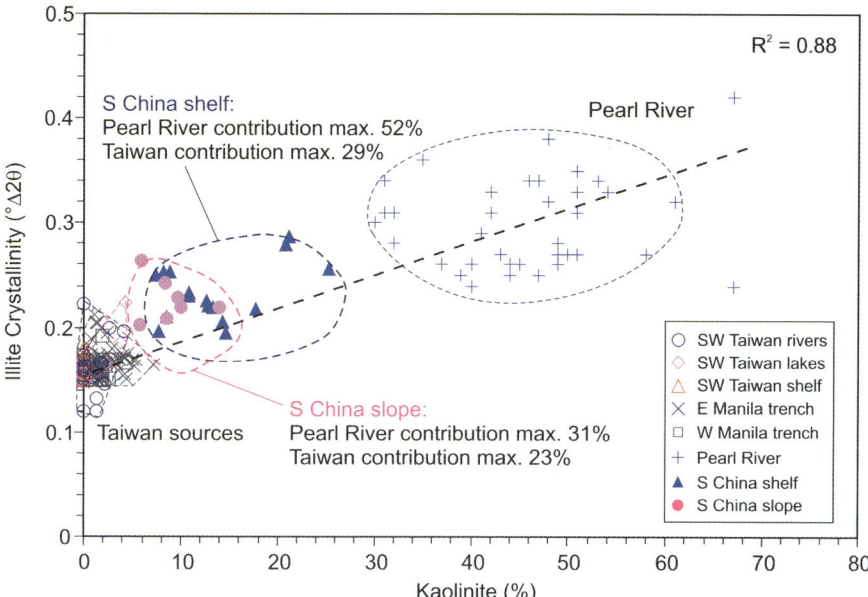

Fig. 4.14 Illite crystallinity and kaolinite (%) in surface sediments from southwestern Taiwan, the Pearl River, and the northern SCS shows a good linear correlation, as indicated by coefficient $R^2 = 0.88$ (*coarse dash line*) (Liu Z. et al. 2008)

Therefore, the similar clay mineral assemblages basin-wide in all of these river basins suggest that the difference in clay mineral distribution can be attributed to the intensity of chemical weathering, tectonic activities, and tectonic-induced river incision, rather than their dissimilarity in parent lithology. The clay mineralogy of sediments hence can be used to distinguish terrigenous sediments from land-source areas. Taking the northern SCS surface sediments as an example and assuming that kaolinite is provided completely from the Pearl River, a linear correlation of illite crystallinity with kaolinite (%) is observed for all surface sediments, with two end-members of the Pearl River and Taiwan sources and transitional South China shelf and slope sediments (Fig. 4.14). A semi-quantitative evaluation indicates that the maximum contribution of clay minerals from Taiwan is 29% to the South China shelf and 23% to the South China slope. Accordingly, the maximum contribution of clay minerals from the Pearl River to the South China shelf and slope is 52% and 31%, respectively. The Luzon Arc accounts for the rest of clay mineral components in the northern SCS mainly by providing smectite (Fig. 4.15) (Liu Z. et al. 2008).

Geochemistry

Geochemical studies have been carried out on drainage basins of the Pearl, Red, and Mekong rivers, and southwestern Taiwan (Selvaraj and Chen 2006; Liu Z. et al. 2007a). The bulk carbonate-free surface sediments from the Pearl, Red, and Mekong basins are characterized by high contents of SiO_2, Al_2O_3, Fe_2O_3, and K_2O, and by

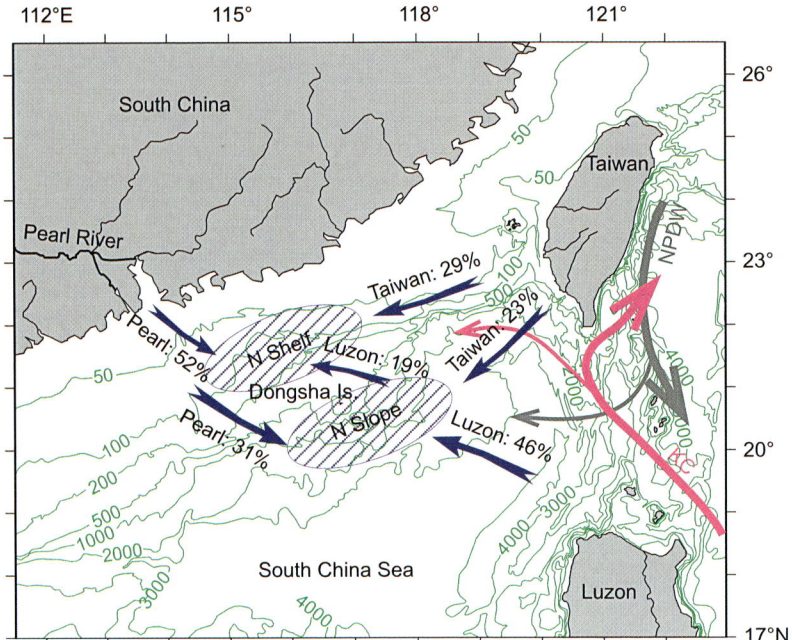

Fig. 4.15 Detrital sediment contributions to the northern SCS from Taiwan and the Pearl River were semi-quantitatively estimated (Liu Z. et al. 2008). Also shown are the North Pacific Deep Water (NPDW) (after Lüdmann et al. 2005) and the surface Kuroshio Current (KC) (after Caruso et al. 2006)

low concentrations of MgO, TiO$_2$, CaO, Na$_2$O, P$_2$O$_5$, and MnO (Fig. 4.16) (Liu Z. et al. 2007a). A moderate to strong negative correlation between SiO$_2$ and Al$_2$O$_3$ and Fe$_2$O$_3$ indicates potential grain-size control on SiO$_2$ content because quartz-rich mineral associations often produce a higher SiO$_2$ concentration. There is a great dispersion in diagrams of K$_2$O versus SiO$_2$ + Al$_2$O$_3$. K$_2$O/Al$_2$O$_3$ ratios are low in smectite and kaolinite, and high in minerals such as illite, muscovite and biotite. Therefore, the major element results imply significant differences in mineralogical compositions between the three rivers.

The degree of chemical weathering can be estimated using the chemical index of alteration (CIA) (Nesbitt and Young 1982). For primary minerals (non altered minerals), all feldspars have CIA value of 50 and the mafic minerals biotite, hornblende, and pyroxenes have CIA values of 50–55, 10–30, and 0–10, respectively. Feldspar and mica weathering to smectite and kaolinite result in a net loss of K and Na in weathering profiles, whereas Al is resistant and therefore enriched in weathering products (Nesbitt and Young 1982). This induces an increase of CIA values by about 100 for kaolinite and 70–85 for smectite. The CIA value is thought to quantify the state of chemical weathering of the rocks by referring to the loss of labile elements such as Na, Ca, and K. CIA values are 75–88 for the Pearl River sediments, 72–83 for the Mekong River sediments, 65–78 for the Red River sediments, and 66–77 for Taiwan samples (Fig. 4.17). Such CIA values are in agreement with the clay

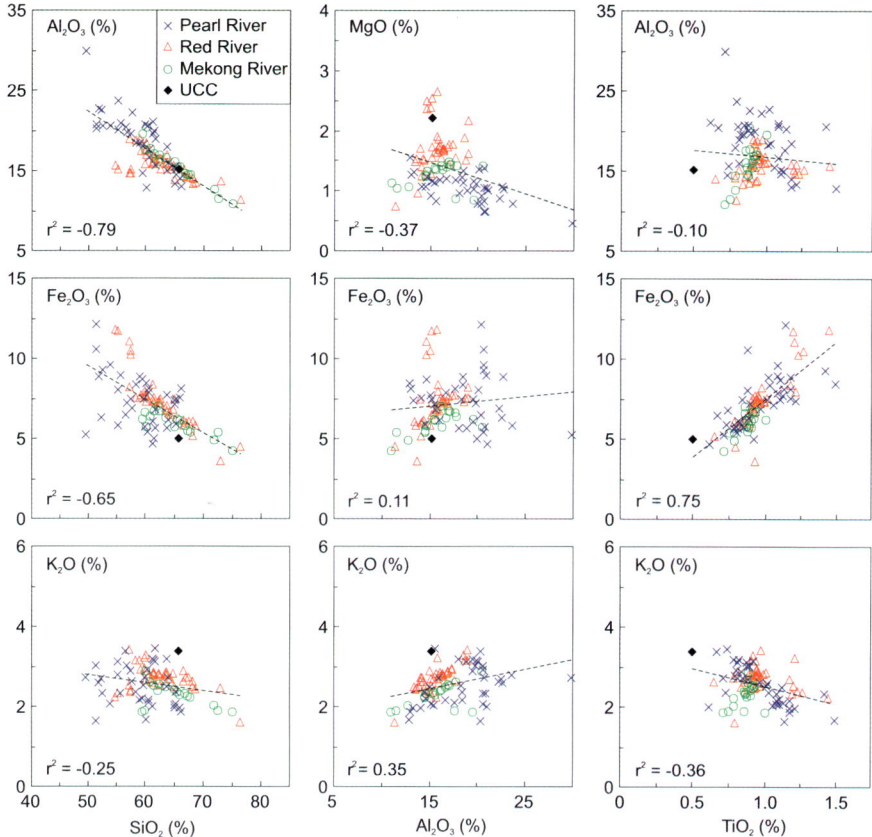

Fig. 4.16 Variation diagrams of selected major elements in the Pearl, Red, and Mekong river basins were plotted against SiO₂, Al₂O₃, and TiO₂, with upper continental crust data (UCC, filled diamond) (Taylor and McLennan 1985) as a reference (Liu Z. et al. 2007a). *Dashed lines* indicate linear correlations marked by a coefficient (r^2)

mineralogical results, which indicate higher kaolinite/illite ratio for the Pearl River sediments and lower values for the Red River and Taiwan sediments.

The CIA values of these river sediments are well correlated with the $K_2O/(Na_2O + CaO)$ molar ratio from river samples (Fig. 4.17A). The range of the $K_2O/(Na_2O + CaO)$ molar ratio strongly suggests mineralogic changes between an end member rich in plagioclase (high content of Na and Ca) and another rich in K-feldspar, micas and illite. Increase in the $K_2O/(Na_2O + CaO)$ molar ratio is associated with CIA increase, suggesting a preferential hydrolysis of plagioclase (enrichment of Na) relative to K-feldspar and micas (enrichment in K) during the silicate weathering process of river sediments. The much higher mobility in elements Ca and Na than in element K from all the three river basins and more plagioclase in Red and Mekong river samples than in Pearl River samples by bulk XRD results support this hypothesis. A linear correlation of CIA with the illite chemical index (Fig. 4.17B) also well demonstrate that the degree of chemical weathering

Fig. 4.17 Comparison between chemical index of alteration (CIA) and (**A**) K$_2$O/(Na$_2$O + CaO) molar ratio and (**B**) illite chemistry index in surface sediments of the Pearl, Red, and Mekong river basins shows linear correlations of mineralogical and element geochemical proxies of chemical weathering (Liu Z. et al. 2007a). Also plotted for comparison are their mean values in clayey sediments from the Amazon River (Vital and Stattegger 2000), the Yellow River and the Yangtze River (Yang et al. 2004)

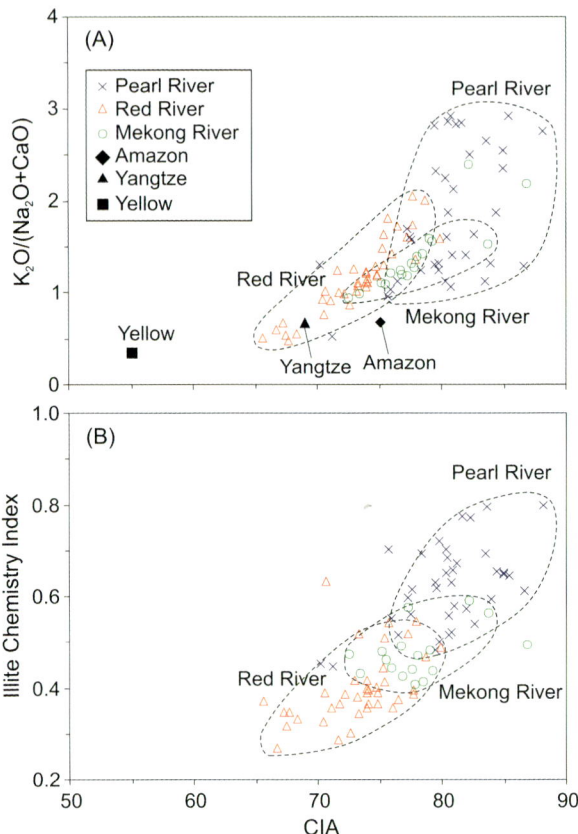

is strongest in the Pearl River basin, moderate in the Mekong River basin, and relatively weak in the Red River basin. In comparison, chemical weathering in the Amazon River (Vital and Stattegger 2000), the Yangtze River and the Yellow River (Yang et al. 2004) are much weaker (Fig. 4.17A,B).

The weathering intensity often influences the Sr isotopic composition in river sediments. On the contrary, neither chemical weathering on land (Borg and Banner 1996) nor grain-size sorting during transport (Tütken et al. 2002; Goldstein and Jacobsen 1988) can significantly modify εNd(0) values in sediments subsequently deposited in river basins. The confinement of all εNd(0) values from the Pearl, Red, and Mekong rivers to a narrow range also conveys a similar message (Fig. 4.18). Because all the Mekong samples come from the delta region and the features of their isotopically distinct sources (if there are any) must have been mixed during the confluence, their Nd isotope range considerably narrower than those from the Pearl and the Red is thus reasonable. As εNd(0) values are independent from the weathering process, therefore, they can be used as a reference to petrological settings. For each individual river, no relationship between the εNd(0) values and the ^{87}Sr/^{86}Sr ratios has been observed, suggesting that variations in Sr isotopic composition are not caused by major changes of petrology in the source region.

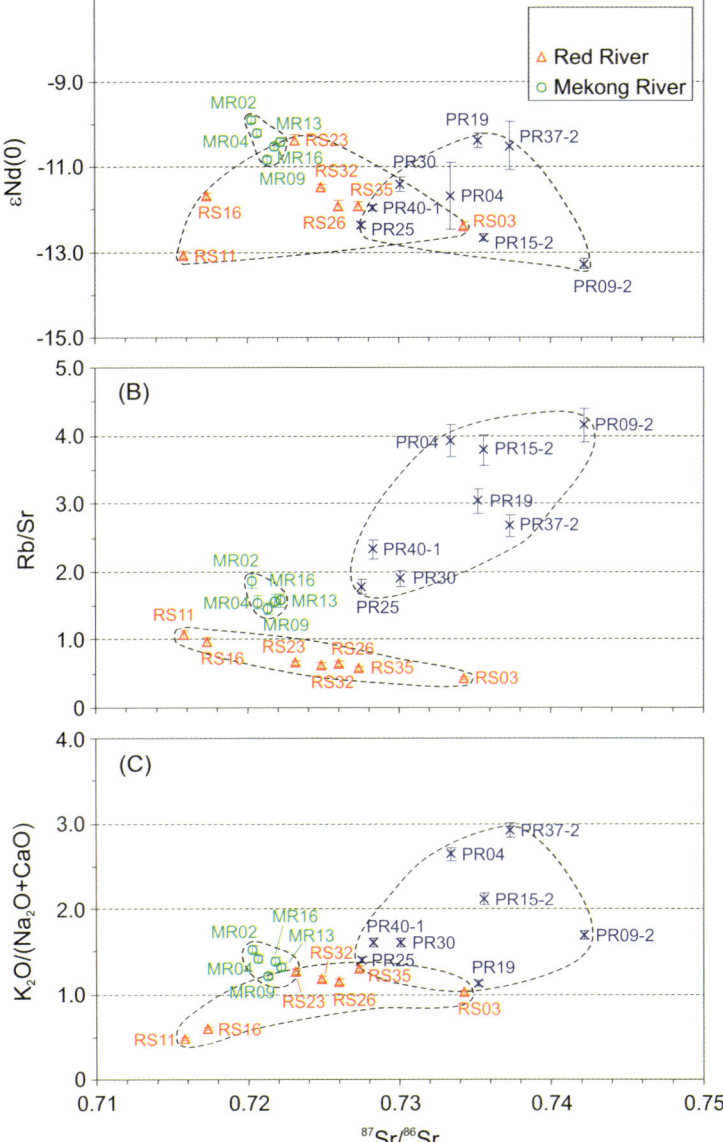

Fig. 4.18 Correlations between (**A**) $^{87}Sr/^{86}Sr$ and $\varepsilon Nd(0)$, (**B**) Rb/Sr, and (**C**) $K_2O/(Na_2O + CaO)$ molar ratio reveal their distribution patterns in argillaceous sediments from the Pearl, Red, and Mekong basins (Liu Z. et al. 2007a). Error bars refer to the measurement uncertainty

A good correlation between $^{87}Sr/^{86}Sr$ and Rb/Sr ratio and $K_2O/(Na_2O + CaO)$ molar ratio for the three rivers (Fig. 4.18B,C), however, denotes a major control on the $^{87}Sr/^{86}Sr$ ratio by Rb/Sr and mineralogical composition in their sediments. High $^{87}Sr/^{86}Sr$ ratios are associated to high $K_2O/(Na_2O + CaO)$ molar ratios, implying the presence of high-Rb, low-Sr minerals such as potassium feldspar and biotite in the river sediments. In contrast, low $^{87}Sr/^{86}Sr$ ratios are associated to low $K_2O/(Na_2O+CaO)$ molar ratios, implying the presence of low-Rb, high-Sr minerals such as plagioclase (rich in Na and Ca). As the Sr concentration in plagioclase is about 10 times higher than biotite, a low proportion of this mineral can strongly decrease the $^{87}Sr/^{86}Sr$ ratios of the river sediments. Because the $K_2O/(Na_2O+CaO)$ molar ratio is also correlated to other indices of chemical weathering (mineralogical proxies and CIA), $^{87}Sr/^{86}Sr$ ratios in river sediments are likely controlled mainly by the state of chemical weathering but not by the presence of basic rocks (such as in the Mekong and Red rivers). Strong physical erosion in the Red River and to a less extent in the Mekong River would produce sediments with high contents of unaltered plagioclase that induce the nonradiogenic Sr composition.

Terrigenous Sediment Supply in Glacial Cycles

The terrigenous sediment supply in the SCS was strongly affected by interglacial/ glacial variations. During interglacial periods, the coastline is approximately similar to present and southwesterly surface currents driven by summer-monsoon winds prevailed, while during glacial periods, the coastline followed the approximate position of the present-day 100 m isobath and the counter-clockwise surface circulation driven by strengthened winter-monsoon winds dominated the semi-enclosed SCS (Wang P. 1995; Wang P. et al. 1995; Huang et al. 1997a) (Fig. 4.11). Here we analyze terrigenous sediment variations in some well-documented records from late Quaternary glacial cycles.

Terrigenous Input in the Northern SCS

Glacial-cyclic terrigenous input is explicit in many sediment cores from the northern SCS, including SONNE95-17940, ODP Sites 1144, 1145, and 1146 (Fig. 4.11). Over the last million years, a strong response in terrigenous grain size and mass accumulation rates to glacial-interglacial cycles has been recorded in all published works, which show coarser grains and higher accumulation rates during glacials, and finer grains and lower accumulation rates during interglacials (Fig. 4.19) (Wang L. et al. 1999; Boulay et al. 2007). The power density spectrum presents significant 100 kyr periodicities due to the influence of the Earth's eccentricity orbit (Boulay et al. 2007), suggesting sea-level related variations for the terrigenous input.

Among the terrigenous components in the SCS, fine fractions mainly as clay minerals are most dominant (Chen P. 1978; Liu Z. et al. 2004). The clay mineral concent results from Site 1146 indicate strong glacial-interglacial cyclicities throughout the past 2 myr (Liu Z. et al. 2003a) (Fig. 4.20). Generally, illite (22–43%) and chlorite (10–30%) show a similar cyclic pattern, with high values reaching 35–43% and 20–30% respectively during glacials. On the contrary, smectite (12–48%) shows a variation range of over 30% with high values during interglacials. The clay mineral ratio, smectite/(illite+chlorite), varies from 0.3 to 1.8, with an average of 0.9. The ratio mean reaches a maximum in the 1,200–400 ka interval. Overall, the record

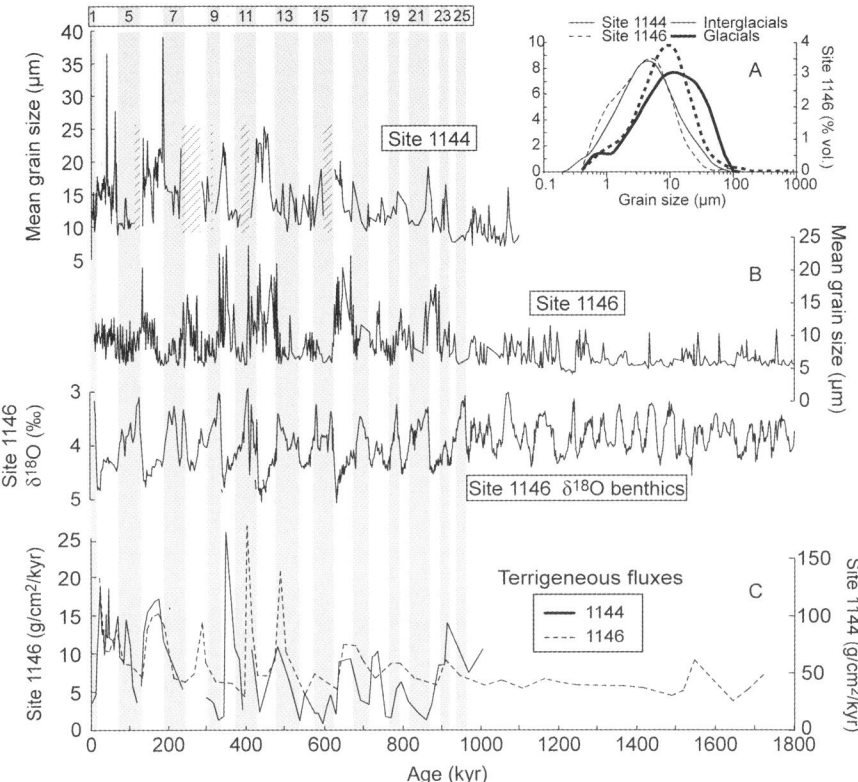

Fig. 4.19 Last 1.8 myr terrigenous grain-size and mass accumulation rate variations recorded at ODP Sites 1144 and 1146 show strong fluxes mainly in interglacials (*shaded*) before 400 ka and mainly in glacials after 400 ka (after Boulay et al. 2007)

of clay mineral ratio is strongly correlated with the oxygen isotope record (higher values during interglacials) prior to about 1,000 ka, moderately from 1,000 ka to 400 ka, and poorly after 400 ka.

However, clay minerals at Site 1145 do not present significant glacial cycles (Boulay et al. 2005). The clay mineral assemblages are dominated by illite (24–44%) and smectite (20–58%), with lesser abundance of chlorite (13–27%) and kaolinite (3–14%) (Fig. 4.21). The smectite/(illite+chlorite) ratio does not change with glacial/interglacial oscillations but varies with the September solar insolation curve calculated for 20 °N of latitude (Fig. 4.22) (Boulay et al. 2005). Each insolation maximum corresponds to an increase in the smectite/(illite+chlorite) ratio, suggesting that summer monsoon intensity has a direct control on the clay mineral composition. Both signals are roughly in phase, indicating rapid response of contineneal weathering and clay mineral deposition to summer monsoon forcing. The $^{87}Sr/^{86}Sr$ and εNd(0) isotopic data, combined with the smectite/(illite+chlorite) ratio, indicate that the Pearl River is the main contributor of detrital material to the site, with variable continental input of volcanic material derived from the erosion of the Luzon

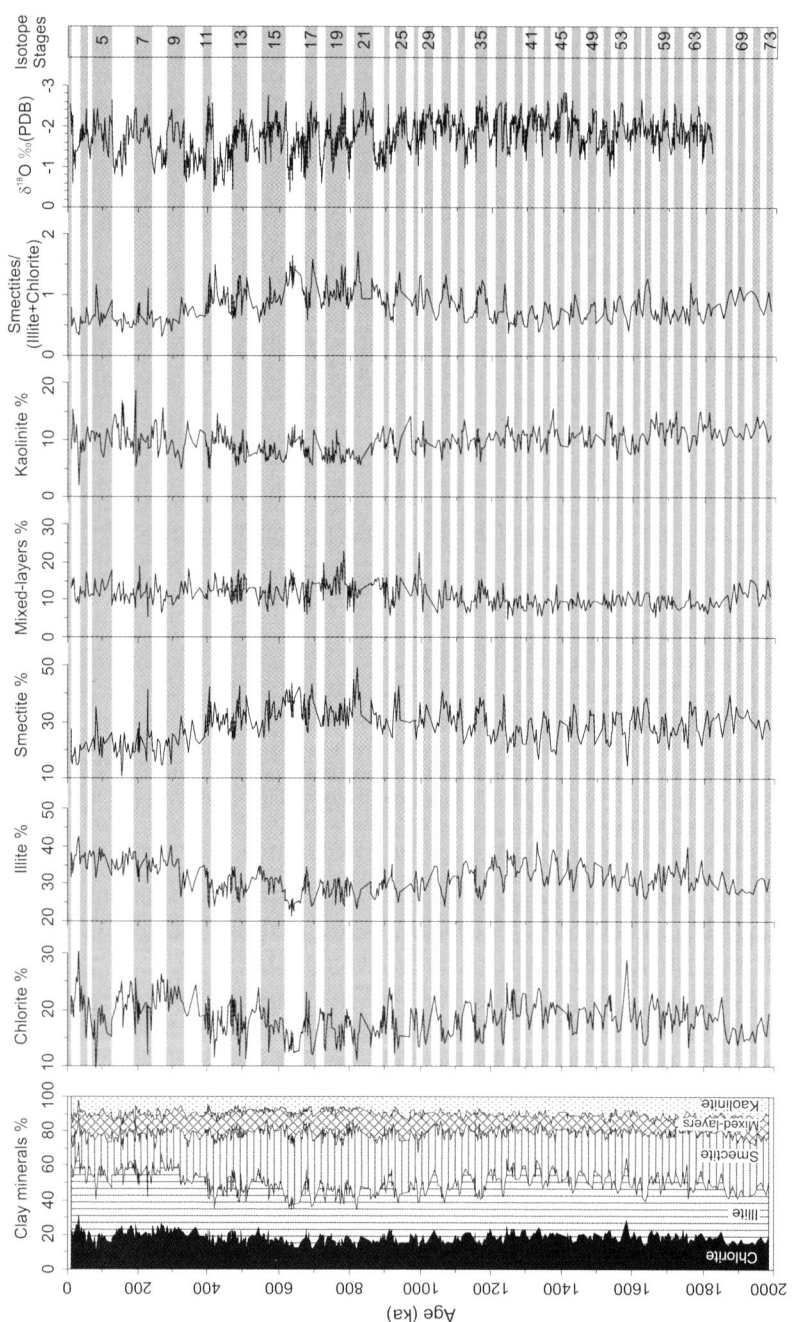

Fig. 4.20 Variations of clay mineral assemblages of ODP Site 1146 over the past 2 myr show a close relationship with glacial cycles (modified from Liu Z. et al. 2003a)

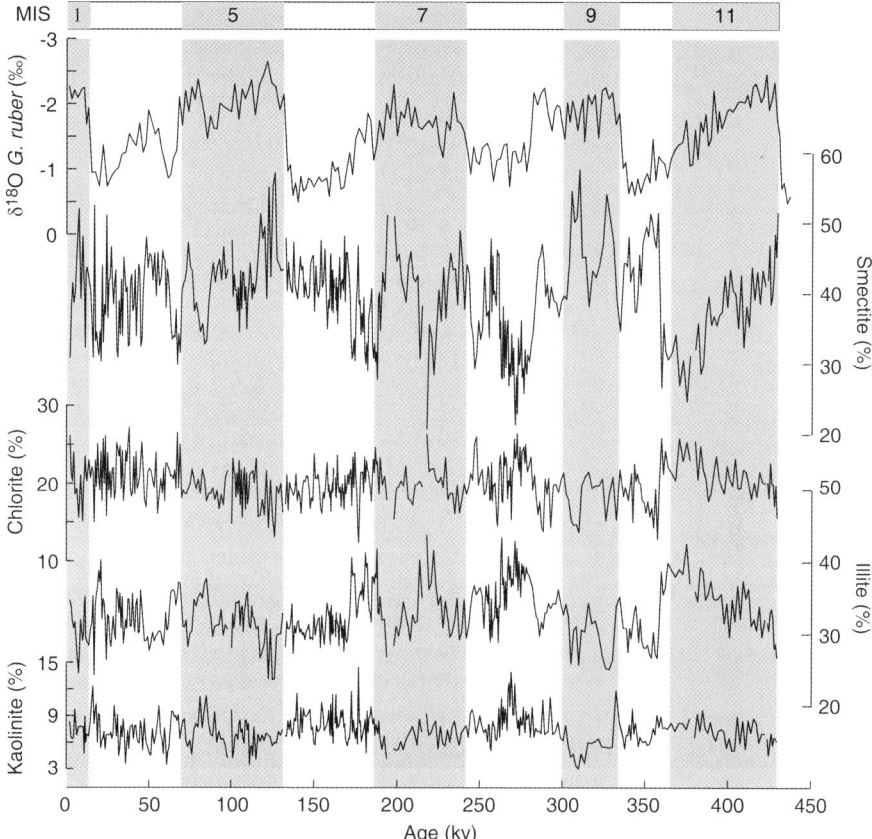

Fig. 4.21 Variations of clay mineral assemblages from ODP Site 1145 show very weak correlation with glacial cycles over the past 450 kyr (from Boulay et al. 2005)

Arc. These input patterns follow the low-latitude solar insolation with a 23 kyr periodicity, as recorded by a periodic change of the clay mineralogy (Fig. 4.22).

Debates exist on provenances of terrigenous materials in the northern SCS not only for the material balance between different drainage basins but also for fluvial or eolian inputs. The northern SCS receives terrigenous materials mainly from the Pearl River in South China, from southwestern Taiwan, and from Luzon (Fig. 4.1). Previous studies of Sites 1144, 1145, and 1148 suggested that clay minerals were mainly derived from the Pearl River on the basis of elemental geochemistry and mineralogy (Clift et al. 2002; Wehausen and Brumsack 2002; Tamburini et al. 2003; Li X. et al. 2003; Boulay et al. 2003, 2005). However, based on high kaolinite contents (>50%) in sediments of the Lingdingyang in the Pearl River estuary, Liu Z. et al. (2003a) concluded that clay minerals at Site 1146 derive mainly from Taiwan, a conclusion supported also by a revisit of Site 1146 material (Wan et al. 2007). As discussed in the previous section, among the total clay fractions in surface sediments of the northeastern SCS, 31% were derived from the Pearl River, 23% from Taiwan, and 46% from Luzon (Fig. 4.5) (Liu Z. et al. 2008).

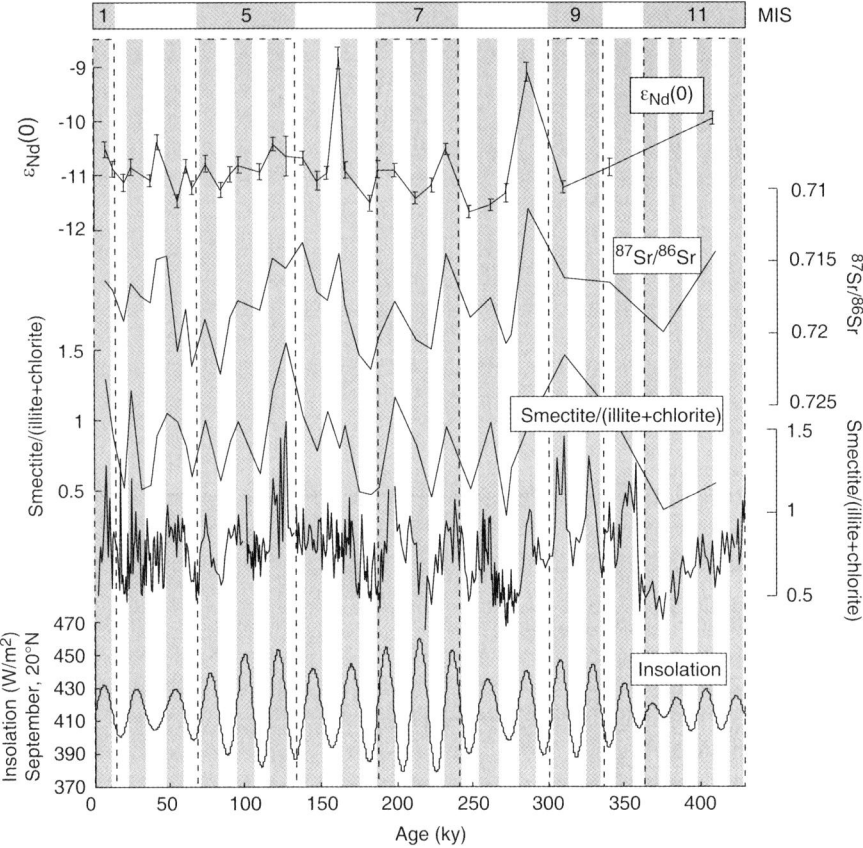

Fig. 4.22 εNd(0) values, $^{87}Sr/^{86}Sr$ ratio and smectite/(illite+chlorite) ratio measured on the carbonate-free fraction of ODP Site 1145 sediments over the last 450 kyr are compared to the insolation curve calculated for September at 20 °N (after Boulay et al. 2005)

Terrigenous Input in the Southern SCS

The Mekong River is the largest terrigenous source in the southern SCS (Fig. 4.1) and must have played a significant role in terrigenous input. Studies of two IMAGES cores, MD01-2393 and MD97-2150, offshore from the Mekong River have shed some insights on its role during the late Quaternary (Liu Z. et al. 2004, 2005).

Core MD01-2393 (10°30.15′N, 110°03.68′E, 1230 m water depth) provides a continuous sediment record down to MIS 6 (about 190 ka). In the <2−μm size fraction, illite (21–40%) and smectite (22–58%) are the dominant clay minerals (Fig. 4.23). Clay minerals present in lesser abundance include chlorite (10–25%) and kaolinite (11–25%). Bulk kaolinite proportions are 6–18% in detrital minerals and bulk quartz contents vary between 11% and 24%, with an average of 16%. These mineral distributions indicate strong glacial-interglacial cyclicity. In general, illite, chlorite and kaolinite concentrations are similar and inversely correlated with

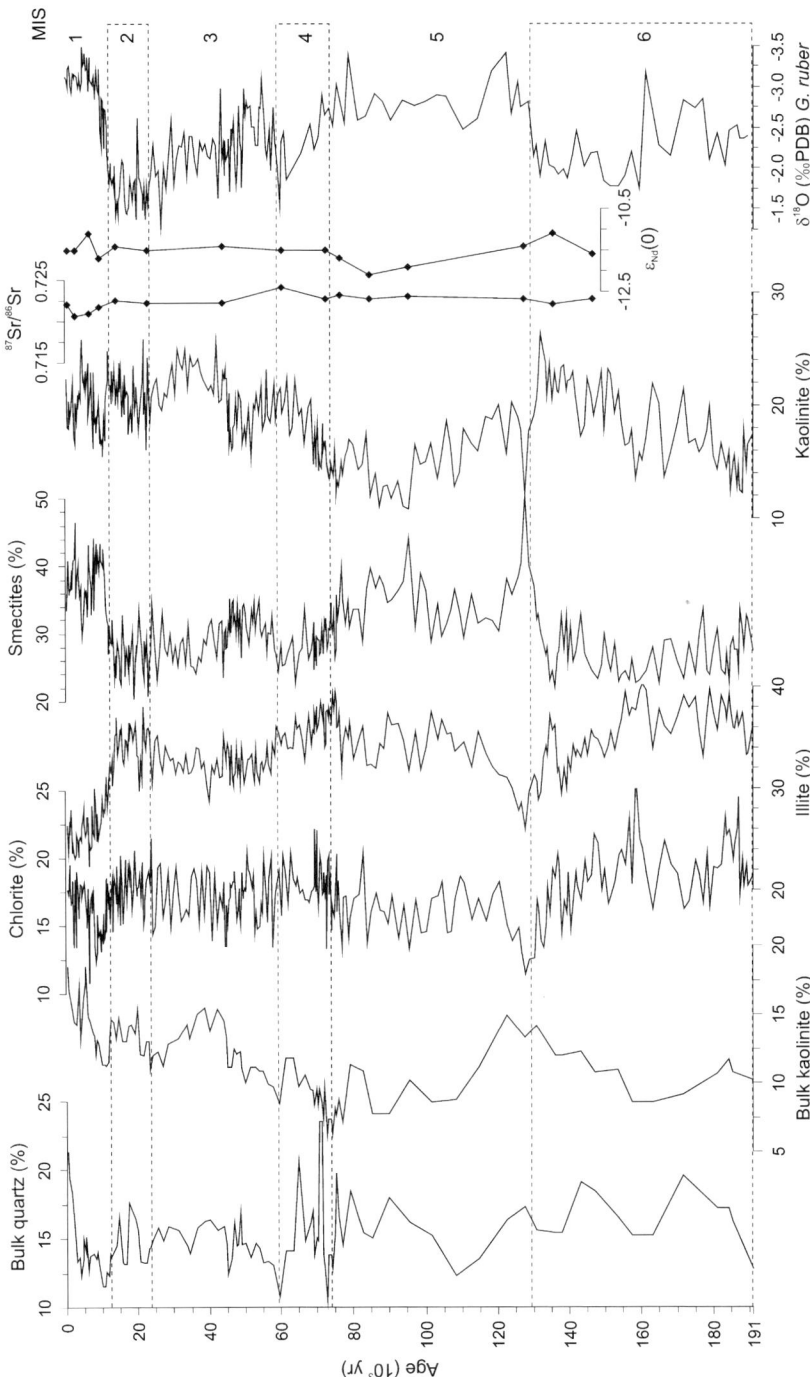

Fig. 4.23 Variations in clay-size, bulk mineral proportions, Sr and Nd isotopes, and planktonic δ¹⁸O in core MD01-2393 show a good correlation (Liu Z. et al. 2005)

the smectite concent (Fig. 4.23). Glacial MIS 2, 3, 4, and 6 are characterized by higher contents of illite (35–40%) than interglacial MIS 1 and 5 (20–35%), whereas smectite shows higher values during interglacials (30–40%) than during glacials (20–35%). Core MD97-2150 (10°11.76′N, 119°31.51′E, 292 m water depth) contains a similar clay mineral contents but provides a shorter record extending only to MIS 4 (about 74 ka) (Liu Z. et al. 2004).

The siliciclastic grain size measurement displays a definite correlation between the curves of the 15–55 μm size fraction and the mean grain size, indicating that the silt population, 15–55 μm, is the main factor controlling mean grain size variations for this period (Liu Z. et al. 2005). On the other hand, the grain size in the clay population (2.5–6.5 μm) has an inverse distribution compared to that of the silt population. Therefore, the grain size ratio of 2.5–6.5 μm/15–55 μm was adopted to represent the sediment discharge from the Mekong River (Fig. 4.24). The Mekong's discharge is derived mainly from the East Asian summer monsoon rainfall between May and October and partly from summer snowmelt in the eastern Tibetan Plateau.

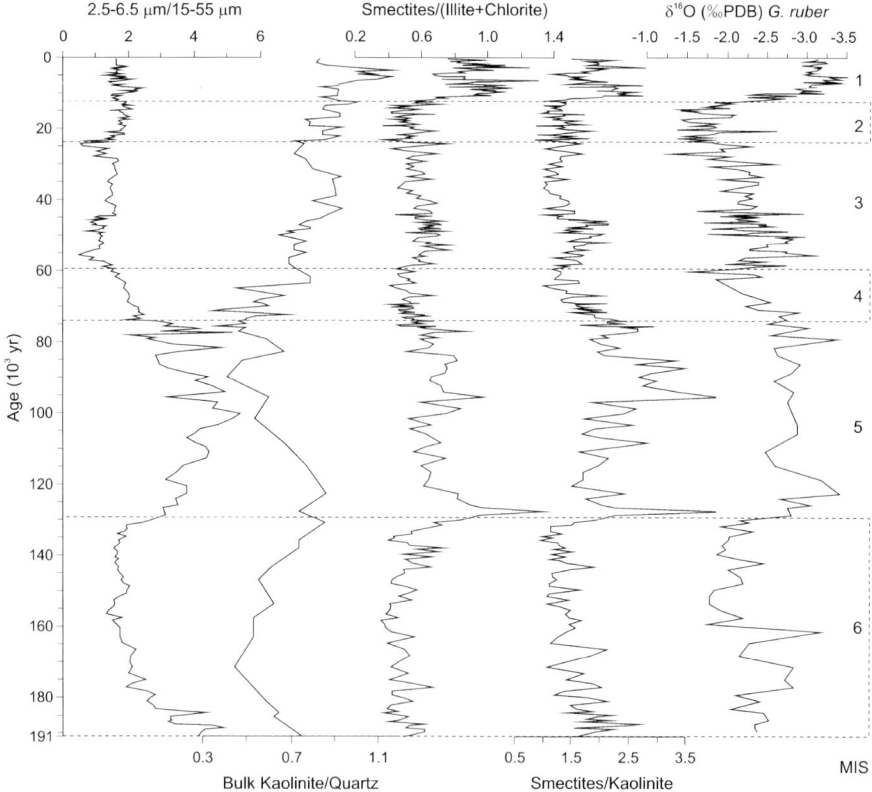

Fig. 4.24 Variations in grain size population (2.5–6.5 μm)/(15–55 μm), bulk kaolinite/quartz, clay-size fraction smectite/(illite+chlorite), and smectite/kaolinite ratios are compared with planktonic $\delta^{18}O$ record from core MD01-2393 (Liu Z. et al. 2005)

Kaolinite seems to be derived mainly from active erosion of inherited clays from reworked sediments in the middle part of the Mekong Basin (Liu Z. et al. 2004). The reworked kaolinitic weathered materials in the landscapes may seriously alter the paleoclimatic signal of kaolinite in the sedimentary record (Thiry 2000). Accordingly, the kaolinite contents, especially the bulk kaolinite/quartz ratio from the Mekong basin can be used to indicate the intensity of physical erosion that occurred in the eastern Tibetan Plateau and in the Mekong basin.

Along with the chemical weathering history obtained by Liu Z. et al. (2004), these results indicate that sediment discharge (revealed by the 2.5–6.5 μm/ 15–55 μm grain size ratio) is well correlated with chemical weathering during the last 190 kyr, as revealed by the smectites/(illite+chlorite) and smectites/kaolinite ratios (Fig. 4.24). Enhanced summer monsoon precipitation, i.e., increased chemical weathering and runoff, was likely the main cause for the increase in sediment discharge. Similarly, variations in the physical erosion (revealed by the bulk kaolinite/quartz ratio) have an inverse correlation with those of the chemical weathering, with the exception of the Holocene (Fig. 4.24). This probably implies a correspondence of intensified physical erosion to weaker summer monsoon rainfall. The results provide strong evidence of monsoon-controlled erosion and weathering processes in the eastern Tibetan Plateau and the Mekong Basin over the past 190 kyr.

Long-Term Changes of Terrigenous Sediment Supply

The Cenozoic is a period of major tectonic activity in SE Asia, where the collision of India with Asia created the Tibetan Plateau by crustal shortening and thickening (i.e. Houseman and England 1993). The collision and continued northward convergence of India led to the southeastward crustal extrusion that contributed at least partly to the opening of the SCS (Tapponnier et al. 1982) (see Chapter 2). Uplift of the Tibetan Plateau most likely strengthened the Asian monsoon, thereby increasing the erosion rate and the flux of fluvial and eolian sediment into the SCS (Clift et al. 2002; Clift 2006). The long-term terrigenous sediment supply to the SCS has been studied using material from ODP Sites 1146 and 1148 (Fig. 4.1), with sediment age ranging from the early Oligocene and from the early Miocene to the present, respectively (Wang P. et al. 2000).

Terrigenous sediment supply since early Miocene at Site 1146 shows some important long-term features of the East Asian monsoon evolution (Wan et al. 2007). The proportion and mass accumulation rate (MAR) of the coarsest end-member EM1 (interpreted as eolian dust), ratios of (illite+chlorite)/smectite, (quartz+feldspar)%, and mean grain-size of terrigenous materials at this site were adopted as proxies for East Asian monsoon evolution (Fig. 4.25). The consistent variation of these independent proxies since 20 Ma shows three profound shifts in the intensity of East Asian winter monsoon relative to summer monsoon, as well as aridity of the Asian continent (Ding et al. 2001; Guo et al. 2002), occurred at ~15 Ma, ~8 Ma and the youngest at about 3 Ma (Fig. 4.25). In comparison, the summer monsoon intensified contemporaneously with the winter monsoon at 3 Ma.

Terrigenous sediment supply since the early Oligocene is best demostrated at ODP Site 1148. Clay mineral input indicates a clear shift from a smectite-dominated mineralogy during the early stages of basin evolution to an illite-dominated character since the middle Miocene (Fig. 4.26) (Clift et al. 2002). Although the total clay abundance increases up-section, there is no sharp change in the total clay which can be correlated with the rapid mineralogy changes. The transition from smectite-dominance to illite-dominance appears to have completed by 15.5 Ma. Chlorite also increases up-section, although to a lesser extent compared with illite. Combined with decreasing εNd values (Fig. 4.26), Clift et al. (2002) interpreted that the drainage basin was progressively eroding more ancient crust within the block following rifting in the SCS. The results indicate the local erosion of the continental arc followed by headward erosion of the Pearl River system into the continental interior since that time (Clift et al. 2002).

However, a parallel geochemical study on Site 1148 sediments found that variations in the abundance of most elements show a remarkable jump in the interval of ~26−23 Ma (477–455 mcd) with Ti, Rb, Zr, Nb, Ce, Sr, and Th all increasing by a factor of 2–7, regardless of their different geochemical behaviour (Li X. et al. 2003). Less pronounced shifts in the contents of Ga, Ti, Rb, and Sr occurred at about 29.5 Ma (~600 mcd). Elemental ratios of Al/Ti, Al/K, Rb/Sr, La/Lu, and Th/La display two remarkable shifts at these two intervals (Fig. 4.27). These two elemental shifts generally coincide with two significant changes in the mass accumulation rate (MAR). The MAR remained high and constant (about $20 \, g/cm^2/kyr$) between 29.3 and 28.5 Ma (589.497 mcd), and then decreased to a negligible value in a slump unit during 26.23 Ma. All these rapid elemental changes also coincide with marked changes in lithology, geochemistry, and Nd isotopes in the slump unit (Fig. 4.27). Similar significant changes also occurred in physical properties of the sediments, particularly color reflectance, natural γ-radiation, magnetic susceptibility, porosity, and density (Wang P. et al. 2000). Along with a double seismic reflector, all these characterize a 'Late Oligocene Unconformity' at Site 1148 (Wang P. et al. 2000). By comparing εNd values of the Pearl River, Mekong River, Sundaland, Borneo, and southern Taiwan, Li X. et al. (2003) considered these geochemical discontinuities during 26–23 Ma at Site 1148 as a result of provenance changes, different from the interpretation by Clift et al. (2002). Pre-27 Ma sediments were proposed

---→

Fig. 4.25 Marine and terrestrial records since 20 Ma from the SCS Site 1146, the north Pacific and Chinese Loess Plateau were well correlated (after Wan et al. 2007). Three dashed lines mark the three profound shifts of the East Asian monsoon intensity at 15 Ma, 8 Ma and 3 Ma. The time series are: (**a–e**) the coarsest end-member EM1%, (illite+chlorite)/smectite, (quartz+feldspar)%, mean grain size and EM1 MAR at Site 1146 (Wan et al. 2007); (**f**) eolian MAR from north Pacific Sites 885/886 (Rea et al. 1998); (**g**) dust accumulation rates of QA-I at Qinan on the Loess Plateau (Guo et al. 2002); (**h**) *Neogloboquadrina dutertrei%* at Site 1146 (Wang P. et al. 2003; Zheng et al. 2004); (**i**) isotopic composition of black carbon at Site 1148 (Jia et al. 2003); (**j**) magnetic susceptibility from the Lingtai section on the Loess Plateau (Ding et al. 2001); (**k**) benthic $\delta^{18}O$ at Site 1148 (Cheng et al. 2004). Some data (fine line) were smoothed by a three-point moving average (coarse line)

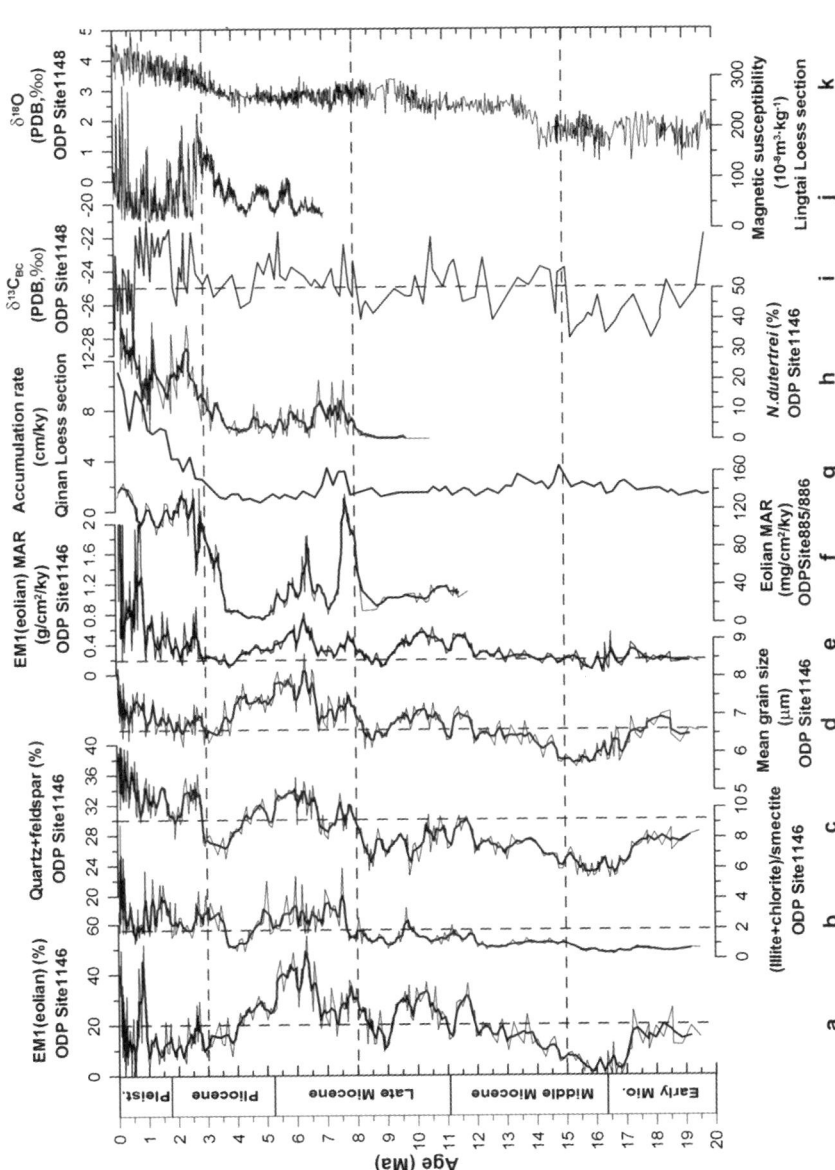

Fig. 4.25 (continued)

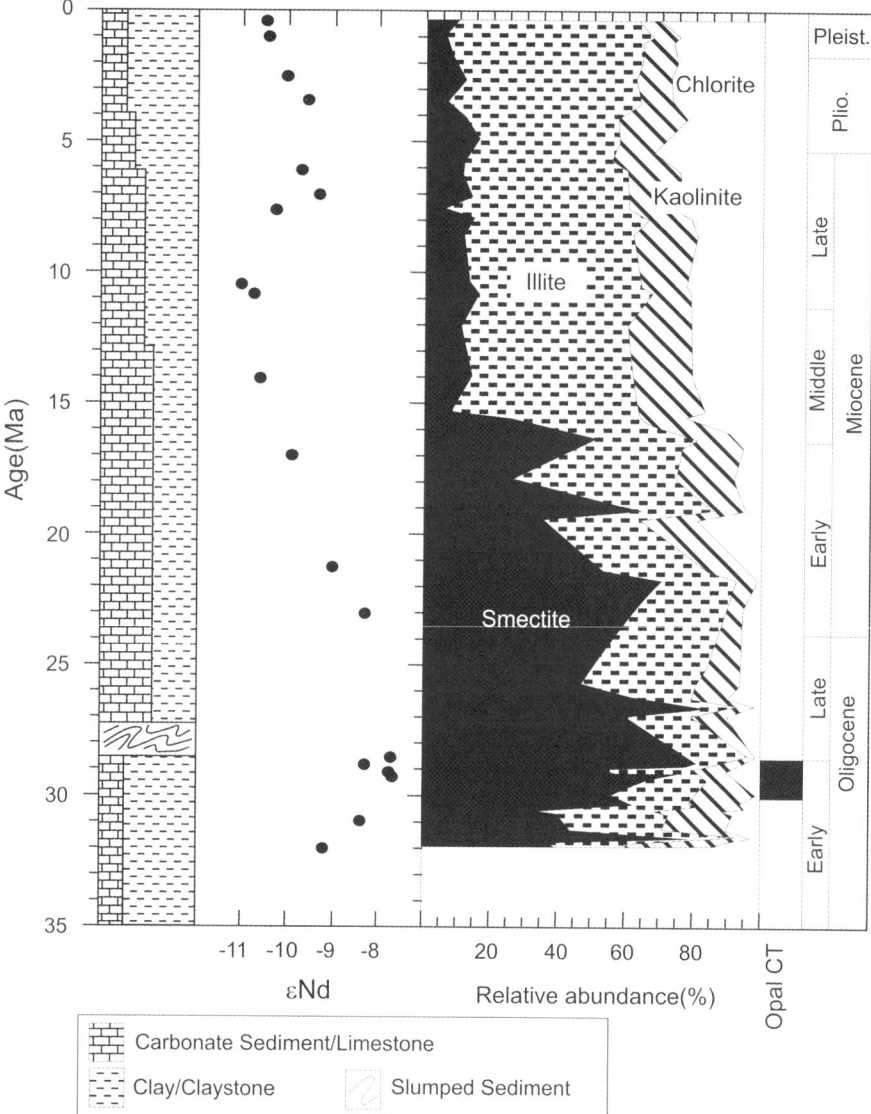

Fig. 4.26 Stratigraphic variations in clay mineralogy and in εNd at ODP Sites 1148 show a decreasing trend from the late Oligocene to middle Miocene (after Clift et al. 2002)

by Li X. et al. (2003) as derived mainly from a southwestern provenance, whereas post-23 Ma sediments were derived from a northern provenance. This provenance change of sediments at ca. 26–23 Ma appears to have been largely caused by a magnificent tectonic event probably relating to a spreading ridge jump from the north to the southwest, associating with a southwestward expansion of the ocean basin and a rise in sea level (Li X. et al. 2003). However, how the southwestern

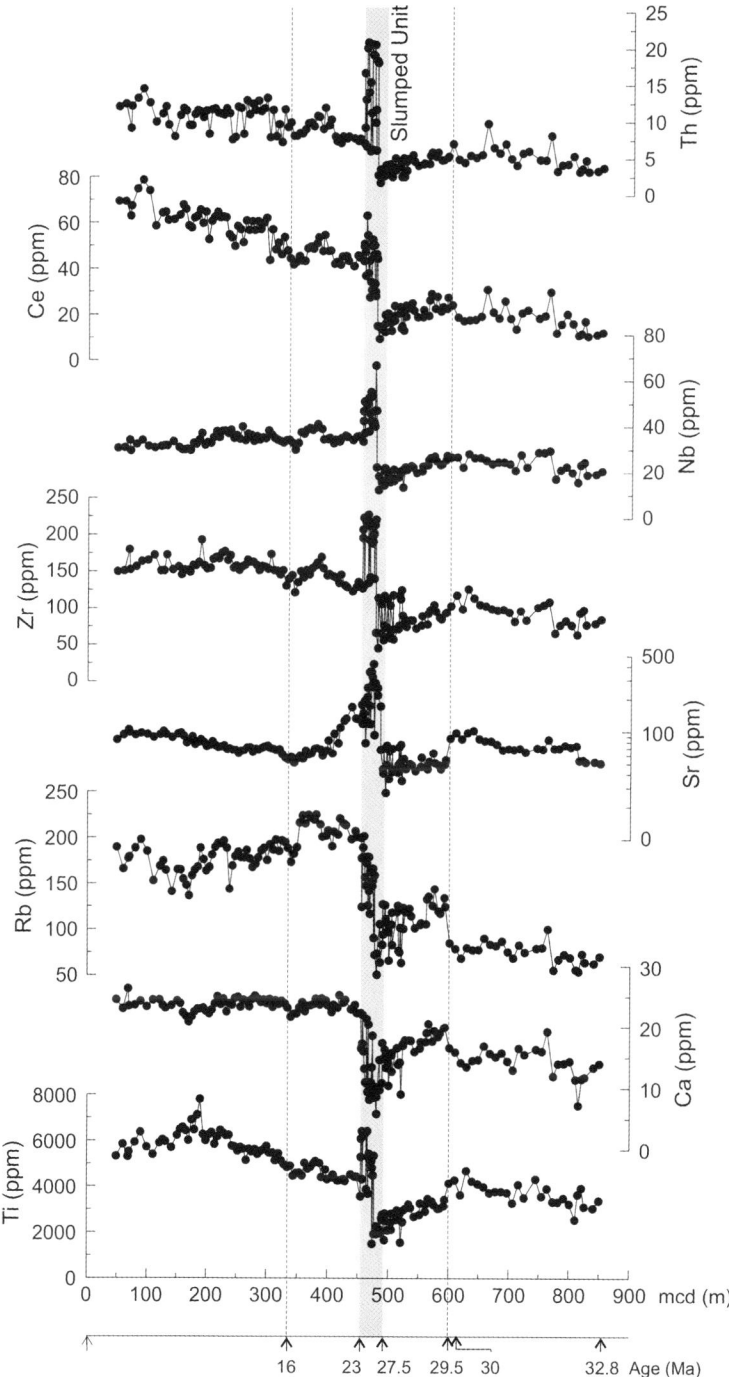

Fig. 4.27 Long-term variations of selected element ratios and Nd values as well as the mass accumulation rate (MAR) from ODP Site 1148 reflect a great tectonic event in the late Oligocene (after Li X. et al. 2003)

provenance influenced sediment transport to the northeastern corner of the SCS and what mechanism caused this transport and the subsequent change still remain open.

4.3 Biogenic Deposition

In the SCS hemipelagic sediments, biogenic components in abundance are inferior to terrigenous components, yet they appear to be much more sensitive to environmental change. Variations in the abundance and composition of biogenic sediments are mainly indicative of surface productivity and oceanography (Barron and Baldauf 1989; Thunell et al. 1992; Wang P. et al. 1995; Wells et al. 1999; Ragueneau et al. 2000), in particular monsoon-induced upwelling in this region (Huang C. et al. 1997b; Huang B. et al. 2000; Jian et al. 1999; Jia et al. 2000). Biogenic deposits in the SCS comprise three major components: carbonate, opal and organic matter. Carbonate is by far the most important biogenic material in the SCS sediment (Fig. 4.5), mostly originating from calcareous microfossils, such as foraminifera and nannoplankton, and from coral reefs. Coral will be discussed in the next section because of its specific nature, and organic matters will be covered in Chapter 7. The focus of this section, therefore, is on pelagic carbonate and biogenic opal, to generalize their distribution patterns in surface sediments and in sediment cores.

Carbonate (Li J. and Wang P.)

Modern Pattern of Carbonate Distribution

Carbonate abundance is the most commonly used proxy data available in sediments. It should not be a mere coincident that paleoceanographic studies in the SCS started from carbonate cycles (e.g., Wang P. et al. 1986). The distribution of carbonate in marine sediments is resulted from a balance between its biogenic production in the surface ocean and its dissolution in deeper waters. With its warm surface water and relatively warm bottom water ($>2\,°C$), the SCS is favorable for carbonate production and preservation. When the four major western Pacific marginal seas are compared, the SCS has the highest and the Sea of Japan the lowest carbonate percentages in surface sediments below $200\,m$ water depths. The northward decrease in $CaCO_3\%$ is ascribed to the declining temperature of bottom water with higher latitudes (Wang P. 1999). The distribution of $CaCO_3$ in the surface sediments of the SCS is shown in Fig. 4.5.

Carbonate is oversaturated in the upper layer of the SCS. Thus, the aragonite saturation (Ωa) reaches 450–550% and calcite saturation (Ωc) measures 250–350% in the surface water on the northern slope, but the saturation declines with depth as shown in Fig. 4.28. Accordingly, the aragonite and calcite saturation horizons in the northern SCS are respectively at about $435\,m$ and $1,630\,m$ (Han and Ma 1988). This is close to the low-latitude Pacific, but much shallower than in the Atlantic.

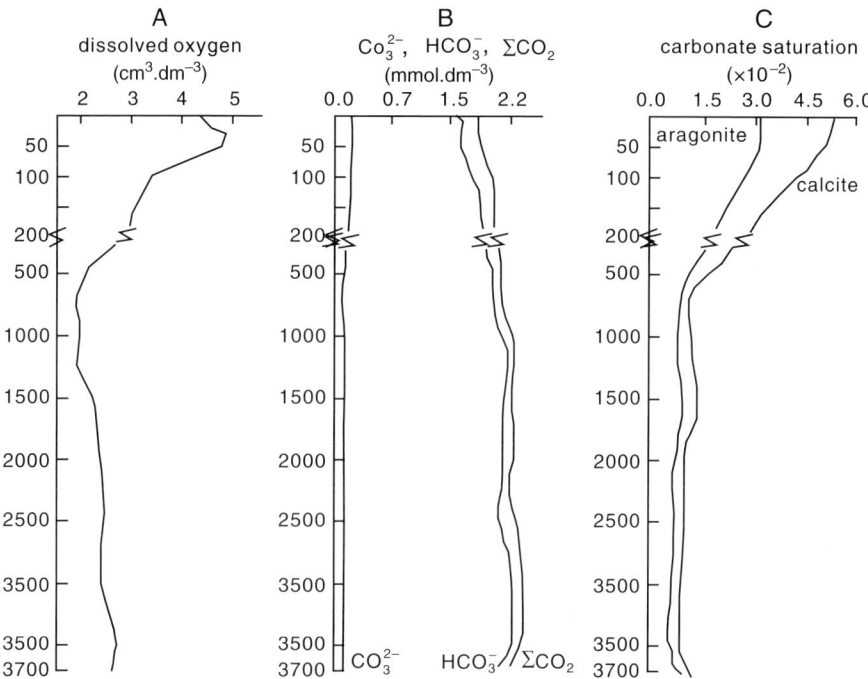

Fig. 4.28 Water chemistry profiles show (**A**) dissolved oxygen, (**B**) CO_3^{2-}, and ΣCO_2^{-}, and (**C**) $CaCO_3$ satuaration in the northern SCS, at Station 461 (118 °E, 19 °N, water depth >3, 700 m) (based on South China Sea Institute of Oceanology, 1985)

According to Chen C. et al. (2006), the typical vertical distributions of aragonite and calcite saturation horizon should be respectively at 600 m and 2,500 m, higher than in the Sulu Sea but lower than in the Western Philippine Sea (Fig. 4.29). Since the pioneering work on foraminiferal preservation in the SCS by Rottman (1979), many authors have contributed to carbonate dissolution in the region. Summarizing the results, the modern aragonite compensation depth in the SCS is estimated at 1,000–1,400 m, and the calcite compensation depth at 3,500–3,800 m, with lysocline lying at about 3,000 m (Wang P. 1995). During past glacial cycles, however, all the depth levels of carbonate saturation changed significantly.

Noteworthy is the newly discovered chemoherm carbonate on the northern slope of the SCS. In June 2004 during the joint Chinese–German RV SONNE Cruise 177, areas of methane seepage were discovered in the northern SCS with buildups of carbonate caps on several ridges (Suess 2005). Since 2004, methane-derived carbonates have either been dredged or had previously been collected from various locations implying the existence of a basin-wide methane seepage at upper to middle slope depths in the SCS (Chen D. et al. 2005; Lu et al. 2005; Chen Z. et al. 2006). Buildups occur at three ridge-crest segments collectively covering about 400 km^2 around 22°05′ N, 118°50′ E, 500 to 770 m water depths. Lithologically, the carbonates are micritic, containing peloids, clasts and clam fragments, dominated by

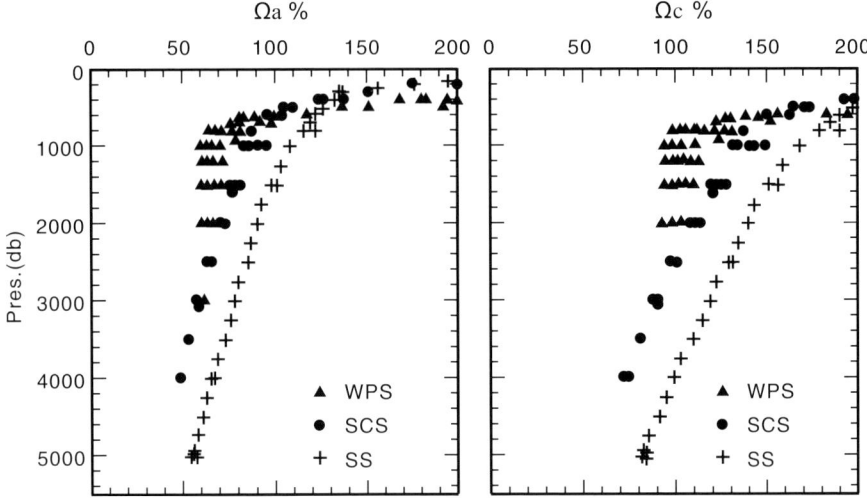

Fig. 4.29 Vertical profiles show the distribution of aragonite saturation (Ωa) and calcite saturation (Ωc) in the Western Philippine Sea (WPS), SCS and Sulu Sea (SS) (Chen C. et al. 2006)

high magnesium calcites (HMC) and aragonite. All of the carbonates are strongly depleted in δ^{13}C (-35.7 to $-57.5‰$ PDB) and enriched in δ^{18}O ($+4.0$ to $+5.3‰$ PDB). These isotopic characteristics, together with the observed microbial structure, are diagnostic of carbonates derived from anaerobic oxidation of methane mediated by microbes. At one site, a carbonate edifice, about 30 m high and at least 100 m in diameter above the base, stands above the surrounding seafloor. Still weakly active, this site has been named Jiulong Methane Reef (Han et al. 2008). Because of the widespread occurrences of methane and the successful recovery of gas hydrates in the northern slope in 2007, many more discoveries of similar chemoherm carbonates in the SCS are expected.

Carbonate Cycles in Quaternary

Although most sediment cores taken from the SCS penetrated only the late Quaternary deposits, the correspondence of carbonate cycles to glacial cycles is evident. It was found, however, that $CaCO_3$% curves from sites above the present lysocline are basically concordant with the δ^{18}O curves, showing glacial decreases and interglacial increases, while those below the lysocline are out of phase with the δ^{18}O, with high carbonate during the last glaciation and low carbonate in the Holocene (Wang P. et al. 1995). Twenty three cores with δ^{18}O-based chronology and relatively high resolution are selected to illustrate the two types of carbonate cycles in the SCS (Table 4.3).

From the 23 cores, 16 reached MIS 1 to 2 or 3 (Fig. 4.30), while 8 penetrated the last 200 kyr, including the last two glacial cycles (Fig. 4.31), with sampling resolution as high as 150–200 years at cores MD97-2148 and -2151 (Chen Y. et al.

Table 4.3 Sediment cores from the SCS with different carbonate cycles are used to compile Figs. 4.30 and 4.31

Core	Location	Water depth. (m)	References
GGC-13	10°36′N, 118°17′E	990	Miao et al. (1994)
SO49-8KL	19°11′N, 114°12′E	1040	Wang L. and Wang (1990)
MD01-2393	10°30′N, 110°03′E	1230	Liu Z. et al. (2004); Michel et al. (1997)
GGC-9	11°38′N, 118°38′E	1455	Miao et al. (1994)
SO17964-3	06°10′N, 112°13′E	1556	Sarnthein et al. (1994)
MD97-2142	12°42′N, 119°28′E	1557	Michel et al. (1997); Chen M.T. et al. (2003a)
MD97-2151	08°44′N, 109°52′E	1589	Huang C. et al. (1999); Michel et al. (1997)
SO17962-2	07°11′N, 112°05′E	1970	Wang P. et al. (1995)
SO49-37KL	17°49′N, 112°47′E	2004	Qian (1999)
ODP1144	20°03′N, 117°25′E	2036	Wang P. et al. (2000)
GGC-11	11°53′N, 118°20′E	2165	Miao et al. (1994)
GGC-12	11°56′N, 118°13′E	2495	Miao et al. (1994)
ODP1143	09°22′N, 113°71′E	2772	Liu Z. et al. (2003b)
MD97-2148	19°05′N, 117°32′E	2830	Michel et al. (1997); Chen Y. et al. (1999)
GGC-6	12°09′N, 118°04′E	2975	Miao et al. (1994)
ODP1148	18°50′N, 116°34′E	3297	Chen X. et al. (2002)
SO17956	13°51′N, 112°35′E	3387	Sarnthein et al. (1994)
SO17937-2	19°30′N, 117°40′E	3428	Sarnthein et al. (1994)
GGC-4	12°39′N, 117°56′E	3530	Miao et al. (1994)
GGC-3	13°16′N, 117°48′E	3725	Miao et al. (1994)
SO50-29KL	18°20′N, 115°59′E	3766	Wang P. et al. (1995)
GGC-2	13°37′N, 117°41′E	4010	Miao et al. (1994)
GGC-1	14°00′N, 117°30′E	4203	Miao et al. (1994)

1999; Huang C. et al. 1999). The striking contrast between carbonates curves from sites below and above the lysocline around 3000 m reminds those two types from the two oceans: the Atlantic and the Pacific. When comparing $CaCO_3$% curves from the high latitude North Atlantic and from the equatorial Pacific, Luz and Shackleton (1975) noticed two different rends nearly counter to each other: the Atlantic $CaCO_3$% curves run roughly parallel to the $\delta^{18}O$ curves ("Atlantic cycles"), whereas the Pacific curves display a different trend with the minimal values from MIS 4 to MIS 5d ("Pacific cycles"). The two types of carbonate cycles are observed in the SCS, the Atlantic type above the modern lysocline and the Pacific type from below it (Figs. 4.30 and 4.31).

In searching for an explanation on the difference between the two oceans, Seibold and Berger (1982) attributed the Atlantic type of carbonate cycles to dilution cycles and the Pacific type to dissolution cycles. Later studies, however, have shown more complex variations in carbonate cycles in the two oceans, and the cause of these variations must have been deeply rooted in changes of deep-water circulation and chemical reorganization in the world ocean (e.g., Crowley 1985; Zahn et al. 1991). Along with carbonate percentages, foraminiferal and pteropod preservation, and carbonate and terrigenous clastic accumulation rates are often used to evaluate the role of dilution and dissolution in carbonate cycles. In the SCS, foraminiferal dissolution index (FDI) and the ratio between benthic and planktonic foraminifera

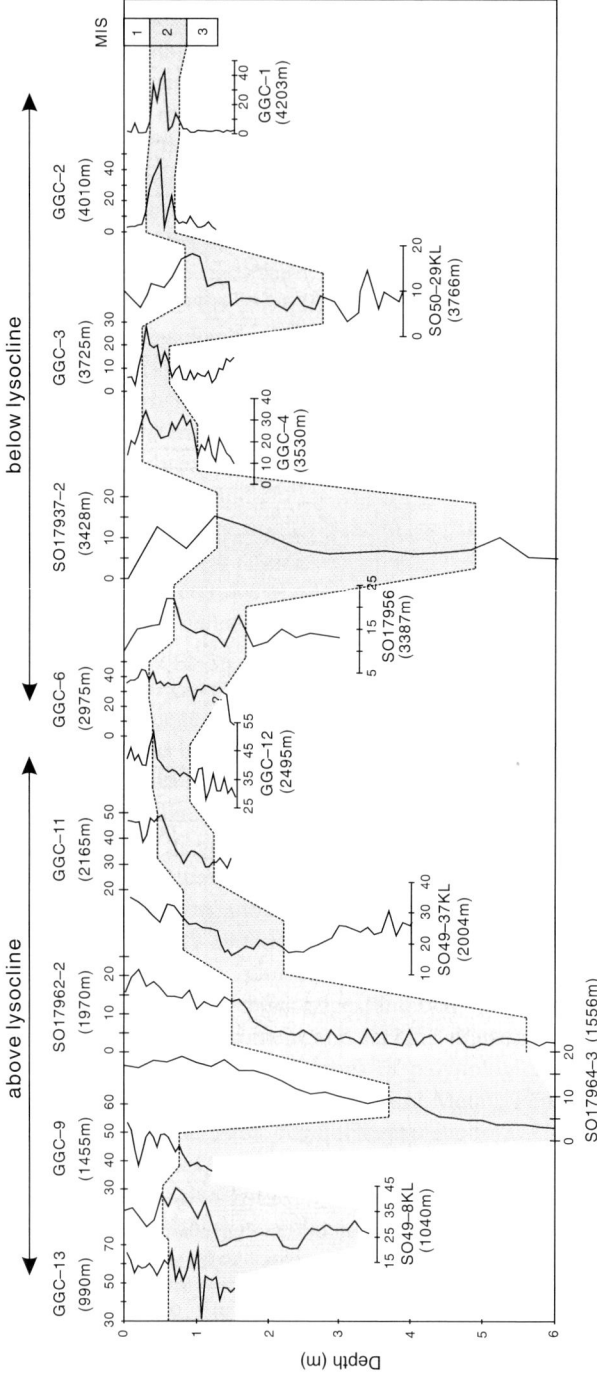

Fig. 4.30 Downcore variations in carbonate% are shown for 16 short cores from above lysocline (*left*) and at or below lysocline (*right*) in the SCS. Refer to Table 4.3 for locations and references

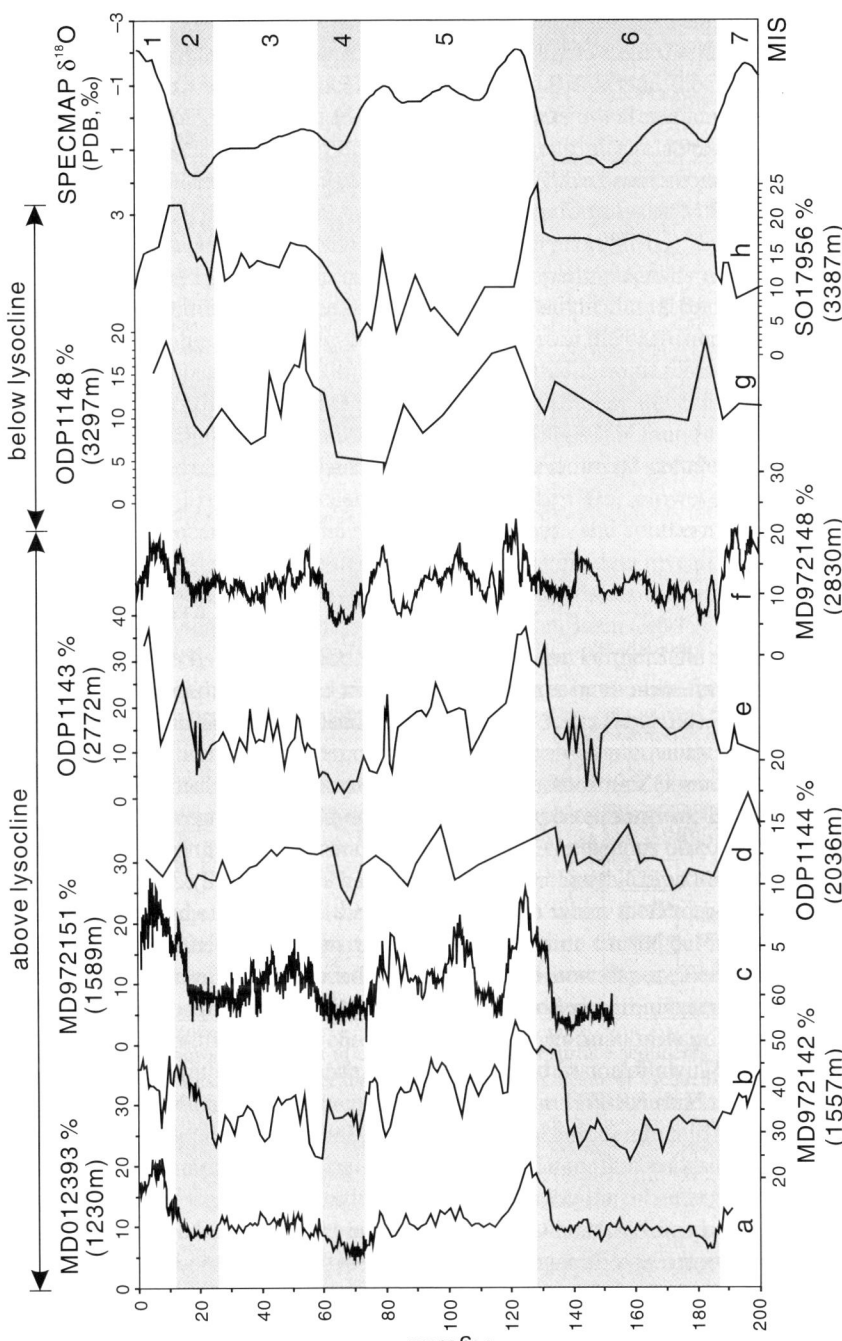

Fig. 4.31 Downcore variations in carbonate% are shown for 8 longer cores from above lysocline (**a–f**, "Atlantic-type") and below lysocline (**g–h**, "Pacific-type") in the SCS. Refer to Table 4.3 for locations and references

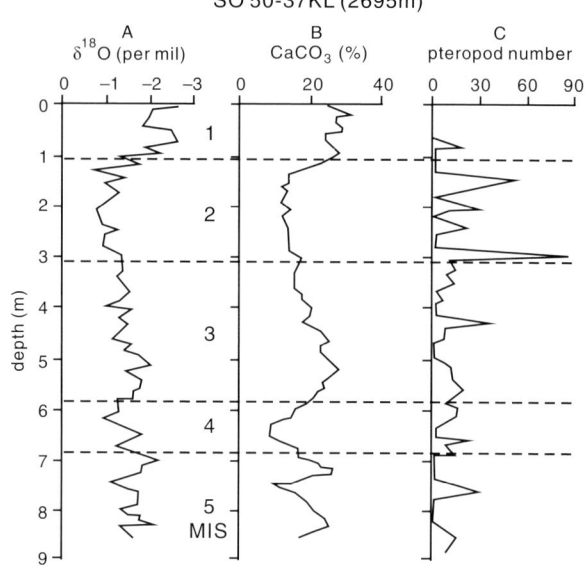

Fig. 4.32 Variations in $\delta^{18}O$ (**A**), carbonate% (**B**) and aragonite preservation (pteropod number per gram sample, **C**) in core SO 50-37 are compared to show an "Atlantic" type carbonate cycle and a Pacific type aragonite preservation (Wang L. et al. 1997)

are applied to indicate foraminiferal preservation. The results show that these fauna-based curves below lysocline swing against the carbonate curve and peak at MIS 1, 3 and 5, implying strengthened dissolution during interglacials (Bian et al. 1992). At sites above lysocline, the deposition rate of non-carbonate material is about two times higher in glacials than in interglacials, while the deposition rate of carbonate displays no glacial/interglacial contrast, implying the "dilution" effect in carbonate cycles (Wang P. et al. 1995).

However, carbonate dissolution may occur at sites above lysocline as well. This applies particularly to aragonite with its modern compensation depth at only about 1,000 m. Core SO 50-37KL (18.9°N, 115.8°E, water depth 2,695 m) from the northern SCS, for example, the carbonate curve shows "Atlantic" type, but the pteropod abundance drastically increases in MIS 2, displaying a typical "Pacific" type of carbonate cycle (Fig. 4.32) (Wang L. et al. 1997).

Therefore, the late Quaternary carbonate cycles in the SCS are superimposed by dissolution and dilution cycles. During glacial sea level lowstands, increased input of terrigenous clasts raises sedimentation rate and dilutes carbonate, generating the apparent "Atlantic cycles". Meanwhile, the continuous connection between the SCS and the Pacific maintains similar water chemistry in these two basins. This explains why all the studied cores in the SCS share the same carbonate dissolution signals as in the Pacific. Regardless of depth position below or above the modern lysocline, dissolution of both calcite and aragonite enhanced in MIS 4 and 5a–d in the SCS, as well as in the western equatorial Pacific, attributed to an uplifted lysocline (Wang P. et al. 1995). Of course, there can be additional local factors to cause unusual carbonate curves. An example is the abnormally high input of terrigenous material after the last glaciation at core SO 17940 in the northeastern SCS, with drastic decrease in CaCO$_3$% from MIS 2 to MIS 1 (Wang P. et al. 1995).

A more complete record of carbonate cycles over the Quaternary from ODP Site 1143 (9°22′N, 113°17′E, water depth 2,772 m) shows clear cycles in carbonate deposition and preservation well correlated to glacial cycles defined by benthic δ^{18}O (Fig. 4.33). All the proxy curves for carbonate preservation (carbonate MAR, CaCO$_3$%, absolute abundance of foraminifers, and coarse fraction) run nearly parallel to that of benthic δ^{18}O, but they peak at glacial/interglacial transitions, or terminations, whereas thedissolution index (fragmentation%) varies in an opposite way (Liu Z. et al. 2003b; Xu 2004). The Site 1143 record reconfirms that the above discussed carbonate preservation model over the last 200 kyr is appliable to the entire Quaternary in the region.

Remarkable are variations in carbonate content on time scales longer than last several glacial cycles, as noticed decades ago in the equatorial Pacific (e.g., Hays et al. 1969). An example is the "Mid-Brunhes Event" of carbonate dissolution starting from MIS 11 (Crowley 1985; Jansen et al. 1986). A similar long-term trend of carbonate preservation occurred also in the SCS (e.g., Li B. et al. 2001). The three major time intervals of low carbonate and intensive dissolution at MIS 13, 27–29 and 53–57 (Fig. 4.33; Xu et al. 2005) are comparable with those in the Indian Ocean (Bassinot et al. 1994), and they may correspond to the three major carbon isotope events discussed in Chapter 7.

Long-Term Changes in Carbonate Deposition

The Cenozoic ocean has experienced significant decrease in carbonate compensation depth (Delaney and Boyle 1988) and carbon isotope value (Shackleton 1985), particularly for the last 23.5 myr, during the Neogene. Lyle (2003) compiled CaCO$_3$ MAR data over the last 25 myr from 144 DSDP and ODP sites in the Pacific, and found no CaCO$_3$ buried below 4,700 m water depth at anytime in the Neogene. However, very little coherence exists between CaCO$_3$ MAR time series from different Pacific regions. A transition from high to low CaCO$_3$ MAR from 23–20 Ma is the only event common to the entire Pacific Ocean. Unfortunately, "the NW Pacific marginal seas suffer from complex topography and a lack of paleoceanographic drilling emphasis" (Lyle 2003) until ODP Leg 184 to the SCS in 1999, which offers the first dataset for the region.

ODP Site 1148 recovered the longest record from the northern slope of the SCS, covering the last 33 myr (Wang P. et al. 2000). When the 24-myr low-resolution carbonate MAR profile from Site 1148 is compared with the stacked Neogene time series from western and central equatorial Pacific, remarkable differences emerge between the three curves (Fig. 4.34). The only general patterns in all the three low-latitude Pacific curves are carbonate MAR decreases at 21−20 Ma and at 5−3 Ma. The 21−20 Ma carbonate MAR decrease corresponded to the early Miocene CCD rise due to a general warming and sea level rise by ∼50 m, resulting in shallow sequestration of CaCO$_3$ as shallow seas expanded and warmed (Lyle 2003). At Site

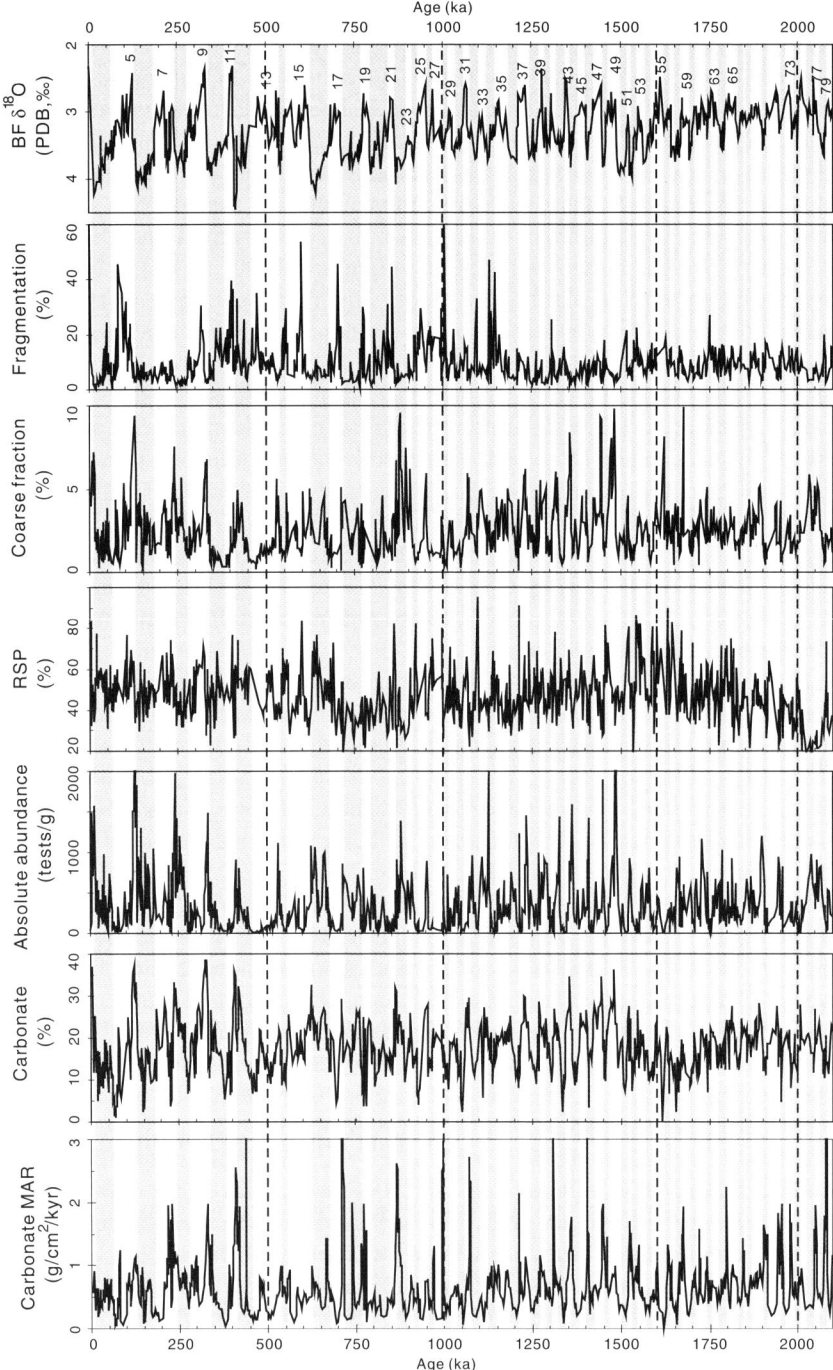

Fig. 4.33 (continued)

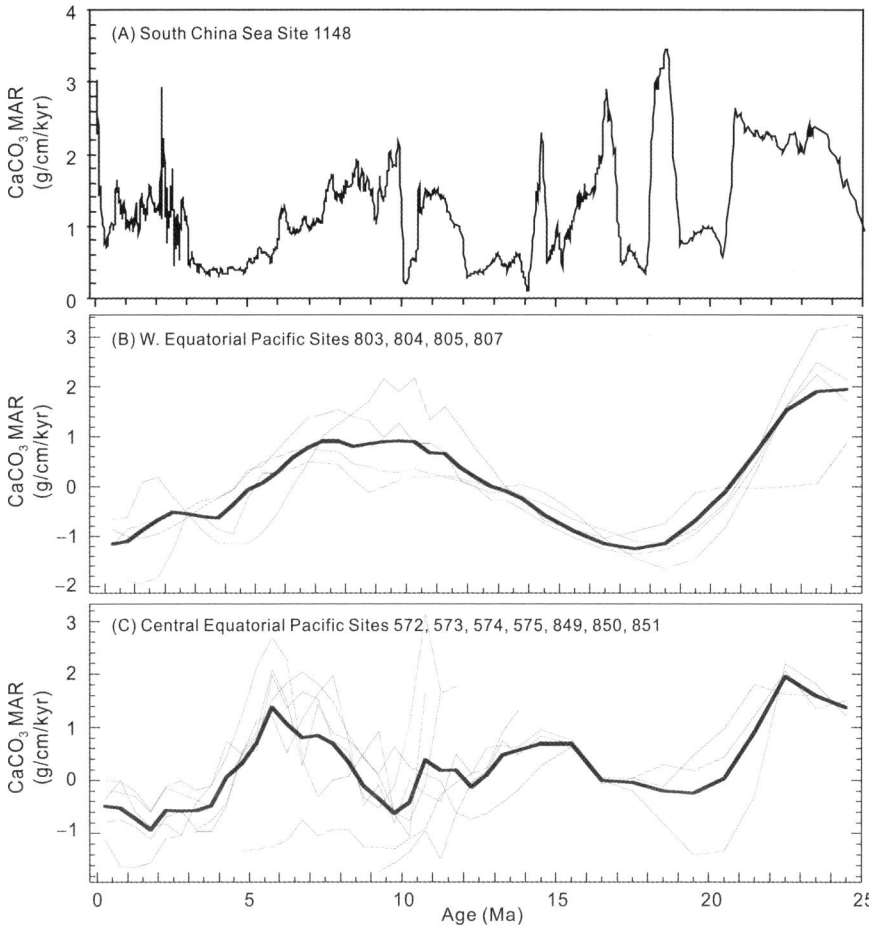

Fig. 4.34 Carbonate MAR time series from (**A**) Site 1148 from the SCS (Chen X. et al. 2002) are compared to sites from (**B**) western and (**C**) central equatorial Pacific (Lyle 2003). The *thick lines* in **D** and **E** are stacks while the thin lines are time series for individual sites

1148, CaCO$_3$%, CaCO$_3$ MAR, fragmentation and benthic foraminifer abundance together convey a clear message that 5 periods of intensive dissolution had occurred in the Miocene–Pliocene SCS. Respectively at ~21 Ma (D-1), 16–15 Ma (D-2), 13–11 Ma (D-3), 10–9 Ma (D-4), and 5–3 Ma (D-5) (Fig. 4.35C), they indicate major steps in the change of SCS deeper waters (Li Q. et al. 2006).

←

Fig. 4.33 A two million years record of carbonate accumulation and dissolution at Site 1143 shows variations in carbonate mass accumulation rate (MAR), carbonate%, absolute abundance of foraminifera (tests per gram sediment), dissolution-resistant species (DRP, %), coarse fraction (% of grains >63 μm), planktonic fragmentation (%), and benthic δ^{18}O (from Xu 2004)

Along with Site 1148, two other sites with time series trace back to the Miocene: an 18-myr record from Site 1146 and a 12-myr record from Site 1143 (Wang P. et al. 2000). The carbonate deposition and dissolution curves from these 3 sites exhibit a similar trend (Fig. 4.35), with a stage of increased CaCO$_3$% from ~10 Ma to ~5 Ma followed by a decline from 5 to 3 Ma. The late Miocene high CaCO$_3$% can be broadly correlated with the "biogenic bloom" in the equatorial Pacific, whereas the decrease of CaCO$_3$% since the late Pliocene is largely associated with enhanced supply of terrigenous clasts as a result of frequent sea-level variations in glacial cycles.

"Biogenic Bloom" was initially reported from the eastern equatorial Pacific (e.g., Theyer et al. 1985; Farrell et al. 1995) for a time interval with enhanced carbonate deposition centred at 6.5−3.5 Ma, shortly after the ~11−8 Ma "carbonate crush" in the same region (Lyle et al. 1995). As the "Biogenic Bloom" is not restricted to high carbonate content but also displays a biogenic silica peak, it has been ascribed to raised productivity in the tropics (Farrell et al. 1995). The event was subsequently found to have occurred not only in the eastern and central Pacific but in the entire Indo-Pacific Ocean, although the timing differs slightly from region to region (Dickens and Owen 1999). Recent studies have found the "Biogenic Bloom" also in the Atlantic, suggesting its global scale (Diester-Haass et al. 2005, 2006). In the SCS, increased carbonate MARs occur during ~8−5 Ma at all the 3 ODP sites, accompanied also by high opal abundance (such as at Site 1143, Fig. 4.35). The carbonate content was low before the event, but no true "carbonate crush" has been found.

Contrary to the decline of CaCO$_3$% over the last 3–5 myr, a significant rise in the carbonate CCD took place in the low-latitude Pacific after the "Biogenic Bloom". However, this is not obvious in the Site 1148 record, in which the percentage of foraminiferal frangmentation and the benthic/planktonic ratio both decline upward instead (Fig. 4.35C). An opposite trend occurred at Site 1143 in the southern SCS with a upward increase in both frangmentation and the benthic/planktonic ratio over the last 5 myr (Fig. 4.35A). In the late Pliocene and Pleistocene, an increase in these two proxy curves is recorded at Site 1146 (Fig. 4.35B), supporting enhanced carbonate dissolution. Therefore, the divergent behavior of carbonate preservation at various sites suggests that the CaCO$_3$% drop over the last 3–5 myr was probably caused by dilution rather than preservation effects.

Recently, a new inversion technique was used to estimate CaCO$_3$% based on high- resolution seismic profiles. When applied to the northeastern SCS where ODP Sites 1146 and 1148 are located (Fig. 4.36) (Xiong et al. 2006), it reveals a declining trend in CaCO$_3$% since the latest Miocene. Therefore, CaCO$_3$ reduction was the main feature in sedimentation at least along the northern SCS slope over the last 5 myr.

Thanks to its low latitude location and a sill depth of about 2,400 m, the SCS enjoys the most favorable conditions for carbonate accumulation among the northwestern Pacific marginal seas. With its modern compensation depth at 1,000–1,400 m for aragonite and at 3,500–3,800 m for calcite, the carbonate content exceeds 10% everywhere in the SCS, except in the central basin below 3,500 m. Because of its better carbonate preservation and higher accumulation rates than most

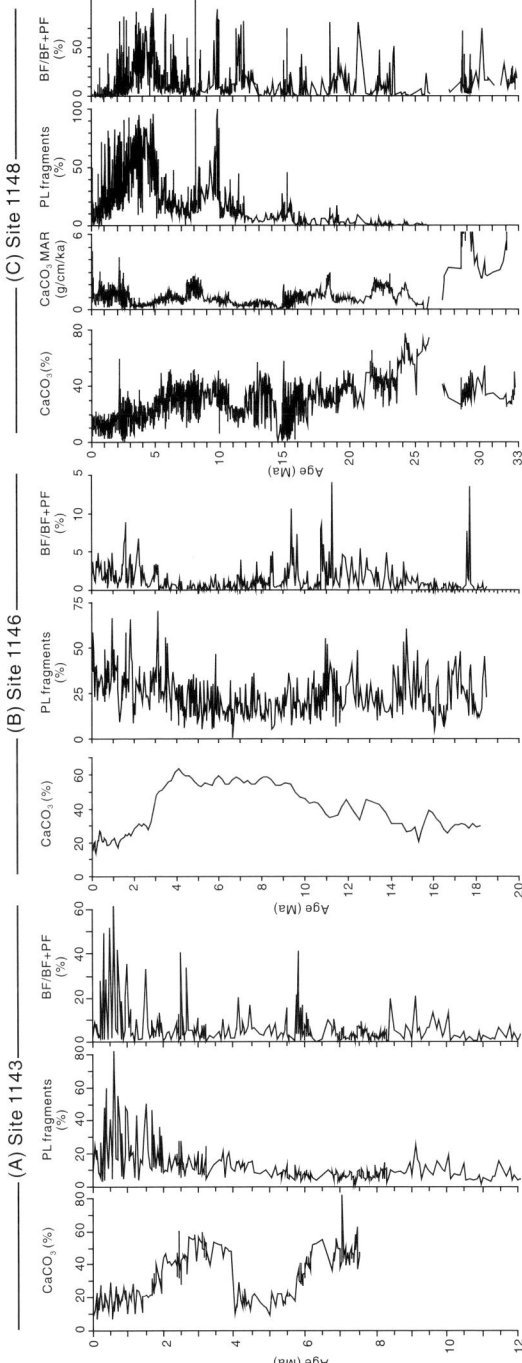

Fig. 4.35 Variations in carbonate deposition and preservation proxies, including CaCO₃%, CaCO₃ MAR (g/cm₂/ka), fragmentation of planktonic foraminifera (PF fragmentation%), and ratio of benthic to total foraminiferal abundance (BF/BF+PF) were shown for Sites 1143 (A, Li B. et al. 2005), 1146 (B, Wang J. 2001), and 1148 (C, Li Q. et al. 2006). Note different vertical scales

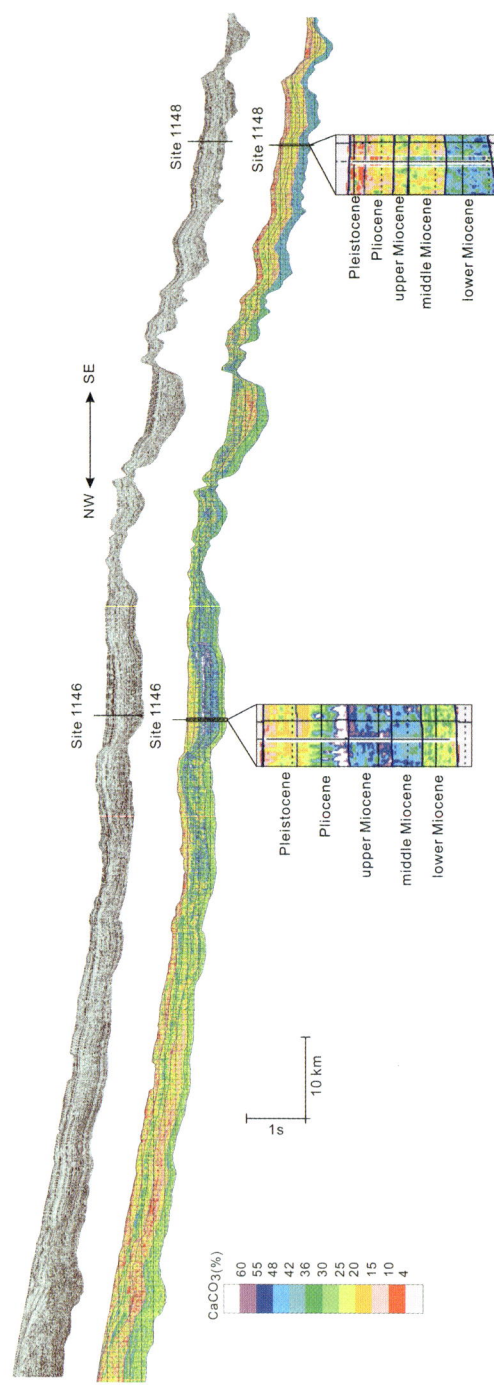

Fig. 4.36 Variations of carbonate content in the Neogene deposits were displayed along the northern slope passing through ODP Sites 1146 and 1148. Upper panel: seismic profile; lower panel: inversed carbonate content profile (modified from Xiong et al. 2006)

western Pacific localities, the SCS hemipelagic deposits provide valuable records for paleoceanographic studies. The Quaternary carbonate cycles in the SCS are basically determined by two major factors: deep-sea dissolution and dilution by terrigenous deposits. Increase in terrigenous inputs during glacials dilutes carbonate, generating the apparent "Atlantic cycles" of carbonate with the $CaCO_3\%$ curve running parallel to that of $\delta^{18}O$. On the other hand, the SCS carbonate record embodies the signals of the Pacific water chemistry because of a continuing connection between the two basins. Even at the sites with the Atlantic-type $CaCO_3\%$ curve, the proxies of carbonate dissolution exhibit downcore variations of the Pacific-type, as seen from the preservation of aragonite pteropod shells from the upper slope (Wang L. et al. 1997). This Pacific imprint in the long-term carbonate records of the SCS is also perceived, such as the early Miocene CCD rise at 20–21 Ma and the "Biogenic Bloom" between 5 and 10 Ma. At Site 1148, five major dissolution events (D-1 to D-5) over the Neogene were found linking to the stepwise development of deep water masses in the deepsea basin (Li Q. et al., 2006).

Opal (Wang R.)

Modern Opal Accumulation and Distribution

In the SCS, biogenic opal is produced mainly by diatoms and radiolarians, and its accumulation is closely related to surface productivity, particularly in monsoon-driven upwelling areas. As discussed in Chapter 2, winter monsoon is stronger and lasts for nearly six months (November to April), while summer monsoon is relatively weaker and lasts for about 4 months (June to August). Accordingly, accumulation of biogenic particulates, including opal, displays strong seasonal variations, as confirmed by sediment trap records from September 1987 to March 1988, at the northern SCS Site SCS-N (Wiesner et al. 1996; Chen J. 2005) and from 1993 to 1996 at the central SCS Site SCS-C (Chen R. et al. 2006) (Fig. 4.37).

A sediment trap deployed at the ODP Site 1143 locality (9°23.2′N, 113°13.9′E), (or Site SCS-S) during May 2004–April 2005 reveals high particulate fluxes and primary productivity during the summer monsoon period but an opposite trend during the winter monsoon period. The positive correlation between opal and organic carbon flux and primary productivity implies that biogenic opal can be used as a surface productivity proxy (Wang R. et al. 2007).

Abundance of biogenic opals in a total of 162 surface sediment samples varies between 9% and 0.33%, averaging 4% (Fig. 4.38). Higher opal contents, >6%, occur in the central part of the southern SCS. Samples with 3–6% opal contents are mainly from the southern SCS, the northern slope, the central basin and offshore from Vietnam, while those with 3% or less are limited to the northern and southern shelves and the northeastern corner near the Bashi Strait (Fig. 4.38). Thus opal distribution is closely associated with local nutrient and productivity levels. Shallow water samples contain lesser opal probably because opal is often diluted on shelf by terrigenous particulates and dissolved by pore waters once buried in sediments

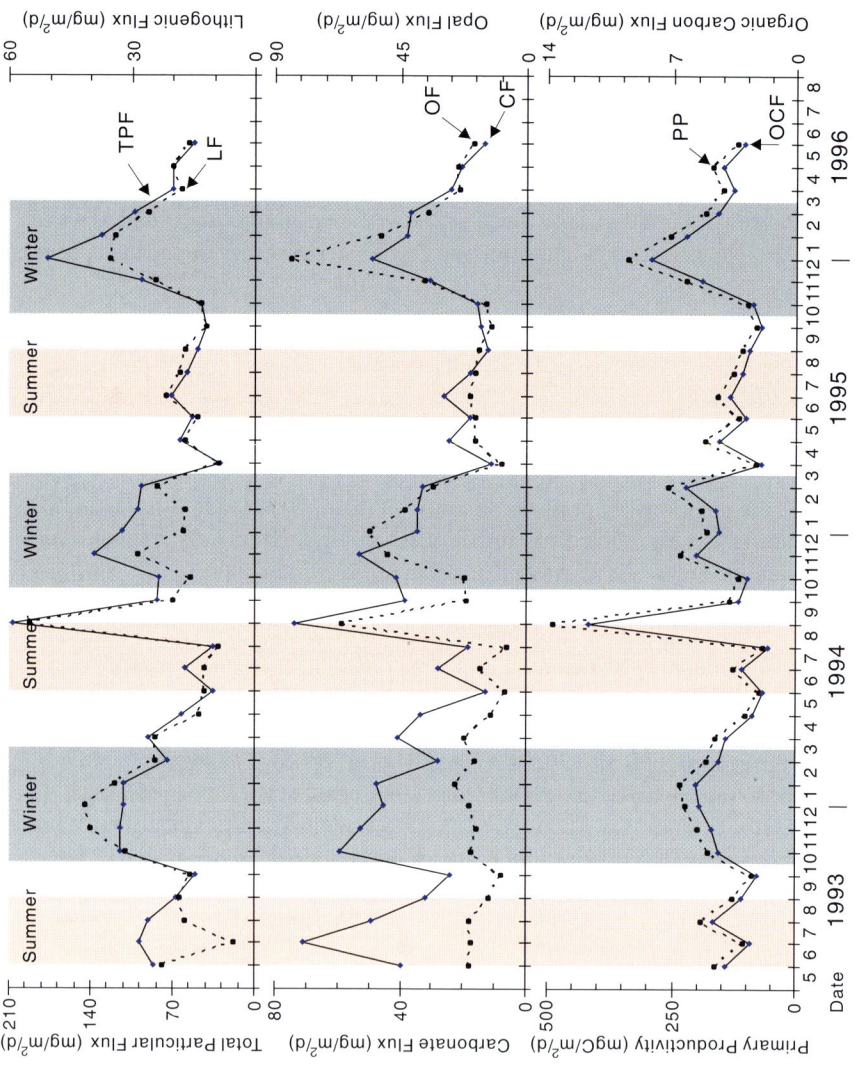

Fig. 4.37 Seasonal changes in particulate fluxes were recorded in a sediment trap deployed at 1,200 m water depth in the central SCS (14°36.2′N, 115°07.1′E) during 1993–1996 (Chen R. et al. 2006)

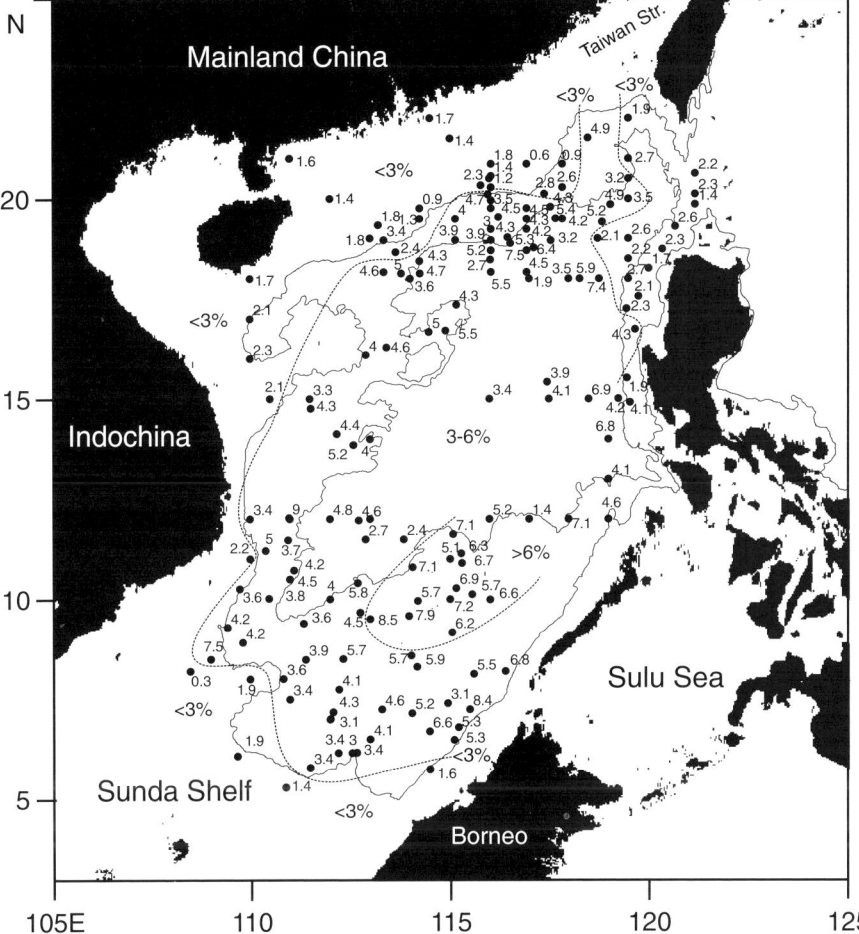

Fig. 4.38 Distribution of opal % in the surface sediment of the SCS was based on over 162 samples. Dashed lines sketch 3% and 6% abundance contours respectively. The 1,000 m and 3,000 m isobaths is marked by *solid lines*

(Dixit et al. 2001). With a low terrigenous input and less organic matters, the main cause to opal dissolution (Ragueneau et al. 2000), deep waters may preserve more biogenic opal.

Quaternary Opal Deposition

The ratio of relative opal abundance (%) vs opal mass accumulation rate (MAR, g cm^{-2} ka^{-1}) between MIS 1 and 2 is low, <1, in late Quaternary samples at northern sites, and slightly higher, >1, at southern and western SCS sites except MD05-2897 and SO 17950-2 (Table 4.4). These statistics indicate opal % and opal-MAR decreasing from MIS 2 to MIS 1 in the northern SCS, but both increasing from MIS 2 to MIS 1 in the southern SCS. This seesaw pattern of opal % and opal-MAR between the northern and southern SCS also occurred in other Quaternary glacial-interglacial intervals (Wang R. et al. 2007).

Variations of opal % and opal-MAR in longer Pleistocene records are shown in Figs. 4.39 and 4.40. At northern sites, such as Site 1144, both opal% and opal-MAR increased markedly near the MIS 22/23 boundary ($\sim$900 ka), at MIS 16 ($\sim$680 ka) and MIS 12 ($\sim$450 ka) (Fig. 4.39) (Wang R. et al. 2007), although the Site 1145 record is not so obvious probably due to low sampling resolution. At southern sites, such as Site 1143, both opal % and opal-MAR started to increase from MIS 11 ($\sim$420 ka) (Fig. 4.40) after a long period with $\sim$1.8% abundance from MIS 48 to MIS 12. Apparently all their variations closely responded to glacial-interglacial cycles, as indicated by the $\delta^{18}O$ records. While opal % and opal-MAR increased stepwise since $\sim$900 ka in the northern SCS with some very high values always in glacial periods, their significant increases started only since 450$-$420 ka in the southern SCS with higher values mostly during interglacial periods (Fig. 4.40).

The opal-MAR increase in glacial northern SCS from 900 ka to 680 ka is considered as linking to the mid-Pleistocene climate transition (Fig. 4.39). An intensified East Asian monsoon, low sea level and increased terrigenous input together caused high surface productivity particularly in the northern SCS (Wang L. et al. 1999; Wang R. et al. 2003, 2004, 2007; Liu K. et al. 2002; Li J. and Wang 2004). Although summer monsoon during interglacial periods could also bring about abundant rainfall discharge, its influence on the northern sites appears to have been very weak when sea level was high like today.

Driven by winter monsoon, abundant eolian dust from Asia continent was deposited in sediments of marginal sea and western Pacific Ocean, causing high surface productivity there (e.g., Rea et al. 1998). Eolian dust transported by winter monsoons to the SCS supplied the nutrient elements, mainly Si and Fe, which stimulate phyto- and zooplankton blooms in the oligotrophic SCS (Chen Y. 2005). Among others, Fe is an important stimulator of high primary productivity in surface waters (Boyd et al. 2000). During the glacial, the North Pacific polar front migrated southwards and brought nutrient-rich temperate Pacific water into the SCS through the Bashi Strait (Wang L. and Wang 1990), forming an alongshore current (Chapter 2). The current and the monsoon-induced upwelling area offshore Luzon are rich in nutrients, also resulting in high surface productivity (Li J. and Wang, 2004).

Paleoproductivity records of nitrogen isotopes, chlorine content and other mineralogical and geochemical results from Site 1144 indicate intensified winter monsoon activity, high terrigenous and eolan inputs and high primary productivity when sea level was low during glacials, and vice versa during interglacals (Higginson et al. 2003; Tamburini et al. 2003). In contrast to these northern SCS records, however, opal% and opal-MAR at southern SCS sites increased significantly from $\sim$450 ka (Wang R. and Li 2003; Wang R. et al. 2007) (Fig. 4.40), although radiolarian abundance at Site 1143 also increased slightly earlier, at $\sim$470 ka (Yang et al. 2002). According to Wiesner et al. (1996), the Site 1143 area was affected by summer monsoon-induced upwelling during interglacial periods.

Table 4.4 Biogenic opal % and opal mass accumulation rate (MAR, g cm^{-2} ka^{-1}) in MIS 1–2 samples are shown for various SCS sites

Site	Age (ka)	MIS 1		MIS 2		Opal-MAR MIS 1/MIS 2	References
		Opal %	Opal-MAR	Opal %	Opal-MAR		
Northern SCS							
Site 1144	0–1,050	3.4	1.25	5.7	6.7	0.19	Wang R. et al. (2007)
Site 1145	0–1,365	3.7	0.3	6.6	0.73	0.41	idem
Site 1146	0–1,800	4.5	0.57	5.7	1.06	0.54	idem
Site 1148	0–2,000	3.9	0.78	5.6	1.16	0.67	idem
MD05-2904	0–264	3.6	1.28	4.8	5.33	0.24	This study
SO 17940-2	0–38	2.9	0.98	3.3	0.58	1.69	Lin et al. (1999)
Western SCS							
SO 17954-3	0–213	3.6	0.14	2.9	0.1	1.4	This study
SO 17950-2	0–200	3.5	2.5	3.4	3.5	0.71	Lin et al. (1999)
MD05-2901	0–448	2.0	0.39	2.3	0.13	3	This study
Southern SCS							
Site 1143	0–1,600	4.5	0.10	2.4	0.09	1.11	Wang R. et al. (2007)
SO 17957-2	0–1,500	17.2	0.24	14.3	0.13	1.85	idem
SO 17959-2	0–800	5.9	0.08	5.1	0.07	1.14	This study
MD05-2897	0–468	2.3	0.03	2.6	0.05	0.6	idem
MD01-2392	0–440	3.1	0.85	1.8	0.45	1.89	idem

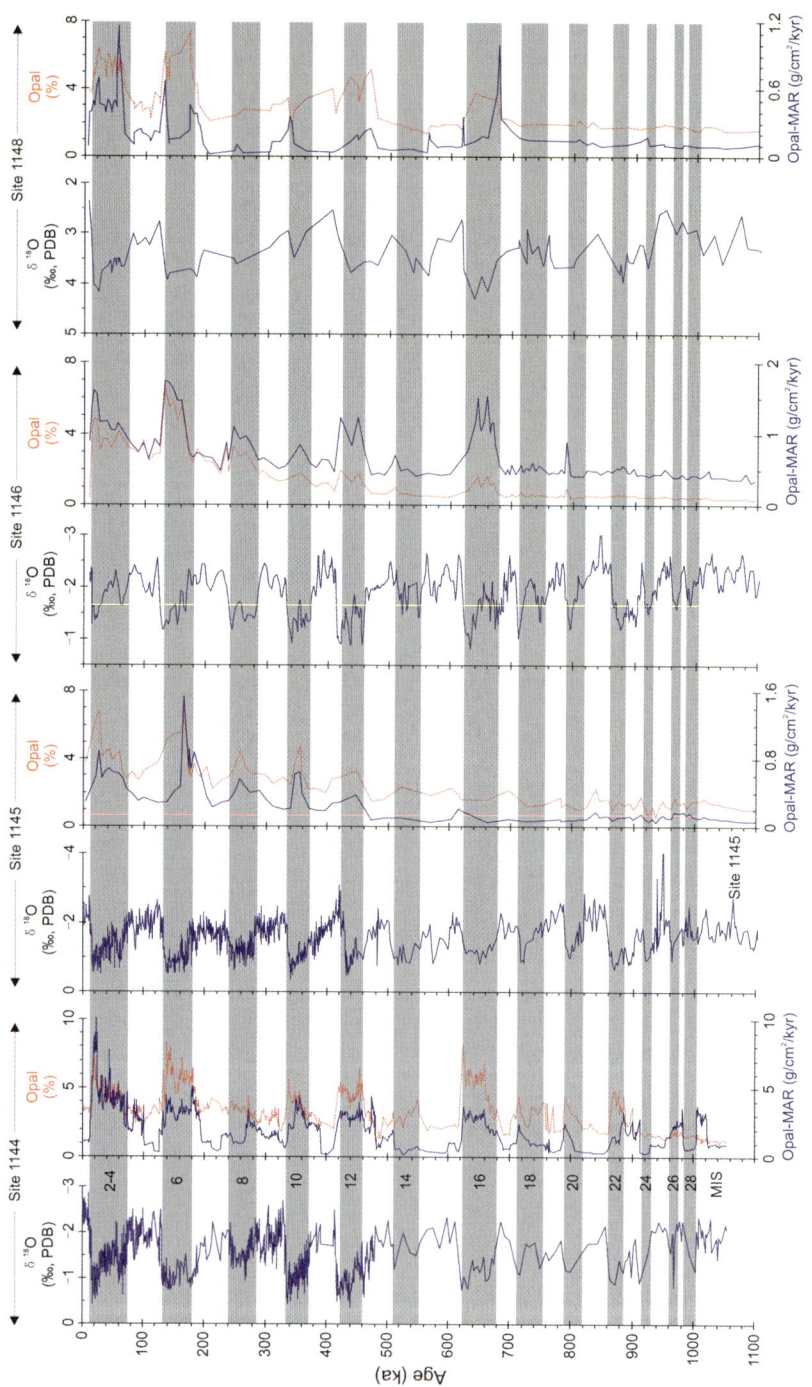

Fig. 4.39 Variations in Pleistocene opal % (*red dashed lines*) and opal-MAR (*solid blue lines*) are correlated to δ^{18}O records at ODP Sites 1144, 1145, 1146 and 1148 from the northern SCS (Wang R. et al. 2007)

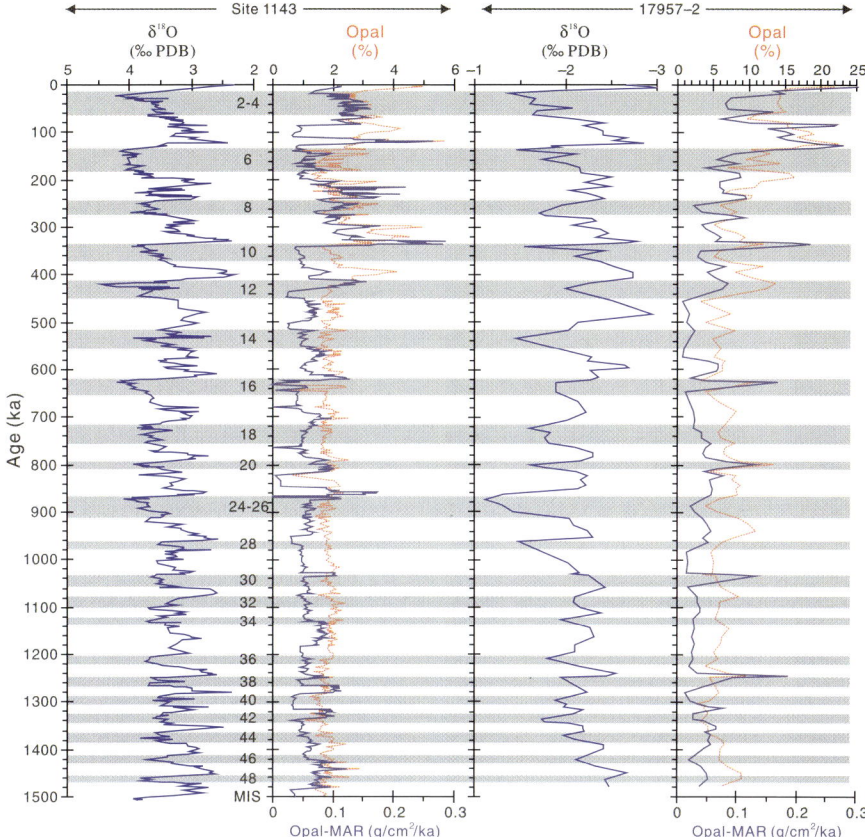

Fig. 4.40 Variations in Pleistocene opal % (*red dashed lines*) and opal-MAR (*solid blue lines*) are correlated to δ¹⁸O records at ODP Site 1143 and core 17957-2 from the southern SCS (Wang R. et al. 2007)

Correlation between thermocline surface radiolarian indexes (TSR) from southern SCS sites and magnetic susceptibility data from the Chinese Loess Plateau confirms that higher sea surface productivity and shoaled thermocline were caused by stronger upwelling and nutrient supply due to prevailing summer monsoon during interglacials (Wang R. and Abelmann 2002). A similar message has also been revealed by mineralogical, geochemical and foraminiferal records from Site 1143 (Tamburini et al. 2003; Xu 2004).

Modern particulate flux data, shown in Fig. 4.37, mirror the Quaternary seesaw pattern of opal-MAR during glacial-interglacial cycles (Wang R. et al. 2007). In the northern SCS, opal-MAR increased during glacials but decreased during interglacials, while the opposite is true for the southern SCS (Figs. 4.39 and 4.40). The glacial- interglacial differences between the northern and southern SCS were obviously caused by monsoons: strenghtened winter monsoon during glacials influenced more on the north while the strengthened summer monsoon during interglacials affected more on the south.

Cross spectrum analyses of opal % at Sites 1143 and 1144 (productivity signal), the benthic $\delta^{18}O$ records (global ice volume signal) and the Earth's orbital parameters (ETP) show a high coefficient between opal % and $\delta^{18}O$ at Site 1144 with distinct 100 kyr and 41 kyr cycles, indicating orbital forcing and ice volume impact on monsoon-induced high productivity in the northern SCS (Fig. 4.41) (Wang R. et al. 2007). Similar results are from pollen records (Herb and *Pinus* contents) at Site 1144, although pollen and ETP spectra show their high coefficient at the 23 kyr precession band (Sun et al. 2003; Tian et al. 2005). All these are consistent with loess and paleosol records from the Chinese Mainland (Ding et al. 1995; An et al. 2001) and eolian sediments in the North Pacific (Rea et al. 1998), confirming an orbital driven monsoon influence.

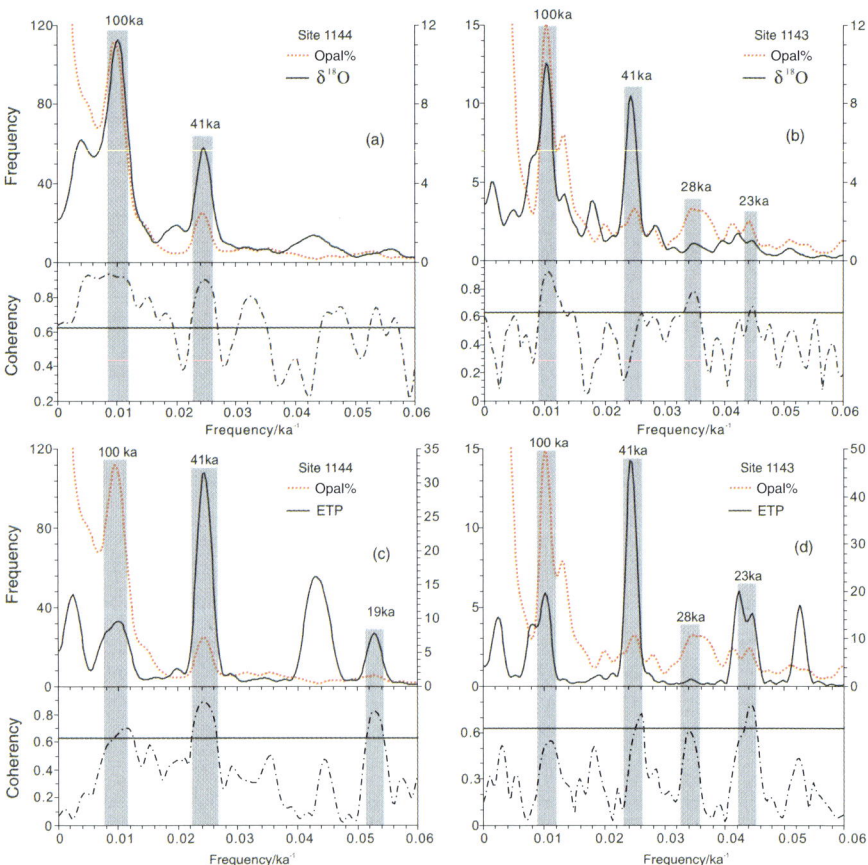

Fig. 4.41 Cross spectral analyses show periodicities and coherencies between opal % and $\delta^{18}O$ at Site 1144 (**a**) and at Site 1143 (**b**), and between ETP and opal % at Site 1144 (**c**) and at Site 1143 (**d**). The *horizontal solid lines* denote coherencies above 80 % standard level. The *grey bars* mark various periodicities (Wang R. et al. 2007)

Winter monsoon strength is controlled by the polar high pressure system in the Northern Hemisphere (Ding et al. 1995; Xiao et al. 1995). As high opal % and high organic carbon fluxes are related to high wind speeds in winter (Chen R. et al. 2006), their similar records from Pleistocene glacial periods in the northern SCS may have been forced by strengthened winter monsoon as global ice volume expansed, as implied also by other studies (Wang L. et al. 1999; Ding et al. 1995; Chen M. and Huang 1998). While the Site 1144 opal records show distinct 100 kyr and 41 kyr cycles, the Site 1143 records exhibit distinct 100 kyr, 28 kyr, and 23 kyr cycles, with the highest correlative coefficient with $\delta^{18}O$ on the 100 kyr band (Fig. 4.41). A phase lead in opal % by $\sim32°$ over $\delta^{18}O$ with a coefficient exceeding 80% may imply a close link between the summer monsoon and the $\delta^{18}O$ at the eccentricity band (Tian et al. 2005), as found also in magnetic susceptibility records (Ding et al. 1995; Kukla et al. 1990). For the 28 kyr periodicity, it probably represents a heterodyne frequency of the primary orbital cycles in a nonlinear nature, which often occurs in the tropical Indian and Pacific Oceans (Prell et al. 1992; Wang P. et al. 2005). The nonlinear response to the 23 kyr (ENSO) and 100 kyr (glacial and interglacial cyclicity) cycles led to the energy transfer from 23 ka cycle to 30 ka and 19 ka cycles (Thevenon et al. 2004). This heterodyne frequency in Site 1143 records may have been due to monsoon changes over the upwelling area in the southern SCS while acting as an important component of the summer monsoon (Jian and Huang 2001). The 28–30 ka heterodyne frequency has also been found in late Quaternary productivity and natural fire records from the Sulu Sea and from the equatorial West Pacific, likely as a long-term ENSO-like forcing on the summer monsoon in glacial cycles (Beaufort et al. 2001, 2003). A relative higher coefficient between opal% and the ETP at the 23 ka band than at the 41 ka band (Fig. 4.41d) also reveals the domination of precession cycle in the low latitude sea areas (Tian et al. 2005), indicating the response of the summer monsoon to orbital forcing mostly at the precession band which controls summer monsoon intensity. More likely, the summer monsoon was driven by a joint force of orbital cycles over the 23 ka band, which is related to solar radiation, and over the 100 ka band, which is associated with the global ice volume (Jian and Huang 2001).

Opal Deposition in the Neogene

Increases at ~3.2 Ma in opal % and TOC % and their MARs at Site 1143 from the southern SCS (Fig. 4.42) may have responded to the formation of ice sheet on the Northern Hemisphere, while their high values from 5.8 Ma to 12.3 Ma may indicate the contemporary Biogenic Bloom Event (Li J. et al. 2002a; Wang R. et al. 2004).

Very high biogenic silica accumulation rates first found in the central equatorial Pacific from the late Miocene to early Pliocene (Leinen 1979) were subsequently confirmed and designated as the Biogenic Bloom Event by scientists of DSDP Leg 85 (Theyer et al. 1985). ODP Leg 138 from the eastern equatorial Pacific also reveals similar results, showing opal-MAR peak values from 7 to 4.5 Ma at Sites 849, 850 and 851 and relative high values from 11.5 to 7.5 Ma at Sites 850 and 851 (Mayer et al. 1992; Farrell et al. 1995) (Fig. 4.43). The concurrence of high opal-MAR with high carbonate-MAR signals a biogenic bloom (Farrell et al. 1995). At ODP

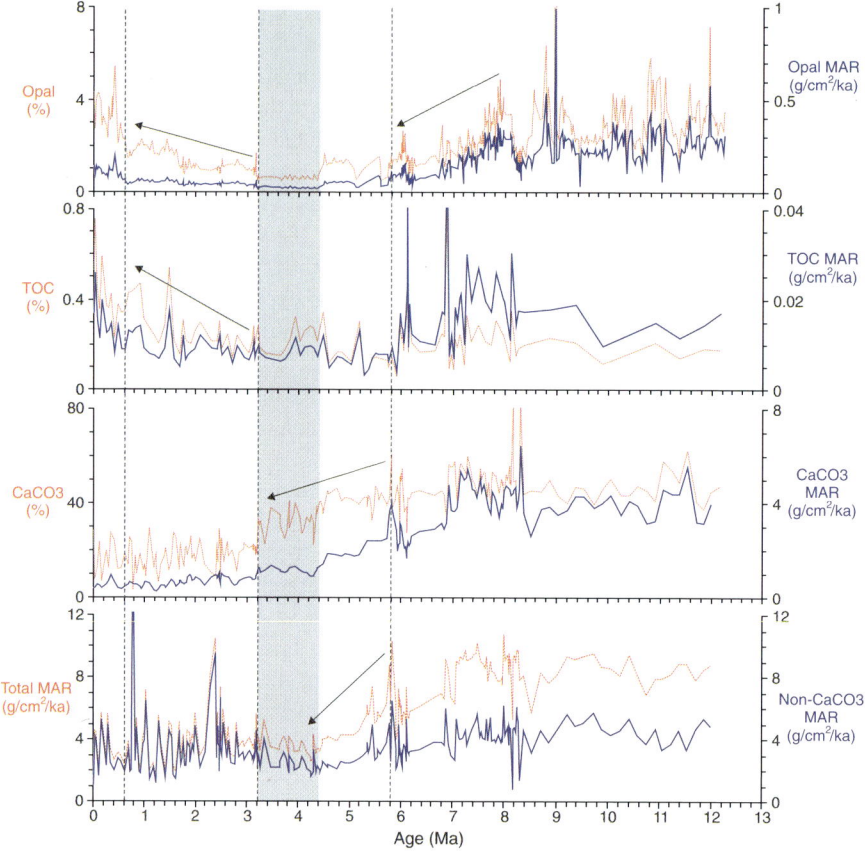

Fig. 4.42 Variations of opal %, opal-MAR and other biogenic components (TOC and CaCO₃) at Site 1143 from the southern SCS indicate high productivity during the period of 12–5 Ma (Wang R. et al. 2004). *Vertical grey bar* marks the period with low opal % and opal-MAR values. Arrows indicate the general trends in different parameters

Site 1021 from the northeastern Pacific, opal-MAR was very high from 10.2 to 7.3 Ma and remained relatively high from ~7 to 4.5 Ma (Lyle et al. 1997). In the western Pacific, high carbonate accumulation rates during 9.4–4.1 Ma at ODP Leg 130 sites on the Ontong Java Plateau may also indicate high biogenic productivity (Berger et al. 1993), although opal-MAR at DSDP Site 586 from the region dropped gradually during 8.8–4.5 Ma (Moberly and Schlanger 1986).

Dickens and co-workers (Dickens and Owen 1996, 1999; Dickens and Barron 1997) found the bloom occurring from the late Miocene to the early Pliocene (~9 to 3.5 Ma) in the Indian-Pacific, especially in upwelling areas with very high productivity. A compilation of the last 15 myr opal deposition data from the world ocean shows that opal main sink areas were in the equatorial eastern Pacific, northeastern and northwestern Pacific and southern South Pacific from 15 to 10 Ma, but shifted to

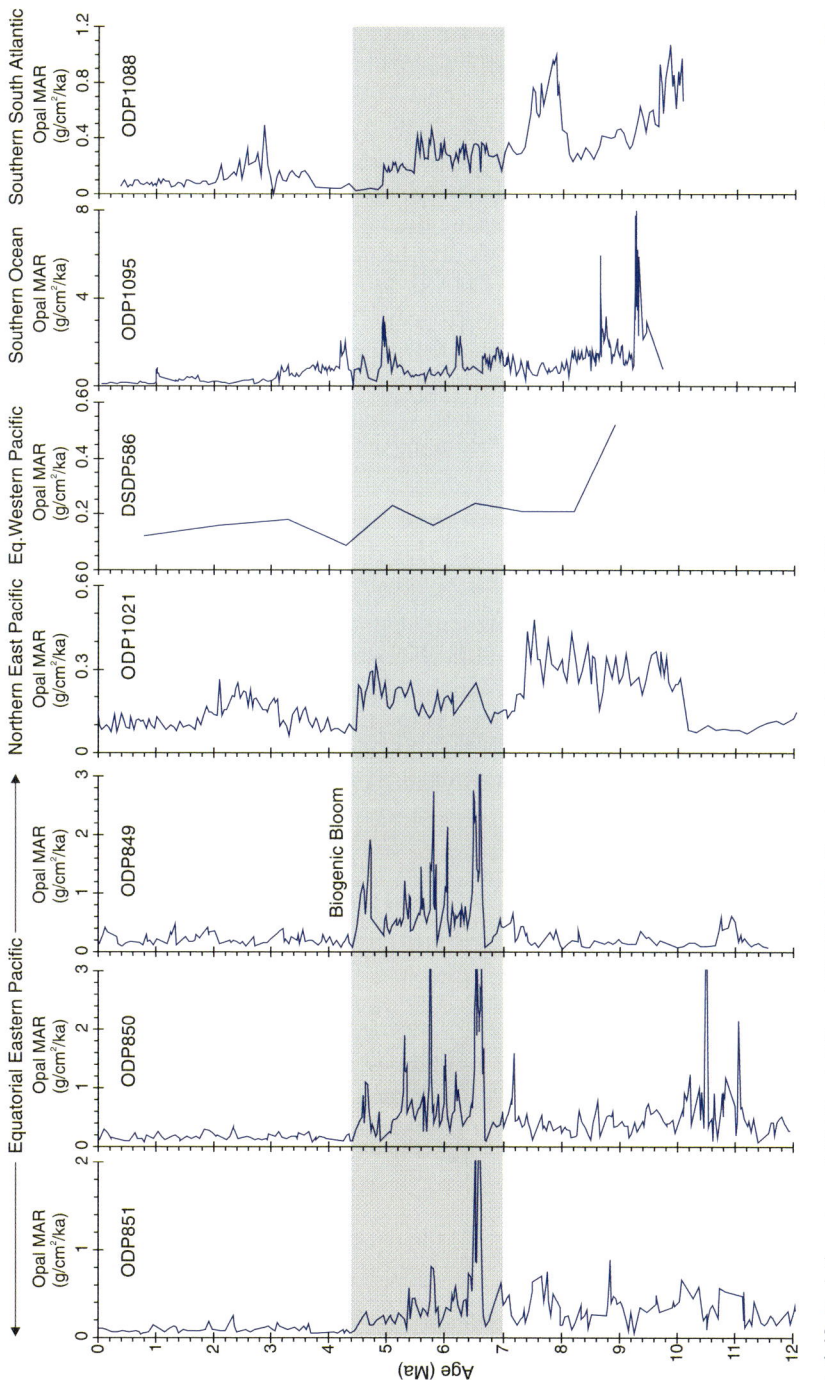

Fig. 4.43 Variations in opal MARs at different DSDP-ODP sites show the originally definded Biogenic Bloom Event (*shaded interval*), compiled from Mayer et al. (1992), Lyle et al. (1997), Moberly and Schlanger (1986), Hillenbrand and Ehrmann (2005) and Diekmann et al. (2003)

the equatorial East and West Pacific and northeastern Pacific from 9 to 3.5 Ma. This shift in opal deposition was obviously linked to reorganization of oceanic circulation and other factors involved in the complex global climatic processes (Cortese et al. 2004). However, a gradual decline in the opal-MAR at ODP Site 1095 in the Southern Ocean from 9.8 to 3.2 Ma and at Site 1088 in the southern South Atlantic from 10 to 4.9 Ma could be simply due to a strong reduction of sea-ice cover in relative warm conditions on Antarctica (Hillenbrand and Ehrmann 2005; Diekmann et al. 2003), if it was not related to the Biogenic Bloom in the Pacific and Indian Oceans.

Therefore, the high productivity record at Site 1143, as indicated by coeval high values in biogenic deposits and their MARs from 12.3 Ma to 5.8 Ma (Fig. 4.42), is considered as corresponding to the Biogenic Bloom Event reported from elsewhere (Li J. et al. 2002a; Wang R. et al. 2004). The significant increase in (pyloniid) radiolarian flux at ∼12 − 11 Ma, close to the middle/late Miocene boundary at Site 1143 may be unrelated but probably linked to the earliest enhencement of the East Asian summer monsoon, which reached a maximum strength at ∼8 Ma (Chen M. et al., 2003b).

Opal Deposition in the Oligocene

Abundant skeletal opal, mainly radiolarians, diatoms and sponge spicules, occurred in samples from 475–600 m at Site 1148 (Fig. 4.44), in an interval corresponding to the middle part of the Oligocene, about 29.5–27.5 Ma. Together with very high H_4SiO_4 values in sediment pore-waters (Fig. 4.45) (Wang R. et al. 2001), high biogenic silica in sediments indicates high biogenic silica productivity and better preservation conditions across the early/late Oligocene boundary in the northern SCS.

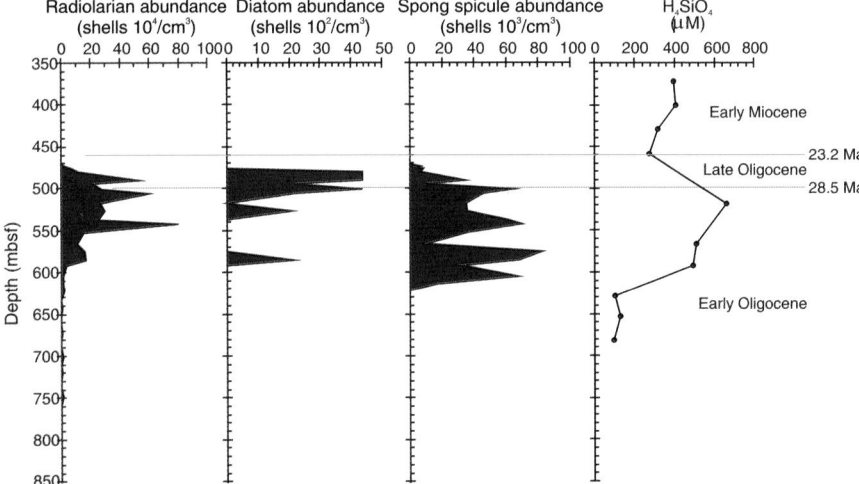

Fig. 4.44 Radiolarians, diatoms and sponge spicules and H_4SiO_4 (in pore-waters) all show very high concentrations between 475 and 600 m at ODP Site 1148 (Wang R. et al. 2001)

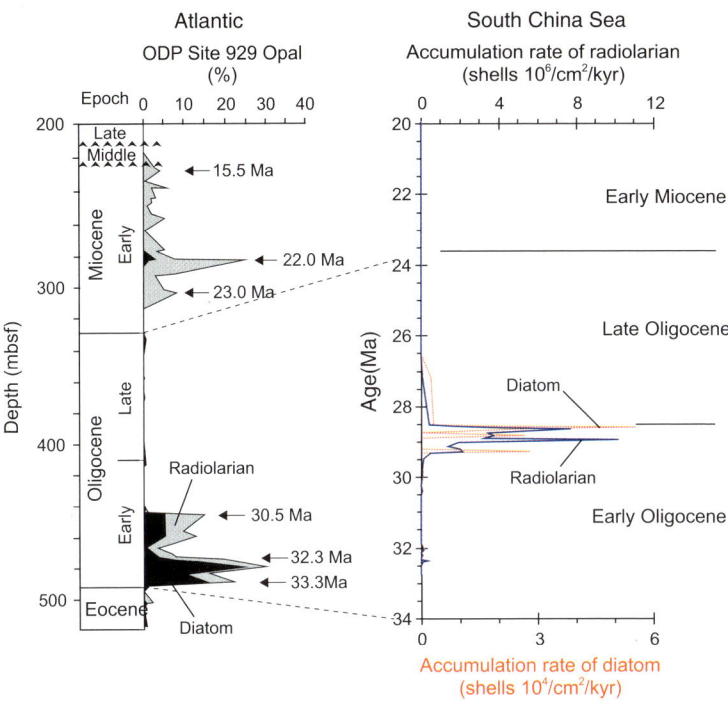

Fig. 4.45 Biogenic opal accumulation at Atlantic Site 929 (Mikkelsen and Barron 1997) and at ODP Site 1148 (Wang R. et al. 2001) shows a contrasting pattern during the Oligocene-Miocene

4.4 Coral Reefs (Yu K. and Zhao J.)

Coral reefs have attracted great attention over the last few decades because they are valuable natural resources and heritage (e.g. habitat, food, medicine, recreation, coastal protection, etc.), and also thought to play a role in modulating global climate and carbon cycle through biological processes. For instance, the "Coral Reef Hypothesis" (Berger 1982) suggests the increase of atmospheric CO_2 during deglaciation resulted from (1) precipitation of carbonate sediments mainly in the coral reef environment, (2) lowering of ocean water pH, (3) reduction in the solubility of CO_2 and (4) subsequent release of CO_2 to the atmosphere.

Recent model also suggests coral reefs play a partial role as a sink of present anthropogenic CO_2 emission (Kinsey and Hopley 1991). However, the importance of coral reefs in global carbon cycle is still a matter of debate. It was considered that the highly-productive coral reefs contribute only about a small portion of the estimated net CO_2 fixation rate in global oceans (Crossland et al. 1991), but the reliability of this estimate is questionable. In addition, the debate on the role of coral reefs in global carbon cycle may also be partially due to the great difference among published estimates of global coral reef areas, varying from 1,440,000 km^2 (Milliman 1974), 617,000 km^2 (Smith 1978), 112,000 km^2 (De Vooys 1979), 1,500,000 km^2 (Copper 1994), 255,000 km^2 (Spalding and Grenfell

1997), to 304,000–345,000 km^2 (Vecsei 2004), with the value of 617,000 km^2 from Smith (1978) being most widely quoted. The huge difference in coral reef area estimation may partly result from the difference on coral reef definition, but is mostly related to the poor knowledge on regional coral reef areas, with the SCS as a typical example. Although coral reefs are widely distributed across the SCS, with reef area around 7974 km^2, matching the Great Barrier Reef in size, latitudinal range and biodiversity, they are rarely mentioned by any existing publications, even including those on global topics (Byran Jr 1953; Milliman 1993; Smith 1978; Smith and Kinsey 1976; Spalding and Grenfell 1997; Spalding et al. 2001). Thus, obtaining accurate information for both regional and global coral reefs becomes essential, particularly when it is more evident that "tropical forcing" plays a significant role in global climate change.

Based on field investigations and literature studies, here we present a summary of the distribution, modern sedimentation, calcium carbonate production and the developmental history of coral reefs in the SCS. Major points derived from this section may not be definitive, mainly because the present understanding of coral reefs in this region is still quite limited.

Modern Coral Reef Distribution

Coral reefs in the studied area of the SCS can be divided into the following seven geographical regions (Fig. 4.46): South China coast and its offshore islands, Hainan Island and its offshore islands, Taiwan Island and its offshore islands, Dongsha islands, Zhongsha Islands, Xisha Islands and Nansha Islands. The former three regions are dominated by fringing reefs, while the latter four islands are mainly composed of atolls.

The estimated total area of modern coral reefs in all these regions is about 7974 km^2 (Table 4.5), including some fore-reef areas estimated by pulsing 6 % of reef flat area for steep-type slopes (Vecsei 2004). Fore-reef is a very important area or zone especially for living corals, but it was seldom included in the coral reef area estimation because of limited knowledge about this zone. Recently, Vecsei (2004) developed a new method for estimating the fore-reef area based the degree of slope.

Fringing Reefs

Fringing reefs are distributed on Leizhou Peninsula, Weizhou Island, Hainan Island and Taiwan Island (Fig. 4.46).The fringing reef on Leizhou Peninsula is the only developed and well-preserved coral reef on mainland China. The reef flat at the Dengloujiao site is about 10 km long and 500–1000 m wide (2 km maximum width) and around 10 km^2 in reef area, and is dominated by *Porites* and *Goniopora* species. Because fringing reef usually develops in relatively shallow water (<10 m), the minor area of its fore-reef is neglected in our reef area estimation. From sea to land, five major biogeomorphologic zones were recognized from this fringing reef (Yu

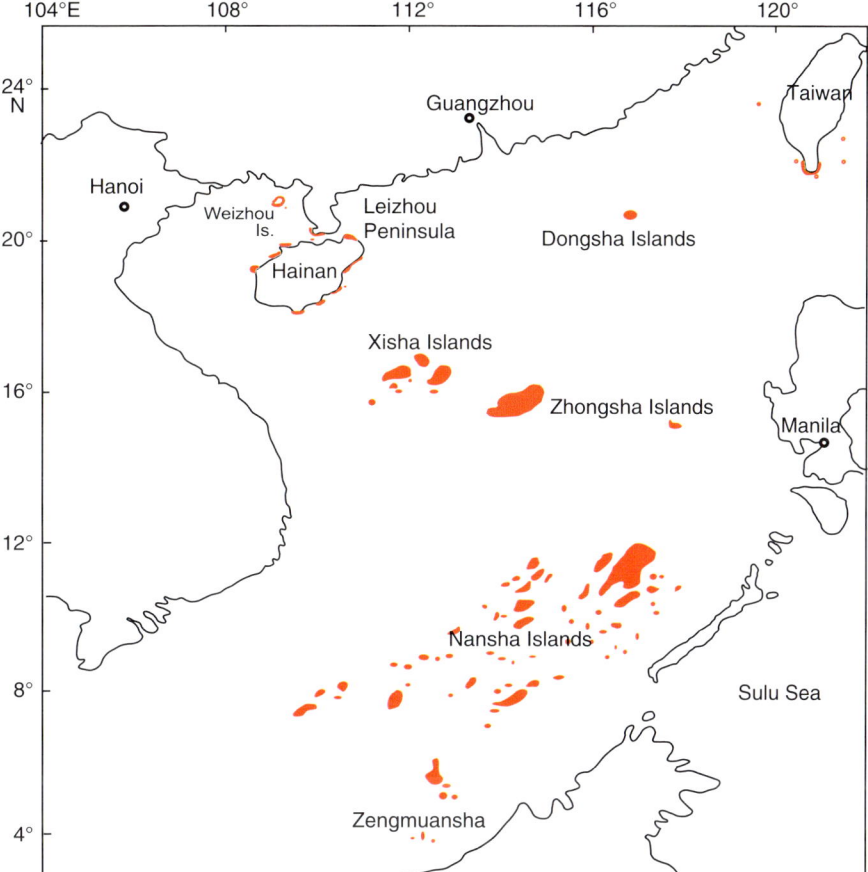

Fig. 4.46 Map shows coral reef distribution in the SCS (after Zhao and Yu 1999)

et al. 2002c). Evidences accumulated since 1993 reveal that the coral reef ecosystem is in its natural recovery. 25 genera and 39 species of reef-building corals have been recorded (Yu 2005). This relatively high-latitude coral reef is particularly good for paleo-environmental research, because it is very sensitive to environmental change. Fossil corals collected from this reef have been used to reconstruct multiple sea level high-stands since mid-Holocene (Nie et al. 1997b; Yu et al. 2002c; Zhao and Yu 2002), high frequency, large amplitude, abrupt cold events ("Leizhou Events") (Yu et al. 2002a) and cold SST-induced coral mortality (cold- bleaching) (Yu et al. 2004c) during the mid-Holocene (7.0–7.5 ka BP).

The coral reef on Weizhou Island (Fig. 4.47) is about 300–400 m wide with branching corals such as *Acropora* as the dominant species. Its water depth is about 3–4 m. About 58 reef-building species from 23 genera have been recorded (Yu et al. 2004a; Zou et al. 1988). The coral reef area for this island is about 6–8 km^2. It represents part of the northmost belt of coral reefs in the SCS. According to Yu et al.

Table 4.5 Estimated area of coral reefs is shown for various regions of the SCS

Region	Reef flat (km²)	Lagoon (km²)	Fore-reef[1] (km²)	Total area (km²)	References
I. S China coast and offshore islands	17	0	0	17.0	This work
II. Hainan and offshore islands	180	0	0	180.0	This work
III. Taiwan and offshore islands	940	0	0	940.0	Spalding et al. (2001)
IV. Dongsha (Pratas) Islands	125	292	7.5	424.5	Zhao and Yu (1999)
V. Huangyandao, Zhongsha Is.	53	77	3.2	133.2	idem
V. Zhongsha Is. (Macclesfield Bank)	1495			1495.0	idem
VI. Xisha (Paracel) Islands	221.6	1614.8	13.3	1849.7	idem
VII. Nansha (Spratly) Islands	507.5	2396.8	30.5	2934.8	idem
Total (km²)	3539.1	4380.6	54.5	7974.2	idem

[1] Fore-reef area was calculated using the method of Vecsei (2001) by pulsing 6 % of reef flat area.

(2004a), sea surface temperature (SST) is an important factor in affecting the coral reef ecosystem development on this island. The age distribution from the data of Liang and Li (2002) suggests that the reef development history is quite similar to that on Leizhou Peninsula, i.e., within Holocene (Yu et al. 2002c; Zhao and Yu 2002).

The coral reef length along Hainan Island is about 200 km, which is about 11.5% of the total coastline length of Hainan Island and its 132 offshore islands. The width of the reef flats varies from 500 m to 2 km (4 km at most). The estimated coral reef

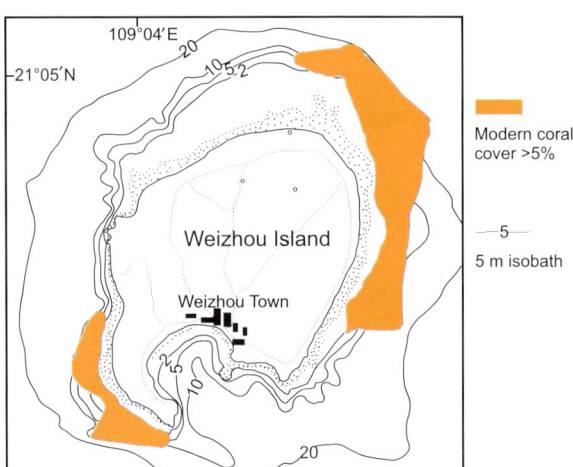

Fig. 4.47 Map shows coral reef distribution on Weizhou Island, northern SCS (after Yu et al. 2004a)

area for this region is about $180\,\mathrm{km}^2$. About 110 reef-building species from 34 genera were recorded for this region (Nie et al. 1997c). Drilled cores reveal that the coral reef thickness is less than 10 m, and dating suggests these coral reefs developed in the Holocene (see review of Zhao and Yu 1999). The most thoroughly-studied reef in this region is the fringing reef at Luhuitou (Fig. 4.48), Sanya, southern Hainan Island. The Luhuitou fringing reef is about 3500 m long and 250–500 m wide, and consists of about 70% coral species so far reported for Hainan Island and its offshore islands. This reef was subdivided into the following major biogeomorphological zones: sandy beach (10–25 m wide), inner reef flat (38–128 m wide), outer reef flat (136–198 m) and reef slope (Zhang 2001). Long-term monitoring suggests that the live coral cover and biodiversity at this site have declined seriously over the last 50 years (Zhao M. et al. in press).

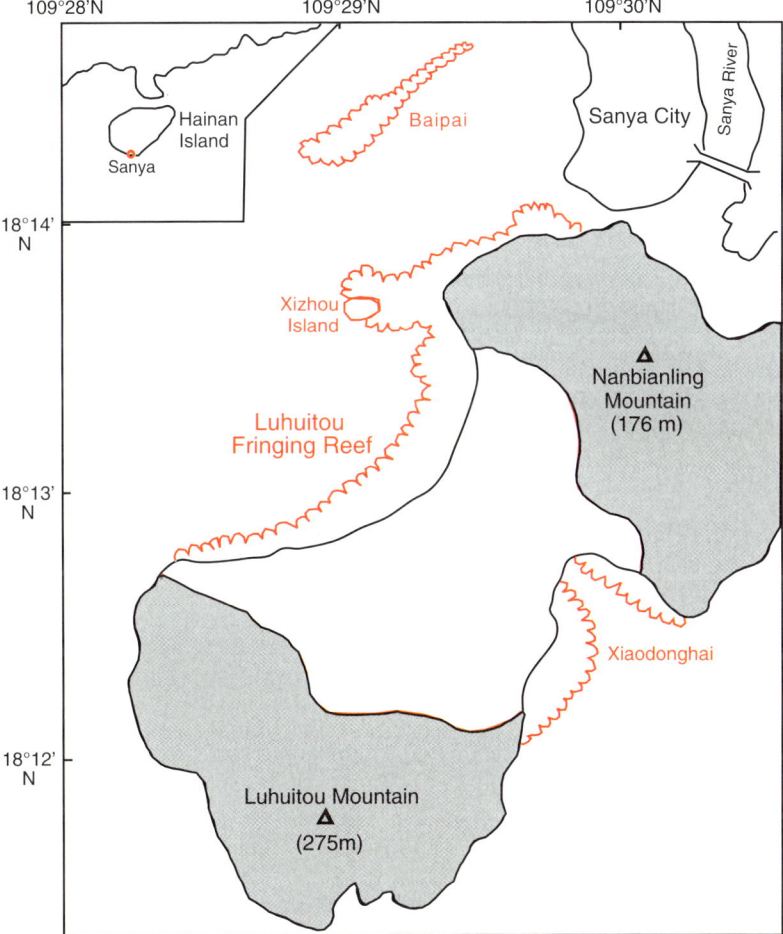

Fig. 4.48 A map shows coral reef distribution on Luhuitou, Sanya, southern Hainan Island

Taiwan Island lies in relatively high-latitude area of the SCS, but it is affected along its southern and eastern edges by the warm Kuroshio Current, resulting in well- developed coral reefs. The best-known and developed reefs of this region are the fringing reefs on Hengchun Peninsula. These reefs occur along a coastline of about 100 km with their reef flats varying from several meters to 250 m wide. Because the coastline along this Peninsula is often broken up by sand channels, the fringing reefs form a discontinuous structure. About 300 reef-building corals representing over 55 genera have been recorded in this reef region (Dai 1991; Nie et al. 1997c; Spalding et al. 2001). The estimated coral reef area in this region is about 940 km^2 (Spalding et al. 2001). Significant progress has been made over the last decade in the study of coral reefs in Taiwan, especially in the field of biology and ecology. For example, Soong and Chen (2003) investigated coral generation using fragments of *Acropora* corals in southern Taiwan; other researchers (Fan et al. 2005; Liu P.J. et al. 2005; Twan et al. 2006) studied coral mass spawning in southern Taiwan; Tsai et al. (2004) investigated the macroalgal assemblage structure on a coral reef in Nanwan Bay (southern Taiwan); and Wallace and Dai (1997) reviewed *Acropora* coral genus across the entire Taiwan Island. Soft corals in Taiwan were also well studied (Benayahu et al. 2004; Benayahu and Perkol-Finkel 2004; El-Gamal et al. 2004).

Atolls

Most coral reefs developed on Dongsha Islands, Zhongsha Islands, Xisha Islands and Nansha Islands (Fig. 4.46) are atolls.

Coral reefs on Dongsha Islands mainly cluster around Dongsha Atoll, which is a circular-shaped atoll with a length of 25.6 km in the NE-SW direction and a width of 20.4 km in the NW-SW direction, and is enclosed by discontinuous reef flats (1.6 to 5.0 km in width). A lime-sand island with an area about 1.7 km^2 is standing on its northwest reef flat. The total coral reef area for this region is 424.5 km^2, including about 292 km^2 as lagoon, 125 km^2 as reef flat and 7.5 km^2 as the fore-reef area (Table 4.5). Forty-four reef-building coral genera have been reported for this region (see review of Nie et al. 1997c).

The coral reefs in Zhongsha Islands include coral reefs on Zhongsha Atoll and those of Huangyan Island. Zhongsha Atoll is a huge ellipse-shaped submerged atoll, with a length of 140 km in the NE-SW direction and a width of 61 km in the NW-SE direction and covers an area about 8540 km^2. The enclosed lagoon has a water depth of 75–85 m. The patched reefs in the lagoon have summits of 15–20 m tall and are well covered by reef-building corals. The submerged reef flats (rim) are 13–20 m deep, and dominated by massive reef-building corals (*Porites*, *Favia*, *Goniastrea*, *Cyphastrea*) at water depth less than 60 m. The area >60 m water depth on the rims is dominated by coralline algae (Corallinaceae), followed by reef-building corals (*Favia*, *Montipora*, *Galaxea*, *Alveopora*, *Leptastrea*, *Lobophyllia*, *Echinophyllia*, *Leptoseris*, *Acropora*, *Pocillopora*, *Seriatopora*, *Acrhelia*), and a few other organisms including non-reef building corals (*Caryophyllia* and *dendrophyllia*) (Nie et al. 1992). A large area of well-developed coral reefs at water depth of 10–20 m was

also discovered in May 2004. The coral reef area for this atoll has never been estimated, although the coral reefs are well developed. Considering the fact that lagoons
often make up a large proportion of the coral reef systems worldwide (Kennedy and
Woodroffe 2000) (e.g., making up 90% of Enewetak Atoll (Yamano et al. 2001))
and ~82.5% of the coral reef systems in Nansha area (Zhao and Yu 1999), we
assume 17.5% of the whole area, or 1495 km^2, as shallow-water reef flat area. The
fore-reef and the lagoon areas cannot be estimated as they are submerged. Huangyan
Island is an isosceles triangle-shaped atoll with a perimeter about 46 km. The reef
flat emerges at low tide, with a width varying from tens to hundreds meters, and
covers about 53 km^2. The enclosed lagoon has a water depth around 9–11 m (20 m
at most) and covers about 77 km^2. The coral reef area on Huangyan Island is about
133.2 km^2 including ~3.2 km^2 fore-reef area. Therefore, the total reef flat area in
this region, including Dongsha Atoll and Huangyan island, is about 1628.2 km^2.
Thirty-four reef-building coral genera were reported for this region (see review
of Nie et al. 1997c). Xisha Islands consist of 29 lime-sand islands, with Yongxing Island (15°50′N, 112°20′E) (Fig. 4.49a), ~1.8 km^2 of exposed area, being the
largest. Some atolls were described by Wang G. (2001). The total coral reef area
(including low tide emerged reef flat and the enclosed shallow water lagoon) in
this region is about 1836.4 km^2. 127 species from 38 reef-building coral genera

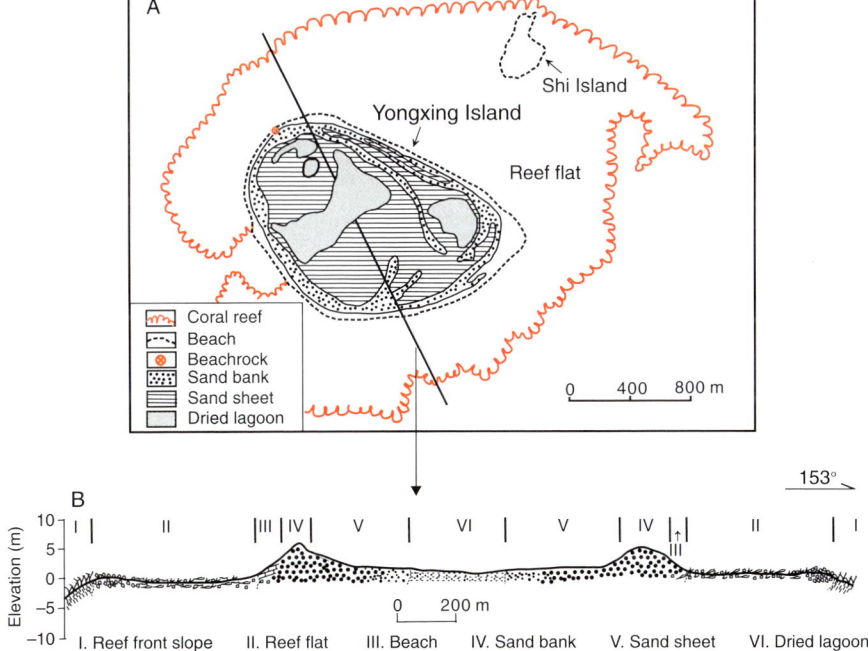

Fig. 4.49 (**A**) Map and (**B**) cross section show geomorphological zonation of Yongxing Island,
Xisha Islands

were reported for this region (Nie et al. 1997c; Zou et al. 1983). Five geomorphological zones were identified on Yongxing Island as reef front slope, reef flat, sandy beach, sand bank, sand sheet and low-lying dried lagoon (Fig. 4.49B) with beach-rock developed on the northwest beach (Yu et al. 1995). Radiocarbon dates of 2680 ± 95 yr BP for the beach-rock (Yu et al. 1995) and 6790 ± 90 yr BP for a sample from 10 m depth of a drill-core (Ye et al. 1985) suggest that the sand island started accumulation from at least mid-Holocene and exposed in late Holocene. A survey carried out in 2002 on Yongxing Island (Li Y. et al. 2004) suggested that the coral cover and coral biodiversity were still high but showing an obvious ecological decline if comparing with the observation in 1978, although this area is several hundred kilometers away from mainland and is rarely influenced by human activities. Because no systemic ecological survey has been carried in this area since early 1980s, little is known about the coral reef's response to global climate change. No information is available either on high temperature-induced coral bleaching that occurred worldwide as the result of global warming. Annual growth rates of massive *Porites lutea* corals range from 7 to 15 mm/yr with an average of 11 mm/yr, and show strong positive correlation with instrumental SST (Nie et al. 1997a). Using the coral growth rate as a SST proxy, Nie et al. (1999) reconstructed the SST variation for the last 220 years in the Xisha waters and found an overall warming trend.

Nansha Islands (Fig. 4.46) consist of 117 reefs, among which 64 are atolls (with 43 low-tide emerged atolls and 21 submerged atolls), and 23 are lime-sand islands, with the largest one, Taiping Island, having an exposed area of $0.43 \, \text{km}^2$. The total area of emerged reef flats and their enclosed shallow water lagoons in this region is about $2904 \, \text{km}^2$. None of the 64 atolls was shown in the world's list of 425 atolls (Byran Jr 1953; Stoddart 1965). 44 reef-building coral genera were reported for this region (Nie et al. 1997c). The best studied reef in this region is Yongshu Reef ($9°32'–42'$N, $112°52'–113°04'$E, Yu et al. 2006a), which is an open spindle-shaped atoll, about 25 km long in the NEE-SWW direction and 6 km wide in the NW-SE direction, and with a total area about $110 \, \text{km}^2$. A closed lagoon (380 m long, 150 m wide and maximum 12 m deep) is situated in the center of the southwest reef flat. The area around the lagoon is termed as the "small atoll". Based on systematic field investigations, six biogeological and sedimentary zones are recognized on the small atoll (Yu et al. 2004b). Detailed studies of drilled cores (Yu et al. 2006b; Zhao et al. 1992; Zhu et al. 1997) from the reef flat and the lagoon show that the coral reef started to develop at a depth of 17–18 m about 7350 to 8000 years ago, and experienced sustained subsidence since its development. The living coral growth rate of *Porites lutea* varies from 10 to 16 mm per year. High-resolution skeletal $\delta^{18}O$ and $\delta^{13}C$ (covering a period from 1950 to 1999) of a living *Porites lutea* from Yongshu Reef were reported by Yu et al. (2001, 2002b), which reveals a general warming trend in SST, consistent with the global trend.

Carbonate Platform Sediments and Calcium Carbonate Production

Coral Reef Zonation and Sediments

In modern coral reefs, carbonate is deposited in two distinct manners: as reef framework or as unconsolidated sediments (Milliman 1974). In the SCS, the reef framework is constructed mainly by massive reef-building corals (such as *Porites*, *Diploastrea*, *Platygyra*, *Favia* and *Favites*), and is further cemented by coralline algae (red algae) (Nie et al. 1997c; Zhao et al. 1992; Zhao and Yu 1999). Both surface investigations and drill cores reveal that the framework composition usually takes <30–40% of the whole reef materials although they play a very important role in the reef-building process. The bulk of the coral reef materials (60–70%) are composed of coral branches, gravels and the unconsolidated sands originated from reef organisms (such as corals, coralline algae, mollusks, *Halimeda*, foraminifera, bryozoans, echinoderms, brachiopods, spongs and sea urchin). Most coral reefs in the SCS can be subdivided into different sedimentation zones (facies), such as reef front slope, reef flat, lagoon slope and lagoon. Characteristics of different zones were used in paleo-facies reconstruction in the analysis of Quaternary coral reef cores (Zhao et al. 1992; Zhu et al. 1997). The distribution of these carbonate sediments or organic components is largely dependent upon the regional hydrological, ecological and sedimentary environments. It was noticed that there is great zonal discrimination in calcification rates (Kinsey 1985) and production (Vecsei 2004), with higher rates present in zones of high coral cover. In order to obtain a more detailed understanding of the geomorphology, biology, ecology and sedimentation of coral reefs, some sub-zones were further divided on the basis of traditional zones.

Zonation of fringing reefs: Zonation is typical of most fringing reefs around the SCS. Here we use the fringing reef on Leizhou Peninsula as an example to illustrate the distribution of different zones. From sea to land, five major biogeomorphologic zones were recognized from this reef (Fig. 4.50) (Nie et al. 1997b; Yu 2005; Yu et al. 2002a,c, 2004c), but characteristics of the loose sediments on this reef have never been reported.

I. Reef-front living coral zone: This zone varies from 10 to 120 m in width and is dominated by *Acropora* corals, which have reached the low spring tide level. The living coral coverage is usually <20%, with the exception of some well-developed area (~0.5 km^2) where the cover coverage is up to 85%. Living *Porites* microatolls, with a diameter about 0.6–1.2 m, are developed sporadically around the inner (landward) part of this zone. During a field investigation in April 2004, we noticed the catastrophic impact of massive growth of macro algae over the living coral surface, which resulted in the decline of living reef coral population. The loose sediments are mainly composed of terrestrial quartz, some organisms, coral fragments, mollusks and coralline algae.

II. Outer reef-flat massive Porites zone: This zone varies from 100 to 150 m in width and is dominated by dead, large massive *Porites* corals (5–6 m in

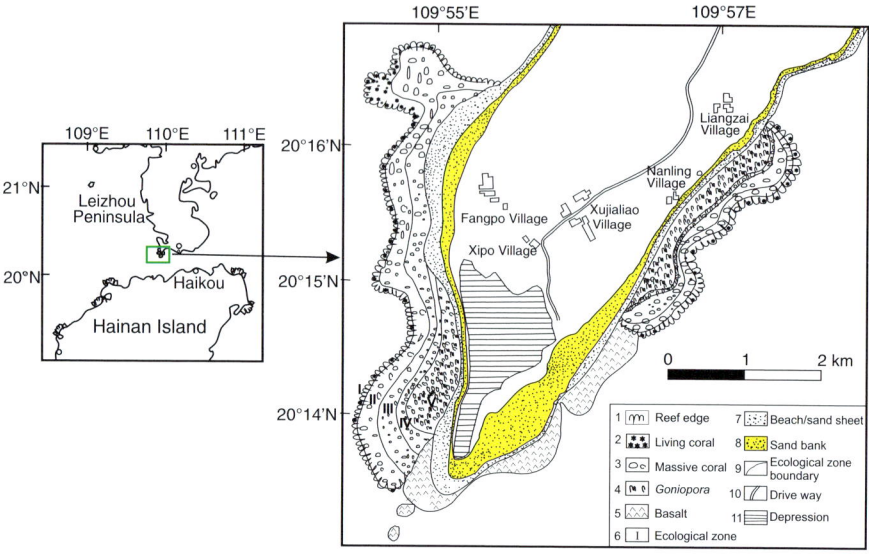

Fig. 4.50 Coral reef distribution and zonation on Dengloujiao, Leizhou Peninsula, show typical fringing reef characters in the SCS

diameter) with elevation about 1–2 m above the reef flat surface. Some of these massive corals occur as microatolls with diameter up to 9.6 m, which multiple rims probably reflecting sea level fluctuations over the period of growth. In the depressions of this zone, there are some small living corals of dominantly massive species, such as *Porites lutea, Galaxea, platygyra*.

III. Within-reef-flat mixed massive coral zone: This zone, ~200 m wide, consists of large dead *Porites, Pavona* and *Favia* corals with some as microatolls. The huge *Pavona* corals are usually 8–10 m in diameter with elevation about 30–70 cm above the reef flat surface.

Detailed elevation survey suggests the dead microatolls in Zones II and III are about 179 to 219 cm above the modern *Porites* microatoll surfaces in Zone 1. TIMS U-series dating of the rims of the dead microatolls show that they developed between 7104±32 and 6657±41 yr BP, suggesting that the sea level was about 179–219 cm above the present level in the mid-Holocene (Yu et al. 2009a; Zhao and Yu 2007).

IV. Within-reef-flat mixed massive-coral/*Goniopora* zone: This zone, 0–120 m wide, contains smaller massive corals mixed with branching *Goniopora* corals.

TIMS U-series ages for 57 in situ *Porites* and 2 *Favia* corals from the above three fossil coral zones show that the majority (~65%) of the corals formed during the period of 7100-6500 years before present (yr BP). The remaining ~35% corals grew periodically since 6300 yr BP, with three major groupings at 6245±15 to 5676±14, 4156±23 to 3675±23 and 2795±14 to 2321±23 yr BP, and two individual dates at 5009±54 and 1511±23 yr BP, respectively. The data suggest that multiple sea-level highstands of at least ~2 m above the present

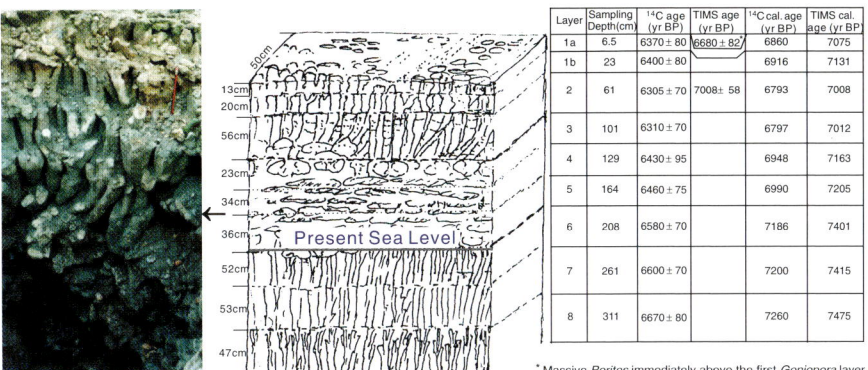

Layer	Sampling Depth(cm)	¹⁴C age (yr BP)	TIMS age (yr BP)	¹⁴C cal. age (yr BP)	TIMS cal. age (yr BP)
1a	6.5	6370 ± 80	6680 ± 82*	6860	7075
1b	23	6400 ± 80		6916	7131
2	61	6305 ± 70	7008 ± 58	6793	7008
3	101	6310 ± 70		6797	7012
4	129	6430 ± 95		6948	7163
5	164	6460 ± 75		6990	7205
6	208	6580 ± 70		7186	7401
7	261	6600 ± 70		7200	7415
8	311	6670 ± 80		7260	7475

* Massive *Porites* immediately above the first *Goniopora* layer.

Fig. 4.51 Cross section of *Goniopora* reef profile on Leizhou Peninsula shows 9 layers in its development, with hiatuses between layers clearly visible. *Goniopora* coral heads on the top surfaces of layers 3, 4 and 5 were eroded. Also shown are measured ¹⁴C (1δ error) and TIMS U-series (2δ error) dates, and their relative depths in the profile. Age values in ¹⁴C cal. ages (aBP) column refer to calibrated calendar years. Age values in the TIMS cal. age (aBP) column are further corrected ages based on TIMS U-series date of 7008±58 aBP for the *Favia* coral sample from layer 2 (after Yu et al. 2004c)

level occurred when the above corals formed (Yu et al. 2002c; Zhao and Yu 2002, 2006, 2007; Zhao et al. submitted).

V. Inner reef-flat *Goniopora* zone: This is a special zone among all reef sites published. It is 100–500 m wide and is dominated by branching *Goniopora* corals with a spatial coverage of >95%. Four wells (referred to as NLH, NLW, DLH and DLW), each 3–4 m in depth, were excavated since 2001 (Yu et al. 2002a,c, 2004c) in the *Goniopora* reef flats in 1998 and 2000. Detailed field observations suggest that the coral profiles in all the wells had almost identical ecological characteristics, and their stratigraphic sequences can be correlated to each other (Yu et al. 2002a, 2004c). From the surface downward, the *Goniopora* reef profile of DLW can be divided into at least 8 layers (Fig. 4.51). Growth discontinuities between individual layers are clearly defined, which suggests multiple stops and recovery in the process of coral reef development. Further study reveals that this reef profile developed during 7.5–7.0 ka BP, recording at least nine abrupt massive *Goniopora* stress and mortality events ("cold-bleaching") during the winter time (within the Holocene climatic optimum) (Yu et al. 2004c). This study provides the first pre-historic evidence for cold-bleaching of reef corals at higher latitudes and adds new dimension to the understanding of coral bleaching. The results also show that it took about 20–25 years for a bleached *Goniopora* coral reef to recover. During this period, sea level rose by ∼3.4 m, with present sea-level reached at ∼7.3 ka BP and a sea-level highstand of at least ∼1.8 m occurred at ∼7.0 ka BP.

Zonation of atolls: Atoll is the most widespread type of coral reefs on Nansha Islands, exemplified by Yongshu Reef of Nansha area as described below. Here

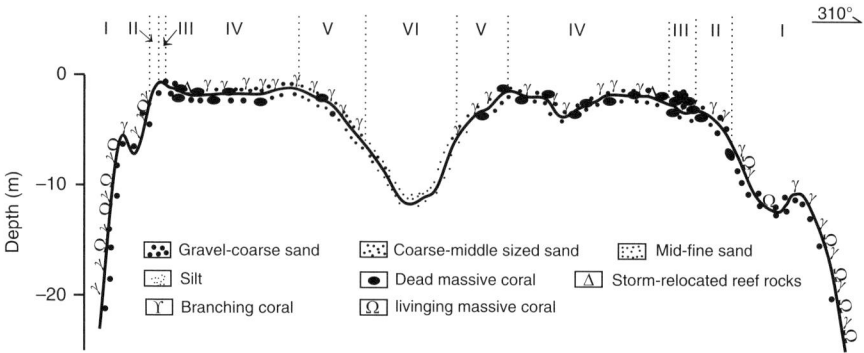

Fig. 4.52 Sketch profile shows biogeological and sedimentary zones (I to VI) for a small atoll on Yongshu Reef (after Yu et al. 2004b)

six biogeological and sedimentary zones have been recognized for a small atoll of Yongshu Reef (e.g. Yu et al. 2004b; Zhao et al. 1992). From outer to inner zones (Fig. 4.52), these are (I) reef-front living coral zone, (II) outer reef-flat coral zone, (III) reef-ridge coral-branch-cemented zone, (IV) inner reef-flat branching-coral/sand zone, (V) lagoon slope branching-coral/fine-sand zone, and (VI) lagoon basin-floor silt zone. Based on detailed field investigation on another seven atolls, this zonation pattern is very similar among all atolls in this region.

   I. Reef-front living coral zone: This zone, or called "outer slope" in some cases (Guilcher 1988), surrounds the whole reef with differing slopes at different sites, which is interrupted by a terrace at a depth of 17–18 m. Other terraces were also reported at water depths from 40 to 1800–1850 m (Fig. 4.53; Zhao and Yu 1999). The presence of submerged reef terraces reflects the response of coral reef development to environmental changes, showing evidence for multiple exposure, erosion and re-development during the entire reef-building

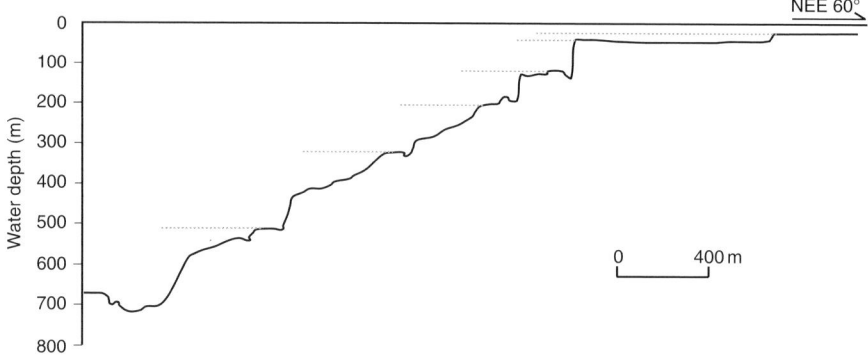

Fig. 4.53 Sketch profile shows terraces developed on the reef-front slope, southeast Yongshu Reef, Nansha Islands (modified from Zhao et al. 1992)

history. This zone consists of numerous spur-and-groove systems, each with 4–5 m width and 2–3 m depth, covered with living corals of dominantly branching *Pocillopora* and *Acropora*. Large massive *Porites lutea* corals (>1.5 m in diameter) are also present, but not widespread, growing on the spurs 3–4 m underwater. Two cores, each about 1.4 m long, were drilled in 1999 and the upper 50-year section of one core was analyzed for C & O isotopes for SST reconstruction (Yu et al. 2001). Other massive corals such as *Favia*, *Goniastera* and *Galaxea* can also be found, but are not as large as *Porites* corals. This zone shows great biodiversity, with almost all species in the Nansha area being found. Living coral coverage within this zone is up to 90%. The sandy sediments are mainly composed of the debris of corals, coralline algae and molluscs.

II. Outer reef-flat coral zone: The surface of this zone, about 10 m wide, is cemented by algae and it is saw-shaped toward the sea, as a result of cutting by the groove and spur systems. Wave-reef interaction keeps this zone wet, allowing many massive and branching coral species to grow, although the size is usually very small. Massive corals, such as *Porites lutea*, are no more than 30 cm in diameter. Some branches and coral gravels can be found in depressions of this zone.

III. Reef-ridge coral-branch-cemented zone: This zone is composed of semi-cemented broken coral branches, rubbles and gravels. It is about 8–10 m wide and 20 cm higher than the reef flat surface. No living coral can be found here. During daily low-tide period, this zone is exposed like a white ring, which is also the highest part of the atoll.

IV. Inner reef-flat branching-coral and sand zone: This is the widest part of the atoll with width varying from about 500 m along the short axis to about 1600 m along the long axis. The usual water depth is about 1–2 m, increasing toward the lagoon. During low spring tides, this zone is generally exposed to the atmosphere. However, numerous pools and long water-filled stripes of various sizes exist, providing habitats for living branching corals and some small massive corals. Towards the lagoon, the coverage of living corals, dominated by *Acropora*, increases (up to 60%), but they may not have a long growth history due to bleaching or other causes. Mass mortality of these branching corals frequently occurs in this zone. Some small living massive *Porites* (<35 cm in diameter) can be found within the *Acropora* colonies. This zone is covered with many in situ dead *Porites* corals and microatolls (with diameter 1–2.5 m), some being surrounded by live *Porites* and forming new microatolls. All 59 cores from these in situ dead *Porites* corals indicate that they are less than 70 cm in depth, implying very short growth histories (<50 years). U-series dating of the death massive corals or microatolls suggest most of them died over the past 200 years, and their mortality ages correlate well with historic El Nino events, with the frequency of mortality increased since 1940s (Yu et al. 2006a). In contrast, storm-transported blocks, most composed of massive *Porites* corals, are 1–2.5 m in heights and 1–3 m^3 in sizes, which are significantly larger than the in situ corals (Yu et al. 2004b). These coral blocks are littered mainly in the

Table 4.6 Biogenic compositions (%) vary considerably in surface sediments from different geomorphological zones, Nansha Islands

Grain size (mm)	Coral	Molluscs	*Halimeda*	Coralline algae	Foraminifera	Others
Reef flat						
5.0–2.0	44.45	12.82	7.73	29.90	3.28	1.84
2.0–1.6	45.72	9.10	16.72	20.12	3.62	4.74
1.6–1.25	55.25	9.46	15.33	15.97	1.66	2.00
1.25–1.0	54.10	10.93	14.11	13.98	2.74	1.65
1.0–0.80	55.34	8.30	10.12	18.79	4.86	2.61
0.80–0.71	63.99	8.29	7.49	12.19	4.65	3.47
0.71–0.63	66.78	7.84	8.54	9.30	4.61	2.94
0.63–0.50	63.34	9.21	6.85	9.55	8.22	2.35
Lagoon slope						
5.0–2.0	55.43	11.54	16.09	8.17	4.99	3.78
2.0–1.6	54.39	17.36	17.96	9.91	0.16	0.23
1.6–1.25	54.35	14.31	24.44	4.66	1.03	0.66
1.25–1.0	64.91	13.50	12.14	6.07	2.27	1.11
1.0–0.80	54.41	7.88	15.31	17.51	4.90	
0.80–0.71	60.34	6.32	5.96	19.55	4.20	3.11
0.71–0.63	68.80	3.17	4.27	18.66	4.27	0.32
0.63–0.50	63.42	6.42	5.99	16.15	5.88	2.14
Lagoon basin floor						
5.0–2.0		9.98	25.85	63.01	1.16	
2.0–1.6	19.54	23.18	19.20		3.97	34.10
1.6–1.25	17.80	14.83	36.36	9.75	2.54	18.22
1.25–1.0	35.93	19.77	25.86	3.99	7.79	6.65
1.0–0.80	28.82	13.65	21.14	16.59	19.81	
0.80–0.71	25.97	10.10	21.64	17.65	20.20	4.44
0.71–0.63	22.47	12.78	24.64	11.60	27.02	1.17
0.63–0.50	19.69	12.56	16.31	15.28	30.73	4.92

southwest part of the zone. All such samples analyzed by Yu et al. (2004b) were from this area. The size and population of storm-transported blocks decrease towards the lagoon. The sands are mainly originated from corals (50–60%), followed by coralline algae (15–20%), mollusc (10%), *Halimeda* (10%) (Table 4.6). Table 4.7 outlines the mineral and grain-size parameters of sandy sediments from this and the next two zones.

V. Lagoon slope branching-coral and fine-sand zone: This zone has a water depth of about 3–5 m, which is totally floored by white fine sand. *Acropora*

Table 4.7 Mineral and grain size compositions (%) of surface sediments are shown for different geomorphological zones, Nansha Islands

	W.d. (m)	Aragonite	Mg Calcite	Calcite	Gravel	Coarse sand	Med. sand	Fine sand	Silt
Reef front slope	10	48.27	51.73	0	4.72	30.48	58.68	5.89	0.23
Reef flat	1.5	46.31	49.55	4.14	9.06	54.09	29.94	4.27	2.01
Lagoon slope	3	50.84	49.16	0	5.91	52.96	33.17	5.87	2.10
Lagoon floor	8	52.43	13.80	0	0.53	3.89	20.47	62.79	12.32

communities with diameter about 1–1.5 m may be found on the sandy floor. Very few massive corals, such as *Porites lutea*, can be found. The sands are mainly originated from corals (60%), followed by coralline algae (12%), *Halimeda* (12%) and mollusc (10%) (Table 4.6).

VI. Lagoon basin-floor silt zone: This zone is about 5–12 m deep, and is entirely covered with white silts, except for a few living *Acropora*. No living massive corals such as *Porites* were found within this zone. This ecological feature is largely controlled by the enclosed lagoon, which does not have any channel for water exchange with the open ocean. Coral debris (25%), *Halimeda* (24%) and foraminifera (18%) are the main sources of the lagoon sediments, following by molluscs (14%) and coralline algae (13%) (Table 4.6). A total of 62 species from 48 genera of foraminifera were identified from the lagoon sediments (Wen et al. 2001).

In open lagoons, which have channels for rapid exchange of lagoon water with the adjacent ocean, the habitat situation is very different. For instance, patched reefs (Fig. 4.54) are well developed in the lagoon (8.6 km × 6.5 km in size, water depth 30 m) of Meiji Reef, and the lagoon floor is usually covered by ramose and foliose corals. However, the organism composition in the lagoon sediments is quite uniform (Yu et al. 1997) regardless of lagoon types. Large, enclosed lagoons tend to contain large quantities of benthic foraminifera and *Halimeda*. Figure 4.55 shows the distribution of *Halimeda* debris in the surface sediments from different geomorphological zones of eight atolls of Nansha Islands. Usually *Halimeda* contributes up to 25–75% of the lagoon sediments if the lagoon is large, deep and well enclosed.

Estimation of Reef Calcium Carbonate Production

There is still some debate on the role of coral reefs as a sink or source of atmospheric CO_2. The observations on the water pCO_2 (comparing to the non-reef area water) implied that coral reef lagoons act as a source for the atmospheric CO_2 (Kawahata et al. 1997, 2000; Suzuki and Kawahata 2003; Suzuki et al. 1997, 2001),

Fig. 4.54 Photo shows various corals growing in the lagoon of Meiji Reef, Nansha Islands

Fig. 4.55 Sketch maps show *Halimeda* distribution in surface sediments from different geomorphological zones of eight atolls, Nansha Islands. Shaded area represents *Halimeda* concentration, with whole circle representing 100% and half circle 50% (Yu et al. 1998)

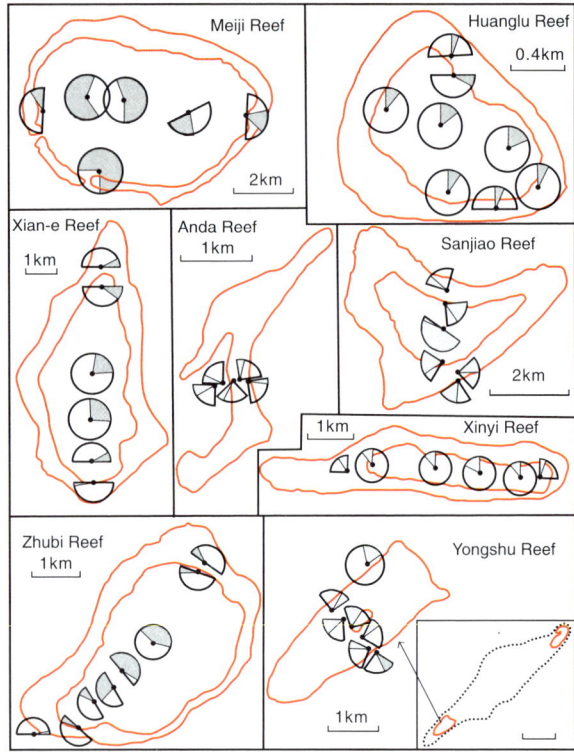

whilst Kayanne et al. (1995) suggested the reefs act as a CO_2 sink based on the diurnal change of pCO_2 in fringing reef water of Ryukyu Islands. The sink/source behavior is controlled by the balance between photosynthesis and calcification, the most important biogeochemical processes (Suzuki and Kawahata 2003). The two processes are shown as:

$$\text{Photosynthesis}: H_2O + CO_2 \rightarrow CH_2O + O_2 \uparrow$$
$$\text{Calcification}: Ca^{2+} + 2HCO_3^- \rightarrow Ca_3\,CO_3 \downarrow + H_2O + CO_2 \uparrow$$

Photosynthesis decreases pCO_2, whilst calcium carbonate production raises pCO_2. Therefore, coral reef calcium carbonate ($CaCO_3$) production is an important component of its participation in the carbon cycle. Coral reef $CaCO_3$ production was conveniently categorized as potential production, gross production and ness production (Chave et al. 1972). Four calculation methods have been used to estimate reefal carbonate production on local to global scales (see review by Vecsei 2004), including the use of hydrochemistry (Milliman 1993), census data (Chave et al. 1972), accumulated reef sediments (Ryan et al. 2001), and numerical modeling (Kleypas et al. 1999).

Hydrochemistry method is based on changes in the hydrochemistry (pH and total alkalinity, or pH and O_2) of the water moving above the reefs sums carbon precipitation and dissolution. Milliman (1993) estimated about 9×10^{11} kg $CaCO_3$ produced by global coral reefs complex based on hydrochemical measurements, with 20% from reef flat and 80% from lagoon. Smith (1978) estimated 6×10^{12} mol (or 6×10^{11} kg) $CaCO_3$ annually.

Census method uses biota censuses (mainly coral cover) and growth rates (linear skeletal extension), measured in the field and sometimes in the laboratory, to calculate gross production before dissolution and erosion. It is time-consuming, but was considered by Vecsei (2001) as the best among all the methods. The annual $CaCO_3$ production by hermatypic coral *Diploria labyrinthiformis* in Florida reef tract is 11.8 ± 0.3 kg/m^2 of the reef space occupied by the corals (Ghiold and Enos 1982).

Accumulation method calculates net production based on vertical accumulation rate of reefal sediments, and its content of carbonate mineral aragonite precipitated by corals. Such estimates yield production after syndepositional dissolution and erosion. With this method, Ryan et al (2001) used data from Great Barrier Reef and estimated global $CaCO_3$ accumulation to be 1.3×10^{12} kg/yr during the Holocene period (7 to 4 ka BP).

The numeric modelling method (Bosscher and Schlager 1992) is based on the fact that light is a major control on reef growth (and carbonate production). Light decreases with water depth, and simulates the growth of coral reefs. With such a method, Kleypas (1997) suggests that global coral reef area is $584–746 \times 10^3$ km^2 and global carbonate production is 1.00×10^{12} kg/year.

Among the four methods described above, accumulation method is probably the best for calculating the coral reef $CaCO_3$ production in the SCS, because there is relatively detailed information of Holocene reef cores and age sequences of fringing reef, reef flat and lagoon of atolls in the region available for this calculation.

Previous studies suggested that the Holocene reef thickness on Leizhou Peninsula is about 5 m with a bottom age around 7500 yr BP (Yu et al. 2004c) and 10 m on Hainan Island with a bottom age of 7910 yr BP (Zhao and Yu 1999), corresponding to reef growth rates of 0.07 cm/yr and 0.13 cm/yr for Leizhou Peninsula and Hainan Island, respectively. The latter can be extensively used for the fringing reefs around Taiwan Island. Using the reef carbonate density of 2.9 g cm^{-3} and a porosity of 50% in reef flat materials (Opdyke and Walker 1992), the annual $CaCO_3$ productions are calculated and listed in Table 4.8.

Three cores (Fig. 4.56), respectively from the outer reef flat (core Nanyong- 2, reaching a depth of 413.69 m), the inner reef flat (core Nanyong-1, reaching a depth of 152.07 m) and the lagoon (core Nanyong-4, reaching a depth of 15.4 m), were drilled from Yongshu Reef, Nansha Islands. Detailed dating (Yu et al. 2006b) of Nanyong-4 shows that all the 26 high-precision TIMS U-series dates (for 25 coral branches) and 5 AMS ^{14}C dates (for foraminifera) for the 15.4-m-long lagoon core are in the correct stratigraphical sequence, although the smallest sampling interval is only 13 cm, and they reveal a ~4000-yr continuous deposition history. The results indicate the deposition rate varied in the range of 0.8 and 24.6 mm/yr, with

Table 4.8 CaCO$_3$ production was estimated for different reef regions of the SCS

Region	Reef flat (km^2)	lagoon (km^2)	Rore-reef (km^2)	Reef flat + fore-reef CaCO$_3$ production ($\times 10^9$ kg)	Lagoon CaCO$_3$ production ($\times 10^9$ kg)	Total ($\times 10^9$ kg)
I. S China coast and offshore islands	17			0.0164		0.0164
II. Hainan and offshore islands	180			0.3300		0.3300
III. Taiwan and offshore islands	940			1.7231		1.7231
IV. Dongsha Islands	125	292	7.50	0.4823	0.7884	1.2707
V. Huangyan Island, Zhongsha Islands	53	77	3.18	0.2045	0.2079	0.4124
V. Zhongsha Atoll, Zhongsha Islands	1495			4.0365		4.0365
VI. Xisha Islands	221.6	1614.8	13.30	0.8550	4.3600	5.2150
VII. Nansha Islands	507.5	2396.8	30.45	1.9581	6.4714	8.4295
Total ($\times 10^9$ kg)				9.6059	11.8277	21.4336

an average of 3.85 mm/yr. Based on the sedimentation rates, the calculated calcium carbonate production rates in the lagoon ($=$ sedimentation rates $\times$ 0.7 g cm^{-3}, which is the average density of the dry loose lagoon sediments and pores in the area) vary from 547 to 17,215 g CaCO$_3$ m^{-2} yr^{-1}, with an average of $\sim$2,700 g CaCO$_3$ m^{-2} yr^{-1} for the last 4000 years. This average value of $\sim$2,700 g CaCO3 m^{-2} yr^{-1} is used in the calculations of the CaCO$_3$ productions in the coral reef lagoons of Nansha, Xisha and Dongsha islands (Table 4.8).

Both Nanyong-1 from inner reef flat and Nanyong-2 from outer reef flat reveal that the Holocene coral reef flat started to develop about 7350 to 8000 years ago on weathered Pleistocene coral reef limestone at a depth of 17–18 m (Yu et al. 2004b, 2006b; Zhu et al. 1997). Radiocarbon ages indicate that the average reef growth rate over this period is $\sim$2.51 mm/yr. Considering the reef carbonate density of 2.9 g cm^{-3} and a porosity of 50% in reef flat materials (Opdyke and Walker 1992), the average carbonate accumulation in reef flats is about 3,640 g CaCO$_3$ m^{-2}yr^{-1}, close to the reef flat production of 4.0±0.3 kg CaCO$_3$ m^{-2}yr^{-1} used by Smith (1978) for the study of global carbonate budget.

Drill cores on Yongxing Island (core Xiyong-2, reaching a depth of 600.02 m), Shi Island (core Xishi-1, reaching a depth of 200.63 m) and Chenhang Island (core Xichen-1, reaching a depth of 802.17 m) of Xisha Islands (Zhang et al. 1989) had a remarkably similar stratigraphy to that of Yongshu Reef, Nansha Islands. A layer of 17–18 m thick un-cemented reef sediments of Holocene age overlies the weathered surfaces of later Pleistocene coral reef limestone. The weathered surfaces of later Pleistocene coral reef limestone are at depth of 16.91 m in core Xichen-1 and 17.72 m in core Xiyong-2. Therefore, the average CaCO$_3$ production rate of 3,640 g CaCO$_3$ m^{-2}yr^{-1} from Yongshu Reef should also apply to coral reefs of

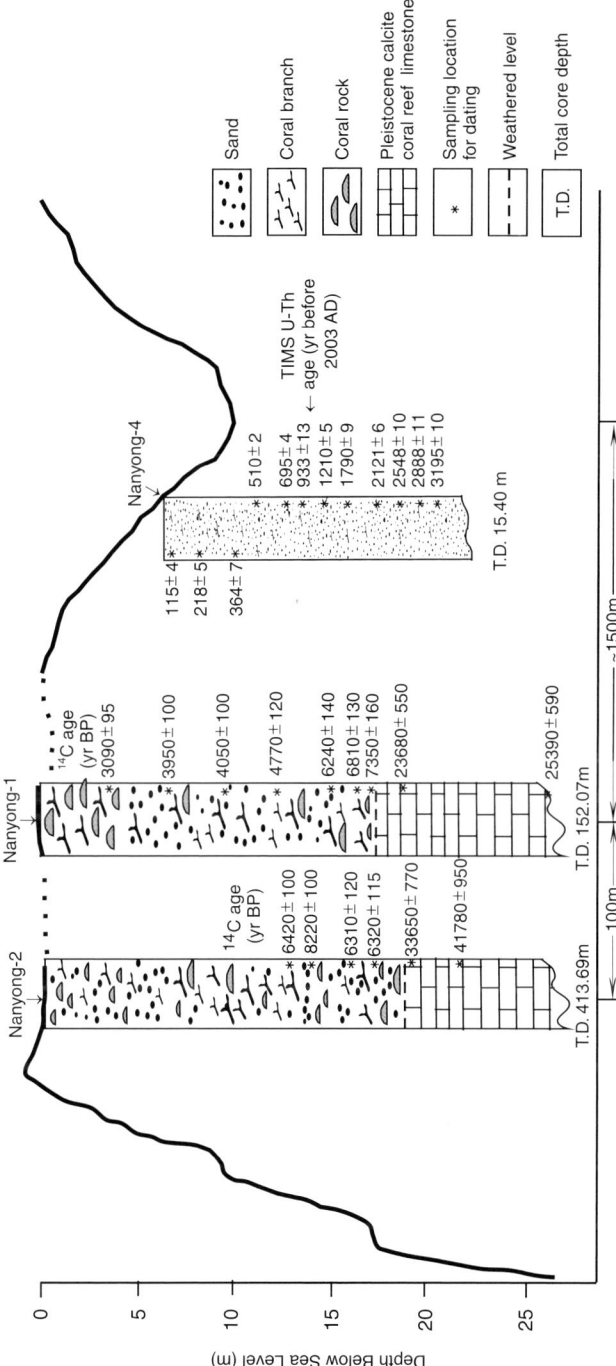

Fig. 4.56 Stratigraphical columns of three cores Nanyang-1, Nanyang-2 and Nanyang-4 show Holocene coral reef structure of Yongshu Reef, Nansha Islands, SCS (after Yu et al. 2004b, 2006b)

Xisha Islands. This value is also used in the annual $CaCO_3$ production calculation for coral reefs on Zhongsha and Dongsha Islands, where no Holocene core data are available. Because the estimated fore-reef area is only about $43.75\,km^2$ and no accurate fore-reef production information is provided in this region, the same $CaCO_3$ production rate is used in the calculation of its fore-reef area. Table 4.7 lists the calculated $CaCO_3$ production for coral reefs of these regions.

The total annual $CaCO_3$ production for all these seven regions is 2.14×10^{10} kg/yr, which is about 1.6–3.3% of global reefal carbonate production estimated by Milliman (1993), Kleypas (1997), Ryan et al. (2001) and Vecsei (2004), respectively. Therefore, the carbonate production of the coral reefs in the SCS should not be neglected in the global estimation of carbonate production.

Cenozoic carbonate production in the SCS might be even higher than present, because many pre-Quaternary reef complexes or carbonate platforms were successively identified outside the modern coral reef areas over the last two decades (Hu and Wang 1996; Hu and Xie 1987; Qiu and Wang 2001; Story et al. 2000; Wiedicke 1987). For example, the carbonate platform area in the Zhujiangkou (Pearl River Mouth) Basin is up to $56,750\,km^2$ (Hu and Wang 1996) and the reef area is about $2,000\,km^2$ (Hu and Xie 1987). Here the reef started to develop from early Miocene, and is dominated by algae and large benthic foraminifera. Oil exploration drilling revealed that the thickness of this reef varies from 23.5 to 560 m. The reef complexes or carbonate platforms are mainly located in basins in the northern, southern and western continental shelves of the SCS, and they were controlled by the paleostructure and river system (Qiu and Wang 2001). The reef complexes are thought to be important targets for oil (gas) exploration because of their good reservoir properties. Sedimentary facies and seismic data derived from the Liuhua 11-1 oilfield, which was developed in the Zhujiangkou Basin, were widely reported (Story et al. 2000; Yue et al. 2005).

Reef History

Reef Thickness and Origin

Over twenty drillholes from fringing reefs on Hainan Island (Zhao and Yu 1999) indicated that the fringing reef thickness in this region is about 10 m and the development of the reef started about 8000 yr ago. The well-studied hole from Luhuitou (Sanya) presents a continuous reef stratigraphy since 7910 yr BP without obvious change in reef-building organisms. Subsequent detailed investigations of the fringing reef on Leizhou Peninsula (Yu et al. 2002c, 2004c; Zhao and Yu 2002, 2006, 2007; Zhao et al. submitted) suggest that the coral reef there also started at about 7500 years ago but the reef development was episodic, with 7500–6500 yr BP as the most flourishing period. Other reef developing periods on Leizhou Peninsula include 6245 ± 15 to 5676 ± 14, 5009 ± 54, 4156 ± 23 to 3675 ± 23, 2795 ± 14 to 2321 ± 23, and 1511 ± 23 yr BP.

More than ten deep cores ($>150m$ in core depth) have been drilled on atolls of the SCS over the past few decades. Two of the deepest holes, Sampaguita-1 and

Xiyong-1, were respectively drilled by petroleum companies on Xisha and Nansha Islands for the purpose of oil exploration in 1970s. Unfortunately, they were not well studied for the purpose of understanding coral reef development. Nevertheless, they do provide us very important information on the basement, reef thickness and origins of the coral reefs in the SCS.

Sampaguita-1 (Du Bois 1981) was drilled to a total depth of 4124 m on northern Nansha Islands, which reveals that the upper sequence from 0 to 2160 m is composed of the upper Oligocene to recent sediments of white to buff limestone, comprising a major long-lasting reef complex. This suggests that around 2000 m thick reef profile in the Nansha Islands was accumulated since upper Oligocene, underlain by marine clastic deposits of Eocene-Oligocene age.

Xiyong-1, with a depth of 1384.68 m, reveals that the reef thickness is 1251 m on Yongxing Island of Xisha Islands. Evidence from micropaleontology analysis indicates that the reef was originated during the Miocene (Zhu et al. 1997). The reef limestone overlies unconformably on metamorphic complex of Precambrian to Paleozoic age (Zhao and Yu 1999).

Reef Stratigraphy and Structure

In order to understand the reef development process, three deep cores (Xiyong-2, 600.02 m; Xishi-1, 200.63 m; and Xichen-1, 802.17 m) were drilled on coral reef islands of Xisha Islands in the 1980s, and two cores (Nanyong-1, 152.07 m and Nanyong-2, 413.69 m) were drilled from Yongshu Reef of Nansha Islands in the 1990s. None of these cores penetrated through the reef complex. Lithological studies (Zhang et al. 1989; Zhao et al. 1992; Zhu et al. 1997) suggest that the cores from both Xisha and Nansha Islands have almost identical stratigraphic characteristics and their stratigraphic sequences can be correlated to each other. The Miocene to Holocene reef history summarized below is based mainly on published results from cores Nanyong-1 and Nanyong-2.

Figure 4.57 shows simplified stratigraphy, lithology and sedimentary facies of core Nanyong-2. Five discontinuous boundaries were identified at core depths of 369.83 m, 268.00 m, 141.71 m, 91.25 m and 17.71 m, respectively, which were formed during periods of sea level lowstands, indicating six large-scale sea level rises and five sea level falls (up to 120 m rise and fall). The mineral compositions include aragonite and high-Mg calcite from core depth 0–17 m, low-Mg calcite from 17 to 142 m and dolomite below 142 m. Corals are the dominant reef-building organisms for the whole 413.69 m core, followed by coralline algae. A total of 31 genera of Scleractinia corals and 1 genera of Octocorallia corals were identified in core Nanyong-2 (Zhu et al. 1997).

Based on the combination of coral genera, foraminifera, lithology, paleomagnetic characteristics, elemental geochemistry and deposition hiatus, five stratigraphic units have been recognized with different facies shown in Fig. 4.57 (Zhu et al. 1997).

  I. Miocene dolomite (413.7–369.8 m) consists of coral gravels and coral sands (debris of corals, coralline algae, *Halimeda*, foraminifera, molluscs, and other

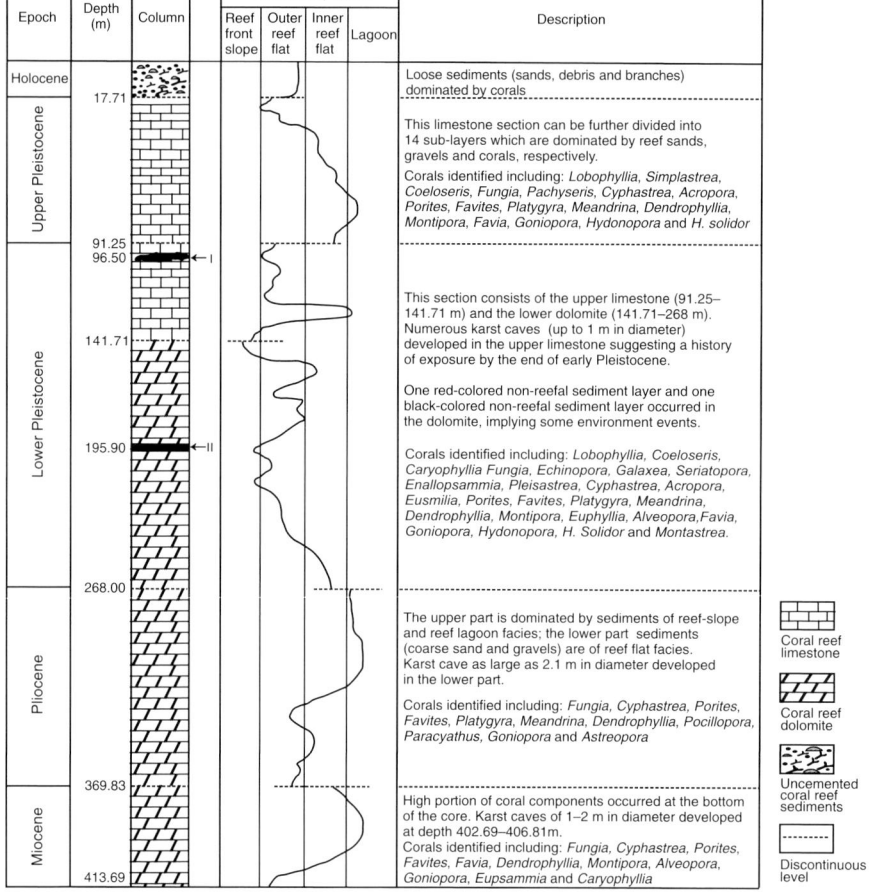

Fig. 4.57 Simplified stratigraphical column of core Nanyong-2 is shown together with projected sedimentary facies. Horizon I indicates the occurrence of the "Black-colored sedimentation event", and horizon II indicates the occurrence of the "Red-colored sedimentation event". See details in the text (after Zhu et al. 1997)

organisms). The foraminifer *Nephrolepidina* indicates mid-Oligocene to late Miocene in age.

II. Pliocene dolomite (369.8–268.0 m) consists of coral sands. The chronology of this unit is based on the following age markers: (1) the deepest vertical distribution of *Lobophyllia*, a characteristic coral genus representing the Quaternary period, was found at 255 m; (2) *Sphaeroidinellopsis*, a foraminifera with an upper-limit age of Pliocene, occurred from 278 to 280 m; (3) the palaeomagnetic boundary between Matsuyama reversed epoch (M) and Gaussian normal epoch (G) was identified at the depth of 256.3 m; (4) elemental concentrations, including S, Mg, Ni, Na, Ba, Mn, Al, Cd, As, Y, Cr and Rb, show distinct variations at around 250–260 m; and (5) a stratigraphical discontinuity exists at 268 m.

III. Lower Pleistocene (268.0–91.3 m) includes a reef dolomite (268–141 m) and a reef calcite limestone (141–91.3 m). A red-colored, Fe-rich non-reefal sediment layer (marked as "Red-colored sedimentation event") occurs at 96.59–99.69 m, and a black-colored, Mn-rich non-reefal sediment layer (marked as "Black-colored sedimentation event") occurs at 195.96–196.26 m (Fig. 4.56). Foraminifera in the black-colored sediments clearly indicate that they came from relatively calm and deepwater environments rather than shallow coral reefs. The enrichment of Mn, Al and Fe implies they were probably products of volcanic eruptions. High-valence Fe in the red-colored sediments also indicates a strong oxidizing environment (Zhu et al. 1997).

IV. Upper Pleistocene (91.3–17.7 m) is represented by a calcite reefal limestone.

V. Holocene consists of un-cemented reef aragonite and high-Mg calcite sediments (17.7–0 m). Radiocarbon dates show that the sediments above the discontinuity at ∼17.7 m were younger than 8220 years, implying that the Holocene reef initiated at about 8.2 ka.

The stratigraphy for all three cores from Xisha Islands, Xiyong-2 (600.02 m), Xishi-1 (200.63 m) and Xichen-1 (802.17 m), is summaried in Fig. 4.58 (Zhang et al. 1989) showing a similar development history to that of Yongshu Reef, Nansha Islands (Nanyong-2). Zhang et al. (1989) subdivided the stratigraphical sequence into five periods: Holocene (10–18 m thick), upper Pleistocene (45 m thick), middle Pleistocene (74 m thick), lower Pleistocene (152 m thick), Pliocene (123 m thick) and Miocene (up to 900 m thick). There appears to be one major difference between cores from Xisha Islands and those from Nansha Islands. In Nansha Islands, reef-building corals were found to have played a major role throughout the entire reef development history. However, in Xisha Islands, the presence of corals has never been mentioned by Zhang et al. (1989), although they concluded that algae were dominant prior to mid-Pleistocene. This apparent difference needs to be verified through further detailed investigations of cores from Xisha Islands.

Coral Reef Records of Recent Environmental Changes

Coral reefs have experienced numerous environmental changes on different time scales throughout their development history, and such changes have been well documented in reef carbonates.

Over the past few years, corals in the SCS have been widely used for U-series dating and high-resolution $\delta^{18}O$, Sr/Ca, Mg/Ca and other proxy analyses for the purpose of paleoenvironmental reconstruction (including sea-level, SST and salinity, ocean circulation patterns) mainly since the Holocene (He et al. 2002; Ma et al. 1999, 2003; Peng et al. 2002, 2003; Shen et al. 1996, 2003, 2004a,b; Sun et al. 2004; Sun D. et al. 2005; Sun Y. et al. 2005; Wei et al. 2000, 2004, 2007; Yu et al. 2001, 2004c, 2005a,b, 2006a,b, submitted; Zhao and Yu 2002, 2006, 2007). Highlights of recent research in these aspects are summarized below:

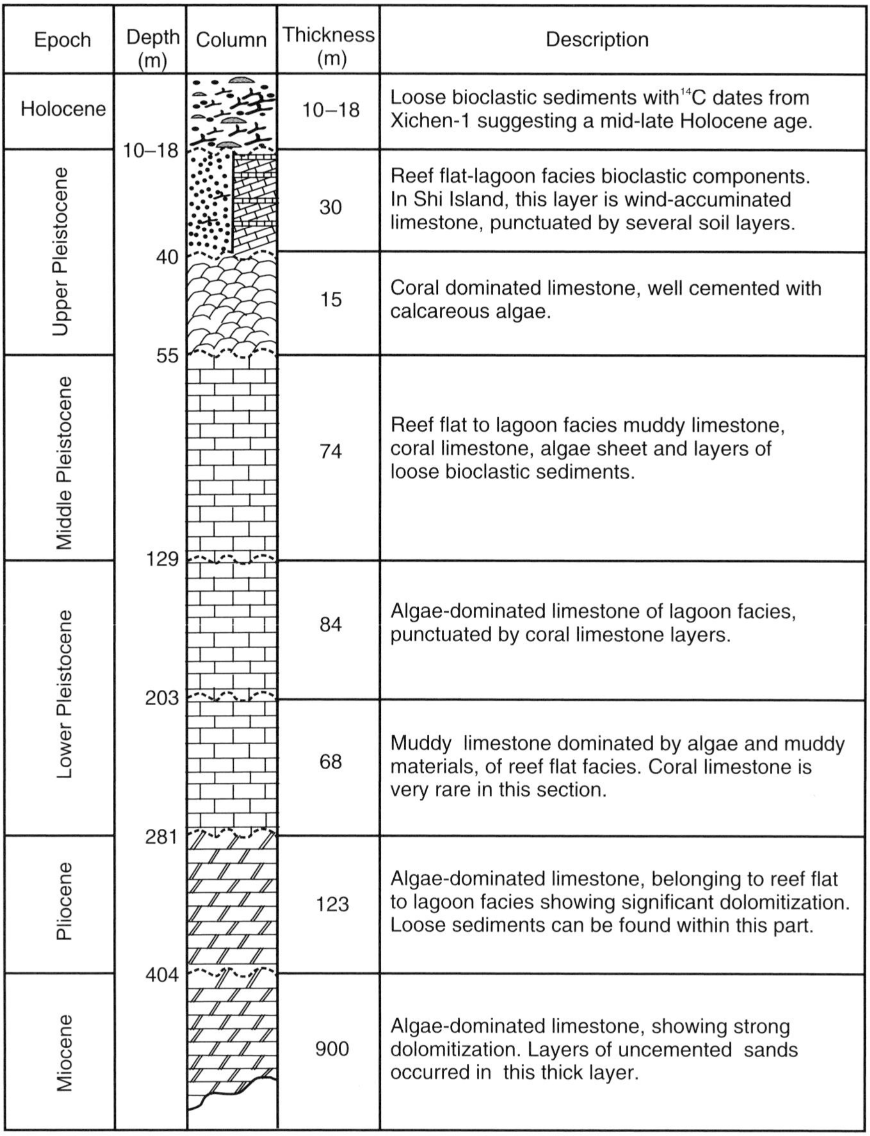

Epoch	Depth (m)	Column	Thickness (m)	Description
Holocene	10–18		10–18	Loose bioclastic sediments with ^{14}C dates from Xichen-1 suggesting a mid-late Holocene age.
Upper Pleistocene	40		30	Reef flat-lagoon facies bioclastic components. In Shi Island, this layer is wind-accumulated limestone, punctuated by several soil layers.
Upper Pleistocene	55		15	Coral dominated limestone, well cemented with calcareous algae.
Middle Pleistocene	129		74	Reef flat to lagoon facies muddy limestone, coral limestone, algae sheet and layers of loose bioclastic sediments.
Lower Pleistocene	203		84	Algae-dominated limestone of lagoon facies, punctuated by coral limestone layers.
Lower Pleistocene	281		68	Muddy limestone dominated by algae and muddy materials, of reef flat facies. Coral limestone is very rare in this section.
Pliocene	404		123	Algae-dominated limestone, belonging to reef flat to lagoon facies showing significant dolomitization. Loose sediments can be found within this part.
Miocene			900	Algae-dominated limestone, showing strong dolomitization. Layers of uncemented sands occurred in this thick layer.

Fig. 4.58 Synthesized stratigraphical column for cores from Xisha Islands is adopted from Zhang et al. (1989)

On Leizhou Peninsula, Yu et al. (2005a) reported a combined seasonal- to monthly- resolution coral skeletal $\delta^{18}O$, Sr/Ca and Mg/Ca record for a modern coral (covering 1989–2000 AD) from Dengloujiao reef. Further combined analyses of coral Sr/Ca and $\delta^{18}O$ for five Holocene *Porites lutea* samples collected from this reef reveal a general decreasing trend in SST from ~6, 800 to 1,500 yr, despite shorter climatic cycles (Yu et al. 2005b). Compared with the mean Sr/Ca-SST in the

1990s (24.8 °C), 10-year mean Sr/Ca-SSTs were 0.9–0.5 °C higher between 6,800 and 5,000 yr BP, dropped to the present level by ~2, 500 yr BP, and reached a low of 22.6 °C (2.2 °C lower) by ~1, 500 yr BP, and the trend was comfirmed by the summer Sr/Ca-SST maxima, which are more reliable due to faster summer-time growth rates and higher sampling resolution. Such a decline in SST is accompanied by a similar decrease in the amount of monsoon moisture transported out of the SCS, resulting in a general decrease in the seawater $\delta^{18}O$ values, reflected by offsets of mean $\delta^{18}O$ relative to that in the 1990s.

Apart from *Porites*, Sr/Ca ratio was also shown to be excellent SST proxy based on branching *Goniopora* corals from Dengloujiao reef, Leizhou Peninsula (Yu et al. 2004c). Here Sr/Ca data for five Holocene *Goniopora* corals suggest that large-amplitude winter cooling frequently occurred in the mid-Holocene warm period. The results show, whilst the annual summer SST (average ~30.5 °C) during the period of 7.5–7.0 cal. ka BP was identical to that of the 1990s (average ~30.3 °C), the winter SST was ~1.5 to 3 °C lower than present, showing a much larger seasonality than today.

On Hainan Island, Wei et al (2000) reported high-resolution Mg/Ca, Sr/Ca and U/Ca analyses of modern *Porites* corals from Sanya, and suggests that these ratios are not significantly affected by coral growth rate and therefore are reliable proxies for historical SST. Further Sr/Ca analyses of three mid-Holocene corals (covering 6,500 to 6,100 yr BP) from the same site (Wei et al. 2007) suggest that the average minimum monthly winter, summer and annual SSTs were about 0.5 to 1.4 °C, 0 to 2.0 °C, and 0.2 to 1.5 °C higher than today's (1970–1994), respectively. Meanwhile, these records imply weaker ENSO variability in the mid-Holocene than present. Similarly, $\delta^{18}O$ study of a mid-Holocene coral from Hainan Island suggests, whereas reconstructed winter SST appears to show an ENSO-like cycle of 6.7 yr (longer than the modern cycle of 3.6 yr), the reconstructed summer SSS (a proxy for monsoon rainfall) variations display no affinity to ENSO (Sun D. et al. 2005). However, Peng et al (2003) reported a good relationship between wind velocities and coral [18]O in a modern *Porites lutea* sample from Longwan (northern Hainan Island) and proposed that coral [18]O was closely correlated with ENSO variations in the modern time.

Along the coast of Taiwan, Fang and Chou (1992) studied fulvic acid concentrations in the growth bands of the coral *Porites lutea* and their relationship with local annual precipitation, and concluded that fluorescence banding in the coral skeleton was a reasonable proxy for local environment. Shen et al. (1996) studied the relationship between Sr/Ca of *Porites* corals and SST from Kenting (southern Taiwan) and reconstructed historical SST. Chen C.T. et al. (2001) reported $\delta^{18}O$ and $\delta^{13}C$ data for a coral collected around a nuclear power plant and interpreted their relationships with temperature and sunshine hours.

On Xisha Islands, Sun et al. (2004) used Sr concentrations of a *Porites* coral (covering 1906–1994) from Yongxing Island (Xisha Islands) as a proxy to reconstruct SST for the last century. They discovered that the late 20th century was about 1 °C warmer than the early 20th century. They also identified two cooling periods (1915/1916 and 1947/1948) and three warming periods (1935/1936, 1960/1961, and

1976/1977) in the last century. As a result, they concluded that the SCS climate is coherent with climatic regime of the tropical western Pacific.

On Nansha Islands, Yu et al. (2001, 2002b) first calibrated the relationship between $\delta^{18}O$ variations in a *Porites lutea* sample and the instrumental temperature (back to 1988), and reconstructed SST back to 1950 (Yu et al. 2001, 2002b). They also showed that coral $\delta^{13}C$ is well correlated with sunshine duration, total cloud cast and rainfall. The reconstructed winter SST shows a good relationship with winter monsoon intensity, providing a useful tool for analysis of high-resolution paleoclimate.

Along with the SST changes, reef corals in the SCS were used for sea-level reconstruction. For instance, through radiocarbon dating of corals and shells from Penghu islands, Chen and Liu (1996) suggested the highest Holocene relative sea level occurred about 4, 700^{14}C yr ago with a height about 2.4 m above the present level. The same dating method was also applied to corals from Leizhou Peninsula, showing the Holocene sea-level highstands (Nie et al. 1997b). This has been subsequently confirmed by extensive high-precision TIMS U-series dating of corals from the same site (Ma et al. 2003; Yu et al. 2009a Zhao and Yu 2002, 2006, 2007; Zhao et al. 2009a). The systematic TIMS U/Th dates and elevations of the corals and microatolls suggest multiple Holocene sea level highstands ($\sim$2 m) occurred since mid-Holocene, and the initiation times of these highstands appear to correlate with millennial Bond cycles in the North Atlantic (Bond et al. 1997) as well as the termination times of short weaker summer monsoon episodes in SE China (Wang Y. et al., 2005).

Reef corals and lagoon deposits in the SCS were also used for past storm reconstruction. For instance, based on high-precision TIMS U-series dating of large wave-transported *Porites* coral blocks on the southwest reef flat of Yongshu Reef (Nansha Islands), Yu et al. (2004b) suggested that during the last 1000 years, at least six strong storms or giant wave events occurred around 1064$\pm$30, 1210$\pm$5$\sim$1201$\pm$4, 1336$\pm$9, 1443$\pm$9, 1685$\pm$8$\sim$1680$\pm$6, 1872$\pm$15 A.D., respectively, with an average 160-yr cycle (110–240 yr). These findings were confirmed by anomalously high sedimentation rates (Yu et al. 2006b) and grain-size distributions (Yu et al. 2009b) in a well-dated core from Yongshu lagoon. Interestingly, the last two events (1685$\pm$8 to 1680$\pm$6, 1872$\pm$15 A.D.) are well correlated with two Krakatau eruptions in 1680 and 1883 AD, respectively. It has been reported that the 1883 Krakatau eruption was historically the most violent on Earth that produced giant waves reaching heights of 40 m above sea level and distances over 7,000 km, devastating everything in their path and hurling ashore coral blocks weighing as much as 600 tons (Fiske and Simkin 1983). Research as such provides possible clues for the understanding of past catastrophic tsunami events.

In addition, high-precision U-series dating of dead in situ massive *Porites* corals on the reef flats of Yongshu and Meiji Reefs, Nansha Islands reveals many massive coral mortality events over the past 200 years (Yu et al. 2006a), most of which appear to correlate in time with historic El Niño events. Despite contrasting local environmental conditions, at least five massive coral mortality events occurred simultaneously on both reefs (e.g., in 1861–1880, 1911–1921, 1951–1963, 1966–1974, 1996–2002 AD), reflecting large-scale regional events. Considering various potential causes of coral mortality (such as human disturbance, high-temperature bleaching, low-temperature bleaching, low-tide emergence, low salinity, freshwater input) and the fact that the reef stands over 2000 m above the sea-floor, is far away from mainland, and is located close to the equator, Yu et al. (2006a) suggest that most such mortality events are probably due to El Niño-related high temperature bleaching.

Recently, with the advancement in analytical techniques, trace element study of coral reefs starts to gain momentum (Alibert et al. 2003; McCulloch et al. 2003). Similarly, in the SCS, Liu X. et al. (2006) studied the concentrations of P and trace metals (Zn, Cu, Cd, Pb and Hg) in faeces, bones, eggshells and feathers of seabirds and in plants, soils and sediments with and without seabird influence on Dongdao Island of Xisha Islands, and concluded that the levels of P, Zn, Cu, Cd and Hg in the materials influenced by seabird droppings are significantly higher than those without such influence. Xie et al. (2005) reported some element concentrations in groundwater samples from other islands of this area and reported evidence of contamination, implying potential influence of human activities on island environment.

4.5 Volcanic Deposition (Liu Z.)

Volcanic materials make up a minor component of sediments in the SCS, generally <1% of the total sediment budget but locally reaching 30% on the eastern slope. The volcanic deposits are significant for understanding the evolution of sedimentation and tectonism surrounding and in the SCS. Volcanic ash is also one of best stratigraphic markers for basin-wide correlation (Bühring et al. 2000; Song et al. 2000).

Volcanic Rock Distribution

In the SCS region, volcanic rocks are widely distributed (Fig. 4.59) and they provide volcanic sediments to the SCS by erosion apart from volcanic eruption. The most developed volcanism in the region is along the Luzon Arc, which was formed by collision of the Eurasian Plate with the Philippine Arc since middle Miocene. From southeast Taiwan to Luzon, the arc chain is composed mainly of andesite and basalt in ages younger than 10 Ma (Yang et al. 1996). These volcanic rocks are suggested to have contributed most of smectite to the northern SCS by chemical weathering (Liu Z. et al. 2008). The second largest distribution of volcanic rocks mainly as basalt of Neogene age is in central and southern Indochina. They erupted from fissures or discrete centers and became stronger after 5 Ma (Hoang and Flower 1998). South China and Hainan Island also have limited volcanic rock distributions, mainly with Paleogene basalt and rhyolite in South China and Pliocene-Quaternary basalt in northern Hainan. A few basaltic trachy-andesite locations were found along the Ailao Shan-Red River Fault Zone (ASRRF) (Wang J. et al. 2001).

Within the SCS, boreholes and geophysical surveys have revealed quite wider distribution of volcanic rocks (Fig. 4.59) (Yan et al. 2006). In the northern part,

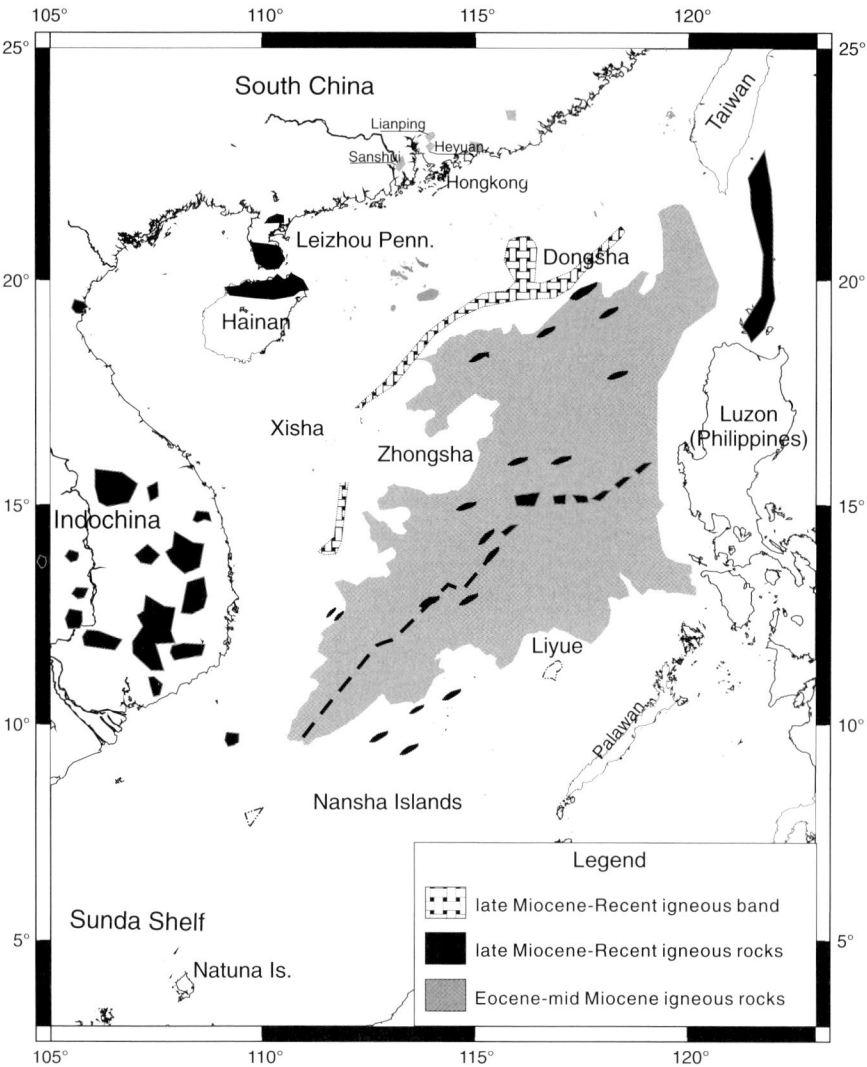

Fig. 4.59 Sketch map shows Cenozoic basalt distribution in and surrounding the SCS (after Yan et al. 2006)

Paleocene-Eocene intermediate-acidic extrusives, mainly andesite, dacite, and rhyolite, formed domes in the Zhujiangkou (Pearl River Mouth) Basin. Oligocene-Miocene basalt and intermediate eruptives occurred along fissures or faults in depressions. In the western Zhujiangkou Basin, the volcanic rock layers are totally 400 m thick with large amounts of tuff and breccia (Yan et al. 2006). Miocene-Quaternary volcanic layers are abundant along the lower slope of the northern SCS margin as shown in seismic profiles (e.g., Lüdmann and Wong 1999; Lüdmann et al. 2001). However, four sites of ODP Leg 184 on the northern slope do not recover large volcanic layers, except some thin ash layers generally <5 cm thick (Wang P. et al. 2000) probably due to their greater water depths. In the central basin, there is a long chain of high seamounts festooning the spreading ridge (Pautot et al. 1990; Li et al. 2002b). The seamounts, consisting of basalt dated as 14–3.5 Ma, rise up to 4,000 m above the abyssal plain (Yan et al. 2006). In the southern part, early Miocene rhyolite and Pleistocene basalt flow were found on the northwest Palawan shelf (Holloway 1982). In the Nansha region, young volcanoes covered by thin Quaternary deposits occurred as high seamounts above the seabed (Yan and Liu 2004).

Volcanic Ash Records

Surface Distribution

Chen Z. et al. (2005) summarized the volcanic ash distribution throughout the surface SCS (Fig. 4.60). Generally, volcanic glasses dispersedly occur in most locations with low average concentration of <5% in the west and relatively high content of about 60% in the east (Yang and Fan 1990; Chen Z. et al. 2005). From north to south, volcanic glasses increase at the lower part of northern slope and then decrease on the Sunda Shelf, indicating eastward increase toward Philippine islands.

The main sources of volcanic ash in surface sediments include Philippine islands (mainly the Luzon Arc), which is the area with most active andesitic and basaltic eruptions (Fig. 4.59), and potential submarine volcanism within the SCS. In despite of the occurrence of Neogene basalt in Indochina and South China, no recent volcanism is observed. Thus, modern volcanic ash contribution from the north and the west to the surface sediment of the SCS should be excluded. Modern volcanic eruptions in Philippine islands, such as the 1991 eruption of Mount Pinatubo, have left discernible records in deep-sea surface sediments in a wide westward elongated lobe covering 30% of the bathypelagic SCS. The thickness of the ash layer reaches 1–2 cm in the area off Luzon (Wiesner et al. 1995).

Late Pleistocene Occurrence

The occurrence of discrete tephra layers of late Pleistocene age has been reported throughout the SCS (e.g., Wang H. et al. 1992; Chen and Zhou 1993; Bühring et al. 2000; Song et al. 2000; Liang et al. 2001; Lee et al. 2004; Chen Z. et al. 2005; Liu Z. et al. 2006). These ash layers evidence their potential as basin-wide stratigraphic markers and provide clues of geological records of past volcanism. The most complete record of late Pleistocene volcanic ash layers was recovered at core MD97-2142, which was raised on the continental slope off the northwestern coast of Palawan. More than 18 distinct, well-defined ash-rich layers with thickness varying from 0.5 to 10 cm were recognized (Fig. 4.61) (Wei et al. 1998). These ash layers consist of vitric, microlite-bearing and crystalline components. Age correlation suggested that these ash-rich layers were deposited at 10 ka, 26–42 ka, 64–79 ka, 110–290 ka, 390–430 ka, and 788 ka (Wei et al. 1998; Lee et al. 2004).

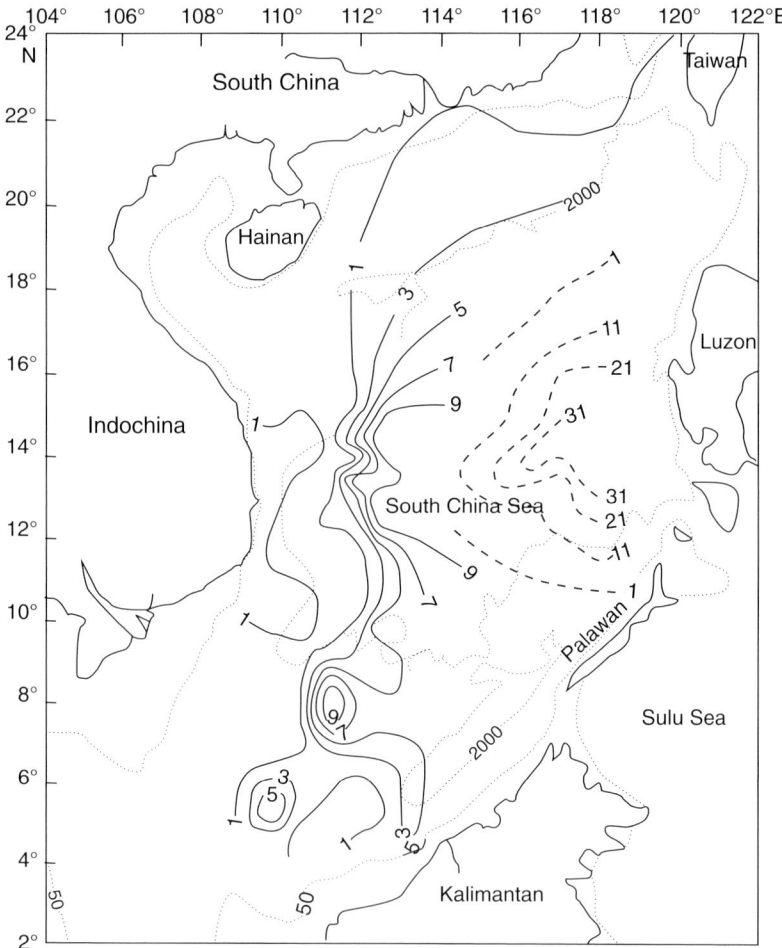

Fig. 4.60 Sketch map shows surface volcanic glass distribution in the SCS. Solid line denotes weight percentage of glass vs total components in size ranges 63–125 μm. Dashed line denotes grain percentage of volcanic clast vs total components (after Chen Z. et al. 2005)

Case Studies: Pinatubo, Toba

Pinatubo

The June 1991 eruption of Mount Pinatubo, Philippines, was one of the largest volcanic eruptions of the twentieth century, emplacing 5–6 km³ of pyroclastic-flow material on the flanks of the volcano (Gran and Montgomery 2005). The magmatic activity at Mount Pinatubo commenced on June 12, 1991, with a series of vertical and lateral blasts, which culminated in a paroxysmal explosion on June 15 (Koyaguchi and Tokuno 1993). Ash was ejected to a maximum altitude of 35–40 km, and subsequently the downwind part of the ash cloud was advected

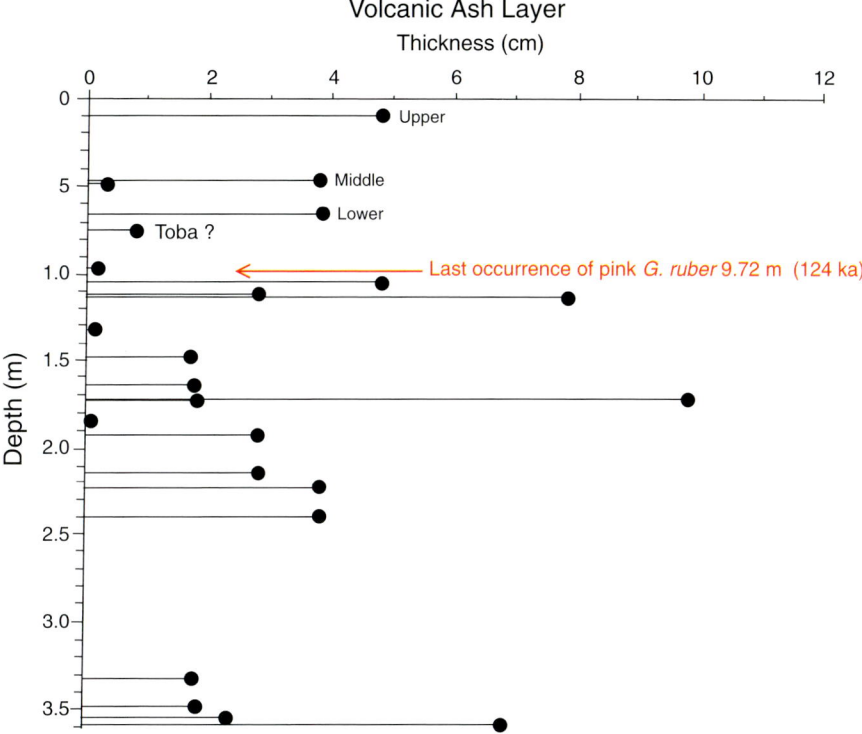

Fig. 4.61 Positions and thickness of *ash layers* in core MD97-2142 show frequent volcanic activities along the Luzon Arc during the late Pleistocene (modified from Wei et al. 1998)

southwestward by high-velocity stratospheric winds. By June 16, the ash cloud passed the SCS and, on June 18, it was centered over the Bay of Bengal (Bluth et al. 1992).

The fallout of Pinatubo tephra was recorded by fully automated sediment traps moored in 1,190 m and 3,730 m water depths at a distance of 586 km to the west of the volcano in the central SCS (Wiesner et al. 1995, 2004) (Fig. 4.62). Within three days after the release of the major eruption plume, ash accumulation (9 kg/m^2) was simultaneously recorded by traps. A numerical simulation of the fallout reveals that the vertical trajectories of the pyroclasts in both the atmosphere and ocean were controlled by particle aggregation. The aggregation process caused a premature subaerial fallout of fine-grained ash. After crossing the air-sea interface, vertical settling of the ash clusters was enhanced by absorption of water leading to settling rates of more than 1670 m/d. Aggregates were observed at all depths, and their rapid settling is reflected in identical pyroclast spectra in intermediate and deep water. This implies that particle sorting must have been complete in the upper water column and that the fallout was not perturbed by oceanic currents (Wiesner et al. 1995). The tephra covers about 30% of the bathypelagic SCS (37 × 10^4 km^2) in a westward elongated lobe (Fig. 4.62), reflecting the prevailing direction of the upper-level winds. It is obvious that settling velocities were sufficiently high to

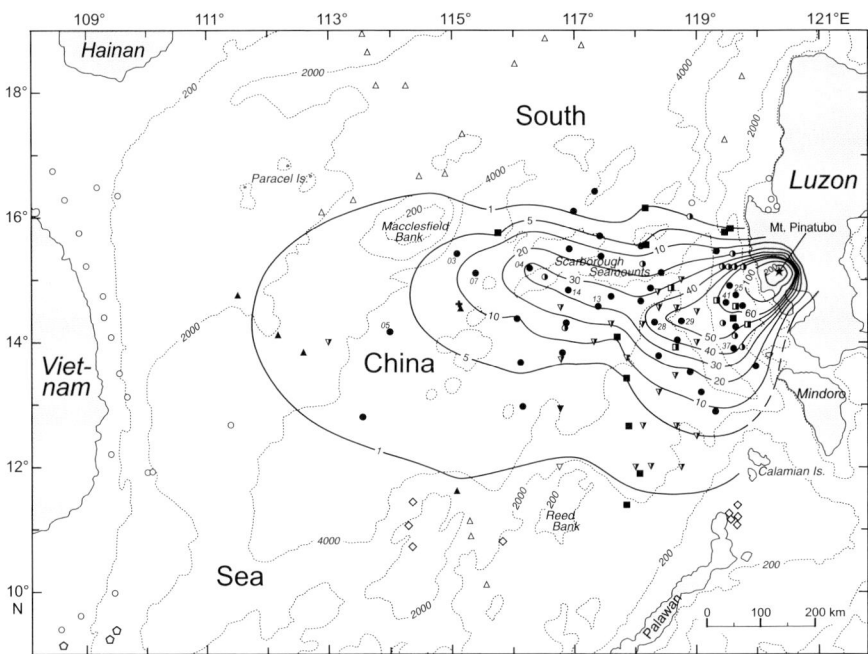

Fig. 4.62 Thickness contours (in mm) of the June 15, 1991 Mt. Pinatubo ash fall deposit in the SCS and on Luzon Island were determined in cores from various cruises (after Wiesner et al. 2004). Fully and partly filled symbols indicate cores with ash layers measured (including those labeled with core numbers) or estimated due to core disturbance. *Unfilled symbols* are core sites without ash. Isobaths are in m

carry the ash rapidly through the strong surface water currents in the central and eastern SCS. Pyroclasts with fine ash in the Pinatubo tephra is common in all of the samples, showing pumice fragment with vesicle fillings largely consisting of tightly interlocked glass shards (Fig. 4.63).

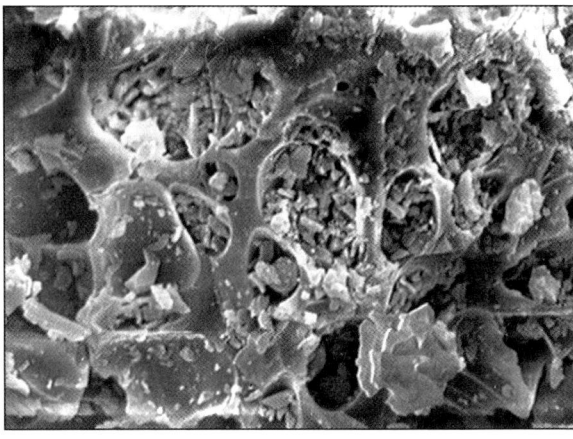

Fig. 4.63 Scanning electron microscopy (SEM) image reveals detailed structure of Pinatubo pyroclasts in a sample taken on May 21 to June 18, 1991, at 3,730 m water depth. Width of field of view is 50 μm (after Wiesner et al. 1995)

Youngest Toba Tephra

The youngest Toba eruption occurred at ~74 ka and may have been the largest single volcanic eruption during the Quaternary (Chesner et al. 1991). The impact of this eruption on global climate and human evolution has been widely discussed (Oppenheimer 2002, and references therein). Dispersed rhyolitic glass and pumice of the Youngest Toba Tephra (YTT) from Toba in northern Sumatra were first found in deep-sea cores in the northeastern Indian Ocean, the Bay of Bengal, the Andaman Sea, and on land in Malaysia (Ninkovich et al. 1978). Similar deposits were then reported from the Indian subcontinent (Acharyya and Basu 1993), the Central Indian Basin (Pattan et al. 1999), the SCS (Bühring et al. 2000; Song et al. 2000; Liang et al. 2001; Liu Z. et al. 2006), and the Arabian Sea (Schulz et al. 2002) (Fig. 4.64). Sediment cores show that the dispersal of the YTT extended to 14 °S in the southern hemisphere, westwards as far as the longitude 64 °E, and northeastwards ~1, 800 km in the SCS, i.e. at least 2% of the Earth surface. Thus, geochemical compositions of glass shards and minerals from deposits of the YTT are usually considered for tephrostratigraphic correlation (Oppenheimer 2002). Despite its exceptional magnitude, however, the youngest Toba eruption had only a minor impact on the evolution of low-latitude monsoonal climate on centennial to millennial time scales (Schulz et al. 2002).

The optical microscope investigation indicates that glass shards and pumice appear fresh, shining, and colorless with different sizes (mainly 20–500 μm) and

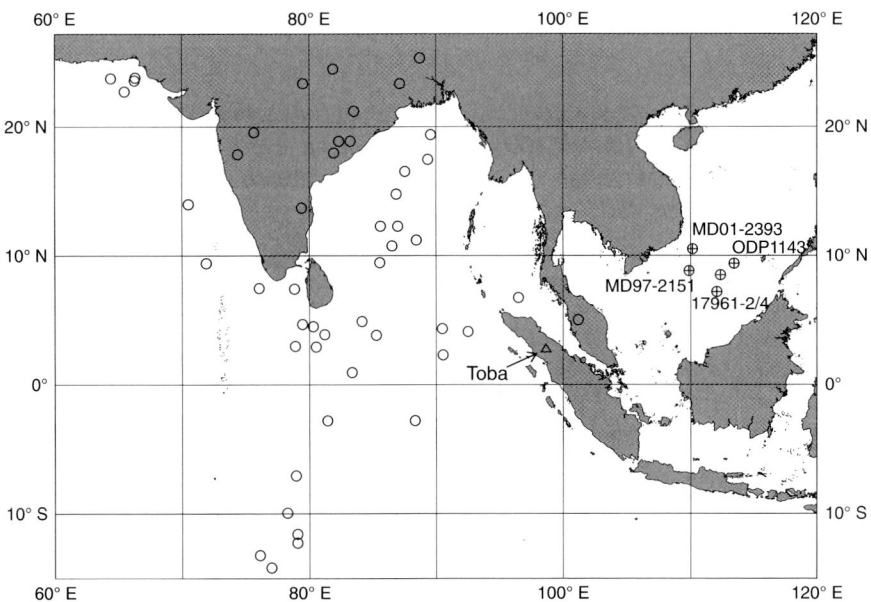

Fig. 4.64 Sketch map shows distribution of known Youngest Toba Tephra deposits (*open circles*) and their known records in the SCS (*crossed circles*) (after Liu Z. et al. 2006)

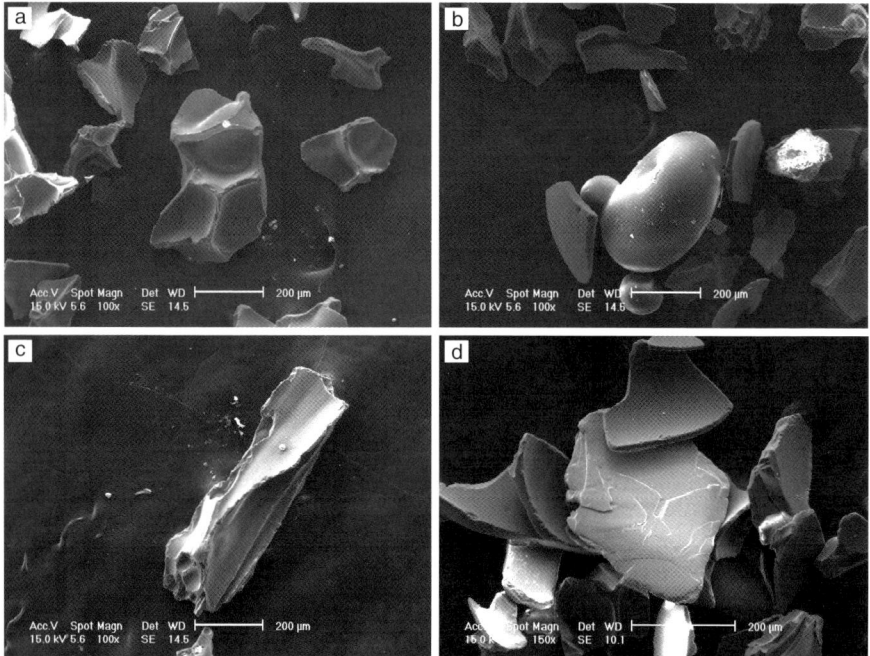

Fig. 4.65 SEM images show morphological variations of youngest Toba glass shards from Core MD01-2393. (**a**) Bubble-wall shards with smooth and slightly curved surfaces; (**b**) elongated bubble with two small coalescent ones; (**c**) pumice shard with elongated vesicles; (**d**) platy and angular biotite mineral (after Liu Z. et al. 2006)

shapes with no signs of alteration (Liu Z. et al. 2006). Minor biotite grains, with sizes ranging between 100 and 400 μm in diameter, are dark brown. Scanning electron microscopy (SEM) studies on the morphology of the glass shards and minerals show that most of the shards are of the bubble-wall variety with smooth and slightly curved surfaces (Fig. 4.65). Planktonic foraminifer *G. ruber* (white) $\delta^{18}O$ record indicates the YTT layer with 0% carbonate content at MIS 4–5 transition, providing an excellent stratigraphic correlation marker in the southern SCS (Song et al. 2000; Liu Z. et al. 2004).

4.6 Estimation of Deposit Mass Since the Oligocene (Huang W. and Wang P.)

As the largest marginal sea in the west Pacific, the SCS is the sink of sediments eroded from catchments of the Asian continent and numerous surrounding islands, which have been affected by tectonoclimatic changes over the last 50 myr or more (Clift 2006). The SCS sediment not only preserves a large amount of resource including petroleum and gas hydrate, but also provides excellent records for studying the history of regional tectonic and paleoceanographic changes. Quantitative reconstruction of the overall sedimentation history in the region is therefore becoming a great interest for both academia and industry, although statistic difficulties still exist especially in answering important questions such as how much sediment has been deposited in the SCS. Nevertheless, the great amount of data generated in last several decades warrants an attempt to quantify the overall deposit mass for the region.

Since 1990s, many international cruises to the SCS have recovered hundreds of sediment cores, resulting in the publication of many monographs and theses (Chapter 1). Using some of these sediment cores, Huang and Wang (1998) and Wang P. et al. (1999) discussed the general distribution of deposit mass in several northern SCS basins. High-resolution seismic stratigraphy also provides a reliable means for reconstructing the sedimentological history in the region (Lüdmann et al. 2001; Zhong et al. 2004). This section summarizes the results from a preliminary study of deposit mass in the SCS in order to shed some light on basin evolution and sedimentation patterns in the SCS.

Data Sources and Analyses

Over its 33 myr history, the SCS has experienced fundamental tectonomorphological changes, and thus its present-day area cannot be used to represent its size in the past (Chapter 2). Since the data are not sufficient for discussing the entire SCS, our estimation of sediment mass covers the area between $0°–24°$N and $104°–121°$E, totally 3.34×10^6 km^2, but excludes most part of the Gulf of Thailand and Sunda Shelf (Fig. 4.66). From this area, 94 seismic profiles, 136 boreholes, 34 industry wells, and many isopach maps and stratigraphic columns were collected as a data bank for this exercise (Fig. 4.66) (Huang 2004).

The 6 ODP sites provide high quality references for stratal analyses and correlation of seismic profiles since the Oligocene (Wang P. et al. 2000, 2003; Li Q. et al. 2005). Other sediment cores, however, are relatively short, mostly reaching only MIS 5 in the late Quaternary. Seismic profiles from the northern SCS provided by the Guangzhou Marine Geological Survey are mostly with a relatively low resolution. However, some large-scale unconformites can be distinguished as representing stratal boundaries, such as Paleocene/Eocene, Oligocene/Miocene, middle/late Miocene, Miocene/Pliocene boundaries and the top of basement. Data from the literature (Sarnthein et al. 1994; The Multidisciplinary Oceanographic Expedition Team of Academia Sinica to Nansha Islands 1989, 1996; Bai et al. 1996; The Second Marine Geological Investigation Brigade of the Ministry of Geology and Resources 1987, 1990, 1992; Qiu et al. 1999; Yao and Zeng 1994; Zhan et al. 2003; Liu H. et al. 2002; Gong and Li 1997) are re-analysed and correlated to the new stratigraphy (Huang 2004; Huang and Wang 2007a).

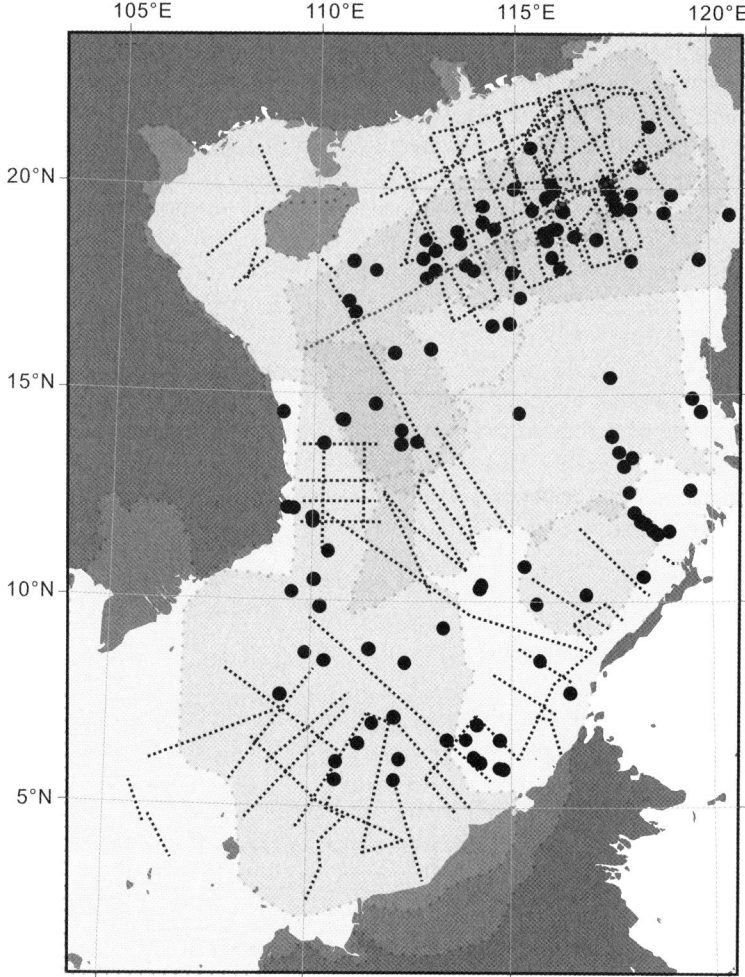

Fig. 4.66 Map shows location of the original data sites and types: cores (*filled circles*), seismic profiles (*bold dotted lines*), and sediment isopachs (*shaded light gray*) (from Huang and Wang 2006)

It is noteworthy that the data used in the present study vary significantly in quality. Although borehole data often provide accurate constraints on stratigraphic boundaries and sediment thickness, core sites are limited in number and unevenly distributed with different spatiotemporal coverages. In order to minimize the discrepancy in stratal correlation, a low resolution method using "epoch" or "period" units was adopted. Accordingly, stratal intervals of Oligocene (E_3), lower Miocene (N_1^1), middle Miocene (N_1^2), upper Miocene (N_1^3), Pliocene (N_2), and Quaternary (Q) were assembled and their sediment masses were individually calculated. Detailed modeling methodology is given in Huang and Wang (2007a).

Sediment Distribution and Mass

Shown in Fig. 4.67A is the calculated total sediment thickness in the SCS since the Oligocene (Huang and Wang 2007a). It confirms a thick sediment cover in shelf-slope basins and a very thin sediment cover in the central Deep Basin, Nansha area and eastern part of the SCS. The ~20 km deposit succession in the Yinggehai and Zengmu basins is the thickest, compared to ~10 km in other basins, as revealed in drilling wells (Chapter 3). If the post-Eocene strata were "removed", a contrast between the "pre-Oligocene base" (Fig. 4.67B) and the modern seabed topography (Fig. 4.67C) is tremendous, with basin areas such as the Yinggehai and Zengmu being the most striking.

The pre-Oligocene base map was generated from a simple calculation of geo-topographic location and sediment thickness. Because the topographic map used for calculation was in much higher resolution than the targeted one especially for areas of Nansha and Zhongsha islands where limited data are available, the outcome is a visible resemblance between the "pre-Oligocene base map" and the modern topography map for these areas.

On the basis of these isopach maps (Fig. 4.67A), the total volume and mass of post-Eocene deposits in the SCS can be calculated to be $7.01 \pm 1.48 \times 10^6$ km^3 and $14.45 \pm 4.06 \times 10^{15}$ t, respectively, with an average accumulation rate of 12.8 g/cm^2/ kyr. The average sedimentation rate is 6.22 cm/kyr, which is significantly higher than the world ocean record of 1 cm/kyr (Kennett 1982). Hay et al. (1988) estimated that about 26.2×10^{16} t of sediment had accumulated on the global ocean floor during the Oligocene-Quaternary. Taken this as a face value, our results may convey a message that nearly 5.5% of the global sediment was deposited in the SCS. As its area of 3.62×10^8 km^2 represents only 0.9% of the global ocean surface, the SCS has been playing a very important role in material transport and deposition in the marginal western Pacific since at least the Oligocene time.

The Central Basin accumulates a sediment mass of 0.68×10^{15} t, or less than 5% of the total, while shelf and slope basins receive the major part of deposition. Using the 2000 m isopach for basin boundary, we have a total basin area of nearly 1.15×10^6 km^2, or 34% of the entire SCS. The deposit mass in all the basin area is 11.9×10^{15} t, accounting for 82% of the total mass. When the pre-Oligocene base is compared with the present-day topography (Fig. 4.67B,C), it is clear that shelf-slope basins surrounding the central area hold the most terrigenous deposit mass. Although river discharges into the SCS could be very high, abyssal fans are rarely formed in the Central Basin. This character is quite different from the Atlantic or Indian oceans where abyssal fans (such as Bengal Fan) are well developed along passive continental margins (Huang and Wang 2007a).

In this preliminary study, the tectonic influence on deposite mass has not been evaluated because to determine how much deposit mass had been lost to plate sub-duction is yet viable. Given the enclosed nature of the SCS basin, local denudation caused by tectonic activities (e.g., Yao and Zeng 1994) may have little influence on calculating deposit mass but inevitably affect more on accumulation rates or other proxies relating to sedimentary process. At Site 1148, the greatest denudation

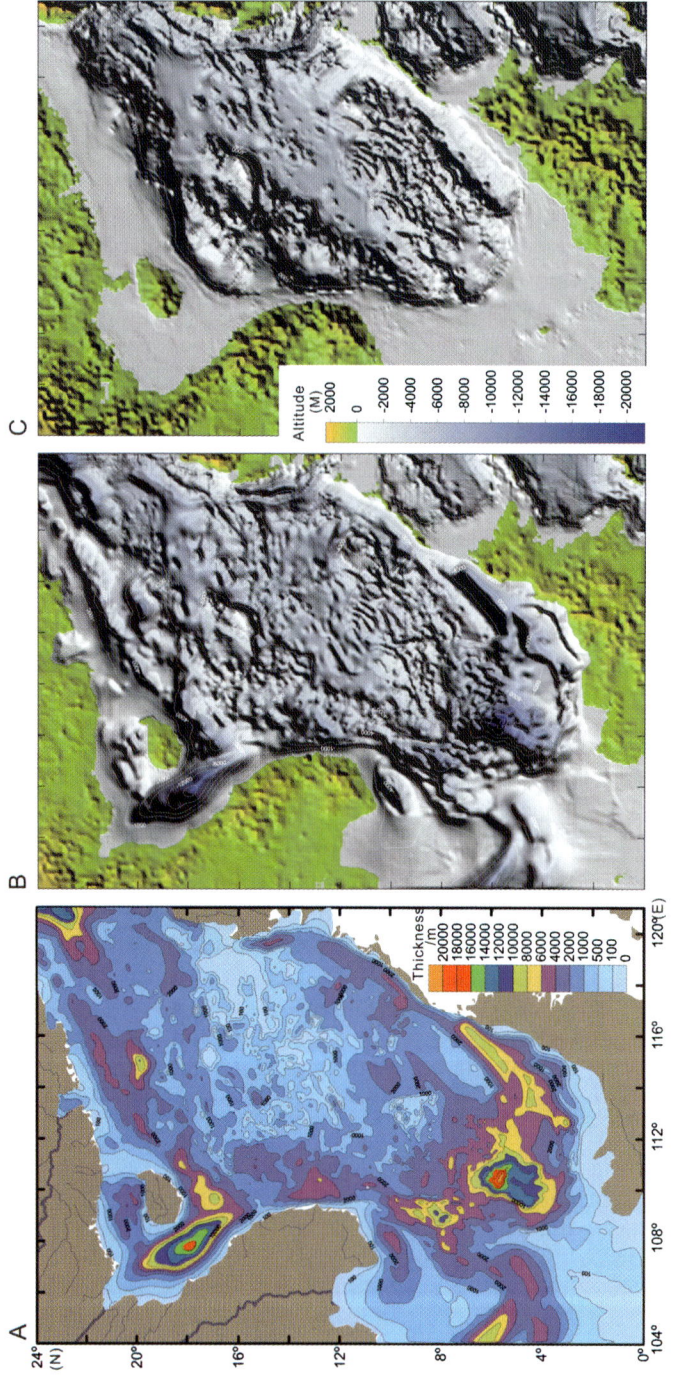

Fig. 4.67 Maps show (**A**) sediment thickness isopaths, (**B**) pre-Oligocene base topography after younger deposits were "removed", and (**C**) modern topographic map of the South China Sea (Huang and Wang 2007a)

occurred during the late Oligocene with several unconformities eroding off a depositional record of about 3 myr (Li X. et al. 2003; Li Q. et al. 2005).

Estimation of Terrigenous and Carbonate Masses

The proportion between terrigenous and carbonate components in the total deposit mass is an important but very complex issue. Terrigenous sediments are common to dominant, while carbonate components increase toward the deep sea. Coral reefs and other calcareous biotic skeletons are the two main kinds of carbonate in the SCS.

Coral reefs distribute over a large area of Dongsha, Xisha, Zhongsha and Nansha Islands (see "Coral reefs" section above). The total area for the largest 5 reefs in Xisha Islands is about $2.4 \times 10^3 \, \mathrm{km}^2$, according to Xu et al. (2000). In Nansha Islands, the total reef area may reach $2.4 \times 10^4 \, \mathrm{km}^2$, including $0.9 \times 10^4 \, \mathrm{km}^2$ for the largest reef Reed Bank (Liu B. 2001). The largest reef in Zhongsha Islands has a similar size as Reed Bank, but reefs in Dongsha Islands are much smaller. Based on these, we estimated the total coral reef area in the SCS to be about $4.0 \times 10^4 \, \mathrm{km}^2$, or only 1% of the entire SCS.

As most coral reefs are steep conic or cylinder in shape, their volume can be calculated based on their base depth and top size to be nearly $7 \times 10^4 \, \mathrm{km}^3$. If a density of $2.7 \, \mathrm{g/cm}^3$ is considered, the total coral reef mass is nearly $0.19 \times 10^{15} \, \mathrm{t}$, or less than 1% of the total deposit mass in the SCS.

The main carbonate components are calcareous biotic skeletons, especially those of foraminifera and nannoplankton. Cores drilled during ODP Leg 184 provide detailed information about long-term carbonate accumulation (Wang P. et al. 2000). When calculating carbonate deposite mass, we allocated at least one reference section for each of the 16 designated depositional provinces (A to P) (Fig. 4.68). The carbonate mass and terrigenous mass in provinces A-H and P were then calculated for each stratigraphic interval by reference to the records of Sites 1146 and 1148, and those in other areas to Site 1143 or Site 1148 for deposits older than the middle Miocene. The preliminary results show that, in average, the total mass consists of 63% terrigenous and 37% carbonate components, confirming the predominant role of terrigenous deposition in the SCS.

The highest total deposit mass is $6.4 \times 10^{15} \, \mathrm{t}$ for the Oligocene, while the Quaternary and the Pliocene are the least, $1.07 \times 10^{15} \, \mathrm{t}$ and $0.9 \times 10^{15} \, \mathrm{t}$, respectively (Huang 2004). These estimates may have some biase for the inclusion of Site 1148 with over 400 m Oligocene strata from the northern lower slope as the key reference section. To understand better the sedimentary history, however, accumulation rate variations calculated for different time periods (Fig. 4.68) may be more relevant than deposit mass. Accumulation rate calculation requires a consideration of sequential size change in the sea basin caused by tectonics, so the total SCS area in different periods can be corrected. At present, however, it is difficult to reconstruct paleogeographic maps for different periods because the extent of deformation and offset of

Fig. 4.68 Map shows depositional provinces A to P (*shaded gray*) and their average accumulation rates (*dark bars* for terrigenous and white bars for carbonate). The ordinates for each bar chart from bottom to top are stratigraphic intervals: Oligocene, early Miocene, middle Miocene, late Miocene, Pliocene and Quaternary, respectively (from Huang and Wang 2006)

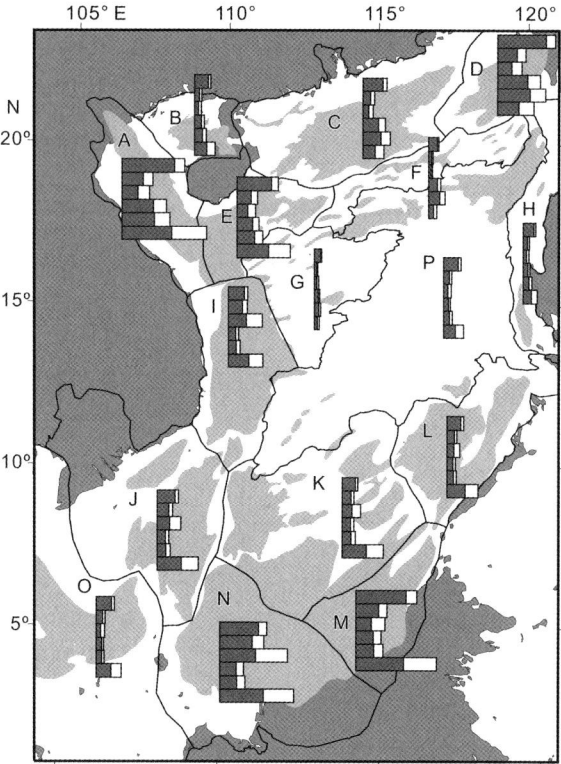

the continental crust in different seafloor spreading stages are not yet fully known. Therefore, we assumed a constant area for the shelf and slope and adjusted only the Central Basin area by reference to magnetic anomaly changes from 30 Ma to 16.7 Ma (Briais et al. 1993) using the timescale of Cande and Kent (1992, 1995). This means only the basin area for the late Oligocene, early Miocene and middle Miocene would be adjusted. The Central Basin area was set at zero at the beginning of seafloor spreading, and increased stepwise to $0.22 \times 10^6 \, \text{km}^2$ at the end of the Oligocene, $0.48 \times 10^6 \, \text{km}^2$ at the end of the early Miocene and $0.52 \times 10^6 \, \text{km}^2$ at the end of the seafloor spreading (16.7 Ma) in the middle Miocene (Huang and Wang 2006).

The average accumulation rates charted in Fig. 4.68 show highest during the Oligocene for most areas, and relative high during the late Miocene for the southern areas and during the Quaternary for the northern areas. The average accumulation rate is $\sim 22 \, \text{g} \cdot \text{cm}^{-2} \cdot \text{kyr}^{-1}$ for the Oligocene, $9–10 \, \text{g} \cdot \text{cm}^{-2} \cdot \text{kyr}^{-1}$ for the early and middle Miocene, $11.34 \, \text{g} \cdot \text{cm}^{-2} \cdot \text{kyr}^{-1}$ for the late Miocene, $\sim 9 \, \text{g} \cdot \text{cm}^{-2} \cdot \text{kyr}^{-1}$ for the Pliocene, and $\sim 15 \, \text{g} \cdot \text{cm}^{-2} \cdot \text{kyr}^{-1}$ for the Quaternary (Fig. 4.69). Today, silt discharge from all rivers into the SCS is nearly $400 \times 10^6 \, \text{t}$, a value corresponding to $\sim 18 \, \text{g} \cdot \text{cm}^{-2} \cdot \text{kyr}^{-1}$ if an area of $3.34 \times 10^6 \, \text{km}^2$ is set for the SCS. Therefore, the

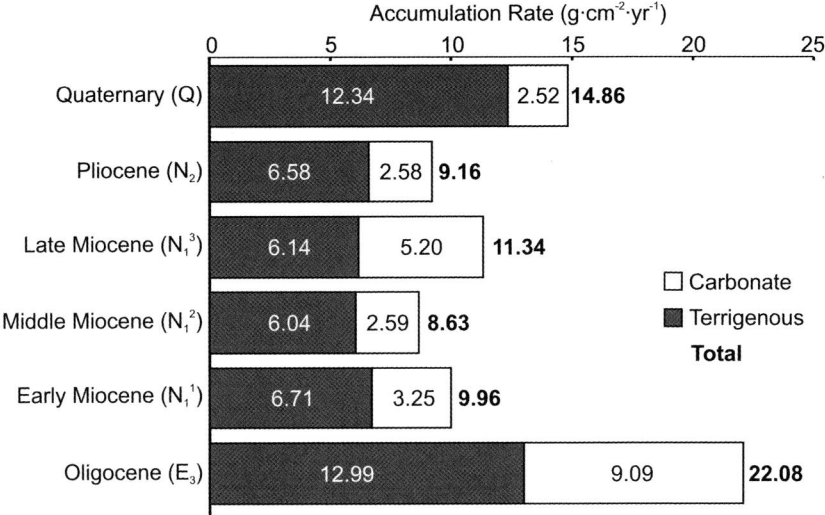

Fig. 4.69 Average carbonate (*unfilled*) and terrigenous (*filled*) accumulation rates were estimated for different time periods

accumulation rate in the present-day SCS is extremely high, largely due to intensive denudation in the modern drainage areas.

Depositional Patterns

Late Oligocene

SCS spreading began in the late Oligocene, and depositional activities at the time were largely limited to the present slope and shelf areas. As the oceanic crust was yet to be formed, the present Central Basin area presumably had received little sediments older than the late Oligocene (Fig. 4.70). However, the northern boundary of the proto-SCS remains unknown, which hampers a detailed evaluation of depositional patterns.

Rapid deposition during the Oligocene in shelf-slope area was related to rifting, with accumulation rates higher than $10\,g\cdot cm^{-2}\cdot kyr^{-1}$ in many parts of the SCS. The central Yinggehai Basin registered highest accumulation rate of $200\,g\cdot cm^{-2}\cdot kyr^{-1}$ with Oligocene deposits of about 8,500 m thick, while the Zengmu Basin from the south had $150\,g\cdot cm^{-2}\cdot kyr^{-1}$ or more. Accumulation rates in other southern basins were also high, $50\,g\cdot cm^{-2}\cdot kyr^{-1}$, and Oligocene strata are generally more than 1,000 m in thickness, or even more than 4,000 m in the central Wan'an, Zengmu and Brunei-Sabah basins. In the north, Oligocene deposits in the Qiongdongnan and Zhujiangkou basins reached a thickness of 2,000 m or more, but often 500 m or less in the Beibuwan and Taixinan basins. Therefore, differences in sediment accumulation between the northern and southern SCS existed already in the Oligocene (Fig. 4.70).

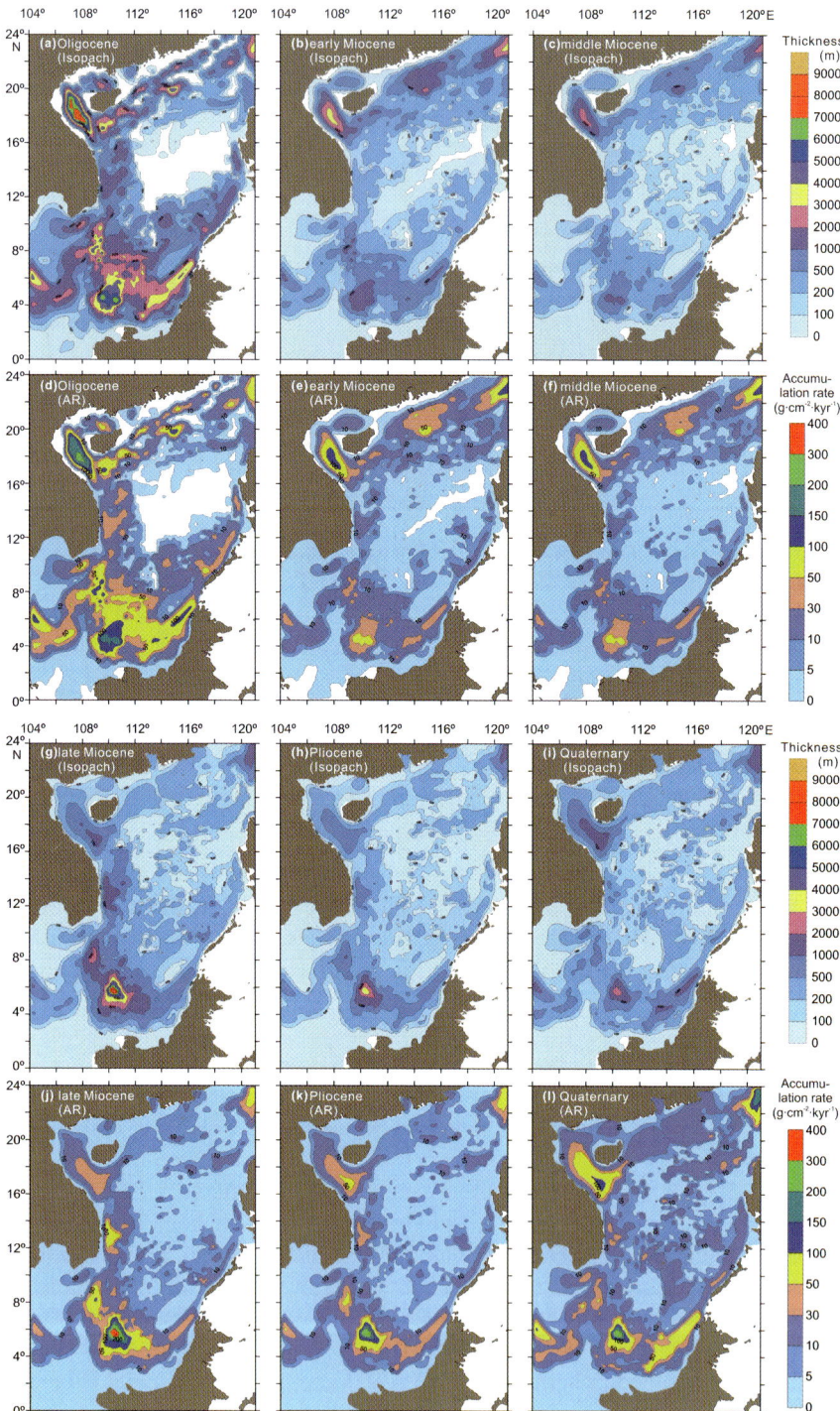

Fig. 4.70 Maps show estimated sediment thickness and accumulation rates (AR) in various periods

Neogene

Intensive tectonic activities in the late Oligocene SCS reconfigured the regional depositional environments (Wang P. et al. 2003). In the early and middle Miocene, average accumulation rates decreased when seafloor spreading rapidly widened the Central Basin, leading to the accumulation of relatively thin sediment layers (Fig. 4.70). Spreading continued at a steady rate until the middle Miocene (Briais et al. 1993), as evidenced also by relative stable depositional record without obvious large-scale variations (Wang P. et al. 2003). During these periods, accumulation rates in most areas decreased to about $10\,\mathrm{g\cdot cm^{-2}\cdot kyr^{-1}}$. Again, accumulation rates in the Yinggehai and Zengmu basins remained very high with $\sim100\,\mathrm{g\cdot cm^{-2}\cdot kyra^{-1}}$ and $\sim50\,\mathrm{g\cdot cm^{-2}\cdot kyr^{-1}}$ respectively (Fig. 4.70). In the Yinggehai Basin, $\sim3,200\,\mathrm{m}$ lower Miocene and $\sim2,200\,\mathrm{m}$ middle Miocene sequences were deposited in the southeastward-moving depocenter. The Zhujiangkou Basin and Taixinan Basin now joined together and accumulated up to $\sim2,000\,\mathrm{m}$ lower Miocene and $\sim1,500\,\mathrm{m}$ middle Miocene sediments (Fig. 4.70), indicating intensive basement subsidence (Gong 1997).

Comparatively low accumulation rates in the early and middle Miocene SCS were in a sharp contrast with rapid denudation in the Tibetan Plateau at these times (Harrison et al. 1992). One reasonable explanation is that deposition in the SCS was then not controlled by uplift and weathering of the (southern) Tibetan, which probably contributed more to the rapid buildup of the Bengal Fan in the northern Indian Ocean (Johnson 1994). Another reason is insufficient data on sediments lost to the end middle Miocene unconformity, a northern SCS-wide event caused by the "Dongsha Movement" about 12 myr ago (Yao and Zeng 1994).

Depositional patterns in the late Miocene between the northern and southern SCS appear to reverse from earlier periods, with higher accumulation rates now occurring in the south (Fig. 4.70g,j). In the Zengmu Basin, $>9,000\,\mathrm{m}$ thick strata were deposited in the northward-moving depocenter with an average accumulation rate of $>300\,\mathrm{g\cdot cm^{-2}\cdot kyr^{-1}}$. An upper Miocene sequence of over $2,000\,\mathrm{m}$ occurred in the Zhongjiannan and Wan'an basins with accumulation rates higher than earlier periods. In the northern SCS, however, accumulation rates decreased to lower than $10\,\mathrm{g\cdot cm^{-2}\cdot kyr^{-1}}$. The Yinggehai Basin and Qiongdongnan Basin now joined together and the new depocenter, on the previous boundary between these two basins, accumulated an upper Miocene sequence of $\sim1,000\,\mathrm{m}$ with maximum accumulation rates of $50\,\mathrm{g\cdot cm^{-2}\cdot kyr^{-1}}$ or lower. The Zhujiangkou Basin and Taixinan Basin became flat without obvious depocenters and deposits grew like thin sheets. All these may indicate stronger tectonic activities in the south than in the north during the late Miocene, probably as a result of further Asia-Australia collison (Cane and Molnar 2001) and the rotation and land-arc collision of the Philippine Plate (Huang C.Y. et al. 2000; Packham 1996). Due to increased isolation, the SCS received more terrigenous sediments along its newly active margin especially in the south.

Depositional patterns in the Pliocene continued on the trend of the late Miocene, indicating similar states in tectonic subsidence and sedimentation as before. Deposition of thin sediment sequences again characterize many basins although the

Pliocene section in the Zengmu Basin may reach nearly 3,800 m thick and accumulation rates in the Yinggehai Basin also increased (Fig. 4.70h,k).

Quaternary

Accumulation rates increased suddenly in the northern SCS during the Quaternary, peaking at $>100 \, \text{g} \cdot \text{cm}^{-2} \cdot \text{kyr}^{-1}$ in the Yinggehai Basin ($\sim 1,000$ m) and $\sim 20 \, \text{g} \cdot \text{cm}^{-2} \cdot \text{kyr}^{-1}$ in the Zhujiangkou and Taixinan basins and some parts of the Central Basin (Fig. 4.70i,l). In contrast, no significant accumulation rate changes are found in the southern SCS although the Zengmu Basin still received sediments of $\sim 2,600$ m thick.

Accelerated sediment deposition since ~ 3 Ma was a global phenomenon, with accumulation rate increases by two to as much as 10 times in all active and passive margins due mainly to global climate change (Zhang et al. 2001). Abrupt changes in temperature, precipitation and vegetation since global cooling and formation of northern hemisphere ice sheet ~ 3 myr ago contributed to intensified deposition world-wide, in a time the glacial regime prevailed. In addition to all these, the SCS record also manifests the intensified East Asian Monsoon (Chapter 5).

The Last Glacial Cycle

Intensive land-sea interactions over the last 20 kyr characterize the late Quaternary glacial-interglacial cycles. To quantify depositional characteristics for this youngest geological period will not only help understand the climate-deposition dynamics in glacial cycles better but may also provide insight into the regional depositional history on a longer time scale. As the shallow shelf sediment has been subject to intensive modifications by frequent sea level changes, we limit our analyses to areas deeper than 100 m water depths based primarily on data from 136 sediment cores (Fig. 4.71) (Huang and Wang 2007b).

Deposit mass and accumulation rates: Sediment isopach maps for MIS 1 (Fig. 4.71A) and MIS 2 (Fig. 4.71B) were reconstructed before their deposit mass and accumulation rates were estimated (Table 4.9). The timescale of Martinson et al. (1987) was followed, with a base age for MIS 1 at 12.05 ka and MIS 2 at 24.11 ka.

For a better interpretation of the data, 7 deposition areas (I to VII) were designated based on the sediment thickness results (Fig. 4.71C). Both from the northern slope, area I is located to the east of the Pearl River estuary, while area II is to its west. Others are from the western and southern slopes and the deep basin: area III between the Vietnam coast and Zhongsha Islands, area IV near the Mekong River estuary, area V for the southern continent slope, area VI for the Nansha Islands region, and area VII for the Central Basin. The average accumulation rates for these areas are presented in Fig. 4.72.

Sediment thickness varies considerably between the 7 deposition areas although MIS 2 deposits are always thicker than the Holocene (Figs. 4.71 and 4.72). Areas I, IV and V show characteristics of newer sedimentary centers with thicker MIS 1 deposits compared to other areas. The total deposit mass for all areas is 1.84 ×

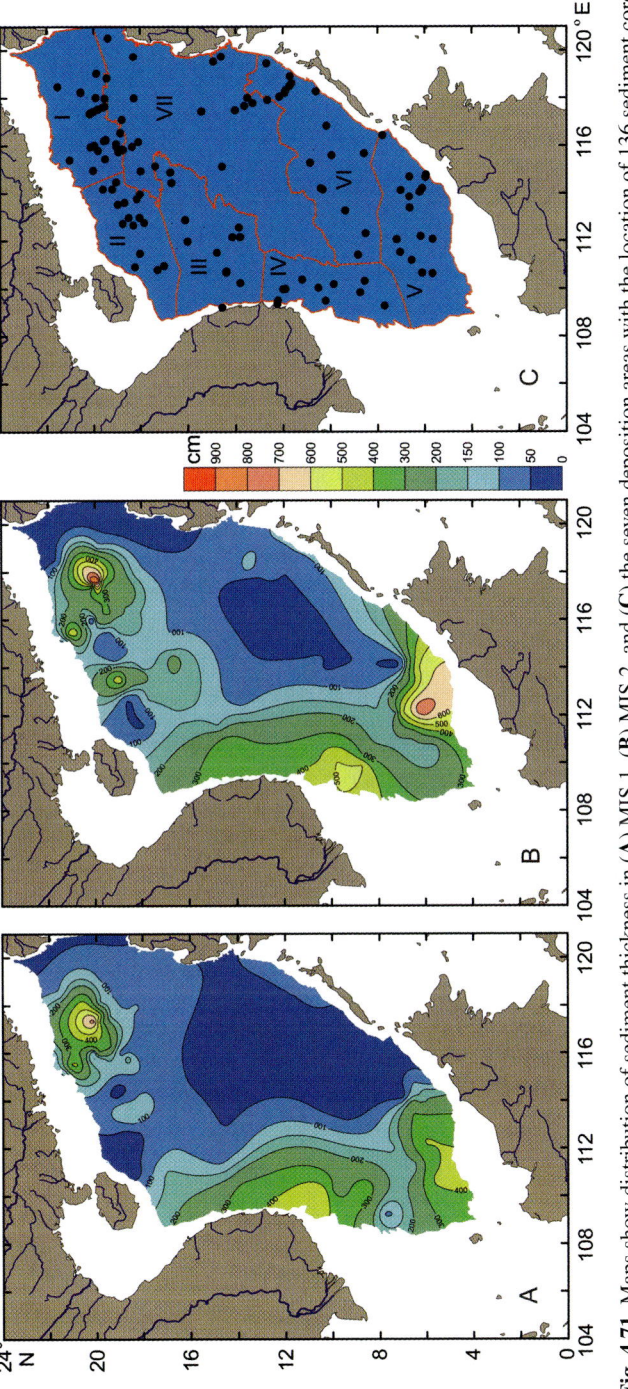

Fig. 4.71 Maps show distribution of sediment thickness in (**A**) MIS 1, (**B**) MIS 2, and (**C**) the seven deposition areas with the location of 136 sediment cores

Table 4.9 Deposit mass ($\times 10^9$ t) and accumulation rates ($g \cdot cm^{-2} \cdot kyr^{-1}$) of terrigenous, carbonate and opal components show large differences between MIS 1 and MIS 2

	Total deposits		Terrigenous		Carbonate		Opal	
	MIS 1	MIS 2	MIS 1	MIS 2	MIS 1	MIS 2	MIS 1	MIS 2
Deposit mass	1836.3	2592.3	1407.7	2146.7	365.4	347.2	63.2	98.5
			76.7%	82.8%	19.9%	13.4%	3.4%	3.8%
Accum. rate	8.17	11.53	6.26	9.55	1.63	1.54	0.28	0.44

10^{12} t for MIS 1 and 2.56×10^{12} t for MIS 2 (Table 4.9), with the majority being terrigenous, or 77% for MIS 1 and 83% for MIS 2. The biogenic component is dominated by carbonate with similar percentages (20% or less) in both stages while the content of opal is very low, between 3.4 and 3.8% (Table 4.9). The average accumulation rate for MIS 2 is $11.53 \, g \cdot cm^{-2} \cdot kyr^{-1}$ (terrigenous 9.55), which is much higher than $8.17 \, g \cdot cm^{-2} \cdot kyr^{-1}$ (terrigenous 6.26) for MIS 1 largely due to enhanced terrigenous supply during the LGM (Table 4.9).

In shallow shelf areas, such as the Sunda Shelf, downslope or lateral transports caused by frequent rapid sea-level fluctuations often resulted in a thinner sediment cover (Hanebuth et al. 2000; Steinke et al. 2003). Accordingly, deposition at middle

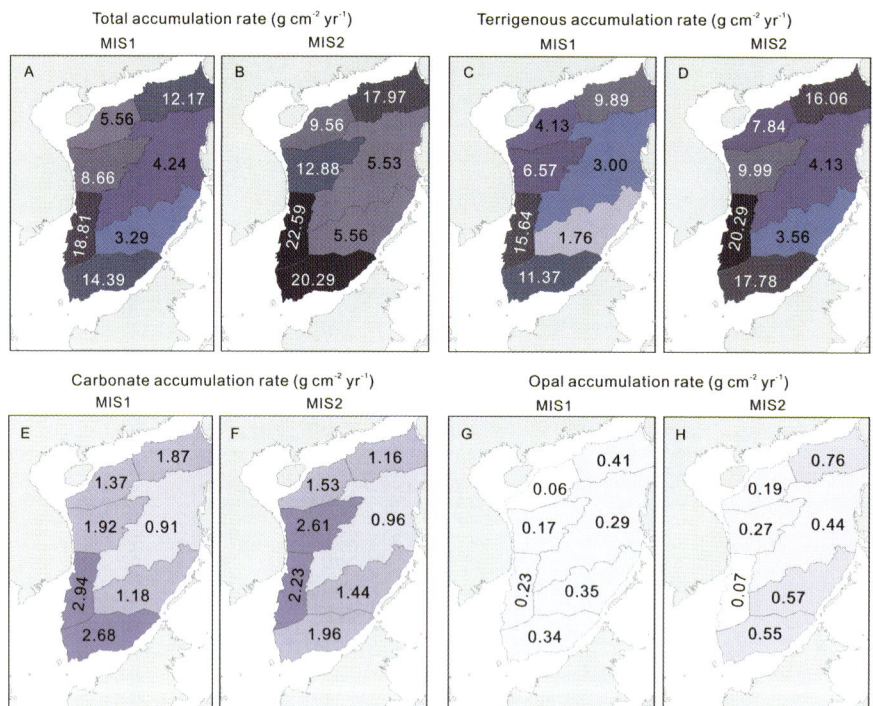

Fig. 4.72 Maps show the calculated average accumulation rates ($g \cdot cm^{-2} \cdot kyr^{-1}$) of various sediment components in different off-shelf areas of the SCS during MIS 1 and MIS 2

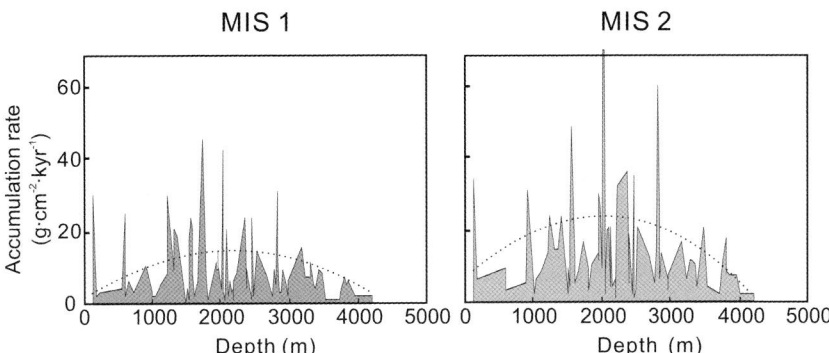

Fig. 4.73 Variations in sediment accumulation rates in off-shelf areas of the SCS during MIS 1 and MIS 2 are compared

slope settings reached a maximum, forming the main feature of SCS sedimentation (Fig. 4.73).

Biogenic components comprise roughly one fifth of the total deposit in the late Quaternary SCS (Table 4.9). Unlike carbonate, opal appears to have been unaffected by dissolution during glacial-interglacial cycles but mainly responded to productivity variations (see "Biogenic opal" section above). A high productivity during MIS 2 can be inferred by average opal accumulation rate of $0.44\,\text{g}\cdot\text{cm}^{-2}\cdot\text{kyr}^{-1}$, compared to $0.28\,\text{g}\cdot\text{cm}^{-2}\cdot\text{kyr}^{-1}$ for MIS 1.

Two types of carbonate cycles existed in the late Quaternary SCS: an "Atlantic type" from above the lysocline with a low carbonate content during glacials, and a "Pacific type" from below the lysocline with a high carbonate content during glacials (see "Carbonate" section above) (Bian et al. 1992; Wang P. et al. 1995, 1997; Zhao and Wang 1999). As shown in Table 4.9, the average carbonate content for MIS 1 $(365 \times 10^9\,\text{t})$ is only slightly higher than MIS 2 $(347 \times 10^9\,\text{t})$, with a small difference within the measurement errors. For individual regions, however, carbonate accumulation rates for MIS 2 were higher than MIS 1 in areas II, III, VI and VII, but lower in areas I, IV and V (Fig. 4.72E,F). Low carbonate accumulation rates in MIS 1 occur in most areas except areas IV and V off the Suna Shelf and Meikong River estuary, indicating the influence of enhanced "Pacific-type" dissolution cycle. During the last glacial maximum, MIS 2, carbonate accumulation rates in areas IV and V remained high, but area III now shown the highest rate due to weaker carbonate dissolution.

Terrigenous components form the main part of sediments in MIS 1 and MIS 2 in both the total mass and average accumulation rates (Fig. 4.72). Area IV from offshore of the Mekong River mouth registered highest terrigenous accumulation rates of $15–20\,\text{g}\cdot\text{cm}^{-2}\cdot\text{kyr}^{-1}$ during MIS 1 and MIS 2, followed by $11–18\,\text{g}\cdot\text{cm}^{-2}\cdot\text{kyr}^{-1}$ in area V from offshore Borneo affected by the paleo-Sunda River and by $10–16\,\text{g}\cdot\text{cm}^{-2}\cdot\text{kyr}^{-1}$ in area I near the Pearl River mouth and toward Taiwan Island. Geochemical analyses indicate that a main part of terrigenous sediments in area I was sourced from Taiwan (Shao et al. 2001). West of the Pearl River estuary, in area II without any Taiwan influence, accumulation rates were almost 50% lower (Fig. 4.72C,D).

Mainly from river discharges, the SCS today receives nearly 400×10^6 t silts annually from the Asia continent and 135×10^6 t from neighboring islands (see also Table 4.1). The total amount of terrigenous deposition in the SCS may reach nearly 600×10^6 t per year if contributions from other island rivers not listed in Table 4.1 are included. Assuming silt discharge from rivers has been constant, terrigenous sediments deposited in the Holocene SCS should be as much as 7.2×10^{12} t, a value much higher than our estimated 1.4×10^{12} t shown in Table 4.9. Even if the rate of sediment delivery from rivers may have been only half as it is today as a result of landscape modifications by human activities is considered (Hay 1998; McLennan 1993), terrigenous sediments for MIS 1 would still be 3.6×10^{12} t, or nearly 3 times high than our calculation. Probably, the difference was caused by the exclusion in our calculation of deposits on shelf, where the main Holocene depositional activities were taken place. Compared with older time intervals, however, the average terrigenous accumulation rate for MIS 2 is still higher than the average value for the Neogene but similar to the Quaternary or Oligocene averages. All this indicates a very active present-day river discharge, which is much higher than any stages before.

Major Characteristics of SCS Sedimentation

The evolution of Cenozoic accumulation rate in the global ocean has been analyzed mainly based on DSDP/ODP data over decades. Figure 4.74 compares the mass/age distribution of oceanic sediment at 5 myr intervals from 0 to 80 Ma (Hay et al. 1988) with other regional studies of Cenozoic sediments, including Asian

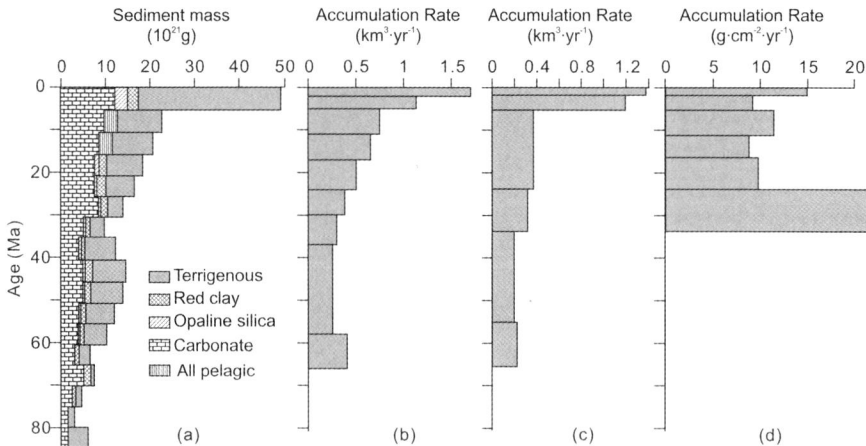

Fig. 4.74 Accumulation rate changes are comparied.between global and regional estimates: (**a**) mass/age distribution in the global ocean (modified from Hay et al. 1988), (**b**) Average solid-phase accumulation rates integrated over all basins of Asia (modified from Métivier et al. 1999), (**c**) sediment–filling estimates for Chinese Cenozoic basins (modified from Liao 1995), and (**d**) Accumulation rates of the SCS since the Oligocene (this study)

basins (Métivier et al. 1999), China and surrounding areas (Liu Z. et al. 2001), and Chinese basin filling estimation (Liao 1995). A similar trend revealed by these studies is that accumulation rates increased step by step from the Oligocene to the Quaternary. In contrast, our sediment mass study for the SCS (Fig. 4.74d) shows high accumulation rates in the Oligocene and Quaternary but low in the Miocene and Pliocene. On the global and regional averages, the Oligocene is characterized by very low sedimentation rates, so the exceptionally high sedimentation rate from the Oligocene SCS is rather unique. Therefore, we can conclude that deposition in marginal seas was primarily controlled by local tectonics. Similarly, the relatively high accumulation rates in Miocene stages may have also resulted from renewal tectonic activities during and after sea-floor spreading in the SCS (Li Q. et al. 2006).

The overall sedimentation rate in the SCS is about 10% higher than the estimated value of $1 \, \text{cm} \cdot \text{kyr}^{-1}$ for the Pacific (Kennett 1982; Lisitzin 1972). As the largest western Pacific marginal sea, the SCS is surrounded by lands and islands which supply ample terrigenous material. Consequently, even the abyssal part of the SCS receives a large amount of terrigenous clastic sediments mixed with various proportions of carbonate and other biogenetic sediments. These sedimentation patterns characterize the SCS as a favorite site locality for the research of paleoceanography, as well as for sea-land interaction studies.

The early Paleogene rifting and the subsequent spreading of the SCS was associated with extension and thinning of the continental margin (Briais et al. 1993), leading to the formation of numerous sediment basins (Chapters 2, 3). As these basins near big estuaries were filled, the slope and shelf became wide and broad as seen today, and abyssal fans are rarely formed due to a lack of large channels or constant sediment flux. Furthermore, the existence of the SCS and other marginal seas block the transport of terrigenous sediments from lands of Asia to bathyal West Pacific, and the fan-lacking western Pacific margin is quite different from the fan-rich Atlantic and Indian margins. The Bengal Fan and Indus Fan in the Indian Ocean both resulted from uplift of the Himalayas since the Eocene, while the Amazon Fan represents a series of deep-water fans along the strongest drained Atlantic margin (Johnson 1994). As the largest deepwater fan in the world, the Bengal Fan consists of fine grain-size turbidite of at least 2,000 m thick at ~4,000 m water depth at the southern far end of the fan (Stow et al. 1990). The absence of large abyssal fans in the SCS and in the West Pacific indicates that balance between tectonic subsidence and sedimentation rates in confined deposition centers and basins on the shelf and slope in this region is similar to abyssal fans in other oceanic margins without typical marginal seas.

Some back-arc basins in the West Pacific, such as the Sulu Sea, are surrounded by island arcs with a steep slope and a narrow shelf, from where sediments are moved directly into the deep basin by turbidity currents (Marsaglia et al. 1995), resulting in Holocene accumulation rates higher than 60 cm/kyr at 4,000 m depths (Kudrass et al. 1991). This is in a sharp contrast with the SCS, in which highest Holocene accumulation rates occur on the middle slope rather than in the deep basin. Therefore, it can be concluded that the basin size and topography of marginal seas determine their sedimentation patterns.

In summary, the analyses of deposit mass and accumulation rate of the SCS since the Oligocene reveal the following macro-sedimentary features that characterize the regional sedimentological history:

1. The total deposit mass in the SCS from the beginning of seafloor spreading in the Oligocene is about 1.44×10^{16} t. The major part of this sediment is terrigenous, and most of it was deposited during the Oligocene.
2. Highest accumulation rates occurred in the Oligocene ($\sim 22\,\mathrm{g \cdot cm^{-2} \cdot kyr^{-1}}$) and Quaternary ($\sim 15\,\mathrm{g \cdot cm^{-2} \cdot kyr^{-1}}$), compared to relatively low accumulation rates ($\sim 10\,\mathrm{g \cdot cm^{-2} \cdot ka^{-1}}$) in the Miocene-Pliocene. The high Oligocene accumulation rate in the SCS is in opposite to a low rate trend in the global sedimentation patterns, indicating the primary control on deposition in marginal seas by local tectonic movements and topography.
3. More than 80% of the SCS sediments were deposited in basins along its continental margin. As the largest marginal sea in the west Pacific, the SCS is characterized by its broad and flat shelf and relative gentle slope where numerous sediment basins were formed and subsequently covered mainly by terrigenous sediments. The lack of abyssal fans in the SCS deep basin easily distinguishes it from the back-arc basins along the western Pacific margin and the passive continental margin of the Indian and Atlantic oceans.
4. Highest accumulation rates occur close to areas of active tectonics but not necessarily near large river mouths, indicating that uplift rate in the source area but not the drainage size has been critical in determining river discharge and sediment deposition. Much higher silt discharge from present-day rivers than accumulation rates estimated for earlier time intervals also reveals current denudation in drainage regions more intensive than ever before.
5. Sediment distribution patterns changed in different evolution stages of the SCS. Tectonics in the late Miocene not only caused the northward shift of the Zhengmu Basin depocenter but also a southward movement of the depocenter of the Yinggehai and Qingdongnan basins after their amalgamation. During the late Quaternary, the accumulation rate in MIS 2 is 1.5 times higher than MIS 1, providing evidence of more active downslope transport at sea-level lowstands during the last glacial maximum.

Acknowledgments This work was supported by the Ministry of Science and Technology of China (NKBRSF Grants 2006BAB19B03 and 2007CB815900), the National Natural Science Foundation of China (Grants 40776027, 40621063, 40506014, 40706023, 40876024), the Shanghai Rising-star Program (07QH14014), the Fok Ying Tung Education Foundation (101018), the Chinese Academy of Sciences (Grant kzcx2-yw-318 to Yu) and the Australian Research Council (Grant DP0773081 to Zhao).

References

Acharyya S.K. and Basu P.K. 1993. Toba ash on the Indian subcontinent and its implications for correlation of late Pleistocene alluvium. Quat. Res. 40: 10–19.

Alibert C., Kinsley L., Fallon S.J., McCulloch M.T., Berkelmans R. and McAllister F. 2003. Source of trace element variability in Great Barrier Reef corals affected by the Burdekin flood plumes. Geochim. Cosmochim. Acta 67: 231–246.

An Z., Kutzbach J.E., Prell W.L. and Porter S.C. 2001. Evolution of Asian monsoons and phased uplift of the Himalaya-Tibetan Plateau since late Miocene time. Nature 411: 62–66.

Bai Z., Zhang G. and Zeng X. 1996. The Geological and Geophysical Comprehensive Research Symposium of the Southeastern Nansha Sea Area. China Univ. Geosci. Press, Wuhan, 89pp (in Chinese).

Barron J.A. and Baldauf J.G. 1989. Tertiary cooling steps and paleoproductivity as reflected by diatoms and biosiliceous sediments. In: Berger W.H., Smetacek V.S. and Wefer G. (eds.), Productivity of the Ocean: Present and Past. John Wiley & Sons Limited., S. Bernhard, Dahlem Konferenzen, pp. 341–354.

Bassinot F.C., Beaufort L., Vincent E. and Labeyrie L.D. 1994. Coarse fraction fluctuatiuons in pelagic carbonate sediments from the tropical Indian Ocean: A 1500-kyr record of carbonate dissolution. Paleoceanography 9: 579–600.

Beaufort L., de Garidel-Thoron T., Linsley B., Oppo D. and Buchet N. 2003. Biomass burning and oceanic primary production estimates in the Sulu Sea area over the last 380 kyr and the East Asian monsoon dynamics. Mar. Geol. 201: 53–65.

Beaufort L., de Garidel-Thoron T., Mix A.C. and Pisias N.G. 2001. ENSO-like forcing on oceanic primary production during the late Pleistocene. Science 293: 2440–2444.

Benayahu Y. and Perkol-Finkel S. 2004. Soft corals (Octocorallia: Alcyonacea) from southern Taiwan. I. *Sarcophyton nanwanensis* sp. nov. (Octocorallia: Alcyonacea). Zool. Studies 43: 537–547.

Benayahu Y., Jeng M.S., Perkol-Finkel S. and Dai C.F. 2004. Soft corals (Octocorallia: Alcyonacea) from southern Taiwan. II. Species diversity and distributional patterns. Zool. Studies 43: 548–560.

Berger W.H. 1982. Increase of Carbon dioxide in the atmosphere during deglaciation: the coral reef hypothesis. Naturwissenschaften 69(2): 87–88.

Berger W.H., Leckie R.M., Janecek T.R., Stax R. and Takayama T. 1993. Neogene carbonate sedimentation on Ontong Java Plateau: highlights and open questions. Proc. ODP Sci. Results 130: 711–744.

Bian Y., Wang P. and Zheng L. 1992. Deep-water dissolution cycles of late Quaternary planktonic foraminifera in the South China Sea. In: Ye Z. and Wang P. (eds.), Contributions to Late Quaternary Paleoceanography of the South China Sea. Qingdao Ocean Univ. Press, Qingdao, pp. 261–273 (in Chinese).

Bluth G.J.S., Doiron S.D., Schnetzler C.C., Krueger A.J. and Walter L.S. 1992. Global tracking of the SO_2 clouds from the June 1991 Mount Pinatubo eruptions. Geophys. Res. Lett. 19: 151–154.

Bond G., Showers W., Cheseby M., Lotti R., Almasi P., de Menocal P., Priore P., Cullen H., Hajdas I. and Bonani G. 1997. A pervasive millennial-scale cycle in North Atlantic Holocene and glacial climates. Science 278: 1257–1266.

Borg L.E. and Banner J.L. 1996. Neodymium and strontium isotopic constraints on soil sources in Barbados, West Indies. Geochim. Cosmochim. Acta 60: 4193–4206.

Bosscher H. and Schlager W. 1992. Computer-simulation of reef growth. Sedimentology 39: 503–512.

Boulay S., Colin C., Trentesaux A., Clain S., Liu Z. and Lauer-Leredde C. 2007. Sedimentary responses to the Pleistocene climatic variations recorded in the South China Sea. Quat. Res. 68: 162–172.

Boulay S., Colin C., Trentesaux A., Frank N. and Liu Z. 2005. Sediment sources and East Asian monsoon intensity over the last 450 ky: Mineralogical and geochemical investigations on South China Sea sediments. Palaeogeogr. Palaeoclimatol. Palaeoecol. 228: 260–277.

Boulay S., Colin C., Trentesaux A., Pluquet F., Bertaux J., Blamart D., Buehring C. and Wang P. 2003. Mineralogy and sedimentology of Pleistocene sediments in the South China Sea (ODP

Site 1144). In: Prell W.L., Wang P., Blum P., Rea D.K. and Clemens S.C. (eds.), Proc. ODP Sci. Result 184: 1–21 [Online].

Boyd P.W., Watson A.J. and Law C.S. 2000. A mesoscale phytoplankton bloom in the polar Southern Ocean stimulated by iron fertilization. Nature 407: 695–702.

Briais A., Patriat P. and Tapponnier P. 1993. Update interpretation of magnetic anomalies and seafloor spreading stages in the South China Sea: Implications for the Tertiary tectonics of southeast Asia. J. Geophys. Res. 98(B4): 6299–6328.

Bühring C., Sarnthein M. and Erlenkeuser H. 2004. Toward a high resolution stable isotope stratigraphy of the last 1.1 m.y.: Site 1144, South China Sea. In: Prell W.L., Wang P., Blum P., Rea D.K. and Clemens S.C. (eds.), Proc. ODP, Sci. Results 184: 1–29 [Online].

Bühring C., Sarnthein M. and Leg 184 Shipboard Scientific Party 2000. Toba ash layers in the South China Sea: Evidence of contrasting wind directions during eruption ca. 74 ka. Geology 28: 275–278.

Byran Jr E.H. 1953. Check list of atolls. Atoll Res. Bull. 19: 1–38.

Cande S.C. and Kent D.V. 1992. A new geomagnetic polarity time scale for the late Cretaceous and Cenozoic. J. Geophys. Res. 97 (B10): 13917–13951.

Cande S.C. and Kent D.V. 1995. Revised calibration of the geomagnetic polarity timescale for the Late Cretaceous and Cenozoic. J. Geophys. Res. 100(B4): 6093–6095.

Cane M.A. and Molnar P. 2001. Closing of the Indonesian seaway as a precursor to east African aridifiation around 3–4 million years ago. Nature 411: 157–162.

Caruso M.J., Gawarkiewicz G.G. and Beardsley R.C. 2006. Interannual variability of the Kuroshio intrusion in the South China Sea. J. Oceanogr. 62: 559–575.

Chamley H. 1989. Clay Sedimentology. Springer, New York, 623pp.

Chave K.E., Smith S.V. and Roy K.J. 1972. Carbonate Production by Coral Reefs. Mar. Geol. 12: 123.

Chen J. 2005. Biogeochemistry of Settling Particles in the South China Sea and Its Significance on Paleo-environment Studies. PhD thesis, Tongji Univ., Shanghai, 136pp.

Chen C.T.A., Hou W.P., Gamo T. and Wang S.L. 2006. Carbonate-related parameters of subsurface waters in the West Philippine, South China and Sulu Seas. Mar. Chem. 99: 151–161.

Chen C.T., Wang C.H., Soong K.Y. and Wang B.J. 2001. Water temperature records from corals near the nuclear power plant in southern Taiwan. Sci. China (D) 44(4): 356–362.

Chen D.F., Huang Y.Y., Yuan X.L. and Cathles III L.M. 2005. Seep carbonates and preserved methane oxidizing bacteria and sulfur reducing bacteria fossils suggest recent gas venting the seafloor in the northeastern South China Sea. Mar. Pet. Geol. 22: 613–621.

Chen M., Wang R., Yang L., Han J. and Lu J. 2003b. Development of east Asian summer monsoon environments in the late Miocene: radiolarian evidence from Site 1143 of ODP Leg 184. Mar. Geol. 201: 169–177.

Chen M.T. and Huang C.Y. 1998. Ice-volume forcing of winter monsoon climate in the South China Sea. Paleoceanography 13: 622–633.

Chen M.T., Shiau L.J., Yu P.S., Chiu T.C., Chen Y.G. and Wei G.Y. 2003a. 500,000-year records of carbonate, organic carbon, and foraminiferal sea-surface temperature from the southeastern South China Sea (near Palawan Island). Palaeogeogr. Palaeoclimatol. Palaeoecol. 197: 113–131.

Chen P.Y. 1978. Minerals in bottom sediments of the South China Sea. GSA Bull. 89: 211–222.

Chen R., Zheng Y., Wiesner M.G., Jin H., Zhao Q., Zheng L. and Chen J. 2006. Seasonal and annual variations of marine sinking particulate flux during 1993–1996 in the central South China Sea. Acta Oceanol. Sinica 28(3): 72–80 (in Chinese).

Chen W. and Zhou F. 1993. A study of volcanic glass in northern South China Sea during the last 100 ka. In: Zheng L. and Chen W. (eds.), Contributions to Sedimentation Process and Geochemistry of the South China Sea. China Ocean Press, Beijing, pp. 39–45 (in Chinese).

Chen X., Zhao Q. and Jian Z. 2002. Carbonate content changes since the Miocene and paleoenvironmental implications, ODP Site 1148, northern South China Sea. Mar. Geol. Quat. Geol. 22: 69–74 (in Chinese).

Chen Y.L. 2005. Spatial and seasonal variations of nitrate-based new production and primary production in the South China Sea. Deep-Sea Res. I 52: 319–340.

Chen Y.Y., Chen M.T. and Fang T.S. 1999. Biogenic sedimentation patterns in the northern South China Sea: an ultrahigh-resolution record MD972148 of the past 150,000 years from the IMAGES III-IPHIS Cruise. Terr. Atmos. Ocean. Sci. (TAO) Taipei 10: 215–224.

Chen Y.G. and Liu T.K. 1996. Sea level changes in the last several thousand years, Penghu Islands, Taiwan Strait. Quat. Res. 45: 254–262.

Chen Z., Xia B., Yan W., Chen M., Yang H., Gu S. and Li Y. 2005. Distribution, chemical characteristics and source area of volcanic glass in the South China Sea. Acta Oceanol. Sinica 27(5): 73–81 (in Chinese).

Chen Z., Yan W., Chen M., Wang S., Lu J., Zhen F., Xiang R., Xiao S., Yan P. and Gu S. 2006. Discovery of seep carbonate nodules as new evidence for gas venting on the northern continental slope of South China Sea. Chinese Sci. Bull. 51(10): 1228–1237.

Cheng X., Zhao Q., Wang J., Jian Z., Xia P., Huang B., Fang D., Xu J., Zhou Z. and Wang P. 2004. Data report: Stable Isotopes from Sites 1147 and 1148. In: Prell W.L., Wang P., Blum P., Rea D.K. and Clemens S.C. (eds.), Proc. ODP, Sci. Results 184: 1–12 [Online].

Chesner C.A., Rose W.I., Deino A., Drake R. and Westgate J.A. 1991. Eruptive history of Earth's largest Quaternary caldera (Toba, Indonesia) clarified. Geology 19: 200–203.

Clift P. 2006. Controls on the erosion of Cenozoic Asia and the flux of clastic sediment to the ocean. Earth Planet. Sci. Lett. 241: 571–580.

Clift P., Lee J. I., Clark M.K. and Blusztajn J. 2002. Erosional response of South China to arc rifting and monsoonal strengthening; a record from the South China Sea. Mar. Geol. 184: 207–226.

Copper P. 1994. Ancient reef ecosystem expansion and collapse. Coral Reefs 13: 3–11.

Cortese G., Gersonde R., Hillenbrand C. and Kuhn G. 2004. Opal sedimentation shifts in the World Ocean over the last 15 Myr. Earth Planet. Sci. Lett. 224: 509–527.

Crossland C.J., Hatcher B.G. and Smith S.V. 1991. Role of Coral Reefs in Global Ocean Production. Coral Reefs 10(2): 55–64.

Crowley T.J. 1985. Late Quaternary carbonate changes in the North Atlantic and Atlantic/Pacific comparison. AGU Geophys. Monogr. 32: 271–284.

Dadson S.J., Hovius N., Chen H., Dade W.B., Hsieh M.L., Willett S.D., Hu J.C., Horng M.J., Chen M.C., Stark C.P., Lague D. and Lin J.C. 2003. Links between erosion, runoff variability and seismicity in the Taiwan orogen. Nature 426: 648–651.

Dai C.F. 1991. Reef environment and coral fauna of southern Taiwan. Atoll Res. Bull. 354(3): 1–28.

Damuth J.E. 1979. Migrating sediment waves created by turbidity currents in the northern South China Basin. Geology 7: 520–523.

Damuth J.E. 1980. Quaternary sedimentation process in the South China Sea basin as revealed by echo-character mapping and piston-core studies. In: Hayes D.E. (ed.), The Tectonics and Geophysical Evolution of Southeast Asian Seas and Islands. AGU Geophys. Monogr. 23: 105–125.

Delaney M.L. and Boyle E.A. 1988. Tertiary Paleogene chemical variability: Unintended consequences of simple geochemical models. Paleoceanography 3: 137–156.

De Vooys C.G.N. 1979. Primary production in aquatic environments. In: Bolin B, Degens E.T., Kempe S. and Ketner P. (eds.), SCOPE, 13: The Global Carbon Cycle. John Wiley and Sons, Chichester, UK, pp. 259–292.

Dickens G.R. and Barron J.A. 1997. A rapid deposited pennate diatom ooze in upper Miocene-lower Pliocene sediment beneath the North Pacific polar front. Mar. Micropaleontol. 31: 177–182.

Dickens G.R. and Owen R.M. 1996. Sediment geochemical evidence for an Early-Middle Gilbert (Early Pliocene) productivity peak in the North Pacific Red Clay Province. Mar. Micropaleontol. 27: 107–120.

Dickens G.R. and Owen R.M. 1999. The latest Miocene—early Pliocene biogenic bloom: a revised Indian Ocean perspective. Mar. Geol. 161: 75–91.

Diekmann B., Fälker M. and Kuhn G. 2003. Environmental history of the southeastern South Atlantic since the middle Miocene: Evidence from the sedimentological records of ODP Sites 1088 and 1092. Sedimentology 50: 511–529.

Diester-Haass L., Billups K. and Emeis K.C. 2005. In search of the late Miocene-early Pliocene "biogenic bloom" in the Atlantic Ocean (Ocean Drilling Program Sites 982, 925, and 1088). Paleoceanography 20: PA4001, doi:10.1029/2005PA001139.

Diester-Haass L., Billups K. and Emeis K.C. 2006. Late Miocene carbon isotope records and marine biological productivity: Was there a (dusty) link? Paleoceanography 21: PA4216, doi:10.1029/2006PA001267.

Ding Z.L., Liu T. and Rutter N.W. 1995. Ice-volume forcing of East Asian winter monsoon variations in the past 800,000 years. Quat. Res. 44: 149–159.

Ding Z.L., Yang S.L., Sun J.M. and Liu T.S. 2001. Iron geochemistry of loess and red clay deposits in the Chinese Loess Plateau and implications for long-term Asian monsoon evolution in the last 7.0 Ma. Earth Planet. Sci. Lett. 185: 99–109.

Dixit S., Van Cappellen P. and van Bennekom A. 2001. Processes controlling solubility of biogenic silica and pore water build-up of silicic acid in marine sediments. Mar. Chem. 73: 333–352.

Du Bois E.P. 1981. Review of principal hydrocarbon-bearings of the South China Sea. Energy 6(11): 1113–1140.

El-Gamal A.A.H., Wang S.K., Dai C.F. and Duh C.Y. 2004. New nardosinanes and 19-oxygenated ergosterols from the soft coral Nephthea armata collected in Taiwan. J. Natural Products 67(9): 1455–1458.

Fan T.Y., Chou Y.H. and Dai C.F. 2005. Sexual reproduction of the alcyonacean coral Lobophytum pauciflorum in southern Taiwan. Bull. Mar. Sci. 76(1): 143–154.

Fang L.S. and Chou Y.C. 1992. Concentration of fulvic-acid in the growth bands of hermatypic corals in relation to local precipitation. Coral Reefs 11(4): 187–191.

Farrell J.W., Raffi I., Janecek T.R., Murray D.W., Levitan M., Dadey K.A., Emeis K.C., Lyle M., Flores J.A. and Hovan S. 1995. Late Neogene sedimentation patterns in the eastern equatorial Pacific Ocean. Proc. ODP Sci. Results 138: 717–756.

Fiske R.S. and Simkin T. 1983. Krakatau, 1883: the Volcanic Eruption and Its Effects. Smithsonian Inst. Press, Washington, D.C., 464pp.

Ghiold J. and Enos P. 1982. Carbonate production of the coral Diploria-Labyrinthiformis in South Florida patch reefs. Mar. Geol. 45: 281–296.

Gingele F.X., Deckker P.D. and Hillenbrand C.D. 2001. Clay mineral distribution in surface sediments between Indonesia and NW Australia – source and transport by ocean currents. Mar. Geol. 179: 135–146.

Goldstein S.J. and Jacobsen D.B. 1988. Nd and Sr isotopic systematic of river water suspended material: Implications for crustal evolution. Earth Planet. Sci. Lett. 87: 215–221.

Gong Z. 1997. The Major Oil and Gas Fields of China Offshore. Petroleum Industrial Press, Beijing, 223pp (in Chinese).

Gong Z. and Li S. (eds.). 1997. Continental Margin Basin Analysis and Hydrocarbon Accumulation of the Northern South China Sea. China Sci. Press, Beijing, 510pp (in Chinese).

Gran K.B. and Montgomery D.R. 2005. Spatial and temporal pattern in fluvial recovery following volcanic eruptions: Channel response to basin-wide sediment loading at Mount Pinatubo, Philippines. GSA Bull. 117: 195–211.

Griffin J.J., Windom H. and Goldberg E.D. 1968. The distribution of clay minerals in the World Ocean. Deep-Sea Res. 15: 433–459.

Guilcher A. 1988. Coral Reef Geomorphology. John Wiley & Sons, New York, pp. 40–44.

Guo Z.T., Ruddiman W.F., Hao Q.Z., Wu H.B., Qiao Y.S., Zhu R.X., Peng S.Z., Wei J.J., Yuan B.Y. and Liu T.S. 2002. Onset of Asian desertification by 22 Myr ago inferred from loess deposits in China. Nature 416: 159–163.

Han W. and Ma K. 1988. Carbonate compensation depth, saturation horizon and lysocline in the northeast region of South China Sea. Tropical Oceanol. 7(3): 84–89 (in Chinese).

Han X.Q., Suess E., Huang Y.Y., Wu N.Y., Bohrmann G., Su X., Eisenhauer A., Rehder G. and Fang Y.X. 2008. Jiulong methane reef: Microbial mediation of seep carbonates in the South China Sea. Mar. Geol. 249: 243–256.

Hanebuth T., Stattegger K. and Groote P.M. 2000. Rapid flooding of the Sunda Shelf: a late-glacial sea-level record. Science 288: 1033–1035.

Harrison T.M., Copeland P., Kidd W.S.F. and Yin A. 1992. Raising Tibet. Science 255: 1663–1670.

Hay W.W. 1998. Detrital sediment fluxes from continents to oceans. Chem. Geol. 145: 287–323.

Hay W.W., Sloan J.L.II and Wold C.N. 1988. Mass/age distribution and composition of sediments on the ocean floor and the global rate of sediment subduction. J. Geophys. Res. 93 (B12): 14933–14940.

Hays J.D., Saito T., Opdyke N.D. and Burckle L.H. 1969. Pliocene-Pleistocene sediments of the equatorial Pacific: their paleomagnetic, biostratigraphic and climatic record. GSA Bull. 80: 1481–1514.

He X.X., Liu D.Y., Peng Z.C. and Liu W.Q. 2002. Monthly sea surface temperature records reconstructed by delta O-18 of reef-building coral in the east of Hainan Island, South China Sea. Sci. China (B) 45: 130–136.

Higginson M., Maxwell J.R. and Altabet M.A. 2003. Nitrogen isotope and chlorin paleoproductivity records from the Northern South China Sea: remote vs. local forcing of millennial- and orbital-scale variability. Mar. Geol. 201: 223–250.

Hillenbrand C. and Ehrmann W. 2005. Late Neogene to Quaternary environmental changes in the Antarctic Peninsula region: evidence from drift sediments. Global Planet. Change 45: 165–191.

Hiscott R.N. 2001. Depositional sequences controlled by high rates of sediment supply, sea-level variations, and growth faulting: the Quaternary Baram Delta of northwestern Borneo. Mar. Geol. 175: 67–102.

Hoang N. and Flower M. 1998. Petrogenesis of Cenozoic basalts from Vietnam: implication for origins of a diffuse igneous province. J. Petroleum 30: 569–595.

Holloway N.H. 1982. North Palawan Block, Philippines, its relation to Asian mainland and role in evolution of South China Sea. AAPG Bull. 66: 1355–1383.

Houseman G. and England P. 1993. Crustal thickening versus lateral expulsion in the Indian-Asian continental collision. J. Geophys. Res. 98: 12233–12249.

Hu P.Z. and Wang J.Z. 1996. Tertiary organic reef in the Pearl River Mouth Basin, organic reef and oil of China. China Ocean Press, Beijing, pp. 294–323 (in Chinese).

Hu P.Z. and Xie Y.X. 1987. Tertiary reef complexes and their relationship with hydrocarbon accumulation in Pearl River Mouth Basin. In: Society G.P. (ed.), International Conference on Prtroleum Geology of the Northern Continental Shelf of the South China Sea, pp. 505–529.

Huang B., Jian Z. and Lin H. 2000. Late Quaternary changes of paleoproductivity in the northeastern South China Sea. Mar. Geol. Quat. Geol. 20(2): 65–68 (in Chinese).

Huang C.Y., Liew P.M., Zhao M., Chang T.C., Kuo C.M., Chen M.T., Wang C.H. and Zhang L.F. 1997a. Deep sea and lake records of the Southeast Asian paleomonsoons for the last 25 thousands years. Earth Planet. Sci. Lett. 146: 59–72.

Huang C.Y., Wang C.C. and Zhao M.X. 1999. High-resolution carbonate stratigraphy of IMAGES core MD972151 from South China Sea. Terr. Atmos. Ocean. Sci. (TAO) Taipei 10: 225–238.

Huang C.Y., Wu S.F., Zhao M., Chen M.T., Wang C.H., Tu X. and Yuan P.B. 1997b. Surface ocean and monsoon climate variability in the South China Sea since the last glaciation. Mar. Micropaleotol. 32: 71–94.

Huang C.Y., Yuan P.B., Lin C.W., Wang T.K. and Chang C.P. 2000. Geodynamic processes of Taiwan arc-continent collision and comparison with analogs in Timor, Papua New Guinea, Ural and Corsia. Tectonophysics 325: 1–21.

Huang W. 2004. Sediment distributional patterns and evolution in the South China Sea since the Oligocene. PhD thesis, Tongji Univ., Shanghai. 114pp (in Chinese).

Huang W. and Wang P. 1998. A quantitative approach to deep-water sedimentation in the South China Sea: changes since the last glaciation. Sci. China (D) 41(2): 195–201.

Huang W. and Wang P. 2006. Sediment mass and distribution in the South China Sea since the Oligocene. Sci. China (D) 49(11): 1147–1155.

Huang W. and Wang P. 2007a. Statistics of sediment mass in the South China Sea: method and results. Fronts. Earth Sci. China 1: 88–96 (in Chinese).

Huang W. and Wang P. 2007b. Accumulation rate characterstics of deep-water sedimentation in the South China Sea during the last glaciation and the Holocene. Acta Oceanol. Sinica 29(5): 69–73 (in Chinese).

Jagodziński R. 2005. Petrography and geochemistry of surface sediments from Sunda and Vietnamese shelves (South China Sea). Adam Mickiewicz Univ. Press, Poznań, 144pp.

Jansen J.H.F., Kuijpers A. and Troelstra S.R. 1986. A Mid-Brunhes climatic event: Long-term changes in global atmosphere and ocean circulation. Science 232: 619–622.

Jia G., Jian Z., Peng P., Wang P. and Fu J. 2000. Biogenic silica records in core 17962 from southern South China Sea and their relation to paleoceanographical events. Geochimica 29(3): 293–296 (in Chinese).

Jia G., Peng P., Zhao Q. and Jian Z. 2003. Changes in terrestrial ecosystem since 30 Ma in East Asia: Stable isotope evidence from black carbon in the South China Sea. Geology 31: 1093–1096.

Jian Z. and Huang B. 2001. Late Quaternary upwelling intensity and East Asian Monsoon forcing in the South China Sea. Quat. Res. 55: 363–370.

Jian Z., Wang L. and Kienast K. 1999. Late Quaternary surface paleoproductivity and variations of the East Asian Monsoon in the South China Sea. Quat. Sci. 1: 32–40 (in Chinese).

Johnson M.R.W. 1994. Volume balance of erosional loss and sediment deposition related to Himalayan uplift. J. Geol. Soc. 151: 217–220.

Kawahata H., Suzuki A. and Goto K. 1997. Coral reef ecosystems as a source of atmospheric CO_2: evidence from PCO_2 measurements of surface waters. Coral Reefs 16(4): 261–266.

Kawahata H., Suzuki A., Ayukai T. and Goto K. 2000. Distribution of the fugacity of carbon dioxide in the surface seawater of the Great Barrier Reef. Mar. Chem. 72: 257–272.

Kayanne H., Suzuki A. and Saito H. 1995. Diurnal changes in the partial-pressure of carbon-dioxide in coral-reef water. Science 269: 214–216.

Kennedy D.M. and Woodroffe C.D. 2000. Holocene lagoonal sedimentation at the latitudinal limits of reef growth, Lord Howe Island, Tasman Sea. Mar. Geol. 169: 287–304.

Kennett J.P. 1982. Marine Geology. Prentice-Hall, Englewood Cliffs, New Jersey, 813pp.

Kinsey D.W. 1985. Metabolism, calcification and carbonate production. I. Systems level studies. In: Gabrie C. and Salvat B. (eds.), Proc. 5th Coral Reef Symposium, Tahiti, pp. 505–526.

Kinsey D.W. and Hopley D. 1991. The significance of coral reefs as global carbon sinks – Response to greenhouse. Global Planet. Change 89: 363–377.

Kleypas J.A. 1997. Modeled estimates of global reef habitat and carbonate production since the last glacial maximum. Paleoceanography 12: 533–545.

Kleypas J.A., Buddemeier R.W., Archer D., Gattuso J.P., Langdon C. and Opdyke B.N. 1999. Geochemical consequences of increased atmospheric carbon dioxide on coral reefs. Science 284: 118–120.

Koyaguchi T. and Tokuno M. 1993. Origin of the giant eruption cloud of Pinatubo, June 15, 1991. J. Volcanol. Geotherm. Res. 55: 85–96.

Kudrass H.R., Erienkeuser H., Vollbrecht R. and Weiss W. 1991. Global nature of the Younger Dryas cooling event inferred from oxygen isotope data from Sulu Sea cores. Nature 349: 406–408.

Kukla G., An Z., Melice J.L., Gavin J. and Xiao J. 1990. Magnetic susceptibility record of Chinese loess. Trans. R. Soc. Edinburgh Earth Sci. 81: 263–288.

Laj C., Wang P. and Balut Y. 2005. IPEV les rapports de campagnes à la mer. MD147/MARCO POLO- IMAGES XII à bord du "Marion Dufresne", 59pp.

Lambiase J.J., bin Abdul Rahim A.A. and Peng C.Y. 2002. Facies distribution and sedimentary processes on the modern Baram Delta: implications for the reservoir sandstones on NW Borneo. Mar. Petroleum Geol. 19: 69–78.

Lee M.-Y., Chen C.-H., Wei K.-Y., Iizuka Y. and Carey S. 2004. First Toba supereruption revival. Geology 32:61–64.

Leinen M. 1979. Biogenic silica accumulation in the central equatorial Pacific and its implications for Cenozoic paleoceanography. Geol. Soc. Am. Bull. 90: 1310–1376.

Li B., Jian Z., Li Q., Tian J. and Wang P. 2005. Paleoceanography of the South China Sea since the middle Miocene: evidence from planktonic foraminifera. Mar. Micropaleontol. 54: 49–62.

Li B., Zhao Q., Chen M.-P., Jian Z. and Wang P. 2001. Carbonate dissolution and deep-water paleoceanography of the South China Sea since the Middle Pleistocene. Chinese Sci. Bull. 46: 1908–1912.

Li J. and Wang R. 2004. Paleoproductivity variability of the northern South China Sea during the past 1 Ma: The opal record from ODP Site 1144. Acta Geol. Sinica 78(2): 228–233 (in Chinese).

Li J., Jin X. and Gao J. 2002b. Morpho-tectonic study on late-stage spreading of the Eastern Subbasin of South China Sea. Sci. China (D) 45: 978–989.

Li J., Wang R. and Li B. 2002a. Variations of opal accumulation rates and paleoproductivity over the past 12 Ma at ODP Site 1143, southern South China Sea. Chinese Sci. Bull. 47: 596–598.

Li Q., Jian Z. and Su X. 2005. Late Oligocene rapid transformations in the South China Sea. Mar. Micropaleontol. 54: 5–25.

Li Q., Wang P., Zhao Q., Shao L., Zhong G., Tian J., Cheng X., Jian Z. and Su X. 2006. A 33 Ma lithostratigraphic record of tectonic and paleoceanographic evolution of the South China Sea. Mar. Geol. 230: 217–235.

Li X. 2005. Sedimentary Characteristics of the western South China Sea and their variations since the late Pleistocene. PhD thesis, Tongji Univ., Shanghai, 91pp (in Chinese).

Li X., Wei G., Shao L., Liu Y., Liang X., Jian Z., Sun M. and Wang P. 2003. Geochemical and Nd isotopic variations in sediments of the South China Sea: a response to Cenozoic tectonism in SE Asia. Earth Planet. Sci. Lett. 211: 207–220.

Li Y.H., Huang X.P., Yue W.Z., Lin Y.T., Zou R.L. and Huang H. 2004. Ecological study on coral reef and intertidal benthos around Yongxing Island, South China Sea. Oceanol. Limnol. Sinica 35(2): 176–182 (in Chinese).

Li Y. and Yang L. 1997. Distributive features of basal matter types of South China Sea. In: Xu D., Liu X. and Zhang X. (eds.), China Offshore Geology. Geol. Publ. House, Beijing, pp. 103–123 (in Chinese).

Liang W. and Li G.Z. 2002. Preliminary study on characteristics of coral reef distribution and environmental protection in Weizhou Island. Res. Environm. Sci. 15(6): 5–7 (in Chinese).

Liang X., Wei G., Shao L., Li X. and Wang R. 2001. Records of Toba eruptions in the South China Sea – Chemical charateristics of the glass shards from ODP 1143A. Sci. China (D) 44(10): 871–878.

Liao Z. 1995. Preliminary estimations of the sediment filling in Chinese Meso-Cenozoic basins. J. Tongji Univ., 23(Suppl.): 162–164 (in Chinese).

Lisitzin A.P. 1972. Sedimentation in the World Ocean. SEPM Paper No. 17, 218pp.

Liu B. 2001. Features Analysis and Measurement of Nansha Islands Remote Sensing Integration Information. China Ocean Press, Beijing, 173pp (in Chinese).

Liu H., Guo L., Sun Y., Zhou D., Su L., Yang S., Zhang Y. and Zhang B. 2002. Study on fault system in Nansha Block (South China Sea) and the block's lithospheric dynamics. China Sci. Press, Beijing, 123pp (in Chinese).

Lin H., Lai C., Ting H., Wang L., Sarnthein M. and Huang J. 1999. Late Pleistocene nutrients and sea surface productivity in the South China Sea: a record of teleconnections with northern hemisphere events. Mar. Geol. 156: 197–210.

Liu K.-K., Chao S.-Y., Shaw P.-T., Gong G.-C., Chen C.-C. and Tang T.Y. 2002. Monsoon-forced chlorophyll distribution and primary production in the South China Sea: observations and a numerical study. Deep-Sea Res. I 49: 1387–1412.

Liu P.J., Fan T.Y. and Dai C.F. 2005. Timing of larval release by the blue coral, *Heliopora coerulea*, in southern Taiwan. Coral Reefs 24: 30–30.

Liu X.D., Zhao S.P., Sun L.G., Yin X.B., Xie Z.Q., Honghao L. and Wang Y.H. 2006. P and trace metal contents in biomaterials, soils, sediments and plants in colony of red-footed booby (Sula sula) in the Dongdao Island of South China Sea. Chemosphere 65: 707–715.

Liu Z., Colin C. and Trentesaux A. 2006. Major element geochemistry of glass shards and minerals of the Youngest Toba Tephra in the southwestern South China Sea. J. Asian Earth Sci. 27: 99–107.

Liu Z., Colin C., Huang W., Chen Z., Trentesaux A. and Chen J. 2007b. Clay minerals in surface sediments of the Pearl River drainage basin and their contribution to the South China Sea. Chinese Sci. Bull. 52(8): 1101–1111.

Liu Z., Colin C., Huang W., Le K.P., Tong S., Chen Z. and Trentesaux A. 2007a. Climatic and tectonic controls on weathering in South China and the Indochina Peninsula: clay mineralogical and geochemical investigations from the Pearl, Red, and Mekong drainage basins. Geochem. Geophys. Geosyst. 8: Q05005, doi:10.1029/2006GC001490.

Liu Z., Colin C., Trentesaux A., Blamart D., Bassinot F., Siani G. and Sicre M.-A. 2004. Erosional history of the eastern Tibetan Plateau since 190 kyr ago: clay mineralogical and geochemical investigations from the southwestern South China Sea. Mar. Geol. 209: 1–18.

Liu Z., Colin C., Trentesaux A., Siani G., Frank N., Blamart D. and Farid S. 2005. Late Quaternary climatic control on erosion and weathering in the eastern Tibetan Plateau and the Mekong Basin. Quat. Res. 63: 316–328.

Liu Z., Trentesaux A., Clemens S.C., Colin C., Wang P., Huang B. and Boulay S. 2003a. Clay mineral assemblages in the northern South China Sea: implications for East Asian monsoon evolution over the past 2 million years. Mar. Geol. 201: 133–146.

Liu Z., Trentesaux A., Clemens S.C. and Wang P. 2003b. Quaternary clay mineralogy in the northern South China Sea (ODP Site 1146) -Implications for oceanic current transport and East Asian monsoon evolution. Sci. China (D) 46(12): 1123–1235.

Liu Z., Wang P., Wang C., Shao L. and Huang W. 2001. Paleotopography of China during the Cenozoic: a preliminary study. Geol. Rev. 47(5): 467–475 (in Chinese).

Liu Z., Tuo S., Colin C., Liu J.T., Huang C.-Y., Selvaraj K., Chen C.-T.A., Zhao Y., Siringan F.P., Boulay S. and Chen Z. 2008. Detrital fine-grained sediment contribution from Taiwan to the northern South China Sea and its relation to regional ocean circulation. Mar. Geol. 255: 149–155.

Liu Z., Zhao Y., Li J. and Colin C. 2007c. Late Quaternary clay minerals off middle Vietnam in the western South China Sea: implications for source analysis and East Asian monsoon evolution. Sci. China (D) 50(11): 1674–1684.

Lu H., Liu J., Chen F., Liao Z., Sun X. and Su X. 2005. Mineralogy and stable isotope composition of authigenic carbonates in bottom sediments on the offshore area of southwest Taiwan, South China Sea: evidence for gas hydrates occurrence. Earth Sci. Frontiers 12: 268–276 (in Chinese).

Lüdmann T. and Wong H.K. 1999. Neotectonic regime on the passive continental margin of the northern South China Sea. Tectonophysics 311: 113–138.

Lüdmann T., Wong H.K. and Berglar K. 2005. Upward flow of North Pacific Deep Water in the northern South China Sea as deduced from the occurrence of drift sediments. Geophys. Res. Lett. 32: L05614, doi:10.1029/2004GL021967.

Lüdmann T., Wong H.K. and Wang P. 2001. Plio-Quaternary sedimentation processes and neotectonics of the northern continental margin of the South China Sea. Mar. Geol. 172: 331–358.

Luz B. and Shackleton N.J. 1975. CaCO3% solution in the tropical east Pacific during the past 130,000 years. Cushman Found. Foraminiferal Res., Special Publ. 13: 142–150.

Lyle M. 2003. Neogene carbonate burial in the Pacific Ocean. Paleoceanography 18: 1059, doi: 10.1029/2002PA000777.

Lyle M., Dadey K.A. and Farrell J.W. 1995. The late Miocene (11–8 Ma) eastern Pacific carbonate crash: evidence for reorganization of deep-water circulation by the closure of the Panama Gateway. Proc. ODP Sci. Results 138: 821–838.

Lyle M., Koizumi I., Richter C., Behl R.J., Boden P., Caulet J.-P., Delaney M.L., deMenocal, P., Desmet M., Fornaciari, E., Hayashida A., Heider F., Hood J., Hovan S.A., Janecek T.R., Janik A.G., Kennett J., Lund D., Machain Castillo M.L., Maruyama T., Merrill R.B., Mossman D.J., Pike J., Ravelo A.C., Rozo Vera G.A., Stax R., Tada R., Thurow J. and Yamamoto M. 1997. Proc. ODP Initial Reports 167. College Station, Texas (Ocean Drilling Program), 1378pp.

Ma Z.B., Xia M., Zhang C.H., Pen Z.C., Wang Z.R., Sun W.D. and An Z.S. 1999. High-precision U-series dating of Holocene corals from South China Sea by thermal ionization mass spectrometry (TIMS). Chinese Sci. Bull. 44(10): 937–941.

Ma Z.B., Xiao J.L., Zhao X.T., Peng Z.C., Xia M., Zhang G.P., Wang Z.R. and An Z.S. 2003. Precise U-series dating of coral reefs from the South China Sea and the high sea level during the Holocene. J. Coast. Res. 19: 296–303.

Marsaglia K.M., Boggs J.S., Clift P., Seyedolali A. and Smith R. 1995. Sedimentation in western Pacific backarc basins: new insights from recent ODP drilling. In: Taylor B. and Natland J. (eds.), Active Margins and Marginal Basins of the Western Pacific, pp. 291–314.

Martinson D.G., Pisias N.G., Hays J.D., Imbrie J., Moore T.C. Jr. and Shackleton N.J. 1987. Age dating and the orbital theory of the ice age: development of a high resolution 0 to 300,000 year chronostratigraphy. Quat. Res. 27: 1–29.

Mayer L., Pisias N. and Janecek T. 1992. Proc. ODP, Initial Reports. College Station, Texas (Ocean Drilling Program), vol. 138, 1462pp.

McCulloch M., Fallon S., Wyndham T., Hendy E., Lough J. and Barnes D. 2003. Coral record of increased sediment flux to the inner Great Barrier Reef since European settlement. Nature 421: 727–730.

McLennan S.M. 1993. Weathering and global denudation. J. Geol. 101: 295–303.

Métivier F., Gaudemer Y., Tapponnier P. and Klein M. 1999. Mass accumulation rates in Asia during the Cenozoic. Geophys. J. Int. 137: 280–318.

Miao Q., Thunell R.C. and Andersen D.M. 1994. Glacial-Holocene carbonate dissolution and sea surface temperatures in the South China and Sulu seas. Paleoceanography 9: 269–290.

Michel E.J., Turon L. and Beaufort B. 1997. Les Rapport de Campagne à la Mer à Bord du Marion Dufresne MD106, IPHIS I and II. Institut Francais pour la Recherche et la Technologie Polaires, L, Brest.

Mikkelsen N. and Barron J.A. 1997. Early Oligocene diatoms on the Ceara Rise and the Cenozoic evolution of biogenic silica accumulation in the low-latitude Atlantic. Proc. ODP Sci. Results 154: 483–490.

Milliman J.D. 1974. Marine Carbonate. Springer-Verlag, New York, 375pp.

Milliman J.D. 1993. Production and accumulation of calcium carbonate in the ocean: budget of a nonsteady state. Global Biogeochem. Cycles 7(4): 927–957.

Milliman J.D. and Syvitski J.P.M. 1992. Geomorphic/tectonic control of sediment discharge to the ocean: the importance of small mountainous rivers. J. Geol. 100: 525–544.

Moberly R. and Schlanger S. 1986. Init. Repts. DSDP. U.S. Govt. Printing Office, Washington D.C., vol. 89, 678pp.

Nesbitt H.W. and Young G.M. 1982. Early Proterozoic climates and plate motions inferred from major element chemistry of lutites. Nature 299: 715–717.

Nie B.F., Chen T.G., Liang M.T., Wang Y.Q., Zhong J.L. and Zhu Y.Z. 1997a. Relationship between coral growth rate and sea surface temperature in the northern part of South China Sea during the past 100 a. Sci. China (D) 40(2): 173–182.

Nie B.F., Chen T.G., Liang M.T., Zhong J.L. and Yu K.F. 1997b. Coral reefs from Leizhou Peninsula and Holocene sea level highstands. Chinese Sci. Bull. 42(5): 511–514 (in Chinese).

Nie B.F., Chen T.G., Liang M.T., Zhong J.L. and Yu K.F. 1997c. The relationship between reefs coral and environmental changes of Nansha Islands and adjacent regions. China Sci. Press, Beijing, 101pp (in Chinese).

Nie B.F., Chen T.G. and Peng Z.C. 1999. Reconstruction of sea surface temperature series in the last 220 years by use of reef corals in Xisha waters, South China Sea. Chinese Sci. Bull. 44(22): 2094–2098.

Nie B.F., Guo L.F., Zhu Y.Z. and Zhong J.L. 1992. Modern sediments of Zhongsha atoll. Nanhai Studia Marina Sinica 10: 1–14 (in Chinese).

Ninkovich D., Shackleton N.J., Abdel-Monem A.A., Obradovich J.D. and Izett G. 1978. K-Ar age of the late Pleistocene eruption of Toba, north Sumatra. Nature 276: 574–577.

Opdyke B.N. and Walker J.C.G. 1992. Return of the coral-reef hypothesis – Basin to shelf partitioning of $CaCO_3$ and iIts effect on atmospheric CO_2. Geology 20: 733–736.

Oppenheimer C. 2002. Limited global change due to the largest known Quaternary eruption, Toba ~74 kyr BP. Quat. Sci. Rev. 21: 1593–1609.

Packham G. 1996. Cenozoic SE Asia: reconstructing its aggregation and reorganization. In: Hall R. and Blundell D. (eds), Tectonic Evolution of Southeast Asia. Geol. Soc. Spec. Publ., London, pp. 123–152.

Pang X., Chen C., Wu M., He M. and Wu X. 2006. The Pearl River deep-water fan systems and significant geological events. Adv. Earth Sci. 21: 793–799 (in Chinese).

Pattan J.N., Shane P. and Banakar V.K. 1999. New occurrence of Youngest Toba Tuff in abyssal sediments of the Central Indian Basin. Mar. Geol. 155: 243–248.

Pautot G., Rangin C., Briais A., Wu J., Han S., Li H., Lu Y. and Zhao J. 1990. The axial ridge of the South China Sea: a seabeam and geophysical survey. Oceanol. Acta 13: 129–143.

Peng Z.C., Chen T.G., Nie B.F., Head M.J., He X.X. and Zhou W.J. 2003. Coral delta O-18 records as an indicator of winter monsoon intensity in the South China Sea. Quat. Res. 59: 285–292.

Peng Z.C., He X.X., Zhang Z.F., Zhou J., Sheng L.S. and Gao H. 2002. Correlation of coral fluorescence with nearshore rainfall and runoff in Hainan Island, South China Sea. Progr. Natural Sci. 12: 41–44 (in Chinese).

Prell W.L., Murray D.W., Clemens S.C. and Anderson D.M. 1992. Evolution and variability of the Indian Ocean summer monsoon: Evidence from the western Arabian Sea drilling program. In: Duncan R.A., Rea D.K., Kidd R.B., von Rad U. and Weissel J.K. (eds.), The Indian Ocean: A Synthesis of Results from the Ocean Drilling Program. Geophys. Monogr. 70, AGU, pp. 447–469.

Qian J. 1999. Paleooceanoraphy for the Late Quaternary in the South China Sea. China Sci. Press, Beijing, 156pp (in Chinese).

Qiu Y. and Wang Y.-M. 2001. Reefs and paleostructure and paleoenvironment in the South China Sea. Mar. Geol. Quat. Geol. 21: 65–73 (in Chinese).

Qiu Y., Yao B., Li T., Biao C. and Gong Y. 1999. Geological structure characteristics of the Zhongjiannan Basin, the Western of the South China Sea. In: Yao B., Qiu Y. and Wu N. (eds.), Geological and Tectonic Characteristics and Cenozoic Sedimentation of the Western South China Sea. Geol. Publ. House, Beijing, pp. 56–70 (in Chinese).

Ragueneau O., Treguer P., Leynaert A., Anderson R.F., Brzezinski M.A., DeMaster D.J., Dugdale R.C., Dymond J., Fischer G., François R., Heinze C., Maier-Reimer E., Martin-Jézéquel V., Nelson D. M. and Quéguiner B. 2000. A review of the Si cycle in the modern ocean: recent progress and missing gaps in the application of biogenic opal as a paleoproductivity proxy. Global Planet. Change 26: 317–365.

Rea D.K., Snoeck H. and Joseph L.H. 1998. Late Cenozoic eolian deposition in the North Pacific: Asian drying, Tibetan uplift, and cooling of the Northern Hemisphere. Paleoceanography 13: 215–224.

Rottman M.C. 1979. Distribution of planktonic foraminifera and pteropods in South China Sea sediments. J Foraminiferal Res. 9: 41–49.

Ryan D.A., Opdyke B.N. and Jell J.S. 2001. Holocene sediments of Wistari Reef: towards a global quantification of coral reef related neritic sedimentation in the Holocene. Palaeogeogr. Palaeoclimatol. Palaeoecol. 175: 173–184.

Sarnthein M., Pflaumann U., Wang P. and Wong H.K. (eds.). 1994. Preliminary Report on SONNE-95 Cruise "Monitor Monsoon" to the South China Sea. Berichte-Reports, Geol.-Palaont. Inst. Univ. Kiel, 48, pp. 1–225.

Schulz H., Emeis K.-C., Erlenkeuser H., von Rad U. and Rolf C. 2002. The Toba volcanic event and interstadial/stadial climates at the Marine Isotopic Stage 5 to 4 transition in the northern Indian Ocean. Quat. Res. 57: 22–31.

Ségalen P. 1995. Les Sols Ferrallitiques et Leur Répartition Géographique. ORSTOM ed., Coll. Etudes et Thèses, Inst. De Rech. Pour le Dév., Paris, vol. 3, 201pp.

Seibold E.B. and Berger W.H. 1982. The Sea Floor. An Introduction to Marine Geology. Springer-Verlag, 288pp.

Selvaraj K. and Chen C.-T.A. 2006. Moderate chemical weathering of subtropical Taiwan: constraints from solid-phase geochemistry of sediments and sedimentary rocks. J. Geol. 114: 101–116.

Shackleton N.J. 1985. Oceanic carbon isotope constraints on oxygen and carbon dioxide in the Cenozoic atmosphere. AGU Geophys. Monogr. 32: 412–418.

Shao L., Li X., Geng J., Pang X., Lei Y., Qiao P., Wang L. and Wang H. 2007. Deep water bottom current deposition in the northern South China Sea. Sci. China (D) 50(7): 1060–1066.

Shao L., Li X., Wei G., Liu Y. and Fang D. 2001. Provenance of a prominent sediment drift on the northern slope of the South China Sea. Sci. China (D) 44: 919–925.

Shen C.C., Lee T., Chen C.Y., Wang C.H., Dai C.F. and Li L.A. 1996. The calibration of D[Sr/Ca] versus sea surface temperature relationship for Porites corals. Geochim. Cosmochim. Acta 60: 3849–3858.

Shen C.D., Yi W.X., Yu K.F., Sun Y.M., Liu T.S., Beer J., Hajdas I. and Bonani G. 2004a. Holocene megathermal abrupt environmental changes derived from C-14 dating of a coral reef at Leizhou Peninsula, South China Sea. Nuclear Instruments and Methods in Physics Res. B 223–24: 416–419.

Shen C.D., Yi W.X., Yu K.F., Sun Y.M., Yang Y. and Zhou B. 2004b. Interannual C-14 variations during 1977–1998 recorded in coral from Daya Bay, South China Sea. Radiocarbon 46(2): 595–601.

Shen C.D., Yu K.F., Sun Y.M., Yi W.X., Yang Y. and Zhou B. 2003. Interannual variations of bomb radiocarbon during 1977–1998 recorded in coral from Daya Bay, South China Sea. Sci. China (D) 46(10): 1040–1048.

Smith S.V. 1978. Coral-reef area and contributions of reefs to processes and resources of world's oceans. Nature 273: 225–226.

Smith S.V. and Kinsey D.W. 1976. Calcium-carbonate production, coral-reef growth, and sea-level change. Science 194: 937–939.

Song S.R., Chen C.H., Lee M.Y., Yang T.F., Iizuka Y. and Wei K.Y. 2000. Newly discovered eastern dispersal of the youngest Toba Tuff. Mar. Geol. 167: 303–312.

Soong K. and Chen T.A. 2003. Coral transplantation: Regeneration and growth of *Acropora* fragments in a nursery. Restoration Ecol.11: 62–71.

South China Sea Institute of Oceanology 1985. Reports on Complex Survey in the South China Sea. Sci. Press, Beijing, Vol. 2, 432 pp. (in Chinese).

Spalding M.D. and Grenfell A.M. 1997. New estimates of global and regional coral reef areas. Coral Reefs 16: 225–230.

Spalding M.D., Ravilious C. and Green E.P. 2001. World Atlas of Coral Reefs. Univ. California Press, Berkeley, 424pp.

Steinke S., Kienast M. and Hanebuth T. 2003. On the significance of sea-level variations and shelf paleo-morphology in governing sedimentation in the southern South China Sea during the last deglaciation. Mar. Geol. 201: 179–206.

Stoddart D.R. 1965. The shape of atolls. Mar. Geol. 3: 369–383.

Story C., Peng P. and Lin J.D. 2000. Liuhua 11–1 field, South China Sea: A shallow carbonate reservoir developed using untrahigh-resolution 3-D seismic, inversion, and attribute-based reservoir modeling. The Leading Edge 19: 834–844.

Stow D.A.V., Anano K., Balson P.S., Brass G.W., Corrigan J., Raman C.V., Tiercelin J.J., Townsend M. and Wjiazanansa N.P. 1990. Sediment facies and processes on the distal Bengal Fan, Leg 116. Proc. ODP Sci. Results 116: 377–396.

Su G. and Wang T. 1994. Basic characteristics of modern sedimentation in the South China Sea. In: Zhou D., Liang Y.B. and Zheng C.K. (eds.), Oceanology of China Seas. Kluwer, New York, pp. 407–418.

Su G., Fan Q., Chen S. 1989. Sediment Atlas of the Central and Northern South China Sea. Guangdong Sci. Tech. Press, Guangzhou, 68pp (in Chinese).

Suess E. 2005. RV SONNE cruise report SO 177, Sino–German cooperative project, South China Sea Continental Margin: geological methane budget and environmental effects of methane emissions and gashydrates. IFM-GEOMAR Reports, 133pp.

Sun D.H., Gagan M.K., Cheng H., Scott-Gagan H., Dykoski C.A., Edwards R.L. and Sua R.X. 2005. Seasonal and interannual variability of the Mid-Holocene East Asian monsoon in coral delta O-18 records from the South China Sea. Earth Planet. Sci. Lett. 237: 69–84.

Sun X., Luo Y., Huang F., Tian J. and Wang P. 2003. Deep-sea pollen from the South China Sea: Pleistocene indicators of East Asian monsoon. Mar. Geol. 201: 97–118.

Sun Y., Sun M., Lee T. and Nie B. 2005. Influence of seawater Sr content on coral Sr/Ca and Sr thermometry. Coral Reefs 24: 23–29.

Sun Y.L., Sun M., Wei G.J., Lee T., Nie B.F. and Yu Z.W. 2004. Strontium contents of a *Porites* coral from Xisha Island, South China Sea: A proxy for sea-surface temperature of the 20th century. Paleoceanography 19: PA2004, doi:10.1029/2003PA000959.

Suzuki A. and Kawahata H. 2003. Carbonate budget of coral reef systems: an overview of observations in fringing reefs, barrier reefs and atolls in the Indo-pacific regions. Tellus. 55B(2): 428–444.

Suzuki A., Kawahata H. and Goto K. 1997. Reef water CO_2 system and carbon cycle in Mahuro Atoll, the Marshall Islands in the central Pacific. In: Lessios H.A. and Macintyre I.G. (eds.), Proc. 8th Int. Coral Reef Symposium, pp. 971–976.

Suzuki A., Kawahata H., Ayukai T. and Goto K. 2001. The oceanic CO_2 system and carbon budget in the Great Barrier Reef, Australia. Geophys. Res. Lett. 28: 1243–1246.

Tamburini F., Adatte T., Föllmi K., Bernasconi S.M. and Steinmann P. 2003. Investigating the history of East Asian monsoon and climate during the last glacial-interglacial period (0–140 000 years): mineralogy and geochemistry of ODP Sites 1143 and 1144, South China Sea. Mar. Geol. 201: 147–168.

Tapponnier P., Peltzer G., Le Dain A.Y., Armijo R. and Cobbold P. 1982. Propagating extrusion tectonics in Asia: new insights from simple experiments with plasticine. Geology 10: 611–616.

Taylor S.R. and McLennan S.M. 1985. The Continental Crust: Its Composition and Evolution. Blackwell, Malden, MA, 312pp.

The Multidisciplinary Oceanographic Expedition Team of Academia Sinica to Nansha Islands 1996. Geology, Geophysics and Natural Resources of Nansha Islands and Adjacent Sea Areas. Sci. Press, Beijing, 252 pp. (in Chinese).

The Multidisciplinary Oceanographic Expedition Team of Academia Sinica to Nansha Islands. 1989. Report of Multidisciplinary Survey on Nansha Islands and Adjacent Sea Areas (1). China Sci. Press, Beijing, 294pp (in Chinese).

The Multidisciplinary Oceanography Expedition Team of Academia Sinica to Nansha Islands. 1993. Sedimentary Atlas of Nansha Islands and Adjacent Sea Area. Hubei Sci. Tech. Press, Wuhan, 94pp (in Chinese).

The Second Marine Geological Investigation Brigade of the Ministry of Geology and Marine Resources 1987. Atlas of Geology and Geophysics of South China Sea. Map Publish House of Guangdong Province, Guangzhou (in Chinese).

The Second Marine Geological Investigation Brigade of the Ministry of Geology and Resources 1990. The Summary Report of Multidisciplinary Geophysical Survey on the Northern Slope of the South China Sea (in Chinese).

The Second Marine Geological Investigation Brigade of the Ministry of Geology and Resources 1992. Chinese Report on China-USA Joint Marine Geological Survey and Research on the South China Sea (Second Stage) (in Chinese).

Thevenon F., Bard E. and Williamson D. 2004. A biomass burning record from the West Equatorial Pacific over the last 360ky: methodological, climatic and anthropic implications. Palaeogeogr. Palaeoclimatol. Palaeoecol. 213: 83–99.

Theyer F., Mayer L.A., Barron J.A. and Thomas E. 1985. The equatorial Pacific high-productivity belt: elements for a synthesis of Deep Sea Drilling Project Leg 85 results. Proc. DSDP Init. Repts. 85: 971–985.

Thiry M. 2000. Palaeoclimatic interpretation of clay minerals in marine deposits: an outlook from the continental origin. Earth-Sci. Rev. 49: 201–221.

Thunell R., Miao Q., Calvert S., Calvert S. and Pedersen T. 1992. Glacial-Holocene biogenic sedimentation patterns in the South China Sea: productivity variations and surface water pCO_2. Paleoceanography 7: 143–162.

Tian J., Wang P., Cheng X., Wang R. and Sun X. 2005. Forcing mechanism of the Pleistocene east Asian monsoon variations in a phase perspective. Sci. China (D) 48(10): 1708–1717.

Tsai C.C., Wong S.L., Chang J.S., Hwang R.L., Dai C.F., Yu Y.C., Shyu Y.T., Sheu F. and Lee T.M. 2004. Macroalgal assemblage structure on a coral reef in Nanwan Bay in southern Taiwan. Botanica Marina 47: 439–453.

Tütken T., Eisenhauer A., Wiegand B. and Hansen B.T. 2002. Glacial-interglacial cycles in Sr and Nd isotopic composition of Arctic marine sediments triggered by the Svalbard/Barents Sea ice sheet. Mar. Geol. 182: 351–372.

Twan W.H., Hwang J.S., Lee Y.H., Wu H.F., Tung Y.H. and Chang C.F. 2006. Hormones and reproduction in scleractinian corals. Comparative Biochem. Physiol. A–Molecular Integr. Physiol. 144: 247–253.

Vecsei A. 2001. Fore-reef carbonate production: development of a regional census-based method and first estimates. Palaeogeogr. Palaeoclimatol. Palaeoecol. 175: 185–200.

Vecsei A. 2004. A new estimate of global reefal carbonate production including the fore-reefs. Global Planet. Change 43: 1–18.

Vital H. and Stattegger K. 2000. Major and trace elements of stream sediments from the lowermost Amazon River. Chem. Geol. 168: 151–168.

Wan S., Li A., Clift P.D. and Stuut J.-B.W. 2007. Development of the East Asian monsoon: mineralogical and sedimentologic records in the northern South China Sea since 20 Ma. Palaeogeogr. Palaeoclimatol. Palaeoecol. 254: 561–582.

Wallace C.C. and Dai, C.F. 1997. Scleractinia of Taiwan 4. Review of the coral genus *Acropora* from Taiwan. Zoological Studies 36: 288–324.

Wang G.Z. 2001. Coral reef sedimentology of the South China Sea. China Ocean Press, Beijing, 313pp (in Chinese).

Wang H., Zhou F. and Jian J. 1992. Volcanic clasts in the periplatform carbonate ooze near Zhongsha Islands and their bearing on paleo-evnironment. In: Ye Z. and Wang P. (eds.), Contributions to Late Quaternary Paleoceanography of the South China Sea. Qingdao Ocean Univ. Press, Qingdao (in Chinese).

Wang J. 2001. Planktonic foraminiferal assemblages and paleoceanography during the last 18 Ma. PhD thesis, Tongji Univ., Shanghai (in Chinese).

Wang J., Yin A., Harrison T.M., Grove M., Zhang Y. and Xie G. 2001. A tectonic model for Cenozoic igneous activities in the eastern Indo-Asian collision zone. Earth Planet Sci. Lett. 188: 13–133.

Wang L., Jian Z. and Chen J. 1997. Late Quaternary pteropods in the South China Sea: carbonate preservation and paleoenviromental variation. Mar. Micropaleotol. 32: 115–126.

Wang L., Sarenthein M., Erlenkeuser H., Grimalt J., Grootes P., Heilig S., Ivanova E., Kienast M., Pelejero C. and Pflaumann U. 1999. East Asian monsoon Climate during the late Pleistocene: high- resolution sediment records from the South China Sea. Mar. Geol. 156: 245–284.

Wang L. and Wang P. 1990. Late Quaternary paleoceano-graphy of the South China Sea: glacial-interglacial contrasts in an enclosed basin. Paleoceanography 5: 77–90.

Wang P. (ed.). 1995. The South China Sea Since 150 ka. Tongji Univ. Press, Shanghai, 184pp (in Chinese).

Wang P. 1999. Response of Western Pacific marginal seas to glacial cycles: paleoceanographic and sedimentological features. Mar. Geol. 156: 5–39.

Wang P., Clemens S., Beaufort L., Braconnot P., Ganssen G., Jian Z., Kershaw P. and Sarnthein M. 2005. Evolution and variability of the Asian monsoon system: state of the art and outstanding issues. Quat. Sci. Rev. 24: 595–629.

Wang P., Min Q., Bian Y. and Feng W. 1986. Planktonic foraminifera in the continental slope of the northern South China Sea during the last 130,000 years and their paleoceanographic implications. Acta Geol. Sinica (Trial English Edition) 60: 1–11.

Wang P., Prell W., Blum P. and the Leg 184 Shipboard Scientific Party 1999. Exploring the Asian monsoon through drilling in the South China Sea. JOIDES J. 25(2): 8–13.

Wang P., Prell W.L., Blum P. (eds.). 2000. Proc. ODP, Init. Repts, Vol. 184 [CD-ROM]. Ocean Drilling Program, Texas A&M University, College Station TX 77845–9547, USA.

Wang P., Wang L., Bian Y. and Jian Z. 1995. Late Qauternary paleoceanography of the South China Sea: surface circulation and carbonate cycles. Mar. Geol. 127: 145–165.

Wang P., Bian Y. and Jian Z. 1997. Late Quaternary carbonate cycles in the Nansha Islands area, South China Sea. Quat. Sci. 17 (4): 293–300 (in Chinese).

Wang P., Zhao Q., Jian Z., Cheng X., Huang W., Tian J., Wang J., Li Q., Li B. and Su X. 2003. Thirty million year deep-sea records in the South China Sea. Chinese Sci. Bull. 48(23): 2524–2535.

Wang R. and Abelmann A. 2002. Radiolarian responses to paleoceanographic events of the southern South China Sea during the Pleistocene. Mar. Micropaleontol. 46: 25–44.

Wang R., Clemens S., Huang B. and Chen M. 2003. Late Quaternary paleoceanographic changes in the northern South China Sea (ODP Site 1146): radiolarian evidence. J. Quat. Sci. 18(8): 745–756.

Wang R. and Li J. 2003. Quaternary high resolution opal record and its paleoproductivity implication at ODP Site 1143, southern South China Sea. Chinese Sci. Bull. 48(4): 363–367.

Wang R., Fang D., Shao L., Chen M., Xia P. and Qi J. 2001. Oligocene biogenetic siliceous deposits on the slope of the northern South China Sea. Sci. China (D) 44(10): 912–918.

Wang R., Jian Z., Xiao W., Tian J., Li J., Chen R., Zheng L. and Chen J. 2007. Quaternary biogenic opal records in the South China Sea: linkages to East Asian monsoon, global ice volume and orbital forcing. Sci. China (D) 50(5): 710–724.

Wang R., Li J. and Li B. 2004. Data report: Late Miocene–Quaternary biogenic opal accumulation at ODP Site 1143, southern South China Sea. In: Prell W.L., Wang P., Blum P., Rea D.K. and Clemens S.C. (eds.), Proc. ODP, Sci. Results 184 [Online].

Wang Y.J., Cheng H., Edwards R.L., He Y.Q., Kong X.G., An Z.S., Wu J.Y., Kelly M.J., Dykoski C.A. and Li X.D. 2005. The Holocene Asian monsoon: Links to solar changes and North Atlantic climate. Science 308: 854–857.

Wehausen R. and Brumsack H.J. 2002. Astronomical forcing of the East Asian monsoon mirrored by the composition of Pliocene South China Sea sediments. Earth Planet. Sci. Lett. 201: 621–636.

Wei G.J., Yu K.F. and Zhao J.X. 2004. Sea surface temperature variations recorded on coralline Sr/Ca ratios during Mid-Late Holocene in Leizhou Peninsula. Chinese Sci. Bull. 49: 1876–1881.

Wei G.J., Deng W.F., Yu K.F., Li X.H., Sun W.D. and Zhao J.X. 2007. Sea surface temperature records in the northern South China Sea from mid-Holocene coral Sr/Ca ratios. Paleoceanography 22: PA3206, doi: 10.1029/2006PA001270.

Wei G.J., Sun M., Li X.H. and Nie B.F. 2000. Mg/Ca, Sr/Ca and U/Ca ratios of a porites coral from Sanya Bay, Hainan Island, South China Sea and their relationships to sea surface temperature. Palaeogeogr. Palaeoclimatol. Palaeoecol. 162: 59–74.

Wei K.Y., Lee T.Q. and Shipboard Scientific Party of IMAGES III/MD106-IPHIS Cruise (Leg II) 1998. Late Pleistocene volcanic ash layers in core MD972142, offshore from northwestern Palawan, South China Sea: A preliminary report. Terr. Atmos. Ocean. Sci. (TAO) Taipei 9(1): 143–152.

Wells M., Vallis G. and Silver E. 1999. Tectonic processes in Papua New Guinea and past productivity in the eastern equatorial Pacific Ocean. Nature 398: 601–604.

Wen X.S., Tu X., Qin G.Q., Zheng F. and Zhao H.T. 2001. Foraminiferal fauna and deponsional environment of core of Nanyong-3 well in lagoon of Yongshu atoll of Nansha Islands. Tropical Oceanol. 20(4): 14–22 (in Chinese).

Wiedicke M. 1987. Biostratigraphy, Microfacies and Diagenesis of Tertiary carbonates from the South China Sea (Dangerous Grounds - Palawan, Philippinen). Facies 16: 195–302.

Wiesner M.G., Wang Y. and Zheng L. 1995. Fallout of volcanic ash to the deep South China Sea induced by the 1991 eruption of Mount Pinatubo (Philippines). Geology 23: 885–888.

Wiesner M.G., Wetzel A., Catane S.G., Listanco E.L. and Mirabueno H.T. 2004. Grain size, areal thickness distribution and controls on sedimentation of the 1991 Mount Pinotubo tephra layer in the South China Sea. Bull. Volcanol. 66: 226–242.

Wiesner M.G., Zheng L. and Wong H.K. 1996. Fluxes of particulate matter in the South China Sea. In: Ittekkot V., Schäfer P., Honjo S. and Depetris P. (eds.), Particle Flux in the Ocean. John Wiley and Sons, New York, pp. 91–154.

Xie Z.Q., Sun L.G., Zhang P.F., Zhao S.P., Yin X.B., Liu X.D. and Cheng B.B. 2005. Preliminary geochemical evidence of groundwater contamination in coral islands of Xi-Sha, South China Sea. Applied Geochem. 20: 1848–1856.

Xiao J., Porter S.C., An Z., Kumai H. and Yoshikawa S. 1995. Grain size of quartz as an indicator of winter monsoon strength on the loess plateau of central China during the last 130,000 yr. Quat. Res. 43: 22–29.

Xiong Y., Zhong G., Li Q., Wu N., Li X. and Ma Z. 2006. Inversion of stratal corbonate content using seismic data. Earth Sci.–J. China Univ. Geosci. 31: 851–856 (in Chinese).

Xu D., Liu X. and Zhang X. (eds.). 1997. China Offshore Geology. Geol. Publ. House, Beijing, 310pp (in Chinese).

Xu H., Wang Y., Cai F., Gou Y., Sun P., Zhang B., Gong J. and Zhang Z. 2000. The Effect of Forming Reef of Biological Stratun in Miocene of Xisha Islands and Algae and the Evolutional Characteristics of Bioherm. China Sci. Press. Beijing, 144pp (in Chinese).

Xu J. 2004. Quaternary planktonic foraminiferal assemblages in the southern South China Sea and paleoclimatic variations. PhD thesis, Tongji Univ., Shanghai (in Chinese).

Xu J., Wang P., Huang B., Li Q. and Jian Z. 2005. Response of planktonic foraminifera to glacial cycles: Mid-Pleistocene change in the southern South China Sea. Mar. Micropaleontol. 54: 89–105.

Yamano H., Kayanne H. and Yonekura N. 2001. Anatomy of a modern coral reef flat: A recorder of storms and uplift in the late Holocene. J. Sedimentary Res. 71: 295–304.

Yan P. and Liu H. 2004. Tectonic-stratigraphic division and blind fold structures in Nansha waters, South China Sea. J. Asian Earth Sci. 24: 337–348.

Yan P., Deng H., Liu H., Zhang Z. and Jiang Y. 2006. The temporal and spatial distribution of volcanism in the South China Sea region. J. Asian Earth Sci. 27: 647–659.

Yang L., Chen M., Wang R. and Zhen F. 2002. Radiolarian record to paleoecological environment change events over the past 1.2 Ma BP in the southern South China Sea. Chinese Sci. Bull. 47(17): 1478–1483.

Yang M., Wu J., Yang R. and Duan W. 1996. Stratigraphic division and nomenclature of the southwestern Nansha sea area. Geol. Res. South China Sea 8: 37–46 (in Chinese).

Yang S., Jung H.-S. and Li C. 2004. Two unique weathering regimes in the Changjjiang and Huanghe drainage basins: geochemical evidence from river sediments. Sedimentary Geol. 164: 19–34.

Yang Y. and Fan S. 1990. Research on volcanic sediments and origin of volcanic substance in the South China Sea during late Quaternary. Tropical Oceanol. 9(1): 52–60 (in Chinese).

Yao B. and Zeng W. 1994. The Geological Memoir of South China Sea Surveyed Jointly by China and USA. China Univ. Geosci. Press, Wuhan, 204pp (in Chinese).

Ye Z.Z., He Q.X. and Zhang M.S. 1985. Study on classfication and characteristic of the islands at Xisha. Mar. Geol. Quat. Geol. 5: 1–13 (in Chinese).

Yu K.F. 2005. The coral reef at Dengloujiao, Leizhou Peninsula, northern coast of the South China Sea – its ecology and sustainable development as a resource. Acta Ecol. Sinica 25: 669–675 (in Chinese).

Yu K.F., Chen T.G., Huang D.C., Zhao H.T., Zhong J.L. and Liu D.S. 2001. The high-resolution climate recorded in the delta O-18 of Porites lutea from the Nansha Islands of China. Chinese Sci. Bull. 46: 2097–2102.

Yu K.F., Liu D.S., Shen C.D., Zhao J.X., Chen T.G., Zhong J.L., Zhao H.T. and Song C.J. 2002a. High-frequency climatic oscillations recorded in a Holocene coral reef at Leizhou Peninsula, South China Sea. Sci. China (D) 45: 1057–1067.

Yu K.F., Liu T.S., Chen T.G., Zhong J.L. and Zhao H.T. 2002b. High-resolution climate recorded in the delta C-13 of Porites lutea from Nansha Islands of China. Progr. Natural Sci. 12: 284–288.

Yu K.F., Jiang M.X., Chen Z.Q. and Chen T.G. 2004a. The latest 42a sea surface temperature changes in Weizhou Island and its influence on the coral reef ecosystem. Appl. Ecol. Acta 15: 506–510 (in Chinese).

Yu K.F., Song C.J. and Zhao H.T. 1995. The characters of geomorphology and modern sediments of Yongxing Island, Xisha Islands. Tropical Oceanol. 14: 24–31 (in Chinese).

Yu K.F., Zhao J.X., Collerson K.D., Shi Q., Chen T.G., Wang P.X. and Liu T.S. 2004b. Storm cycles in the last millennium recorded in Yongshu Reef, southern South China Sea. Palaeogeogr. Palaeoclimatol. Palaeoecol. 210: 89–100.

Yu K.F., Zhao J.X., Done T. and Chen T.G. 2009a. Microatoll record for large century-scale sea level fluctuations in the mid-Holocene. Quat. Res., Doi:10.1016/j.yqres.2009.02.003.

Yu K.F., Zhao J.X., Liu T.S., Wei G.H., Wang P.X. and Collerson K.D. 2004c. High-frequency winter cooling and reef coral mortality during the Holocene climatic optimum. Earth Planet. Sci. Lett. 224: 143–155.

Yu K.F., Zhao J.X., Shi Q. and Meng Q.S. 2009b. Reconstruction of storm/tsunami records over the last 4000 years using transported coral blocks and lagoon deposits in the southern South China Sea. Quatern. Intern., 195: 128–137.

Yu K.F., Zhao J.X., Shi Q., Chen T.G., Wang P.X., Collerson K.D. and Liu T.S. 2006a. U-series dating of dead Porites corals in the South China Sea: Evidence for episodic coral mortality over the past two centuries. Quat. Geochronol. 1: 129–141.

Yu K.F., Zhao J.X., Wang P.X., Shi Q., Meng Q.S., Collerson K.D. and Liu T.S. 2006b. High-precision TIMS U-series and AMS ^{14}C dating of a coral reef lagoon sediment core from southern South China Sea. Quat. Sci. Rev. 25: 2420–2430.

Yu K.F., Zhao J.X., Wei G.J., Cheng X.R. and Wang P.X. 2005b. Mid-late Holocene monsoon climate retrieved from seasonal Sr/Ca and d^{18}O records of Porites lutea corals at Leizhou Peninsula, northern coast of the South China Sea. Global Planet. Change 47: 301–316.

Yu K.F., Zhao J.X., Wei G.J., Cheng X.R., Chen T.G., Wang P.X. and Liu T.S. 2005a. d^{18}O, Sr/Ca and Mg/Ca records of Porites lutea corals from Leizhou Peninsula, northern South China Sea and their applicability as paleothermometers. Palaeogeogr. Palaeoclimatol. Palaeoecol. 218: 57–73.

Yu K.F., Zhao H.T. and Zhu Y.Z. 1998. Modern sedimentary characteristics of Halimeda on the coral reefs of Nansha islands. Acta Sedimentol. Sinica 16: 20–24 (in Chinese).

Yu K.F., Zhong J.L., Zhao J.X., Shen C.D., Chen T.G. and Liu D.S. 2002c. Biological-geomorphological zones in a coral reef area at southwest Leizhou Peninsula unveil multiple sea level high-stands in the Holocene. Mar. Geol. Quat. Geol. 22(2): 27–33 (in Chinese).

Yu K.F., Zhu Y.Z. and Zhao H.T. 1997. Modern clastic sediments of atoll reefs (Xinyi Reef and other 3 reefs) in Nansha Islands. Nanhai Studia Marina Sinica 12: 119–147 (in Chinese).

Yue D.L., Wu S.H., Lin C.Y., Wang Q.R., Heng L.Q. and Li Y. 2005. Sedimentary and diagenetic evolution pattern of reef limestone reservoirs in Liuhua 11–1 oilfield. Oil Gas Geol. 26: 518–529.

Zahn R., Rushdi A., Pisias N.G., Bornhold B.D., Blaise B. and Karlin R. 1991. Carbonate deposition and benthic δ^{13}C in the subarctic Pacific: implication for changes of the oceanic carbonate system during the past 750,000 years. Earth Planet. Sci. Lett. 103: 116–132.

Zhan W., Qiu X., Sun Z., Zu J. and Tang C. 2003. Red river active fault zone in northwestern South China Sea. Tropical Oceanol. 22(2): 10–16 (in Chinese).

Zhang P., Molnar P. and Downs W.R. 2001. Increased sedimentation rates and grain sizes 2–4 Myr ago due to the influence of climate change on erosion rates. Nature 410: 891–897.

Zhang Q.M. 2001. On biogeomorphology of Luhuitou fringing reef of Sanya city, Hainan Island, China. Chinese Sci. Bull. 46(Suppl.): 97–102.

Zhang M., He Q., Ye Z., Han C., Li H., Wu J. and Ju L. 1989. Sedimentary Geology of Xisha Reef Carbonates. China Sci. Press, Beijing, 117pp (in Chinese).

Zhao H.T. and Yu K.F. 1999. Coral reefs of the South China Sea. In: Zhao H.T. (ed.), Geomorphology and Eenvironment of the South China Coast and the South China Sea Islands. China Sci. Press, Beijing, pp. 370–453 (in Chinese).

Zhao H.T., Sha Q.A. and Zhu Y.Z. 1992. Quaternary Coral Reef: Geology of Yongshu Reef, Nansha Islands. China Ocean Press, Beijing, 264pp (in Chinese).

Zhao J.X. and Yu K.F. 2002. Timing of Holocene sea-level highstands by mass spectrometric U-series ages of a coral reef from Leizhou Peninsula, South China Sea. Chinese Sci. Bull. 47: 348–352.

Zhao J.X. and Yu K.F. 2006. U-series dating of coral reefs from the South China Sea. Geochim. Cosmochim. Acta. 70(Suppl. 1): A741.

Zhao J.X. and Yu K.F. 2007. Millennial-, century- and decadal-scale oscillations of Holocene sea-level recorded in a coral reef in the northern South China Sea. Quat. Int. 167–168(Suppl.): 473.

Zhao J.X., Yu K.F. and Chen T.G. Holocene multiple sea-level highstands recorded in a coral reef in the northern South China Sea. Earth Planet. Sci. Lett. (submitted).

Zhao M.X., Yu K.-F., Zhang Q.M. and Shi Q. Long-term dynamics of coral cover in Luhuitou Fring Reef, Sanya. Mar. Environ. Sci. (in press).

Zhao Q. and Wang P. 1999. Pregress in Quaternary paleoceanography of the South China Sea: a review. Quat. Sci. 6: 481–501 (in Chinese).

Zheng H.B., Powell C.M., Rea D.K., Wang J.L. and Wang P.X. 2004. Late Miocene and mid-Pliocene enhancement of the East Asian monsoon as viewed from the land and sea. Global Planet. Change 41: 147–155.

Zhong G., Geng J., Wong H.K., Ma Z. and Wu N. 2004. A semi-quantitative method for the reconstruction of eustatic sea level history from seismic profiles and its application to the southern South China Sea. Earth Planet. Sci. Lett. 223: 443–459.

Zhong G., Li Q., Hao H. and Wang L. 2007. Current status of deep-water sediment wave studies and the South China Sea perspectives. Adv. Earth Sci. 22: 907–913 (in Chinese).

Zhu Y.Z., Sha Q.A., Guo L.F., Yu K.F. and Zhao H.T. 1997. Cenozoic Coral Reef Geology of Yongshu Reef, Nansha Islands. China Sci. Press, Beijing, 134pp (in Chinese).

Zou R.L., Meng Z.M. and Guan X.L. 1983. Ecological analysis of ahermatypic corals from the northern shelf of South China Sea. Tropical Oceanol. 2: 1–26 (in Chinese).

Zou R.L., Zhang Y.L. and Xie Y. 1988. An ecological study of reef corals around Weizhou Island. In: Xu G.Z. and Mortor B. (eds.), Proceedings on Marine Biology of the South China Sea. China Ocean Press, Beijing, pp. 201–211 (in Chinese).

.

Chapter 5
Upper Water Structure and Paleo-Monsoon

Zhimin Jian, Jun Tian and Xiangjun Sun

Introduction

Hemipelagic sediments in the SCS often register higher depositional rates. Carbonate compensation depth (CCD) in this marginal ocean basin is generally deeper than neighboring sea basins. Thus the SCS offers an ideal locality for paleoceanographic studies in the low-latitude western Pacific (Wang P. 1999). The first several cores in the world ocean used for high-resolution stratigraphy using accelerator mass spectrometer (AMS) ^{14}C dating technique were from the southern part of this basin (Andree et al. 1986; Broecker et al. 1988). This is particularly significant for the western Pacific region where high-resolution paleo-studies are hampered by poor carbonate preservation and low sedimentation rates.

Climate and hydrography in the SCS are largely controlled by the monsoonal wind system which is characterized by its pronounced seasonality. Strong southwesterly winds during summer and northeasterly winds during winter drive a semi-annual reversal in surface-water circulation from approximately clockwise to anti-clockwise (Wyrtki 1961). The oceanographic features of the upper ocean such as sea surface temperatures (SST) and the thermocline are driven by the monsoon circulation and hence display conspicuous seasonality (see Chapter 2). Since the 1990s, numerous experiments and expeditions have been organized to study the East Asian monsoon in the SCS, such as the "South China Sea Monsoon Experiment (SCSMEX)" (1996–2001) by meteorologists (Ding et al. 2004), "Monitor Monsoon" expedition in 1994 (Sarnthein et al. 1994) and ODP Leg 184 (Wang P.

Z. Jian (✉)
State Key Laboratory of Marine Geology, Tongji University, Shanghai 200092, China
e-mail: jian@tongji.edu.cn

J. Tian
State Key Laboratory of Marine Geology, Tongji University, Shanghai 200092, China
e-mail: tianjun@tongji.edu.cn

X. Sun
Institute of Botany, Chinese Academy of Sciences, Beijing 100093, China; State Key Laboratory of Marine Geology, Tongji University, Shanghai 200092, China
e-mail: sunxj@tongji.edu.cn

P. Wang, Q. Li (eds.), *The South China Sea*, Developments in Paleoenvironmental Research 13, DOI 10.1007/978-1-4020-9745-4_5,
© Springer Science+Business Media B.V. 2009

et al. 2000) by earth scientists. Now the SCS and the Arabian Sea respectively have become the foci for the East and South Asian monsoon studies in both modern and paleo-climatology (Wang B. et al. 2003).

The upper ocean and monsoon of the SCS in the geological past were highly sensitive to changes in climate boundary conditions. During glacial periods, the SCS became a semi-enclosed basin connected to the western Pacific through the Bashi Strait and to the Sulu Sea through the Balabac and Mindoro Straits due to low sea level, inevitably resulting in a pronounced difference in surface and deep circulation (Wang P. et al. 1995; Jian and Wang 1997) and hence altering the upper water structure and deep water conditions. Since the 1990s, a large number of papers have been published on the monsoonal evolution and variability using proxy measurements relating to changes in SST, vertical thermal structure, productivity, wind-blown (e.g., eolian dust and specific pollens) and river-discharged grains (Wang P. et al. 2005 and references herewith). The present chapter provides a synthesis of the upper ocean structure and East Asian monsoon history in the SCS, beginning with the SST and thermocline depth history, then the vegetation history in the deep-sea pollen record, and finally an outline of the monsoon history.

5.1 Sea Surface Temperature History (Jian Z. and Tian J.)

SST Proxies

Sea surface temperature (SST) is not only one of the main subjects for international paleoceanographic studies since the 1970s like CLIMAP, but also the starting point of the East Asian paleo-monsoon study since the 1990s in the SCS. So far, three methods including paleoecology of planktonic foraminiferal transfer function, alkenone unsaturation index ($U^{K'}_{37}$) of sediments, and Mg/Ca ratios of planktonic foraminiferal tests have been applied to reconstruct the past SST variations in the SCS. Numerous paleo-SST sequences have been generated, in particular for the late Quaternary, although the use of some SST proxies remains debatable.

Foraminiferal Transfer Function

The first attempt to reconstruct paleo-SST in the SCS was in 1990, when transfer function FP-12 was applied to foraminiferal census over the last glacial cycle at 3 core sites in the northern SCS (Wang and Wang 1990). It was found that the glacial/interglacial SST contrast was much more significant in the SCS than in the adjacent open Pacific, indicating an amplifier effect of glacial signals in the marginal sea. However, the transfer function FP-12E initially employed was a regional version of Imbrie-Kipp's transfer function, based on census data of planktonic foraminifers in 165 coretops from the open western Pacific Ocean, with the standard errors of 2.48 °C for winter SST and 1.46 °C for summer SST (Thompson 1981). The SST reconstruction using FP-12E has inevitable biases in the marginal seas where upper ocean hydrology is different from that of the open Pacific.

To avoid these drawbacks, Pflaumann and Jian (1999) developed a calibrated SIMMAX-28 transfer function to estimate past SST in the SCS. SIMMAX method used modern analog technique with a similarity index (Pflaumann et al. 1996). SIMMAX-28 was based on 30 new planktonic foraminiferal census data of surface sediment samples recovered between 630 and 2883 m water depths in the SCS, together with the 131 data sets published from the western Pacific. The standard errors of the transfer function SIMMAX-28 are 1.27 °C for winter SST and 0.45 °C for summer SST. This regional SIMMAX method offers a slightly better understanding of the marginal sea conditions of the SCS than FP-12E, but is also biased toward the tropical temperature regime, like FP-12E, because of the very limited data from temperate and subpolar regions. Application of SIMMAX-28 transfer function on sediment core 17940 from the northern SCS (Pflaumann and Jian 1999) revealed nearly unchanged summer temperatures around 28 °C for the last 30 kyr, while winter temperatures varied between 19.5 °C in the last glacial maximum (LGM) and 26 °C during the Holocene, and during Termination 1A the winter estimates show a Younger Dryas cooling by 3 °C subsequent to a temperature optimum of 24 °C during the Bølling/Allerød period.

The different results from the two methods can be illustrated with an example from the southern SCS, namely the SST-sequence of core 17957-2 over the past 1.5 myr (Fig. 5.1) (Jian et al. 2000b). Estimates of the winter and summer SSTs for the core top are 27.2 °C and 29.4 °C using FP-12E, and 27.0 °C and 28.9 °C

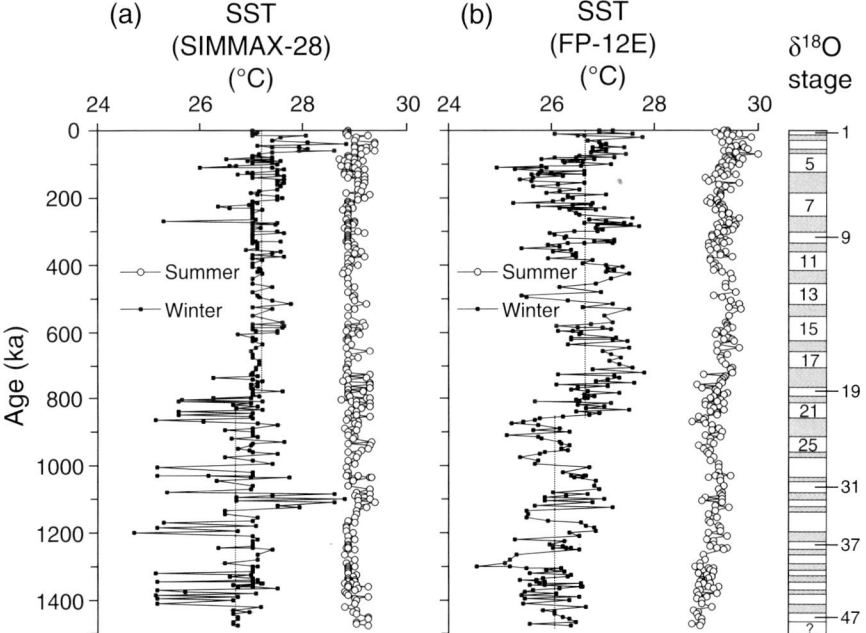

Fig. 5.1 Down-core variations in the estimated winter and summer SSTs in core 17957-2 using (**a**) SIMMAX-28 and (**b**) FP-12E methods show slightly different results (from Jian et al. 2000b). *Dashed lines* mark the average winter SST before and after marine isotope stages (MIS) 21–20

using SIMMAX-28, respectively. For the whole core, the SIMMAX-28 derived winter SST varies from 24.7 to 27.8 °C, while the summer SST varies from 28.8 to 29.4 °C. Both FP-12E and SIMMAX-28 derived winter and summer SSTs are slightly warmer with a narrow range of seasonality in core 17957-2 when compared to previous estimates using transfer function FP-12E (Wang and Wang 1990; Miao et al. 1994; Wang P. et al. 1995). Glacial/interglacial cycles are well presented in the FP-12E derived SST records but are somewhat obscured in the SIMMAX-28 derived SST records (Jian et al. 2000b).

For core MD972151 from the southern SCS, five transfer functions including the Imbrie-Kipp method (IKM), modern analog technique (MAT), modern analog technique with similarity index (SIMMAX), revised analog method (RAM), and the artificial neural network technique (ANN) were used to evaluate the magnitude of cooling during the LGM (Chen M. et al. 2005). The MAT, SIMMAX, RAM and ANN showed similar SST estimation results with ≤ 1 °C uncertainties in coretop SST calibrations. The IKM performed significantly worst in the calibration exercise, producing glacial SST estimates similar to present-day values, whereas all other four techniques indicated ~ 1 °C cooler glacial SST in the tropical western Pacific.

Alkenone Unsaturation Index

Pelejero and Grimalt (1997) established a calibrated linear relationship between $U^{K'}_{37}$ in surface sediment samples mainly consisting of coccolithophorids and SST for the SCS. The relationship between C_{37} alkenones composition and water temperature is based on the degree of unsaturation of the ketones, which is measured by means of several indices: $U^{K}_{37} = (C_{37:2} - C_{37:4})/(C_{37:2} + C_{37:3} + C_{37:4})$ and $U^{K'}_{37} = (C_{37:2})/(C_{37:2} + C_{37:3})$, where $C_{37:4}$, $C_{37:3}$ and $C_{37:2}$ are the concentrations of heptatriaconta-(8E, 15E, 22E, 29E)- tetraen-2-one, heptatriaconta-(8E, 15E, 22E)-trien-2-one and heptatriaconta-(15E, 22E)-dien-2-one, respectively. Even in the warm waters (25–29 °C) of the SCS, $U^{K'}_{37}$ and SST show a good linearity, indicating that the alkenone unsaturation index ($U^{K'}_{37}$) is a good SST proxy. This linear relationship is expressed as an equation, $U^{K'}_{37} = 0.031T + 0.092$, which is very similar to the frequently used equation ($U^{K'}_{37} = 0.033T + 0.043$) established for *Emiliania huxleyi* in culture and column water samples (Prahl and Wakeham 1987). The equation of $U^{K'}_{37}$ is valid for temperatures of the annually averaged water mass between 0 and 30 m in the SCS. SST estimates using this equation for core 17961-2 from the southern SCS show significant SST increases during Termination I. In sediment cores 18252-3 and 18287-3 from the southwestern and southern SCS, $U^{K'}_{37}$-derived SST increased abruptly by 1 °C or more at the end of Termination IA (Kienast et al. 2001, 2003), concurrent with the pre-Bølling warming about 14.6 ka ago observed in the Greenland Ice Sheet Project (GISP) 2 ice core. In IMAGES core MD972151 from the southern SCS, Zhao et al. (2006) used $U^{K'}_{37}$ to reconstruct the SST history for the past 150 kyr on a millennial scale, suggesting glacial/interglacial SST changes by 4 °C and 5 °C for Termination I and Termination II, respectively.

However, Bentaleb et al. (2002) questioned the usage of $U^{K'}_{37}$ as a SST proxy after calibrating water samples collected from the western Pacific Ocean. Their results showed a constant value of unity (1.0) for the $U^{K'}_{37}$ index when the surface water temperature reached 26.4 °C, indicating that paleo-SSTs above 26.4 °C may not be accurately calculated by using the $U^{K'}_{37}$ index. Pelejero and Calvo (2003), however, pointed out that Bentaleb et al. (2002) might have been made errors in designating a value of unity to the $U^{K'}_{37}$ index when the $C_{37:3}$ alkenone was not detected. Zhao et al. (2006) detected both $C_{37:2}$ and $_{37:3}$ alkenones in all samples from the southern SCS, enabling them to estimate SST up to 29 °C during MIS 5.5, which supports the use of the $U^{K'}_{37}$ index as a paleothermometer in the region.

Mg/Ca Ratios of Foraminifera

The Mg/Ca ratio of planktonic foraminiferal shells has also been widely used as useful proxy of past SST reconstruction in the SCS. This method has been proved to be valid in a number of oceanographic settings (Hasting et al. 1998; Lea et al. 2000). Mg paleothermometry has certain unique advantages over other proxies (Lea et al. 2000), the most important of which is that Mg/Ca is measured in foraminiferal shells, which are by themselves a vital archive of past climate and the carrier phase for oxygen isotopes. Measuring both Mg/Ca and $\delta^{18}O$ in foraminiferal shells from the same sample makes it possible to separate the magnitude and timing of SST and $\delta^{18}O_{water}$ changes (Elderfield and Ganssen 2000). Hasting et al. (2001) developed an empirical equation based on core top sample calibrations, i.e., Mg/Ca (mmol.mol^{-1}) = 0.38 × exp(0.089 × SST(°C)), and then compared three independent paleotemperature estimates from an AMS ^{14}C-dated core 18287-3 from the southern SCS: Mg/Ca ratio from planktonic foraminifera (*Globigerinoides ruber* and *Globigerinoides sacculifer*), alkenone thermometry ($U^{K'}_{37}$) and different foraminiferal transfer functions (SIMMAX28, RAM and FP-12E). Results from these three different methods show similar average glacial/interglacial temperature differences of about 2.5 °C. The most important shift that appears in all the three records is an abrupt SST increase by about 1.3–1.8 °C near the end of the last glacial period, synchronous with the 14.6 ka warming event observed in the GISP2 ice core. At two depth intervals with enhanced preservation, the Mg/Ca-deduced SSTs were 1 to 3.5 °C higher than predicted, which led Hasting et al. (2001) to conclude that even in a shallow core expected to have no significant carbonate dissolution, enhanced preservation events may lead to more positive Mg/Ca values and, therefore, higher inferred SST.

Not only dissolution but also planktonic foraminiferal species variations affect the validity of Mg/Ca as a paleo-SST proxy. Core top Mg/Ca measurements of samples from the tropical Pacific and Atlantic indicate *G. ruber* being the most accurate SST recorder and *G. sacculifer* a subsurface temperature register at 20–30 m (Dekens et al. 2002). When considering water depth of a core as a dissolution correction parameter and ΔCO_3^{2-} a dissolution parameter, a calibrated equation for *G. ruber* Mg/Ca in the Pacific can be used: Mg/Ca = 0.38exp[SST-0.61(core depth km)−1.6 °C] (Dekens et al. 2002). All the results from using this equation are plausible, as seen from SST reconstructions for Site 1143 between 3.3 and 2.5 Ma (Tian et al. 2006) and for Site 1145 for the past 145 kyr (Oppo and Sun 2005).

Wei et al. (2007) measured Mg/Ca in *G. sacculifer* for a past ~260 kyr SST reconstruction at Site 1144. Because there is no calibration for *G. sacculifer* in the

SCS, Wei et al. (2007) established the relationship between SST and Mg/Ca ratios on *G. sacculifer* shells without a sac chamber by five calibrations, including shells from culture (Nürnberg et al. 1996), from core-top sediments (Dekens et al. 2002; Rosenthal and Lohmann 2002), and from sediment traps (Anand et al. 2003). Each calibration was evaluated by two calculated results: SST of the topmost sample and SST change from LGM to the Holocene. Evaluation shows that SST calculations based on *G. sacculifer* in the SCS from Nürnberg et al. (2000) (T = (log Mg/Ca − log 0.491)/0.033), Anand et al. (2003) (Mg/Ca = 0.347 exp(0.090T)) and Dekens et al. (2002) for the Atlantic (Mg/Ca = 0.37 exp[0.09(T − 0.36 core depth km)]) agree quite well with each other with average temperature offsets less than 0.1 °C.

The planktonic foraminifer *G. ruber* (white) has two morphotypes, *G. ruber* sensu stricto (s.s.) living in the top 30 m of the water column and *G. ruber* sensu lato (s.l.) living at depths below 30 m (Wang L. 2000). Steinke et al. (2005) showed that *G. ruber* s.s. often register significantly higher Mg/Ca ratios than *G. ruber* s.l. as the latter precipitated its shells in slightly colder waters. Increased temperature differences between the two morphotypes occur mostly during periods of low salinity, especially during the Bølling/Allerød and the early Holocene. Their findings are supported by $\delta^{18}O$ and $\delta^{13}C$ differences between the two morphotypes and by stable oxygen isotope studies on core-top material from the SCS (Wang L. 2000; Löwemark et al. 2005) and by plankton tow and pumping studies from the Pacific (Kuroyanagi and Kawahata 2004). However, even the author themselves (Steinke et al. 2005, 2008a) accept that specimens of the two morphotypes of *G. ruber* are usually insufficient in high-resolution core samples for separate Mg/Ca analyses, thus a SST bias using both morphotypes appears to be unavoidable.

Intercomparison

SST reconstructions derived from planktonic foraminiferal transfer functions usually produce significant discrepancies with those derived from alkenone unsaturation and Mg/Ca ratios (Huang et al. 1997b; Steinke et al. 2001). In addition, SST records from different foraminifer transfer functions do not always agree with each other (Jian et al. 2000b; Steinke et al. 2001; Chen M. et al. 2005). To test the reliability of SST reconstructions, Steinke et al. (2008b) compared SST estimates using five different transfer function techniques on planktonic foraminiferal faunal census data from core MD01-2390 in the tropical southern SCS. All SST records derived from transfer functions indicate nearly unchanged or unusually higher temperatures during the LGM relative to modern temperatures, in contrast to substantial cooling of 2–5 °C inferred by $U^{K'}_{37}$ or Mg/Ca ratios from the same core (Steinke et al. 2006). The glacial planktonic foraminiferal assemblage from the southern SCS that has no modern analog is the main reason why the transfer function derived SST estimates for the LGM are inaccurate (Steinke et al. 2008b). The abundances of *Pulleniatina obliquiloculata* and *Neogloboquadrina pachyderma* (dextral) are abnormally high in glacial assemblages. The glacial high abundance of *P. obliquiloculata* probably resulted from a process involving stronger mixing and/or enhanced upwelling due to an intensified winter monsoon, which prevented shallow-dwelling, warm indicators to establish larger populations. Similarly, the glacial high abundance of *N. pachyderma* (dextral) was likely caused by a winter inflow of cold surface water from the northeast via the Bashi Strait due to the combined effects of an intensified winter monsoon, a southward shift of the polar front and the eastward migration of the Kuroshio Current during the LGM (Steinke et al. 2008b). As reported also by Jian et al. (1999), Huang et al. (2002), Steinke and Chen (2003) and Xu et al. (2005), *P. obliquiloculata* during glacials was consistently more abundant at southern than northern SCS localities, implying a unique phenomenon from the southern SCS. The abnormal glacial planktonic foraminiferal assemblage often causes warm SST estimates for the southern SCS using transfer functions (Chen M. et al. 2005; Steinke et al. 2008b). Comparatively, therefore, SST estimates based on Mg/Ca ratios and alkenone unsaturation index are more realistic and reliable (Steinke et al. 2008a,b).

In core 17940 from the northern SCS, the estimated annual SST curve (average of summer and winter SST) based on $U^{K'}_{37}$ measurements shows a smoother outline without the obvious short-term fluctuations revealed by the SIMMAX-derived SST curve during the Holocene and the LGM (Pelejero et al. 1999a). The discrepancy between the planktonic foraminiferal- and $U^{K'}_{37}$ (coccolithophorids)-derived SST may be attributed to the disparity between these two organisms. While coccolithophorids are nannoplankton living in the uppermost water, planktonic foraminifera live in water column from the sea surface down to several hundred meters and different species reach peak abundances at different times throughout the year. In addition to temperature, planktonic foraminiferal assemblage may have been affected by other water mass variables such as salinity and by fertility (Pflaumann and Jian 1999).

Paleo-SST Reconstruction

Most of the continuous paleo-SST records from the SCS, especially those derived from geochemical methods ($U^{K'}_{37}$ and Mg/Ca ratios), cover only the late Pleistocene. Although faunal transfer functions are biased in SST estimates for the glacial southern SCS (Chen M. et al. 2005; Steinke et al. 2008b), they are still the main technique used to reveal the long-term trend of SST variability in the SCS on the tectonic timescale.

Plio-Pleistocene

The longest paleo-SST sequences from the SCS are those derived from ODP Leg 184 cores. Late Pliocene-Pleistocene SST records have been obtained using PF-12E transfer functions at Site 1143 in the southern SCS (2.5–0 Ma), and at Site 1146 (4–0 Ma) and Site 1144 (1.1–0 Ma) in the northern SCS (Li B. et al. 2004; Zheng et al. 2005) (Fig. 5.2). In these studies, the extinct species were replaced by their modern counterparts for SST calculations (Anderson 1997). The results show that winter SST gradually decreased at Site 1146 since 3.1 Ma, superimposed by large amplitude fluctuations. The decrease by more than 7 °C on average was in parallel with an upcore reduction in the abundance of mixed-layer species influenced by the

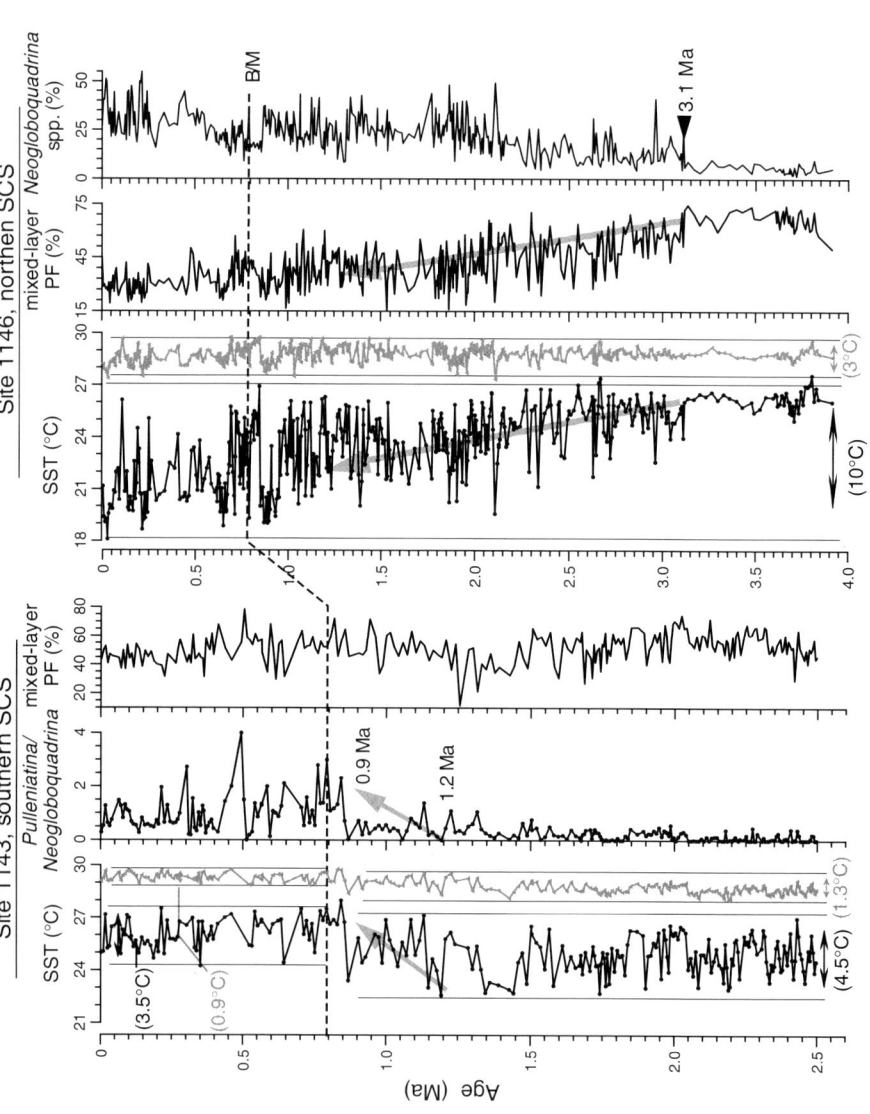

Fig. 5.2 Comparison of transfer-function estimated winter (*black*) and summer (*gray*) SST variations between Site 1143 and Site 1146 shows different trends since the late Pliocene (after Li B. et al. 2004 with Site 1146 data from Huang B. 2002). B/M is Brunhes/Matuyama paleomagnetic boundary at 0.78 Ma

East Asian winter monsoon variability. The decreased SST at Site 1146, therefore, was probably responding to intensified East Asian winter monsoon especially during the period of 3.1–2.0 Ma, as documented also in terrestrial records from the loess plateau (An 2000) and marine isotopic records from the southern SCS (Tian et al. 2004, 2006).

At Site 1143 from the southern SCS, the amplitudes of the SST fluctuations were 1.3 °C for summer and 4.5 °C for winter between 2.5 and 0.9 Ma but changed to 0.9 °C for summer and 3.5 °C for winter after 0.9 Ma, consistent with SST estimates from other southern SCS sites (e.g., Fig. 5.1). Therefore, the southern SCS has maintained relatively warm and stable SST since the late Pliocene, while the northern SCS experienced larger amplitude SST fluctuations especially during the Pleistocene.

Of particular interest are rapid SST changes in such periods as 3.1–2.5 Ma and 1.2–0.9 Ma. For the period from 3.3 to 2.5 Ma, the estimated SST at ODP Site 1143 using Mg/Ca ratio of *G. ruber* varies between 31 and 26.4 °C, showing a pattern of stepwise decrease corresponding to the onset of the significant Northern Hemisphere Glaciation (Fig. 5.3) (Tian et al. 2006). Three steps (periods 1–3) of SST changes can be recognized: MIS M2-G18 (3.3–2.97 Ma), G18-G6 (2.97–2.7 Ma), and G6-99 (2.7–2.5 Ma), with SST decreasing by ~5 °C from 3.3 to ~2.7 Ma before increasing by ~1.5 °C until ~2.5 Ma. Moreover, the amplitude of the millennial- or orbital-scale variability of SST also decreased (Tian et al. 2006).

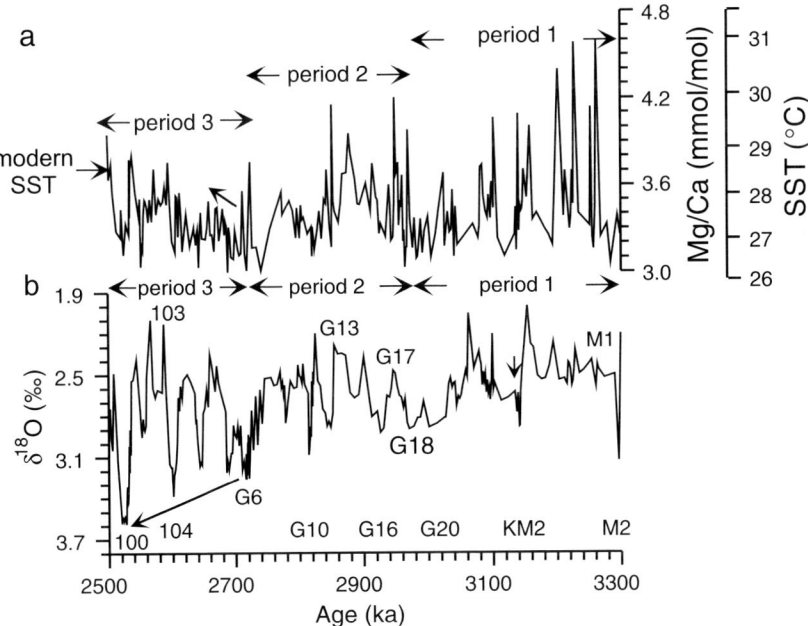

Fig. 5.3 Paleo-SST reconstruction for 3.3–2.5 Ma at ODP Site 1143, southern SCS, based on Mg/Ca ratio of *G. ruber* shows stepwise SST decreases from period 1 to period 3 (modified from Tian et al. 2006). (**a**) Mg/Ca ratio and Mg/Ca derived sea surface temperature; (**b**) *Cibicidoides* $\delta^{18}O$. The letters and numbers denote Marine Isotope Stages (MISs)

For the period from 1.2 to 0.9 Ma, the estimated SST at ODP Site 1143 using transfer function PF-12E shows winter SST increases by ~2 °C (Fig. 5.2a; Li B. et al. 2004). Because *P. obliquiloculata* contributes 80% to a tropical dissolution-resistant factor in PF-12E (Thompson 1981), its relatively high abundance since 0.9 Ma probably contributed to estimated SST higher than in earlier periods, consistent also with an increase in the *Pulleniatina/Neogloboquadrina* ratio (Fig. 5.2b). As *Pulleniatina* and *Neogloboquadrina* are both deep-dwelling species but respectively indicate warmer and cooler water temperatures (Hemleben et al. 1989; Chaisson 1995), their ratio increase at 0.9 Ma may indicate a warm subsurface water mass in the southern SCS at the time. The SST variations, *Pulleniatina* abundance, and *Pulleniatina/Neogloboquadrina* ratio at Site 1143 all indicate relatively warm surface waters in the southern SCS during the Pleistocene with enhanced warm subsurface waters right after the mid-Pleistocene climate transition between 1.2 and 0.9 Ma. A similar finding was reported from core 17957-2 also in the southern SCS, although the SST estimates using FP-12E and SIMMAX-28 show slightly different results (Fig. 5.1) (Jian et al. 2000b). The SIMMAX-28-derived winter SST exhibit larger amplitude fluctuations before the Brunhes/Matuyama (B/M) boundary, whereas the PF-12E-derived winter SST show greater changes across the MIS 22/21 transition (Fig. 5.1).

Late Pleistocene

Late Pleistocene SST variations have been studied in numerous cores in the SCS with higher time resolutions than those Pliocene-Pleistocene records. Wang L. et al. (1999a) reported a millennial-scale $U^{K'}_{37}$ SST record for the past 35 kyr from core 17940, northern SCS, by calibrating $U^{K'}_{37}$ to annual mean SST in 0–30 m water depth, together with the FP-12E-based summer and winter SST estimates (Fig. 5.4). While the summer SST remained almost constant over the entire last glacial cycle, the winter SST decreased with a number of short-term negative oscillations by ~2 °C during the MIS 3, paralleling some "warm" $\delta^{18}O$ minima. The MIS 2 SST minima match well with heavy $\delta^{18}O$ values toward the end of the LGM. Prominent declines in winter SST also occur in the late Holocene.

In the southern SCS, Steinke et al. (2006) reported a high resolution SST record for the past 22 kyr based on planktonic foraminiferal Mg/Ca ratios from core MD01-2390 (Fig. 5.5), showing an average temperature change of 3.1 ± 0.6 °C. This temperature change is relatively larger than previous reconstructions. This could be related to the fact that the entire *G. ruber* population instead of "morphospecific" samples of *G. ruber* s.s. from the glacial period was used for Mg/Ca measurements. As mentioned above, *G. ruber* s.l. specimens generally yield lower Mg/Ca ratios and hence lower SST estimates compared to G. *ruber* s.s. In core MD01-2390, the abundance of *G. ruber* in LGM samples is very low, and the use of both *G. ruber* s.l. and *G. ruber* s.s. for Mg/Ca analyses contributes to lower SST-estimations for the LGM (Steinke et al. 2006, 2008a).

An abrupt temperature increase by at least 1 °C at the end of Heinrich 1 (H1) event (Termination IA) is recorded in two short $U^{K'}_{37}$ SST records spanning the

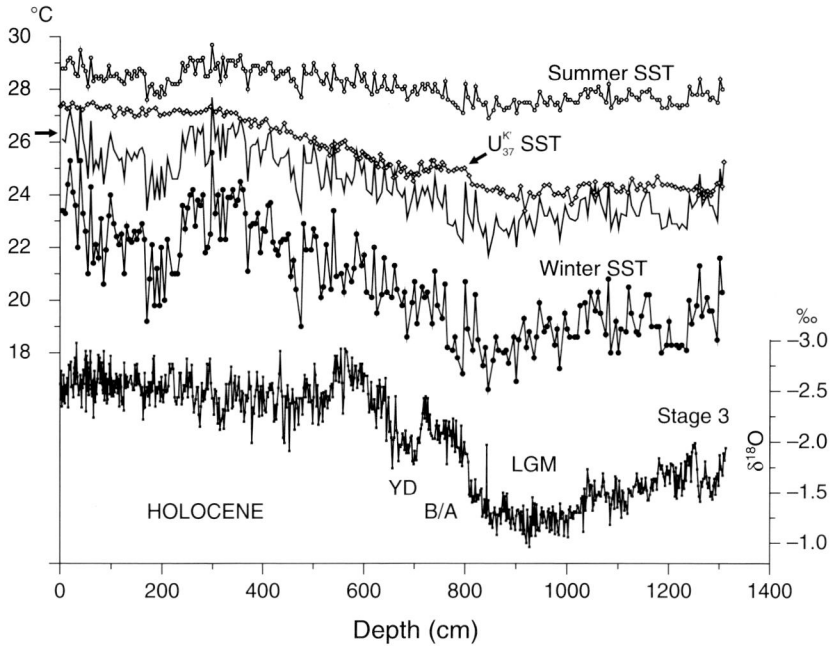

Fig. 5.4 SST estimates for core 17940 in the northern SCS are based on FP-12E transfer function (*circles* for summer, *dots* for winter and *coarse line* for the annual mean) and based on the $U^{K'}_{37}$ (*diamonds* for annual mean SST), as compared with *G. ruber* $\delta^{18}O$ curves (modified from Wang L. et al. (1999a). YD = Younger Dryas; B/A = Bølling-Allerød. Horizontal arrow marks modern annual mean SST value at 0–30 m

last 13,000 yr from cores 18287-3 and 18252-3 in the southern SCS (Fig. 5.6) (Kienast et al. 2001). Within the recognized dating uncertainties, this SST increase was synchronous with the Bølling warming 14.6 ka ago as recorded in the GISP2 ice core.

SST Variations at Millennial Scales

In recent years, remarkable progresses have been made on SST reconstructions on millennial time scales in the SCS. Representing these are the alkenone derived SST record in IMAGES core MD972151 from the southern SCS (Zhao et al. 2006) and the Mg/Ca derived SST record at ODP Site 1145 from the northern SCS (Oppo and Sun 2005).

In the southern SCS, the alkenone data from core MD972151 generate a SST record over the past 150 kyr, with sampling resolution of 150 to 200 years (Fig. 5.7) (Zhao et al. 2006). It shows glacial SST variations from 23.5 to 26 °C and interglacial SST variations from 27 to 28.9 °C, in general agreement with previously reported records from the southern SCS (Pelejero et al. 1999a,b; Wang P.1999; Kienast et al. 2001; Steinke et al. 2001; Chen M. et al. 2005) and from the tropical western Pacific ocean (Lea et al. 2000; Stott et al. 2002; Rosenthal et al. 2003; Visser et al. 2003).

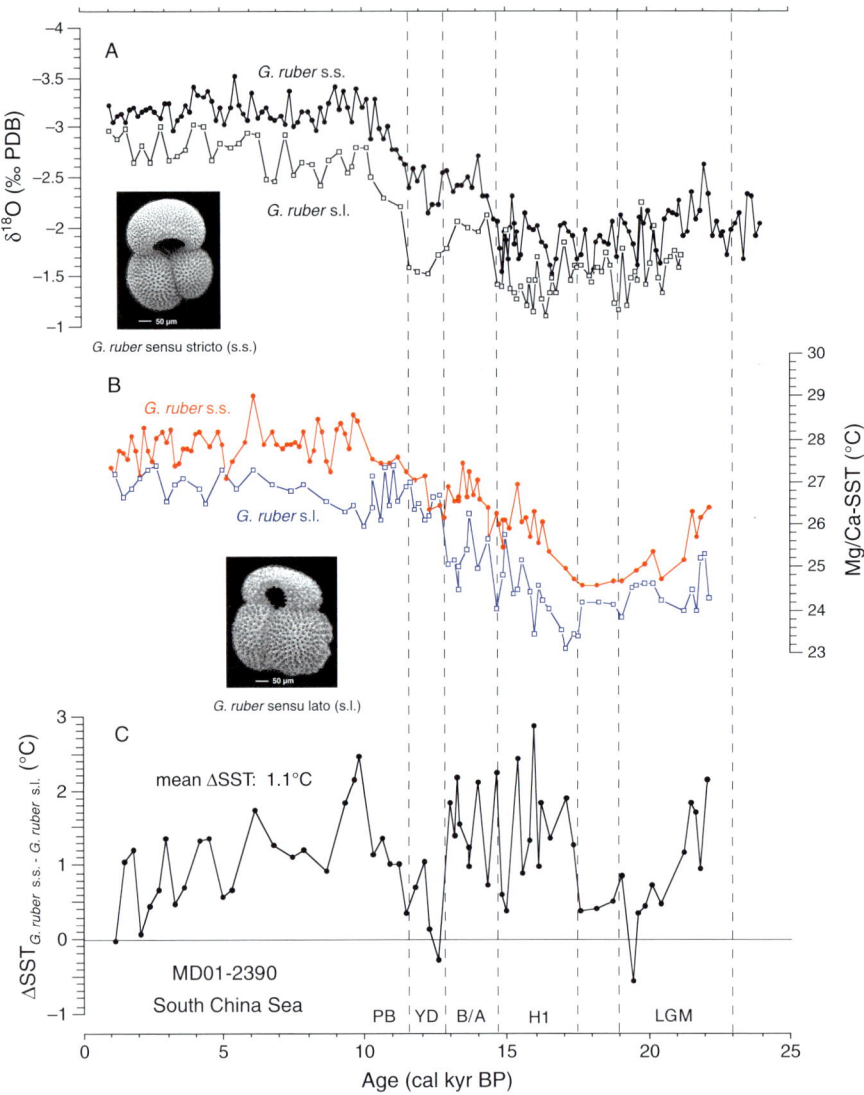

Fig. 5.5 Paleo-records from two morphotypes of *G. ruber* in core MD01-2390 are compared: (A) δ¹⁸O of *G. ruber* s.s. and *G. ruber* s.l., (B) inferred SSTs using Mg/Ca ratios in *G. ruber* s.s. (*filled circles*) and *G. ruber* s.l. (*open circles*), and (C) SST differences between these two morphotypes (from Steinke et al. 2006, 2008a). PB = Preboreal; YD = Younger Dryas; B/A = Bølling–Allerød; H1 = Heinrich 1 event; LGM = Last Glacial Maximum

SST increased by ∼5 °C during Termination II and by ∼4 °C during Termination I. For Termination I, the lowest SST (23.8–24.0 °C) was recorded at 15.2–14.1 ka, just prior to the Bølling transition instead of during the LGM between 23 and 19 ka. Following this was an increase by 3 °C (to 27 °C) within 100 years during

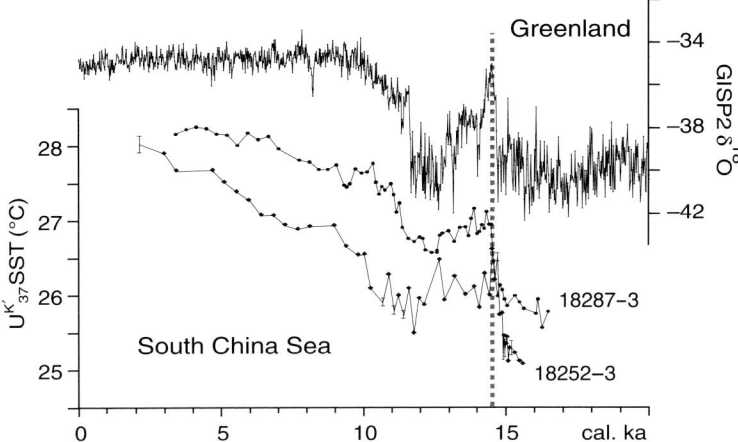

Fig. 5.6 $U^{K'}_{37}$ SST records in cores 18252-3 and 18287-3 from the southern SCS on independent calendar time scales are compared with the GISP2 $\delta^{18}O$ record (Stuiver and Grootes 2000), showing contemporary SST increase at 14.6 ka marked by the vertical dashed line (Kienast et al. 2001)

the Bølling/Allerød and IS1 warm event. A similar SST pattern occurred also in Termination II, with a very cold event (23.5–23.7 °C) lasting 2700 years (between 137 and 134 ka) in late MIS 6 (Fig. 5.7). Frequent SST oscillations were observed within MIS 3, with frequencies similar to the D-O cycles observed in Greenland ice cores. SST was around 26 °C during typical warmer interstadial events and dropped to ~24.4 °C during colder stadial events.

The SST curve from core MD972151 is comparable with the $\delta^{18}O$ record of the Hulu stalagmite from eastern China (Wang Y. et al. 2001) with some minor differences (Fig. 5.7). Between 65 and 10 ka, the SST oscillated within a range of 24–26 °C, while the Hulu $\delta^{18}O$ fluctuated between −6.5 and −9.5‰ during the period of 60–30 ka before increasing to a maximum of −4.5‰ at 16 ka. These millennial-scale oscillations in the two records correlate well with each other and with the Greenland GISP2 record. Within the error of the age model, for example, most of the positive SST excursions on the order of 1.0–1.5 °C over a few hundred years can be correlated with the interstadials in the GISP2 record and with more negative $\delta^{18}O$ values in the Hulu $\delta^{18}O$ record, indicative of stronger summer monsoons. Similarly, several significant cool intervals (24.4–24.2 °C) between these interstadial events possibly correspond to the Heinrich events reported from the North Atlantic Ocean (Bond et al. 1993; Chapman et al. 2000) (Fig. 5.7). Therefore, Zhao et al. (2006) concluded that millennial-scale SST changes in the SCS were mainly caused by the variability of the winter monsoon. Moreover, the simultaneous rise in sea level and in $U^{K'}_{37}$ SST during Termination I indicates an important role of sea-level change on the regional SST variability by influencing the exchange of tropical ocean warm surface water with the SCS water through the Sunda Shelf region (sill depth 30–50 m). When sea-level drop was less than 30 m relative to

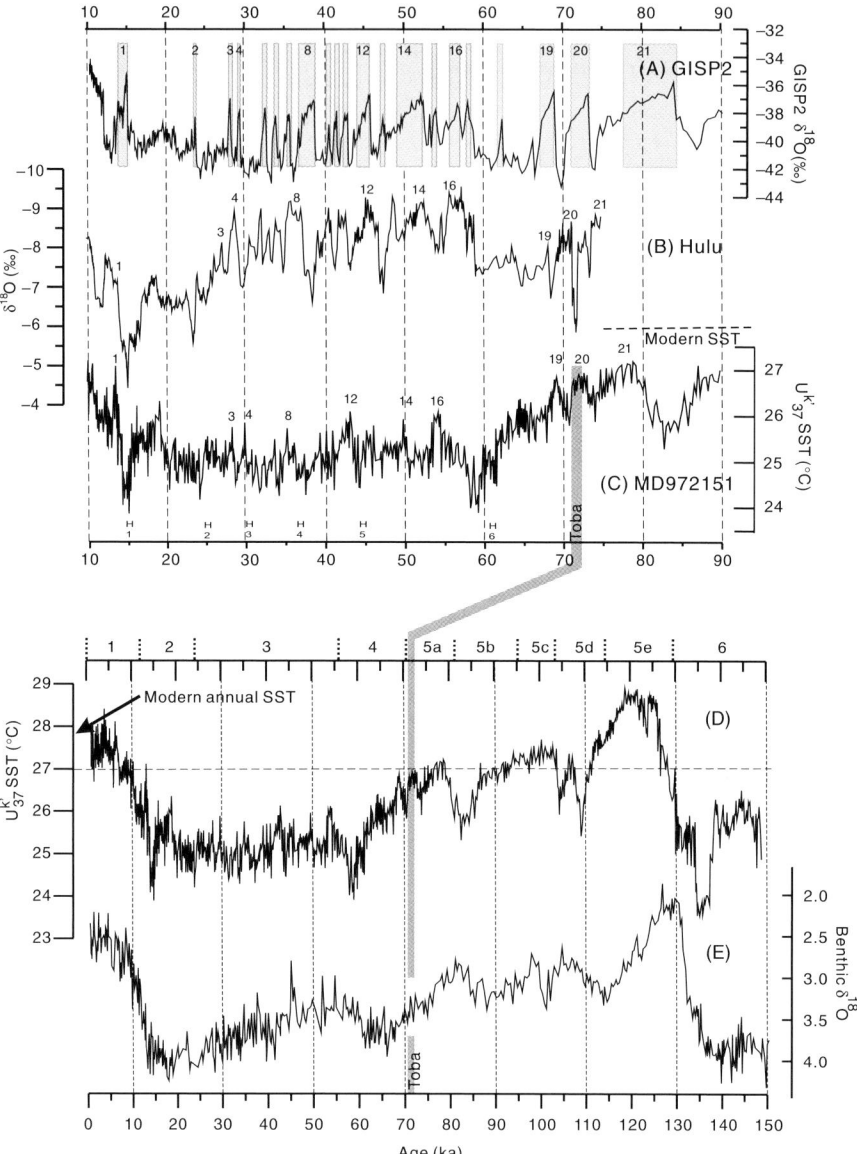

Fig. 5.7 Comparison between $U^{K'}_{37}$ SST and benthic $\delta^{18}O$ records of core MD972151 (C, D, E) and the $\delta^{18}O$ record of the Hulu Cave (B) (Wang Y. et al. 2001) and the $\delta^{18}O$ record of the GISP2 ice core from Greenland (A) (Dansgaard et al. 1993) shows a good correlation of interstadial events and Heinrich events (H1 to H61) (from Zhao et al. 2006). The vertical bar indicates the youngest Toba tephra layer. Below 40 ka, GISP2 has turned out to deviate by 2500 yrs from new annual-layer counts (Svensson et al. 2008)

today's, SST at MD972151 was constantly above 27 °C, a situation which occurred during MIS 5e, part of MIS5c and 5a, and over the last 11 kyr (Zhao et al. 2006).

In the northern SCS, Oppo and Sun (2005) reconstructed SST variations at ODP Site 1145 for the past 140 kyr (Fig. 5.8) using an equation developed for *G. ruber* Mg/Ca ratios in the Pacific (Dekens et al. 2002). The estimated SST for the late Holocene was ~28 °C, corresponding to an abundance peak of *G. ruber* in late summer. The maximum SST of the previous interglacial (MIS 5e) was ~1 °C warmer than the late Holocene, in a good agreement with alkenone-based SST estimates (Pelejero et al. 1999a). Glacial SST was ~4 °C colder than the late Holocene, also consistent with previous estimates for the northern SCS (Wang P. et al. 1995; Pelejero et al. 1999a). After the MIS 5e temperature peak, SST within the last interglacial changed several times by ~3 °C marking other MIS 5 substages (Fig. 5.8).

Suborbital SST changes at Site 1145 are best developed since ~80 kyr. Based on a visual correlation between the SST record and the Hulu δ^{18}O record, Oppo and Sun (2005) suggested that SST changes in the northern SCS were synchronous with abrupt events in the seasonality of monsoon precipitation and in temperature changes over Greenland, and that the distinct cold events likely corresponded to the Heinrich events from episodic massive iceberg discharges into the subpolar North

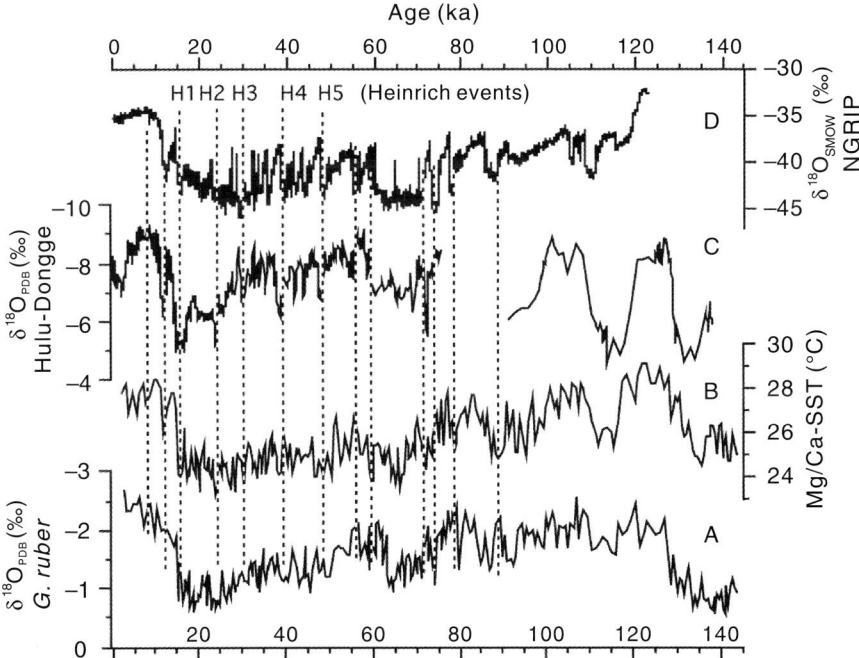

Fig. 5.8 Planktonic δ^{18}O (A) and Mg/Ca-based SST (B) from ODP Site 1145 are compared with various δ^{18}O records: (C) speleothem δ^{18}O from Hulu Cave (Wang Y. et al. 2001) and Dongge Cave (Yuan et al. 2004), and (D) Greenland ice core δ^{18}O (NGRIP Members 2004). *Dashed lines* mark likely correlations between marine and terrestrial records (from Oppo and Sun 2005)

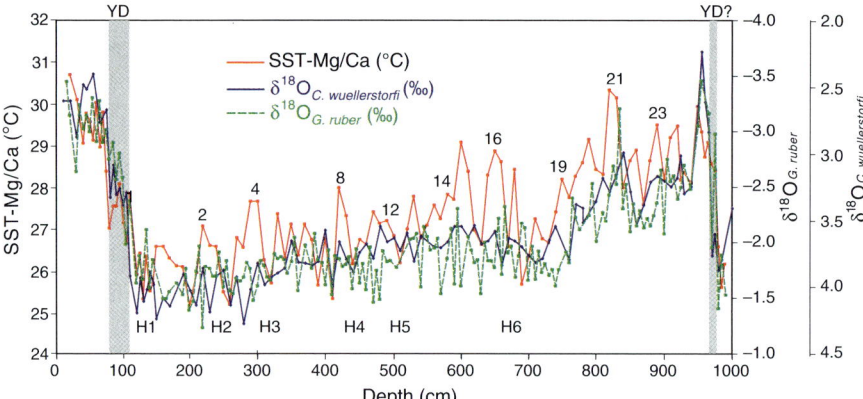

Fig. 5.9 Comparison between Mg/Ca SST estimates and $\delta^{18}O$ records of benthic (*C. wuellerstorfi*) and planktonic foraminifera (*G. ruber*) in core 17961 from the southern SCS (Jian et al. 2008). Also shown are interstadial events (2 to 23), Heinrich events (H1 to H61) and the Younger Dryas (YD) or similar events (*gray bars*)

Atlantic (Fig. 5.8). A good match between the Mg/Ca SST record and the benthic $\delta^{18}O$ from the same site also implies that the demise of ice sheets during the last deglaciation did not precede Northern Hemisphere summer insolation increase (Oppo and Sun 2005).

The outlined SST changes are confirmed by our new Mg/Ca-derived SST record from core 17961 in the southern SCS (Fig. 5.9). During the last deglaciation, SST changes were apparently synchronous with changes in both benthic (*C. wuellerstorfi*) and planktonic (*G. ruber*) $\delta^{18}O$, without any noticeable phase lead or lag between them (Jian et al. 2008). Together with the recent published SST records (Kienast et al. 2001; Oppo and Sun 2005), our new data support the inference that, in the East Asian monsoon region, tropical SST changes during the last deglaciation were synchronous with warming events in Greenland. These SCS records are in conflict with postulates from the equatorial western Pacific Ocean (Lea et al. 2000; Visser et al. 2003), although the SCS, particularly its southern part, belongs to the tropical Pacific. Therefore, the precise phase relationship between tropical and high-latitude climates during past abrupt climate events require further studies on higher time resolution before these inconsistencies can be resolved.

Paleo-SST Patterns

Paleo-SST in the SCS

As described in Chapter 2, the modern SST distribution in the SCS is largely controlled by the seasonally reversing monsoon-driven surface circulation. With the seasonal alternation of prevailing winds, i.e. the southwest monsoon in summer and northeast monsoon in winter, the SCS displays a trans-basinal pattern of surface currents with opposite directions during summer and winter. In summer, surface water of the tropical Indian Ocean flows northward into the SCS, while in winter the northeast wind drives the tropical and subtropical Pacific waters together with the cooler coastal waters into the SCS, leading to a steeper south-north (S-N) temperature gradient in the winter SCS. According to Levitus and Boyer (1994), the SST of the SCS ranges from 22 to 28.8 °C during winter, with steep gradients toward the coast of China, while during summer, the SST varies only between 27 and 29 °C.

During the LGM, however, when sea level dropped by 100–120 m, all the southern connections to the open ocean closed, and the SCS changed to a semi-closed basin with Bashi Strait as its only water passage way to the open ocean. Winter and summer SST estimates for the LGM from ten cores (Fig. 5.10) show distinct S-N contrast (Wang P. et al. 1995). The winter SST displayed a south-north (S-N) trend, from 17.5 °C in the north to 23 °C in the south, whereas the summer SST showed

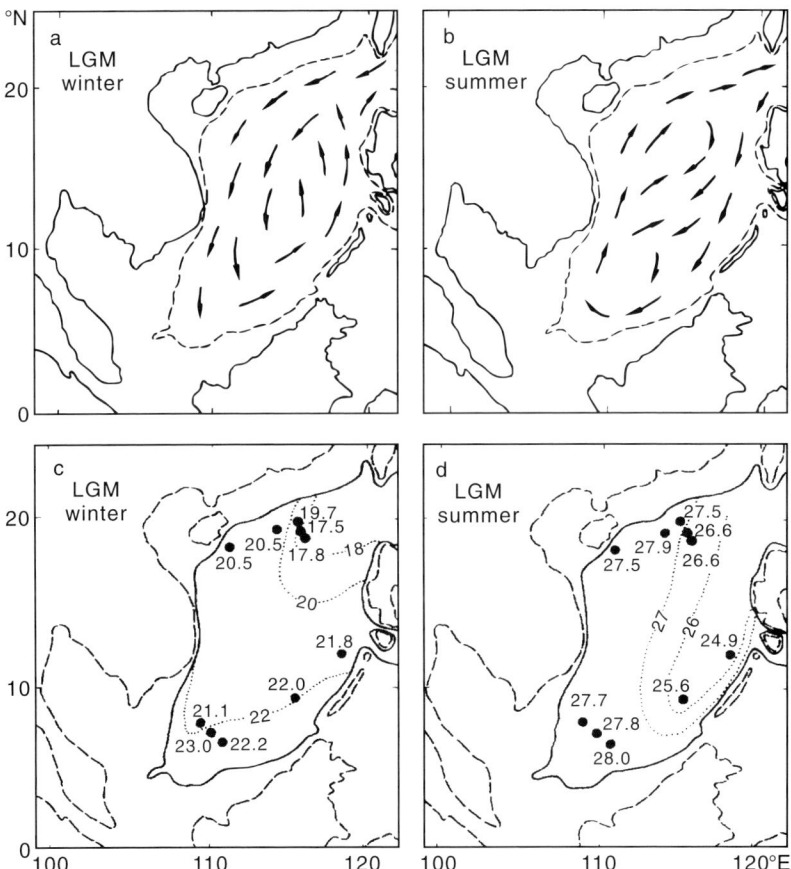

Fig. 5.10 Surface circulation (**a**, winter; **b**, summer) and average SST (**c**, winter; **d**, summer) in the SCS during the LGM show different distribution patterns (Wang P. et al. 1995). *Dotted lines* denote isotherms (°C). Filled circles with numbers indicate the position of cores and average SST during the LGM, respectively

an east-west trend, from <25 °C in the east to 27–28 °C in the west, indicating basic changes in surface circulation.

These SST patterns during the LGM are also supported by numerical simulations. As shown in Fig. 5.10, when sea level was low, the glacial surface circulation was counter-clockwise for winter and basically clockwise for summer. Therefore, the earlier interglacial (and postglacial) trans-basinal patters of circulation must have been replaced by semi-enclosed patterns as schematically shown in Fig. 5.10 (Wang P. et al. 1995).

North-South Comparison of SST

The S-N temperature gradient in the winter SCS seems to exist throughout the entire Quaternary and can be used to indicate changes in the strength of the East Asian winter monsoon. Reconstructions of SST at ODP Sites 1146 and 1143 reveal increased SST gradient between the northern and southern SCS during glacials of the late Pleistocene due to strengthened winter monsoon, consistent with other winter monsoon proxy records (Fig. 5.11). For example, the S-N SST difference (ΔSST) between core 17964 (112°12.8′E, 06°09.5′N) from the south and core 17940 (117°23.0′E, 20°07.0′N) from the north displays increased values during the last glacial stage and decreased values during the Holocene, respectively, indicating intervals of strengthened and weakened winter monsoons (Fig. 5.12).

The SST estimates in cores 17940 and 17964 are based on planktonic foraminiferal transfer function and possibly biased by age control and ecological effects. Recently, Oppo and Sun (2005) published the first SST record based on Mg/Ca ratio measurements at ODP Site 1145. This northern SCS record is compared with a southern SCS record in core 17961 (112°19.9′E, 08°30.4′) derived also from the Mg/Ca ratio (Jian et al. 2008). The S-N SST and planktonic δ^{18}O differences between Site 1145 and core 17961 were calculated after interpolating the two sets of data to 1 kyr interval. The results show that the S-N SST gradient in the SCS fluctuated frequently, although the time resolution for the lower part of core 17961 is lower than that at Site 1145 (Fig. 5.13). Spectral analyses of the ΔSST reveal a precessional cycle (~23.2 kyr) and millennial variabilities mainly at 2.0–2.4 kyr and 1.4 kyr (Fig. 5.14). Particularly, during the D-O interstadials in the last glacial stage, the S-N SST differences decreased correspondingly, indicating weakened winter monsoons.

Paleo-SST in the Western Pacific

On the basis of micropaleontological data, the CLIMAP studies concluded that the "ice-age ocean was strikingly similar to the present ocean in at least one respect:

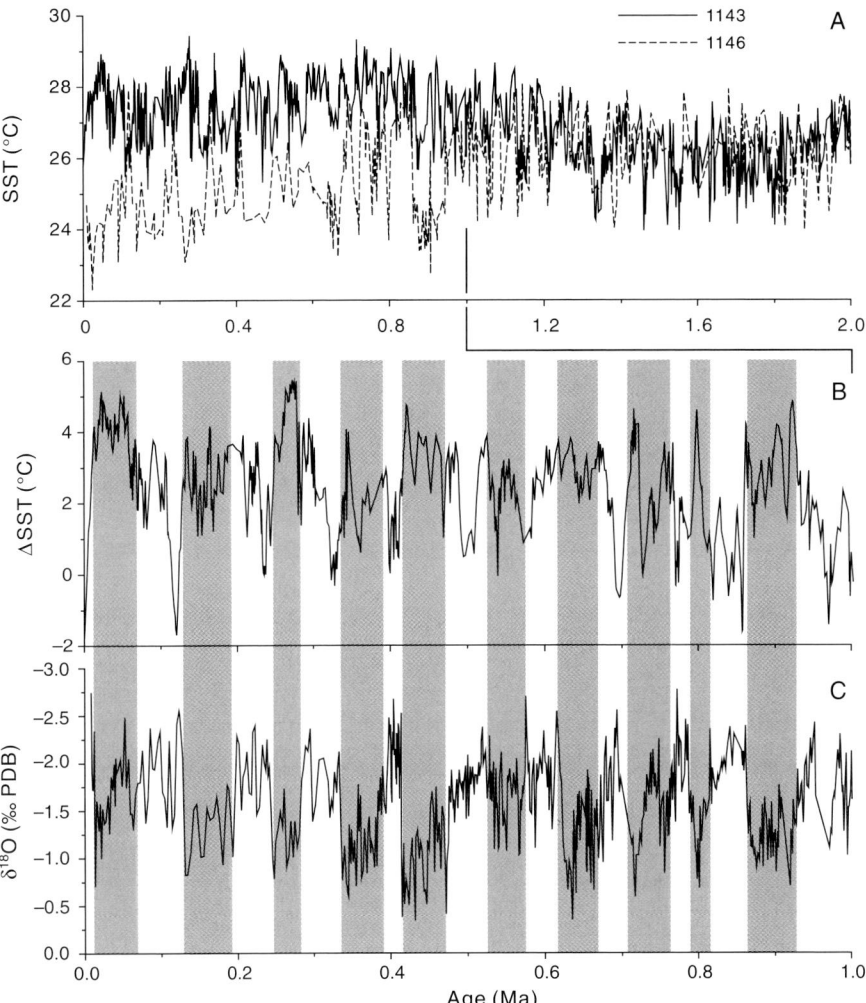

Fig. 5.11 SST variations are revealed at (**A**) the southern Site 1143 (*dashed line*, Xu et al. 2005) and the northern Site 1146 (*solid line*, Huang et al. 2003) derived from foraminiferal transfer function, (**B**) SST difference (ΔSST) between these two sites, and (C) *G. ruber* δ¹⁸O at Site 1146. *Vertical gray bars* indicate glacial stages

large areas of the tropics and subtropics within all oceans had sea-surface temperature as warm as, or slightly warmer than, today" (CLIMAP 1981). Recent studies, however, have different conclusions (e.g., Hostetler et al. 2006).

Wang P. (1999) summarized the available paleo-SST data from the SCS and East China Sea, Sulu Sea and the adjacent western Pacific during the LGM, which were based on planktonic foraminiferal census using the transfer function FP-12E. As seen from Fig. 5.15, the LGM summer SST for the South China Sea and Sulu Sea

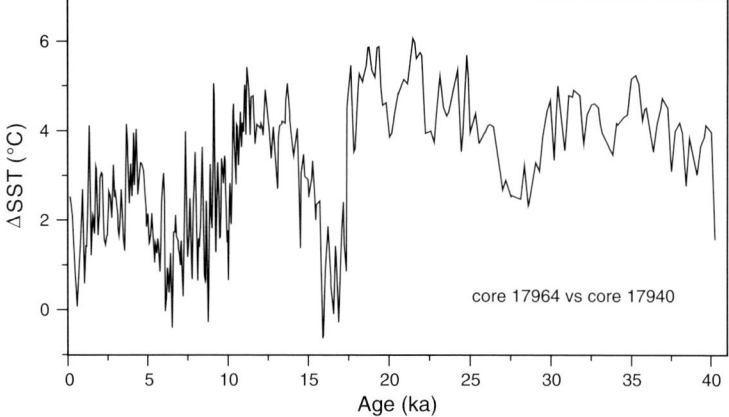

Fig. 5.12 S-N SST difference (ΔSST) between core 17964 from the southern and core 17940 (Pflaumann and Jian 1999) from the northern SCS was based on inferred SSTs using foraminiferal transfer function

between 5 and 20 °N ranges from 25.6 to 29.0 °C, averaging 27.8 °C, while in the open western Pacific at the same latitudes it ranges from 27.1 to 29.6 °C with an average of 28.7 °C. The LGM winter SST varies from 16.0 to 24.0 °C in the South China and Sulu Seas, averaging 21.1 °C, and from 23.8 to 28.0 °C in the open ocean, averaging 26.0 °C, or 4.9 °C higher than that in the marginal seas. Thus, the winter SST at the LGM was much cooler in the western Pacific marginal seas than in the open ocean, whereas in summer the SST was similar in the marginal seas and ocean, resulting in a much more intensive seasonality during the LGM in the marginal seas. In general, the winter SST at the LGM was at least 3–4 °C lower in the SCS and Sulu Sea than in the open Pacific, and the LGM seasonality is about 4 °C higher in the open ocean, supporting the early finding of the amplifying effect of glacial signals in the marginal seas (Wang and Wang 1990).

These different results of paleo-SST reconstructions led to new efforts to improve the paleoecological transfer function technique and to debate whether the technique is applicable to the whole tropics (Anderson and Webb 1994). Up to now, nearly ten cores in the SCS have been analyzed for $U^{K'}_{37}$ measurements, confirming LGM-Holocene SST contrast by 4–4.5 °C in the north and by 2.5 °C in the south (Huang et al. 1997a; Pelejero et al. 1999a; Kienast et al. 2001; Zhao et al. 2006), all slightly exceeding that in the open Pacific (Lea et al. 2000).

Oxygen isotope data for the late Quaternary are now available from many sites in the western Pacific marginal seas, in particular the SCS. As seen from LGM-Holocene changes in the $\delta^{18}O$ of shallow-dwelling planktonic foraminifers (*G. sacculifer* or *G. ruber*) (Fig. 5.15d), the $\delta^{18}O$ contrast is again much more significant in the marginal seas than in the open ocean, with <1.7‰ in the open ocean south of 30 °N but >1.7‰ in the marginal seas at the same latitudes. Although the greater $\delta^{18}O$ difference in the marginal seas might be partly caused by salinity changes and

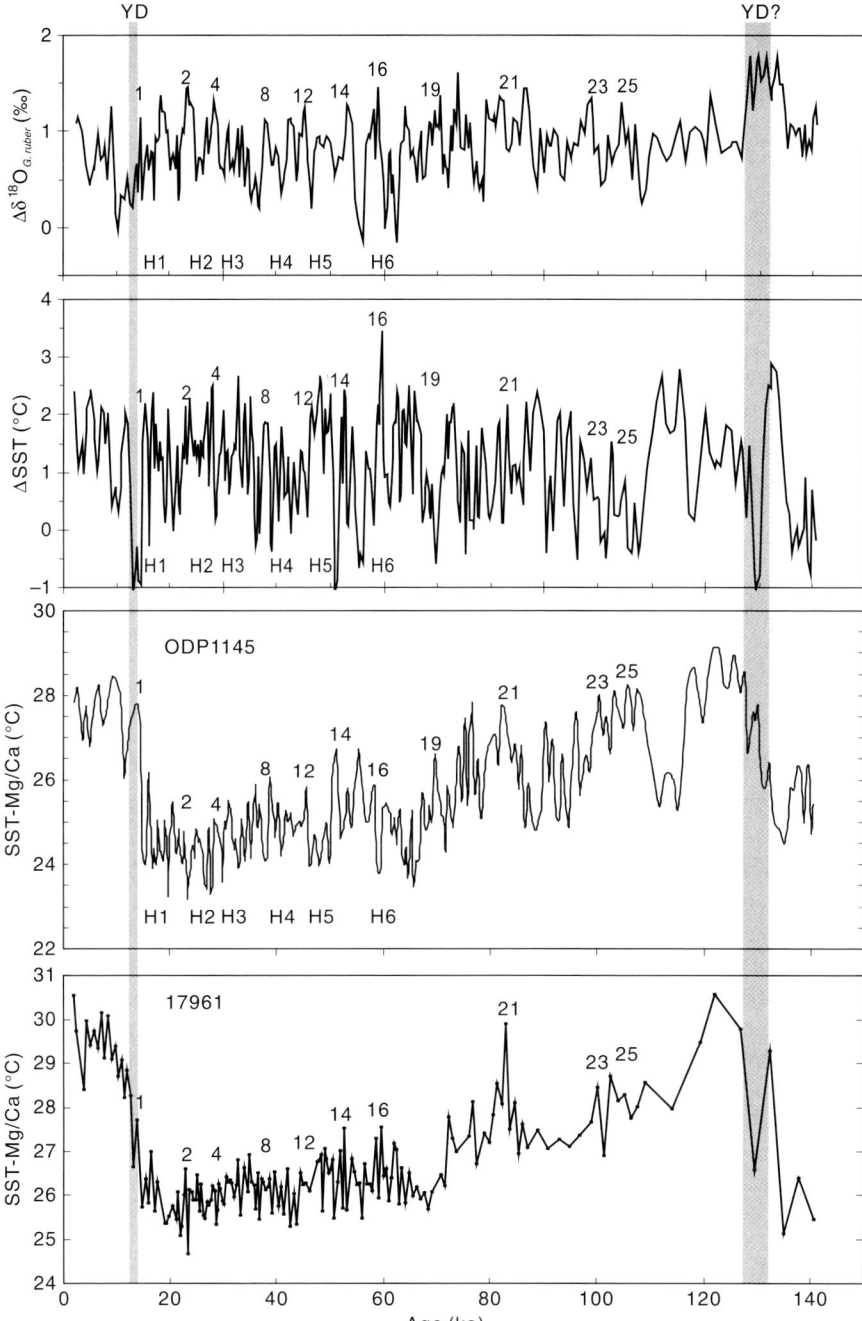

Fig. 5.13 S-N differences are shown in SST (ΔSST) and *G. ruber* δ^{18}O ($\Delta\delta^{18}$O$_{G.\ ruber}$) between cores 17961 from the southern (Jian et al. 2008) and ODP Site 1145 from the northern SCS (Oppo and Sun 2005). *Gray bars* show the Younger Dryas (YD) or similar events. Numbers denote the D-O events, and H1 to H6 indicate Heinrich events

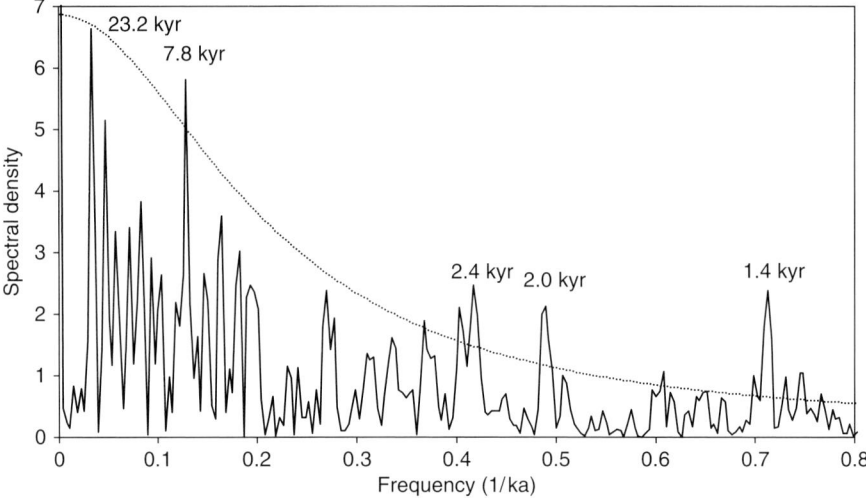

Fig. 5.14 Results of spectral analyses on the S-N SST difference (ΔSST) between core 17961 and Site 1145 reveal several precessional and millennial periods (numbers) exceeding the upper limit of red noise at 90% confidence level (*dotted line*)

in part controlled by sedimentation rates and bioturbation, there is no contradiction between the δ^{18}O data and the trend of paleo-SST change discussed above (Wang P. 1999).

5.2 Thermocline Depth History (Tian J. and Jian Z.)

The upper-water structure, particularly the thermocline depth, is of particular importance for the low-latitude Pacific and the SCS because of its close ties with El Nino-Southern Oscillation and with the East Asian monsoon. Therefore, past thermocline changes in the SCS can shed light on the history of both the East Asian monsoon (Jian et al. 2000a; Huang et al. 2002; Tian et al. 2005b) and the Western Pacific Warm Pool (Li B. et al. 2004; Li Q. et al. 2006). Seasonal variations of thermocline depth in the modern SCS are discussed in Chapter 2 (Thermocline and upwelling), here we review the history of thermocline depth changes starting from its proxies.

Proxies of Thermocline Depth

Foraminiferal Assemblage

Planktonic foraminiferal assemblages, transfer functions, and δ^{18}O gradient between surface and sub-surface planktonic foraminifers have been applied for reconstructing past thermocline changes in the SCS. Quantitative analyses of the modern planktonic foraminiferal distribution indicate that shallow dwelling species such as

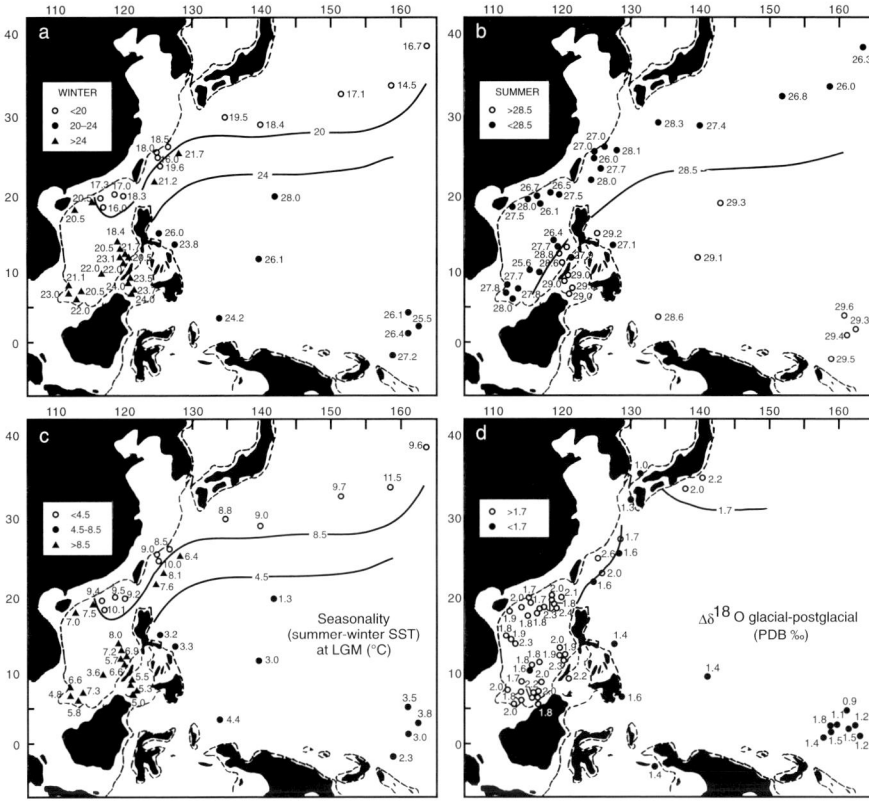

Fig. 5.15 SST estimates in cores from the low- and mid-latitude western Pacific and marginal seas at the LGM based on transfer function FP-12E show variations in (**a**) winter SST, (**b**) summer SST and (**c**) seasonality in SST (summer SST minus winter SST), as compared with (**d**) LGM-Holocene planktonic $\delta^{18}O$ differences (Wang P. 1999)

G. ruber and *G. sacculifer* dominate the assemblage when the surface mixed layer is deep and the thermocline is depressed to below the photic zone, and vice versa for the deep thermocline dwelling species such as *P. obliquiloculata* and *Globorotalia tumida* (Bé et al. 1985; Ravelo and Fairbanks 1992; Jian et al. 2000a). The vertical depth habitats of tropical planktonic foraminifers provide a primary tool for reconstructing upper water structure in paleo-oceans (Fairbanks et al. 1982; Thunell et al. 1983; Bé et al. 1985), so paleo-thermocline depth can be reconstructed either using the abundance ratio between shallow and deep dwelling species or transfer functions, as in the case of SST reconstruction. While the former may indicate relative changes, the latter provides quantitative estimates of the thermocline depth. In the tropical Pacific Ocean, two transfer function techniques, the Imbrie-Kipp Method (IKM) (Imbrie and Kipp 1971) and the modern analog technique (MAT) (Hutson 1980) have been widely used and the results show that subsurface hydrography of the tropical Pacific during the LGM was only slightly different from the present-day's structure there (Andreasen and Ravelo 1997).

$\delta^{18}O$ Gradient Between Surface and Subsurface Planktonic Foraminifera

Another useful proxy of the paleo-thermocline depth in the western Pacific is the $\delta^{18}O$ difference between surface and subsurface dwelling planktonic foraminiferal species (Ravelo and Shackleton 1995; Jian et al. 2001). Larger $\delta^{18}O$ differences between subsurface and surface species usually indicate a shallow thermocline with a large temperature range in the photic zone with colder intermediate waters getting close to the surface, whereas smaller $\delta^{18}O$ differences often result from a deepened thermocline with a narrow temperature range (Ravelo and Shackleton 1995). For example, in the modern tropical Pacific, the thermocline is deep in the west but shallow in the east. The subsurface to surface foraminiferal $\delta^{18}O$ differences in core top samples are smaller (1.0‰) in the western Pacific but larger (1.9‰) in the eastern Pacific, corresponding to the deep and shallow thermoclines in these two Pacific sectors, respectively (Billups et al. 1999). Similarly, in the equatorial Atlantic the subsurface to surface foraminiferal $\delta^{18}O$ differences in core top samples are smaller (0.9‰) in the west with a deep thermocline but larger (1.4‰) in the east with a shallow thermocline (Billups et al. 1999). By using $\delta^{18}O$ differences between the subsurface *G. tumida* and the near-surface dwelling species *G. sacculifer*, Billups et al. (1999) constructed the upper ocean thermal gradient variations at the open west Pacific ODP Site 806 between 5 and 3 Ma. The consistently small $\delta^{18}O$ differences led the authors to suggest that the thermal gradient in the photic zone remained small and the mixed layer remained deep during the early Pliocene at Site 806.

The same has been observed in the modern SCS. As shown in Fig. 5.16, the core top $\delta^{18}O$ differences ($\Delta\delta^{18}O_{(P-G)}$) between *P. obliquiloculata* and *G. ruber* in the southern SCS ranges from 0.76 to 1.38‰ in the summer monsoon upwelling area off Vietnam, but from 1.68 to 1.83‰ outside this area. Smaller $\delta^{18}O$ differences between subsurface and surface species mean shallower thermocline in the upwelling area (Tian et al. 2005b). Therefore, the $\delta^{18}O$ difference $\Delta\delta^{18}O_{(P-G)}$, can be used to indicate thermal gradient variations in the SCS, with large $\Delta\delta^{18}O_{(P-G)}$ values implying decreased mixed layer depth, and vice versa for smaller $\Delta\delta^{18}O_{(P-G)}$ values.

Paleo-Thermocline Depth

Thermocline Evolution on Tectonic Timescales Since the Late Miocene

Abundance variations of deep-dwelling planktonic foraminiferal species have been used to estimate relative thermocline changes in the tropical and subtropical Pacific (Kennett et al. 1985) as well as in the SCS. Based on the relative abundances of the total deep-dwelling species and such species groups as *Neogloboquadrina* spp., *Pulleniatina* spp., and *Globoquadrina dehiscens* at Sites 1143 and 1146, Li B. et al. (2004) revealed the relative thermocline evolution in the northern and southern SCS for the past 12 myr.

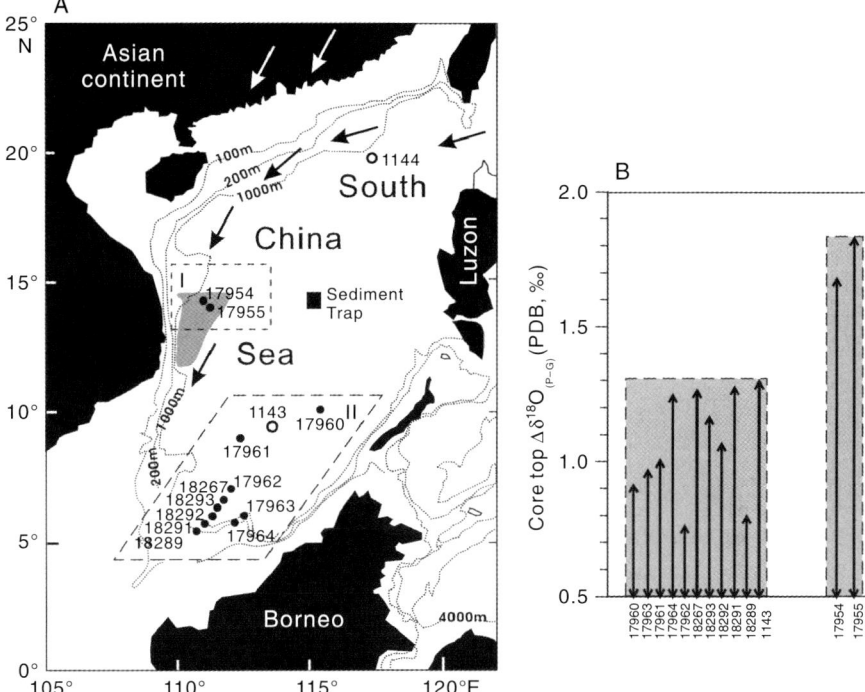

Fig. 5.16 $\delta^{18}O$ differences between subsurface-dwelling *P. obliquiloculata* and surface dwelling *G. ruber* ($\Delta\delta^{18}O_{(P-G)}$) (**B**) in core top samples can be used as an indicator of thermocline depth changes. (**A**) Locations of coretop sites (*black dots* = piston cores, *open circles* = drill holes). *Arrows* denote the direction of the East Asian winter monsoon wind and surface circulation. *Black solid square* denotes the location of the sediment trap in the central SCS. Cores 17954 and 17955 from summer upwelling region off Vietnam have larger $\Delta\delta^{18}O_{(P-G)}$ values (Tian et al. 2005b)

As shown in Fig. 5.17, a major decrease in the abundance of deep-dwelling planktonic foraminifera between 11 and 9 Ma at both sites, centered at 10 Ma, may indicate a deeper thermocline in the SCS after 10 Ma. Lower total abundances of deep-dwelling planktonic foraminiferal species from 10 to 8.2 Ma (Site 1143) and 8.58 Ma (Site 1146) suggest that a deeper thermocline persisted until 8.2 and 8.58 Ma in the southern and northern SCS, respectively. The thermocline in the western Pacific is 50–100 m deeper than in the eastern Pacific (Levitus and Boyer 1994). It has been suggested that the closure of the Indonesian Seaway played an important role in the early formation of the WPWP (Kennett et al. 1985). The closure of the Indonesian Seaway could have contributed greatly to WPWP evolution owing to the piling up of warm surface water in the western equatorial Pacific and the strengthening of the Equatorial Under Current (Maier-Reimer et al. 1990; Hirst and Godfrey 1993). Accordingly, the synchronous thermocline deepening at 10 Ma at sites 1143 and 1146 likely has been caused by initial formation of the WPWP due to the increased closure of the Indonesian Seaway in the early part of the late Miocene.

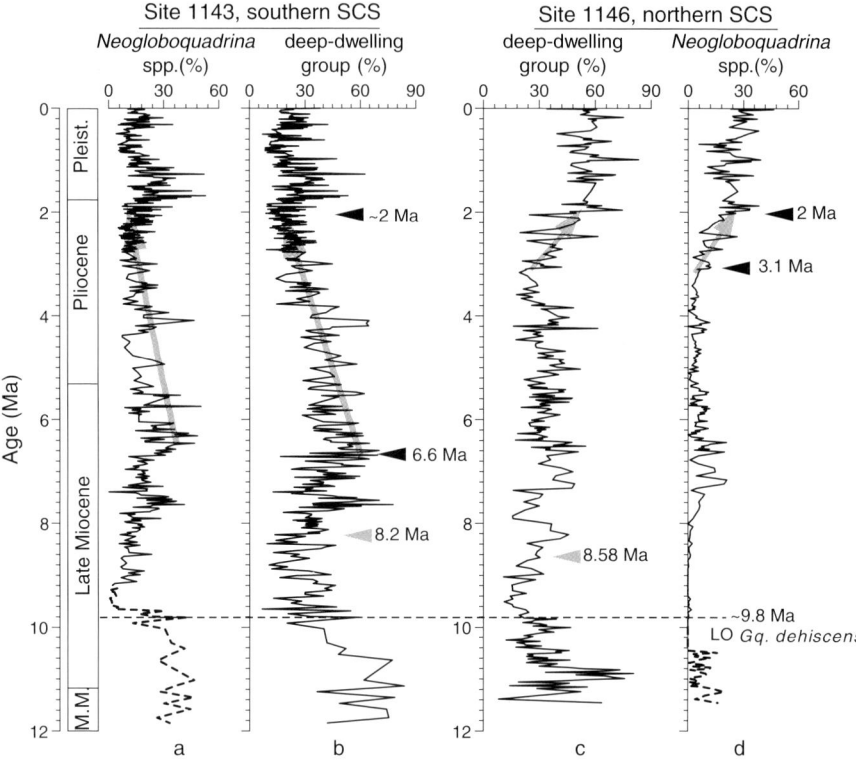

Fig. 5.17 Down-core variations of *Neogloboquadrina* spp. and total deep-dwelling planktonic foraminifera at Site 1143 (**a, b**) and Site 1146 (**c, d**) show different trends (Li B. et al. 2004). *Neogloboquadrina* spp. (*solid line*) includes *Neogloboquadrina dutertrei, Neogloboquadrina pachyderma, Neogloboquadrina humerosa*, and *Neogloboquadrina acostaensis*. *Dashed lines* indicate the abundance of *Globorotalia siakensis-Globorotalia mayeri* complex. M.M. = Middle Miocene; LO = last occurrence

Globoquadrina dehiscens was a typical subsurface species living mainly in warm temperate waters during the Miocene, which became extinct from the world oceans at 5.6 Ma near the Miocene/Pliocene boundary (Kennett et al. 1985; Berggren et al. 1995). In the SCS, however, the faunal data from Sites 1143 and 1146 show that the last occurrence (LO) of *Gq. dehiscens* at core depths of 470 m at Site 1143 and 412 m at Site 1146 corresponds to an age at ∼9.8 Ma (Fig. 5.17). This event has also been reported from numerous petroleum wells on the northern shelf of the SCS, such as BY 7-1-1 (Qin 1996). The LO of *Gq. dehiscens* at these industrial sites is close to the zone N15/N16 boundary, and is accompanied by the first occurrences of *Neogloboquadrina acostaensis* and *Globorotalia merotumida*. The disappearance of *Gq. dehiscens* after 10 Ma from both the southern and northern parts of the SCS probably provides additional evidence for the initial development of the WPWP during the early part of the late Miocene (Li Q. et al. 2006). The pile up of warm water during the formation of the WPWP probably eliminated the temperate water dwellers such as *Gq. dehiscens* that characterized the fauna before 10 Ma in the SCS.

At approximately 6.6 Ma the deep-dwelling planktonic foraminiferal species at Site 1143 constituted more than 60% of the total fauna, reflecting a shallower thermocline at that time. Later, it gradually decreased to the lowest abundance values of 10% at 2 Ma (Fig. 5.17b), a trend also observed in the *Neogloboquadrina* group (Fig. 5.17a). Site 1143 now lies within the modern WPWP, so thermocline variations at this locality should have responded to the warm pool development. These faunal decreases imply a thermocline deepening after 6.6 Ma in the southern SCS, probably related to the evolution of the western Pacific hydrography toward modern "warm pool" conditions after 6.6 Ma.

At Site 1146 in the northern SCS, the total deep dwelling species and species of *Neogloboquadrina* did not change significantly until 3.1 Ma (Fig. 5.17c and d). *Neogloboquadrina* spp. increased markedly over the 3.1–2.0 Ma period, from less than 10% to more than 20% of the fauna, and increased further to 30% in the Pleistocene. The total deep-dwelling species also increased from 25 to 55% over the same period, reaching the highest value of >83% at 1.0 Ma in the middle Pleistocene. These abundance increases at Site 1146 reflect a sudden shoaling of the thermocline after 3.1 Ma that continued into the Pleistocene in the northern SCS. These patterns contrast sharply with those at Site 1143, indicating that different controlling factors had been operating on the upper waters of the southern and northern parts of the SCS since the late Miocene. The faunal data from Site 1146 also suggest that the upper water conditions analogous to the modern northern SCS first occurred in the middle Pliocene. The shoaling thermocline from 3.1 to 2 Ma at Site 1146 thus indicates an intensified East Asian winter monsoon during that period, consistent with the eolian sediment records from central China (An et al. 2001).

Pleistocene Thermocline Variations on Orbital Scales in the Southern SCS

Transfer function derived thermocline change. Compared to the Neogene history discussed above, the Pleistocene thermocline depth can be estimated quantitatively using foraminiferal transfer function. The very deep dwelling species *Globorotalia truncatulinoides* has an unusual life cycle and is very useful in the reconstruction of the upper thermal structure. It reproduces at ∼600 m and from this depth juveniles rapidly travel to the surface then sink slowly through the water column, growing by adding chambers (Bé 1977; Bé et al. 1985; Hemleben et al. 1989). A higher proportion of this species probably indicate a very deep thermocline and/or thick mode water thermostads (Lohmann and Schweitzer 1990; Ravelo and Fairbanks 1992; Martinez 1994, 1997). In core 17957-2 from the southern SCS, the percentage abundance of *G. truncatulinoides* shows a trend of gradual decrease during the Brunhes chronozone which indicates water mixing and the depth of thermocline (DOT) gradually decreased (Fig. 5.18). Especially, the *G. truncatulinoides* left-coiling form, which requires a thermocline much deeper than the right-coiling form, and *Globoquadrina conglomerata* abruptly increased in MIS 5, indicating that the thermal structure of upper water column greatly changed during at that time.

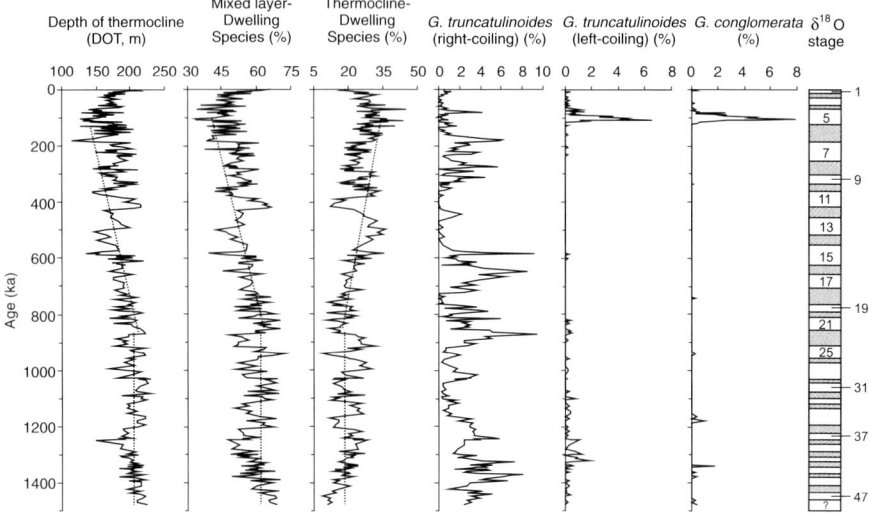

Fig. 5.18 Pleistocene thermocline history in the southern SCS is revealed in faunal proxies from core 17957-2, including the depth of thermocline (DOT) using transfer function of Andreasen and Ravelo (1997), relative abundance of mixed layer dwelling species, thermocline dwelling species, right-coiling and left-coiling *G. truncatulinoides*, and *Globoquadrina conglomerata*. Also indicated are marine isotope stages (numbers) and the general trends of changes before and after MIS 22 – 21 (*dashed lines*) (from Jian et al. 2000a)

The average Holocene DOT estimates in core 17957-2 is ~190 m, agreeing with the modern annual DOT derived from the world ocean atlas for this region (Levitus and Boyer 1994). The estimated downcore DOT ranges from 115 to 230 m, relatively deeper during interglacials than during glacials. Before MIS 22-21 or before the mid-Pleistocene revolution (MPR), DOT changed little around 200 m. Later it gradually decreased during the Brunhes chronozone, with an average of 180 m. The shallowest DOT of ~115 m occurred within MIS 6 before deepening again after abrupt increases in the *G. truncatulinoides* left-coiling form and in *G. conglomerata* during MIS 5.

In responding to DOT shoaling, the abundance of mixed layer dwelling species reduced while thermocline dwelling species increased (Ravelo et al. 1990; Ravelo and Fairbanks 1992). The mixed layer dwelling species in core 17957-2 decreased in abundance since the MPR and reached a minimum during MIS 6-5, while the thermocline dwelling species changed in an opposite trend (Fig. 5.18), reflecting a shoaling DOT, a phenomenon also observed in the western equatorial Pacific during the same time by the $\delta^{18}O$ difference between *G. sacculifer* and *P. obliquiloculata* (Schmidt et al. 1993).

However, the transfer function-derived DOT changes do not always display similar patterns of glacial/interglacial changes in different parts of the SCS. The prevailing monsoon system coupled with Ekman effect results in winter upwelling off the northwestern edge of the Philippines and summer upwelling along the Vietnam coast (Chapter 2). The winter upwelling region centered about 100 km offshore between 16 and 19 °N off the northwestern Philippines had been revealed by the distributions of temperature, salinity and dissolved oxygen concentration, and also by the tracer distribution obtained from a numerical experiment (Chao et al. 1996; Shaw et al. 1996). The summer upwelling is also predicted from a climatology-driven circulation model in response to summer monsoonal winds (see Chapter 2). Late Quaternary records of monsoon-driven coastal upwelling can be highly promising for reconstructing the past thermocline changes and hence the East Asian monsoon variability. For example, shoaled thermocline and enhanced organic carbon flux have been inferred as result of intensified upwelling off eastern Vietnam during interglacials and off the northwestern Philippines during glacials (Jian et al. 2001; Fig. 5.19), showing a seesaw pattern of the DOT changes during

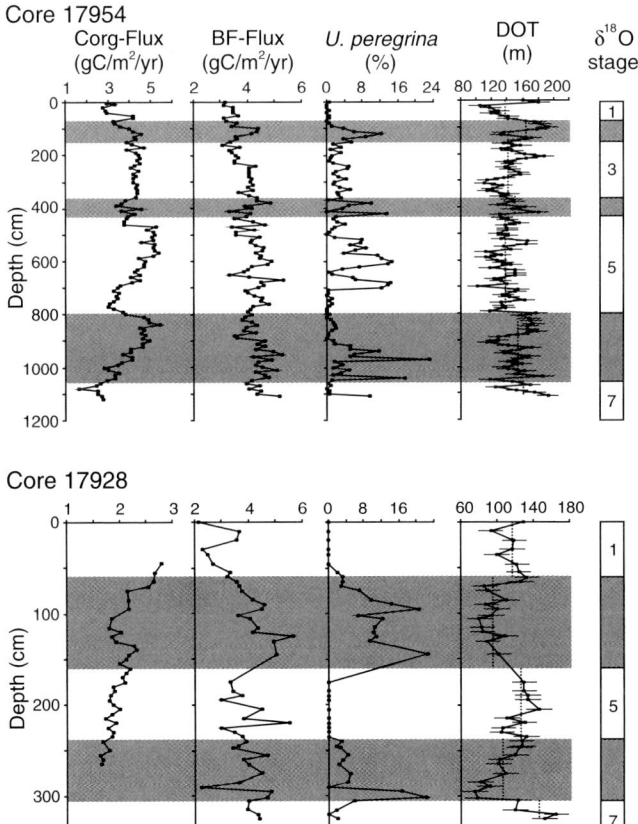

Fig. 5.19 Downcore variations in paleoproductivity indicators are compared between core 17954 off Vietnam and core 17928 off northwestern Luzon: C_{org}-Flux (organic carbon flux to the seafloor based on organic carbon content using the equation of Sarnthein et al. 1992), BF-flux (organic carbon flux to the seafloor based on the relative abundance data of benthic foraminifera using the technique of Kuhnt et al. 1999), the relative abundance of benthic foraminifer *Uvigerina peregrina* indicative of high organic carbon flux, and the depth of thermocline (DOT) with error bars (*short horizontal lines*) and average DOT for each MIS (from Jian et al. 2001)

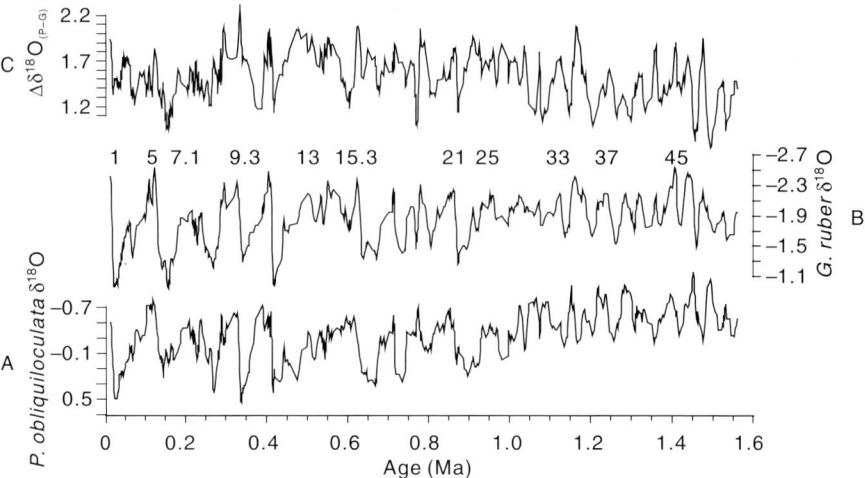

Fig. 5.20 Last 1.56 myr δ^{18}O records at ODP Site 1143 from the southern SCS after 3-point Gaussian smoothing show variations in the upper water structure: (**A**) δ^{18}O of *P. obliquiloculata* (PDB, ‰,), (**B**) *G. ruber* δ^{18}O (PDB, ‰), and (**C**) $\Delta\delta^{18}$O$_{(P-G)}$ (*P. obliquiloculata* δ^{18}O minus *G. ruber* δ^{18}O) (PDB, ‰) (from Tian et al. 2005b)

glacial/interglacial cycles from the east to the west. The seasaw pattern of the DOT changes is probably unique to the SCS due to the monsoon influences in both summer and winter seasons.

δ^{18}O gradient derived thermocline change. An alternative way of thermocline reconstruction is by $\Delta\delta^{18}$O$_{(P-G)}$. Its values at ODP Site 1143 were reduced by 0.5–1.0‰ during glacials or stadials compared to the adjacent interglacials or interstadials, suggesting that the thermocline depth was deeper during glacial and stadial periods over the last 1.56 myr (Fig. 5.20) (Tian et al. 2004). On the contrary, during interglacial or interstadial periods, large $\Delta\delta^{18}$O$_{(P-G)}$ values imply a shallower thermocline. Exceptions to this general pattern are rare, including relatively high $\Delta\delta^{18}$O$_{(P-G)}$ values for MIS 16 and relatively low values for MIS 17 (Fig. 5.20).

However, the estimated DOT changes using these two methods show two opposite patterns for the early Pleistocene glacial/interglacial cycles in the southern SCS. The δ^{18}O gradient between *P. obliquiloculata* and *G. ruber* at Site 1143 exhibits deeper thermocline during glacials and shallower thermocline during interglacials throughout the Pleistocene (Fig. 5.20C), whereas transfer function derived DOT shows entirely different trends before and after the mid-Pleistocene revolution (MPR) around 0.9Ma mostly due to the abnormally high abundance of *P. obliquiloculata* in post-0.9 Ma glacial times. *P. obliquiloculata* is a typical thermocline dweller and its abundance heavily influences the accuracy of planktonic foraminiferal transfer functions. Its abundance shows higher values in interglacials before MPR but higher values in glacials after MPR in the southern SCS, such as at ODP Site 1143 (Xu et al. 2005), SONNE cores 17957-2 (Jian et al. 2000b) and 17962 (Fang et al.

2000). Although the transfer function techniques of Andreasen and Ravelo (1997) were successfully applied to reconstruct past thermocline depths in the open tropical Pacific, these techniques should be applied with caution in the southern SCS because of the potential no-analog behavior of planktonic foraminiferal species, especially of *P. obliquiloculata*.

Trans-Pacific Comparison and the Western Pacific Warm Pool

The upper ocean structure across the tropical Pacific is characterized by marked gradients of SST and thermocline depth, which are affected by ENSO (El Niño-Southern Oscillation) on different time scales. During normal conditions, the tropical eastern Pacific is bathed with cold surface water and a shallow thermocline, and the western Pacific with warm surface water and a deep thermocline. However, during El Niño years, the slopes of the seesaws of SST and thermocline across the equatorial Pacific were reduced, as a result of thermocline deepening and surface water warming in the east and thermocline shoaling and surface water cooling in the west. The SST records derived from the planktonic Mg/Ca ratio at ODP Site 806 in the west and at ODP Site 847 in the east well document the evolution of the trans-Pacific equatorial SST gradient over the last 5 myr and reveal permanent El Niño-like conditions in the late Pliocene period (Ravelo et al. 2004; Wara et al. 2005). The SST records show that the SST gradient across the equatorial Pacific was only 1.5 °C before ~2 Ma, very similar to the situation during a modern El Niño event. However, the SST gradient reached as high as 4–5 °C just after 2 Ma. In addition, the proxy records of thermocline changes using the planktonic $\delta^{18}O$ difference at ODP Site 847 reveal significant thermocline shoaling or water cooling in the eastern equatorial Pacific at ~3.5 Ma, much earlier than the abrupt increase in the SST gradient. Similar changes in planktonic $\delta^{18}O$ differences are also reported from ODP Site 851, another site from the eastern equatorial Pacific (Cannariato and Ravelo 1997).

The 28 °C isotherm, which constrains the northern boundary of the Western Pacific Warm Pool (WPWP), separates the SCS into two parts from the northeast to the southwest. The thermocline in the modern SCS shoals from ~175 m in the south to ~125 m in the north (Levitus and Boyer 1994), a seesaw pattern similar to that across the West and East Pacific. Variations in the relative abundances of the deep-dwelling planktonic foraminifera at ODP Site 1146 and at ODP Site 1143 reflect the evolution of the thermocline depth gradient across the northern and southern SCS since the late Miocene (Jian et al. 2006) (Fig. 5.21). The thermocline gradient between the northern and southern SCS probably enhanced for the first time during the 11.5–10.6 Ma, as indicated by an opposite change in the relative abundance of deep-dwelling planktonic foraminifera between Site 1146 and Site 1143. Between 10.6 and 4.0 Ma was a period with weakened thermocline gradient between the two sites. More significant increases in the thermocline gradient occurred at about 4.0–3.2 Ma, as marked by a jump in the abundance of deep-dwelling species at Site 1146 but a major decrease at Site 1143 (Jian et al. 2006). The faunal evidence of thermocline deepening from ODP Site 1143 at about

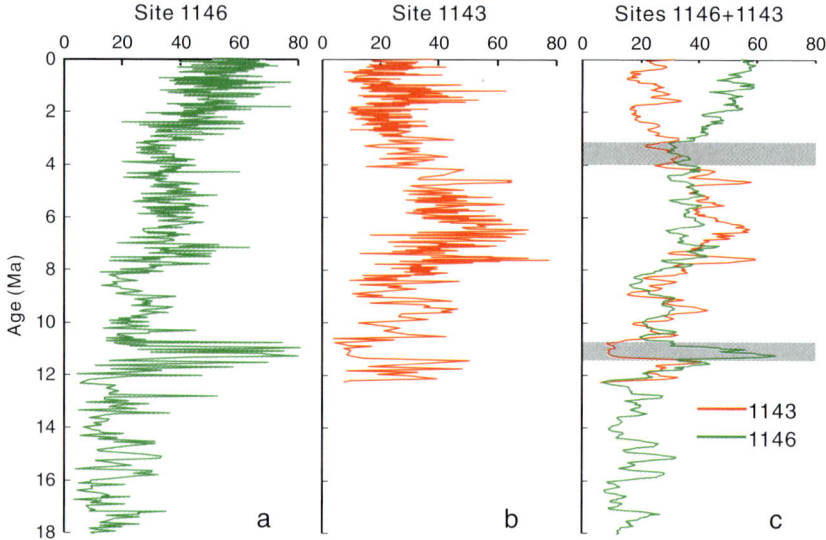

Fig. 5.21 Changes in the abundances (%) of planktonic foraminiferal deep-dwelling species at ODP Site 1146 (**a**) and Site 1146 (**b**) show increased south-north thermocline gradient (**c**, 5-point moving average) at ∼11 Ma and since ∼3 Ma (*horizontal bars*) in the SCS (Jian et al. 2006)

4.0–3.2 Ma matches well with the record thermocline shoaling in the east equatorial Pacific around ∼3.5 Ma (Wara et al. 2005). The thermocline variations between the eastern and western equatorial Pacific reflect the long-term evolution of climate conditions in the pan-Pacific region. For example, the similarly decreased thermocline gradients across the east-west equatorial Pacific in the late Pliocene and in the modern El Niño years probably reflect the El Niño-like climate conditions during the late Pliocene and subsequent weakening. The development of a S-N thermocline gradient in the SCS since the Pliocene is comparable to the trans-Pacific east-west thermocline variations, indicating the long-term evolution of the WPWP.

5.3 Vegetation History in Deep-Sea Record (Sun X.)

Pollen grains in marine sediments are originating from vegetation on land and then buried in the sea after being transported by winds or currents. Deep-sea palynology yields terrestrial climate signals among the ocean records and bridges the paleoenvironmental studies across from land to sea. Pollen grains derived from the deep-sea, however, are typically transported over long distance and they integrate palynological information over a large area. Therefore, the use of pollen as climate and vegetation proxy depends on adequate knowledge of its modern distribution. A survey of modern pollen in surface sediments is a prerequisite for paleoenvironment interpretation of the SCS sediment records.

Pollen Distribution in Surface Sediments

A total of 40 surface sediment samples from the SCS were investigated for pollen (Fig. 5.22) (Sun et al. 1999). The results are calculated and summarized by isopoll maps of pollen concentration (grains/g) and pollen percentage, which display distribution patterns of different pollen types related to their dispersal routes and mechanisms in the SCS. The concentrations of total pollen, total tree pollen, pine and fern spores bear very similar distribution patterns. Their highest concentrations occur in the northern part of the SCS, adjacent to Taiwan and Bashi Straits and decrease towards southwest along the direction of winter monsoon and surface currents. Their concentration isopoll figures are stretched as a tongue, extending from the Bashi trait towards the southwest. This can be illustrated with *Pinus* concentration isopoll map (Fig. 5.22A), and is most likely resulting from the northeastern winter monsoon which brings the conifer pollen from the southeast part of mainland China and Taiwan to the sea before being transported further afield by surface currents (Sun et al. 1999). This interpretation is confirmed by the depositional process recorded by sedimentation traps in the northern SCS (Su and Wang 1994). On the other hand, pollen concentrations of tropical/subtropical trees and herbs are very low and their values decrease from the continental shelf to the deep sea, implying their dispersal from the coastal areas of south China and Taiwan mainly by river discharges (Fig. 5.22B).

The pollen concentrations in the southern SCS are much lower, only one tenth of those of the northern part, and decrease from the coastal areas of islands (mainly Borneo) to the deep-sea. This pollen distribution pattern indicates that the pollen source areas in the southern SCS are the islands east of the Sunda Shelf, mainly Borneo, and pollen grains reach there chiefly by river transport (Sun et al. 1999).

Pollen percentage results display that, along the continental shelves of the northern and southern SCS, the fern spores are the dominant component, respectively reaching 80% and 70% of total pollen and spores sum. The subdominant components are trees, reaching 40% in the northern, southern, and eastern margins of the SCS but diminishing seaward. Percentages of herb pollen are quite low, contributing less than 10% of the total in near 90% of samples.

Among tree pollen, *Pinus* is found to be prevalent in the surface sediments from the northern SCS, contributing 80% in most of samples, but it is only less than 10% in the southern SCS (Fig. 5.22C). Its maximal values of concentration occur in the northeast SCS and stretch as a saddle from NE to SW, being consistent with the direction of the NE winter monsoon and the surface current. Pollen of tropical and subtropical complex reaches 60% in the southern part, but only <1% in the north (Fig. 5.22D). Mangrove pollen is abundant in the southern part, reaching a maximum of 20%, but only ~1% in the northern part (Fig. 5.22E).

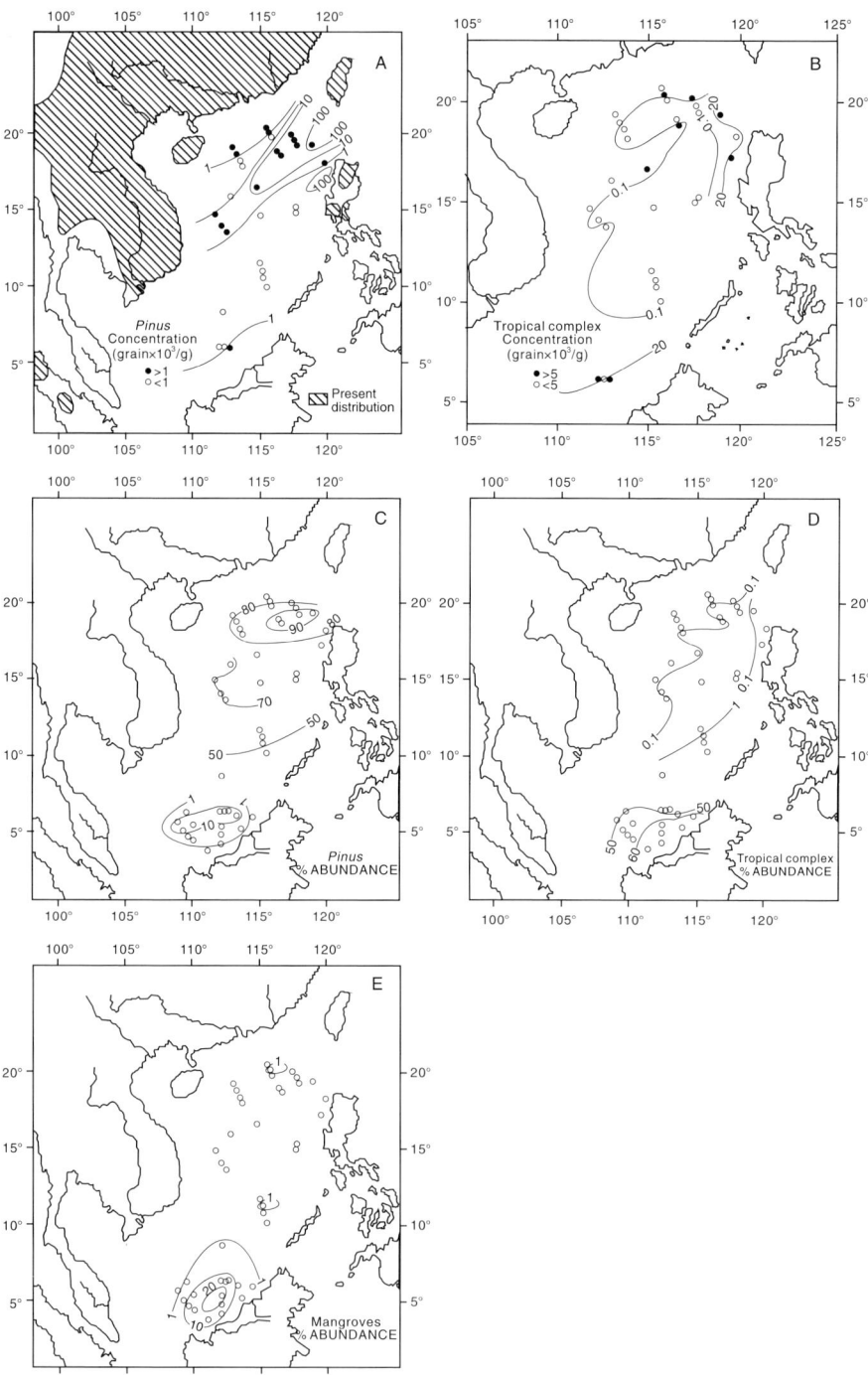

Fig. 5.22 Maps show concentration isopolls (10^3/g.dry sediment) and relative abundance isopoll (%) of (**A**) *Pinus* concentration, (**B**) tropical complex concentration, (**C**) *Pinus*%, (**D**) tropical complex %, and (**E**) mangroves% in the modern SCS (modified from Sun et al. 1999)

Long-Term Evolution

All of the pollen sequences discussed here were based on samples retrieved during cruises "Sonne 95" in 1994 (Sarnthein et al. 1994), "Sonne 115" in 1996 (Stattegger et al. 1997) and "ODP Leg 184" in 1999 (Wang P. et al. 2000) (see Table 5.1, Fig. 1.1). Of those only the ODP cores enabled us to study the long-term vegetation history before the late Pleistocene. Because of the extremely low pollen concentrations in a number of stratigraphic intervals, only the early-to-middle Oligocene, late Miocene-Pliocene, and Pleistocene pollen sequences are discussed here.

Oligocene

The longest deep-sea record from the SCS was recovered at ODP Site 1148 from the northern lower continental slope (at modern water depth of 3,294 m), but from the entire profile only the Oligocene section yielded sufficient quantity of pollen for a statistically meaningful study. As seen from Fig. 5.23, the Oligocene assemblage is dominated by montane conifer tree pollen. The broad-leave tree pollen groups include mainly tropical-subtropical components. A distinct change in pollen assemblage occurred at around 32.0 Ma when the temperate montane conifer and cool and drought-enduring deciduous tree taxa remarkably increased. Therefore, the pollen data suggest that tropical montane rainforest and lowland rainforest developed in the neighboring areas before 32.0 Ma, but the climate turned to be cooler and drier after 32.0 Ma. This climate change is well correlated with previous observations during oil exploration in the Zhujiangkou (Pearl River Mouth) Basin (Wu et al. 2003).

Noteworthy is the occurrence of abundant coastal and neritic dinoflagellate cysts together with the Oligocene pollen (Mao et al. 2007), which seem to be incompatible with the deep-water benthic fauna from the same samples (see Chapter 6). This unusual combination of coastal dinoflagellates with deepwater ostracoda and foraminifera is most probably associated with the narrow gulf shape of the early SCS basin at the early stage of seafloor spreading. Since then, the depth of the SCS basin increased during its further opening, as evidenced by the changes in benthic microfauna, accompanied by increased abundance of oceanic dinoflagellate species upward in the profile (Mao et al. 2007).

Late Miocene-Pliocene

The late Miocene-Pliocene pollen sequence is based on 380 samples from the depth interval of 76–512 mcd (meter composite depth) at ODP Site 1143, southern SCS. Four pollen zones are defined from the pollen diagrams (Fig. 5.24) (Luo and Sun 2007):

Table 5.1 Location, water depth, core length, chronology, sampling interval, time resolution and references of studied pollen sites are listed

Site	Location	Water Depth (m)	Core Length (m)	Age	No. of samples	Sampling Interval (cm)	Resolution (y)	Reference
ODP1144	20°3.18'N 117°23.0'E	2037	520	1.03 Ma	1250		820 (154–1160)	Sun and Luo 2001 Luo et al. 2001 Sun et al. 2003 Luo et al. 2005
17940	20°07'N 117°23.0'E	1727	13.30	37 ka	102	10	360	Sun and Li 1999 Sun et al. 1999 Sun et al. 2000b
ODP 1148	18°50'N 116°34'E	3294	850	33 Ma	169		43000	Wu et al. 2003 Mao et al. 2007
ODP1143	9°22'N 113°17'E	2772	510	~12 Ma	380	28–220	7000–52500	Luo and Sun 2007
17962	7°11'N 112°5'E	1970	8.0	30 ka	82	10	365	Sun et al. 2002
17964	06°09.5'N 112°2.8'E	1556	13.03	26 ka	62	20	420	Li and Sun 1999
18287	5°39'N 110°39'E	598	5.66	16.5 ^{14}C ka	112	5	120	Wang X. et al. 2007
18300	4°21'N 108°39'E	91	8.85	39.2 ^{14}C ka	52	20	800	Wang X. 2006
18302	4°09'N 108°34'E	83	5.98	20.16 ^{14}C ka	41	20	500	Wang X. 2006
18313	3°52'N 108°52'N	98	6.2	No record	12	40		Wang X. 2006
18323	2°47'N 107°53'E	92	5.4	31.27 ^{14}C ka	50	20	600	Wang X. 2006

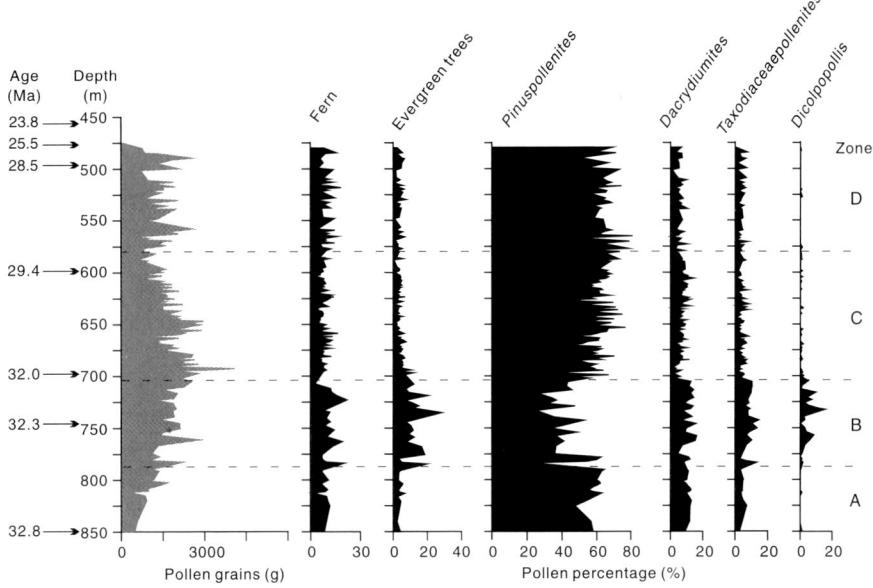

Fig. 5.23 Pollen concentration and percentage diagrams of selected palynomorph components in the Oligocene section of ODP Site 1148 show a major change at ∼32 Ma (based on Wu et al. 2003)

P1 (521–413 mcd, 11.9–8.15 Ma) is characterized by very low pollen influx (48 grains/cm^2/kyr in average). *Pinus* (52.1%) is the dominant taxon, and the tropical lowland rainforest pollen (16.5%) comes second. The percentages of fern spores are very high, occupying 107.9% of the total land seed plants.

P2 (413–171 mcd, 8.15–4.29 Ma) is characterized by abruptly increased pollen influx (1648 grains/cm^2/kyr in average), which is nearly 35 times more than those in P1. Cool-tolerant pollen groups increased during this stage, at the cost of warm-tolerant groups. But pollen of the tropical lowland rainforest (6.5%) and fern spores (62.6%) dropped sharply.

P3 (171–122 mcd, 4.29–2.63 Ma) is characterized by decreased pollen influx (85 grains/cm^2/kyr) and dramatically increased fern spores (329.4%). During this stage, pollen of the temperate forest fell sharply (2.1%), and the mangrove pollen appeared in very low percentages.

P4 (122–76 mcd, 2.63–1.58 Ma) is characterized by a second rise in pollen influx (194 grains/cm^2/kyr) with strong fluctuations. The lowland rainforest pollen and especially the ferns (329.4%) increased in percentages. More mangrove pollen (1.3%) were found during this stage.

A distinct boundary occurs at 8.1 Ma when pollen influx values increased abruptly. The cool-tolerant montane and temperate groups also increased while percentages of tropical lowland group and fern declined. These dramatic changes in pollen influx presumably resulted from exposure of the Sunda Shelf from 8.1 to 4.3 Ma because of the sea level drop.

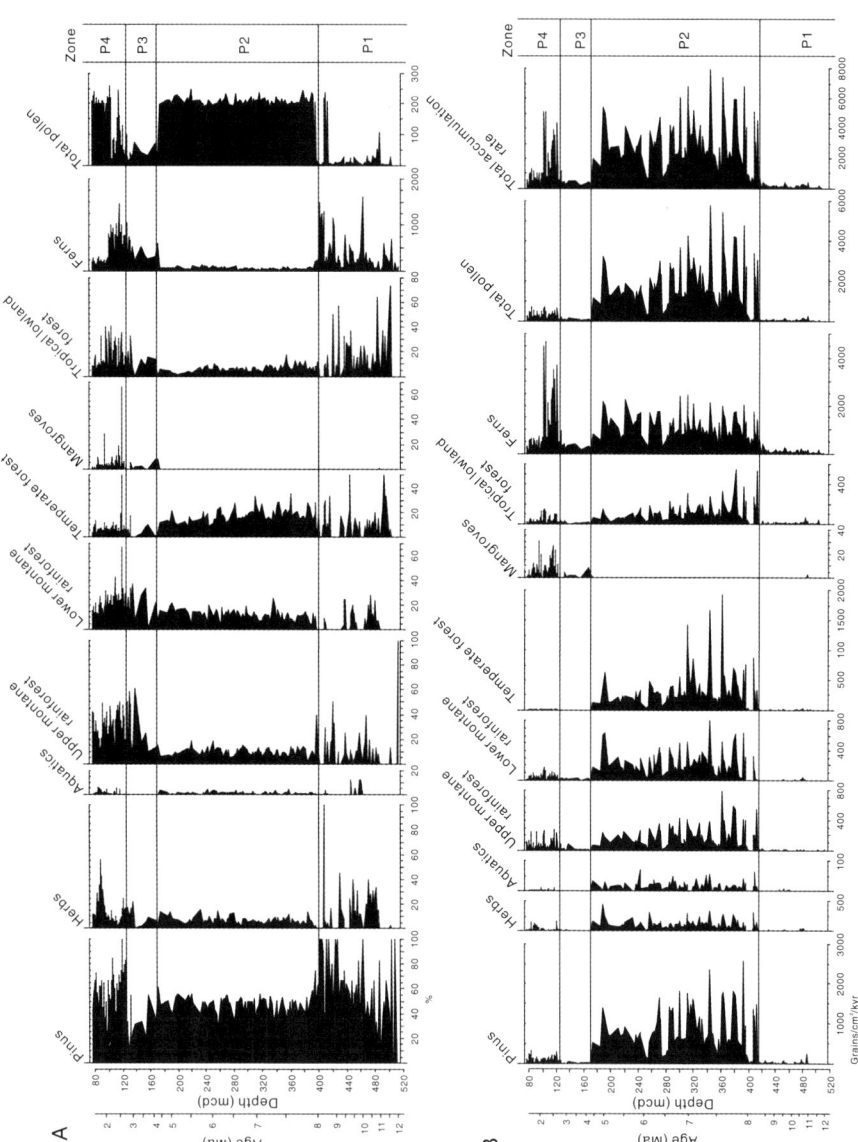

Fig. 5.24 (continued)

The increase of cool-tolerant group probably indicates a climate cooling since ~8 Ma, in consistent with loess sequence results and other records. The loess records demonstrate a massive eolian deposition since 7–8 Ma, implying arid climate and desert environment developed in the late Miocene (Sun and An 2001). The terrestrial pollen records demonstrate that a bloom of grass vegetation occurred during time intervals of 8.5–6.0 Ma in the Jiuxi Basin (Ma et al. 2004), 8.6–8.4 Ma and 6.9–6.6 Ma in the Linxia Basin (Ma et al. 1998) from the northwestern China, indicating dry and cold climate. The pollen influx at Site 1143 decreased from 4.3 to 2.6 Ma, probably related to sea level rise and submergence of the Sunda Shelf. Since ~2.6 Ma, the pollen influx increased again and fluctuated in a wide range. A rough comparison of pollen influx values with $\delta^{18}O$ data indicates that the time intervals with high influx values correspond to glacial periods with heavy $\delta^{18}O$ values and the low influx values are correlated with interglacial periods with lighter $\delta^{18}O$ values (Luo and Sun 2007).

Pleistocene

The best palynological record of the Pleistocene in the SCS is from ODP Site 1144, northern SCS. It provides a long pollen sequence spanning the past 1.3 myr, with an average time resolution of about 820 years. A total of 1250 pollen samples were analyzed for the 504 m long sequence. Chronology of the sequence is based on micropaleontology and magnetostratigraphy (Wang P. et al. 2000), as well as the $\delta^{18}O$ record of *G. ruber* (Bühring et al. 2001). Sediments of the profile are almost continuous except for a hiatus at 196.64 m where MIS 8 is almost completely missing, and two short-term hiatuses respectively within MIS 5.5 and MIS 11.31. Often, lighter $\delta^{18}O$ stages are correlated to pine-dominant pollen zones assigned to interglacials, and heavier $\delta^{18}O$ stages correspond to herb-predominant pollen zones belonging to glacials. On this basis, a total of 29 pollen zones have been recognized which almost completely coincide with isotopic stages MISs 1–29 (Fig. 5.25).

A total of 174 pollen types have been identified, but except for pine and some herbs, most of the pollen types contain very few grains, in particular for some tropical and subtropical taxa (Fig. 5.26). *Pinus* pollen dominates pollen assemblages throughout the profile, followed by herbs in percentages. These two types together can reach up to 80% of the total pollen sum of land seed plants, though the downcore variations are very significant. *Pinus* almost always shows higher percentages in the interglacial (varying from 39 to 81%) than in the adjacent glacial stage (30–54%). Herbs pollen includes 29 pollen components, with the most important components from *Artemisia*, Poaceae and Cyperaceae. Pollen from Chenopodiaceae and other Asteraceae are moderate, and all other taxa are very low in percentage. In contrast

Fig. 5.24 Late Miocene–Pliocene pollen diagrams (**A**, pollen percentages; **B**, pollen influx) from ODP Site 1143 can be divided into 4 pollen zones, indicating major changes at 8.1, 4.3 and ~2.6 Ma (from Luo and Sun 2007). Pollen percentages are based on total pollen numbers of seed land plants, while pollen influx represents numbers of pollen grains/cm^2/kyr

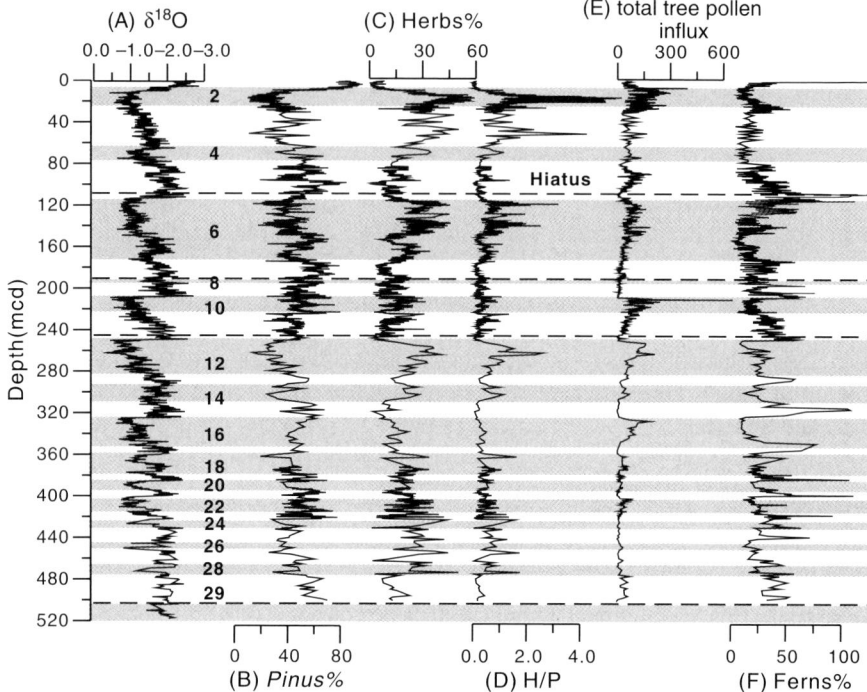

Fig. 5.25 Glacial/interglacial contrasts in pollen assemblages (B to F) are compared to *G. ruber* $\delta^{18}O$ record (A) at ODP Site 1144, northern SCS (from Sun et al. 2003). Numbers denote the marine isotope stages. Dashed horizontal lines mark the presence of hiatus

to *Pinus*, the percentages of herbs and their main taxa are low during interglacial periods (4.9–25.5%), but considerably higher during glacials (12.8–32.7%) (Sun et al. 2003).

The tropical and subtropical group includes a large number of taxa, but only a few grains of each taxon can be encountered per sample, except *Quercus* (evergreen type). This group ranks third, inferior to *Pinus* and herbs in pollen percentages, and ranges from 6.3 to 20%. No distinct glacial/interglacial variations in its percentage can be found, although the middle part of the profile (from MIS 19 to 12) bears

Fig. 5.26 Pollen diagram of ecological groups and selected taxa from ODP Site 1144 are shown by (**A**) pollen percentages, calculated on the total pollen sum of land seed plants, and (**B**) pollen influx (grains/cm^2/yr) (from Sun et al. 2003). Ecological groups shown in the figure are: Herbs (*Artemisia*, Poaceae and Cyperaceae as the main components, with sparse Asteraceae, Chenopodiaceae etc.); Boreal conifers (*Picea, Abies*, and *Tsuga*); Tropical upper mountain group (*Podocarpus, Dacrycarpus, Dacrydium* and *Phyllocladus*); Temperate deciduous group (*Betula, Alnus, Carpinus, Juglans, Ulmus* etc.); Tropical and subtropical evergreen group (*Quercus, Altingia, Ilex, Castanopsis/Lithocarpus, Mallotus/Macaranga*, Euphorbiaceae, Palmae, Melastomataceae, Meliaceae, Euphorbiaceae, Moraceae); Mangroves (mainly *Rhizophora* and *Sonneratia*) and Aquatics (*Typha, Myriophyllum, Nymphoides*)

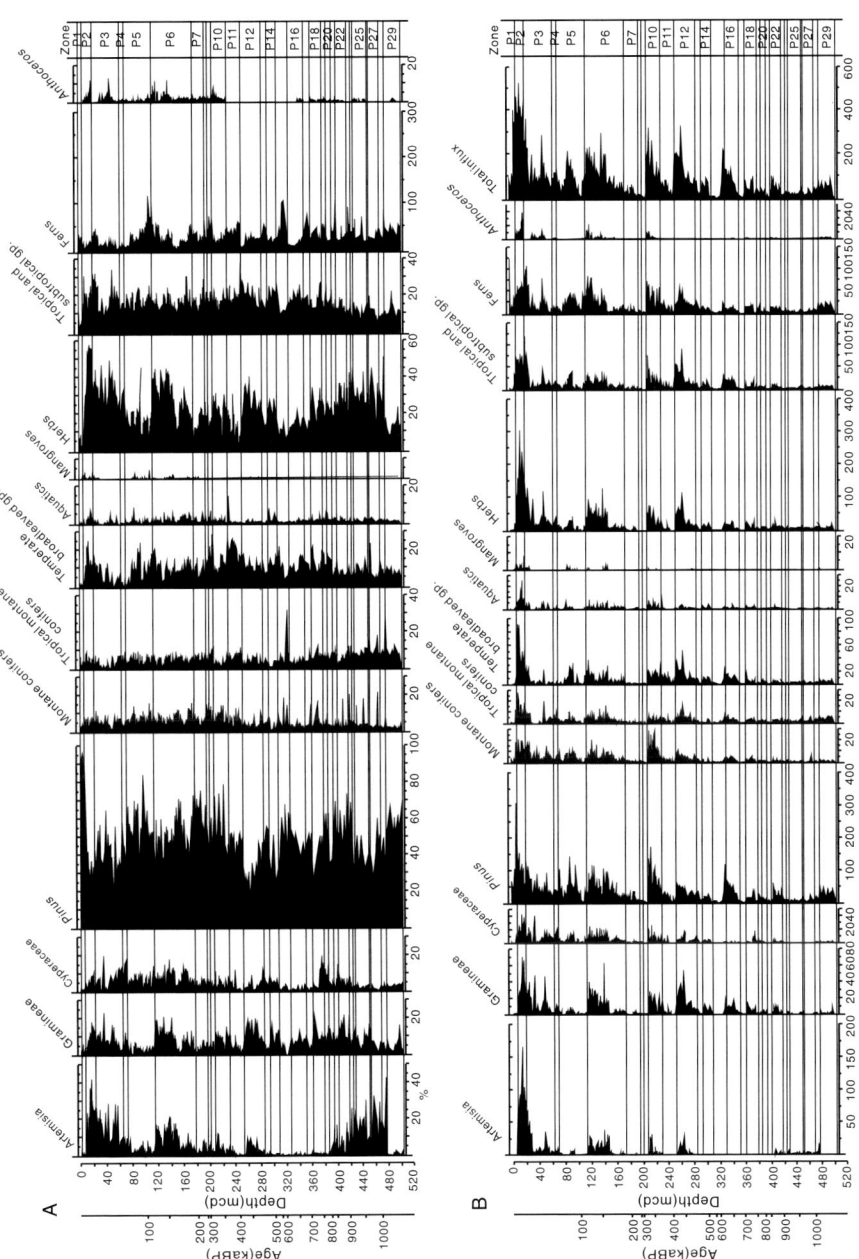

Fig. 5.26 (continued)

higher percentages than the upper and lower parts. Mangroves, of which *Rhizophora* is the most important, is very low in percentage, ranging from 0 to 0.3%. Mangrove pollen is almost completely absent before MIS 12, occur only in trace numbers since MIS 12, but increase to maximum values in the LGM and the Holocene (MIS 2 and 1). *Anthoceros*, the dwarf liverwort genus is also recorded, with insignificant numbers in interglacial intervals. Aquatic plants occur in low proportions (less than 2%) throughout the profile. Fern spores are found in large numbers ranging from 11.7 to 58.2% of total land seed plants and represent a variety of types with spores of *Cyathea* and *Gleichenia* dominating. Generally, fern spores are more frequent in interglacials, and achieve highest percentages in MIS 1, 5e and 15 (Fig. 5.26A).

Pollen influx (Fig. 5.26B) is expressed as numbers of pollen grains accumulated on one square centimeter during one year. In general, influx values of total pollen and pollen of each ecological group or taxon are very low (4 grains/cm^2/yr) in the lower part of the profile (from the bottom of the profile to MIS 17), then begin to increase upward and reach the maximum (864 grains/cm^2/yr) in MIS 2 (LGM). The range of pollen influx variations can exceed two orders of magnitude (Fig. 5.26B). In addition, the influx values of total pollen and pollen of each ecological group or taxon are clearly greater in glacials than those in neighboring interglacials.

The pollen sequence at Site 1144 provides a vegetation and monsoon history of the Pleistocene. During interglacials, low amounts of pollen are found in the sediments, particularly from tropical/subtropical plants and herbs, thereby providing very limited information about the terrestrial vegetation. In contrast, a large amount of pollen in glacial sediments at Site 1144 could have been brought in by a strengthened northeast winter monsoon from the Asian mainland and Taiwan Island. In addition, pollen grains could also have been transported from the exposed continental shelf by water flow and wind at glacial sea-level lowstands (Sun et al. 2003).

Pollen assemblages at Site 1144 have revealed the changes in evergreen forest composition over the last million years, indicating that the evergreen broadleaved forests still survived in the southern coastal areas of China in glacials, probably broader in its distribution than today. The early period of the vegetation history before 900 ka is marked by relatively high percentages of tropical montane pollen and *Altingia* within the tropical and subtropical pollen, implying cooler climate than today. The period of 900–360 ka saw the occurrence of Dipterocarpaceae, Celastraceae, *Cycas, Eugenia, Mallotus/Macaranga* and *Trema*, indicating a relatively warm climate. Since ~360 ka trees from Fagaceae began to expand and became absolutely dominant in the forest, probably implying strengthened seasonality and cooler climate than before.

The pollen data from core 17940, a site near Site 1144, have already showed that, during the LGM, the emerged part of the continental shelf was mainly covered by grassland inferred from the high percentages of herb pollen (Sun et al. 1999). This is further confirmed by the pollen records from Site 1144. The long-term trends in the pollen record also bear information about evolution of the shelf and its vegetation during the glacial emergence. Before 900 ka, the narrowly emerged shelf was probably covered by grassland mainly of *Artemisia* during glacials, and then mainly of Poaceae and Cyperaceae during glacials between 900 and 160 ka. The *Artemisia* population increased again and occupied most part of the extensive emerged continental shelf during the last glacial stage. There were also sedge swamps and wetlands on the continental shelf during glacials, as indicated by the pollen of Cyperaceae, water plants like *Typha* and *Myriophyllum*, and small liverwort (*Anthoceros*). The existence of mangrove pollen since MIS 10 (Fig. 5.26), even in very low percentages, indicates that mangrove survived along the northern coast of the SCS during some glacial periods within the last 360 ka.

The frequent alternations between pine and herbaceous pollen at Site 1144 correspond to sea level changes during the last million years (Sun et al. 2003). The pollen ratio between shore plants (represented by herbaceous pollen, H) and upland ones (indicated by pine pollen, P) may indicate the relative distance of the studied site from the coast (Traverse 1988), with high ratio of H/P implying shorter distance to the coast and low ratio vice versa. The palynologically inferred changes in the distance from the coast indicated by H/P ratios at Site 1144 should be ascribed to eustatic sea-level changes during glacial cycles (Fig. 5.25). H/P ratios reach the maximum (5.5) during the LGM, and the minimum (0.01) in the Holocene. The large amplitude variations of H/P values in MIS 3 may suggest significant sea-level changes in the area, as recorded in the coastal zone of China (Wang P. et al. 1981).

However, a broader emerged shelf can also result in a higher H/P ratio than a narrow one. Changes in the size or width of the exposed continental shelf in the northern SCS are probably other factors influencing H/P variations. The small amplitudes in variations and low values of H/P before MIS 6 suggests a narrow and steep continental shelf, with only a limited area exposed at the lower sea-level stand, and the broadening of the continental shelf around MIS 6 was probably caused by neotectonics of the China continent related to the uplift of the Tibetan plateau (Fig. 5.27) (Li J. 1991; Li and Fang 1996).

Pollen records at Site 1144 serve as direct proxies of the past variability of the East Asian winter and summer monsoons. The downcore variations of the tree pollen influx (Fig. 5.26) show that the higher values occur in glacial periods compared with lower values in neighboring interglacials, indicating intensification of the winter monsoon in glacial periods and weakening in interglacials. Moreover, in most cases the pollen influx values are very low at the beginning of a glacial cycle and then increase gradually, reaching the maximum at its end and then abruptly falling down to the minimum at the beginning of the next interglacial. Similar to

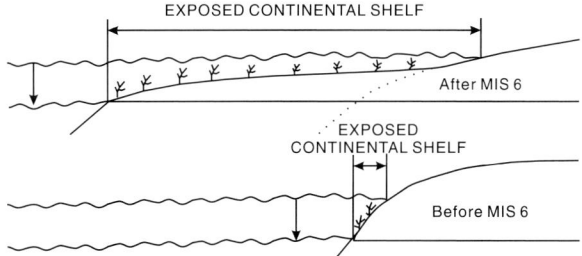

Fig. 5.27 Schematic diagrams show changes in width of the northern continental shelf of the SCS during glacial stages before and after MIS 6, as inferred from the pollen record (from Sun et al. 2003)

foraminiferal $\delta^{18}O$ variations, they indicate a causal relationship between global ice volume and monsoon variations.

Over the past 1 myr, the tree pollen influx value noticeably increased since MIS 16, indicating that the Asian winter monsoon began to intensify since ~670 ka ago when the European Alps experienced the first major glacial, the Günz glaciation. Grain size analysis of a number of loess sequences from the Loess Plateau in central China also shows the intensification of the winter monsoon during the last ~0.6–0.45 myr. Significant advance of the Mu Us desert (north of the Loess Plateau) and better development of paleosols than those formed before imply enhancement of both winter and summer monsoons during the last 0.6 myr (Ding et al. 1999). Similarly, increased dust influx in Lingtai and Xifen loess profiles is regarded as an increase of aridity and, hence, of winter monsoon intensity during the last ~0.6 myr (Hovan et al. 1989; Sun and An 2001).

Other pollen records such as herbs demonstrate similar glacial cycles as the tree pollen influx. Fox example, the high values of tree pollen influx at MIS 2, 5, 12 are accompanied by similar peaks of *Artemisia* and herbs in general (Fig. 5.25), all displaying sawtooth-like curves. However, tree pollen influx also shows high values at MIS 5b and two spikes within MIS 3, probably related to intensive and variable winter monsoon during these interglacial periods. The unstable climate of MIS 3 is also evidenced by the high H/P ratio and high percentage of *Anthoceros* at Site 1144.

For the pollen sequence at Site 1144, there are humidity-indicative palynomorphs closely related to the summer monsoon intensity. Ferns usually grow under humid conditions, with higher fern percentages suggesting wetter climate (Van der Kaars 1991; Van der Kaars et al. 2000). In addition, fern spores are produced in enormous numbers and hence suitable for quantitative analyses. Therefore, the fern spore proportions to the total pollen sum of land seed plants may serve as summer monsoon proxy. Downcore variations of fern percentages show higher values in interglacials and lower values in glacials, suggesting strong summer monsoons during warm periods (Fig. 5.25F). The significantly strong summer monsoon have occurred during late MIS 1, 5e and 15, evidenced by very high values of fern spore percentage. Fern spore percentage maintained high and constant values before MIS 16, probably reflecting relatively strong and stable summer monsoon before MIS 16.

Spectral analyses of herbs and pine percentages at Site 1144 discover a set of Milankovich cycles for the last 1 myr, the strong 100 kyr eccentricity cycle, the weak 41 kyr obliquity cycle and a more distinct 23 kyr precession cycle, as well as some significant semi-precessional cycles (~11, 10, and 9 kyr) (Fig. 5.28), revealing the orbital forcing on the vegetation changes in the source regions that delivered pollen to the northern SCS. Cross-spectral analyses of herb and pine pollen percentages with planktonic foraminiferal $\delta^{18}O$ records show very high non-zero coherences (>90%) at the 100 kyr and 23 kyr bands (Fig. 5.29). Phase relationships between pollen and $\delta^{18}O$ records show that the vegetation changes are nearly in phase with or slightly lag the global ice volume maxima at the 100 kyr and the 23 kyr bands, respectively (Sun et al. 2003). Spectral and cross-spectral analyses reveal close relationship of the vegetation changes in the northern SCS with those changes in global ice volume, sea level and monsoons.

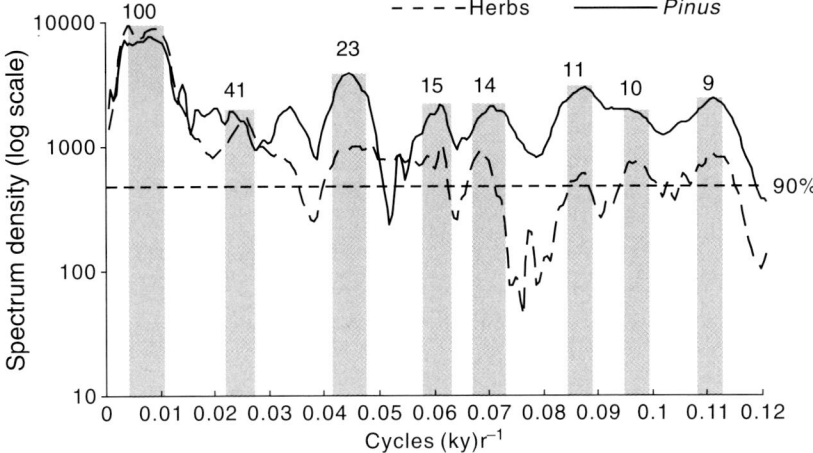

Fig. 5.28 Spectral analysis of pollen percentage values of herbs (*dashed line*) and pine (*solid line*) at ODP Site 1144 shows the presence of various Milankovich and suborbital cyclicities for the last 1.0 myr (from Sun et al. 2003). The *horizontal dashed line* indicates the 90% level of significance

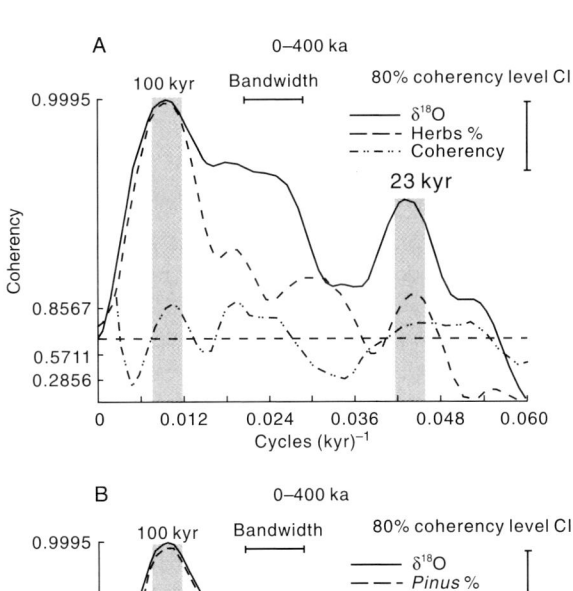

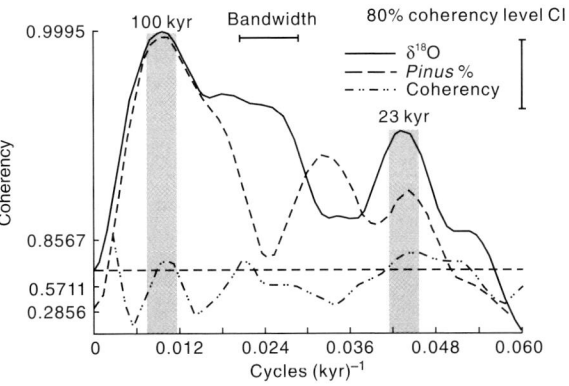

Fig. 5.29 Cross spectral analyses of $\delta^{18}O$ (*solid line*) with herbs% (**A**, *dashed line*) and *Pinus*% (**B**, *dashed line*) in the last 400 kyr at ODP Site 1144 show strong coherence over the 100 kyr eccentricity band and the 23 kyr precessional band (from Sun et al. 2003). Spectral densities are normalized and plotted on a log scale. The coherency spectra are plotted on a hyperbolic arctangent scale. The *horizontal dashed line* denotes 80% coherency level. Point-dashed lines indicate coherency

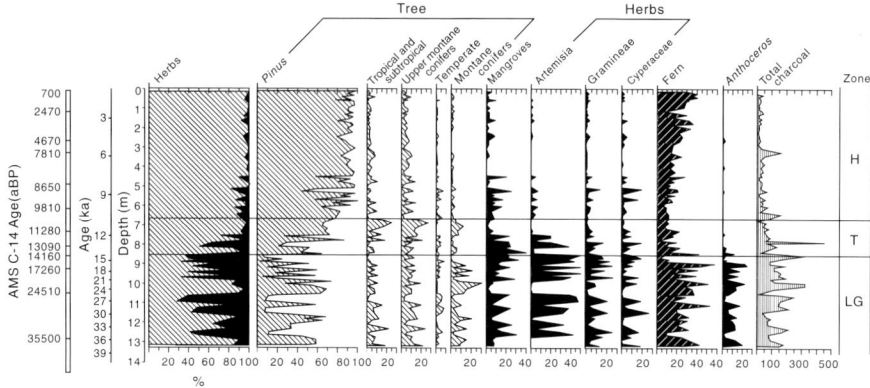

Fig. 5.30 Pollen percentage profiles and charcoal concentration curve (grains/cm^3) from core 17940, the northern SCS, show changes in three pollen zones (from Sun and Li 1999). Ecological groups include: tropical and subtropical broadleaved taxa (mainly *Castanopsis, Quercus, Ilex, Altingia, Elaeocarpus*, Palmae, Sapindaceae, Araliaceae, Gesneriaceae, etc.), temperate broadleaved taxa (mainly *Betula, Carpinus, Alnus, Juglans*, etc.), montane conifers (*Picea, Abies, Tsuga*) and upper montane rain forest taxa (*Podocarpus, Dacrydium, Phyllocladus, Dacrycarpus*)

Last Glacial Pollen Records: North-South Differences

Northern SCS

The best pollen record over the last glacial from the northern SCS was generated from core 17940, in which a total of 161 pollen types were identified, with 13 AMS ^{14}C dates for time constraint. A large number of tree pollen taxa from this core, except the predominant *Pinus*, were grouped into several categories according to their ecology (Fig. 5.30) (Sun and Li 1999). On the basis of downcore variation in different ecological groups it can be clearly divided into three zones, namely Zone LG, Zone T and Zone H (Table 5.2).

During MIS 3 and MIS 2 (the LGM), the pollen assemblages are marked by alternating dominance of montane conifers and upper montane rainforest taxa with herb taxa, denoting cooler climate than the present day. Today the montane conifers and montane rainforest gymnosperm incline to inhabit cool and humid environments. They are distributed either in montane areas of northern China and Taiwan (such as Picea, Abies) or in montane areas of tropical subtropical areas (*Dacrydium, Podocarpus, Dacrycarpus, Phyllocladus*). *Anthoceros* is a small liverwort, recorded

Table 5.2 Division of pollen zones in core 17940 is based on Sun and Li (1999)

Zone	Depth (cm)	Interval/Period	Age (ka)
H	0–660	Holocene (MIS 1)	0–10
T	660–723	Younger Dryas	10–11.3
	723–870	Bølling-Allerød	11.3–15
LG	870–1050	Last Glacial Maximum	15–25.3
	1050–1306	MIS 3	25.3–37

almost only during the last glaciation. The living *Anthoceros* prefers to grow in deforested areas, on the edges of forests, or on moist soil in northeastern China (Gao and Zhang 1981). High percentages of these pollen and spores denote much cooler climate than in the present day.

Large amount of herb pollens was found during the glacial period. As the dominant component in the herb group, *Artemisia* is a herb or a small shrub widely distributed in temperate grasslands in the northern Hemisphere. In the pollen assemblages from the surface sediments of the SCS, only a few pollen grains of this taxon occur in rare samples adjacent to the northern continent. During the LGM, high frequencies of *Artemisia* occurred in northern China and declined eastward and southward. In southeastern China, close to core 17940, *Artemisia* grains were sporadic or even absent at 18 ka (Sun and Li 1999; Sun et al. 2000b). So, where did such a great amount of *Artemisia* pollen in the glacial SCS come from? A plausible answer is that they came from the exposed continental shelf of the SCS which was occupied by grassland, dominated by *Artemisia*, under a comparatively dry and temperate condition. Judging from the above mentioned pollen data, the exposed continental shelf during the last glacial time was probably occupied by grassland, with montane conifers growing on nearby mountains. Therefore, the northern part of the SCS had experienced remarkable decline in temperature and humidity. Moreover, the amount of charcoal during the last glacial was much higher than during the Holocene, suggesting frequent fire due to the dry climate (Sun et al. 2000a). However, the frequent alternating predominance of the herb-dominant and montane confers-dominant groups also implies unstable conditions with frequent alternations between relatively cool and humid and comparatively dry and warm stages.

During Termination I, the vegetation and climate experienced remarkable fluctuations. Those include a sudden expansion of tropical and subtropical vegetation and mangroves, which indicates abrupt warming during the Bølling and Allerød, and a rapid increase of montane conifers, which indicates abrupt cooling during the Younger Dryas. These changes exhibit the main futures of climatic oscillations of the same time in the northern Hemisphere.

The Holocene is distinguished by absolute dominance of pine pollen (~90%), gradually increasing of fern spores and disappearance of *Anthocerus* and mangroves. The pollen records are quite similar to those of the surface sediments from the northern SCS, implying that the vegetation and climate was similar to that of the present day, with gradual warming of the climate and continuing submerging of the continental shelf. The sudden rise of spore *Dicranopteris* about 2 kyr ago might have resulted from intensification of human activities. *Dicranopteris* is atypical tropical or subtropical fern, often distributing in deforested places after human interferences (Guangdong Institute of Botany 1976).

Southern SCS

Pollen types are different between the southern and northern parts of the SCS. Four main ecological groups of pollen were discovered in pollen assemblages from the south (Li and Sun 1999). The first group includes tropical montane rainforest, reflecting a cool and humid climate. The second group is low montane rainforest taxa. The third group is rainforest taxa with more than one hundred pollen morphs identified, denoting hot and humid climate. The fourth group includes mangroves, *Rhizaphora, Sonneratia* and *Liuminizera* growing in brackish water along tropical coasts and serving as an indicator of tropical coastal line. The rise of mangrove percentage in pollen diagram is closely related to sea level rise (Ellison 1993; Grindrod and Rhodes 1984). The last glacial pollen data of the southern SCS come from two areas: the southern continental slope and the Sunda Shelf.

Southern slope. Detailed pollen data are available from 3 cores: 17964, 17962 and 18287 (see Fig. 1.1 and Table 5.1 for locations). The bottom of core 17964 was dated at 26 ka, and the pollen assemblages from this core are dominated by tree pollen, with less herb pollen (<20%). Remarkable from the first glance is the absence of the glacial/interglacial contrast frequently observed in the northern SCS. Nevertheless, four pollen zones (P1-P4) were recognized that represent different stages of vegetation and climatic changes (Fig. 5.31) (Li and Sun 1999).

The glacial time (P1) is dominated by tree pollen taxa (>80%), excluding considerable amount of pine, composed mainly of upper montane rainforest and lowland rainforest and mangroves. Herb pollen percentages are quite stable through the profile (around 20%), while fern spores are very high in percentages. Pollen assemblage from this stage probably indicates that the climate was gradually getting cooler from MIS 3 towards the LGM. Mangroves pollen percentages increase upwards the profile as well, suggesting gradually exposure of the Sunda Shelf with sea level drop.

The striking feature of the deglaciation (P2) is that the total pollen influx (grains/cm^2/yr) and influxes of each group and taxa are in their maximum values, probably resulting from the rapid migration of the coastline and accompanied erosion. The pollen record at this stage denotes that the Sunda Land was covered mainly by lowland rainforest. The pollen percentages of mangroves increased during the later stage of deglaciation (P3).

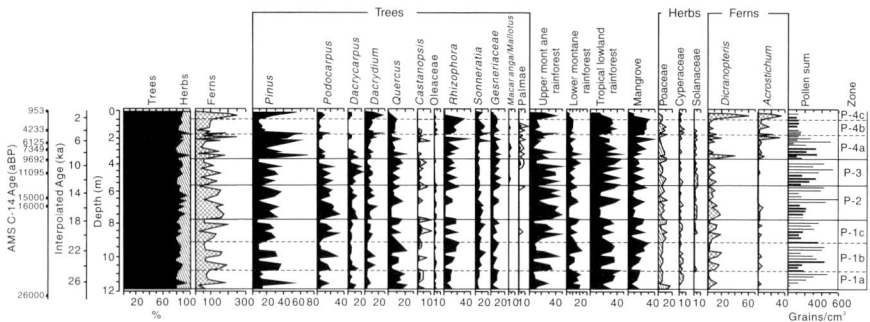

Fig. 5.31 Pollen percentage diagram of core 17964 from the southern South China Sea with four pollen zones (P1-P4) recognized: P1 (718–1178 cm, 18.3–26.5 ka), P2 (478–718 cm, 13.8–18.3 ka), P3 (318–478 cm, 9.9–13.8 ka), and P4 (0–318 cm, 0–9.9ka) (modified from Li and Sun 1999)

The Holocene (P4) can be divided into three substages. The early part (9.9–5.2 ka) is characterized by abruptly increased pine pollen and mangroves, and a slight decline of lowland rainforest. In the middle part of P4 (5.2–2.3 ka), pollen percentages of the lowland rainforest and mangroves (especially *Rhizophora* and *Sonneratia*) increased their pollen percentages, but those of the montane rainforests continuously declined and almost disappeared, indicating the warmest climate developed in the middle Holocene. In the most recent part (2.3 ka-present), pollen percentages of the lowland rainforest and mangroves declined, but those of *Pinus*, upper montane rainforest and fern spores (spores of *Dicranopteris* and *Acrostichum* in particular) increased considerably, implying a cooler climate and human disturbances during the late Holocene (Li and Sun 1999).

Core 17962 is quite similar to core 17964 in pollen assemblages, especially during the Holocene (Sun et al. 2002). However, the pollen assemblages of core 17962 changed much more smoothly (Fig. 5.32) and did not display clear millennial-scale oscillations in different ecological groups of pollen taxa. In addition, the lowland rainforest dominated the whole pollen profile, except the early Holocene. The pollen influx values in core 17962 were high during the glacial stage with maximum values in the LGM, but they progressively decrease during the Termination I and to a minimum in the Holocene (Fig. 5.32).

Core 18287 recovered from the upper slope close to the Sunda Shelf covers only Termination I and the Holocene, and the pollen profile is dominant by tree pollen taxa (Fig. 5.33; see Figs. 1.1 and 5.35 for location). The upper montane rainforest pollen of this core is low in percentages throughout the whole profile, but the

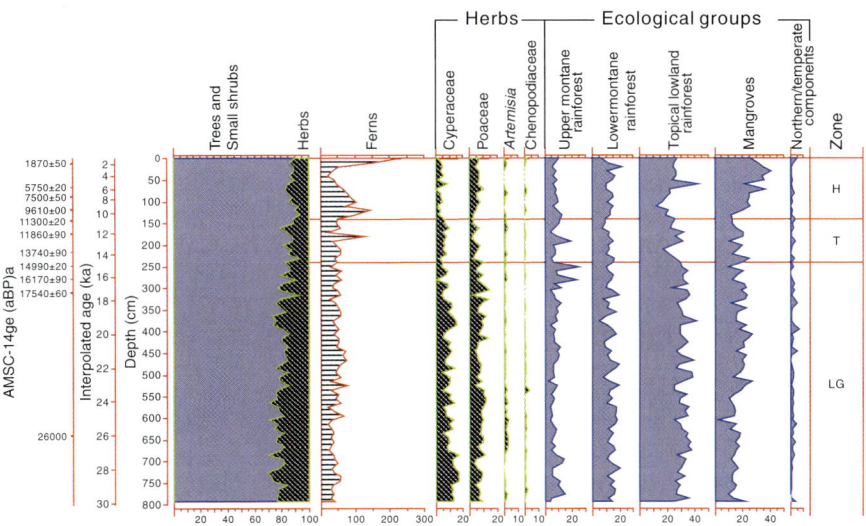

Fig. 5.32 Pollen percentage diagram of selected taxa from core 17962 from the southern SCS shows weaker variations compared to other site localities (Sun et al. 2002). LG = last glacial; T = Termination; H = Holocene

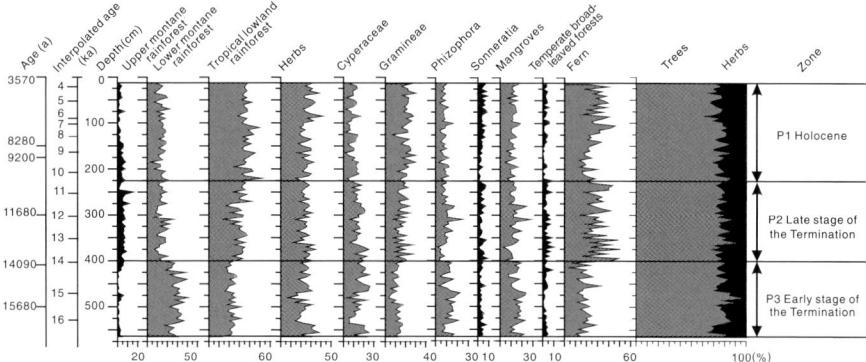

Fig. 5.33 Pollen percentage diagram of core 18287 from the southern SCS shows variations from Termination I to the Holocene (from Wang X. et al. 2007)

lower montane rainforest pollen serving as cool climate indicator is quite abundant. During the Termination the lowland rainforest and fern expanded their distributions and the lower montane rainforest progressively moved up to the mountains and its distribution area on the shelf shrank, probably caused by climate warming during this period (Wang X. 2006; Wang X. et al. 2007).

The three cores discussed above (17962, 17964, 18287) range from north to south. A comparison of their pollen influx profiles reveals a southward migration of its maximum value: at LGM in the northern core 17962 but at Termination I in the southern core 18287 (Fig. 5.34). This pattern confirms that the pollen was transported northward from the Sunda Shelf and the maximum influx moved to the

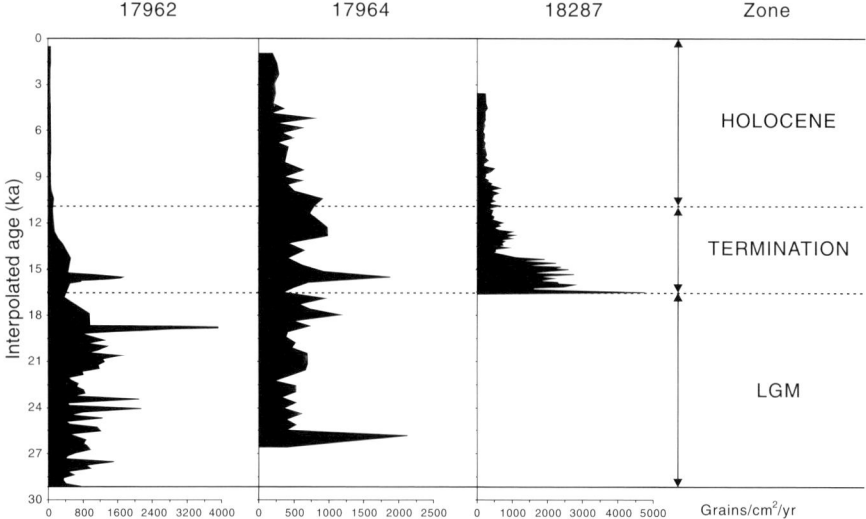

Fig. 5.34 Total pollen influx (grains/cm²/yr) curves for cores 17962 (Sun et al. 2002), 17964 (Li and Sun 1999) and 18287 (Wang X. 2006) indicate a shift of pollen deposition center southward with sea level rise from the LGM to Termination I

south with the postglacial retreat of the coastline. In the Holocene, extremely low pollen influx at these cores indicates that the sea level rose and the Sunda Shelf submerged into the seawater. Therefore, the neighboring islands, mainly Borneo, became the only pollen source area. Noticeable is the middle Holocene, when both sea level rise and the submerged area of the shelf reached its maximum, and the climate was warmer and more humid than today, as evidenced by maximum pollen percentages of mangroves. During the late Holocene, pine pollen increased again but lowland rainforest and mangroves sharply decreased, probably related to climatic cooling. The sudden increase of fern spores in the late Holocene as recorded in core 17962 (Fig. 5.32) suggests intensive deforestation by human activities (Li and Sun 1999; Sun et al. 2002).

Sunda Shelf. During the LGM, the Sunda Shelf was widely subaerially exposed as a large low-gradient coastal plain, while the modern Malayan Peninsula, Borneo and Sumatra Islands formed highlands to the west and south (Tjia 1980; Hanebuth et al. 2000). Several rivers originating here drained the coast lowland during the LGM. Pollen assemblages from three cores were studied in detail (Wang X. 2006; Wang X. et al. 2007, 2008): cores 18300 and 18302 from the outer shelf in modern water depths of 92–98 m and core 18323 from the inner shelf but in a paleo-river channel with modern water depth of 87 m (Fig. 5.35). The pollen records from the three cores are shown in Fig. 5.36.

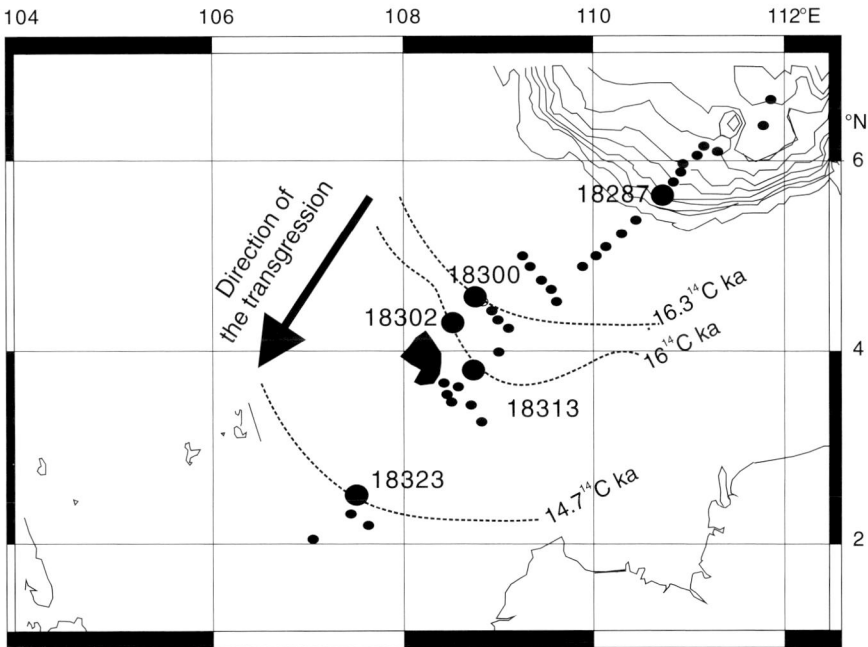

Fig. 5.35 Location map shows the palynologically studied sites (*large filled circles*) revealing the rising trend of sea level and the submergence of the Sunda Shelf after the LGM (from Wang X. 2006). *Dotted lines* show paleo-coastline. *Small dots* are other sites cored during SONNE Cruise 115 (Stattegger et al. 1997)

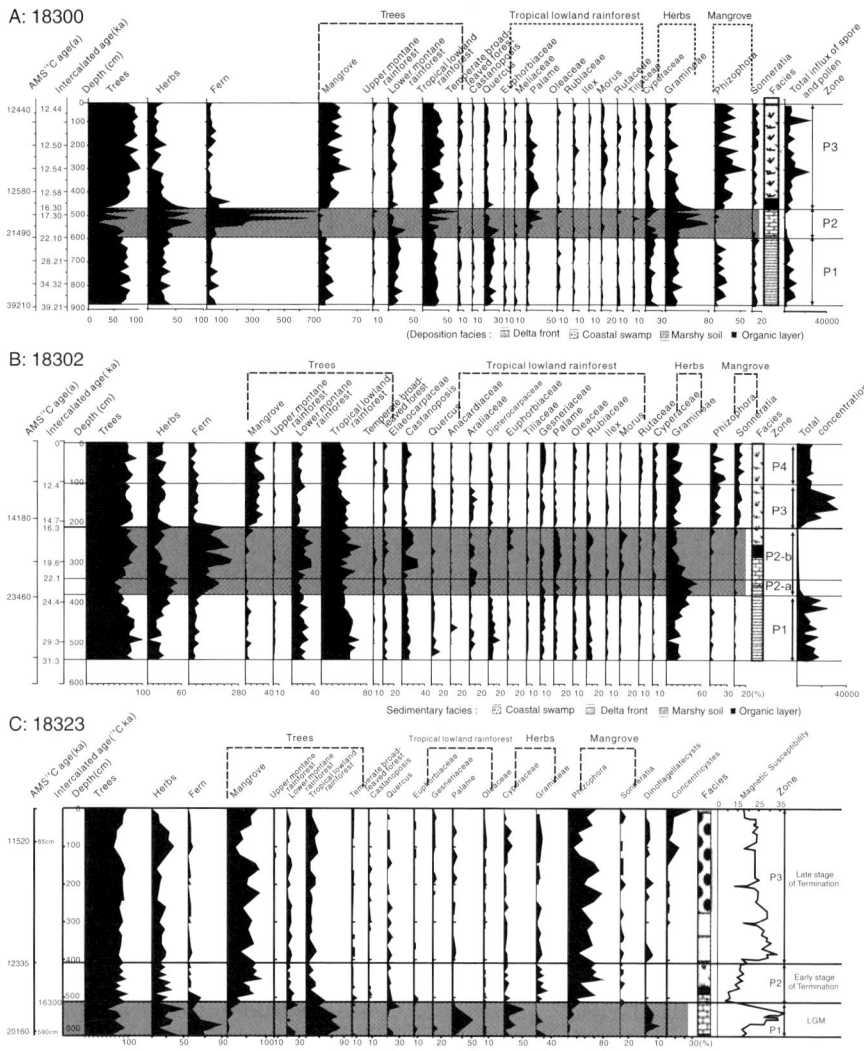

Fig. 5.36 Pollen percentage diagrams of cores 18300 (**A**), 18302 (**B**), and 18323 (**C**) from the Sunda Shelf indicate a major change during the LGM (*shaded interval*) (modified from Wang X. (2006)

All the 3 cores reached the last glacial. Most interesting is the pollen record of the LGM (including early H1 event; shaded zone in Fig. 5.36), although ages slightly differ between cores (~22.1–16.3 ka for P2 at 18300, ~20.16–16.3 ka for P1 at 18302, and ~23.46–16.3 ka for P2 at 18323). Before the LGM (such as P1 in 18300), tree pollen is prevailing with considerable amount of mangrove pollen. The LGM and early H1 times, however, are characterized by dominance of herb pollen (mainly Poaceae) and fern spores (mainly *Cyathea*) and by very low pollen influx and occurrence of *Phagmites* (sedge, a water or swamp plant). Tree pollen dropped to ~50% and mangrove pollen almost disappeared during the interval. This is very different from the pollen assemblages of MIS 3 and subsequent Termination I (P3 in core 18300), which features a sharp decrease in herb pollen and fern spores but increase in tree pollen. Mangroves increased remarkably, suggesting that the area was submerged with the coastline approaching again (Fig. 5.36) (Wang X. et al. 2008).

Summarizing the pollen records from cores 18300, 18323, 17962 and 17964, regional environmental changes since MIS 3 in the southern SCS can be reconstructed. In the course of sea level drop, the inner shelf was already subaerially exposed and covered mainly by lowland rainforest, but the outer shelf was still covered by shallow seawater (Fig. 5.37A). The lower montane rainforest probably migrated from montane areas of the southern island down to the shelf and became an important part of the glacial vegetation. The climate at that time was cooler than the present day, but could be still very humid. During the LGM and early H1, the pollen assemblages from the continental slope were dominated by lowland rainforest and mangroves, whereas those from the shelf were prevailed by high percentages of herb pollen and fern spores without mangroves. One source area of the pollen was from the marshy plants growing along the Sunda River, and the other was from the vegetation distributed on the Sunda Land (Fig. 5.37B). Lowland rain forest and lower montane rainforest covered the exposed Sunda Land during the LGM and H1, and the upper montane rainforest periodically migrated down along the montane slopes of the southern islands. The climate should be cooler than the present day, but still humid. During the Termination I after ~16 ka, pollen assemblages are dominated by lowland rainforest and mangrove (Fig. 5.37C). Mangrove pollen reappeared and increased quickly upwards the profiles on the shelf, indicating sea level rise and shelf submerging again. These changes indicate postglacial warming and lowland rainforest covering the exposed part of the shelf during the Termination.

North-South Comparison of the Vegetation During the LGM

The pollen data summarized above have outlined a distinct picture of north-south contrast of vegetation in the late Quaternary SCS, specifically for peak glacial and H1 times. Grassland vegetation, mainly composed of *Artemisia*, then covered the exposed northern continental shelf of the SCS, indicating colder and drier climate relative to the present day. But the subaerially exposed southern continental shelf, the Sunda Land, was covered by lowland rainforest and lower montane rainforest; mangroves grew along the coast and montane rainforest migrated down the montane slopes many times during this period. Along the North Sunda River distributed marshy vegetation. This evidence of pollen assemblages indicates cool but humid climate in the southern SCS during the LGM.

The difference in humidity between the northern and southern SCS was probably caused by changes in the East Asian monsoon system (Sun et al. 2000b). The

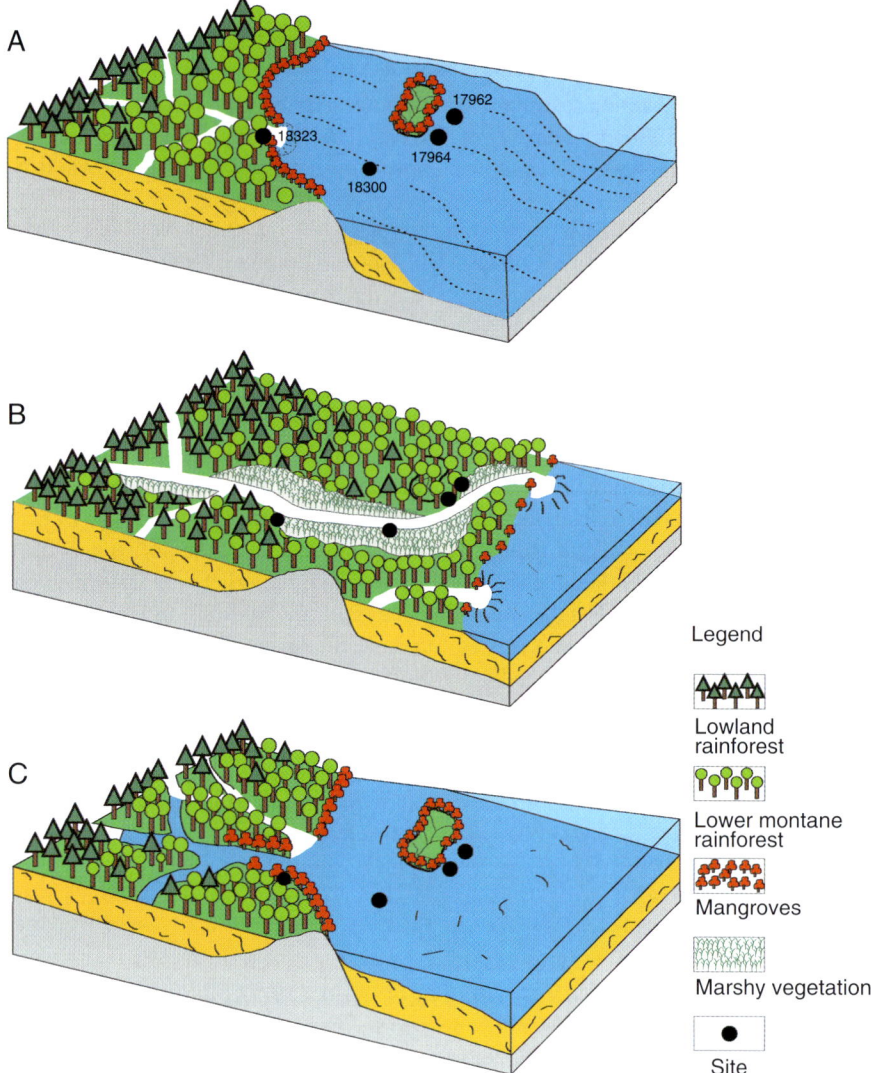

Fig. 5.37 Schematic diagrams show paleovegetation evolution on the Sunda Land from (**A**) MIS 3, to (**B**) LGM and H1 times, and (**C**) Termination I after 16 ka

intensification of Siberian High over the East Asian continent during the glacial period strengthened the East Asian winter monsoon, which in turn led to the lower temperature and reduced humidity in the region including the northern SCS. By contrast, the East Asian winter monsoon becomes Australian summer monsoon when it crosses the Equator, and brings precipitation to the islands of southeastern Asia. During the last glaciation, the strengthened boreal winter monsoon absorbed moisture when crossing the SCS and provided more precipitation to the Sunda

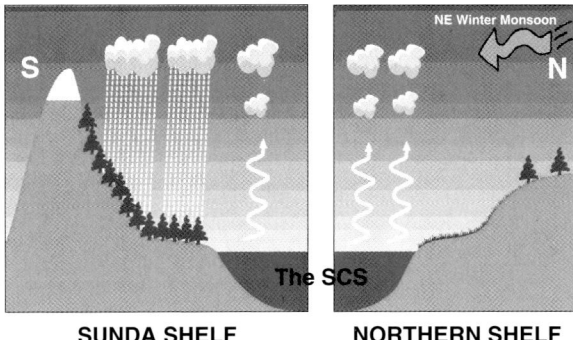

Fig. 5.38 A schematic diagram shows vegetation and rainfall differences between the northern (*right panel*) and southern SCS (*left panel*) during glacial stages. While grassland vegetation covered the exposed continental shelf under strong winter monsoon in the north, more precipitation in the south was provided by the strengthened winter monsoon which absorbed moisture when crossing the sea, enabling lowland rainforest to grow

Land, leading to the continued growth of lowland rainforest and mangroves there. Moreover, the island areas south of the southern SCS without relief obstruction could also receive more rainfall to enable humid rainforest vegetation to grow (Walker and Flenley 1976; Stuijts et al. 1988; Hope and Tulip 1994). However, areas located in the rain-shadow behind mountains may have experienced an arid climate and grown grassland vegetation due to a lack of moisture from the monsoon (Fig. 5.38).

This idea was confirmed by *n*-alkane stable carbon isotopic analysis ($\delta^{13}C$) of core 17962. Accumulation rates of long–chain *n*-alkanes suggest that intensified river flows occurred on Sunda Land due to intensified winter monsoon precipitation during the last glacial period. The isotopic composition ranges from $-27.1‰$ to $-33.9‰$ for C27-C33 n-alkanes in the entire core sequence, indicating an input mainly from C3 higher plants (Hu et al. 2003). This means that rainforest continued growing on the Sunda Shelf in a constant humid climate although the possibility of certain decrease in precipitation cannot be excluded. Since the present climate in the Sunda Shelf is extremely humid, "there could be a decrease in the total precipitation which the area receives, without there necessarily being any recognizable effect on vegetation" (Newsome and Flenley 1988).

5.4 Monsoon History (Jian Z. and Tian J.)

The climate records in the SCS discussed so far in this chapter cover a broad spectrum from land vegetation to upper water structure, but all are focused on the main feature of the region: the East Asian monsoon system. The following is a brief overview of the monsoon history as recorded in the SCS. A correct use of proxies is a prerequisite for paleo-monsoon reconstruction.

Monsoon Proxies

Terrestrial sediment records provide reliable East Asian Monsoon proxies. For example, loess deposits show clear evidences of monsoon changes in the late Cenozoic, with magnetic susceptibility and Rr/Sr ratio indicating summer monsoon change, and grain size and Al flux indicating winter monsoon change (Porter et al. 1992; Ding et al. 1992, 1994; Vandenberghe et al. 1997; An et al. 2001; Guo et al. 2002). Stable oxygen isotopes of cave stalagmites from the southern China document millennial and orbital scale variabilities of precipitation mostly during summer seasons, serving as a perfect East Asian Summer Monsoon proxy (Wang Y. et al. 2001, 2005, 2008; Cheng et al. 2006). Recently, the magnetic properties and the titanium content in sediments of Lake Huguang Maar in coastal southeast China have been found to be East Asian winter monsoon proxies (Yancheva et al. 2007), anticorrelated to the cave stalagmite records of summer monsoon.

Marine deposition in the SCS involves a complex of physical, biological, and chemical processes, making marine proxies associated not only with monsoon changes but also with several other variables, and thus is usually more complex than terrestrial deposition. In addition, there has been much less observation of monsoon-related modern processes in the SCS as compared to the Arabian Sea. Therefore, the use of monsoon proxies in the SCS needs to be employed with caution to avoid misinterpretation. As everywhere, it is better to utilize a multi-proxy method for monsoon study in the SCS, with support of evidences from modern hydrological features and from core top and sediment trap analyses.

In 2005, the SCOR-IMAGES Evolution of Asian MONSoon (SEAMONS) working group provided a summary of geological archives and their proxy data of the Asian monsoon (Wang P. et al. 2005). In general, monsoon proxies from the SCS can be divided into two groups according to the primary aspects of the monsoon that they address: direct proxies related to monsoon winds (direction and strength), and indirect proxies related to monsoon-induced precipitation or upper ocean structure changes such as the thermocline depth (Table 5.3). Many of the monsoon proxies have been already discussed in the previous paragraphs, some additional remarks and a few proxies specific of the SCS, grouped into paleontological, isotopic, mineral and elemental geochemical, and organic geochemical proxies, are briefly introduced below.

Microfossil and Isotopic Proxies

As shown in preceding sections, micropaleontological and palynological proxies are among the most frequently used methods in paleo-monsoon studies, yet the application of a particular proxy has its spatial and temporal limitations. In the Indian Ocean, for example, the census count of the planktonic foraminifera *Globigerina bulloides* serves as a good upwelling indicator and thus an ideal proxy of the Indian monsoon, as evidenced by sediment trap time series and plankton-tow data (Curry et al. 1992) and confirmed by its geographic distribution in core-top sediments (Prell 1984). *G. bulloides* has been widely used in upwelling regions of the Arabian Sea for paleo-monsoon studies on long-term tectonic scales (Kroon et al. 1991) to

Table 5.3 Synthesis of monsoon proxies indicates two groups of proxies (direct and indirect) commonly used in the SCS (modified from Wang P. et al. 2005)

Features and processes	Proxies	References
Wind system (direct proxies)		
Wind directions (surface and mid-tropospheric wind)	Lithogenic tracers	
	Clay minerals	Liu Z. et al. 2003
	Specific pollen types and assemblages	Sun et al. 1999
	Loess-type sediments (>6 μm)	Wang L. et al. 1999a
	Wind-induced coastal upwelling:	Wang L. et al. 1999a
	1. Summer upwelling S.W. of Vietnam	Jian et al. 2001
	2. Winter upwelling N.W. of Luzon	Huang B. et al. 2002
Wind strength	Silt modal grain sizes (>6 μm)	Wang L. et al. 1999a
Wind-driven precipitation (indirect proxies)		
Processes linked with wind strength	Thermocline depth	Jian et al. 2000b
	Planktonic foraminifera and nannofossil indices	Tian et al. 2005b Liu C. et al. 2002
1. Structure of surface ocean	Meridional SST gradient	
	Difference in planktonic $\delta^{18}O$	Wei et al. 2003
2. Upwelling productivity	Micropaleontological proxies:	
	1. Planktonic foraminifera Abundance of *N. dutertrei*	Jian et al. 2003
	2. Benthic foraminifera indicative of high organic-carbon flux	Jian et al. 1999; Kuhnt et al. 1999
	3. Radiolaria Upwelling index	Wang and Abelmann 2002 Chen M. et al. 2003
	4. Nannofossils	Liu C. et al. 2002
	Geochemical proxies	
	1. Organic carbon % and flux	Jian et al. 1999
	2. Opal % and flux	Lin H. et al. 1999
	3. Ba/Al ratio	Wehausen and Brumsack 2002; Lin H. et al. 1999
	4. Cd/Ca ratio in foraminiferal tests	
	5. $\delta^{15}N$	Kienast et al. 2002 Higginson et al. 2003
	6. $\delta^{13}C$ in near-surface dwelling planktonic foraminifera as inverse nutrient signal	Wang L. et al. 1999a Jian et al. 2003
Continental runoff	Sea surface salinity estimates Planktonic $\delta^{18}O$	Wang L. et al. 1999a,b
Precipitation rate	Pollen (vegetation changes)	Sun and Li 1999; Sun et al. 2003
	Charcoal	Luo et al. 2001
Weathering and pedogenesis	Clay minerals	Trentesaux et al. 2004
	K/Si	Wehausen and Brumsack 2002
	Magnetic grain size (ARM/SIRM ratio)	Kissel et al. 2003

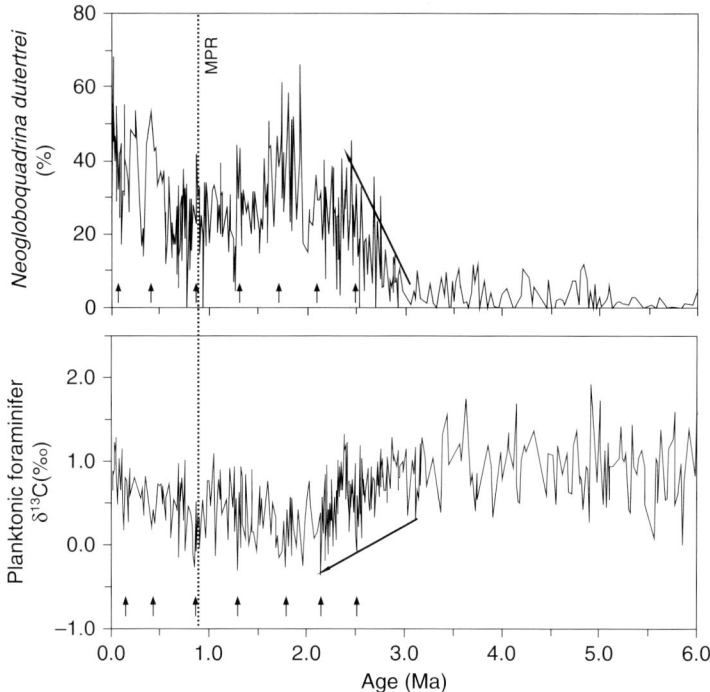

Fig. 5.39 Abundance changes in *Neogloboquadrina dutertrei* (*upper panel*) and planktonic δ^{13}C of *G. ruber* (*lower panel*) at ODP Site 1148 can be used as indirect monsoon proxies. *Oblique arrow* shows the East Asian winter monsoon strengthening in the late Pliocene. *Short arrows* indicate short-term winter monsoon enhancements as indicated by decreased δ^{13}C and increased abundance of *N. dutertrei*. *Vertical dashed line* indicates the mid-Pleistocene revolution (MPR)

high-resolution millennial scales (Anderson et al. 2002). However, its use in the SCS as monsoon indicator is hampered by its rarity because of the semi-enclosed nature of the sea basin. In the upwelling regions of the SCS, *Neogloboquadrina dutertrei*, rather than *G. bulloides*, is dominant (Pflaumann and Jian 1999). Modern observations in sediment traps (Chen et al. 2000) and surface sediments indicate that *N. dutertrei* is a typical winter and high-productivity species and can therefore serve as an indirect proxy for the East Asian winter monsoon in the SCS, as exemplified by a study of the Plio-Pleistocene sequence at ODP Site 1148 from the northern SCS (Fig. 5.39) (Jian et al. 2001, 2003).

δ^{18}O differences between subsurface *P. obliquiloculata* and surface *G. ruber* reflect monsoon-driven changes in the thermocline depth, thus are useful as monsoon indicator (see Fig. 5.20). Similarly, *G. ruber* δ^{13}C may also serve as proxy of the East Asian winter monsoon, as this seasonal reproducer (Chen et al. 2000) displays a gradual decrease in its δ^{13}C towards the Luzon Strait where the inflow of nutrient-enriched surface water is driven by the East Asian winter monsoon (Wang L. et al. 1999a; Jian et al. 2003; Cheng X. et al. 2005). At ODP Site 1148 from the northern SCS, rapid decreases in *G. ruber* δ^{13}C were coupled with increased abundance in *N. dutertrei* during the period of 3.2–2.2 Ma, probably illustrating the trend of gradually enhanced East Asian winter monsoon. After ~2.2 Ma, several conspicuous *G. ruber* δ^{13}C negative excursions associating with increases in relative abundance of *N. dutertrei*, particularly at ~1.7, 1.3, 0.9, 0.45, and 0.15 Ma, or about every 0.4 Ma, may imply that the winter monsoon strengthened at those times (Fig. 5.39) (Jian et al. 2003).

Mineral and Elemental Geochemical Proxies

With the rapid development in technology, some mineral and elemental ratios of sediments have become progressively widely used proxies in paleo-monsoon studies. The following are a few examples from the SCS.

Ratio of smectite vs illite+chlorite is used to estimate the relative strength of the summer vs winter monsoons based on the observation that modern illite and chlorite in the SCS mainly come from the mainland of China and Taiwan Island, while smectite largely originates from Luzon and other islands of the southern SCS (Liu Z. et al. 2003; see "Clay Mineral" section in Chapter 4). The record of this ratio at ODP Site 1146 demonstrates distinctive glacial/interglacial cycles for the past 2 million years, with high values implying strong summer monsoons during interglacials and low values implying strong winter monsoons during glacials (Liu Z. et al. 2003), consistent with the results of palynological analysis at ODP Site 1144 (Fig. 5.40). Also at Sites 1143 and 1146, ratios of (illite+chlorite)/smectite,

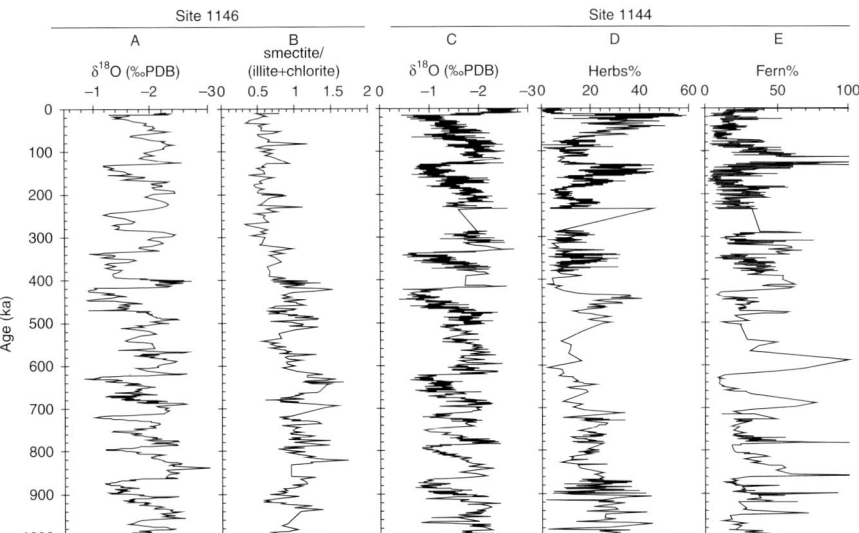

Fig. 5.40 Variations of monsoon proxies during the past one million years are from ODP Site 1146 (**A**, planktonic δ^{18}O; **B**, ratio of smectite/illite+chlorite) and ODP Site 1144 (**C**, planktonic δ^{18}O; **D**, herbs%; **E**, ferns% (percentage in the total pollen sum)) in the northern SCS (Wang P. et al. 2003)

(quartz+feldspar)% and mean grain-size of terrigenous materials have been considered as direct monsoon proxies, revealing a 20 myr long history of the East Asian Monsoon on the tectonic scale (Wan et al. 2006, 2007).

Ratio of hematite vs goethite (Hm/Gt) has been used as a reasonable precipitation proxy over the last 600 kyr in the southern SCS (Zhang et al. 2007). These authors measured the concentrations of hematite and goethite, two climatically significant Fe oxide minerals, using diffuse reflectance spectroscopy in ODP Site 1143 samples, and found a good correlation between variations in the Hm/Gt ratio and in stalagmite oxygen isotopes from South China, indicating that the Hm/Gt ratio provides a unique, long-term record of monsoon precipitation in Southeast Asia.

Ratios of K/Si and Ba/Al also show responses to insolation-driven monsoon variability in the northern SCS (Wehausen and Brumsack 2002). While the fluvial input (K/Si) responds to changes in the summer monsoon, productivity variations, as documented by Ba enrichments, seem to reflect variations in winter monsoon intensity. A stronger winter monsoon may have increased nutrient availability via dust input and/or upwelling activities. Ba/Al peaks, indicating enhanced productivity, occur during K/Si minima at ODP Site 1145 from 3.2 to 2.5 Ma, implying that summer and winter monsoons were approximately 180° out-of-phase (Fig. 5.41).

Chemical index of alteration (CIA) and elemental ratios, such as Ca/Ti, Na/Ti, Al/Ti, Al/Na, Al/K, and La/Sm, are sensitive to weathering which is closely related to East Asian summer monsoon changes (Wei G. et al. 2006). These proxies at ODP Site 1148 from the northern SCS are found to have recorded the chemical weathering history on a tectonic timescale since 23 Ma in South China. The results show that the East Asian summer monsoon has dramatically affected South China in the early Miocene, but its weathering influence decreased continuously since that time, probably because of the intensification of the winter monsoon (Fig. 5.42).

Organic Geochemical and Isotopic Proxies

$\delta^{15}N$ and chlorine accumulation rate of sediments have been employed to reconstruct oceanic nitrate inventory, the balance between denitrification and N fixation, and paleoproductivity since the LGM in the northern SCS. N fixation was locally enhanced during the LGM, coincident with increased inputs of iron-rich aeolian dust raised by invigorated winter monsoon winds from a desiccated continent (Higginson et al. 2003), indicating the possibility of $\delta^{15}N$ to be an East Asian winter monsoon index.

The accumulation rates of C_{37} alkenones and C_{30} alkyl diols have been used as East Asian winter monsoon proxies in the southern SCS (Hu et al. 2002). Pollen records indicate that drought occurred in the northern SCS and South China during the LGM (Sun et al. 1999). However, this is not the case in the southern SCS, where the enhanced winter monsoon brought precipitation of moisture from the West Pacific to the emerged Sunda Land during glacial times (Pelejero et al. 1999a,b), as evidenced in the enhanced accumulation rates of terrestrial biomarkers in core 17962 from the southern SCS (Hu et al. 2002). Both the proxies from core 17962 show enhanced paleoproductivity of coccolithophorids and microalgae such

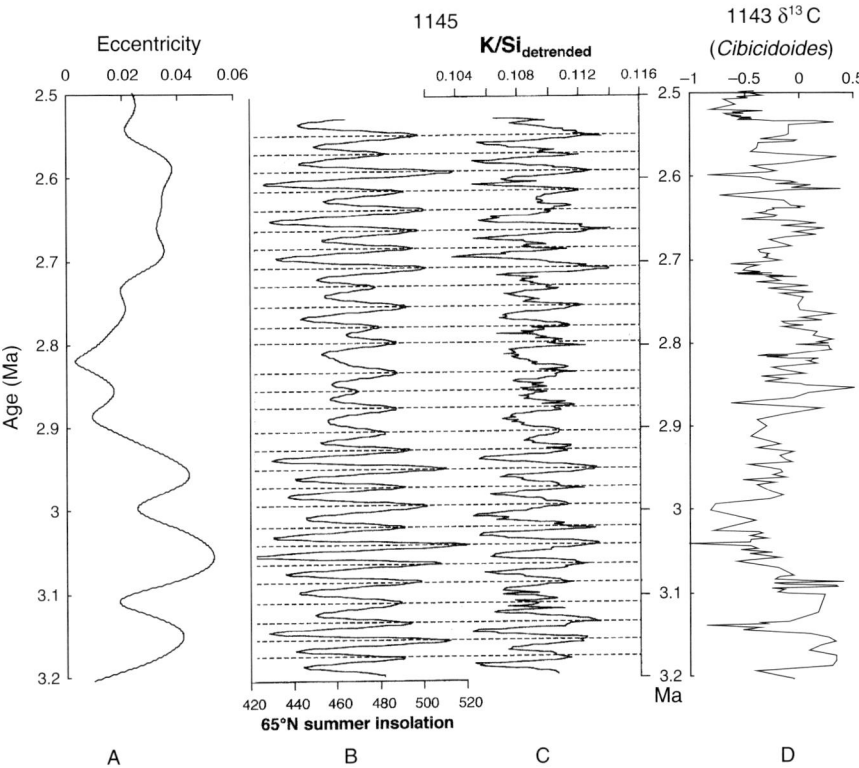

Fig. 5.41 K/Si ratio reflects the orbital forcing of summer monsoon variations in the late Pliocene (3.2–2.5 Ma) at ODP Site 1145 (**A**, eccentricity; **B**, insolation at 65 °N; **C**, K/Si ratio; from Wehausen and Brumsack 2002). Benthic $\delta^{13}C$ at ODP Site 1143 (**D**) is shown for comparison (Wang P. et al. 2003)

as eustigmatophytes during the glacial time, correlating well with those of terrestrial biomarkers (e.g., long-chain n-alkanes, high molecular n-alkanols and long-chain n-alkanoic acids), suggesting that the enhancement of marine productivity during the last glacial period would have been triggered by an increased supply of nutrients from rivers on Sunda Land, which probably had been caused by abundant monsoon rainfall.

Black carbon $\delta^{13}C$ or difference between atmospheric $\delta^{13}C$ and black carbon $\delta^{13}C$ (Δp) has been used to indicate East Asian monsoon, especially the summer monsoon variations since 30 Ma ago on the tectonic timescale (Jia et al. 2003). Because the negative adjustment in Δp of C_3 ecosystems is usually related to moisture deficits, and C_4 photosynthesis is commonly associated with hot, dry environments with warm-season precipitation in a low atmospheric pCO_2 background, secular changes in terrestrial Δp could be reasonably related to the East Asian climate evolution toward a monsoon circulation system. The reconstruction of black carbon $\delta^{13}C$ and Δp at ODP Site 1148 from the northern SCS reveals five events of

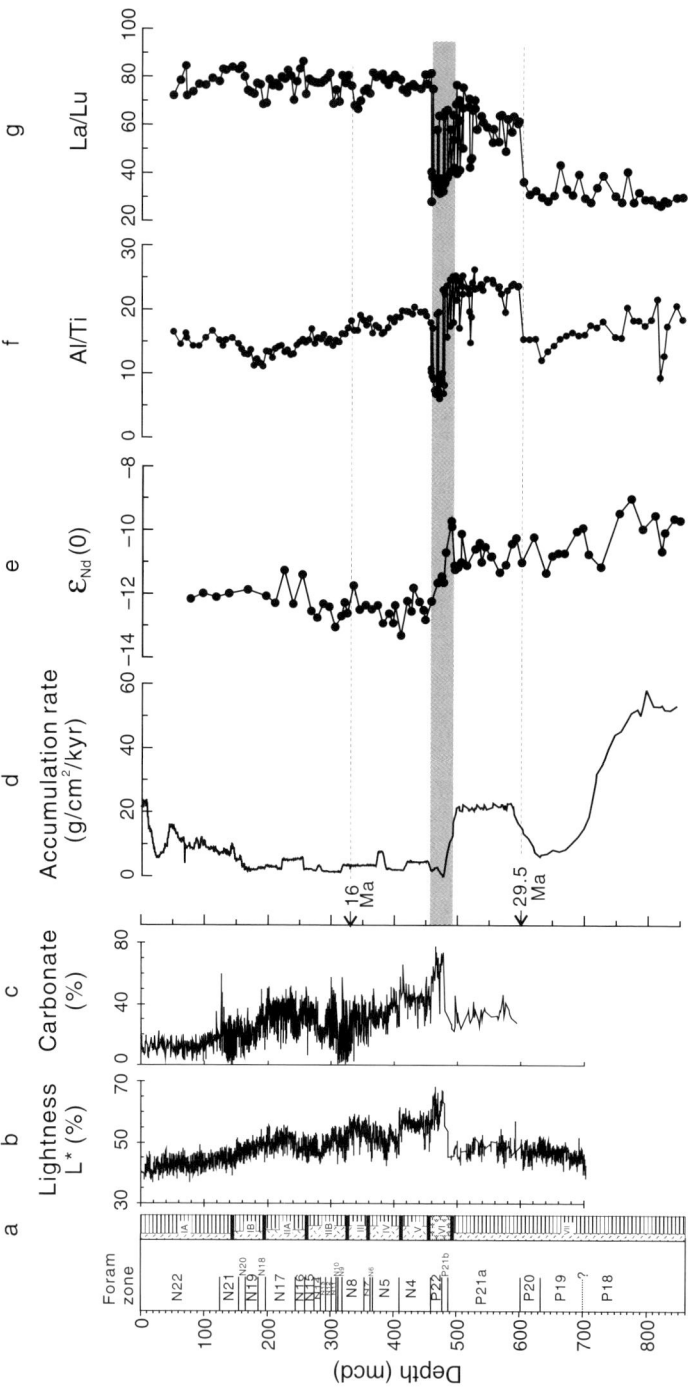

Fig. 5.42 Stratigraphy, lithology and geochemistry at ODP Site 1148 show a sudden change at ~29.5 Ma indicating increased humidity: (**a**) lithological units in Roman numbers, (**b**) color reflectance L*, (**c**) CaCO₃%, (**d**) accumulation rate, (**e**) ε_Nd, (**f**) Al/Ti ratio, (**g**) La/Lu ratio (from Wang P. et al. 2003)

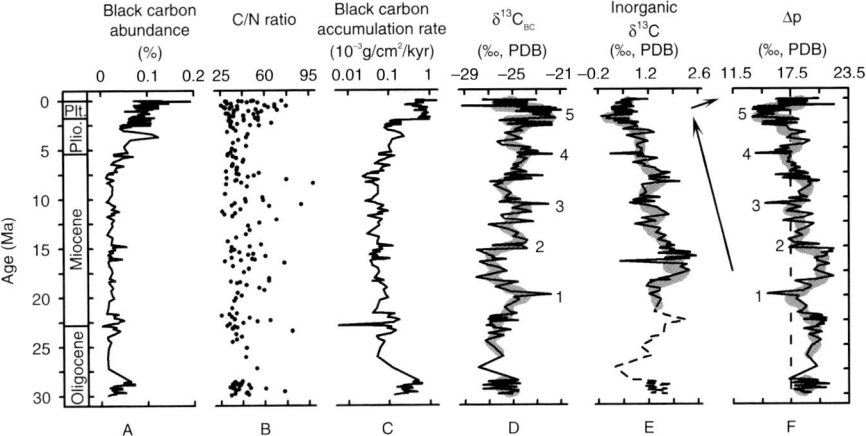

Fig. 5.43 Analytical data from ODP Site 1148 show 5 stages in carbon isotope changes related to monsoon development (Jia et al. 2003). (**A**) Black carbon abundance in dry bulk sediments; (**B**) C/N elemental ratio in treated samples containing black carbon; (**C**) black carbon accumulation rate; (**D**) isotopic composition of black carbon; (**E**) isotopic composition of marine inorganic carbon in planktonic foraminifera (*solid line*) and in bulk carbonate (*dashed line*); (**F**) difference (Δ_P) between atmospheric $\delta^{13}C$ and $\delta^{13}C_{BC}$. *Thick gray lines* in D, E, and F are 5 point running averages. PDB = Peedee belemnite; Plio = Pliocene; Plt = Pleistocene

the East Asian summer monsoon intensification since the early Miocene (Fig. 5.43; Jia et al. 2003).

Noteworthy is that the integrity of each monsoon proxy depends on the extent to which it responds to monsoon forcing only or, if influenced also by non-monsoon processes, the extent to which this additional signal can be identified and removed. Examples include the difficulty in differentiating the monsoon signal from the sea level signal when interpreting the deep-sea pollen record (Sun et al. 2003; Wang P. et al. 2005). For these reasons a multi-proxy approach is often employed, an example is the summer monsoon factor for the northern Arabian Sea based on factor analyses of five proxies: lithogenic grain size, Ba accumulation rate, $\delta^{15}N$, abundance of *G. bulloides* and opal mass accumulation rate (Clemens and Prell 2003). While the multi-proxy approach is clearly useful in that it provides a means of accounting for problems associated with multiple mechanisms influencing individual proxies, it does not replace the need for detailed understanding of the mechanisms influencing individual proxies.

The key to reliable paleoclimate reconstruction is an exact understanding of the various processes contributing to variance in any specific climate proxy. One means of accomplishing this is through long-term sediment trap studies. Careful analyses and applications of the sediment trap data will significantly improve our knowledge of monsoon proxies. However, it must be kept in mind that sediment trap results do not include many processes which take place during and after deposition which can strongly distort the integrity of a climate proxy signal. In addition, the majority of trap deployments is of short duration (two years or

less) and limited to one or two trap locations. Long-duration efforts are required
to record the entire range of climate variability associated, for example, with El
Niño. Short duration efforts of a year or two cannot record the entire range of variability necessary to evaluate proxy response under the full range of modern climate
conditions. Multiple traps are required to record gradients which are often prevalent and important in regions of large variability such as the monsoon influenced
regions.

High-quality core-top data sets are another base to establish and evaluate monsoon proxies. An example is the distribution of planktonic foraminiferal species in
the northern Indian Ocean based on analyses of 251 core-top samples (Cullen and
Prell 1984). This survey not only illustrated the link between *G. bulloides*% and
monsoon-driven upwelling, but also brought to light the alteration of planktonic
foraminiferal assemblages by carbonate dissolution. Similar work has been done
for planktonic foraminifera (Pflaumann and Jian 1999) and pollen (Sun et al. 1999)
in the SCS, but the number of sites is relatively low. In general, there are too few
core-top data-sets published for the Asian monsoon region, and continued work on
systematic core-top analyses to further refine monsoon proxies should be a high
priority.

Tectonic-Scale Long-Term Evolution

Long-term monsoon records have been developed from DSDP/ODP cruises to the
Indian Ocean and the Mediterranean sea (Legs 22, 24, 115, 116, 117, 121), as well
as from studies of the Chinese Loess Plateau. Early monsoon studies in the SCS,
however, were limited to the late Quaternary (Sarnthein et al. 1994; Wang L. et al.
1999a) until ODP Leg 184 in 1999, which has enabled us to study the long-term
evolution of summer and winter monsoons. The first question to be answered is
how far the East Asian monsoon history can be traced back.

Onset of the East Asian Monsoon System

The monsoon upwelling indicator, *Globigerina bulloides*, significantly increased
only about 8.5 Ma at ODP Site 722, Arabian Sea (Kroon et al. 1991; Prell et al.
1992). In age, this is very close to the rapid ecological transition from C_3-dominated
to C_4-dominated vegetation about 7.4–7.0 Ma, as revealed by the $\delta^{13}C$ data of pedo-
genic carbonates from northern Pakistan in the Himalayan foreland, and interpreted
then as marking the origination or intensification of the Asian monsoon system
(Quade et al. 1989). Numerical modeling supports the hypothesis of intensified
uplift of the Tibetan Plateau around 8 Ma causing enhanced aridity in the Asian
interior and the onset of the Indian monsoon (Prell and Kutzbach 1997). The loess-
paleosol profile in China extended the history of the monsoon from 2.6 back to
7–8 Ma (Sun D. et al. 1998; Ding et al. 1998; An et al. 2001). However, the recent
discovery of the Miocene loess-paleosol profile at Qin'an, western Loess Plateau,

indicates that the East Asian monsoon could be traced back to 22 Ma at the least (Guo et al. 2002).

There is yet no deep-sea data to discern the East Asian monsoon history over the entire late Cenozoic because pollen records from the SCS are limited only to the Oligocene and the Quaternary, although the longest sediment record of Leg 184 (Site 1148 on the lowermost northern slope) spans the last 32 myr. At Site 1148, a suite of element ratios such as Al/Ti, Al/K. Rb/Sr and La/L, indicative of the intensity of chemical weathering, increased abruptly around 29.5 Ma, in an early seafloor spreading stage of the SCS (Fig. 5.42). This event implies an increase in humidity, but whether it was related to the first East Asian monsoon is not clear (Wang P. et al. 2003).

A 30 myr long stable isotopic record of marine-deposited black carbon from regional terrestrial biomass burning from ODP Site 1148 of the northern SCS reveals photosynthetic pathway evolution of terrestrial ecosystems in the late Cenozoic. This record revealed 5 positive excursions and indicates that C_3 plants negatively adjusted their isotopic discrimination and C_4 plants appeared gradually as a component of land vegetation in East Asia since the early Miocene (21–22 Ma). This record coincides with the Qin'an Miocene loess profile in geological time (Guo et al. 2002), but significantly predates the 7–8 Ma age of sudden expansion of C_4 plants in Pakistan during the late Miocene and Pliocene (Fig. 5.43). The changes in terrestrial ecosystems with time can be reasonably related to the evolution of East Asian monsoons (Jia et al. 2003), although this evidence cannot be conclusive as the C_4 plants expansion may be caused by factors other than monsoon.

Sun and Wang (2005) compiled available marine records from the South and East China Seas together with profiles from onshore basins of China to approach the Cenozoic evolution of the East Asian monsoon. From 125 off- and on-shore sites with pollen and paleobotanical data collected, plus lithological indicators from all over the country, the distribution patterns of arid versus humid climates were reconstructed for five epochs. The results support the model that a broad arid zone was stretching across China in the Paleogene, but retreated to northwest by the end of the Oligocene, indicating the transition from a planetary to a monsoonal system in atmospheric circulation over the region. A variety of evidence, such as the Qin'an Miocene loess profile (Guo et al. 2002), the monsoonal Miocene mammalian fauna discovered in Southeast Asia (Ducrocq et al. 1994), and paleo-climate modeling (Ramstein et al. 1997), all support the existence of the Asian monsoon before the early Miocene. However, these new data do not support an onset of the Asian monsoon system around 8 Ma (Prell and Kutzbach 1997). Rather, the new data led to a hypothesis that the transition to the monsoon climate system in East Asia occurred in the latest Oligocene. The reorganization of the climate system, the appearance of C_4 plants and the increase in humidity in East Asia around the Oligocene/Miocene boundary provide evidence for the establishment of the modern East Asian monsoon. Since then, the Neogene has witnessed significant variations of the monsoon system, including enhancement of aridity and monsoon intensity around 8 Ma and 3 Ma. Thus, the Asian monsoon system has a longer history than previously thought (Wang P. et al. 2005; Sun and Wang 2005).

Development of the East Asian Monsoon System

No remarkable change in sediment accumulation was found at ODP Leg 184 sites in the SCS around 8 Ma (Wang P. et al. 2000). In oceans, the monsoon-driven upwelling can lead to increased productivity and shoaled thermocline. Important faunal signals of strengthened monsoons include abundance increases in productivity-indicative planktonic foraminifera due to upwelling and decreases in the percentage of planktonic foraminifera living in the mixed layer because of a shoaled thermocline. As already mentioned (see "Monsoon proxies"), the relative abundance of *G. bulloides* is a good proxy of upwelling-related high productivity in the Arabian Sea, but it never shows high abundance in the SCS. Instead, the percentage of *N. dutertrei* better unveils the monsoonal variability. According to previous studies and comparison of various stable isotope measurements, species of *Globigerinoides*, *Globigerinita*, and *Globigerina* comprise the major part of shallow water dwellers in the mixed layer. Figure 5.44 plots the variations of *N. dutertrei*% and shallow water dwellers% at Site 1146 in the northern SCS over the past 12 myr. It shows that *N. dutertrei*% increased abruptly at 7.6 Ma, and further increased from 3.2 to 2.0 Ma. Opposite to this trend are an abrupt decrease in the shallow water dwellers% after 8 Ma and a further decrease from 3.2 to 2.0 Ma, although their abundance has also been affected by carbonate dissolution around 11 Ma (Wang P. et al. 2003). Together, these planktonic foraminiferal results indicate paleo-monsoon enhancements at 8–7 Ma and 3.2–2.0 Ma.

The strengthening of the summer monsoon around 8 Ma is also supported by the increase in Pyloniid radiolarians at Site 1143 (Chen et al. 2003). Considering

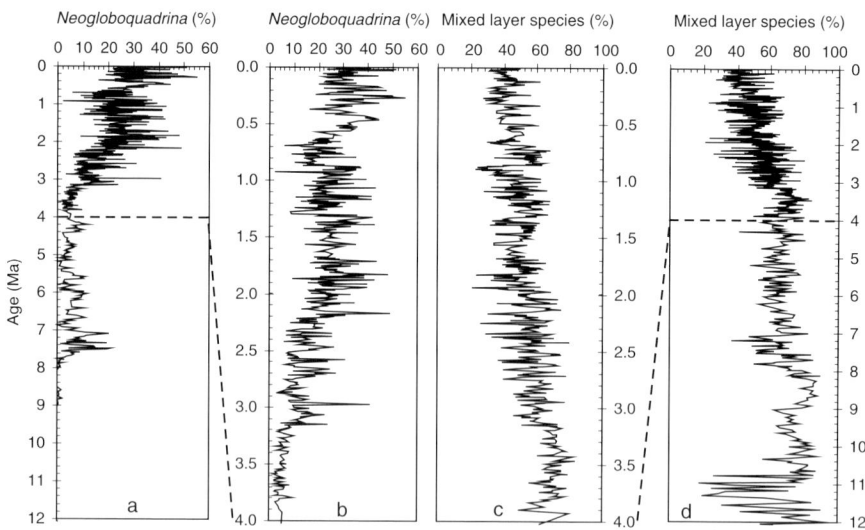

Fig. 5.44 Downcore variations of planktonic foraminiferal percentages at ODP Site 1146 show major changes at ∼8 Ma and since ∼3 Ma: (**a, b**) *Neogloboquadrina dutertrei*%; (**c, d**) mixed layer shallow-water species% (Wang P. et al. 2003)

the first obvious strengthening of the Indian monsoon at 8 Ma broadly concurred with the initiation of the Red Clay eolian deposition on the loess plateau, the SCS results confirm a significant enhancement of the Asian monsoon system around 8 Ma. Further development of monsoons from 3.2 to 2.0 Ma was also manifested by the increase of opal abundance at Site 1143 (Li J. et al. 2002) and by drastic coarsening of terrigenous clastic grain size due to intensified eolian transportation (Ding et al. 1992). Judging from the modern planktonic foraminiferal $\delta^{13}C$ distribution in the region, the prominent decrease in planktonic $\delta^{13}C$ at Site 1148 from 3.1 to 2.2 Ma was likely responding to the intensification of the winter monsoon (Jian et al. 2003). In the modern SCS, winter monsoons exert a major influence on the variation of productivity and thermocline depth. Therefore, the Leg 184 monsoonal proxies discussed above are interpreted as result of the intensification of the winter monsoon caused by the growth of boreal ice sheets.

The enhancements of the East Asian monsoon at 8–7 Ma and 3.2–2.0 Ma correspond well to the Indian monsoon records (Prell et al. 1992). Zheng H. et al. (2004) compared the monsoonal records from the Indian Ocean, Loess Plateau, the SCS and the North Pacific, and found that intensification of Asian aridity and monsoons around 8 Ma and 3 Ma ago has been convincingly documented both on land and in the ocean, which display similar stages in the development of the East and South Asian monsoons (Fig. 5.45), with an enhanced winter monsoon over East Asia being the major difference (Wang P. et al. 2003). Increased aridity around 8–7 Ma was recorded also in the North Pacific as a peak in dust accumulation rate (Rea et al. 1998), but this interval of higher dust accumulation was not sustained, unlike the younger record of a major dust increase since about 3.5 Ma (Fig. 5.45).

The number and geographic coverage of monsoon records decrease with increasing age and thus our knowledge of pre-Quaternary monsoon history remains relatively poor. Only long records will provide the opportunity to test the numerous hypotheses regarding the long-term history of the East Asian monsoon and the relative roles played by uplift, sea-land distribution and oceanic gateways. Therefore, we need more long and high-quality marine records from the SCS, including hemipelagic and lacustrine sediments of Miocene and Paleogene age.

Orbital-Scale Variability

Late Pliocene Monsoon Variations

Orbital-driven monsoon variations in the Pliocene can be demonstrated with isotopic records from ODP sites in the SCS. The period from 3.3 to 2.5 Ma in the late Pliocene is characterized by a continuous increase of global ice volume as recorded by a trend of positive shift in benthic $\delta^{18}O$. The planktonic $\delta^{18}O$ level, however, remains relatively constant, resulting in an increasing difference between benthic and planktonic isotopes (Jian et al. 2003; Tian et al. 2006). To test the connection of the isotopic signals with monsoon variability, the ice volume, water temperature, and salinity effects need to be distinguished first. As already introduced in the section

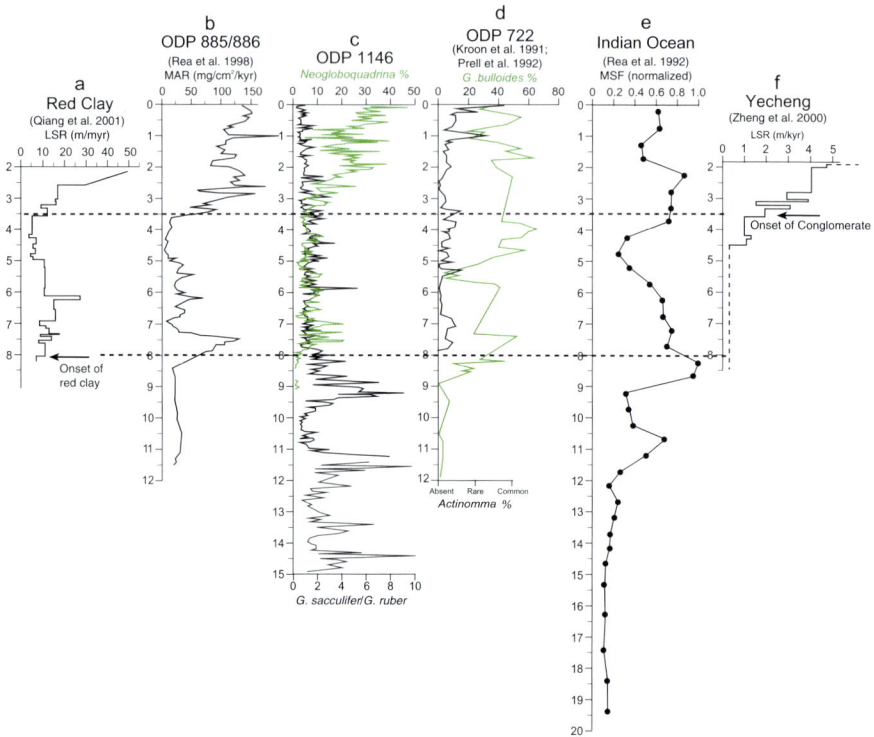

Fig. 5.45 Terrestrial and marine records reveal the monsoon history since the middle Miocene: (**a**) linear sedimentation rate of Jiaxian red clay in the Chinese Loess Plateau; (**b**) mass accumulation rate of eolian flux to the North Pacific (ODP Sites 885/886); (**c**) ratio of abundance of *G. sacculifer* and *G. ruber* (*solid line*) and abundance of *Neogloboquadrina* (*green line*) from the SCS (ODP Site 1146); (**d**) abundance of *G. bulloides* and relative abundance of radiolarian *Actinoma* spp. from the Indian Ocean (ODP Site 722); (**e**) normalized mean sediment flux to the Indian Ocean; (**f**) linear sedimentation rate of Yecheng molass (from Zheng et al. 2004)

"Sea surface temperature history", the paleo-SST at ODP Site 1143 from 3.3 to 2.5 Ma was estimated using the *G. ruber* Mg/Ca ratio (Fig. 5.3). The surface water $\delta^{18}O_{sw}$ can be calculated from the paired-*G. ruber* $\delta^{18}O$ and Mg/Ca ratio-based SST using the *Orbulina* low-light paleotemperature equation (Bemis et al. 1998). The derived $\delta^{18}O_{sw}$ shows a pattern of steady glacial/interglacial cycles from 3.3 to 2.5 Ma, without any obvious long-term trends (Tian et al. 2006). Since the $\delta^{18}O_{sw}$ depends both on global ice volume and water salinity, the regional sea surface salinity (SSS) variations can be estimated by removing the ice volume effects from the $\delta^{18}O_{sw}$, i.e., the residual $\Delta\delta^{18}O_{sw-b}$. The $\Delta\delta^{18}O_{sw-b}$ of Site 1143 was obtained by subtracting the benthic foraminiferal $\delta^{18}O$ from the calculated $\delta^{18}O_{sw}$. The derived $\Delta\delta^{18}O_{sw-b}$ shows a stepwise decrease during the period of significant ice sheet growth 3.3–2.5 Ma ago (Fig. 5.46C) (Tian et al. 2006).

The $\Delta\delta^{18}O_{sw-b}$ reflects SSS variations at Site 1143, which are associated with regional changes of precipitation and fluvial runoff in the southern SCS region. The

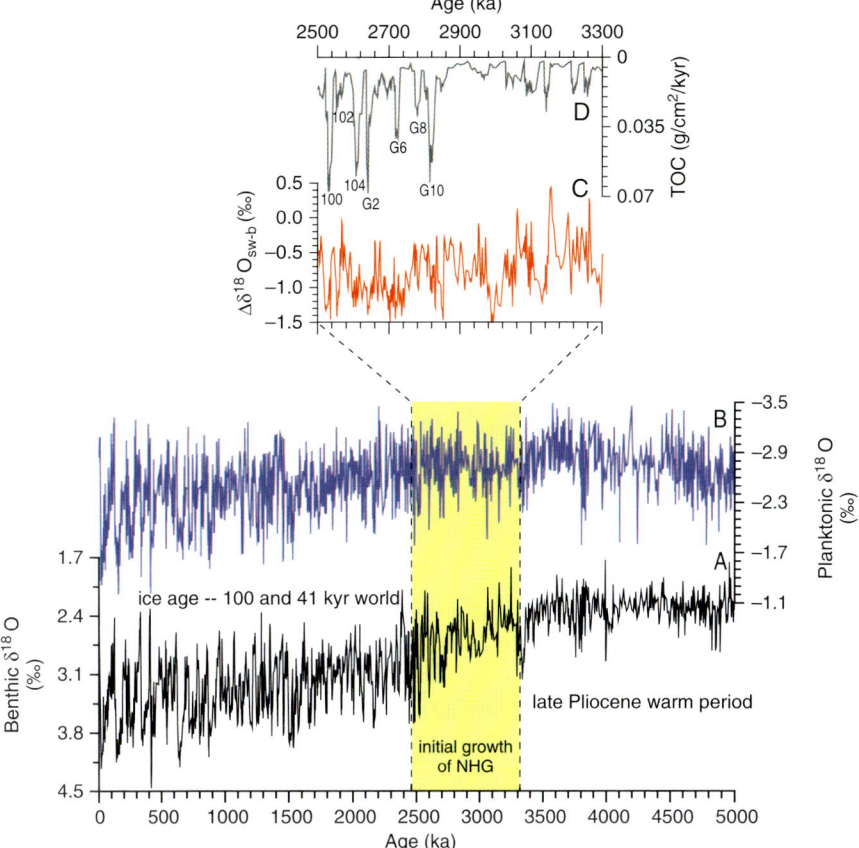

Fig. 5.46 Paleo-monsoon records from ODP Site 1143 show monsoon-driven salinity changes in the period of 2.5–3.3 Ma when Northern Hemisphere Glaciation (NHG) started: (A) *Cibici-doides* $\delta^{18}O$ for the past 5 myr, (B) *G. ruber* $\delta^{18}O$ for the past 5 myr, (C) $\Delta\delta^{18}O_{sw-b}$, the difference between sea water $\delta^{18}O$ and *Cibicidoides* $\delta^{18}O$, (D) total organic carbon mass accumulation rate, with marine isotope stages indicated (from Tian et al. 2006)

ice sheet growth caused an overall long-term sea level lowering of ~43 m (Mudelsee and Raymo 2005), which shortened the distance between Site 1143 and the continent, leading to a local decrease of SSS. Meanwhile, strengthened East Asian summer monsoons increased the annual mean precipitation in and the fluvial runoff to the southern SCS, which in turn resulted in SSS decreases at Site 1143. Thus, the $\Delta\delta^{18}O_{sw-b}$ variations from 3.3 to 2.5 Ma may reflect a stepwise development of the East Asian monsoons (Fig. 5.46C).

The down-core measurements of TOC flux at Site 1143 show clear glacial/interglacial cycles from 3.3 to 2.5 Ma, with higher values during glacial periods and lower values during interglacial periods (Fig. 5.46D). Just after 2.82 Ma (MIS G10), the amplitude of the TOC flux fluctuations within a glacial/interglacial cycle rapidly

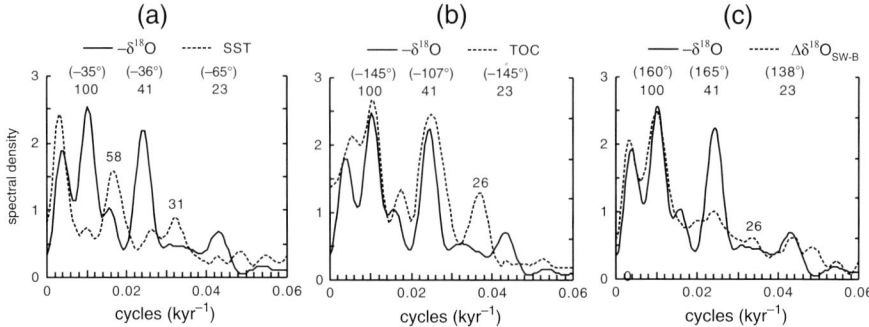

Fig. 5.47 Spectral analyses of *Cibicidoides* $-\delta^{18}O$ and Mg/Ca-derived SST, TOC and $\Delta\delta^{18}O_{sw-b}$ records from ODP Site 1143 reveal major cyclicities for the period of 2.5–3.3 Ma: (**a**) spectrum of $-\delta^{18}O$ (*solid line*) and SST (*dashed line*); (**b**) spectrum of $-\delta^{18}O$ (*solid line*) and TOC (*dashed line*); (**c**) spectrum of $-\delta^{18}O$ (*solid line*) and $\Delta\delta^{18}O_{sw-b}$ (*dashed line*) (Tian et al. 2006). Numbers denote the phases at the 100 kyr, 41 kyr and 23 kyr bands. Negative phases denote SST leads $-\delta^{18}O$, and positive phases denote TOC or $\Delta\delta^{18}O_{sw-b}$ lags $-\delta^{18}O$

increased, reaching a level nearly 3 to 4 times the amplitude observed prior to MIS G10 (Fig. 5.46D). As modern productivity in the southern SCS is associated with summer monsoon driven upwelling, the variations of TOC flux at Site 1143 indicate a rapid strengthening of the East Asian summer monsoon after 2.82 Ma. Both TOC flux and $\Delta\delta^{18}O_{sw-b}$ records reveal a general enhancement of the East Asian monsoons, especially the summer monsoon during the period of the late Pliocene ice sheet growth (Fig. 5.46).

Since SST, TOC and $\Delta\delta^{18}O_{sw-b}$ all vary someway in response to changes in monsoon intensity, cross-spectrum analyses were carried out between the monsoon proxies and benthic foraminiferal $\delta^{18}O$ (Fig. 5.47). Compared to the benthic $\delta^{18}O$, the SST records show weak variations at the primary orbital periodicities of 100 kyr, 41 kyr and 23 kyr, but relatively strong variations at the periodicities of 58 kyr and 31 kyr. Coherencies at the orbital periodicities are all above the 80% alarm test level, indicating a coherent relationship between the benthic $\delta^{18}O$ and the SST records. Phase relations reveal a lead of the SST changes relative to the global ice volume changes at each cycle (Fig. 5.47a), namely a lead of 35° at the 100 kyr band, equivalent to 9.72 kyr; a lead of 36° at the 41 kyr band, equivalent to 4.1 kyr; and a lead of 65° at the 23-kyr band, equivalent to 4.1 kyr. From 3.3 to 2.5 Ma, the 3–4 kyr lead of SST change at ODP Site 1143 relative to global ice volume signal at both the obliquity and precession bands is similar to that in the early Pleistocene, indicating a stationary phase relationship since the Pliocene.

TOC and $\Delta\delta^{18}O_{sw-b}$ are spectrally similar to the benthic $\delta^{18}O$, performing strong 100 kyr, 41 kyr and 23 kyr cycles (Fig. 5.47b,c). In addition, both the TOC and $\Delta\delta^{18}O_{sw-b}$ records show a periodicity of 26 kyr which is within the range of the precession in the Pliocene (Berger et al. 1992). Coherencies above the 80% alarm test level indicate coherent relationship between the benthic $\delta^{18}O$ and these two monsoon-related proxy records at the primary orbital periodicities. However, phase relations reveal that these two records lag the global ice volume change at the 100 kyr, 41 kyr and 23 kyr bands. The phase offsets between the TOC and the benthic $\delta^{18}O$ record are 145° at the 100 kyr band equating with a lag of 40.2 kyr, and 107° at the 41 kyr band equating with a lag of 12.2 kyr, and 145° at the 23 kyr band equating with a lag of 9.2 kyr (Fig. 5.47b). The phase offsets between the $\Delta\delta^{18}O_{sw-b}$ and the benthic foraminiferal $\delta^{18}O$ are 160° at the 100 kyr band equating with a lag of 44.4 kyr, and 165° at the 41 kyr band equating with a lag of 18.8 kyr, and 138° at the 23-kyr band equating with a lag of 8.8 kyr (Fig. 5.47c). The cross spectral analyses thus show that the global ice volume change is a factor internal to the climate system with great influence on the variability of East Asian monsoons at least since the Pliocene. In the Pleistocene, monsoon proxy records in the Arabian Sea indicate that the monsoon maximum usually lags the minimum of global ice volume by 2–4 kyr at both the obliquity and precession bands (Clemens et al. 1991, 1996), much smaller than the lag of 8–18 kyr found at Site 1143. The differences reveal a nonstationary phase of the East Asian monsoon relative to the global ice volume in the Plio-Pleistocene. The numerical climate-model experiment reveals that the evolution of the Asian monsoons is linked to phases of Himalaya-Tibetan plateau uplift and to the boreal ice sheet growth (Prell and Kutzbach 1997). Thus, the decreased phase lag at the obliquity and precession bands is possibly linked to the amplified northern hemisphere ice sheets and the increased contrast between glacial and interglacial periods.

To sum up, the data from the southern SCS show important linkage of the East Asian monsoon with both the tropical and high latitude forcings. The records of ODP Site 1143 from the southern SCS reveal that East Asian monsoons gradually strengthened in response to the phased expansion of ice sheets in the high northern latitudes during the late Pliocene, with increased phase offsets at the obliquity and precession bands relative to the late Pleistocene. This finding suggests that the East Asian monsoons are not only simply driven by northern summer insolation at the precession period but also modulated by global ice volume change in high latitudes (Tian et al. 2006).

Pleistocene Monsoon Variations: Glacial/Interglacial Cycles

Compared to the late Pliocene, the Pleistocene displayed much more significant amplitude changes with glacial/interglacial cycles due to a prevalent monsoon climate. In the modern SCS, winter (from December to March) is the season when the average seasonal mixed layer reaches its maximum depth, and the SST decreased to the annual minimum, while the proxies of productivity, including opal%, *P. obliquiloculata* and *G. ruber* flux, organic carbon flux, all exhibit higher values than other seasons (Fig. 5.48). Accordingly, in the Pleistocene, all the monsoon proxies are expected to vary more strongly in glacial than interglacial cycles. Decreased SST and intensified winter monsoon during glacials strengthened the mixing in the upper water and thickened the mixed layer in the region as indicated by decreased $\Delta\delta^{18}O_{(P-G)}$ values at Site 1143 (Fig. 5.20C).

Fig. 5.48 Sediment trap records from the central SCS indicate seasonal changes in all variables, including (**A**) the average seasonal mixed layer depth for the SCS by integrating data from Levitus (1982); (**B**) SST °C; (**C**) opal%; (**D**) *P. obliquiloculata* (*dashed line*) and *G. ruber* (*solid line*) flux, individuals.m^{-2}.d^{-1}; (**E**) organic carbon flux, mg.m^{-2}.d^{-1}; (**F**) primary productivity, mg.m^{-2}.d^{-1} (from Tian et al. 2005b)

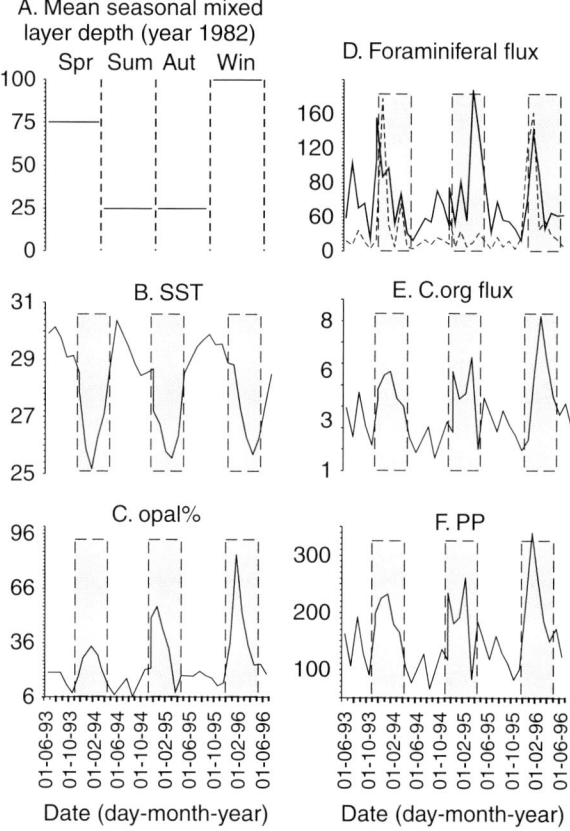

A compilation of some monsoon proxies over the last 800 kyr (Fig. 5.49) shows higher herbaceous pollen%, lower $\Delta\delta^{18}O_{(P-G)}$ values and lower *P. obliquiloculata*% during glacial intervals, indicating intensified winter monsoon in the SCS and southern China. *P. obliquiloculata* is a typical deep dwelling planktonic foraminifer, living in the uppermost part of the thermocline or below the bottom of the mixed layer. Since the sediment trap reveals its high flux (also *G. ruber* flux) corresponding to the deep winter mixed layer, the higher abundances of *P. obliquiloculata* during glacials (Fig. 5.49B) possibly reflect a deeper thermocline associated with cooling climate and intensified winter monsoon (Tian et al. 2005b). However, relating its higher abundances with a deep thermocline in the southern SCS remains to be confirmed because other studies indicate that deep-dwelling species including *P. obliquiloculata* and *Globorotalia tumida* often dominate the planktonic assemblage during times of a shallower surface mixed layer when the thermocline is located within the photic zone (Bé et al. 1985).

Analogous to higher opal concentrations during winter within the annual cycle as revealed in the sediment trap study in the central SCS (Fig. 5.48C), the opal flux at Site 1144 from the northern SCS displays higher concentrations during glacials than

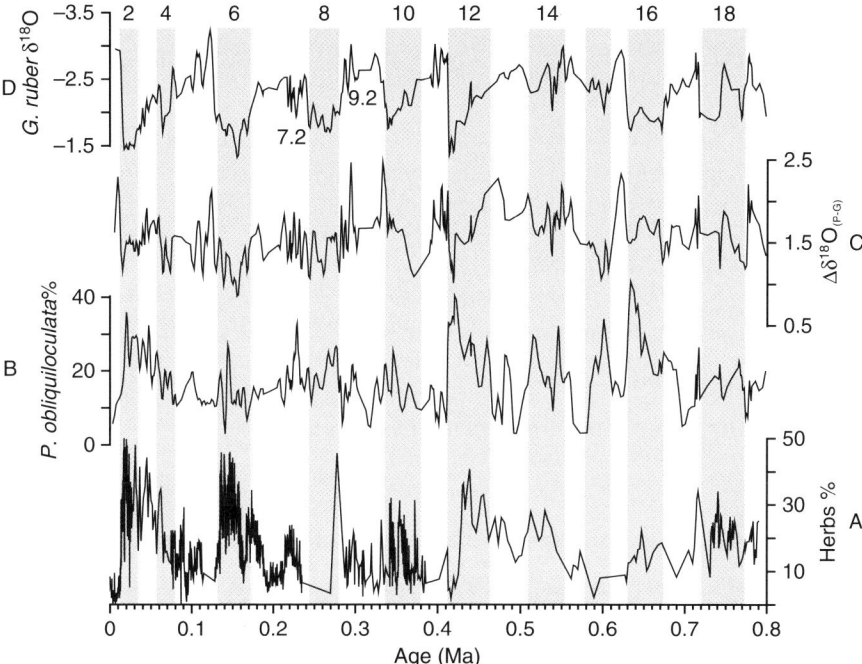

Fig. 5.49 East Asian monsoon proxy records from the SCS show monsoon variations over the last 800 kyr: (A) herbs% of ODP Site 1144 (Sun et al. 2003); (B) *P. obliquiloculata*% of Site 1143 (Xu et al. 2005); (C) $\Delta\delta^{18}O_{(P-G)}$ of ODP Site 1143 (PDB, ‰, 3-point Gaussian smoothing); (D) *G. ruber* $\delta^{18}O$ of Site 1143 (PDB, ‰, 3-point Gaussian smoothing), with numbers and shaded bars marking cold marine isotope stages (MIS) (Tian et al. 2005b)

during interglacials for the past 1.56 myr (Li and Wang 2004). At a nearby locality, Site 1146, siliceous microfossils also show higher abundances and higher accumulation rates during glacials than during interglacials for the past 1.0 myr (Wang R. et al. 2003). The evidence of increased opal flux and siliceous microfossils together implies higher glacial siliceous productivity in the northern SCS caused by strong East Asian winter monsoon winds. The intensified East Asian winter monsoon winds during glacials likely enhanced the inflow of upwelled nutrient-rich water through Bashi Strait and the transportation of more eolian dust, resulting in nutrient increases and higher productivity in surface waters of the northern SCS (Wang L. et al. 1999a).

Previous studies have revealed that most marine proxies are not associated with any particular climatic variable but with the integration of several variables, and only the variance held in common among several proxies can be attributed to the variable of interest. This approach is most useful when the proxies are of sufficiently different origins (chemical, physical, biological, isotopic) and from an array of sites such that the variance not held in common is largely independent (Clemens and Prell 2003). Although the $\Delta\delta^{18}O_{(P-G)}$ and *P. obliquiloculata*% of Site 1143, as well as herbs% and opal flux of Site 1144, show many differences in both long- and short-term variations (Fig. 5.49), they all share common features of strong glacial/interglacial variations that may have been caused by strong winter monsoons during glacials.

Therefore, the main implication of the $\delta^{18}O$ differences between *G. ruber* and *P. obliquiloculata* from Site 1143 is that an intensified East Asian winter monsoon was the primary factor affecting upper ocean thermal gradient variations, especially the mixed layer depth or the thermocline depth in the southern SCS during the Pleistocene. This conclusion differs from faunal-climate records from the East Pacific where the cool sea surface and the shoaled thermocline are driven by wind-induced upwelling. The hydrologic feature of a deep mixed layer or thermocline driven by strong winter monsoons in the SCS, as demonstrated by modern observations and comparisons between paleo-climate proxies, seems to be unique in marginal seas where upwelling is weak but monsoon influence is strong.

Pleistocene Monsoon Variations: Coherence and Phase Relationship

Clemens et al. (1991) found that the Earth's geometry (ETP, eccentricity+obliquity-precession) serves as the external forcing of the Indian Ocean summer monsoon whereas the latent heat across the equator from the southern Indian Ocean to the northern serves as the internal forcing. In addition, they also found that the global ice volume changes have little impact on the evolution of the Indian Ocean summer monsoon, conflicting with the GCM simulations (Kutzbach and Guetter 1986; Prell and Kutzbach 1987). Later, Wang L. et al. (1999a) considered that the East Asian monsoon probably had a similar forcing mechanism to that of the Indian Ocean summer monsoon based on several short and mainly low resolution proxy records from the SCS. Cores from ODP Leg 184 sites provide high quality and long sediment records, which enable a test of this hypothesis by examining all available faunal, floral and geochemical proxies, especially the coherency and phase relationship of the East Asian Monsoon with orbital forcing and global ice volume changes during the Pleistocene.

ODP Site 1143 in the southern SCS. The flux of opal (including radiolarians and diatoms) usually reflects modern ocean productivity (Dickens and Barron 1997), and monsoon-driven upwelling can enhance the siliceous productivity of the upper ocean. In the Indian Ocean, opal flux indicates changes in the intensity of the Indian summer monsoon. Similarly in the SCS, the siliceous productivity indicated by Opal% is also related to the East Asian monsoon changes (Wang and Li 2003; Wang R. et al. 2003). Sediment trap studies in the northern and central SCS show opal flux peaking in both winter and summer monsoon seasons (Tian et al. 2004). At Site 1143, the opal% is lower before 400 ka, ranging from 1.5 to 2.5%, with a smaller glacial/interglacial amplitude; after 400 ka, it increases abruptly from 1.5 to 6.0%, with a larger glacial/interglacial amplitude (Fig. 5.50). In general, the opal% of Site 1143 is lower during glacials but higher during interglacials, indicating stronger summer monsoon during interglacials.

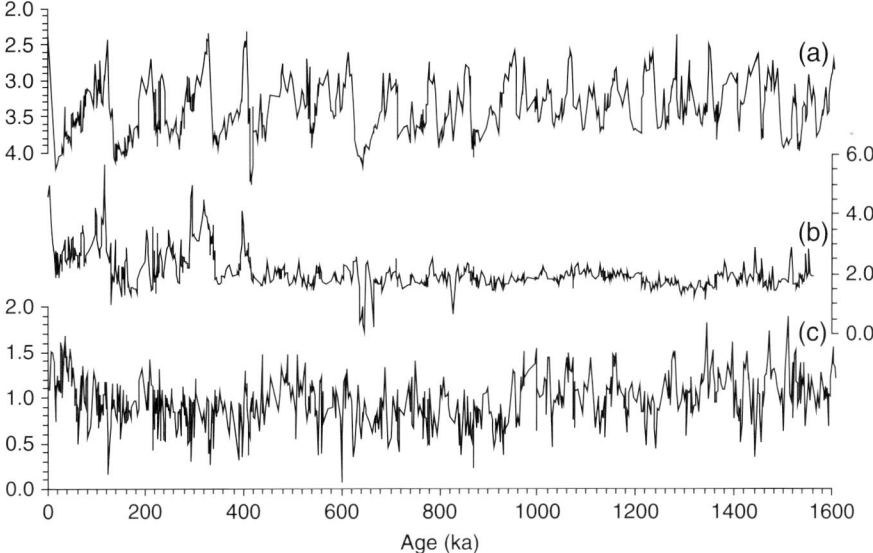

Fig. 5.50 Paleo-climate records of ODP Site 1143 show monsoon variations over the past 1.6 myr in the southern SCS: (**a**) benthic foraminiferal *Cibicidoides* δ^{18}O (‰, PDB) (Tian et al. 2002); (**b**) opal percentage (%) (Wang and Li 2003); (**c**) planktonic foraminiferal *G. ruber* δ^{13}C (‰, PDB) (Tian et al. 2005a)

When the nutrient-rich subsurface water upwells to the surface by the monsoon, the δ^{13}C of the sea surface water will decrease. If the foraminiferal shell equilibriums with the ambient water during the process of its calcification, the changes of the shell's δ^{13}C should record the changes of the monsoon intensity. In the SCS, in the upwelling areas off the northwest coast of the Philippines and off the east coast of Vietnam, the *G. ruber* δ^{13}C and *P. obliquiloculata* δ^{13}C in core-top samples show relatively lower values than in other areas (Wang L. et al. 1999). The planktonic foraminifer *G. ruber* δ^{13}C for the past 1.6 myr varies from 0.2 to 2.0‰ (Fig. 5.50). Although the glacial/interglacial change of *G. ruber* δ^{13}C is not as obvious as that of the benthic δ^{18}O, the high-frequency and large amplitude fluctuations in the *G. ruber* δ^{13}C of Site 1143 may reflect productivity variations related to monsoon variability.

ODP Site 1144 in the northern SCS. The herbs and *Pinus* found in the deep sea sediments of Site 1144 were probably transported by strong winter monsoon-driven currents (Sun et al. 2003), and thus can be used as the proxies of the East Asian winter monsoon. The abundance sum of these two groups exceeds 50% in most samples and they always show a reverse relationship with higher *Pinus*% in interglacials and higher herbs% in glacials (Fig. 5.25), indicating strengthened winter monsoon during glacials and weakened winter monsoon during interglacials (see detailed discussion in section "Vegetation history in deep-sea record" above).

Coherencies and phases of monsoonal proxy records with orbital and global ice volume forcing. Cross spectral analyses indicate that the pollen records of the

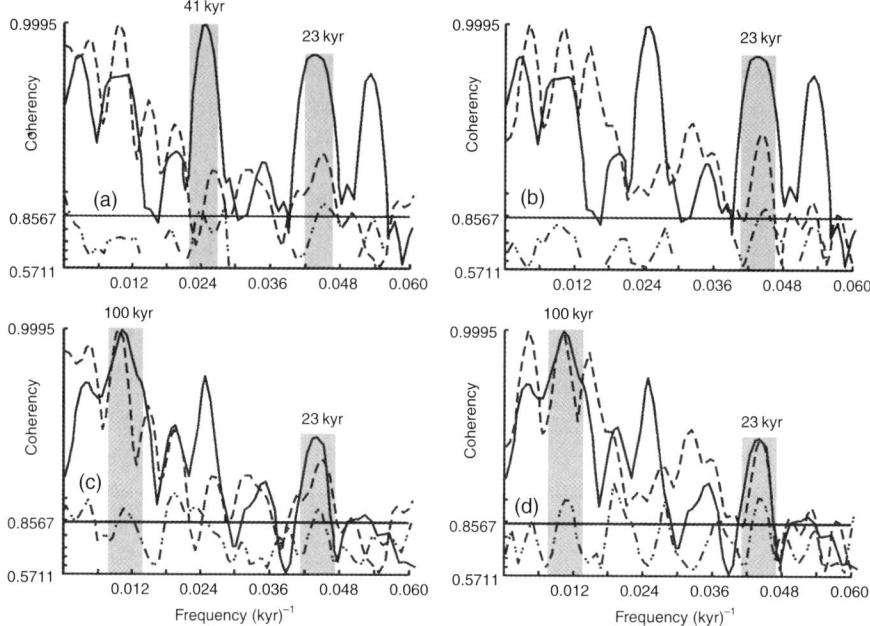

Fig. 5.51 Cross spectral analyses of the pollen records of ODP Site 1144 (0–1 Ma) with the orbital forcing (ETP) and the global ice volume variations (δ^{18}O) show dominant orbital periods at 100 kyr, 41 kyr and 23 kyr (*grey bars*) where high coherencies are present: (**a**) ETP vs herbs%; (**b**) ETP vs *Pinus*%; (**c**) δ^{18}O vs herbs%; (**d**) δ^{18}O vs *Pinus*% (Tian et al. 2005a). *Solid lines* denote the spectrums of ETP or δ^{18}O. *Dashed lines* denote the spectrums of the pollen records. *Dotted dashed line* denotes coherency. The *horizontal solid lines* denote the coherencies above 80% standard level

past 1 myr (Herbs% and *Pinus*%) at ODP Site 1144 demonstrate strong variances at the 100 kyr band and moderate to strong variance at the 41 kyr and 23 kyr bands, but they are coherent with the orbital forcing (ETP) only at the 23 kyr precession band (Fig. 5.51). The upper ocean temperature contrast in the southern SCS is also coherent with the ETP only at the precession band (Tian et al. 2004), as also in the east tropical Pacific (Ravelo and Shackleton 1995). These indicate that the tropical climate is externally controlled by the precessional radiation.

The herbs% and *Pinus*% are highly coherent with the global ice volume change (represented by foraminiferal $-\delta^{18}$O) at the 100 kyr and the 23 kyr bands (Fig. 5.51c,d). Though the cross spectrum between the pollen and the δ^{18}O records show no coherent relationship at the 41 kyr band, a highlighted coherency stands at the 54 kyr band, a heterodyne frequency of the primary orbital periods. A similar coherent relationship at the 54 kyr band also exists between the Indian Ocean monsoon tracers and the concentration of the orbital variance (Clemens et al. 1991). The 100 kyr glacial cycles are dominant in the Earth's Pleistocene climate records (Imbrie et al. 1992), especially in the foraminiferal δ^{18}O records. Both the herbs% and the *Pinus*% of Site 1144 display perfect glacial/interglacial variations for the past 1 myr. The graphic structure of the pollen records, which is identical to that of the $\delta^{18}O$, in conjunction with the cross spectral analyses, indicate that the East Asian winter monsoon has been greatly influenced by and coherent with the global ice volume change. The inference is consistent with the conclusions drawn from the loess-paleosol sequence (An et al. 2001) and from the north Pacific dust concentration (Rea et al. 1998) as well as from numerical simulations (Kutzbach and Guetter 1986). As the other important component of the Asian monsoon system, the Indian Ocean summer monsoon appears to be different, with proxies not coherent with the global ice volume change at the 100 kyr band but coherent at the 41 kyr and the 23 kyr bands. In addition, the fact that the ice-volume minima (maximum effectiveness of sensible heating) lag the Indian Ocean summer monsoon maxima by 33 kyr over the eccentricity band, excludes a forcing-response relationship between the Northern Hemisphere Glaciation and the Indian Ocean summer monsoon changes (Clemens et al. 1991).

Both *G. ruber* $\delta^{13}C$ and opal% of Site 1143 are coherent with the concentration of orbital variances at the three primary orbital cycles, the 100 kyr, the 41 kyr and the 23 kyr bands (Fig. 5.52a,b). Particularly, the coherency of *G. ruber* $\delta^{13}C$ with the ETP at the 23 kyr precession band is the highest, nearly two times the coherency

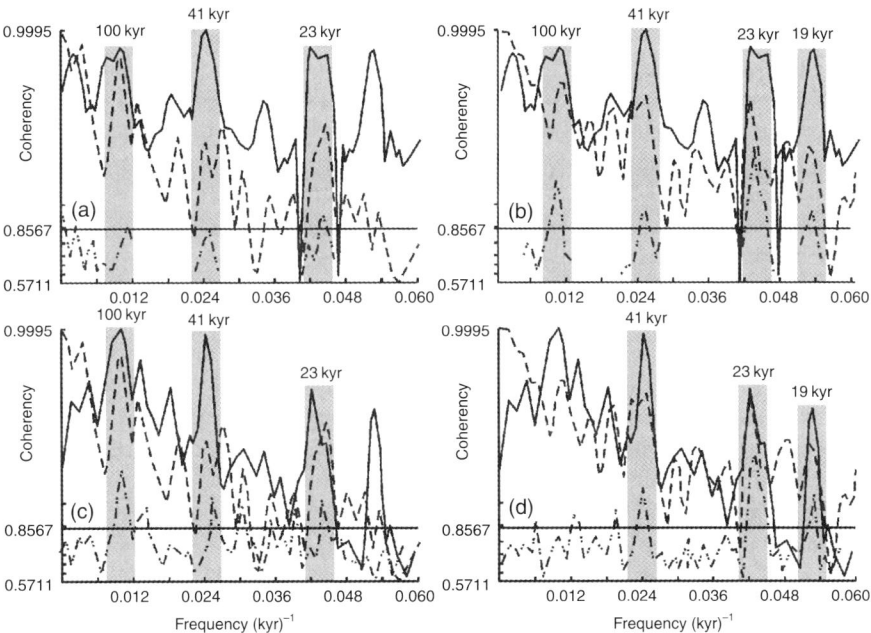

Fig. 5.52 Cross spectral analyses of *G. ruber* $\delta^{13}C$ and opal% of ODP Site 1143 with the orbital forcing (ETP) and the global ice volume variations ($\delta^{18}O$) show dominant orbital periods at 100 kyr, 41 kyr and 23 kyr (*grey bars*) where high coherencies are present over the last 1.6 myr: (**a**) ETP vs opal%; (**b**) ETP vs $\delta^{13}C$; (**c**) $\delta^{18}O$ vs opal%; (**d**) $\delta^{18}O$ vs $\delta^{13}C$ (Tian et al. 2005a). *Solid lines* denote the spectrums of ETP or $\delta^{18}O$. *Dashed lines* denote the spectrums of the proxy records. *Dotted dashed line* denotes coherency. The *horizontal solid lines* denote the coherencies above 80% standard level

at the eccentricity or obliquity bands. Also, the coherency of opal% with the ETP at the 23 kyr precession band is the highest among the three coherencies. These mean that the strongest responses of the East Asian monsoon to the orbital forcing occur at the precession band. This kind of relationship between the East Asian monsoon proxies and the orbital forcing highlights the precession as the primary force driving the tropical climate change. The opal% of Site 1143 is also highly coherent with the $-\delta^{18}O$ at the three primary orbital cycles; in addition, the *G. ruber* $\delta^{13}C$ is also strongly coherent with the $-\delta^{18}O$ at the 41 kyr band and the two precession bands (23 kyr and 19 kyr). Though the coherency of *G. ruber* $\delta^{13}C$ with the $-\delta^{18}O$ at the 100 kyr band does not exceed the 80% statistical level, a strong 100 kyr cycle also occurs in its spectrum. The coherent relationship of the opal% and *G. ruber* $\delta^{13}C$ with the $-\delta^{18}O$ is consistent with that between the pollen records and the $-\delta^{18}O$ at Site 1144 (Fig. 5.51c), revealing a close relationship between the East Asian monsoon variations and global ice volume changes.

Cross spectrum analysis has also revealed the phase relationships of the proxies with the orbital variance (ETP) at the orbital cycles. At the 100 kyr band, the opal% leads the ETP by $2.9° \pm 2.4°$, close to zero, while the pollen records are not coherent with the ETP, but in phase with $-\delta^{18}O$. Again, this indicates that the East Asian winter monsoon is greatly influenced by the global ice volume change. At the 41 kyr band, all monsoon tracers and the global ice volume change are coherent with the concentration of the ETP. Their phases relative to the ETP are close to each other if the phase errors are considered. This means that the orbital forcing has almost the same controls over the East Asia monsoon and the global ice volume change at the 41 kyr band. At the precession band, the phases of the monsoon proxies relative to ETP are overall much closer at the precession band than at the obliquity band, but they departure away from the phase of the $-\delta^{18}O$ relative to the ETP (see Tian et al. 2005a for further details).

Suborbital-Scale Variability

Paleomonsoon research was initiated with efforts to understand links between changes in monsoon intensity and large-scale boundary conditions such as orbital forcing and changes in global ice volume. These remain active areas of research. However, ice cores from Greenland have revealed records of climate change on far shorter time scales as well (Dansgaard et al. 1993; Grootes and Stuiver 1997). Now, it has been revealed that decadal- to millennial-scale climate variability is a global phenomenon, which was not limited to the ice cores and high latitudes of the North Atlantic, but also extended to other regions such as the tropical Pacific and monsoonal Asia. Millennial-to-centennial- scale variations have been extensively reported in the monsoon climate, including in the SCS, and a number of conceptual models are under development in efforts to constrain the forcing mechanisms underlying short-term changes in monsoon climate. Most hypothesize that the short-term changes in monsoon intensity are linked to internal oscillations in thermohaline circulation as well as atmospheric energy and moisture transfer. In some cases, a link to millennial- and centennial-scale variations of solar activity has been made on the basis of variations in ^{10}Be flux and ^{14}C production (Bond et al. 2001; Schulz and Paul 2002; Sarnthein et al. 2002). Exploration of the tidal influences on monsoon variability, like the solar cycles, has also been discussed (e.g. Fairbridge 1986). Because very high-resolution records are requested for reconstructing the sub-orbital variations in monsoon climate, only fragmentary pieces of information from the Pleistocene history are available so far in the SCS.

Millennial-Scale Cycles in the Middle Pleistocene

It has been found that the rapid climate changes occurred not only in the last glacial stage, but also in the Holocene in Asian monsoon regions (Schulz et al. 1998; Wang L. et al. 1999a; von Rad et al. 1999), implying that the decadal- to millennial-scale climate variability operates independently of the glacial/interglacial climate cycle. Similarly, sub-orbital climate fluctuations existed beyond the late Quaternary.

McManus et al. (1999) reported the millennial-scale climate variability throughout the past 500 kyr according to the changes of the ice rafted debris in the North Atlantic. Almost at the same time, Raymo et al. (1998) found that the millennial-scale climate variability can be traced back to 1.2 Ma. They claimed that millennial-scale climate instability may be a pervasive and long-term characteristic of Earth's climate, rather than just a feature of the strong glacial/interglacial cycles of the past 800 kyr. These findings fundamentally altered the way Earth scientists thought about the operation of the Earth's climate system and the relative sensitivity of this system to major climatic shifts.

In the SCS, ODP Site 1144 is distinguished by extremely high sedimentation rate in the last 1 myr (Bühring et al. 2001). Its ultrahigh-resolution color reflectance (L*) data (4 cm interval; time resolution of $\sim$10 yr), as a proxy of carbonate content in deep sea sediments (Sarnthein et al. 1994), provide a unique possibility to study millennial-scale climate variability during the middle Pleistocene climate transition (MPT).

Figure 5.53 shows the results of wavelet spectral analysis of the color reflectance (L*) data of Site 1144 based on the oxygen isotopic stratigraphy of Bühring et al. (2001), which demonstrate that the D/O-like millennial-scale climate fluctuations existed throughout the glacial, deglacial and postglacial stages over the last 1 myr (Jian and Huang 2003). Of particular interest is that after the mid-Pleistocene revolution (MPR) at 900 ka, along with the strengthening of the East Asian winter monsoon, the signal of millennial-scale climate fluctuations became stronger during glacial stages. Before the MPR, the signals of millennial-scale climate fluctuations were relatively stronger during interglacials, but after the MPT they became relatively stronger during glacials. It seems that the MPT was reflected not only by a change in dominant climate periodicities from 41 to 100 kyr, but also by a change in the characteristics of millennial-scale climate fluctuations. Although the new results need further evidence from high-resolution isotope analyses, it illuminates that orbital and millennial/centennial scale climate changes are interwoven between

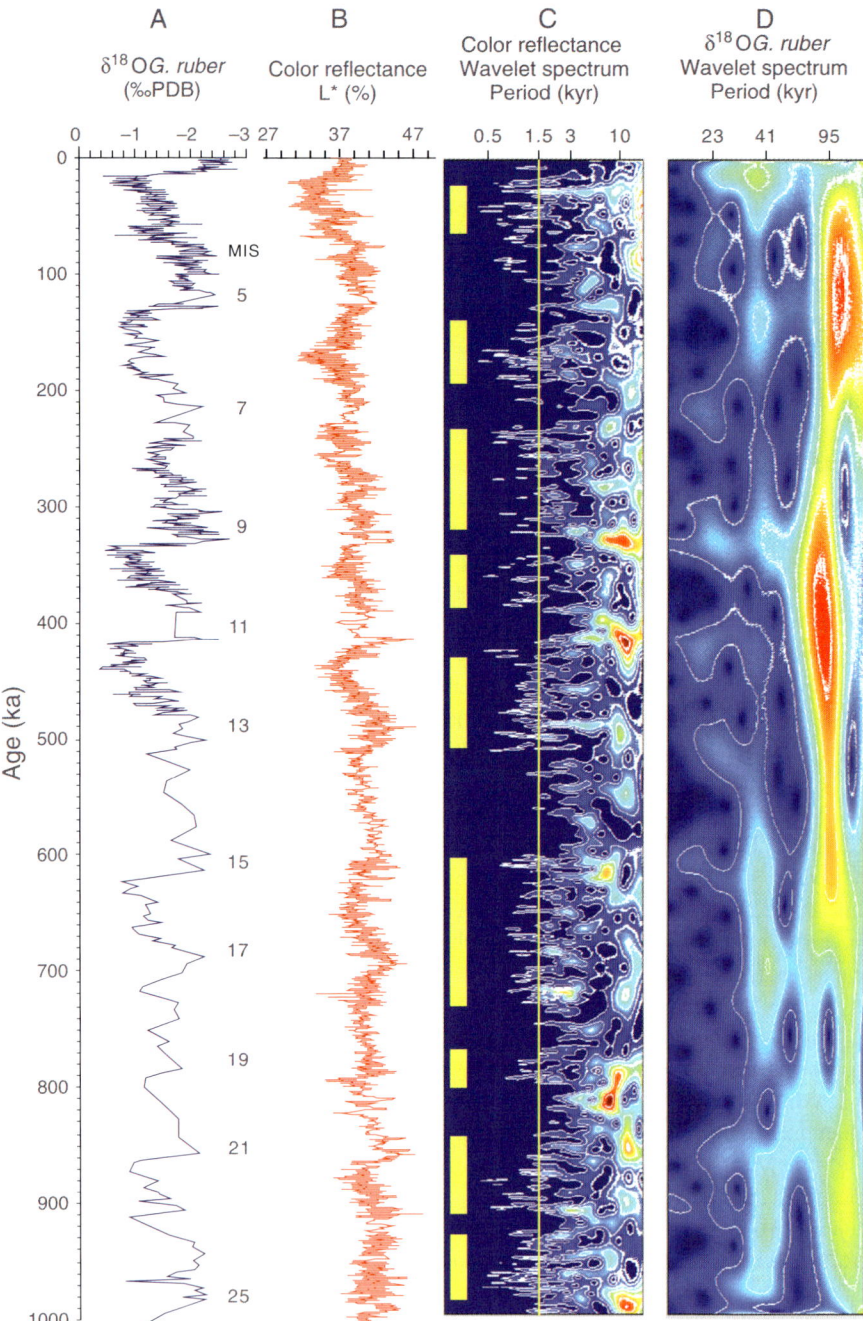

Fig. 5.53 Wavelet spectra of the δ^{18}O (**A, D**) and the color reflectance L* (**B, C**) show orbital and suborbital changes at ODP Site 1144 (Jian and Huang 2003). Number in (**A**) denotes interglacial marine isotopic stages (MIS). *Vertical yellow line* in (**C**) indicates the period of 1.5 kyr, and *short yellow bars* indicate intervals with strong millennial-scale climate fluctuations

them. This is very important for understanding the sub-orbital variability of the Asian monsoon system.

Dansgaard-Oeschger Cycles During the Last Glaciation

Monsoon variations on Dansgaard-Oeschger (D-O) time scales ($\sim$1.5 kyr) were reported from the carved sediment sections off Pakistan (Schulz et al. 1998), from bioturbated high-resolution sediment sections in the Arabian Sea (Sirocko et al. 1996), Bengal Fan (Kudrass et al. 2001), the SCS (Wang L. et al. 1999a; Bühring et al. 2001; Higginson et al. 2003; Zhao et al. 2006), the Sulu Sea (Oppo et al. 2001; de Garidel-Thoron et al. 2001), and the Sea of Japan (Tada et al. 1999). Wang L. et al. (1999a) were the first to report glacial millennial-scale D-O events from the SCS (Fig. 5.54). IMAGES Core MD972151 (Zhao et al. 2006) also offers an opportunity to correlate millennial- to centennial-scale SST variabilities in the SCS during the glacial period with the monsoon record from the Hulu Cave of eastern China (Wang Y. et al. 2001) and the D-O cycles in the GISP2 ice core record (Fig. 5.7). These results support that the glacial millennial-scale D-O events recorded in the SCS and on Chinese mainland are nearly synchronous with those found in Greenland ice cores, suggesting global climatic tele-connections.

As discussed earlier in this chapter, the SST difference between the southern and northern SCS can be used to monitor the strength of the East Asian winter monsoon, as illustrated in Fig. 5.55 with Mg/Ca-derived SST data from core 17961 and ODP Site 1145 (Oppo and Sun 2005). The S-N SST difference displays clear millennial-scale fluctuations at a period of $\sim$1.4 kyr (Figs. 5.13 and 5.14). During interstadials of the D-O events, the S-N SST difference decreased correspondingly, indicating the decreased winter monsoon, and vice versa during stadials (Fig. 5.55). The overriding features of glacial D-O cycles correspond to increased summer monsoons during interstadials and increased winter monsoons during stadials, consistent with the exceptionally well-dated monsoon records of the Hulu Cave (Wang Y. et al. 2001) as well as with loess records from central China (An 2000).

Millennial-To-Centennial-Scale Cycles in the Holocene

Sub-orbital climate fluctuations also occurred in the Holocene, although the D/O-type variability of the last glaciation is no longer dominant (Schulz et al. 1999). Instead, the Holocene interval is characterized by variability with a broad range of periodicities near 890–950 yr, 550 yr, 200 yr, 145 yr, 80–105 yr, 20–24 yr and 11 yr (Wang P. et al. 2005). These scales of variability may be linked to changes in solar activity, which produce very weak variations in incident solar energy (Labitzke 2001; Beer et al. 2000; Haigh 1996).

In the northern SCS, fluctuations in the SSS during the Holocene reflect centennial periods of 775 yr and 102/84 yr which exceed the upper limit of red noise at 80% confidence level (Fig. 5.56). These periods lie far below the established range of Milankovitch orbital forcing, and also differ from the millennial-scale glacial D-O cycles (Wang L. et al. 1999a). Instead, the 102/84-year periodicities possibly reflect the Gleissberg cycle of solar activity.

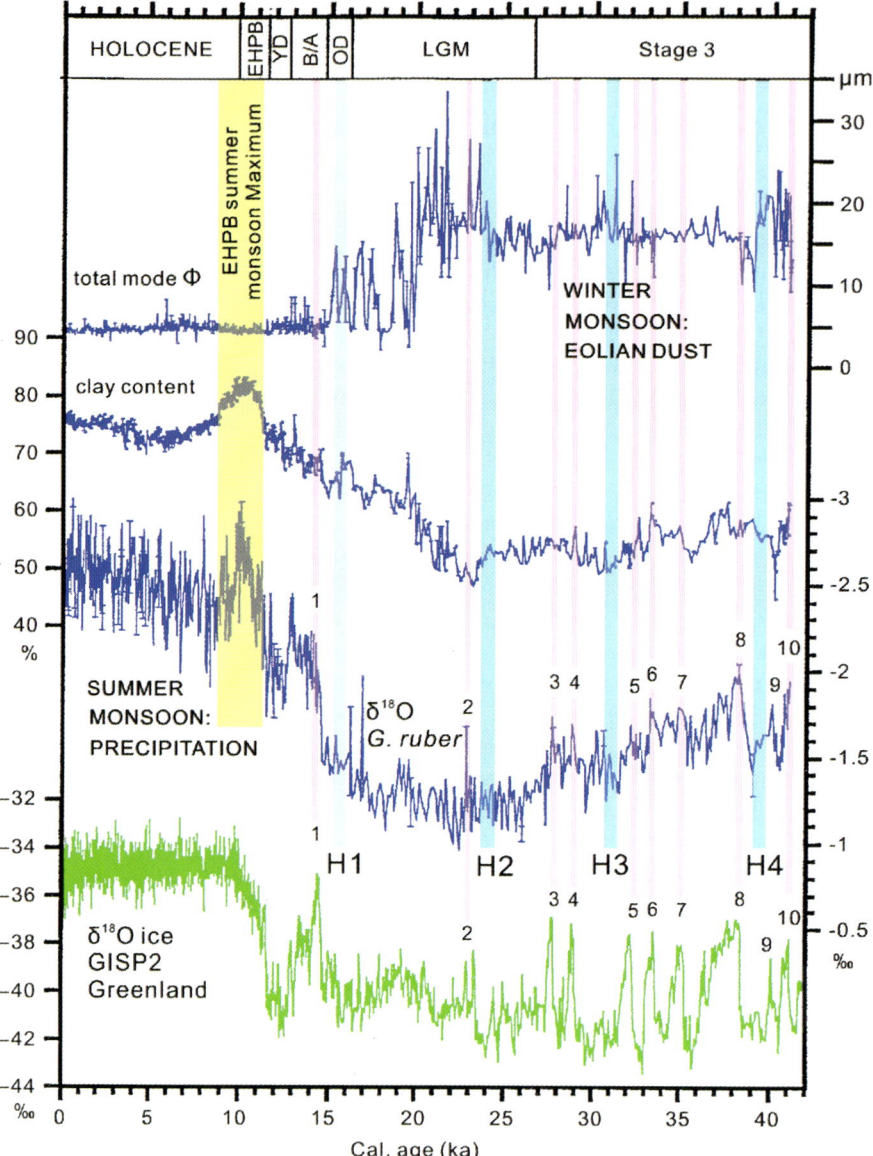

Fig. 5.54 Dansgaard-Oeschger (D-O in numbers) events and Heinrich events (H1 to H4) are recorded in core 17940-2 from the northern SCS (modified from Wang L. et al. 1999a). OD = Oldest Dryas; B/A = Bølling-Allerød; YD = Younger Dryas; EHPB = early Holocene/Preboreal

These new marine records can now be compared with a set of ultrahigh-resolution companion records from terrestrial archives, as obtained from East African and Tibetan lake sediments (Gasse et al. 1996), and Chinese speleothems (Wang Y. et al. 2001; Yuan et al. 2004). Among the millennial-scale climate records generated to

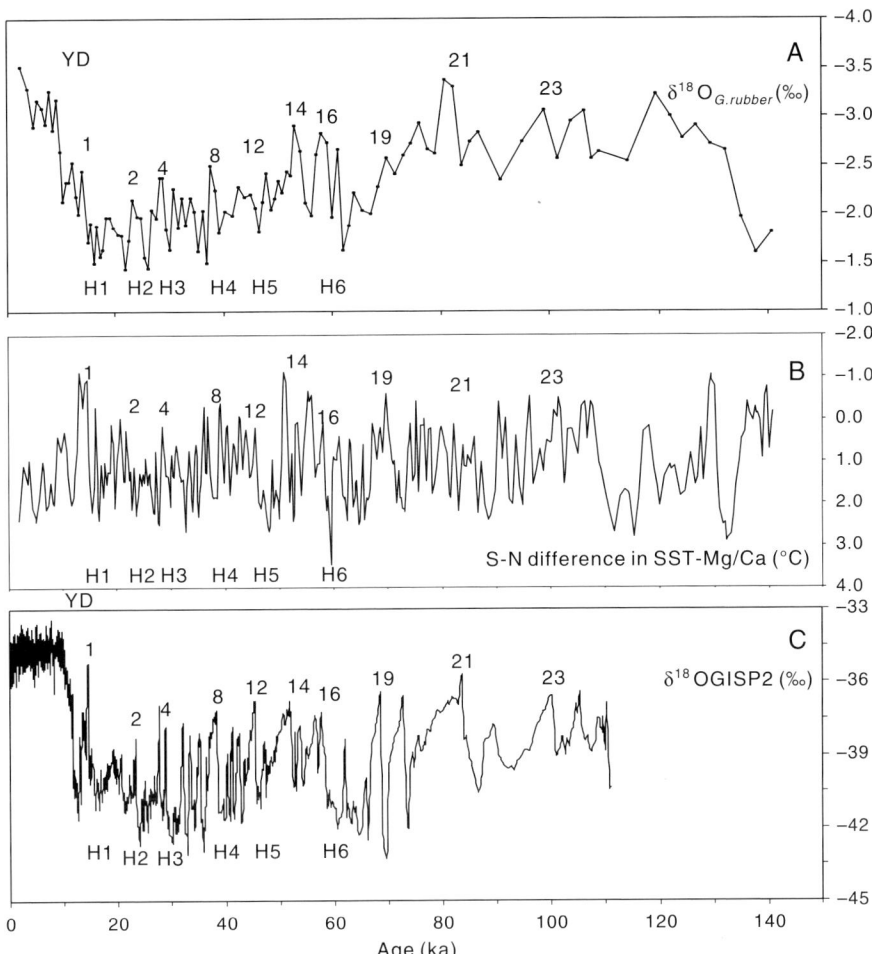

Fig. 5.55 Last glacial climate records from various regions show similar Dansgaard-Oeschger (D-O) events and Heinrich events (H1 to H6): (**A**) core 17961-2, southern SCS; (**B**) SST difference between core 17961 (Jian et al. 2008) from the southern and Site 1145 from northern SCS (Oppo and Sun 2005); (**C**) Greenland ice core $\delta^{18}O$ (Stuiver and Grootes 2000)

date, those from monsoonal regions show extremely strong similarities to those from the Greenland ice cores, indicating the coupled nature of high- and low-latitude abrupt climate change (Figs. 5.55 and 5.56). Despite their nature and mechanism are still vigorously being debated, factors such as tropical forcing and solar activity other than the glacial/interglacial ice cover volume change are most likely responsible for the sub-orbital climate changes. More high quality sediment sequences with high sedimentation rates in the SCS are needed for evaluating the phase relationships of climate signals in the monsoon region as compared to high-latitude climate components. This information will enhance our understanding of the role that monsoon circulation plays in sub-orbital variability.

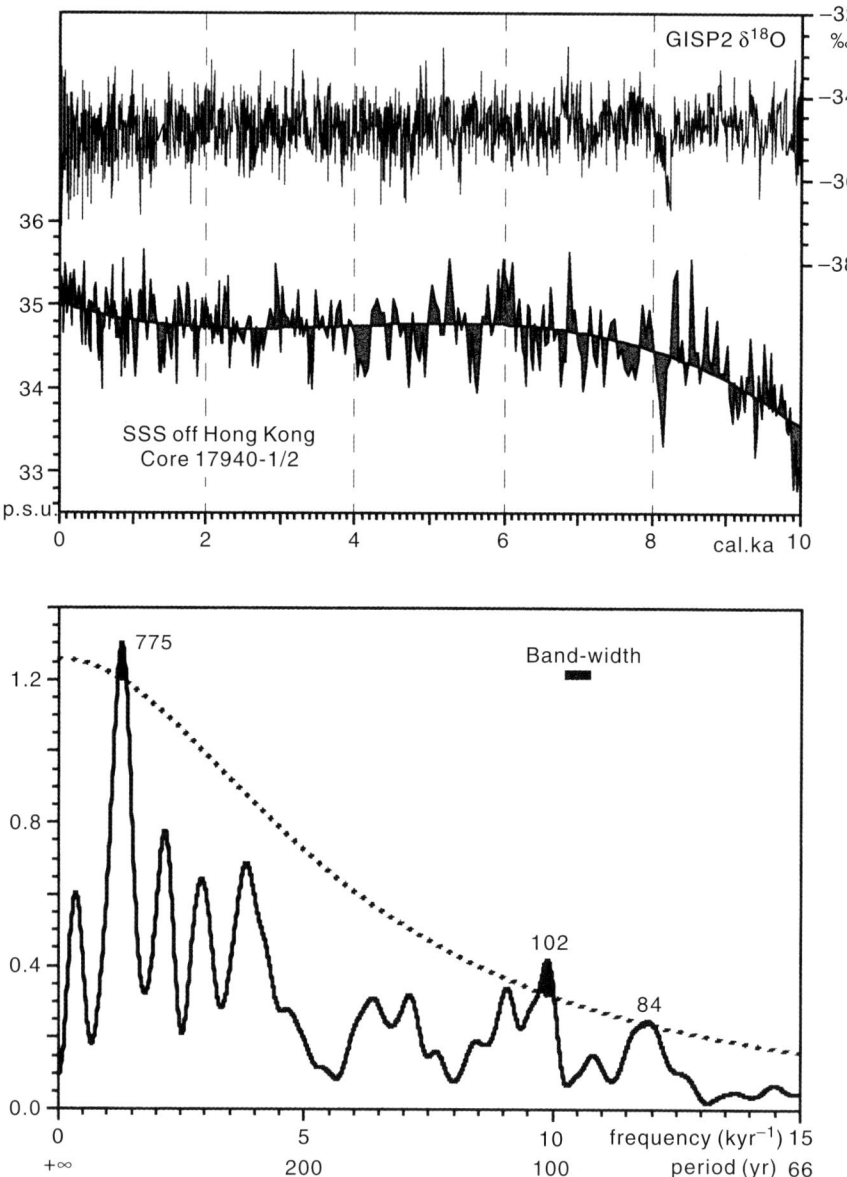

Fig. 5.56 Comparison between the δ^{18}O-based Holocene temperature record in Greenland ice-core GISP2 (Grootes and Stuiver 1997) and sea-surface salinity (SSS) changes in core 17940 off Hong Kong shows sub-orbital variations over the last 10,000 yr (from Wang L. et al. 1999a). Data are presented in the time (*upper panel*) and frequency domain (*lower panel* for SSS). Major temperature lows (δ^{18}O minima) on Greenland summit were coeval with SSS maxima, equal to lows and subsequent highs in monsoon precipitation in subtropical South China. *Horizontal bar* shows band-width. Numbers in the SSS power spectrum are significant periods in years, which exceed the *upper limit* of red noise at 80% confidence level (*dotted line*)

Summary

As the largest of the East Asian monsoon-dominated marginal seas between the largest continent Asia and the largest ocean Pacific, the SCS has witnessed dramatic changes in paleo-sea surface temperature, thermocline depth and paleo-vegetation of the surrounding continents/islands on tectonic, orbital and sub-orbital time scales in the late Cenozoic, which are summarized as follows:

1. *Upper water structure variations.* The upper water structure in the SCS is sensitive to changes in the East Asian monsoon and global ice-volume in the late Cenozoic. The foraminiferal assemblage data reveal that the depth of thermocline (DOT) in the SCS has experienced major changes at 11.5–10.6, 4.2–3.2 and 0.9 Ma, possibly related to the evolution of the WPWP. The Mg/Ca ratio derived SST in the southern SCS had ever decreased by ~5 °C during the formation of the Northern Hemisphere Glaciation. Since then, due to the strengthening of the winter monsoon and reorganization of sea circulation, the estimated SST decreased and the DOT deepened during glacials, and vice versa during interglacials. As shown by the paleo-SST estimations based on the transfer function technique, the average SST in the SCS was 0.9 °C cooler than in the Pacific at similar latitudes in summer, while the difference was as much as 4.9 °C in winter, resulting in an enhanced seasonality in the marginal seas, especially during glacials. The increased glacial/interglacial SST contrast in the marginal seas has been supported by $U^{K'}_{37}$ and Mg/Ca ratio analyses, showing the "amplifying effect" of climate changes in the SCS.
2. *East Asian monsoon evolution and variability.* A multi-proxy method for monsoon study in the SCS is recommended. The monsoon system started in the latest Oligocene and experienced major stages of intensification at ~8.0, 3.2, and 2.0 Ma, in accordance with the terrestrial records from the Loess Plateau. The evolution of East Asian monsoon was similar to the South Asian monsoon in stages, with an enhancement of the winter monsoon in East Asia being the major difference. On the orbital time scale, previous studies on Indian monsoonal upwelling have revealed that the Indian monsoon was externally forced by cyclical changes in solar insolation associated only with the obliquity and precession parameters. However, the SCS-based studies have shown that direct local insolation forcing could be less important in driving the East Asian winter monsoon variability, but the ice volume forcing may be the primary factor in determining their strength and timing. Therefore, intensified winter monsoon and weakened summer monsoon is typical of glacial periods, and vice versa for interglacial periods. The rich spectra of monsoon variability from the southern SCS exhibit characteristic features of orbital forcing in the low latitude ocean ranging from the 400 kyr eccentricity to 10 kyr semi-precessional cycles. The high-resolution Quaternary records have demonstrated that the D/O-like millennial-scale climate fluctuations exist throughout the last 1.0 myr and the rapid climatic events in the SCS were coeval with those found in Greenland ice cores, suggesting global climatic tele-connections.

3. *Land-Sea comparison.* The sediment records (e.g., marine pollen and clay mineral records) provide important information for the land-sea comparison of paleoclimatological changes in the SCS. The increase of cool-tolerant pollen group probably indicates climate cooling since ~8.0 and 2.6 Ma, consistent with the Chinese loess sequences. During the LGM, grassland vegetation mainly composed of *Artemisia* covered the exposed northern continental shelf of the SCS, indicating colder and drier climate relative to the present day. The glacial emergence of the continental shelf and SST decrease in SCS led to a considerable reduction of evaporation, involving a strongly enhanced continental aridity in the subtropical South China. Moreover, the evidence of pollen assemblages indicates cool but humid climate in the southern SCS during glacials. Since the southern SCS is a part of the WPWP, its exposure and cooling must have weakened the role of the warm pool in regional climate during glacial periods.

4. *South-North contrast.* The northern part of the SCS basin is controlled by the monsoon-driven cyclonic gyre, while the southern part belongs to the WPWP with much less monsoon influence. In the past, this S-N contrast became intensified with the growth of the Boreal ice sheet and the strengthening of the winter monsoon. Enhanced also by geological contrast between the landward sides to the NW (continent) and the SE (island arc), the southern and northern SCS are also different in many sedimentological and paleoceanographic aspects, including the different responses of the two parts to glacial cycles in micropaleontology, SST, and monsoon-induced upwelling. Palynological analyses have found grassland on the northern continental shelf, but forest on the southern shelf of the glacial SCS, indicating their different vegetational and climatic response to the LGM. Interestingly, the spectral analysis of the S-N SST difference (ΔSST) in the SCS has revealed precessional cycle and millennial variability, reflecting the changes in the East Asian monsoon on orbital and sub-orbital time scales.

Acknowledgments This work was supported by the Ministry of Science and Technology of China (NKBRSF Grant 2007CB815900), the National Natural Science Foundation of China (Grants 40621063, 40331002, 40476027 and 40776028) and the Ministry of Education of China (FANEDD Grant No. 2005036 to Tian).

References

An Z. 2000. The history and variability of the East Asian paleomonsoon climate. Quat. Sci. Rev. 19: 171–187.

An Z., Kutzbach J.E., Prell W.L. and Porter S.C. 2001. Evolution of Asian monsoons and phased uplift of the Himalaya-Tibetan Plateau since late Miocene time. Nature 411: 62–66.

Anand P., Elderfield H. and Conte M.H. 2003. Calibration of Mg/Ca thermometry in planktonic foraminifera from a sediment trap time series. Paleoceanography 18: 1050, doi:10.1029/2002PA000846.

Anderson C. 1997. Thansfer function vs. modern analog technique for estimating Pliocene sea-surface temperature based on planktonic foraminiferal data, west equatorial Pacific Ocean. J. Foraminiferal Res. 27: 123–132.

Anderson D. and Webb R.S. 1994. Ice-age tropics revisited. Nature 367: 23–24.

Anderson D.M., Overpeck J.T. and Gupta A.K. 2002. Increase in the Asian Southwest monsoon during the past four centuries. Science 297: 596–599.

Andreasen D.J. and Ravelo A.C. 1997. Tropical Pacific Ocean thermocline depth reconstructions for the last glacial maximum. Paleoceanography 12: 395–413.

Andree M., Oeschger H., Broecker W.S., Beavan N., Mix A., Bonani G., Hofmann H.J., Morenzoni E., Nessi M., Suter M. and Wolfli W. 1986. AMS radiocarbon dates on foraminifera from deep sea sediments. Radiocarbon 28(2A): 424–428.

Bé A.W.H. 1977. An ecological, zoogeographical and taxonomic review of recent planktonic foraminifera. In: Ramsay A.T.S. (ed.), Oceanic Micropaleontology. Academic Press, London, 1, pp. 1–100.

Bé A.W.H., Bishop J.K.B., Swerdlove M.S. and Gardner W.D. 1985. Standing stock, vertical distribution and flux of planktonic foraminifera in the Panama Basin. Mar. Micropaleontol. 9: 307–333.

Beer J., Mende W. and Stellmacher R. 2000. The role of the sun in climate forcing. Quat. Sci. Rev. 19: 403–415.

Bemis B., Spero H., Bijma J. and Lea D. 1998. Reevaluation of the oxygen isotopic composition of planktonic foraminifera: Experimental results and revised paleotemperature equations. Paleoceanography 13: 150–160.

Bentaleb I., Fontugne M. and Beaufort L. 2002. Long-chain alkenones and $U_{37}^{K'}$ variability along a south-north transect in the Western Pacific Ocean. Global Planet. Change 34: 173–183.

Berger A., Loutre M.F. and Laskar J. 1992. Stability of the astronomical frequencies over the Earth's history for paleoclimate studies. Science 255: 560–566.

Berggren W.A., Kent D.V., Swisher III C.C. and Aubry M.P. 1995. A revised Cenozoic geochronology and chronostratigraphy. In: Berggren W.A., Kent D.V., Aubry M.P. and Hardenbol J. (eds.), Geochronology, Time Scales and Global Stratigraphic Correlation. SEPM Spec. Publ. 54: 129–212.

Billups K., Ravelo A.C., Zachos J.C. and Norris R.D. 1999. Link between oceanic heat transport, thermohaline circulation, and the intertropical convergence zone in the early Pliocene Atlantic. Geology 27: 319–322.

Bond G., Broecker W., Johnsen S., McManus J., Labeyrie L., Jouzel J. and Bonani G. 1993. Correlations between climate records from North Atlantic sediments and Greenland ice. Nature 365: 143–147.

Bond G., Kromer B., Beer J., Muscheler R., Evans M.N., Showers W., Hoffmann S., Lotti-Bond R., Hajdas I. and Bonani G. 2001. Persistent solar influence on North Atlantic climate during the Holocene. Science 294: 2130–2136.

Broecker W.S., Andree M., Klas M., Bonani G., Wolfli W. and Oeschger H. 1988. New evidence from the South China Sea for an abrupt termination of the last glacial period. Nature 333: 156–158.

Bühring C., Sarnthein M. and Erlenkeuser H. 2001. Toward a high-resolution stable isotope stratigraphy of the last 1.1 million years: Site 1144, South China Sea. In: Prell W.L., Wang P., Blum P., Rea D. and Clemens S. (eds.), Proc. ODP Sci. Res. 184.

Cannariato K.G. and Ravelo A.C. 1997. Pliocene-Pleistocene evolution of eastern tropical Pacific surface water circulation and thermocline depth. Paleoceanography 12: 805–820.

Chaisson W.P. 1995. Planktonic foraminiferal assemblages and paleoceanographic change in the trans-tropical Pacific ocean: a comparison of West (Leg 130) and East (leg 138), latest Miocene to Pleistocene. In: Pisias N.G., Mayer L.A., Jenecek T.R., Palmer-Julson A. and van Andel T.H. (eds.), Proc. ODP Sci. Res. 138: 555–583.

Chao S.V., Shaw P.T. and Wu S.Y. 1996. Deep water ventilation in the South China Sea. Deep-Sea Res. II 43: 445–466.

Chapman M.R., Shackleton N. and Duplessy J.C. 2000. Sea surface temperature variability during the last glacial–interglacial cycles: assessing the magnitude and pattern of climate changes in the North Atlantic. Palaeogeogr. Palaeoclimatol. Palaeoecol. 157: 1–25.

Chen M.T., Huang C.C., Pflaumann U., Waelbroeck C. and Kucera M. 2005. Estimating glacial western Pacific sea-surface temprature: methodological overview and data compilation of surface sediment planktonic foraminifer faunas. Quat. Sci. Rev. 24: 1049–1062.

Chen M., Wang R., Yang L., Han J. and Lu J. 2003. Development of east Asian summer monsoon environments in the late Miocene: radiolarian evidence from Site 1143 of ODP Leg 184. Mar. Geol. 201: 169–177.

Chen R., Jian Z., Zheng Y. and Chen J. 2000. Seasonal variations of the planktonic foraminiferal flux in the central South China Sea. J. Tongji Univ. (Natural Sci.) 28(1): 73–77 (in Chinese).

Cheng H., Edwards R.L., Wang Y.J., Kong X., Ming Y., Gallup C.D., Kelly M.J., Wang X. and Liu W. 2006. A penultimate glacial monsoon record from Hulu Cave and two-phase glacial terminations. Geology 34: 217–220.

Cheng X., Huang B., Jian Z., Zhao Q., Tian J. and Li J. 2005. Foraminiferal isotopic evidence for monsoonal activity in the South China Sea: a present-LGM comparison. Mar. Micropaleontol. 54: 125–139.

Clemens S.C., Murray D.W. and Prell W.L. 1996. Nonstationary phase of the Plio-Pleistocene Asian monsoon. Science 274: 943–948.

Clemens S.C. and Prell W. 2003. A 350,000 year summer-monsoon multi-proxy stack from the Owen Ridge, Northern Arabian Sea. Mar. Geol. 201: 35–51.

Clemens S.C., Prell W., Murray D., Shimmield G. and Weedon G. 1991. Forcing mechanisms of the Indian Ocean monsoon. Nature 353: 720–725.

CLIMAP. 1981. Seasonal Reconstructions of the Earth's Surface at the Last Glacial Maximum. GSA Map. Chart Ser. MC-36.

Cullen J.L. and Prell W.L. 1984. Planktonic Foraminifera of the northern Indian Ocean: Distribution and preservation in surface sediments. Mar. Micropaleontol. 9: 1–52.

Curry W.B., Ostermann D.R., Guptha M.V.S. and Ittekkot V. 1992. Foraminiferal production and monsoonal upwelling in the Arabian Sea: Evidence from sediment traps. In: Summerhays C.P., Prell W.L. and Emeis K.C. (eds.), Upwelling Systems: Evolution Since the Early Miocene. Geological Society of London, Special Publication 63, pp. 93–106.

Dansgaard W., Johnsen S.J., Clausen H.B., Dahl-Jensen D., Gundestrup N.S., Hammer C.U., Hvidberg C.S., Steffensen J.P., Sveinbjornsdottir A.E., Jouzel J. and Bond G. 1993. Evidence for general instability of past climate form a 250-kyr ice-core record. Nature 364: 218–220.

de Garidel-Thoron T., Beaufort L., Linsley B.K. and Dannenmann S. 2001. Millennial-scale dynamics of the East-Asian winter monsoon during the last 200,000 years. Paleoceanography 16: 491–502.

Dekens P.S., Lea D.W., Park D.K. and Spero H.J. 2002. Core top calibration of Mg/Ca in tropical foraminifera: Refining paleotemperature estimation. Geochem. Geophys. Geosyst. 3: 10.1029/2001GC000200.

Dickens G.R. and Barron J.A. 1997. A rapid deposited pennate diatom ooze in upper Miocene-lower Pliocene sediment beneath the North Pacific polar front. Mar. Micropaleontol. 31: 177–182.

Ding Y., Li C. and Liu Y. 2004. Overview of the South China Sea Monsoon Experiment. Adv. Atmospheric Sci. 21: 343–360.

Ding Z.L., Rutter N.W., Han J.T. and Liu T.S. 1992. A coupled environmental system formed at about 2.5 Ma over East Asia. Palaeogeogr. Palaeoclimatol. Palaeoecol. 94: 223–242.

Ding Z.L., Sun J. and Liu D. 1999. Stepwise advance of the Mu Us Desert since Pliocene: Evidence of a red clay-loess record. Chinese Sci. Bull. 44(13): 1211–1214.

Ding Z.L., Sun J.M., Yang S.L. and Liu T.S. 1998. Preliminary magnetostratigraphy of a thick eolian red clay-loess sequence at Lingtai, the Chinese Loess Plateau. Geophys. Res. Lett. 25: 1225–1228.

Ding Z.L., Yu Z.W., Rutter N.W. and Liu T.S. 1994. Towards an orbital time scale for Chinese loess deposits. Quat. Sci. Rev. 13: 39–70.

Ducrocq S., Chaimanee Y., Suteethorn V. and Jaeger J.J. 1994. Ages and paleoenvironment of Miocene mammalian faunas from Thailand. Palaeogeogr. Palaeoclimatol. Palaeoecol. 108: 149–163.

Elderfield H. and Ganssen G. 2000. Past temperature and $\delta^{18}O$ of surface ocean waters inferred from foraminiferal Mg/Ca ratios. Nature 405: 442–445.

Ellison J.C. 1993. Mangrove retreat with rising sea level, Bermuda. Estuar. Coast. Shelf Sci. 37: 75–88.

Fairbanks R.G., Sverdlove M., Free R., Wiebe P.H. and Bé A.W.H. 1982. Vertical distribution and isotopic fractionation of living planktonic foraminifera from the Panama Basin. Nature 298: 841–844.

Fairbridge R.W. 1986. Monsoons and paleomonsoons. Episodes 9: 143–149.

Fang D., Jian Z. and Wang P. 2000. The paleoproductivity recorded in the southern Nansha sea area for about 30 ka. Chinese Sci. Bull. 45: 227–230.

Gao Q. and Zhang Q. 1981. Bryophytes of Northeast China. China Sci. Press, Beijing (in Chinese).

Gasse F., Fontes J.Ch., Van Campo E. and Wei K. 1996. Holocene environmental changes in Bangong Co basin (Western Tibet). Part 4: Discussion and conclusions. Palaeogeogr. Palaeoclimatol. Palaeoecol. 120: 79–92.

Grindrod J. and Rhodes E.G. 1984. Holocene sea-level history of a tropical estuary: Missionary Bay, North Queensland. In: Thom B.G. (ed.), Coastal Geomorphology in Australia. Academic Press, Sydney, pp. 151–178.

Grootes P.M. and Stuiver M. 1997. Oxygen 18/16 variability in Greenland snow and ice with 10^{-3}-to 10^{-5}-year time resolution. J. Geophys. Res. C102: 26455–26470.

Guangdong Institute of Botany 1976. Vegetation of Guangdong. China Sci. Press, Beijing (in Chinese).

Guo Z.T., Ruddiman W.F., Hao Q.Z., Wu H.B., Qiao Y.S., Zhu R.X., Peng S.Z., Wei J.J., Yuan B.Y. and Liu T.S. 2002. Onset of Asian desertification by 22 Myr ago inferred from loess deposits in China. Nature 416: 159–163.

Haigh J.D. 1996. The impact of solar variability on climate. Science 272: 981–984.

Hanebuth T., Stattegger K. and Groote P.M. 2000. Rapid flooding of the Sunda Shelf: a late-glacial sea-level record. Science 288: 1033–1035.

Hastings D., Kienast M., Steinke S. and Whitko A.A. 2001. A Comparison of three independent paleotemperature estimates from a high resolution record of deglacial SST records in the tropical South China Sea. EOS Trans. AGU 82: PP12B-10.

Hastings D., Russell A.D. and Emerson S.R. 1998. Foraminiferal magnesium in *Globigerinoides sacculifer* as a paleotemperature proxy. Paleoceanography 13: 161–169.

Hemleben C., Spindler M. and Anderson O.R. 1989. Modern Planktonic Foraminifera. Springer-Verlag, New York, 363pp.

Higginson M., Maxwell J.R. and Altabet M.A. 2003. Nitrogen isotope and chlorin paleoproductivity records from the Northern South China Sea: remote vs. local forcing of millennial- and orbital-scale variability. Mar. Geol. 201: 223–250.

Hirst A.C. and Godfrey J.S. 1993. The role of Indonesian throughflow in a global ocean GCM. J. Phys. Oceanogr. 23: 1057–1086.

Hope G.S. and Tulip J. 1994. A long vegetation history from lowland Iran Jaya, Indonesia. Paleogeogr. Paleoclimatol. Paleoecol. 109: 385–398.

Hostetler S., Pisias N. and Mix A. 2006. Sensitivity of Last Glacial Maximum climate to uncertainties in tropical and subtropical ocean temperatures. Quat. Sci. Rev. 25: 1168–1185.

Hovan S.A., Rea D.K., Pisias N.G. and Shackleton N.J. 1989. A direct link between the China loess and marine $\delta^{18}O$ records: Aeolian flux to the north Pacific. Nature 340: 296–298.

Hu J., Peng P., Fan D., Jia G., Jian Z. and Wang P. 2003. No aridity in Sunda land during the Last Glaciation: Evidence from molecular-isotopic stratigraphy of long-chain *n*-alkans. Palaeogeogr. Palaeoclinatol. Palaeoecol. 201: 269–281.

Hu J., Peng P., Jia G., Fang D., Fu J. and Wang P. 2002. Biological markers and their carbon isotopes as an approach to the paleoenvironmental reconstruction of Nansha area, South China Sea, during the last 30 ka. Org. Geochem. 33: 1197–1204.

Huang B., Cheng X., Jian Z. and Wang P. 2003. Response of upper ocean structure to the initiation of the North Hemisphere glaciation in the South China Sea. Palaeogeogr. Palaeoclimatol. Palaeoecol. 196: 305–318.

Huang B., Jian Z., Cheng X. and Wang P. 2002. Foraminiferal responses to upwelling variations in the South China Sea over the last 220 000 years. Mar. Micropaleontol. 47: 1–15.

Huang C.Y., Liew P.M., Zhao M., Chang T.C., Kuo C.M., Chen M. T., Wang C.H. and Zheng L.F. 1997a. Deep sea and lake records of the Southeast Asian paleomonsoons for the last 25 thousand years. Earth Planet. Sci. Lett. 146: 59–72.

Huang C.Y., Wu S.F., Zhao M., Chen M.T., Wang C.H., Tu X. and Yuan P.B. 1997b. Surface ocean and monsoon climate variability in the South China Sea since the last glaciation. Mar. Micropaleotol. 32: 71–94.

Hutson W.H. 1980. The Agulhas Current during the Late Pleistocene: Analysis of modern faunal analogs. Science 207: 64–66.

Imbrie J., Boyle E., Clemens S., Duffy A., Howard W.R., Kukla G., Kutzbach J., Martinson D.G., McIntyre A., Mix A.C., Molfino B., Morley J.J., Peterson L.C., Pisias N.G., Prell W.L., Raymo M.E., Shackleton N.J. and Toggweiler J.R. 1992. On the structure and origin of major glaciation cycles, 1, Linear responses to Milankovitch forcing. Paleoceanography 7: 701–738.

Imbrie J. and Kipp N.G. 1971. A new micropaleontological method for quantitative paleoclimatology: Application to a late Pleistocene Carinbbean core. In: Turekian K.K. (ed.), The Late Cenozoic Glacial Ages. Yale Univ. Press, New Haven, Conn., pp. 71–181.

Jia G., Peng P., Zhao Q. and Jian Z. 2003. Changes in terrestrial ecosystem since 30 Ma in East Asia: Stable isotope evidence from black carbon in the South China Sea. Geology 31: 1093–1096.

Jian Z. and Huang W. 2003. Rapid climate change and high resolution deep-sea sedimentary records. Adv. Earth Sci. 18(5): 673–680 (in Chinese).

Jian Z., Huang B., Lin H. and Kuhnt W. 2001. Late Quaternary upwelling intensity and East Asian monsoon forcing in the South China Sea. Quat. Res. 55: 363–370.

Jian Z., Li B., Huang B. and Wang J. 2000a. *Globorotalia truncatulinoides* as indicator of upper-ocean thermal structure during the Quaternary: Evidences from the South China Sea and Okinawa Trough. Palaeogeogr. Palaeoclimatol. Palaeoecol. 162: 287–298.

Jian Z., Wang B. and Qiao P. 2008. Late Quaternary changes of sea surface temperature in the southern South China Sea and their comparison with the paleoclimatic records of polar ice cores. Quat. Sci. 3: 391–398 (in Chinese).

Jian Z. and Wang L. 1997. Late Quaternary benthic foraminifera and deep-water paleoceanography in the South China Sea. Mar. Micropaleotol. 32: 127–154.

Jian Z., Wang L., Kienast M., Sarnthein M., Kuhnt W., Lin H. and Wang P. 1999. Benthic foraminiferal paleoceanography of the South China Sea over the last 40,000 years. Mar. Geol. 156: 159–186.

Jian Z., Wang P., Chen M.P., Li B., Zhao Q., Bühring C., Laj C., Lin H.L., Pflaumann U., Bian Y., Wang R. and Cheng X. 2000b. Foraminiferal response to major Pleistocene paleographic changes in the southern South China Sea. Paleoceanography 15: 229–243.

Jian Z., Yu Y., Li B., Wang J., Zhang X. and Zhou Z. 2006. Phased evolution of the south-north hydrographic gradient in the South China Sea since the middle Miocene. Palaeogeogr. Palaeoclimatol. Palaeoecol. 230: 251–263.

Jian Z., Zhao Q., Cheng X., Wang J., Wang P. and Su X. 2003. Pliocene-Pleistocene stable isotope and paleoceanographic changes in the northern South China Sea. Palaeogeogr. Palaeoclimatol. Palaeoecol. 193: 425–442.

Kennett J.P., Keller G. and Srinivasan M.S. 1985. Miocene planktonic foraminiferal biogeography and paleoceanographic de velopment of the Indo-Pacific region, in The Miocene Ocean: Paleogeography and Biogeography. GSA Memoir 163: 197–236.

Kienast M., Hanebuth T.J.J., Pelejero C. and Steinke S. 2003. Synchroneity of meltwater pulse 1a and the Bølling warming: new evidence from the South China Sea. Geology 31: 67–70.

Kienast M., Steinke S., Stattegger K. and Calvert S.E. 2001. Synchronous tropical South China Sea SST change and Greenland warming during deglaciation. Science 291: 2132–2134.

Kienast S.S., Calvert S.E. and Pedersen T.F. 2002. Nitrogen isotope and productivity variations along the northeast Pacific margin over the last 120 kyr: Surface and subsurface paleoceanography. Paleoceanography 17: 1055, doi: 10.1029/2001PA000650.

Kissel C., Laj C., Clemens S. and Solheid P. 2003. Magnetic signature of environmental changes in the last 1.2 Myr at ODP Site 1146, South China Sea. Mar. Geol. 201: 119–132.

Kroon D., Steens T.N.F. and Troelstra S.R 1991. Onset of monsoonal related upwelling in the western Arabian Sea. Proc. ODP Sci. Results 117: 257–264.

Kudrass H.R., Hofmann A., Doose H., Emeis K. and Erlenkeuser H. 2001. Modulation and amplification of climatic changes in the Northern Hemisphere by the Indian summer monsoon during the past 80 k.y. Geology 29: 63–66.

Kuhnt W., Hess S. and Jian Z. 1999. Quantitative composition of benthic foraminiferal assemblages as a proxy indicator for organic carbon flux rates in the South China Sea. Mar. Geol. 156: 123–157.

Kuroyanagi A. and Kawahata H. 2004. Vertical distribution of living planktonic foraminifera in the seas around Japan. Mar. Micropaleontol. 53: 173–196.

Kutzbach J.E. and Guetter P.J. 1986. The influence of changing orbital parameters and surface boundary conditions on climate simulations for the past 18000 years. J. Atmos. Sci. 43: 1726–1759.

Labitzke K. 2001. The global signal of the 11-year sunspot cycle in the stratosphere: differences between solar maxima and minima. Meteorol. Z. 10: 901–908.

Lea D.W., Pak D.K. and Spero H.J. 2000. Climate impact of Late Quaternary equatorial Pacific sea surface temperature variation. Science 289: 1719–1724.

Levitus S. 1982. Climatological Atlas of the World Ocean. In: NOAA Professional Paper 13. Washington, D.C.: U.S. Government Printing Office.

Levitus S. and Boyer T.P. 1994. World Ocean Atlas, vol. 4: Temperature. In: NOAA Atlas NESDIS 4. U.S. Govt. Printing Office, Washington, D.C.

Li J. 1991. The environmental effects of the uplift of the Qinghai-Xizang Plateau. Quat. Sci. Rev. 10: 479–483.

Li J. and Fang S. 1996. Studies on the uplift and environmental changes of the Qinghai-Xizang Plateau. Chin. Sci. Bull. 41: 316–322.

Li B., Wang J., Huang B., Li Q., Jian Z. and Wang P. 2004. South China Sea surface water evolution over the last 12 Ma: A south-north comparison from ODP Sites 1143 and 1146. Paleoceanography 19: PA1009, doi:10.1029/2003PA000906.

Li J. and Wang R. 2004. Paleoproductivity variability of the northern South China Sea during the past 1 Ma: The opal record from ODP Site 1144. Acta Geol. Sinica 78(2): 228–233 (in Chinese).

Li J., Wang R. and Li B. 2002. Variations of opal accumulation rates and paleoproductivity over the past 12 Ma at ODP Site 1143, southern South China Sea. Chinese Sci. Bull. 47: 596–598.

Li Q., Li B., Zhong G., McGowran B., Zhou Z., Wang J. and Wang P. 2006. Late Miocene development of the western Pacific warm pool: Planktonic foraminifer and oxygen isotopic evidence. Palaeogeogr. Palaeoclimatol. Palaeoecol. 237: 465–482.

Li X. and Sun X. 1999. Palynological records since Last Glacial Maximum from a deep-sea core in the southern South China Sea. Quat. Sci. 4: 526–535 (in Chinese).

Lin H., Lai C., Ting H., Wang L., Sarnthein M. and Huang J. 1999. Late Pleistocene nutrients and sea surface productivity in the South China Sea: a record of teleconnections with northern hemisphere events. Mar. Geol. 156: 197–210.

Liu C., Cheng X., Zhu Y., Tian J. and Xia P. 2002. Oxygen and carbon isotopic records of calcareous nannofossils for the past 1 Ma in the southern South China Sea. Chinese Sci. Bull. 47(10): 798–803.

Liu Z., Trentesaux A., Clemens S.C., Colin C., Wang P., Huang B. and Boulay S. 2003. Clay mineral assemblages in the northern South China Sea: implications for East Asian monsoon evolution over the past 2 million years. Mar. Geol. 201: 133–146.

Lohmann G.P. and Schweitzer P.N. 1990. *Globorotalia truncatulinoides* growth and chemistry as probes of the past thermocline: 1. Shell size. Paleoceanography 5: 55–75.

Löwemark L., Hong W.L., Yui T.F. and Hung G.W. 2005. A test of different factors influencing the isotopic signal of planktonic foraminifers in surface sediments from the northern South China Sea. Mar. Micropaleontol. 55: 49–62.

Luo Y., Cheng H., Wu G. and Sun X. 2001. Records of natural fire and climate history during the last three glacial- interglacial cycles around the South China Sea - Charcoal record from the ODP 1144. Sci. China (D) 44(10): 897–904.

Luo Y. and Sun X. 2007. Deep-sea pollen in the southern South China Sea during 12 ~ 1.6Ma BP and its response to the global climate change. Chinese Sci. Bull. 52(15): 2115–2122.

Luo Y., Sun X. and Jian Z. 2005. Environmental change during the penultimate glacial cycle: a high-resolution pollen record from ODP Site 1144, South China Sea. Mar. Micropaleontol. 54: 107–123.

Ma Y., Fan X., Li J., Wu F. and Zhang J. 2004. Vegetational and environmental changes during late Tertiary- early Quaternary in Jiuxi Basin. Sci. China (D) 34: 107–116 (in Chinese).

Ma Y., Li J. and Fan X. 1998. Pollen- based vegetation and climate records during 30.6 to 5.0 My from Linxia area, Gansu. Chinese Sci. Bull. 43(3): 301–304 (in Chinese).

Maier-Reimer E., Mikolajewicz U. and Crowley T. 1990. Ocean general circulation model sensitivity experiment with an open central American isthmus. Paleoceanography 5: 349–266.

Mao S., Li J., Wu G. and Harland R. 2007. Dinoflagellate cysts and environmental evolution of the Oligocene to Lower Miocene at Site 1148, ODP Leg 184, South China Sea. Palynology 31: 37–52.

Martinez J.I. 1997. Decreasing influence of Subantarctic Mode Water north of the Tasman Front over the past 150 kyr. Palaeogeogr. Palaeoclimatol. Palaeoecol. 131: 355–364.

Martinez J.I. 1994. Late Pleistocene paleoceanography of the Tasman Sea: implications for the dynamics of the warm pool in the western Pacific. Palaeogeogr. Palaeoclimatol. Palaeoecol. 112: 19–62.

McManus J.F., Oppo D.W. and Cullen J.L. 1999. A 0.5 million year record of millennial-scale climate variability in the North Atlantic. Science 283: 971–975.

Miao Q., Thunell R.C. and Andersen D.M. 1994. Glacial-Holocene carbonate dissolution and sea surface temperatures in the South China and Sulu seas. Paleoceanography 9: 269–290.

Mudelsee M. and Raymo M.E. 2005. Slow dynamics of the Northern Hemisphere glaciations. Paleoceanography 20: PA4022, doi:10.1029/2005PA001153.

Newsome J.C. and Flenley J.R. 1988. Late Quaternary vegetation history of the Central Highland of Sumatra II. Palaeopalynology and vegetation history. J. Biogeogr. 15: 555–578.

NGRIP (North Greenland Ice Core Project) Members 2004. High-resolution record of Northern Hemisphere climate extending into the last interglacial period. Nature 431: 147–151.

Nürnberg D., Bijma J. and Hemleben C. 1996. Assessing the reliability of magnesium in foraminiferal calcite as a proxy for water mass temperature. Geochim. Cosmochim. Acta 60: 803–814.

Nürnberg D., Muller A. and Schneider R.R. 2000. Paleo-sea surface temperature calculations in the equatorial east Atlantic from Mg/Ca ratios in planktic foraminifera: a comparison to sea surface temperature estimates from $U^{K'}_{37}$, oxygen isotopes, and foraminiferal transfer function. Paleoceanography 15(1): 124–134.

Oppo D.W., Linsley B.K., Dannemann S. and Beaufort L. 2001. A 400-kyr long planktic $\delta^{18}O$ record from the Sulu Sea: constraints on ice volume and the tropical hydrologic cycle. EOS Trans. AGU 82(47): F737.

Oppo D.W. and Sun Y. 2005. Amplitude and timing of sea-surface temperature change in the northern South China Sea: Dynamic link to the East Asian monsoon. Geology 33: 785–788.

Pelejero C. and Calvo E. 2003. The upper end of the $U_{37}^{K'}$ temperature calibration revisited. Geochem. Geophys. Geosyst. 4: doi:10.1028/2002GC000431.

Pelejero C. and Grimalt J.O. 1997. The correlation between the $U_{37}^{K'}$ index and sea surface temperatures in the warm boundary: the South China Sea. Geochim. Cosmochim. Acta 61: 4789–4797.

Pelejero C., Grimalt J.O., Heilig S., Kienast M. and Wang L. 1999a. High-resolution $U_{37}^{K'}$ temperature reconstructions in the South China Sea over the past 220 kyr. Paleoceanography 14: 224–231.

Pelejero C., Grimalt J.O., Sarnthein M., Wang L. and Flores J.A. 1999b. Molecular biomarker record of sea surface temperature and climatic change in the South China Sea during the last 140,000 years. Mar. Geol. 156: 109–121.

Pflaumann U., Duprat J., Pujol C. and Labeyrie L.D. 1996. SIMMAX: a modern analog technique to deduce Atlantic sea surface temperatures from planktonic foraminifera in deep-sea sediments. Paleoceanography 11: 15–35.

Pflaumann U. and Jian Z. 1999. Modern distribution patterns of planktonic foraminifera in the South China Sea and western Pacific: A new transfer technique to estimate regional sea-surface temperatures. Mar. Geol. 156: 41–83.

Porter S.C., An Z. and Zheng H. 1992. Cyclic Quaternary alleviation and terracing in a nonglaciated Drainage Basin on the north flank of Qingling Shan, central China. Quat. Res. 38: 157–169.

Prahl F.G. and Wakeham S.G. 1987. Calibration of unsaaturation patterns in long-chain ketone composition for paleotemperature assessment. Nature 330: 367–369.

Prell W.L. 1984. Covariance pattern of foraminiferal δ^{18}O: An evaluation of Pliocene ice volume change near 3.2 million years ago. Science 226: 692–693.

Prell W.L. and Kutzbach J.E. 1987. Monsoon variability over the past 150,000 years. J. Geophys. Res. 92: 8411–8425.

Prell W.L. and Kutzbach J.E. 1997. The impact of Tibetan-Himalayan elevation on the sensitivity of the monsoon climate system to changes in solar radiation. In: Ruddiman W.F. (ed.), Tectonic Uplift and Climate Change, Plenum Press, pp. 171–201.

Prell W.L., Murray D.W., Clemens S.C. and Anderson D.M. 1992. Evolution and variability of the Indian Ocean summer monsoon: Evidence from the western Arabian Sea drilling program. In: Duncan R.A., Rea D.K., Kidd R.B., von Rad U. and Weissel J.K. (eds.), The Indian Ocean: A Synthesis of Results from the Ocean Drilling Program. Geophys. Monogr. 70, AGU, pp. 447–469.

Qin G. 1996. Biostratigraphic zonation and correlation of the late Cenozoic planktonic foraminifera in Pearl River Basin. In: Hao Y. (ed.), Research on Micropalaeontology and Paleoceanography in Pearl River Mouth Basin, South China Sea. China Univ. Geosci. Press, Beijing, pp. 19–31 (in Chinese).

Quade J., Cerling T.E. and Bowman J.E. 1989. Development of Asian monsoon revealed by marked ecological shift during the latest Miocene in northern Pakistan. Nature 342: 163–166.

Qiang X.K., Li Z.X., Powell C. and Zheng H. 2001. Magnetostratigraphic record of the Late Miocene onset of the East Asian monsoon, and Pliocene uplift of northern Tibet. Earth Planet. Sci. Lett. 187: 83–93.

Ravelo A.C., Andreasen D.H., Lyle M., Olivarez Lyle A. and Wara M.W. 2004. Regional climate shifts caused by gradual cooling in the Pliocene epoch. Nature 429: 263–267.

Ravelo A.C., Fairbanks R.G. and Philander S.G.H. 1990. Reconstructing tropical Atlantic hydrography using planktonic foraminifera and an ocean model. Paleoceanography 5: 409–431.

Ravelo A.C. and Fairbanks R.G. 1992. Oxygen isotopic composition of multiple species of planktonic foraminifera: Recorders of the modern photic zone temperature gradient. Paleoceanography 7: 815–831.

Ravelo A. and Shackleton N.J. 1995. Evidence for surface-water circulation changes at Site 851 in the eastern tropical Pacific Ocean. Proc. ODP Sci. Res. 138: 503–514.

Raymo M.E., Ganley K., Carter S., Oppo D.W. and McManus J. 1998. Millennial-scale climate instability during the early Pleistocene epoch. Nature 392: 699–702.

Ramstein G., Fluteau F., Besse J. and Joussaume S. 1997. Effect of orogeny, plate motion and land-sea distribution on Eurasian climate change over the past 30 million years. Nature 386: 788–795.

Rea D.K. 1992. Delivery of Himalayan sediment to the northern Indian Ocean and its relation to global climate, sea level, uplift, and seawater strontium. In: Duncan R.A. (ed.), Synthesis of Results from Scientific Drilling in the Indian Ocean. Geophys. Monogr. Ser. 70: 387–402.

Rea D.K., Snoeck H. and Joseph L.H. 1998. Late Cenozoic eolian deposition in the North Pacific: Asian drying, Tibetan uplift, and cooling of the Northern Hemisphere. Paleoceanography 13: 215–224.

Rosenthal Y. and Lohmann G.P. 2002. Accurate estimation of sea surface temperatures using dissolution corrected calibrations for Mg/Ca paleothermometry. Paleoceanography 17: 10.1029/2001PA000749.

Rosenthal Y., Oppo D.W. and Linsley B.K. 2003. The amplitude and phasing of climate change during the last deglaciation in the Sulu Sea, western equatorial Pacific. Geophys. Res. Lett. 30(8): 1428, doi: 10.1029/2002GL016612.

Sarnthein M., Kennett J.P., Allen J.R.M., Beer J., Grootes P., Laj C., McManus J., Ramesh R. and SCOR-IMAGES Working Group 117. 2002. Decadal-to-millennial-scale climate variability - chronology and mechanisms: summary and recommendations. Quat. Sci. Rev. 21: 1121–1128.

Sarnthein M., Pflaumann U., Ross R., Tiedemann R. and Winn K. 1992. Transfer functions to reconstruct ocean paleoproductivity: a comparison. Geol. Soc. Spec. Public. 64:411–427.

Sarnthein M., Pflaumann U., Wang P. and Wong H.K. (eds.). 1994. Preliminary Report on SONNE-95 Cruise "Monitor Monsoon" to the South China Sea. Berichte-Reports, Geol.-Palaont. Inst. Univ. Kiel, 48, pp. 1–225.

Schmidt H., Berger W.H., Bickert T. and Wefer G. 1993. Quaternary carbon isotope record of pelagic foraminifers: Site 806, Ontong Java Plateau. Proc. ODP Sci. Res. 130: 397–409.

Schulz M., Berger W.H., Sarnthein M. and Grootes P. 1999. Amplitude variations of 1470-year climate oscillations during the last 100,00 years linked to fluctuations of continental ice mass. Geophys. Res. Lett. 26: 3385–3388.

Schulz M. and Paul A. 2002. Holocene climate variability on centennial-to- millennial time scales: 1. Climate records from the North-Atlantic realm. In: Wefer G., Berger W.H., Behre K.E. and Jansen E. (eds.), Climate Development and History of the North Atlantic Realm, Springer Verlag, Berlin, pp. 41–54.

Schulz H., von Rad U. and Erlenkeuser H. 1998. Correlation between Arabian Sea and Greenland climate oscillations of the past 110,000 years. Nature 393: 54–57.

Shaw P.T. 1996. Winter upwelling off Luzon in the Northeastern South China Sea. J. Geophys. Res. 101: 16435–16448.

Sirocko F., Garbe-Schönberg C.D., McIntyre A. and Molfino B. 1996. Teleconnections between the subtropical monsoons and high-latitude climates during the last deglaciation. Nature 272: 526–529.

Stattegger K., Kuhnt W., Wong H.K. and Scientific Party 1997. Cruise Report SONNE 115 SUNDAFLUT. Berichte-Report 86, Institüt für Geowissenschaften, Univ. Kiel, 211pp.

Steinke S. and Chen M.T. 2003. The spatial distribution patterns of Pulleniatina obliquiloculata in the South China Sea: implications of latitudinal differences in the strength of the East Asian winter monsoon intensity during the last glaciations. EGS-AGU-EUG Joint Assembly, Scientific Program, 327.

Steinke S., Chiu H.Y, Yu P.S., Shen C.C., Erlenkeuser H., Löwemark L. and Chen M.T. 2006. On the influence of sea level and monsoon climate on the southern South China Sea freshwater budget over the last 22,000 years. Quat. Sci. Rev. 25: 1475–1488.

Steinke S., Chiu H.Y, Yu P.S., Shen C.C., Löwemark L., Mii H.S. and Chen M.T. 2005. Mg/Ca ratios of two Globigerinoides ruber (white) morphotypes: implications for reconstructing past tropical/subtropical surface water conditions. Geochem. Geophys. Geosyst. 6: doi:10.1029/2005GC000926.

Steinke S., Kienast M., Groeneveld J., Lin L.C., Chen M.T. and Rendle-Bühring R. 2008a. Proxy dependence of the tempral patten of deglacial warming in the tropical South China Sea: toward resolving seasonality. Quat. Sci. Rev. 27: 688–700.

Steinke S., Kienast M., Pflaumann U., Weinelt M. and Stattegger K. 2001. A high resolution sea-surface temperature record from the tropical South China Sea (16,500–3000 B.P.). Quat. Res. 5: 353–362.

Steinke S., Yu P.S., Kucera M. and Chen M.T. 2008b. No-analog planktonic foraminiferal faunas in the glacial southern South China Sea: Implications for the magnitude of glacial cooling in the western Pacific warm pool. Mar. Micrpaleontol. 66: 71–90.

Stott L., Poulsen C., Lund S. and Thunell R. 2002. Super ENSO and global climate oscillations at millennial time scales. Science 297: 222–226.

Stuijts I., Newsome J.C. and Flenley J.R. 1988. Evidence for late Quaternary vegetational change in the Sumatran and Javan Highlands. Rev. Palaeobot. Palynol. 55: 207–216.

Stuiver M. and Grootes P.M. 2000. GISP2 oxygen isotope ratios. Quat. Res. 53: 277–284.

Su G. and Wang T. 1994. Basic characteristics of modern sedimentation in the South China Sea. In: Zhou D., Liang Y.B. and Zheng C.K. (eds.), Oceanology of China Seas. Kluwer, New York, pp. 407–418.

Sun Y. and An Z. 2001. History and variability of aridification in Asian Interior as recorded in dust flux of the Loess Plateau over the last 7 ma. Sci. China (D) 31(9): 769–767 (in Chinese).

Sun D.H., Shaw J., An Z., Cheng M. and Yue L. 1998. Magnetostratigraphy and paleoclimatic interpretation of a continuous 7.2 Ma Late Cenozoic eolian sediments from the Chinese Loess Plateau. Geophys. Res. Lett. 25: 85–88.

Sun X. and Li X. 1999. A pollen record of the last 37 ka in deep sea core 17940 from the northern South China Sea. Mar. Geol. 156: 227–244.

Sun X., Li X. and Beug H.J. 1999. Pollen distribution in hemipelagic surface sediments of the South China Sea and its relation to modern vegetation distribution. Mar. Geol. 156: 211–226.

Sun X., Li X. and Chen H. 2000a. Evidence for natural fire and climate history since 37 ka BP in the northern part of the South China Sea. Sci. China (D) 43(5): 487–493.

Sun X., Li X., Luo Y. and Chen X. 2000b. The vegetation and climate at the last glaciation on the emerged continental shelf of the South China Sea. Palaeogeogr. Palaeoclimatol. Palaeoecol. 160: 301–316.

Sun X., Li X. and Luo Y. 2002. Vegetation and climate on the Sunda Shelf of the South China Sea during the Last Glactiation-Pollen results from station 17962. Acta Bot. Sinica 44(6): 746–752.

Sun X. and Luo Y. 2001. Pollen record of the last 280 ka from deep-sea sediments of the northern South China Sea. Sci. China (D) 44(10): 879–888.

Sun X., Luo Y., Huang F., Tian J. and Wang P. 2003. Deep-sea pollen from the South China Sea: Pleistocene indicators of East Asian monsoon. Mar. Geol. 201: 97–118.

Sun X. and Wang P. 2005. How old is the Asian monsoon system? – Palaeobotanical records from China. Palaeogeogr., Palaeoclimatol., Palaeoecol. 222: 181–222.

Svensson A., Andersen K.K., Bigler M., Clausen H.B., Dahl-Jensen D., Davies S.M., Johnsen S.J., Muscheler R., Parrenin F., Rasmussen S.O., Röthlisberger R., Seierstad I., Steffensen J.P. and Vinther B.M. 2008. A 60 000 year Greenland stratigraphic ice core chronology. Clim. Past 4: 47–57.

Tada R., Irino T. and Koizumi I. 1999. Land-ocean linkages over orbital and millennial timescales recorded in late Quaternary sediments of the Japan Sea. Paleoceanography 14: 236–247.

Thompson P.R. 1981. Planktonic foraminifer in the western North Pacific during the past 150,000 years: comparison of modern and fossil assemblages. Palaeogeogr. Palaeoclimatol. Palaeoecol. 35: 441–479.

Thunell R.C., Curry W.B. and Honjo S. 1983. Seasonal variation in flux of planktonic foraminifera time-series sediment trap results from the Panama Basin. Earth Planet. Sci. Lett. 64: 44–55.

Tian J., Pak D.K., Wang P., Lea D., Cheng X. and Zhao Q. 2006. Late Pliocene monsoon linkage in the tropical South China Sea. Earth Planet. Sci. Lett. 252: 72–81.

Tian J., Wang P. and Cheng X. 2004. Development of the East Asian monsoon and Northern Hemisphere glaciation: Oxygen isotope records from the South China Sea. Quat. Sci. Rev. 23: 2007–2016.

Tian J., Wang P., Cheng X., Wang R. and Sun X. 2005a. Forcing mechanism of the Pleistocene east Asian monsoon variations in a phase perspective. Sci. China (D) 48(10): 1708–1717.

Tian J., Wang P., Chen R. and Cheng X. 2005b. Quaternary upper ocean thermal gradient variations in the South China Sea: Implications for east Asian monsoon climate. Paleoceanography 20: PA4007, doi:10.1029/2004PA001115.

Tian J., Wang P., Cheng X. and Li Q. 2002. Astronomically tuned Plio-Pleistocene benthic $\delta^{18}O$ records from South China Sea and Atlantic-Pacific comparison. Earth Planet. Sci. Lett. 203: 1015–1029.

Tija H.D. 1980. The Sunda Shelf, Southeast Asia. Zeitschrift fuer Geomorphologie N.F. 24: 405–427.

Traverse A. 1988. Production, dispersal, and sedimentation of spore/pollen. In: Traverse. A. (eds.), Palynology, Charpter 17. Unwin Hyman, Boston, pp. 375–430.

Trentesaux A., Liu Z., Colin C., Clemens S.C., Boulay S. and Wang P. 2004. Clay mineral assemblages in the northern South China Sea: Implications for the East Asian monsoon evolution over the past 2 million years. In: Prell W.L., Wang P., Blum P., Rea D.K. and Clemens S.C. (eds.), Proc. ODP, Sci. Results 184 [Online].

Vandenberghe J., An Z., Nugteren G., Lu H. and Van Huissteden K. 1997. New absolute time scale for the Quaternary climate in the Chinese loess region by grain-size analysis. Geology 25: 270–273.

Van der Kaars S. 1991. Palynology of eastern Indonesian marine piston-core: a late Quaternary vegetational and climatic record for Australasia. Palaeogeogr. Palaeoclimatol. Palaeoecol. 85: 239–302.

Van der Kaars S., Wang S., Kershaw P., Guichard F. and Setiabudi D.A. 2000. A late Quaternary palaeoecological record from the Banda Sea, Indonesia: patterns of vegetation and biomass burning in Indonesia and northern Australia. Palaeogeogr. Palaeoclimatol. Palaeoecol. 155: 135–153.

Visser K., Thunell R.C. and Stott L. 2003. Magnitude and timing of temperature change in the Indo-Pacific warm pool during deglaciation. Nature 421: 152–155.

von Rad U., Schaaf M., Michels K.H., Schulz H., Berger W.H. and Sirocko F. 1999. A 5000-yr record of climate change in varved sediments from the oxygen minimum zone off Pakistan, northeastern Arabian Sea. Quat. Res. 51: 39–53.

Walker D. and Flenley J.R. 1976. Late Quaternary vegetational history of the Enga District of upland Papua New Guinea. Phil. Trans. R. Soc. B286: 265–344.

Wan S., Li A., Clift P.D. and Jiang H. 2006. Development of the East Asian summer monsoon: Evidence from the sediment record in the South China Sea since 8.5 Ma. Palaeogeogr. Palaeoclimatol. Palaeoecol. 241: 139–159.

Wan S., Li A., Clift P.C. and Stuut J.B.W. 2007. Development of the East Asian monsoon: Mineralogical and sedimentologic records in the northern South China Sea since 20 Ma. Palaeogeogr. Palaeoclimatol. Palaeoecol. 254: 561–582.

Wang B., Clemens S.C. and Liu P. 2003. Contrasting the Indian and East Asian monsoons: implications on geological timescales. Mar. Geol. 201: 5–21.

Wang L. 2000. Isotopic signals in two morphotypes of *Globigerinoides ruber* (white) from the South Chian Sea: implication ns for monsoon climate change during the last glacial cycle. Palaeogeogr. Palaeoclimatol. Palaeoecol. 161: 381–394.

Wang L., Sarenthein M., Erlenkeuser H., Grimalt J., Grootes P., Heilig S., Ivanova E., Kienast M., Pelejero C. and Pflaumann U. 1999a. East Asian monsoon Climate during the late Pleistocene: high- resolution sediment records from the South China Sea. Mar. Geol. 156: 245–284.

Wang L., Sarenthein M., Grootes P. and Erlenkeuser H. 1999b. Millennial reoccurrence of century- scale abrupt events of East Asian monsoon: A possible heat conveyor for the global deglaciation. Paleoceanography 14: 725–731.

Wang L. and Wang P. 1990. Late Quaternary paleoceano-graphy of the South China Sea: glacial-interglacial contrasts in an enclosed basin. Paleoceanography 5: 77–90.

Wang P. 1999. Response of Western Pacific marginal seas to glacial cycles: Paleoceanographic and sedimentological features. Mar. Geol. 156: 5–39.

Wang P., Clemens S., Beaufort L., Braconnot P., Ganssen G., Jian Z., Kershaw P. and Sarnthein M. 2005. Evolution and variability of the Asian monsoon system: state of the art and outstanding issues. Quat. Sci. Rev. 24: 595–629.

Wang P., Jian Z., Zhao Q., Li Q., Wang R., Liu Z., Wu G., Shao L., Wang J., Huang B., Fang D., Tian J., Li J., Li X., Wei G., Sun X., Luo Y., Su X., Mao S. and Chen M. 2003. Evolution of the South China Sea and monsoon history revealed in deep-sea records. Chinese Sci. Bull. 48(23): 2549–2561.

Wang P., Min Q., Bain Y. and Cheng X. 1981. Strata of Quaternary transgressions in East China: a preliminary study. Acta Geol. Sinica 1: 1–13 (in Chinese).

Wang P., Prell W.L., Blum P. (eds.). 2000. Proc. ODP, Init. Repts, Vol. 184 [CD-ROM]. Ocean Drilling Program, Texas A&M University, College Station TX 77845–9547, USA.

Wang P., Wang L., Bian Y. and Jian Z. 1995. Late Qauternary paleoceanography of the South China Sea: surface circulation and carbonate cycles. Mar. Geol. 127: 145–165.

Wang R. and Abelmann A. 2002. Radiolarian responses to paleoceanographic events of the southern South China Sea during the Pleistocene. Mar. Micropaleontol. 46: 25–44.

Wang R.J. and Li J. 2003. Quaternary high-resolution opal record and its paleo-productivity implication at ODP Site 1143, southern South China Sea. Chin. Sci. Bull. 48(4): 363–367.

Wang R., Clemens S., Huang B. and Chen M. 2003. Late Quaternary paleoceanographic changes in the northern South China Sea (ODP Site 1146): radiolarian evidence. J. Quat. Sci. 18(8): 745–756.

Wang Y.J., Cheng H., Edwards R.L., An Z.S., Wu J.Y., Shen C.C. and Dorale J.A. 2001. A high-resolution absolute-dated late Pleistocene monsoon record from Hulu Cave, China. Science 294: 2345–2348.

Wang Y.J., Cheng H., Edwards R.L., He Y.Q., Kong X.G., An Z.S., Wu J.Y., Kelly M.J., Dykoski C.A. and Li X.D. 2005. The Holocene Asian monsoon: Links to solar changes and North Atlantic climate. Science 308: 854–857.

Wang Y.J., Cheng H., Edwards L.R., Kong X., Shao X., Chen S., Wu J., Jiang X., Wang X. and An Z. 2008. Millennial-and orbital-scale changes in the East Asian monsoon over the past 224,000 years. Nature 451: 1090–1093.

Wang X. 2006. Paleovegetation on the Sunda Shelf over the last 40 ka and its paleoenveronment. Ph.D. Thesis, School of Ocean & Earth Sciences, Tongji University (in Chinese).

Wang X., Sun X., Wang P. and Stattegger K. 2007. A high-resolution history of vegetation and climate history on Sunda Shelf since the last glaciation. Sci. China (D) 50: 75–80.

Wang X., Sun X., Wang P. and Stattegger K. 2008. The records of coastline changes reflected by mangroves on the Sunda Shelf since the last 40 ka. Chinese Sci. Bull. 53(13): 2069–2076.

Wara M.W., Ravelo A.C. and Delaney M.L. 2005. Permanent El Niño–like conditions during the Pliocene warm period. Science 309: 758–761.

Wehausen R. and Brumsack H.J. 2002. Astronomical forcing of the East Asian monsoon mirrored by the composition of Pliocene South China Sea sediments. Earth Planet. Sci. Lett. 201: 621–636.

Wei G.J., Deng W.F., Liu Y. and Li X.H. 2007. High-resolution sea surface temperature records derived from foraminiferal Mg/Ca ratios during the last 260ka in the northern South China Sea. Palaeogeogr. Palaeoclimatol. Palaeoecol. 250: 126–138.

Wei G.J., Li X., Liu Y., Shao L. and Liang X. 2006. Geochemical record of chemical weathering and monsoon climate change since the early Miocene in the South China Sea. Paleoceanography 21: PA4214, doi: 10.1029/2006PA001300.

Wei K.Y., Chiu T.C. and Chen Y.G. 2003. Toward establishing a maritime proxy record of the East Asian summer monsoons for the late Quaternary. Mar. Geol. 201: 67–79.

Wu G., Qin J. and Mao S. 2003. Deep-water Oligocene pollen record from South China Sea. Chinese Sci. Bull. 48(22): 2511–2515.

Wyrtki K. 1961. Scientific results of marine investigations of the South China Sea and Gulf of Thailand 1959~1961. Naga Rep. 2, Scripps Institution of Oceanography, University of California, San Diego, pp. 164–169.

Xu J., Wang P., Huang B., Li Q. and Jian Z. 2005. Response of planktonic foraminifera to glacial cycles: Mid-Pleistocene change in the southern South China Sea. Mar. Micropaleontol. 54: 89–105.

Yancheva G., Nowaczyk N.R., Mingram J., Dulsji P., Schetter G., Negendank J.F.W., Liu J., Sigman D.M., Peterson L.C. and Haug G.H. 2007. Influence of the intertropical convergence zone on the East Asian monsoon. Nature 7123: 74–77.

Yuan D., Chen H., Edwards R.L., Dykoski C.A., Kelly M.J., Zhang M., Qing J., Lin Y., Wang Y., Wu J., Dorale J.A., An Z. and Cai Y. 2004. Timing, duration, and transitions of the last interglacial Asian monsoon. Science 304: 575–578.

Zhang Y., Ji J., Baslsam W.L., Liu L. and Chen J. 2007. High resolution hematite and goethite records from ODP 1143, South China Sea: Co-evolution of monsoon precipitation and El Niño over the past 600,000 years. Earth Planet. Sci. Lett. 264: 136–150.

Zheng H.B., Powell C.M., Rea D.K., Wang J.L. and Wang P.X. 2004. Late Miocene and mid-Pliocene enhancement of the East Asian monsoon as viewed from the land and sea. Global Planet. Change 41: 147–155.

Zheng F., Li Q., Li B., Chen M., Tu X., Tian J. and Jian Z. 2005. A millennial scale planktonic foraminifer record of the mid-Pleistocene climate transition from the northern South China Sea. Palaeogeogr. Palaeoclimatol. Palaeoecol. 223: 349–363.

Zheng H., Powell C. McA., An Z., Zhou J. and Dong G. 2000. Pliocene uplift of the northern Tibetan Plateau. Geology 28: 715–718.

Zhao M., Huang C.Y., Wang C.C. and Wei G. 2006. A millennial-scale $U_{37}^{K'}$ sea-surface temperature record from the South China Sea ($8°N$) over the last 150 kyr: Monsoon and sea-level influence. Paleogeogr. Paleoclimat. Paleoecol. 236: 39–55.

Chapter 6
Deep Waters and Oceanic Connection

Quanhong Zhao, Qianyu Li and Zhimin Jian

Introduction

Among the low to mid latitude western Pacific marginal seas are the South China Sea (SCS), Sulu Sea and Sea of Japan, the three major enclosed seas in the region with passages connecting to the open ocean. The passage sill depths are ~130 m for the Sea of Japan, ~420 m for the Sulu Sea, and ~2,400 m for the SCS. Therefore, water exchanges with the deeper ocean are more effective in the SCS than in other sea basins. The last 30 Ma deep water history since seafloor spreading in the SCS is a history of basin evolution relating especially to subsidence and sediment preservation, as well as changes in deep water properties. The long sequence recovered at ODP Site 1148 provides a unique opportunity to trace this history. In this chapter, we overview recent progresses of deep water research in the region by focusing on benthic foraminiferal, ostracod and isotopic geochemistry records from Site 1148.

Q. Li (✉)
State Key Laboratory of Marine Geology, Tongji University, Shanghai 200092, China; School of Earth and Environmental Sciences, The University of Adelaide, South Australia 5005, Australia
e-mail: qli01@tongji.edu.cn; qianyu.li@adelaide.edu.au

Q. Zhao
State Key Laboratory of Marine Geology, Tongji University, Shanghai 200092, China,
e-mail: qhzhaok@online.sh.cn

Z. Jian
State Key Laboratory of Marine Geology, Tongji University, Shanghai 200092, China
e-mail: zjiank@online.sh.cn; jian@tongji.edu.cn

P. Wang, Q. Li (eds.), *The South China Sea*, Developments in Paleoenvironmental Research 13, DOI 10.1007/978-1-4020-9745-4_6,
© Springer Science+Business Media B.V. 2009

6.1 Modern Deep Waters and Their Faunal Features

Marginal Seas in the Western Pacific

More than 75% of world's marginal basins are distributed along the western Pacific continental margin, as a result of late Cenozoic deformation of Asia and subduction of the Pacific Plate (Wang P. 1999; Wang P. 2004). Table 6.1 lists morphological features of the three major enclosed seas from the low and mid latitude western Pacific margin: the SCS, Sulu Sea, and Sea of Japan, all having a central deep basin of over 4,000 m. Unlike other marginal seas, however, they are connected to the open ocean through narrow passages at various sill depths. The deepest passage is the Bashi (Luzon) Strait connecting the SCS with the western Pacific or the open western Philippine Sea at ~2,400 m water depth (Qu et al. 2006). The deepest passage linking the Sulu Sea to the SCS is the Mindoro Strait at 420 m. With extreme shallow sills all <130 m, the Sea of Japan is the most isolated. During glacial sea level lowstands, therefore, the Sea of Japan was completely closed, the Sulu Sea was connected with the SCS only with a shallow Mindoro Strait, and the SCS had the only deep passage Bashi Strait connecting to the open Pacific water. The broad Sunda Shelf and the northern shelf in the glacial SCS were largely exposed, as also the flat and shallow East China Sea (over 75% in <200 m). Therefore, the hydrological system in the region is highly sensible to sea level fluctuations, often leading to large circulation changes that will ultimately impact regional land-sea interaction and climate.

The geometric features of these enclosed basins determine the property and flow patterns of their deep waters. The Japan Sea Proper Water below 200 m is produced from the northern part of the basin in winter and characterized by a low temperature of 0–1 °C. Therefore, the carbonate compensation depth (CCD) is shallow, ~1,400–1,600 m, and preservation of carbonate and calcareous faunas is poor in the Sea of Japan. In contrast, the Sulu Sea is bathed with warm deep water of 10 °C originating from the upper layers of the SCS, whereas the >2,500 m deep water in the SCS is constantly cold (~2 °C) with a deeper CCD at about 3,500 m (Wang P. 1999).

Deep waters from both the Sea of Japan and the SCS have a very short resident time: ~120 years for the Sea of Japan (Chen et al. 1995) and generally less than

Table 6.1 Morphological features of the three enclosed marginal seas in the West Pacific are modified from Wang L. et al. (1999), showing the four main passages connecting the South China Sea. The Bashi Strait sill depth at ~2,400 m (Qu et al. 2006) is used here

Sea	Mean depth (m)	<200 m area (%)	Max depth (m)	Sill depth (m)	Passage (strait)	Passage connection
South China Sea	1,212	52.4	5,377	2,400	Bashi	W Philippine Sea
				70	Taiwan	East China Sea
Sulu Sea	1,570	34.3	5,580	420	Mindoro	South China Sea
				100	Balabac	South China Sea
Sea of Japan	1,361	26.3	4,049	130	Korean	East China Sea
				116	Tsugaru	North Pacific

50 years for the SCS (e.g., Qu et al. 2006) (Chapter 2). A fast flush of the deep water in the SCS precludes full decomposition of sink particulate matters in the water column. Long-term deep water variations occurred in various stages of basin evolution, particularly in responding to changes in basin configuration and in surface circulation, and their feedback may also play an important role in the process of climate change in the region (Wang P. 1999).

Modern Intermediate and Deep Waters in the South China Sea

The property and flow patterns of deeper waters in the SCS are much less studied compared to those of the upper-layer water. Limited data indicate that intermediate and deeper waters in the sea basin are similar to those in the tropical-subtropical western Pacific above ~2,000 m but comparatively slightly less oxygenated (Table 6.2). This is because the intermediate waters between the two basins exchange freely through the ~2,400 m Bashi Strait, but not below the sill depth. As waters are often re-juvenated at the sill depth before entering the SCS (Fig. 6.1), the deeper

Table 6.2 Properties of subsurface water masses in the South China Sea and the tropical western Pacific are similar according to various authors: Wyrtki (1961a,b), SOA (1988), Chen et al. (2001) and Qu et al. (2006)

Water mass	Depth (m)	Temp (°C)	Salinity (%)	Oxygen (ml/l)
South China Sea				
Intermediate water	300–1,000	12.0–5.0	34.4–34.5	2.5–1.6
Upper deep water	1,000–2,500	5.0–2.4	34.5–34.6	1.6–2.5
Lower deep water	>2,500	<2.4	~34.6	~2.5
Tropical western Pacific				
Intermediate water	300–1,000	11.0–5.0	34.1–34.6	3.0–1.2
Upper deep water	1,000–2,500	5.0–3.0	34.5–34.6	2.0–3.0
Lower deep water	>2,500	1.6	34.6–34.7	3.4

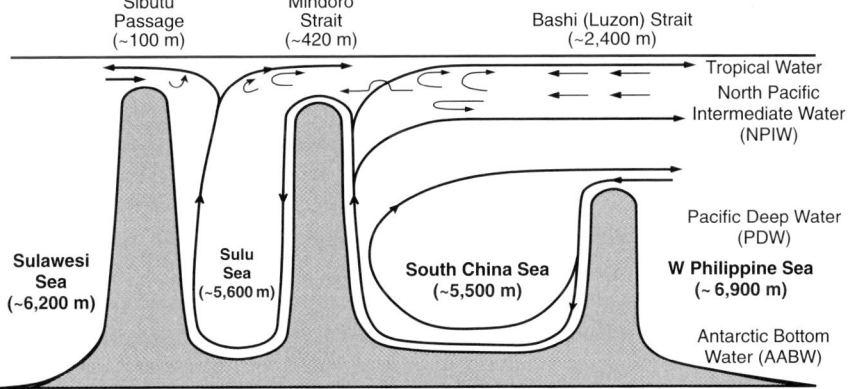

Fig. 6.1 Schematic diagram shows the major flow patterns in the SCS and neighboring sea basins (modified from Chen C. et al. 2006). Note that lateral flows and mixing in deeper layers are less understood

water in the SCS is less corrosive relative to the deep Pacific. Analyses of modern carbonate dissolution data indicate that the aragonite (or pteropod) composition depth (ACD) in the SCS lies at ~1,000 m, the lysocline depth at ~3,000 m, and the CCD at ~3,500 m (Rottman 1979; Li C. 1987; Han and Ma 1988; Chen and Chen 1989; Calvert et al. 1993; Zheng et al. 1993; Miao and Thunell 1993).

Monsoonal upwelling in areas such as offshore Luzon and Vietnam (Chapter 2) causes mixing between intermediate and surface waters. Simultaneous downwelling may also occur off Luzon and Palawan, causing a basin-wide circulation dynamics due to monsoon winds (Chao et al. 1996). As a result, water renewal is achieved and deep water down to 2,000 m or more becomes ventilated. The residence time of SCS waters varies between 40 and 155 years depending on localities (Gong et al. 1992; Chen et al. 2001; Qu et al. 2006), which is considerably less than 1,000 or more years for today's tropical Pacific (e.g., Broecker 1992).

Table 6.2 lists the general characteristics of various layers of water in the SCS. Chapter 2 provides a more detailed account on their flowing patterns.

Modern Deep-Sea Benthic Foraminifera and Ostracods

Benthic microfauna such as Foraminifera and Ostracoda are widely used as proxies for deep-water changes because their sensitivity to environment conditions in bottom waters. Benthic foraminifera and ostracods from modern deep-sea SCS have been studied by Cai (1982, 1991), Cai and Tu (1983), Tu (1983), Zheng (1987), Tu and Zheng (1991), Jian and Zheng (1992, 1995), Miao and Thunell (1993), Whatley and Zhao (1993), Zhao and Zhou (1995), Zhao and Zheng (1996), Hess and Kuhnt (1996), Jian and Wang (1997), Hess (1998), Kuhnt et al. (1999) and Zhao et al. (2000). The results from these studies show a close relationship between faunal assemblages and environmental factors such as depth, temperature, water masses, corrosiveness of bottom water and food supply.

Benthic Foraminifera

Numerous studies on the distribution of benthic foraminifera in deep sea sediments (Murray 1991, 2006) reveal a cosmopolitan pattern for many species. Some, however, prefer living in certain bathymetric depth ranges closely associating with particular deepwater masses. This relationship justifies their use as water mass indicators although such credibility is questioned by some researchers (e.g., Jorissen et al. 2007) because many species appear to respond more to oxygen and productivity levels than to bathymetry changes (Altenbach and Sarnthein 1989; Murray 2006). Table 6.3 lists species commonly found in the SCS with known ecological features which can be used to assist interpreting deep water changes in the region. Although the indication for different water masses is given by the cited references, most listed species appear to occur in all intermediate and deep water masses without any obvious preference (Murray 2006).

In the modern SCS, three benthic foraminiferal assemblages have been recognized (Fig. 6.2): an intermediate water assemblage dominated by *Globocassidulina*

Table 6.3 Common deep-sea benthic foraminifera and their known ecological data compiled from various sources

Taxa	Ecological preference[1]	References[2]
Bulimina	infaunal, high food supply, different oxygen levels, AABW, CPDW, PIW, PDW	12, 14, 15, 18, 20, 23, 26, 31, 32
Buliminella	infaunal, low oxygen with high food supply	29
Cibicidoides	epifaunal to shallow infaunal, oxic to suboxic, low food supply, IW and PDW of Sulu Sea	14, 15, 18
Epistominella exigua	epifaunal, oxic, phytodetritus feeder, seasonally high food supply, all water masses	1, 2, 3, 4, 5, 10, 13 14, 16, 17, 21, 27, 31
Favocassidulina favus	suboxic, PDW	2, 3, 22, 30
Globobulimina	deep infaunal, dysoxic	14, 15, 18, 23, 24, 26, 28, 31
Globocassidulina subglobosa	epifaunal and shallow infaunal, oxic, phyto- detritus feeder, low food supply, strong bottom current, AABW, AAIW, IBA, NADW; suboxic, food supply, IW of the SCS	1, 3, 4, 8, 14, 15, 17, 20, 21, 23, 32; 9, 12, 19, 25
Gyroidinoides	epifaunal to shallow infaunal, suboxic, phyto- detritus feeder, seasonally food supply	14, 15, 18, 23, 33
Hoeglundina elegans	shallow infaunal, suboxic, carbonate well- preserved water, NAIW, PIW	2, 3, 7, 12, 15, 31
Melonis	infaunal, suboxic, high food supply, NADW, PDW, PIW	2, 3, 10, 12, 14, 15, 20, 24, 26, 31
Nuttallides umboniferus	epifaunal, oxic, phytodetritus feeder, seasonally high food supply, deep to bottom waters	1, 2, 3, 4, 8, 26
Oridosalis umbonatus	epifaunal, oxic, low food supply, AABW, CPDW, NADW, PDW, PBW; trasitional, suboxic, oligotrophic, Sulu Sea	2, 3, 8, 9, 12, 14, 15, 20, 25, 31; 18, 23
Osangularia culter	shallow infaunal, oxic, oligotrophic, NADW; transitional infaunal, suboxic, IW and DW of the SCS	26, 29, 31; 23
Cibicidoides wuellerstorfi	epifaunal, oxic, suspension feeder, seasonal food supply, high bottom current, AABW, NADW, PBW, PDW	1, 3, 4, 6, 11, 12, 14, 15, 21, 22, 30, 36
Pullenia quinqueloba	shallow infaunal, suboxic, high food supply, PIW, PDW and Arabian Sea OMZ	2, 3, 37
Rhabammina	epifaunal, oxic, suspension feeder, oligotrophic, strong bottom current, AABW, NADW	11, 34, 32
Uvigerina	shallow infaunal, suboxic, high food supply, NAIW, IBW, IDW, IIW, PIW	2, 3, 9, 12, 14, 15, 18, 21, 23, 25, 26, 34, 35, 37

[1] Water masses: AABW=Antarctic Bottom Water; AAIW=Antarctic Intermediate Water; ABW=Atlantic Bottom Water; CPDW=Circum-Polar Deep Water; IBW=Indian Bottom Water; IDW=Indian Deep Water; IIW=Indian Intermediate Water; IW=Indian Intermediate Water; NADW=North Atlantic Deep Water; NAIW=North Atlantic Intermediate Water; PIW=Pacific Intermediate Water; PDW=Pacific Deep Water; PBW=Pacific Bottom Water.

[2]Reference sources: 1, Corliss (1979); 2. Burke (1981); 3. Douglas and Woodruff (1981); 4, Corliss (1983); 5, Gooday (1988); 6, Lutze and Thiel (1989); 7, Hermelin and Shimmield (1990); 8, Hayward et al. (2007); 9, Jian and Zheng (1992); 10, Burke et al. (1993); 11, Linke and Lutze (1993); 12, Miao and Thunell (1994); 13, Gooday (1993); 14, Gooday (1994); 15, Kaiho (1994a.b); 16, Mackensen et al. (1993); 17, Clark et al. (1994); 18, Rathburn and Corliss (1994); 19, Jian and Zheng (1995); 20, Mackensen et al. (1995); 21, Nomura (1995); 22, McDougall (1996); 23, Rathburn et al. (1996); 24, Fariduddin and Loubere (1997); 25, Jian and Wang (1997); 26, Schmiedl et al. (1997); 27, Smart and Gooday (1997); 28, Jorissen (1999); 29, Gupta (1999); 30, Kaiho (1999); 31, den Dulk et al. (2000); 32, de Mello e Sousa et al. (2006); 33, Schmiedl et al. (2000); 34, Altenbach et al. (2003); 35, Murgese and De Deckker (2005, 2007); 36, Gupta et al. (2006); 37, Schumacher et al. (2007).

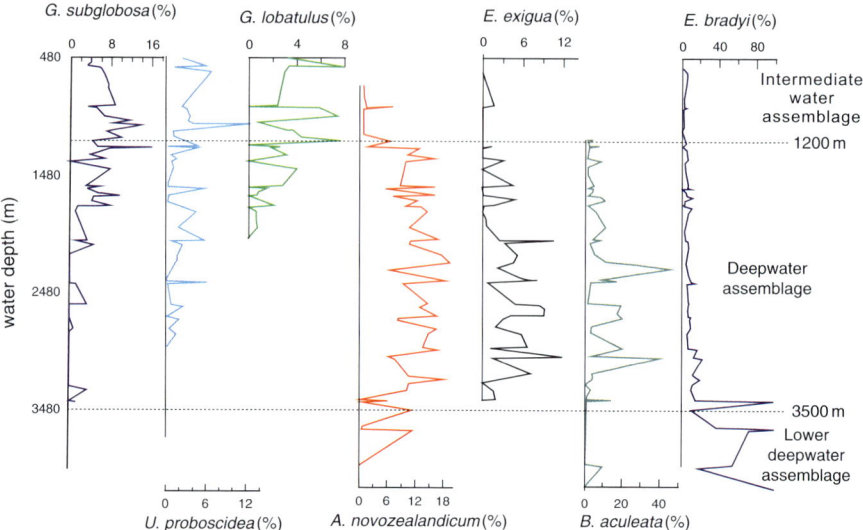

Fig. 6.2 Distribution of selected deep-sea benthic foraminiferal species in surface sediments of the SCS shows three faunal associations relating to intermediate, deep and lower deep waters respectively (modified from Jian and Wang 1997)

subglobosa and *Uvigerina*, a deep water assemblage by *Astrononion* and *Bulimina aculeata* with common occurrence of *Oridorsalis*, and a lower deep water assemblage dominated by *Eggerella bradyi* and other agglutinated taxa from below the CCD.

Clearly, the intermediate and deep water benthic foraminiferal assemblages from the SCS and tropical western Pacific are similar (Table 6.4). However, *Favocassidulina favus* and *Nuttallides umboniferus*, the two typical components of the Pacific Bottom Water (PBW) are absent from the modern SCS, suggesting that the PBW is unable to flow over the Bashi Strait into the SCS (Jian and Wang 1997).

Kuhnt et al. (1999) found some benthic foraminiferal species with high negative scores in Factor 1 of their corresponding analysis mainly relating to high organic carbon flux in the SCS. These species include: *Uvigerina peregrina*, *Bolivina* spp., *Trifarina bradyi*, *Rosalina concinna*, *Cassidulina crassa*, *Epistominella rugosa* and *Globobulimina affinis*. Low organic flux indicators are *Siphotextularia rolshauseni*, *Eggerella bradyi*, *Globocassidulina elegans*, *Astrononion novozealandicum*, *Epistominella exigua*, *Bulimina aculeata* and *Quinqueloculina venusta*.

In a study of benthic foraminiferal recolonization after the eruption of Mt. Pinatubo volcano in 1991, Hess and Kuhnt (1996) found that the eruption caused mass mortality of benthic foraminifera in a vast area of the eastern SCS. Recolonization of the ash substrate was step-wise, mainly by infaunal morphotypes including species of the genus *Reophax*. Three years after the eruption, the benthic foraminiferal community structure is still far from recovery to background levels. After about 6–7 years, post-eruption assemblages with high proportions

Table 6.4 Deep-sea benthic foraminiferal assemblages from the South China Sea and the tropical western Pacific

	Intermediate water	Upper deep water	Lower deep water
South China Sea:			
Jian and Wang (1997)	480–1,200 m *Globocassidulina subglobosa, Uvigerina* spp., *Cibicidoides bradyi, Gavelinopsis lobatulus, Pullenia bulloides*	1,200–2,500 m *Astrononion novozealandicum, Bulimina aculeata, Oridorsalis umbonatus, Epistominella exigua, Sphaeroidina bulloides, Eggerella bradyi*	>2,500 m *Eggerella bradyi*
Miao and Thunell (1993)	<1,500 m *G. subglobosa, Melonis affinis, Uvigerina* spp., *Bolivina robusta, Hoeglundina elegans*	1,500–3,000 m *A. novozealandicum, B. aculeata, O. umbonatus, Cassidulina subcarin ta, C. wuellerstorfi, Bolivina robusta*	>3,000 m *Rhabdammina abyssorrum, Cyclammina cancellata, Ammoglobigerina globigeriniformis, Karreriella bradyi*
Tropical West Pacific:			
Douglas and Woodruff (1981)	<2,500 m *Uvigerina* spp., *Hoeglundina elegans, Gyroidina* spp., *Ehrenbergina* sp.	2,500–3,500 m *Epistominella exigua, Pullenia quinqueloba, Melonis* spp., *Favocassidulina* sp.	>3,500 m *Nuttallides umboniferus, E. exiqua, Pullenia bulloides, Cibicidoides wuellerstorfi*
Burke (1981)	1,200–2,500 m *Siphouvigerina inter- rupta, H. elegans, Triloculina* sp.B	2,500–3,000 m *E. exiqua, Favocassidulina favus, Melonis* spp., *Pullenia quinqueloba*	>3,000 m *Epistominella exiqua, Nuttallides umboniferus*

of epibenthic suspension feeders became established (Hess et al. 2001; Kuhnt et al. 2005).

Ostracods

Zhao and Zhou (1995) recognized 4 ostracod assemblages in the modern SCS (Table 6.5). The subsurface water assemblage (180–300 m) is dominated by *Neonesidea elegans* (33.9%), the intermediate water assemblage (300–1,000 m) by *Krithe* (20.6%) and *Xestoleber* (9.3%), and the deep water assemblage (900–2,500 m) and lower deep water assemblage (>2,500 m) both by *Krithe* (38.7%) and *Argilloecia* (7.4%) but the mean abundance and diversity in the latter are the lowest. The highest mean abundance, or "incidence" representing the number of valves per 10-gram dry sediment, is found in the intermediate water assemblage. Many cosmopolitan deep-sea species such as *Abyssocythereis sulcatoperforata, Henryhowella asperima, Legitimocythere acanthoderma* and *Poseidonamicus* occur frequently in samples from below 1,000 m (see Table 6.5).

Table 6.5 Ostracod assemblages and species mean percentages from bottom surface sediments of the South China Sea (after Zhao and Zhou 1995)

	Subsurface water 180–300 m	Intermediate water 300–1,000 m	Upper deep water 1,000–2,500 m	Lower deep water >2,500 m
Mean incidence (valves/10 g)	103.7	394.7	20.2	6.4
Mean species diversity	20	36	13	2.1
Abrocythere	1.2%	4.6%	0	0
Abyssocythere	0	0	3.7%	1.4%
A. sulcatoperforata	0	0	5.0%	3.6%
Acanthocythereis	13.3%	0	0	0
Ambocythere	1.2%	5.0%	0.8%	0
Argilloecia	1.9%	5.7%	7.4%	10.9%
Bradleya	6.7%	6.8%	3.3%	6.0%
Cytherella	0.7%	2.9%	1.1%	0
Cytherelloidea	12.6%	1.2%	0	0
Cytheropteron	9.9%	8.3%	3.0%	0.4%
Henryhowella	0	0.4%	6.1%	4.0%
Krithe	0	20.6%	38.7%	31.8%
L. acanthoderma	0	0.8%	3.2%	4.8%
Neonesidea	33.9%	2.6%	0	0
Palmoconcha	0	1.7%	0.2%	0
Parakrithe	0.4%	4.5%	6.9%	1.7%
Rugocythereis	0	0	3.4%	4.4%
Xestoleberis	4.9%	9.3%	2.1%	5.0%

6.2 Late Quaternary Deep-Water Faunas and Stable Isotopes

Only a few studies have been carried out on the variations of deep-water faunas and stable isotopes in late Quaternary sediments of the SCS. From a water depth of ~2,100 m, Jian and Wang (1997) reported the typical Pacific Deep Water (PDW) species *F. favus* in core samples immediately below the last glacial maximum (LGM) interval. Because this species marks the upper depth limit of the PDW (Table 6.3), its occurrence in pre-LGM samples may suggest an elevated deep water mass by ~500 m before its depression by the formation of the North Pacific Deep Water (NPDW) during the LGM (Jian and Wang 1997) (Fig. 6.3).

Abundance variations in dominant benthic foraminiferal species in piston cores 17940-2 from the northern SCS and 17964-2/3 from the southern SCS indicate their close links with productivity-induced organic carbon flux (Fig. 6.4) (Jian et al. 1999a,b). High organic carbon flux intervals (>3.5 g/m^2/yr calculated using organic carbon accumulation rates) are dominated by a group of infaunal detritus feeders including *Bulimina aculeata* and *Uvigerina peregrina*, respectively at and before the LGM in the south and at about 10 ka in the northern SCS. As soon as C$_{org}$ flux decreased to 2.5–3.5 g/m^2/yr, suspension feeders such as *Cibicidoides wuellerstorfi* (=*Planulina wuellerstorfi*) and a group of "opportunistic" species including *Oridorsalis umbonatus*, *Melonis barleeanum* and *Chilostomella ovoidea* gradually became more abundant (Fig. 6.4).

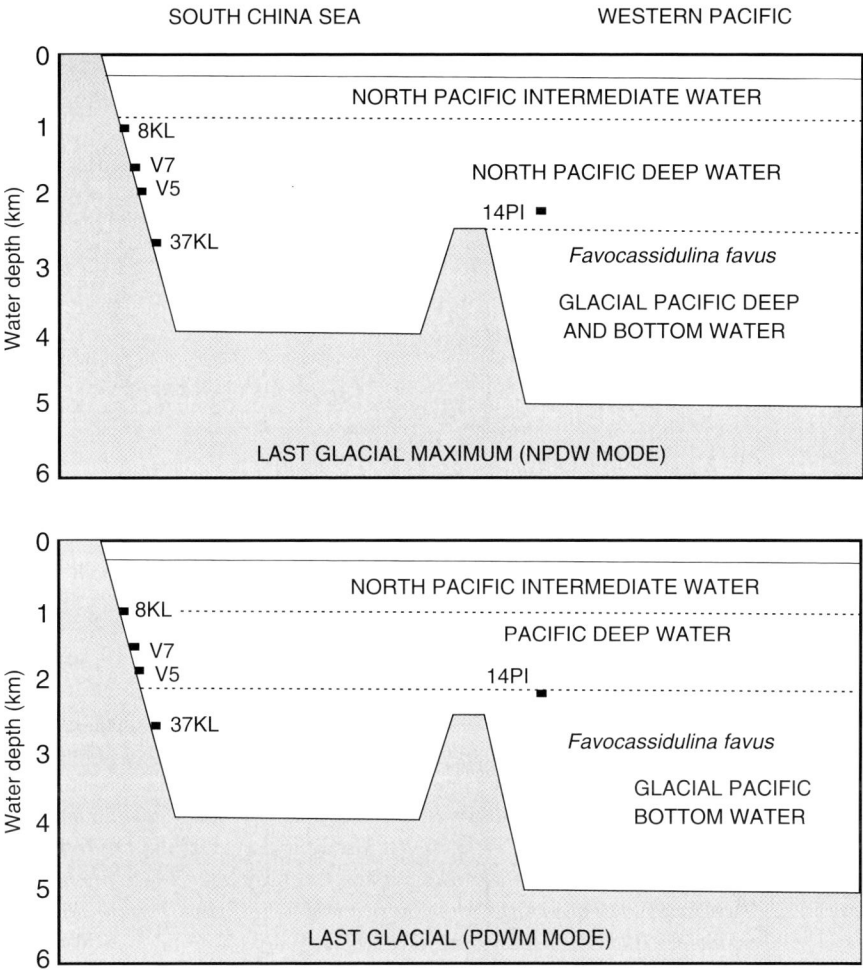

Fig. 6.3 The occurrence of benthic foraminiferal species *F. favus* only in pre-LGM samples from core 37 KL may indicate the influence of Pacific Deep Water Mass (PDWM) (*lower panel*) before the formation of the North Pacific Deep Water (NPDW) (*upper panel*) (Jian and Wang 1997)

Using their high and low organic flux groupings of modern benthic foraminifera, however, Kuhnt et al. (1999) obtained consistently higher C_{org} flux values in cores 17964-2/3: ~6–8 g/m^2/yr during the LGM and Termination I before reducing to 3–4 g/m^2/yr in the Holocene. Therefore, simple groupings of species often help identify productivity and carbon flux trends but may be insufficient for quantifying these trends or their variation patterns. Another factor which also needs attention is that the SCS deep water is more oxic than other oceanic settings and organic matter decay at the sea floor is relatively fast, which could have resulted in significantly underestimate of past carbon fluxes using the Corg accumulation rate.

Oxygen and carbon stable isotopes on benthic foraminifer *C. wuellerstorfi* and planktonic species *Globigerinoides ruber* and *Pulleniatina obliquiloculata* in

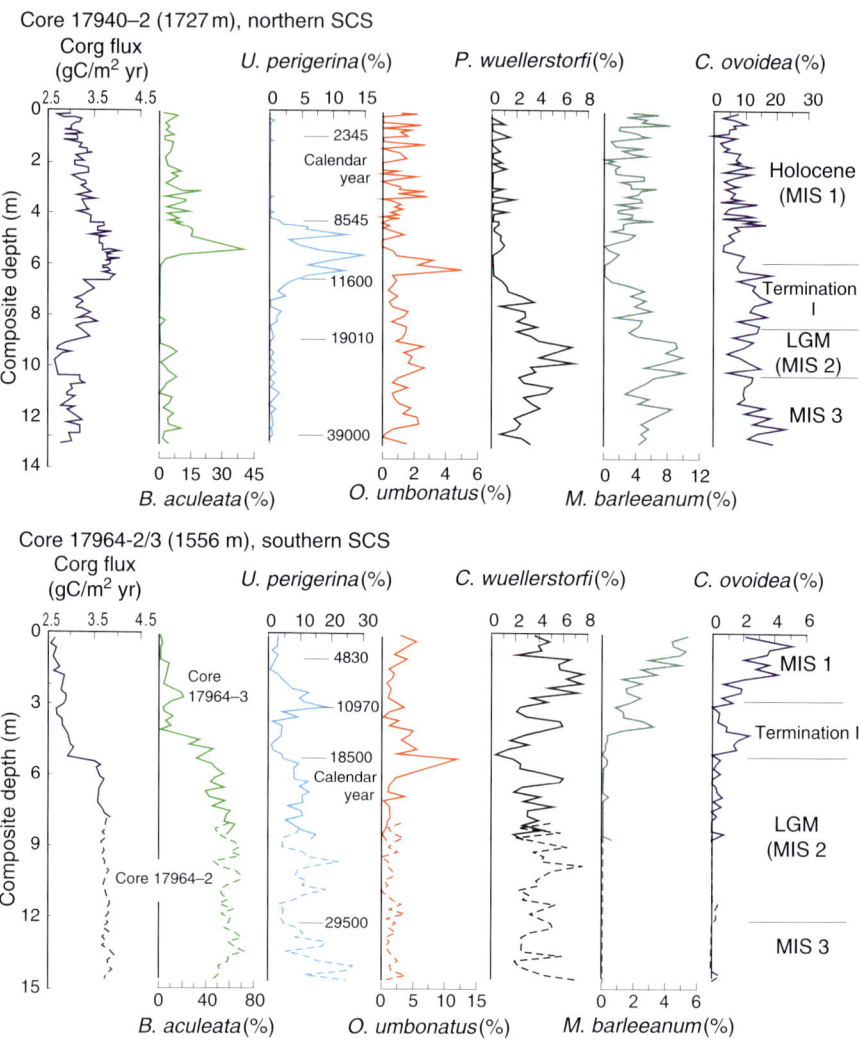

Fig. 6.4 Down-core variations in relative abundances of dominant benthic foraminiferal species in cores 17940-2 and 17964-2/3 show a close relationship with C_{org} flux (Jian et al. 1999b). The calendar ages are from Wang L. et al. (1999)

Holocene and LGM samples from various parts of the SCS were systematically analyzed by Cheng et al. (2005). The results show consistent lighter $\delta^{18}O$ but heavier $\delta^{13}C$ values in LGM samples. Respectively, the average planktonic $\delta^{18}O$ values are $-2.85‰$ for the Holocene (52 samples) and $-1.24‰$ for the LGM (17 samples), and the average benthic $\delta^{18}O$ values are $2.18‰$ for the Holocene (53 samples) and $4.08‰$ for the LGM. The differences between benthic and planktonic $\delta^{18}O$, or $\Delta\delta^{18}O_{b-p}$, are 5.03 for the Holocene and 5.32 for the LGM. Heavier planktonic

$\delta^{18}O$ values, corresponding to low $\Delta\delta^{18}O_{b-p}$, form a tongue in the northeastern SCS, near the Bashi Strait, probably indicating the invasion of North Pacific waters. This tongue is also marked by lighter planktonic and benthic $\delta^{13}C$ values (Cheng et al. 2005) (Fig. 7.17 in Chapter 7).

The distribution of heavier benthic $\delta^{18}O$ values mainly within the deep sea basin indicates the existence of a stable local deep water in the SCS, marked also by lighter $\delta^{13}C$ values. However, LGM samples show lighter planktonic $\delta^{13}C$ values respectively from the northeast and southeast, probably relating more to higher nutrient supply induced by monsoons (cf. Winn et al. 1992). Heavier $\delta^{13}C$ values by $\sim2‰$ in LGM samples than their Holocene counterparts appear to suggest a well-mixed bottom water in the glacial SCS, but the small LGM data set hampers detailed analyses.

6.3 Neogene and Oligocene Deep-Water Benthic Faunas from ODP Leg 184 Sites

Due to limited material available, the distribution of deep-sea benthic foraminifera and ostracods in early Quaternary and older sequences of the SCS was poorly known until ODP Leg 184. Well preserved Oligocene–Quaternary sediments recovered in Leg 184 cores provide unprecedented material for paleoceanographic studies (Wang P. et al. 2003b; Li Q. et al. 2006). Biofacies analyses have generated useful information for reconstructing regional paleoenvironmental changes. For example, a preliminary study of benthic foraminifera from Site 1148 by Kuhnt et al. (2002) identified 5 assemblages: 1 and 2 for the middle and early Miocene and 3 to 5 for the Oligocene. Unlike the younger assemblages with a Pacific origin, the Oligocene assemblages bear a close affinity to those of Paratethys origin (Kuhnt et al. 2002). At the southern Site 1143, Hess and Kuhnt (2005) found increase of carbon flux indicators since the late Pliocene. These authors considered high carbon flux proxies to include positive benthic $\delta^{13}C$ difference between *C. wuellerstorfi* and *U. peregrina*, higher benthic foraminiferal accumulation rates (BFARs), low species diversity, and higher relative abundance of flux indicator species listed in Kuhnt et al. (1999). In the following, we follow the time scale of Gradstein et al. (2004) with some modification by Li Q. et al. (2006) for Site 1148.

Site 1148 Benthic Foraminifera

Benthic foraminifera in the $>150\,\mu m$ fraction of 866 samples from ODP Site 1148 are studied. An earlier study by Kuhnt et al. (2002) analyzed 316 samples from the $>10\,Ma$ lower sediment section. Common deep-water benthic foraminifera are illustrated in Fig. 6.5. The results indicate that the Miocene is the period with relative low absolute abundance, as shown by the mean incidence (individuals per gram dry sediment). The early and middle Oligocene is characterized by high abundance of agglutinated species, mainly *Nothia, Rhabdammina, Saccammina, Ammodiscus,*

(a)   (b)

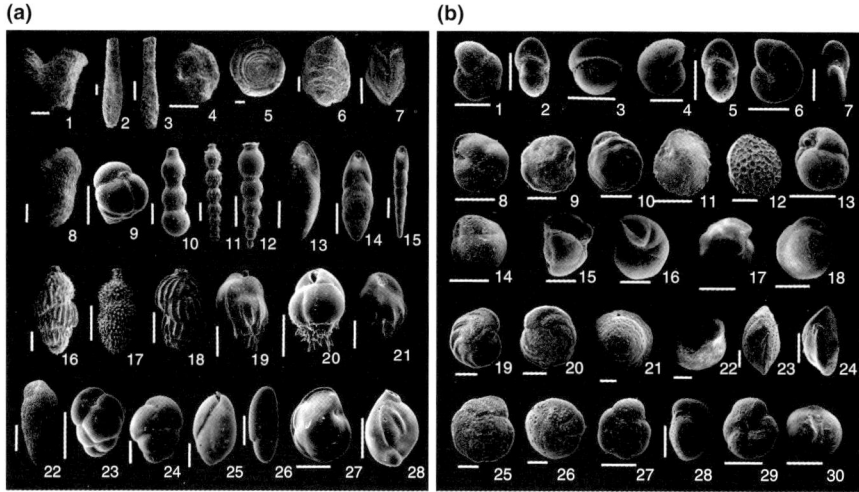

Fig. 6.5 (**a**) Typical deep-sea benthic foraminifera from ODP Hole 1148A are shown with scale bars indicating 200 μm. 1 *Nothia excelsa* (Grzybowski), 714.93 mcd; 2, 3 *Rhabdammina discreta* Brady, 671.97 mcd; 4 *Saccammina* sp., 632.27 mcd; 5 *Ammodiscus* sp., 681.57 mcd; 6 *Vulvulina spinosa* Cushman, 478.75 mcd; 7 *Tritaxia* sp., 545.02 mcd. 8 *Karreriella bradyi* (Cushman), 623.67 mcd; 9 *Eggerella bradyi* (Cushman), 657.07 mcd; 10 *Stilostomella abyssorum* (Brady), 366.59 mcd; 11 *Stilostomella subspinosa* (Cushman), 204.59 mcd; 12 *Stilostomella antillea* (Cushman), 204.59 mcd; 13 *Pleurostomella alternans* Schwager, 699.91 mcd; 14 *Pleurostomella* acuminate Cushman, 608.97 mcd; 15 *Pleurostomella* sp., 396.54 mcd; 16 *Uvigerina peregrina* Cushman, 404.07 mcd; 17 *Uvigerina hispida* Schwager, 374.39 mcd; 18 *Uvigerina havanensis* Cushman and Bermudez, 456.64 mcd; 19 *Uvigerina crassicostata* Schwager, 714.93 mcd; 20 *Bulimina aculeata* d'Orbigny, 478.45 mcd; 21 *Bulimina striata* d'Orbigny, 478.45 mcd; 22 *Bulimina tuxpamensis* Cole, 404.07 mcd; 23, 24 *Buliminella parvula* Brotzen, 392.04 mcd; 25 *Globobulimina affinis* (d'Orbigny), 20.56 mcd; 26 *Chilostomella oolina* Schwager, 20.56 mcd; 27 *Hoeglundina elegans* (d'Orbigny), 31.08 mcd; 28 *Qinqueloculina* sp., 323.97 mcd, (**b**) with scale bars = 200 μm. 1, 2 *Pullenia quinqueloba* (Reuss), 478.25 mcd; 3 *Pullenia bulloides* (d'Orbigny), 204.24 mcd; 4, 5 *Melonis barleeanus* (Williamson), 695.77 mcd; 6, 7 *Melonis affinis* (Reuss), 472.94 mcd; 8 *Epistominella exigua* (Brady), 71.94 mcd; 9, 10 *Nuttallides umboniferus* (Cushman), 86.04 mcd; 11 *Osangularia culter* (Parker and Jones), 358.60 mcd; 12 *Favocassidulina favus* (Brady), 218.04 mcd; 13, 14 *Globocassidulina subglobosa* (Brady), 478.67 mcd; 15, 16 *Gyroidinoides soldanii* (d'Orbigny), 467.47 mcd; 17, 18 *Oridorsalis umbonatus* (Reuss), 517.07 mcd; 19, 20 *Cibicidoides wuellerstorfi* (Schwager), 274.99 mcd; 20–23 *Cibicidoides havanensis* (Cushman and Bermudez), 474.44 mcd; 24–26 *Cibicidoides praemundulus* Berggren and Miller, 539.87 mcd; 27–29 *Cibicidoides bradyi* (Trauth), 331.29 mcd; 30. *Saitoella globosa* Kaiho, 545.02 mcd

Tritaxia, Eggerella bradyi, Karreriella and *Vulvulina* (Fig. 6.6A), although most of them disappeared at about 27.5 Ma. As observed by Kuhnt et al. (2002), the Oligocene assemblages exhibit striking similarities with deep-water assemblages of the Central Paratethys and are significantly different from contemporary assemblages in other oceans. In contrast, Miocene assemblages are cosmopolitan and contain the whole spectrum of taxa known in the Pacific Ocean.

Calcareous species increased substantially in the late Oligocene (Fig. 6.6). The dominant species include *Oridorsalis umbonatus, Cibicidoides, Globocassidulina subglobusa, Gyroidinoides, Stilostomella* and *Uvigerina*. In the early Miocene,

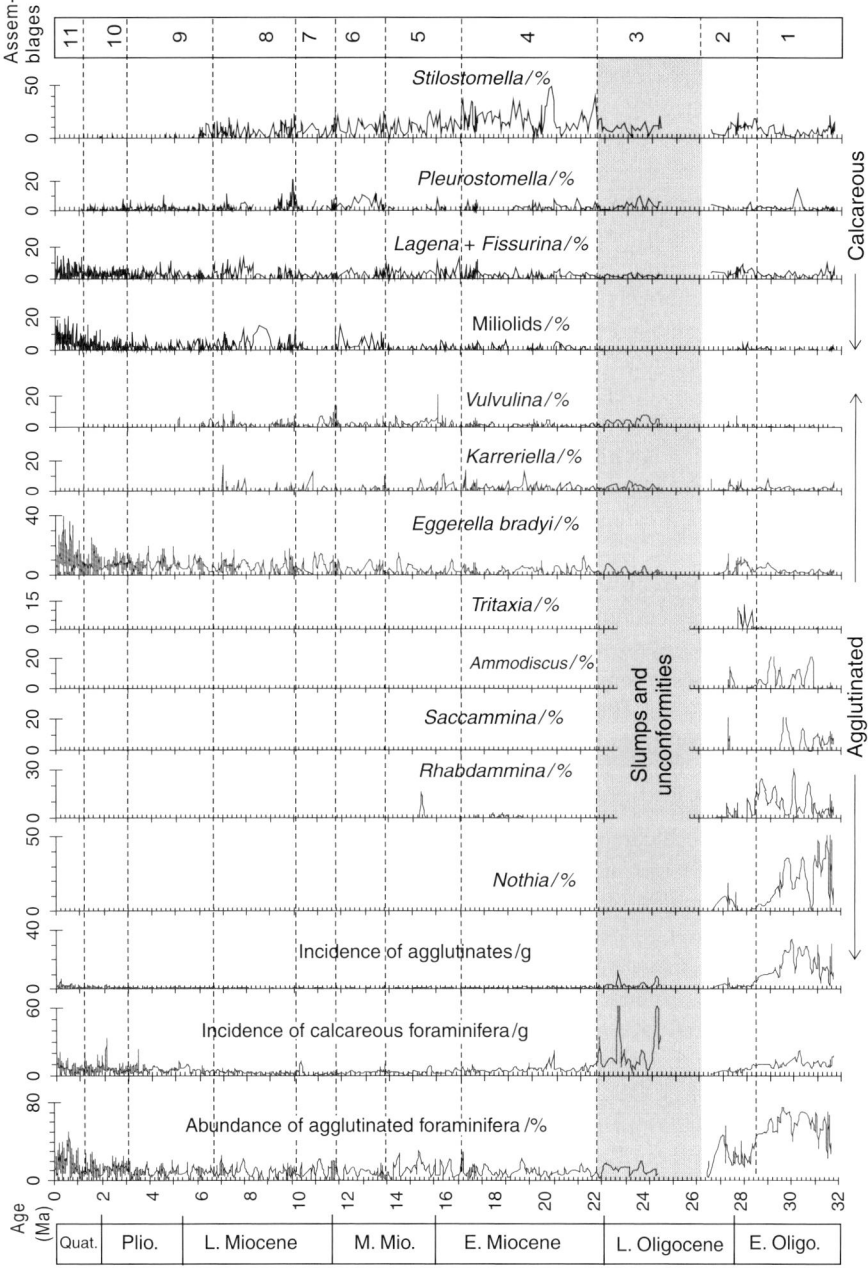

Fig. 6.6A Changes with time in relative abundance and incidence of agglutinated and calcareous benthic foraminiferal groups (individuals per gram of dry sediment) are shown together with percentage abundance of selected taxa (% in total fauna) from Site 1148. Also shown are 1 to 11 assemblages (see Table 6.6)

however, their overall abundance decreased and reached very low values during the middle and late Miocene. At least 4 periods of major turnovers are noted: (1) rapid increase in *C. wuellerstorfi* at 13.8 Ma, (2) sudden decrease in *C. wuellerstorfi* at 10 Ma, accompanied by significant increase in *Pullenia quinqueloba* and *Melonis*, (3) prominent increase in *N. umboniferus*, *F. favus* and *C. wuellerstorfi* but rapid reduction to almost zero in *Stilostominella* and *Buliminella* at ∼6 Ma, and (4) rapid increase in *Uvigerina* since 3 Ma. Among others, *Globobulimina* and *Hoeglundina elegans* became important only in the Quaternary (Fig. 6.6B).

Eleven benthic foraminiferal assemblages can be recognized from Site 1148 (Table 6.6): assemblages 1 to 3 for the Oligocene, assemblages 4 to 8 for the Miocene, and assemblage 9 to 11 for the latest Miocene to Quaternary. The agglutinated species-dominated early Oligocene assemblage 1 was probably associated with strong bottom currents in ∼2,000 m water depths (∼2,500 m given in Kuhnt et al. 2002). Increase in calcareous forms in the late Oligocene indicates a normal deep sea setting but high percentages in epifaunal *Cibicidoides* (18.3%) and *O. umbonatus* (13.4%) and infaunal *Stilostomella* (9.5%) may suggest strong mixing in this interval due to slumped and redeposited sediments (Wang P. et al. 2000; Li Q. et al. 2006).

The early Miocene is characterized by many calcareous forms with abundance alternating between infaunal *Stilostomella* and epifaunal *Cibicidoides, Oridorsalis* and *Gyroidinoides* in assemblage 4 (Table 6.6, Fig. 6.6). As the extinct *Stilostomella* and similar species often dominate assemblages from oxygen-poor, nutrient-rich environments (Kaiho 1991, 1994a; Gupta 1993; Kaiho and Hasegawa 1994; Kawagata et al. 2005), its common occurrence in assemblage 4 may reflect poorly ventilated bottom waters with low dissolved-oxygen levels during the early Miocene (Kuhnt et al. 2002).

Although most calcareous species continued to occur in middle and late Miocene assemblages (5–8), epifaunal species now became dominant. Newly occurring epifaunal species include *Osangularia culter* and *C. wuellerstorfi* (Fig. 6.6). These assemblages appear to denote stepwise increases in deep water ventilation. The rapid increase in *C. wuellerstorfi* in assemblage 6 to maximum values in assemblage 7 probably responded to further sea floor oxygenation after major expansion of the East Antarctic ice cap at ∼14 Ma. Similar assemblages have been reported from Site 289 on the Ontong-Java Plateau (Woodruff and Douglas 1981; Douglas and Woodruff 1981), ODP Sites 757 and 761 from the southeastern Indian Ocean (Singh and Gupta 2004; Holbourn et al. 2004), and the neighboring Site 1146 in the northern SCS (Holbourn et al. 2004). After ∼10 Ma, *C. wuellerstorfi* dramatically declined to ∼1% in assemblage 8 before becoming common again from ∼6 Ma. The decline of *C. wuellerstorfi* between 10 and 6 Ma coincided with a rebound in the abundance of *Gyroidinoides, O. umbonatus* and *G. subglobosa*, as well as a substantial increase in infaunal *Pullenia quinqueloba, Melonis* and *Buliminella*. Judging from ecologic data given in Table 6.3, this association appears to reflect bottom conditions with elevated organic matter flux and low dissolved-oxygen levels.

Following the decline of *Stilostomella, Buliminella* and *Pullenia* at ∼6 Ma, *O. umbonatus C. wuellerstorfi, G. subglobosa* and *Cibicidoides* became common again

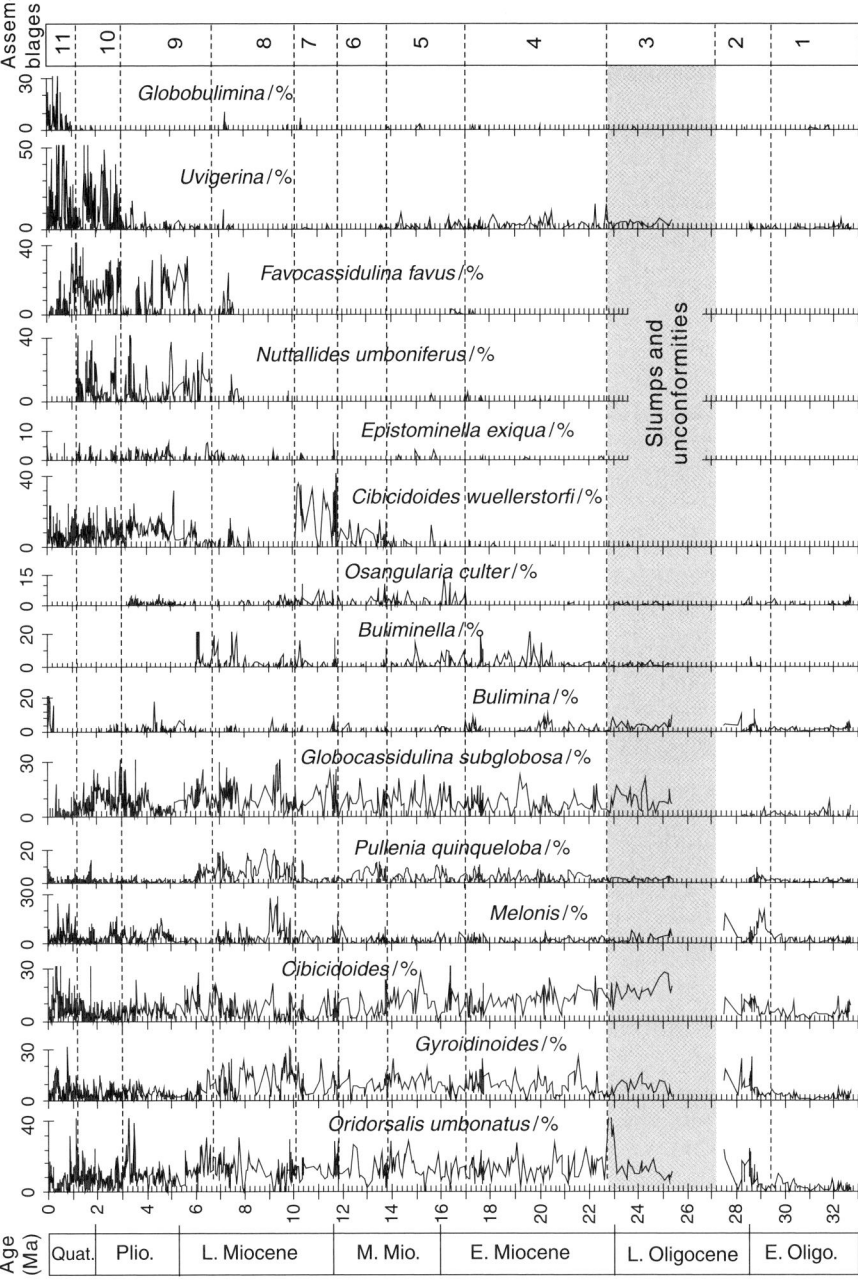

Fig. 6.6B Changes with time in relative abundance of selected calcareous benthic foraminifera from Site 1148 are useful for defining different assemblages (see Table 6.6)

Table 6.6 Benthic foraminiferal assemblages are identified based on the relative abundance of dominant taxa at Site 1148

Assem[1]	indiv./g	Abundance (%)[2]		Dominant taxa (mean relative abundance)	Age (Ma)
		Aggl.	Calc.		
11	8.1	16.7	83.3	*Uvigerina* (12%), *E. bradyi* (10.3%), *Cibicidoides* (9.0%), *C. wuellerstorfi* (7.2%), *F. favus* (7.0%), *O. umbonatus* (6.9%)	1.2–0
10	6.5	11.0	89.0	*F. favus* (11.9%), *Uvigerina* (11.5%), *G. subglobsa* (8.7%), *N. umboniferus* (8.2%), *C. wuellerstorfi* (7.9%), *E. bradyi* (7.3%),	3.0–1.2
9	5.8	8.4	91.6	*O. umbonatus* (10.7%), *C. wuellerstorfi* (8.6%), *G. subglobosa* (8.1%), *Cibicidoides* (7.8%), *N. umboniferus* (7%), *F. favus* (6.3%)	6.5–3.0
8	3.3	9.0	91.0	*O. umbonatus* (12.4%), *Gyroidinoides* (11.6%), *G. subglobosa* (10.9%), *P. quinque- loba* (7.3%), *Stilostomella* (6.7%)	10–65
7	3.1	8.8	91.2	*C. wuellerstorfi* (19.7%), *O. umbonatus* (13%), *G. subglobosa* (9.2%), *Gyroidinoides* (9%), *Stilostomella* (8.4%)	11.8–10
6	3.5	8.1	91.9	*O. umbonatus* (11.9%), *Stilostomella* (10.3%), *Gyroidinoides* (9.7%), *Cibicidoides* (9.1%), *G. subglobosa* (8.4%), *C. wuellerstorfi* (4.7%)	14–11.8
5	3.6	10.1	89.9	*O. umbonatus* (13.2%), *Cibicidoides* (12.8%), *Stilostomella* (10.9%), *Gyroidinoides* (10.0%), *G. subglobosa* (9.1%)	17–14
4	5.9	9.2	90.8	*Stilostomella* (17.5%), *O. umbonatus* (13.3%), *Cibicidoides* (11.8%), *Gyroidinoides* (9.3%), *G. subglobosa* (6.6%)	22.5–17
3	21.9	10.0	90.0	*Cibicidoides* (18.3%), *O. umbonatus* (13.4%), *Stilosomella* (9.5%), *G. subglobosa* (9.1%), *Gyroidinoides* (8.1%)	27–22.5
2	6.4	26.4	73.6	*Stilostomella* (9.6%), *Gyroidinoides* (9.3%), *O. umbonatus* (9.0%), *Cibicidoides* (8.8%), *Melonis* (6.6%)	29.4–27
1	26.9	55.3	44.7	*Nothia* (23.2%), *Rhabdammina* (7.0%), *Stilostomella* (6.6%), *Saccammina* (5.3%), *Cibicidoides* (5.2%)	32.5–29.4

[1] Assemblages 7+6+5 are equivalent to biofacies association 1 of Kuhnt et al. (2002), assemblage 4 to association 2, and assemblages 3 to 1 to associations 3 to 5, respectively.
[2] Aggl=agglutinated group; Calc=calcareous group.

in assemblage 9 (Table 6.6, Fig. 6.6). The most noticeable feature, however, is the consistent abundance in newly evolving species *Nuttallides umboniferus* and *Favocassidulina favus*, indicating a significant change relating to the influx of the Pacific Bottom Water (PBW).

The late Pliocene-early Quaternary assemblage 10 (3.0–1.25 Ma) is characterized by abundant *Uvigerina* (mainly *U. peregrina*), *F. favus*, *N. umboniferus*, *G. subglobosa* and *C. wuellerstorfi*. The *Uvigerina*-dominated assemblage is often associated

Table 6.7 Site 1148 ostracod assemblages are identified based on relative abundance of dominant and common taxa

Assem[1]	Valves/ sample	Diversity (spp/sample)	Dominant and common taxa (average %)	Age (Ma)
7	2.0	1.6	*Krithe* (28.8), *Legitimocythere* (13.5), *Abyssocythereis* (7.4), *Bradleya* (6.5), *Abyssocythere* (6.0), *Henryhowella* (5.2), *Cytheropteron* (4.8), *Rugocythereis* (4.4)	<3.0
6	low	low	Very rare, mainly *Krithe*. *Abyssocythere, Abyssocythereis, Bradleya, Rugocythere*	5.4–3.0
5	2.2	1.5	*Krithe* (45.1), *Abyssocythereis sulcatoperforata* (17.0), *Legitimocythere* (8.6), *Abyssocythere* (8.5)	13.2–5.4
4	5.2	3.8	*Krithe* (46.3), *Pelecocythere* (7.3), *Bradleya* (7.3), *Abyssocythereis sulcatoperforata* (6.3), *Rugocythereis* (5.5), *Henryhowella* (5.3) and *Abyssocythere* (5.3)	23.0–13.2
3	21.6	7.2	*Krithe* (42.6), *Pelecocythere* (7.5), *Rugocythereis* (7.5), *Henryhowella* (7.3) and *Bradleya* (6.6), *Argilloecia* (4.9)	25.5–23.0
2	low	low	Very rare, with only few valves of *Krithe*. *Argilloecia, Aversovalva* and *Eucytherura*	28.5–27
1	13.5	3.8	*Argilloecia* (25.8), *Krithe* (23.2), *Parakrithe* (9.7), bairdiids (7.0), *Xestoleberis* (3.8), *Pelecocythere* (2.1)	32.5–28.5

[1]No ostracods from 13.0–12.6, 10.5–11.5, 9.7–10.1 and 6.3–6.7 Ma intervals.

with high surface productivity and enhanced organic matter fluxes. Its occurrence at Site 1148 is considered as a result of enhanced East Asian monsoon during the course of the northern hemisphere glaciation, as found also at the neighboring Site 1146 (Hess and Kuhnt 2005). Strong fluctuations in the abundance of *Uvigerina* with higher values in glacials and lower values in interglacials over the last 3 Ma further corroborate such a relationship (Fig. 6.6B), similar to the late Quaternary record discussed above (Fig. 6.4) (Jian et al. 1999b).

Apart from *Uvigerina*, *Globobulimina* is also common in the youngest benthic foraminiferal assemblage (11). The abundance of these infaunal taxa in glacials is often replaced in interglacials by epifaunal *Cibicidoides*, *C. wuellerstorfi*, *F. favus*, *O. umbonatus* and *Eggerella bradyi*. Several species are very rare or absent from this assemblage, including *O. culter*, *N. umboniferus*, *Buliminella* spp. and *Pleurostomella* spp. (Fig. 6.6). A significant reduction in the PDW marker *N. umboniferus* at ~1.2 Ma was followed by a period of very low to sporadic occurrence before its complete disappearance at ~0.21 Ma (Fig. 6.6B), likely relating to the periodic shut-down of the inflow of PDW caused by a recent uplift of the Bashi Strait sill depth.

Site 1148 Ostracods

The distribution of ostracods in 885 samples from Site 1148 was analyzed by Zhao (2005). A total of more than 100 species belonging to 54 genera have been identified. The results show that the Oligocene section as a whole contains the highest species number and individual valves per sample (or mean incidence). Among them, *Krithe* is the most abundant, followed by *Argilloeciea, Abyssocythereis sulcatoperforata, Bradleya, Pelecocythere, Abyssocythere, Henryhowella, Legitimocythere, Rugocythereis*, and *Parakrithe* (Table 6.7). Allochthonous shelfal ostracods likely from down-slope transport are common (up to 17%) in samples older than 32 Ma, including *Aurila, Bosasella, Bythoceratina, Cletocythereis, Cytherelloidea, Eucytherura, Hemicytherura, Loxoconcha, Mydionobairdia, Paracytheridea*, and *Uroleberis*. In contrast, most taxa occurring in the late Oligocene and younger intervals are typical psychrospheric deep-sea ostracods common in world oceans. Some selected ostracod taxa are illustrated in Fig. 6.7.

Faunal Indication of Deep-Water Mass Changes

Deep-Water Benthic Faunal Groupings

Although the validity of using benthic foraminifera as water mass indicators is in debate (Jorissen et al. 2007), the results of benthic foraminifera and ostracods from ODP Site 1148 appear to contain information on the development of regional deep water masses in various stages of sea basin evolution in addition to changes in regional productivity and oxygenation discussed above. We consider various deep waters existing in different periods of SCS evolution over the last 30 Ma, which resulted in variations of benthic faunal assemblages. As the linkage between water property, food supply and bathymetry in the SCS is not clear, we only attempt an overall evaluation of faunal inference of these probable factors.

Earlier studies in the Pacific and SCS (Burke 1981; Douglas and Woodruff 1981; Miao and Thunell 1993; Zhao and Zhou 1995; Zhao and Zheng 1996; Jian and Wang 1997) provide vital paleoecological data for many species. Therefore, three foraminiferal bathymetry groups and two ostracod bathymetry groups including most common taxa can be assembled, together with benthic foraminiferal species groups respectively indicating high productivity and high oxygen levels (Table 6.8). According to Burke (1981) and Douglas and Woodruff (1981), the bottom water group (BWG) indicates more than 3,500 m water depths, the deep water group (DWG) about 2,500–3,500 m, and the intermediate water group (IWG) generally from <2,500 m. As shown in Table 6.8 and Fig. 6.8, the distribution of these groups

Fig. 6.7 Typical deep-sea ostracods from Hole 1148A are shown with scale bars = 200 μm. 1 *Bairdoppilata* sp., 667.05 mcd; 2 *Cytherella* sp., 822.30 mcd; 3 *Argilloecia* sp., 367.08 mcd; 4 *Argilloecia* sp., 367.08 mcd; 5 *Krithe* sp., 338.4 mcd; 6 *Krithe* sp., 367.08 mcd; 7 *Krithe* sp., 318.97 mcd; 8 *Parakrithe* sp., 578.06 mcd; 9 *Cytheropteron paracarolinae* Zhao et al., 141.70 mcd; 10 *Cytheropteron syntomoalatum* Whatley and Coles, 24.74 mcd; 11 *Cytheropteron lobatulum* Ayress et al., 28.26 mcd; 12 *Cytheropteron* sp., 585.83 mcd; 13 *Pelecocythere foramen* Whatley and Coles, 400.38 mcd; 14 *Pelecocythere* sp., 156.2 mcd; 15 *Ambocythere* sp., 817.1 mcd; 16 *Ambocythere* sp., 128.16 mcd; 17 *Abyssocythereis sulcatoperforata* (Brady), 405.7 mcd; 18 *Abyssocythere trinidadensis* (van den Bold), 473.18 mcd; 19 *Abyssocythere regalis* Zhao, 270.7 mcd; 20 *Bradleya dictyon* (Brady), 149.18 mcd; 21 *Bradleya* sp., 320.23 mcd; 22 *Henryhowella asperima* (Reuss), 473.18 mcd; 23 *Henryhowella* sp., 212.0 mcd; 24 *Legitimocythere acanthoderma* (Brady), 367.08 mcd; 25 *Pennyella dorsoserrata* (Brady), 125.91 mcd; 26 *Pennyella fortedimorphica* Coles and Whatley, 468.58 mcd; 27 *Poseidonamicus major* Benson, 278.53 mcd; 28 *Poseidonamicus praenudus* Whatley et al., 367.08 mcd; 29, 30 *Dutoitella eocenica* (Benson), 453.18 mcd; 31 *Arcacythere enigmatica* (Whatley et al.), 473.18 mcd; 32 *Xestoleberis* sp., 831.2 mcd

indicates some general trends in paleodepth and productivity changes in various periods of the SCS.

In general, both foraminiferal and ostracod DWGs increased gradually since the early Oligocene and became dominant in the late Miocene-Pliocene with strong fluctuations (Fig. 6.8). This indicates a gradual increase in water depths at the Site 1148 locality from ∼2,000–2,500 m initially to about 3,500–4,000 m in the late Miocene. The later depth range is estimated from significant increases in the abundance of DWG species, especially *C. wuellerstorfi* and *Krithe* spp. Further increase in paleobathymetry to ∼4,500 m in the later late Miocene to early Pleistocene is implied by the occurrence of the BWG marker *N. umboniferus*, although it may signify the influence of Pacific Bottom Water (PBW) as a result of stronger AABW production more than an increase in the local water depth (Woodruff 1992).

Table 6.8 Bathymetry groups of benthic foraminifera and ostracods are listed with their representing species from Site 1148

Faunal group	Benthic foraminifera	Ostracods
Intermediate water group (IWG)	*Ehrenbergina, Globobulimina, H. elegans, Uvigerina*	*Ambocythere, Argilloecia, Cytherella, Cytheropteron, Parakrithe, Pelecocythere*
Deep water group (DWG)	*E. exigua, F. favus, Gyroidinoides, O. umbonatus, C. wuellerstorfi*	*Abyssocythere, Abyssocythereis, Krithe, Legitimocythere, Poseidonamicus*
Bottom water group (BWG)	*N. umboniferus*	
High productivity group (HPG)	*Bulimina, Buliminella, Globobulimina, Uvigerina*	
High oxygen group (HOG)	*E. exigua, F. favus,* milliolids, *Nothia, N. umborniferus, C. wuellerstorfi, Rhabdammina.*	

Evidence of the PBW influence, particularly during 5–2 Ma, is from up to 50–90% planktonic foraminiferal fragmentation, as discussed below. The foraminiferal IWG increased from ~3 Ma and remained high and fluctuating in the youngest two assemblages 10 and 11. Coupled with the sudden drop in the BWG (*N. umboniferus*) across the assemblage 10/11 boundary, the continued high percentages of the IWG since ~1.2 Ma likely reflect either an elevation of the bottom surface by about 500–1,000 m or further oxygenation of the bottom water. In responding to these water depth and water property changes, the average abundance of *Krithe* ostracods in the last 3 Ma also declined to <30% from a maximum of ~45% in the Miocene and early Pliocene (Fig. 6.9).

Food Supply and Oxygen Levels

Listed alongside the water mass groupings in Table 6.8 and Fig. 6.8 are benthic foraminiferal associations, high productivity group (HPG) and high oxygen group (HOG) which can be used to infer food supply and dissolved-oxygen levels induced by surface productivity. High organic carbon fluxes mean high food supply and high productivity, while a high dissolved-oxygen level will often result from low organic carbon fluxes (Gooday 1993, 1994; Rathburn and Corliss 1994; Thomas and Gooday 1996; Schmiedl and Mackensen 1997; Hayward et al. 2006; Kawagata et al. 2006; Smart et al. 2007). Among others, *G. subglobosa* can adapt to a wide range of trophic and dissolved-oxygen levels, and *O. umbonatus* is flourishing in the deep water of the Sulu Sea where both dissolved-oxygen and food supply are low (Miao and Thunell 1993; Rathburn and Corliss 1994).

Clearly, the HPG and HOG show an inverse correlation at Site 1148 (Fig. 6.8), indicating a series of changes in food supply and oxygen level since the early Oligocene. While the HPG curve shows a relative stable long-term food supply except lower values in assemblage 9 from the latest late Miocene and early Pliocene, the highly fluctuating HOG curve may reveal several major steps of bottom water ventilation.

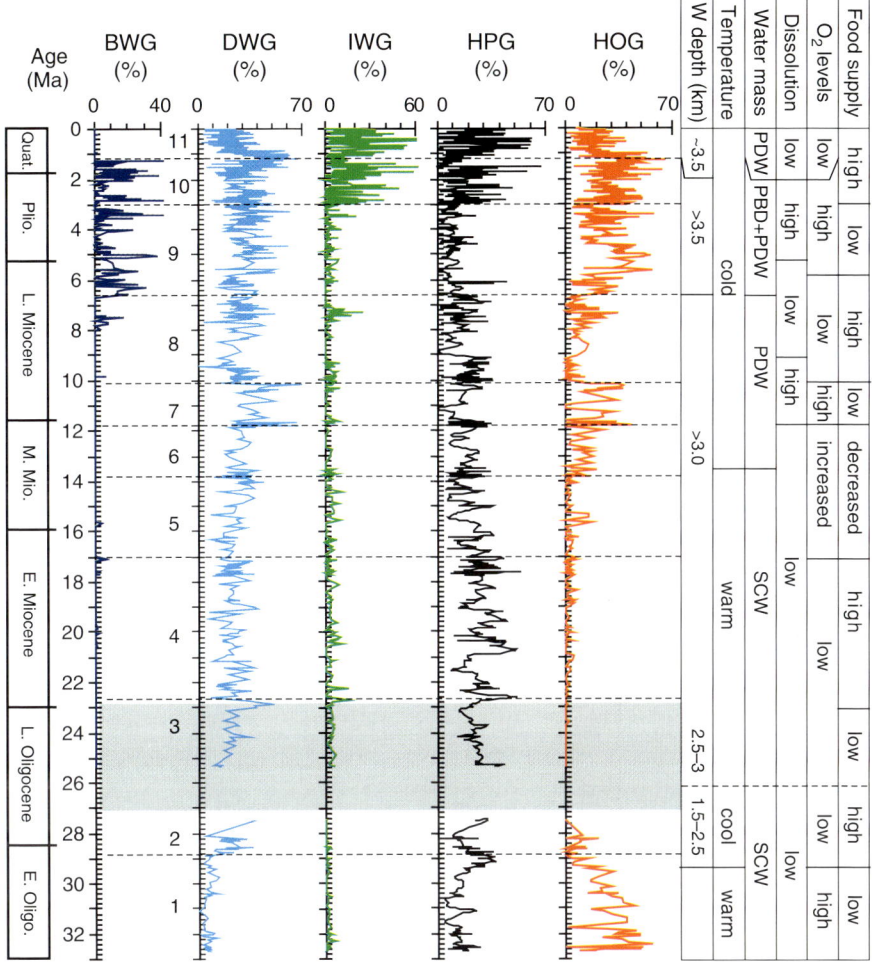

Fig. 6.8 Variations of different benthic foraminiferal groups from ODP Site 1148 (see Table 6.8) are used to interpret change in deep water masses and bathymetry. PDW=Pacific Deep Water; PBW=Pacific Bottom Water; SCW=Southern Component Water

The high HOG values from the early Oligocene probably signal a well oxygenated bottom condition in a relative shallow depth of ~2,000 m in the early stage of seafloor spreading and rapid basin fill. This was followed by a long period of low oxygen level from the late Oligocene to early middle Miocene, as seen also in the greenish lithology from this interval (Wang P. et al. 2000). The period between 14 and 10 Ma saw the first significant increase in HOG values when the site locality had subsided to ~3,500 m in the early middle Miocene. This significant deep water ventilation event was no doubt related to the formation of Pacific Deep Water due to global cooling and ice buildup on East Antarctica since ~14 Ma (see below). The lithology also becomes reddish brown, producing some peak values in the red

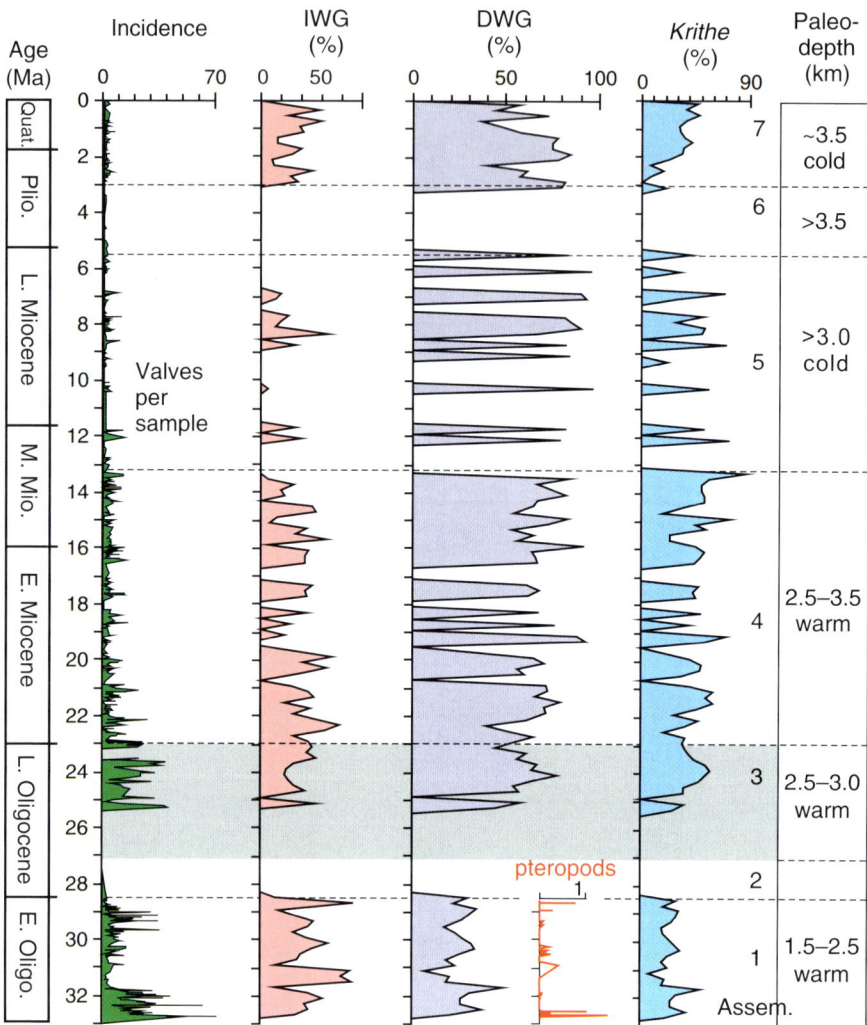

Fig. 6.9 Site 1148 ostracod assemblages are characterized by abundance variations in the incidence, the intermediate water group (IWG), deep water group (DWG) and *Krithe*%. The occurrence of pteropods (per gram of dry sediment) only in the early Oligocene indicates a warm and relative shallow bottom environment at the time

parameter (a*) (Wang P. et al. 2000). Further HOG increases since ~6 Ma were likely related to more oxidized bottom water when the collision along the eastern SCS margin intensified, leading to the closure of the sea basin and ultimately the uplift of Taiwan (Chapter 2). Probably, the influence of the PDW now became more constant with waters mainly from the North Pacific, causing severe dissolution in the deep sea SCS.

Step-wise increases in deep-water ventilation are also recorded at other localities from the northern and southern SCS (Figs. 6.10, 6.11, and 6.12) (Hess and Kuhnt

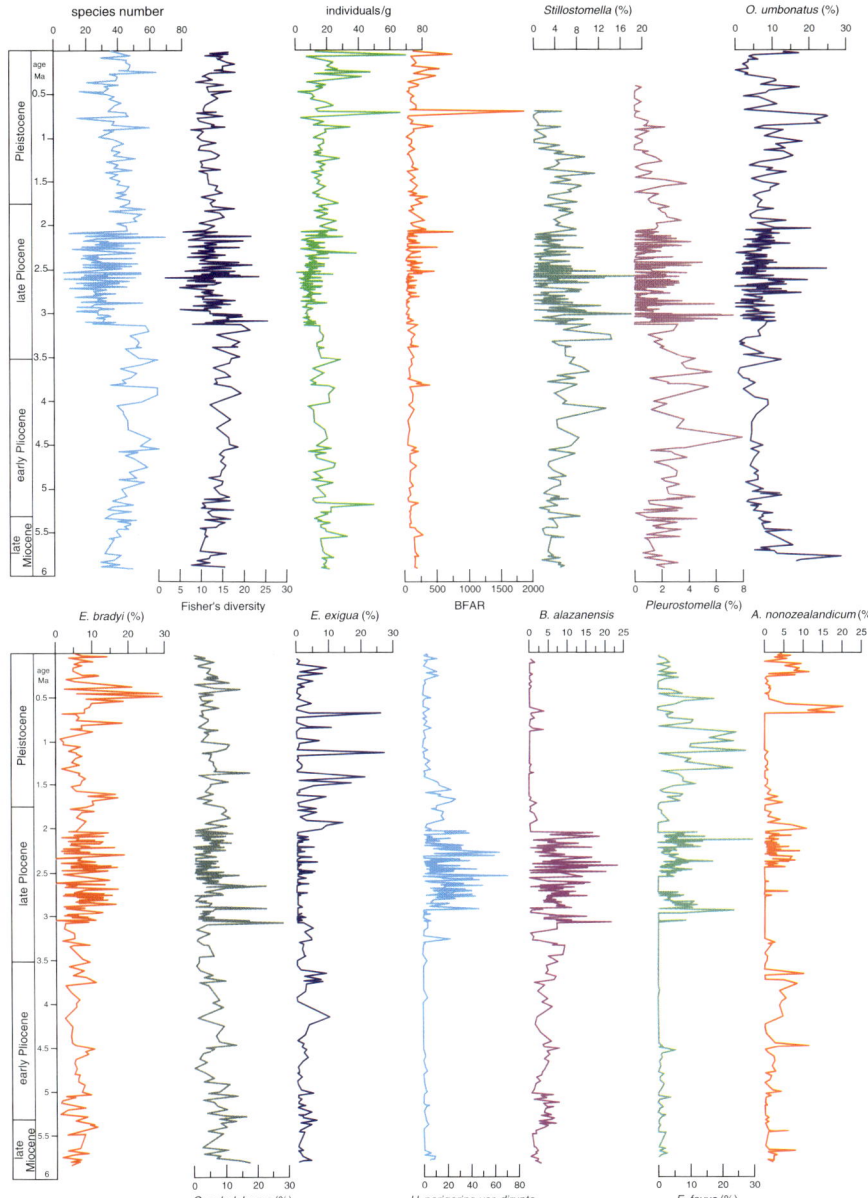

Fig. 6.10 Abundance variations of benthic foraminifera from Site 1143 (Hess and Kuhnt 2005) show important changes in the South China Sea deep water at ~3 Ma, 2.1 Ma and ~0.9 Ma, as recorded also at other site localities shown in Figs. 6.6, 6.11 and 6.12

2005; Huang et al. 2007; Kawagata et al. 2007). Increases in carbon flux indicator species at Site 1143 are coupled with an overall decrease in benthic foraminifer diversity around 3 Ma due to enhanced offshore upwelling and high productivity (Fig. 6.10) (Hess and Kuhnt 2005). At ~2.1 Ma, sudden decline in the abundance

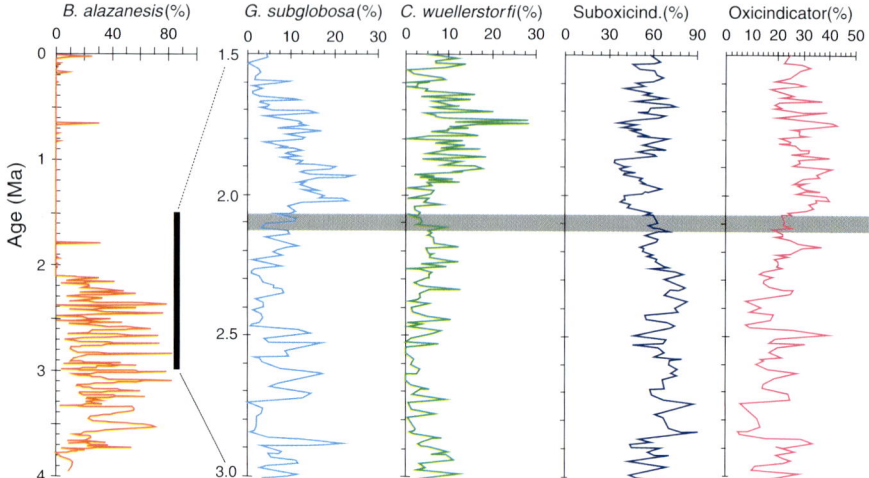

Fig. 6.11 Dramatic reduction in infaunal species was accompanied by increases in oxic and sub-oxic species at ~2.1 Ma from ODP Site 1146, likely responding to increased oxygenation of the deep water (Huang B. et al. 2007)

of *Bulimina alazanensis* and other infaunal species was recorded at both Site 1143 (Fig. 6.10) and Site 1146 (Fig. 6.11) (Huang B. et al. 2007). This 2.1 Ma infaunal decline event was followed by significant increase in the abundance of epifaunal species such as *C. wuellerstorfi* commonly living in oxic to suboxic bottom

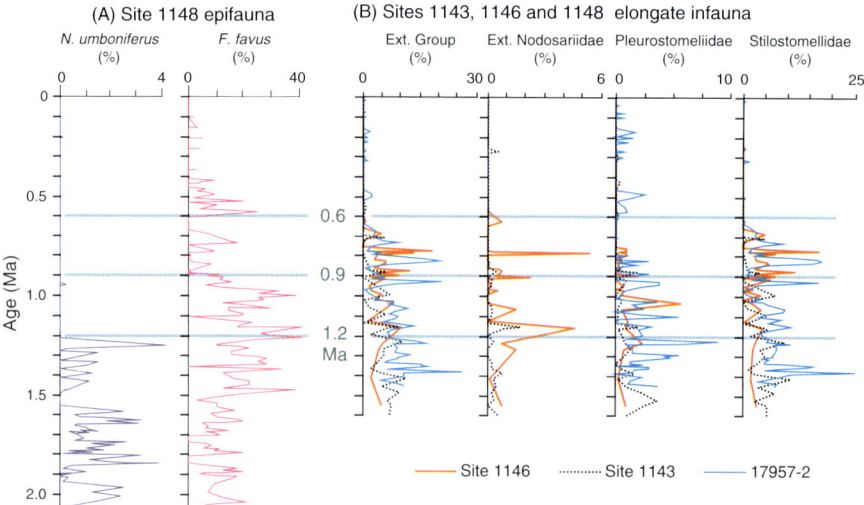

Fig. 6.12 Major changes in benthic foraminifera during the middle Pleistocene between 1.2 and 0.6 Ma are recorded in (**A**) epifaunal and (**B**) infaunal taxa. Site 1148 data are from Zhao et al. (2008 MS), and the elongate infaunal data from Kawagata et al. (2007)

conditions, suggesting increased oxygenation in the deep water (Huang B. et al. 2007). Similarly, during the mid- Pleistocene climate transition between 1.2 and 0.6 Ma, most species of the elongate, cylindrical benthic species (mainly nodosariids, pleurostomellids and stilostomellids) disappeared from Sites 1143, 1146 and 1148 (Fig. 6.12). The extinction of these infaunal species ~0.9 Ma ago was again responding to increased glacial cooling and consequent increased ventilation of the deep-sea water masses (Kawagata et al. 2007), probably also associating with stronger seasonal carbon flux after 0.9 Ma (Hess and Kuhnt 2005). These results may further suggest that North Pacific Deep Water, which many workers infer was formed in the northern Pacific during the last glacial, may have begun forming during MPT glacials (Kawagata et al. 2007).

6.4 Deep Water Evolution: Evidence from Carbonate Preservation and Isotopes

Carbonate Dissolution

Strong variations in the carbonate content and carbonate mass accumulation rate (MAR) have been observed in most samples collected from Site 1148 (Fig. 6.13). With moderate $CaCO_3\%$ and relative high MARs, the early Oligocene appeared to be the only period containing Pteropods, indicating no or very weak dissolution (Fig. 6.9). Dissolution started to affect sediment preservation in the middle Oligocene, accompanied by biogenic silica accumulation (Wang R. et al. 2001). In the Neogene, carbonate peaks (>40%) occurred in the early Miocene and several short interval of middle and late Miocene, while the intervals between 15–14 Ma and 3–0 Ma were characterized by low $CaCO_3$ of 10% or less.

The Site 1148 locality now lies between the modern lysocline (~3,000 m) and CCD (~3,500 m) (Wang P. et al. 2000). All calcareous components in the sediment recovered are affected at least partially by dissolution. The <150 μm residue contains more fragmented planktonic foraminiferal tests than in the coarser fraction. When few tests are left in the >150 μm residue, the entire finer fraction often contains 100% fine-grade shell fragments or even totally barren. The planktonic foraminiferal assemblage is dominated by solution-resistant taxa such as *Sphaeroidinellopsis, Globoquadrina* and *Paragloborotalia* (Li Q. et al. 2004). From the >150 μm residue, 10% fragmentation values of planktonic foraminifera first occur at ~21 Ma, and those with over 20% fragments at 16–15 Ma and 12–11 Ma. From 10 to ~8 Ma and from 5 to 3 Ma and younger are two intervals characterized by assemblages with over 50% fragments (Fig. 6.13). These high fragment values are accompanied by low $CaCO_3\%$ and MAR and high benthic foraminiferal abundance. Therefore, the $CaCO_3\%$, $CaCO_3$MAR, fragmentation and benthic foraminiferal abundance together convey a clear message that 5 periods of intensive dissolution had occurred in the Miocene-Pliocene South China Sea. Respectively, they occurred at ~21 Ma (D1), 16–15 Ma (D2), 13–11 Ma (D3), 10–9 Ma (D4), and 5–3 Ma (D5) (Fig. 6.13) (Li Q. et al. 2006).

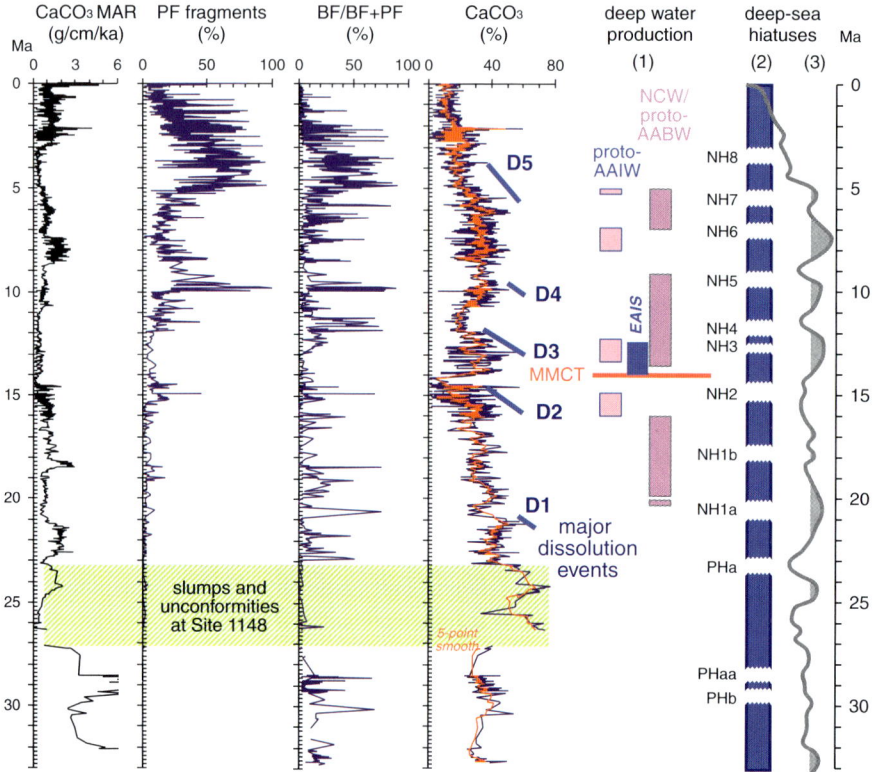

Fig. 6.13 Site 1148 dissolution events D1 to D5 are identified on the basis of CaCO₃%, CaCO₃, mass accumulation rate (MAR), planktonic foraminiferal (PF) fragments, and benthic foraminiferal (BF) abundance. Comparison between the local dissolution events and global deep water production (1, Ramsay et al. 1994), deep-sea hiatuses (2, Keller and Barron 1983, 1987) and continental margin hiatuses (3, Spencer-Cervato 1998) indicates global climate forcing (from Li Q. et al. 2006). MMCT=mid-Miocene climatic transition; EAIS=East Antarctica ice sheet development

These 5 major dissolution events can be correlated respectively to deep-sea hiatuses NH1a, NH2, NH3-4, NH5, and NH7–8 identified from other oceanic localities (e.g., Keller and Barron 1983, 1987; Ramsay et al. 1994) and continental margins (Spencer-Cervato 1998) (Fig. 6.13). The oldest dissolution event D1, for example, can be viewed as a local manifestation of global CaCO₃ reduction at 23–20 Ma (Lyle 2003). Periodic invasion of corrosive bottom waters after a shift of bottom water source from a Paratethys type in the early Oligocene to a Pacific type since the early Miocene most likely caused the dissolution events identified at Site 1148. In particular, post-Oligocene carbonate reduction and dissolution in the SCS would have been largely in responding to the production of the southern component water (SCW) and proto-Antarctic intermediate water (AIW), both influenced by the proto-Antarctic bottom water (AABW) (Ramsay et al. 1994).

Unlike most Site 1148 dissolution events that often fall in periods of increased production of the corrosive AABW, dissolution event D2 at 16–15 Ma appears to correspond mainly to the AAIW or SCW (Fig. 6.13). Since 16–15 Ma, the overall low CaCO$_3$% and more frequent severe dissolution at ODP Site 1148 can be interpreted as a result of local deepwater circulation change or even the initial development of a local bottom water mass after the cessation of seafloor spreading.

Dissolution event D3 from ∼13 Ma to 11 Ma matches well with a temporary carbonate reduction on the Ontong Java Plateau (Berger et al. 1993). This dissolution event D3 was probably affected at least partly by the SCW about 1 Ma after the initial establishing of the East Antarctic ice sheet at about 14 Ma (Flower and Kennett 1994; Zachos et al. 2001a), when global sea level dropped significantly at the middle/late Miocene boundary. From the neighboring Site 1146, Holbourn et al. (2005) identified the importance of orbital forcing and declining atmospheric carbon dioxide levels both acting on the thermal isolated Antarctica as possible driving mechanism for this 14 Ma global event. Noteworthy is that better CaCO$_3$ preservation after 14 Ma has been observed in the eastern subtropical Pacific (Lyle 2003; Holbourn et al. 2005) but not in the central or western Pacific (Lyle 2003), suggesting a more complex cause of deep sea dissolution than a simple factor of deep water production driven by ice sheet growth.

Dissolution event D4 at 10–9 Ma corresponds to the "carbonate crash", a worldwide dissolution event at ∼11 − 10 Ma caused by an intensified NCW production (Lyle 2003). Other consequences include a reorganization of global deep water circulation and the decoupling of Atlantic from Pacific deep water systems (Roth et al. 2000). It was also marked by a significant change in benthic foraminifera featuring a sudden decline in *C. wuellerstorfi* at the assemblages 7/8 boundary (Fig. 6.6). Although the relationship between dissolution events D3 and D4 is not clear, D3 as a precursor to D4 in respectively marking the initial and first maximum deepwater cooling in the post-spreading South China Sea was not impossible (Li Q. et al. 2006).

Since ∼5 Ma, CaCO$_3$ reduction continued in the SCS, and a long period of intensive dissolution is recorded also in the planktonic foraminiferal fragments (Fig. 6.13). Dissolution event D5 at 5–3 Ma was associated with a contemporary CaCO$_3$ decline at other central and western equatorial Pacific localities (Berger et al. 1993) but a more significant increase of CaCO$_3$ accumulation in the eastern Pacific (Lyle 2003), implying another major change in water masses ocean-wide. Probably, it signifies the first formation of modern lysocline (∼3,000 m) and CCD (∼3,500 m) in the Pliocene SCS. Since then, the region has been bathed in corrosive bottom waters with depressed CaCO$_3$%, punctuated also by a sudden decline of the BWG in the mid-Pleistocene (Fig. 6.9) and high swings in the differences between planktonic and benthic δ^{13}C as discussed below.

Comparison between dissolution proxies from Sites 1143, 1146 and 1148 (Fig. 6.14) indicates some general trends in carbonate accumulation and dissolution at different depth localities between the southern and northern SCS. Over the last 5 Ma, Site 1146 has been a site having more carbonate accumulation and lesser dissolution, although a gradual decrease in carbonate content is observed at all

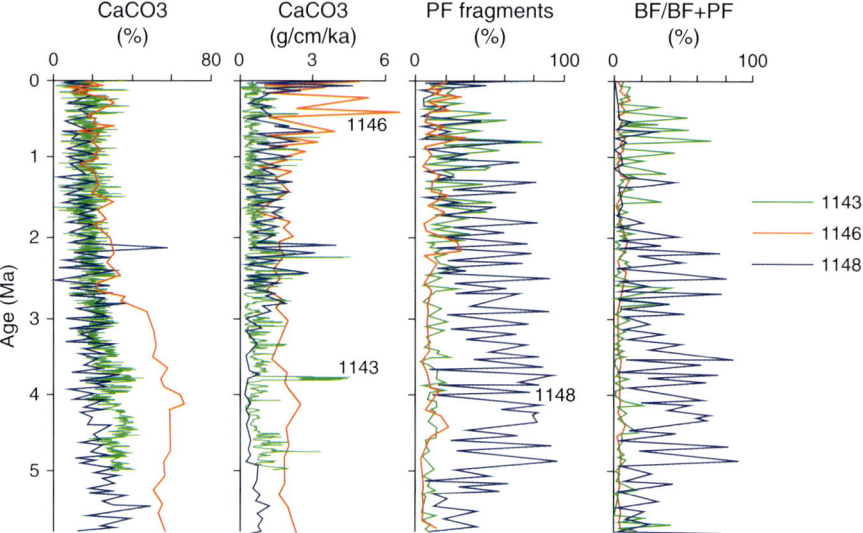

Fig. 6.14 Dissolution proxies are compared to show stronger dissolution at the deeper Site 1148, higher CaCO₃ accumulation before ~3 Ma and after 0.9 Ma at site 1146, and higher benthic foraminifer abundance between 1 and 0.5 Ma at Site 1143. Site 1143 data are from Li B. et al. (2005), Site 1146 data from Wang J. (2001), and Site 1148 data from Chen X et al. (2002)

the 3 sites. Significant variations between these sites occurred during the period of ~1 − 0.5 Ma when productivity as indicated by benthic foraminifera increased at Site 1143 and faster CaCO₃ accumulation occurred at Site 1146, probably linking to the most recent development and differentiation of the upper deep water (Sites 1143 and 1146) from the lower deep water (Site 1148) in the SCS.

Isotopic Records

Benthic foraminifera in 1621 samples and planktonic foraminifera in 978 samples from Site 1148 were measured for stable isotopes (Zhao et al. 2001a,b, this study). Mainly the benthic species *C. wuellerstorfi* and planktonic species *G. ruber* or allied forms were used, and they cover sediment sections of the last 25 Ma and ~20 Ma respectively. However, benthic data from 25 to 23 Ma cannot be used because of the slumped sediment that characterizes the late Oligocene from the site (Li Q. et al. 2005), and planktonic samples are lacking from several intervals (particularly ~11 Ma and ~6 Ma) due to dissolution and/or inadequate specimens (Fig. 6.15). These data are supplemented by isotopic records of Site 1146 (2,092 m) for benthic results from 17 to 11 Ma (Holbourn et al. 2004) and of Site 1143 (2,772 m) for benthic and planktonic results from 5 to 0 Ma (Tian et al. 2006).

   Benthic $\delta^{18}O$ (and $\delta^{13}C$) from Site 1148 show a general increase (decrease) trend since the early Miocene, a trend roughly followed by planktonic $\delta^{13}C$ but not by its

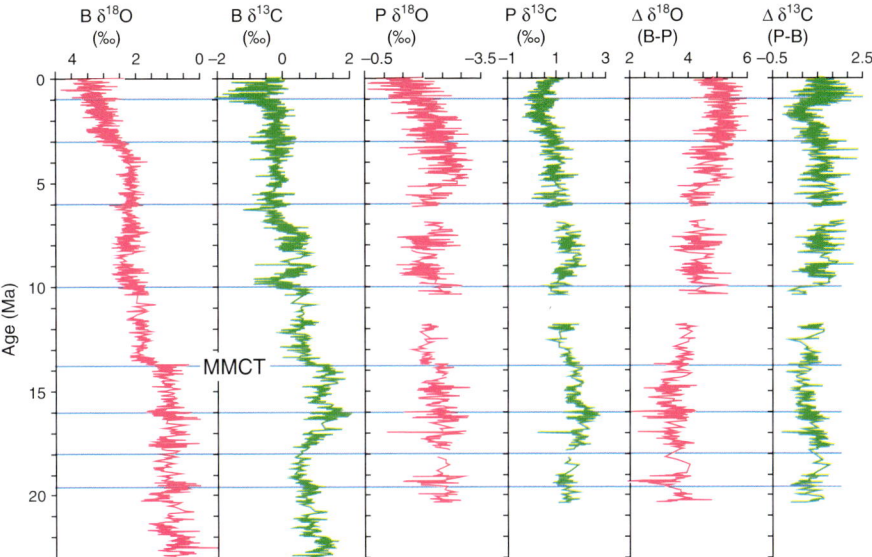

Fig. 6.15 Variations in Site 1148 benthic (B) and planktonic (P) isotopic results indicate major paleoceanographic changes various intervals marked by horizontal lines. MMCT is referred to mid-Miocene climatic transition at ~13.9 Ma

$\delta^{18}O$ which, instead, remained relatively stable from ~20 Ma to ~3 Ma (Fig. 6.14). The implications of this trend may include global cooling and deep water cooling (as indicated by increased benthic $\delta^{18}O$) and reducing deep water ventilation and increasing nutrient flux by intensified monsoons (as indicated by decreased planktonic and benthic $\delta^{13}C$). More likely, the $\delta^{13}C$ changes manifest carbon reservoir variations driven by climatic forcing (Wang P. et al. 2003a) (Chapter 7).

Interrupting this general trend are several major positive or negative shifts in the $\delta^{18}O$ and $\delta^{13}C$ records (Fig. 6.15), likely relating to the development of a regional deep water mass. At ~18 Ma, the near parallel swings in the early Miocene benthic isotopes were followed by a gradual positive excursion in planktonic and benthic $\delta^{13}C$ to maximum values at ~16 Ma, apparently in responding to increased carbon burial during the mid-Miocene climate optimum (Zachos et al. 2001a). However, how much these isotopic signals were related to the end of seafloor spreading at ~16 Ma in the SCS is not clear.

According to Sarnthein and Winn (1990) and Wang L. (1992), lower values in the $\delta^{13}C$ difference between planktonic and benthic foraminifera, or $\Delta\delta^{13}C$ (P-B), can be served as a proxy of low surface productivity and low exported productivity to the sea floor. The relative low $\Delta\delta^{13}C$ (P-B) from 17 to 10 Ma (Fig. 6.15) appears to support the decreasing HPG abundance (Fig. 6.8) in indicating a period with low organic carbon.

A major change in all isotopic values at ~14 Ma was obviously related to global cooling and permanent ice cap buildup on East Antarctica, which resulted in stronger cold water production and increased bottom water ventilation in world oceans (Woodruff and Douglas 1981; Woodruff and Savin 1989; Wright et al. 1992; Flower and Kennett 1994; Pearson and Palmer 2000; Mutti 2000). This mid-Miocene climatic transition, or MMCT, appears to be induced by eccentricity cycles, as revealed in many Pacific and Indian Ocean sites including Site 1146 (Holbourn et al. 2004, 2005). A more significant drop in the benthic $\delta^{13}C$ at $\sim$10 Ma was accompanied by gradual increase in planktonic $\delta^{13}C$, indicating their first diversion in the SCS records probably in relating to increased vertical contrast in the water column. Their contemporary decline at $\sim$6 Ma was followed by slowly increase only in planktonic $\delta^{18}O$. As described above, this 6 Ma event caused many infaunal benthic species to reduce almost completely (Fig. 6.6) probably due to Pacific Bottom Water influx. At $\sim$3 Ma, further declines in benthic $\delta^{18}O$ and planktonic $\delta^{18}O$ and $\delta^{13}C$ were accompanied by large swings in the benthic $\delta^{13}C$ (Fig. 6.15).

Comparison between isotopic records from Sites 1143 and 1148 (Fig. 6.16) reveals sequential changes in glacial cooling, productivity and other regional paleoceanographic events over the last 5 Ma. The positive benthic $\delta^{18}O$ excursions between $\sim$3.5 and 2.5 Ma record the global deep water cooling and ice buildup on high latitude northern hemisphere, while the relative stable to slightly positive excursions in planktonic $\delta^{18}O$ over the same period may have been influenced by intensified East Asian monsoons (Fig. 6.16) (Tian et al. 2004). Rapid increase in benthic foraminifer *Uvigerina* (Fig. 6.6) and deep water ostracod *Krithe* (Fig. 6.9) may also

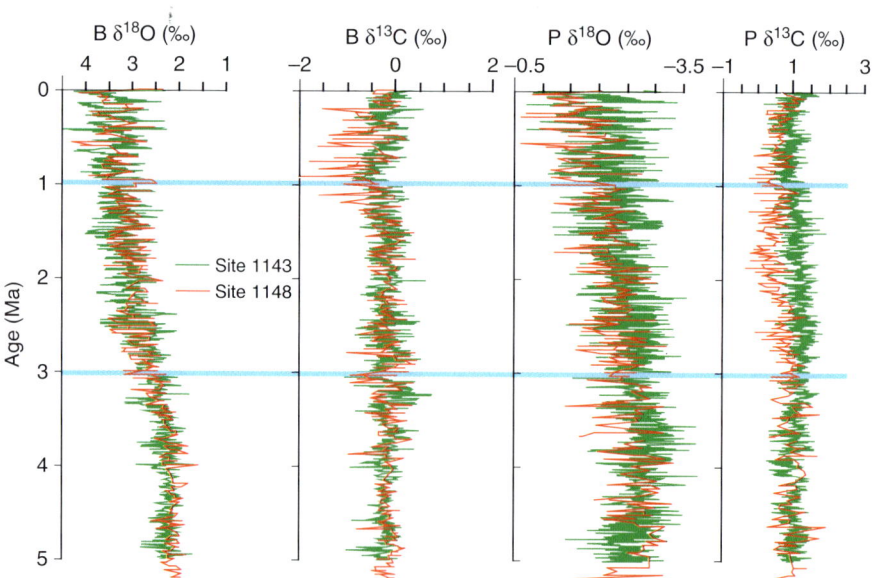

Fig. 6.16 Benthic and planktonic $\delta^{18}O$ and $\delta^{13}C$ records from Sites 1143 and 1148 show parallel variation trends over the last 5 Ma. However, more negative planktonic $\delta^{13}C$ excursions since $\sim$3 and more negative benthic $\delta^{13}C$ excursions since $\sim$1.2 Ma at Site 1148 may imply sequential changes in surface and bottom water property at this lower slope site

convey a similar message. The more negative $\delta^{13}C$ excursions first occurred in the planktonic records from ~3 to ~1 Ma and then in the benthic records since ~1.2 Ma all at the deeper water Site 1148 (Fig. 6.16) reflect enhanced differences between the northern and southern SCS, as well as between upper and lower deep water masses in the region. For example, a period of very low productivity centered at ~1.8 Ma is indicated by a distinct $\Delta\delta^{18}C$ (P-B) trough in the Site 1148 record (Fig. 6.15), coinciding with stronger PBW influx marked by abundant *N. umboniferus* in assemblage 10 (Fig. 6.6). At ~1 Ma, another negative $\delta^{13}C$ excursion occurred on both the planktonic and benthic $\delta^{13}C$ records (Fig. 6.16), and its association with major decline in both *N. umboniferus* and *F. favus* (Fig. 6.12) can be attributed to a reduced PDW influence at the beginning of the mid-Pleistocene climatic transition, probably as a result of uplift of the Bashi Strait sill depth to near the present level (Li Q. et al. 2008).

6.5 Oceanic Connection

When compared to the composite record of Zachos et al. (2001a), the benthic $\delta^{18}O$ from Site 1148 is lighter with stronger fluctuations before ~15 Ma, and the benthic $\delta^{13}C$ is lighter with stronger fluctuations after 15 Ma (Chapter 3). These differences distinguish the first order SCS Neogene isotopic characteristics from the world ocean, implying the influence of regional dynamics especially seafloor spreading which ended at 16–15 Ma and intensification of monsoons.

In Figs. 6.17 and 6.18, the Site 1148 benthic $\delta^{18}O$ and $\delta^{13}C$ curves are further compared with records from North Atlantic, Southern Ocean, tropical Pacific and Indian Ocean. It is clear that the lighter $\delta^{18}O$ before 15 Ma and lighter $\delta^{13}C$ after 15 Ma largely remain for the SCS, but its $\delta^{13}C$ curve matches better with records from the western Pacific Site 289 and northern Indian Ocean Site 709 (Fig. 6.18). The similar $\delta^{13}C$ variations from these three Sites 1148, 289 and 709 may imply a closer connection between their deep waters since at least the early Miocene.

Between 17 and 12 Ma, benthic $\delta^{13}C$ at Site 1146 is lighter than Site 588 (1,533 m) from western Pacific and Site 761 (2,179 m) from eastern Indian Ocean (Fig. 6.19) (Holbourn et al. 2004). The heaviest $\delta^{13}C$ record with a wider range from Site 761 was interpreted as representing a relatively well ventilated deep sea with close proximity to a southern high latitude source. The close coherent $\delta^{13}C$ records between these three sites led Holbourn et al. (2004, 2005) to suggest that $\delta^{13}C$ changes had common global origin (carbon burial) and no open low latitude deep water connection is required to explain these synchronous changes. However, the impact of Southern Component Water (SCW) on these records is not yet fully understood.

As a typical association with the modern Pacific Deep Water (PDW), the *C. wuellerstorfi*-dominated benthic foraminiferal assemblages 6 and 7 from 14 to 10 Ma occurred not only at Site 1146 but also at Sites 757 and 761 (Singh and Gupta 2004; Holbourn et al. 2004). Therefore, the SCW as a precursor to the PDW

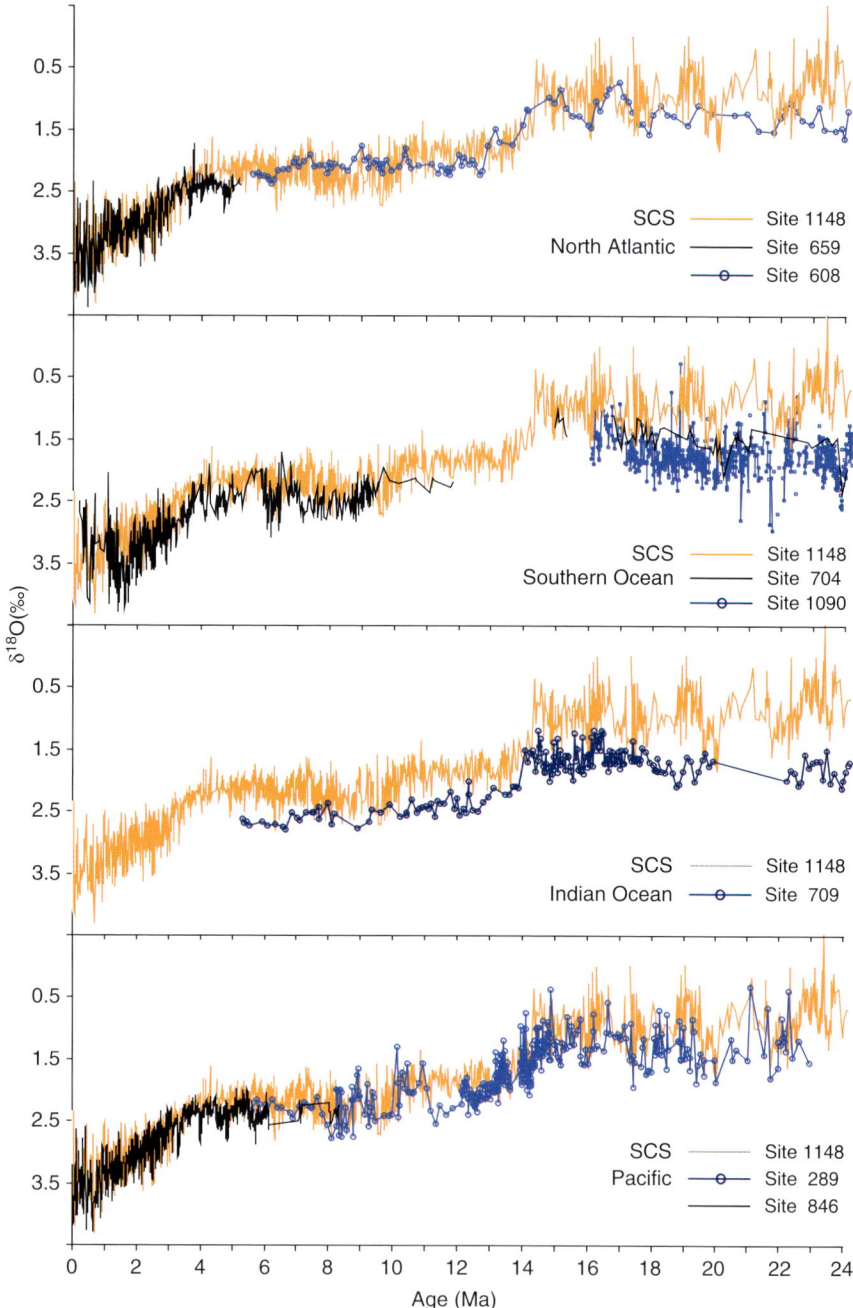

Fig. 6.17 (continued)

remained influential at these localities, and deep water connection between the then SCS and northern Indian Ocean probably still existed. However, this connection appeared to cease at ~10 Ma due to further collision along the eastern SCS margin and in the SE Asian region at large (Fig. 6.20), which led to significant closure of the Indonesian seaway (Chapter 2). The species *C. wuellerstorfi* dramatically decreased at ~10 Ma, and did not fully recover until ~6 Ma (Fig. 6.6).

The short-term decrease by ~0.8‰ in the Site 1148 $\delta^{13}C$ at ~10 Ma is unique (Fig. 6.18). Considering only a minor contemporary drop in other records such as Sites 289 and 709, this 10 Ma event probably marks a complete separation between Indian Ocean and SCS deep waters and relative weak connection between western Pacific and SCS deep waters (Fig. 6.18). Most likely, therefore, this 10 Ma event represents an early attempt in the development of a local deep water mass in the Miocene SCS.

Summary

1. The benthic foraminiferal, ostracod and isotopic records from ODP Sites 1143, 1146 and 1148 and other SCS localities provide important information on the evolution of deep waters in the marginal western Pacific. A general deep sea cooling trend since the Oligocene was punctuated by short-term events of strong carbonate dissolution and bottom water ventilation.
2. Abundant agglutinated benthic foraminifera found in the early Oligocene imply a relative warmer and better ventilated deep water environment due to rapid deposition at ~2,000–2,500 m water depths. Their similarity with Paratethys assemblages suggests a common influence by the Southern Component Water in similarly restricted deep gulf environments.
3. Increase in carbonate dissolution since the early Miocene was in pace with increase in paleobathymetry and stronger influence of the Pacific Deep Water and Pacific Bottom Water. Dissolution events D1 to D5 signify stepwise development of the local deep water mainly controlled by global climate and regional tectonics.
4. Deep sea connection between the SCS and the northern Indian Ocean appears to remain open until ~10 Ma. Sudden declines at ~10 Ma in both $\delta^{13}C$ and *C. wuellerstorfi* from Site 1148 may evince the complete detachment between the two sea basins.

←_____

Fig. 6.17 Site 1148 benthic $\delta^{18}O$ is compared with similar records from other DSDP/ODP sites. Data sources are: Site 659 (3,070 m, Savin et al. 1981; Woodruff and Savin 1991; Hodell and Vayavananda 1993), Site 608 (3,526 m, Wright et al. 1991, 1992), Site 704 (2,532 m, Hodell and Ciesielki 1991; Müller et al. 1991; Wright et al. 1992), Site 1090 (3,699 m, Billups et al. 2002), Site 709 (3,038 m, Woodruff et al. 1990; Shackleton and Hall 1990), Site 289 (2,206 m, Savin et al. 1981; Woodruff and Savin 1991; Hodell and Vayavananda 1993), Site 846 (3,730 m, Shackleton et al. 1995)

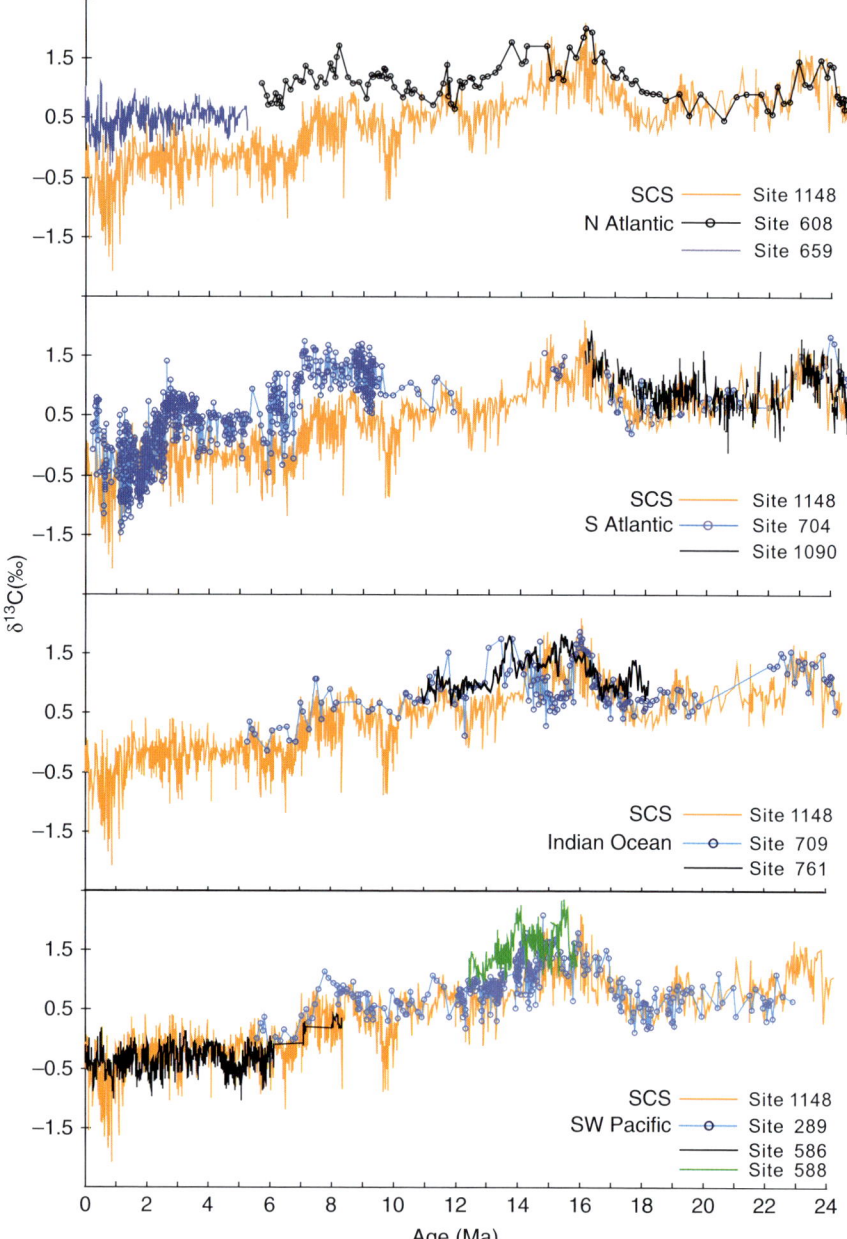

Fig. 6.18 Site 1148 benthic δ^{13}C is compared with contemporary records from the South Atlantic, Indian Ocean and SW Pacific. Data sources see Fig. 6.17 except Site 586 (2,218 m, Whitman and Berger 1993) and Site 588 (1,533 m, Flower and Kennett 1993)

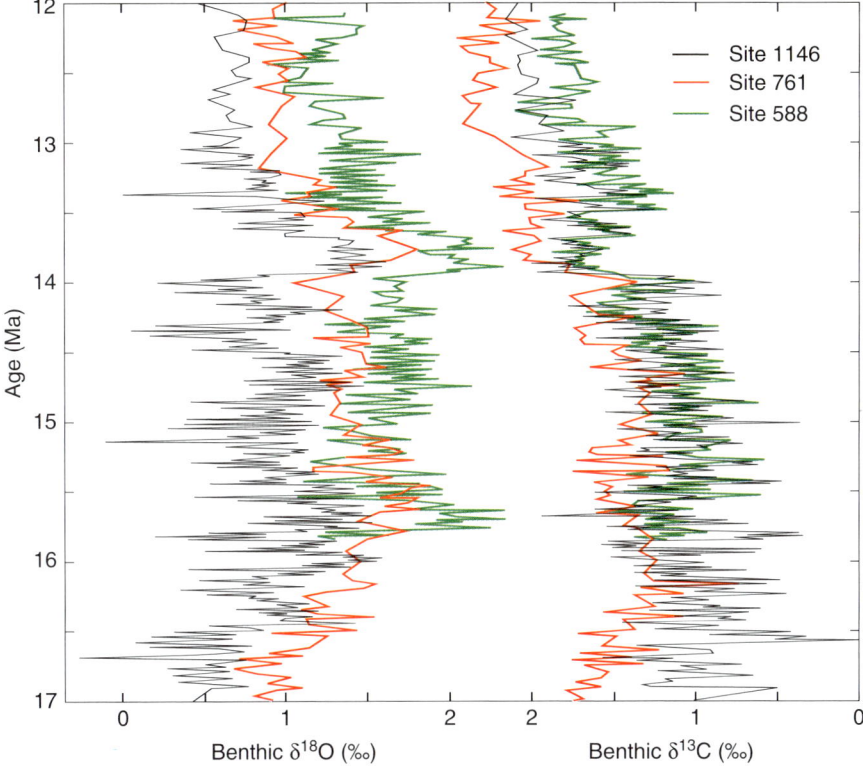

Fig. 6.19 Benthic δ^{13}C and δ^{18}O results from Site 1146 are compared with records from western Pacific Site 588 and eastern Indian Ocean Site 761 (modified from Holbourn et al. 2004)

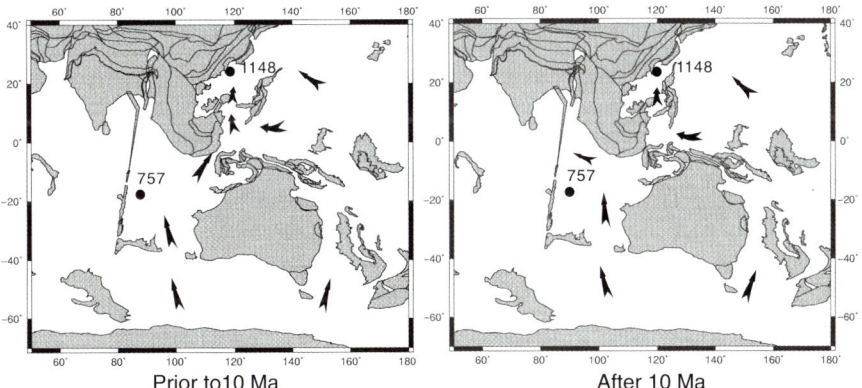

Fig. 6.20 Maps show possible changes in deep-water connection between the western Pacific and Indian Ocean before and after 10 Ma, as indicated by isotopic and microfossil data. Black arrows denote cold and well-oxygenated deep waters. The reconstruction maps were based on data from http://www.odsn.de/odsn/services/paleomap/paleomap.html

5. Deep water ventilation in the SCS was also in steps, mainly responding to global cold water production and local subsidence, as well as changes in its connection to the open ocean. In the late Neogene, significant ventilation events occurred at ∼3 Ma and ∼2.1 Ma, with the former also marked by increased surface productivity probably due to intensified monsoons in the region.

6. A completely isolated SCS deep basin began from ∼1.2 Ma, at the onset of the mid-Pleistocene climatic transition, as indicated by a major decline to almost zero in the abundance of *N. umboniferus*. Since ∼0.9 Ma, the deep sea SCS has been prevailed by the Pacific Intermediate Water.

Acknowledgments This work was supported by the Ministry of Science and Technology of China (NKBRSF Grant 2007CB815900) and the National Natural Science Foundation of China (Grants 40331002, 40576031, 40621063, 40631007 and 40776028).

References

Altenbach A.V. and Sarnthein M. 1989. Productivity record in benthic foraminifera. In: Berger W.H., Smetacek V.S. and Wefer G. (eds.), Productivity of the Ocean: Present and Past. John Wiley & Sons Limited., S. Bernhard, Dahlem Konferenzen, pp. 255–269.

Altenbach A.V., Lutze G.F., Schiebel R. and Schönfeld J. 2003. Impact of interrelated and interdependent ecological controls on benthic foraminifera: an example from the Gulf of Guinea. Palaeogeogr. Palaeoclimatol. Palaeoecol. 197: 213–238.

Berger W.H., Leckie R.M., Janecek T.R., Stax R. and Takayama T. 1993. Neogene carbonate sedimentation on Ontong Java Plateau: highlights and open questions. Proc. ODP Sci. Results 130: 711–744.

Billups K., Channell J. and Zachos J. 2002. Late Oligocene to early Miocene geochronology and paleoceanography from the subantarctic South Atlantic. Paleoceanography 17: 1, doi: 10.1029/2000PA000568.

Broecker W.S. 1992. The Glacial World according to Wally. Lamont-Doherty Geol. Observatory, Columbia Univ., 228pp, Appendices.

Burke S. 1981. Recent benthic foraminifera of the Ontong Java Plateau. J. Foraminiferal Res. 11: 1–19.

Burke S.K., Berger W.H., Coulbourrn W.T. and Vincent E. 1993. Benthic foraminifera in box core ERDC 112, Ontong Java Plateau. J. Foraminiferal Res. 23: 19–39.

Cai H. 1982. Distribution of Ostracoda in the northwester water of the South China Sea. Tropical Oceanol. 8: 19–27 (in Chinese).

Cai H. 1991. Ostracoda from the Nansha Islands and the adjacent sea areas. In: The Multidisciplinary Oceanographic Expedition Team of Academia Sinica to the Nansha Islands (ed.), Quaternary Biological Groups of the Nansha Islands and the Neighbouring Waters. Zhongshan Univ. Press, Guangzhou, pp. 82–128 (in Chinese).

Cai H. and Tu X. 1983. Distribution of Foraminifera and Ostracoda from the surface of the bottom sediments off the Xisha-Zhongsha Islands, South China Sea. Nanhai Studia Marine Sinica 4: 25–64 (in Chinese).

Calvert S.E., Pedersen T.F. and Thunell R.C. 1993. Geochemistry of the surface sediments of the Sulu and South China Seas. Mar. Geol. 114: 207–231.

Chao S.V., Shaw P.T. and Wu S.Y. 1996. Deep water ventilation in the South China Sea. Deep-Sea Res. II 43: 445–466.

Chen M. and Chen S. 1989. On carbonate dissolution and the distribution model of deep sea sediment types in South China Sea. Tropical Oceanol. 15: 20–26 (in Chinese).

Chen C.T.A., Wang S.L. and Bychkov A.S. 1995. Carbonate chemistry of the Sea of Japan, J. Geophys. Res. 100(C7): 13737–13746.

Chen W.S., Ridgway K.D., Hong C.S., Chen Y.K., Shea K.S. and Yeh M.G. 2001. Stratigraphic architecture, magnetostratigraphy, and incised-valley systems of the Pliocene-Pleistocene collisional marine foreland basin of Taiwan. GSA Bull. 113: 1249–1271.

Chen C.T.A., Hou W.P., Gamo T. and Wang S.L. 2006. Carbonate-related parameters of subsurface waters in the West Philippine, South China and Sulu Seas. Mar. Chem. 99: 151–161.

Chen X., Zhao Q. and Jian Z. 2002. Carbonate content changes since the Miocene and paleoenvironmental implications, ODP Site 1148, northern South China Sea. Mar. Geol. Quat. Geol. 22: 69–74 (in Chinese).

Cheng X., Huang B., Jiang Z., Zhao Q., Tian J. and Li J. 2005. Foraminiferal isotopic evidence for monsoonal activity in the South China Sea: a present-LGM comparison. Mar. Micropaleontol. 54: 125–139.

Clark F.E., Oattersom R.T. and Fishbein E. 1994. Distribution of Holocene benthic foraminifera from the tropical southwest Pacific. J. Foraminiferal Res. 24: 241–257.

Corliss B.H. 1979. Recent deep-sea benthic foraminiferal distributions in the southeastern Indian Ocean: Inferred bottom-water routes and ecological implicationos. Mar. Geol. 31: 115–138.

Corliss B.H. 1983. Distribution of Holocene benthic foraminifera in the southwest Indian Ocean. Deep-Sea Res. I 3(2A): 95–117.

de Mello e Sousa S.H., Passos R.F., Fukumoto M., Almeida da Silveira I.C., Figueira R.C.L., Koutsoukos E.A.M., de Mahques M.M. and Rezende C.E. 2006. Mid-low bathyal benthic foraminifera of the Camos basin, Southeast Brazilian margin: Biotopes and controlling ecological factors. Mar. Micropaleontol. 61: 40–57.

den Dulk M., Reichart G.T., van Heyst S., Zachariasse W.J. and Van der Zwaan G.J. 2000. Benthic foraminifera as proxies of organic matter flux and bottom water oxygenation? A case history from the northern Arabian Sea. Palaeogeogr. Palaeoclimatol. Palaeoecol. 161: 337–359.

Douglas R. and Woodruff F. 1981. Deep sea benthic foraminifera. In: Emiliani (ed.), The Ocean Lithosphere. The Sea, vol. 7, J. Wiley & Sons, New York, pp. 1233–1327.

Fariduddin M. and Loubere P. 1997. The surface ocean productivity response of deeper water benthic foraminifera in the Atlantic Ocean. Mar. Micropaleontol. 32: 289–310.

Flower B.P. and Kennett J.P. 1993. Middle Miocene ocean-climate transition: high resolution oxygen and carbon isotopic records from DSDP Site 588A, southwest Pacific. Paleoceanography 8: 811–843.

Flower B.P. and Kennett J.P. 1994. The middle Miocene climatic transition: East Antarctic ice sheet development, deep ocean circulation and global carbon cycling. Palaeogeogr. Palaeoclimatol. Palaeoecol. 108: 537–555.

Gong G.C., Liu K.K. and Liu C.T. 1992. Chemical hydrography of the South China Sea and a comparison with the West Philippine seas. Terr. Atmos. Ocean. Sci. (TAO) Taipei 3: 587–602.

Gooday A.J. 1988. A response by benthic foraminifera to the deposition of phytodetritus in deep sea. Nature 332: 70–73.

Gooday A.J. 1993. Deep-sea benthic foraminiferal species which exploit phytodetritus: characteristic features and controls on distribution. Mar. Micropaleontol. 22: 187–205.

Gooday A.J. 1994. The biology of deep-sea foraminifera: A review of some advances and their applications in paleoceanology. Palaios 9: 14–31.

Gradstein F., Ogg J. and Smith A. (eds.). 2004. A Geologic Time Scale 2004. Cambridge Univ. Press, Cambridge, 589pp.

Gupta A.K. 1993. Biostratigraphic vs. paleoceanographic importance of *Stilostomella lepidula* (Schwager) in the Indian Ocean. Micropaleontology 39(1): 47–51.

Gupta A.K. 1999. Latest Miocene-Pliocene productivity and deep-sea ventilation in the northwestern Indian Ocean (Deep Sea Drilling Project Site 219). Paleoceanography 14: 62–73.

Gupta A.K., Sarkar S. and Mukherjee B. 2006. Paleoceanographic changes during the past 1.9 Myr at DSDP Site 238, Central Indian Ocean Basin: Benthic foraminiferal proxies. Mar. Micropaleontol. 60: 147–166.

Han W. and Ma K. 1988. Carbonate compensation depth, saturation horizon and lysocline in the northeast region of South China Sea. Tropical Oceanol. 7(3): 84–89 (in Chinese).

Hayward B.W., Grenfell H.R., Sabaa A.T. and Nei H.L. 2007. Factors influencing the distribution of subantarctic deep-sea benthic foraminifera, Campbell and Bounty Plateau, New Zealand. Mar. Micropaleontol. 62: 141–166.

Hayward B.W., Kawagata S., Grenfell H.R., Droxler A.W. and Shearer M. 2006. Mid-Pleistocene extinction of bathyal benthic foraminifera in the Caibbean Sea. Micropaleontology 52: 245–266.

Hermelin J.O.R. and Shimmield G.B. 1990. The importance of the oxygen minimum zone and sediment geochemistry in the distribution of Recent foraminifera in the northwest Indian Ocean. Mar. Geol. 91: 1–29.

Hess S. 1998. Vertelungsmuster rezenter brenthischer Foraminiferen im Südchinesischen Meer. Berichte-Reports, Geol.-Palaont. Inst. Univ. Kiel 91: 1–173.

Hess S. and Kuhnt W. 1996. Deep-sea benthic foraminiferal recolonization of the 1991 Mt. Pinatubo ash layer in the South China Sea. Mar. Micropaleontol. 28: 171–197.

Hess S. and Kuhnt W. 2005. Neogene and Quaternary paleoceanographic changes in the southern South China Sea (Site 1143): the benthic foraminiferal record. Mar. Micropaleontol. 54: 63–87.

Hess S., Kuhnt W., Hill S., Kaminski M.A., Holbourn A. and de Leon M. 2001. Monitoring the recolonzation of the Mt Pinatubo 1991 ash layer by benthic foraminifera. Mar. Micropaleontol. 43: 119–142.

Hodell D.A. and Ciesielki F.C. 1991. Stable isotopic carbonate stratigraphy of the late Pliocene and Pleistocene of Hole 704A: eastern subantarctic south Atlantic. Proc. ODP Sci. Results 114: 459–474.

Hodell D.A. and Vayavananda A. 1993. Middle Miocene paleoceanography of the western equatorial Pacific (DSDP Site 289) and the evolution of *Globorotalia* (*Fohsella*). Mar. Micropaleontol. 22: 279–310.

Holbourn A., Kuhnt W. and Schulz M. 2004. Orbitally paced climate variability during the middle Miocene: high resolution benthic stable-isotope records from the tropical western Pacific. In: Clift P.D., Wang P., Hayes D. and Kuhnt W. (eds.), Continent-Ocean Interactions in the East Asian Marginal Seas. AGU Geophys. Monogr. 149, pp. 321–337.

Holbourn A., Kuhnt W., Schulz M. and Erlenkeuser H. 2005. Impacts of orbital forcing and atmospheric carbon dioxide on Miocene ice-sheet expansion. Nature 438: 483–487.

Huang B., Jian Z. and Wang P. 2007. Benthic foraminiferal fauna turnover at 2.1 Ma in the northern South China Sea. Chinese Sci. Bull. 52(6): 839–843.

Jian Z. and Wang L. 1997. Late Quaternary benthic foraminifera and deep-water paleoceanography in the South China Sea. Mar. Micropaleotol. 32: 127–154.

Jian Z. and Zheng L. 1992. Benthic foraminifera in surface sediments and the deeper waters masses of the central South China Sea. In: Ye Z. and Wang P. (eds.), Contributions to Late Quaternary Paleoceanography of the South China Sea. Qingdao Ocean Univ. Press, Qingdao, pp. 119–140 (in Chinese).

Jian Z. and Zheng L. 1995. Deep-sea benthic Foraminifera from the surface sediments of the South China Sea. In: Wang P. (ed.), The South China Sea During the last 150,000 Years. Tongji Univ. Press, Shanghai, pp. 148–158 (in Chinese).

Jian Z., Wang L. and Kienast K. 1999a. Late Quaternary surface paleoproductivity and variations of the East Asian Monsoon in the South China Sea. Quat. Sci. 1: 32–40 (in Chinese).

Jian Z., Wang L., Kienast M., Sarnthein M., Kuhnt W., Lin H. and Wang P. 1999b. Benthic foraminiferal paleoceanography of the South China Sea over the last 40,000 years. Mar. Geol. 156: 159–186.

Jorissen F.J. 1999. Benthic foraminiferal microhabitats between the sediment-water interface. In: Sen Gupta B.K. (ed.), Modern Foraminifera. Kluwer, Dordrecht, pp. 161–179.

Jorissen F.J., Fontanier C. and Thomas E. 2007. Paleoceanographical proxies based on deep-sea benthic foraminiferal assemblage characteristics. In: Hillaire-Marcel C. and de Vernal A. (eds.), Proxies in Late Cenozoic Paleoceanography. Developments in Marine Geology. vol. 1, Elsevier, pp. 263–325.

Kaiho K. 1991. Global changes of Paleocene aerobic/anaerobic benthic foraminifera and deep-sea circulation. Palaeogeogr. Palaeoclimatol. Palaeoecol. 83: 65–85.

Kaiho K. 1994a. Benthic foraminiferal dissolved-oxygen index and dissolved-oxygen levels in modern ocean. Geology 22: 719–722.

Kaiho K. 1994b. Planktonic and benthic foraminiferal extinction events during the last 100 m.y. Palaeogeogr. Palaeoclimatol. Palaeoecol. 111: 45–71.

Kaiho K. 1999. Effect of organic carbon flux and dissolved oxygen on the benthic foraminifral oxygen index (BFOI). Mar. Micropaleontol. 37: 67–76.

Kaiho K. and Hasegawa T. 1994. End-Cenomanian benthic foraminiferal extinctions and oceanic dysoxic events in the northwestern Pacific Ocean. Palaeogeogr. Palaeoclimatol. Palaeoecol. 111: 29–43.

Kawagata S., Hayward B W., Grenfell H R. and Sabaa A. 2005. Mid-Pleistocene extinction of deep-sea foraminifera in the North Atlantic Gatway (ODP Site 980 and 982). Palaeogeogr. Palaeoclimatol. Palaeoecol. 221: 267–291.

Kawagata S., Hayward B.W. and Gupta A.K. 2006. Benthic foraminiferal extinctions linked to late Pliocene-Pleistocene deep-sea circulation changes in the northern Indian Ocean. Mar. Micropaleontol. 58: 219–242.

Kawagata S., Hayward B.W. and Kuhnt W. 2007. Extinction of deep-sea foraminifera as a result of Pliocene-Pleistocene deep-sea circulation changes in the South China Sea (ODP Sites 1143 and 1146). Quat. Sci. Rev. 26: 808–827.

Keller G. and Barron J.A. 1983. Paleoceanographic implication of Miocene deep-sea hiatuses. GSA Bull. 94: 590–613.

Keller G. and Barron J.A. 1987. Paleodepth distribution of Neogene deep-sea hiatuses. Paleoceanography 2: 697–713.

Kuhnt W., Hess S. and Jian Z. 1999. Quantitative composition of benthic foraminiferal assemblages as a proxy indicator for organic carbon flux rates in the South China Sea. Mar. Geol. 156: 123–157.

Kuhnt W., Holbourn A. and Zhao Q. 2002. The early history of the South China Sea: evolution of Oligocene-Miocene deep water environments. Rev. Micropaleontol. 45: 99–159.

Kuhnt W., Hess S., Holbourn A., Paulsen H. and Salomon B. 2005. The impact of the 1991 Mt. Pinatubo eruption on deep-sea foraminiferal communities: A model for the Cretaceous-Tertiary (K/T) boundary? Palaeogeogr. Palaeoclimatol. Palaeoecol. 224: 83–107.

Li B., Jian Z., Li Q., Tian J. and Wang P. 2005. Paleoceanography of the South China Sea since the middle Miocene: evidence from planktonic foraminifera. Mar. Micropaleontol. 54: 49–62.

Li C. 1987. Sediment types and sedimentation of the central South China Sea Basin. Donghai Mar. Sci. 5: 10–18 (in Chinese).

Li Q., Jian Z. and Li B. 2004. Oligocene-Miocene planktonic foraminifer biostratigraphy, Site 1148, northern South China Sea. In: Prell W.L., Wang P., Blum P., Rea D.K. and Clemens S.C. (eds.), Proc. ODP, Sci. Results, 184 [Online].

Li Q., Jian Z. and Su X. 2005. Late Oligocene rapid transformations in the South China Sea. Mar. Micropaleontol. 54: 5–25.

Li Q., Li B., Zhong G., McGowran B., Zhou Z., Wang J. and Wang P. 2006. Late Miocene development of the western Pacific warm pool: Planktonic foraminifer and oxygen isotopic evidence. Palaeogeogr. Palaeoclimat. Palaeoecol. 237: 465–482.

Li Q., Wang P., Zhao Q., Tian J., Cheng X., Jian Z. Zhong G. and Chen M. 2008. Paleoceanography of the mid-Pleistocene South China Sea. Quat. Sci. Rev. 27: 1217–1233.

Linke P. and Lutze G.F. 1993. Microhabitat preferences of benthic foraminifera: a static concept or a dynamic adaptation to optimized food acquisition? Mar. Micropaleontol. 20: 215–233.

Lutze G.F. and Thiel H. 1989. *Cibicidoides wuellerstorfi* and *Planulina ariminensis*, elevated epibenthic foraminifera. J. Foraminiferal Res. 19: 153–158.

Lyle M. 2003. Neogene carbonate burial in the Pacific Ocean. Paleoceanography 18: 1059, doi: 10.1029/2002PA000777.

Mackensen A., Fütterer D.K., Grobe H. and Schmiedl G. 1993. Benthic foraminiferal assemblages from the eastern South Atlantic Polar Front region between 35° and 57 °S: Distribution, ecology and fossilization potential. Mar. Micropaleontol. 22: 33–69.

Mackensen A., Schmiedl G., Harloff J. and Giese M. 1995. Dee-sea foraminifera in the South Atlantic Ocean: Ecology and assemblage generation. Micropaleontology 41: 342–358.

McDougall K. 1996. Benthic foraminiferal response to the emergence of the Isthmus of Panama and coincident paleoceanographic changes. Mar. Micropaleontol. 28: 133–169.

Miao Q. and Thunell R.C. 1993. Recent deep-sea benthic foraminiferal distribution in the South China Sea. Mar. Micropaleontol. 22: 1–32.

Murgese D.S. and De Deckker P. 2005. The distribution of deep-sea benthic foraminifera in core tops from the eastern Indian Ocean. Mar. Micropaleontol. 56: 25–49.

Murgese D.S. and De Deckker P. 2007. The late Quaternary evolution of water masses in the eastern Indian Ocean between Australia and Indonesia, based on benthic foraminifera faunal and carbon isotopes analyses. Palaeogeogr. Palaeoclimatol. Palaeoecol. 247: 382–401.

Murray J. 1991. Ecology and Palaeoecology of Benthic Foraminifera. Longman, Harlow, 397pp.

Murray J. 2006. Ecology and Applications of Benthic Foraminifera. Cambridge Univ. Press, Cambridge, 426pp.

Mutti M. 2000. Bulk $\delta^{18}O$ and $\delta^{13}C$ records from Site 999, Colombian Basin, and Site 1000, Nicaraguan Rise (latest Oligocene to middle Miocene): Diagenesis, link to sediment parameters, and paleoceanography. Proc. ODP Sci. Results 165: 275–283.

Müller D.W., Hodell D.A. and Ciesielski F.C. 1991. Late Miocene to earliest Pliocene (9.8–4.5 Ma) paleoceanography of the subantarctic southeast Atlantic: stable isotopic, sedimentologic, and microfossil evidence. Proc. ODP Sci. Results 114: 459–474.

Nomura R. 1995. Paleogene to Neogene deep-sea paleoceanography in the eastern Indian Ocean: benthic foraminifera from ODP Sites 747, 757 and 758. Micropaleontology 41: 251–290.

Pearson P.N. and Palmer M.R. 2000. Atmospheric carbon dioxide concentrations over the past 60 million years. Nature 406: 695–699.

Qu T., Girton J.B. and Whitehead J.A. 2006. Deepwater overflow through Luzon Strait. J. Geophys. Res. 111: C01002, doi: 10.1029/2005JC003139.

Ramsay A.T.S., Sykes T.J.S. and Kidd R.B. 1994. Waxing (and waning) lyrical on hiatuses: Eocene-Quaternary Indian Ocean hiatuses as proxy indicators of water mass production. Paleoceanography 9: 857–877.

Rathburn A.E. and Corliss B.H. 1994. The ecology of living (stained) deep-sea benthic foraminifera from the Sulu Sea. Paleoceanography 9: 87–150.

Rathburn A.E., Corliss B.H., Tappa K.D. and Lohmann K.C. 1996. Comparisons of the ecology and stable isotopi0063 compositions of living (stained) benthic foraminifera from the Sulu and South China Seas. Deep-Sea Res. I 43: 1617–1646.

Roth J.M., Droxler A.W. and Kameo K. 2000. The Caribbean carbonate crash at the middle to late Miocene transition: linkage to the establishment of the modern global ocean conveyor. Proc. ODP Sci. Results 165: 249–273.

Rottman M.C. 1979. Distribution of planktonic foraminifera and pteropods in South China Sea sediments. J Foraminiferal Res. 9: 41–49.

Sarnthein M. and Winn K. 1990. Reconstruction of low and middle latitude export productivity, 30,000 years B.P. to present: Implication for control of global carbon reservoirs. In: Schlesinger M.E. (ed.), Climate-Ocean Interaction. Kluwer Academic Publishers, Dordrecht, pp. 319–342.

Savin S.M., Douglas R.G., Keller G., Killingley J.S., Shaughnessy L., Sommer M.A., Vincent E. and Woodruff F. 1981. Miocene benthic foraminiferal isotope records: A synthesis. Mar. Micropaleont. 6: 423–450.

Schmiedl G. and Mackensen A. 1997. Late Quaternary paleoproductivity and deepwater circulation in the eastern South Atlantic Ocean: Evidence from benthic foraminifera. Palaeogeogr. Palaeoclimatol. Palaeoecol. 130: 43–80.

Schmiedl G., de Bovèe F., Buscal R., Hemleben C., Medernarch L. and Picon P. 2000. Tropic control of benthic foraminiferal abundance and microhabitat in the bathyal Gulf of Lions, western Mediterranean. Mar. Micropaleontol. 40: 167–188.

Schmiedl G., Mackensen A. and Müller P.J. 1997. Recent benthic foraminifera from eastern
    South Atlantic Ocean: Dependence on supply and water masses. Mar. Micropaleontol. 32:
    249–287.
Schumacher S., Jorissen F.J., Dissard D., Larkin K.E. and Gooday A.J. 2007. Live (Rose Bengal
    stained) and dead benthic foraminifera from the oxygen minimum zone of the Pakistan
    continental margin (Arabian Sea). Mar. Micropaleontol. 62: 45–73.
Shackleton N.J. and Hall M.A. 1990. Pliocene oxygen isotope stratigraphy of Hole 709C. Proc.
    ODP Sci. Results 115: 529–538.
Shackleton N.J., Hall M.A. and Pate D. 1995. Pliocene stable isotope stratigraphy of Site 846.
    Proc. ODP Sci. Results 138: 337–355.
Singh R.K. and Gupta A.K. 2004. Late Oligocene-Miocene paleoceanographic evolution of the
    southeastern Indian Ocean: evidence from deep-sea benthic foraminifera (ODP Site 757). Mar.
    Micropaleontol. 51: 153–170.
Smart C.W. and Gooday A.J. 1997. Recent benthic foraminifera in the abyssal northeast Atlantic
    Ocean: Relation to phytodetrital inputs. J. Foraminiferal Res. 27: 85–92.
Smart C.W., Thomas E. and Ramsay T.S. 2007. Middle-late Miocene benthic foraminifera in a
    western equatorial Indian Ocean depth transect: Paleoceanographic implications. Palaeogeogr.
    Palaeoclimatol. Palaeoecol. 247: 402–420.
Spencer-Cervato C. 1998. Changing depth distribution of hiatuses during the Cenozoic. Paleo-
    ceanography 13: 178–182.
State Oceanic Administration of China (SOA). 1988. Reports of Multidisciplinary Investigations in
    Central part of South China Sea for Resources and Environment. China Ocean Press, Beijing,
    419pp (in Chinese).
Thomas E. and Gooday A.J. 1996. Cenozoic deep-sea benthic foraminifers: Tracers for changes in
    oceanic productivity? Geology 24: 355–358.
Tian J., Wang P. and Cheng X. 2004. Development of the East Asian monsoon and Northern
    Hemisphere glaciation: Oxygen isotope records from the South China Sea. Quat. Sci. Rev.
    23: 2007–2016.
Tian J., Pak D.K., Wang P., Lea D., Cheng X. and Zhao Q. 2006. Late Pliocene monsoon linkage
    in the tropical South China Sea. Earth Planet. Sci. Lett. 252: 72–81.
Tu X. 1983. Distribution and habitats of foraminifera in bottom sediments of the northeastern South
    China Sea. Tropical Oceanol. 2: 11–19 (in Chinese).
Tu X. and Zheng F. 1991. Foraminifera in surface sediments of the Nansha sea area. In: The Mul-
    tidisciplinary Oceanographic Expedition Team of Academia Sinica to the Nansha Islands (ed.),
    Quaternary Biological Groups of the Nansha Islands and the Neighboring Waters. Zhongshan
    Univ. Press, Guangzhou, pp. 129–198 (in Chinese).
Wang J. 2001. Planktonic foraminiferal assemblages and paleoceanography during the last 18 Ma.
    PhD thesis, Tongji Univ., Shanghai (in Chinese).
Wang L. 1992. Late Quaternary carbon isotope records from the South China Sea and their bearing
    on paleoproductivity. In: Ye Z. and Wang P. (eds.), Contributions to Late Quaternary Paleo-
    ceanography of the South China Sea. Qingdao Ocean Univ. Press, Qingdao, pp. 119–226 (in
    Chinese).
Wang L., Sarenthein M., Erlenkeuser H., Grimalt J., Grootes P., Heilig S., Ivanova E., Kien-
    ast M., Pelejero C. and Pflaumann U. 1999. East Asian monsoon Climate during the late
    Pleistocene: high- resolution sediment records from the South China Sea. Mar. Geol. 156:
    245–284.
Wang P. 1999. Response of Western Pacific marginal seas to glacial cycles: Paleoceanographic and
    sedimentological features. Mar. Geol. 156: 5–39.
Wang P. 2004. Cenozoic deformation and the history of sea-land interactions in Asia. In: Clift
    P., Wang P., Kuhnt W. and Hayes D. (eds.), Continent-Ocean Interactions in the East Asian
    Marginal Seas. AGU Geophys. Monogr. 149: 1–22.
Wang P., Prell W.L., Blum P. (eds.). 2000. Proc. ODP, Init. Repts, Vol. 184 [CD-ROM]. Ocean
    Drilling Program, Texas A&M University, College Station TX 77845–9547, USA.
Wang P., Tian J., Cheng X., Liu C. and Xu J. 2003a. Carbon reservoir change preceded major
    ice-sheets expansion at Mid-Brunhes Event. Geology 31: 239–242.

Wang P., Zhao Q., Jian Z., Cheng X., Huang W., Tian J., Wang J., Li Q., Li B. and Su X. 2003b. Thirty million year deep-sea records in the South China Sea. Chinese Sci. Bull. 48(23): 2524–2535.

Wang R., Fang D., Shao L., Chen M., Xia P. and Qi J. 2001. Oligocene biogenetic siliceous deposits on the slope of the northern South China Sea. Sci. China (D) 44(10): 912–918.

Whatley R. and Zhao Q. 1993. The *Krithe* problem: A case history of the distribution of *Krithe* and *Parakrithe* (Crustacea, Ostracoda) in the South China Sea. Paleogeogr. Palaeoclimtol. Palaeoecol. 103: 281–297.

Whitman J.M. and Berger W.H. 1993. Pliocene-Pleistocene carbon isotope record, Site 586, Ontong Java Plateau. Proc. ODP Sci. Results 130: 333–348.

Winn K., Zheng L., Erlenkeuser H. and Stoffers P. 1992. Oxygen/carbon isotopes and paleoproductivity in the South China Sea during the past 110,000 years. In: Jin X., Kudrass H.R. and Pautot G. (eds.), Marine Geology and Geophysics of the South China Sea. China Ocean Press, Beijing, pp. 154–166.

Woodruff F. 1992. Deep-sea benthic foraminifera as indicators of Miocene oceanography. In: Yakayanagi Y. and Saito T. (eds.), Studies in Benthic Foraminifera, Benthos '90. Sendai, Japan, pp. 55–66.

Woodruff F. and Douglas R.G. 1981. Response of deep-sea benthic foraminifera to Miocene paleoclimatic events, DSDP Site 289. Mar. Micropaleontol. 6: 617–632.

Woodruff F. and Savin S.M. 1989. Miocene deepwater oceanography. Paleoceanography 4: 87–140.

Woodruff F. and Savin S.M. 1991. Mid-Miocene isotope stratigraphy in the deep sea: high-resolution correlations, paleoclimatic cycles and sediment preservation. Paleoceanography 6: 755–806.

Woodruff F., Savin S.M. and Abel L. 1990. Miocene benthic foraminifer oxygen and carbon isotopes, Site 709, Indian Ocean. Proc. ODP Sci. Results 115: 519–528.

Wright J.D., Miller K.G. and Fairbanks R.G. 1991. Evolution of modern deepwater circulation: evidence from the late Miocene southern ocean. Paleoceanography 6: 275–290.

Wright J.D., Miller K.G. and Fairbanks R.G. 1992. Early and middle Miocene stable isotopes: implications for deepwater circulation and climate. Paleoceanography 7: 357–389.

Wyrtki K. 1961a. Physical oceanography of the Southeast Asian waters. NAGA, La Jolla, Calif., Rept. 2: 1–195.

Wyrtki K. 1961b. Scientific results of marine investigations of the South China Sea and Gulf of Thailand 1959 ~ 1961. Naga Rep. 2, Scripps Institution of Oceanography, University of California, San Diego, pp. 164–169.

Zachos J., Pagani M., Sloan L., Thomas E. and Billups K. 2001a. Trends, rhythms, and aberrations in global climate 65 Ma to present. Science 292: 686–693.

Zachos J.C., Shackleton N.J., Revenaugh J.S., Pälike H., Flower B.P. 2001b. Climate response to orbital forcing across the Oligocene-Miocene boundary. Science 292: 274–278.

Zhao Q. 2005. Late Cainozoic ostracod faunas and paleoenvironmental changes at ODP Site 1148, South China Sea. Mar. Micropaleontol. 54: 27–47.

Zhao Q. and Zheng L. 1996. Distribution of deep-sea Ostracoda in bottom sediments of the South China Sea. Acta Oceanol. Sinica 18(1): 61–72 (in Chinese).

Zhao Q. and Zhou B. 1995. Deep-sea Ostracoda from the surface sediments of the South China Sea. In: Wang P. (ed.), The South China Sea Since the last 150,000 Years. Tongji Univ. Press, Shanghai, pp. 140–148 (in Chinese).

Zhao Q., Whatley R. and Zhou B. 2000. The taxonomy and distribution of Recent species of the ostracod genus *Cytheropteron* in the South China Sea. Rev. Esp. Micropaleontol. 32: 259–281.

Zhao Q., Jian Z., Wang J., Cheng X., Huang B., Xu J., Zhou Z., Fang D. and Wang P. 2001a. Neogene oxygen isotopic stratigraphy, ODP Site 1148, northern South China Sea. Sci. China (D) 44(10): 934–942.

Zhao Q., Wang P., Cheng X., Wang J., Huang B., Xu J., Zhou Z. and Jian Z. 2001b. A record of Miocene carbon excursions in the South China Sea. Sci. China (D) 44: 943–951.

Zheng L. 1987. A preliminary study on the foraminifera in surface sediments of the central South China Sea. Donghai Mar. Sci. 5: 19–41 (in Chinese).

Zheng L., Ke J., Winn K. and Stoffers P. 1993. Carbonate sedimentation cycles in the northern South China Sea during the late Quaternary. In: Zheng L. and Chen W. (eds.), Contributions to Sedimentation process and Geochemistry of the South China Sea. China Ocean Press, Beijing, pp. 109–123 (in Chinese).

Chapter 7
Biogeochemistry and the Carbon Reservoir

Meixun Zhao, Pinxian Wang, Jun Tian and Jianru Li

Introduction

The biogeochemistry and the carbon reservoir of the South China Sea (SCS) are attractive because of two reasons. First, the monsoon-driven seasonal patterns in bioproductivity and nutrient dynamics distinguish the SCS from other low-latitude waters which, as a rule, are insensitive to seasonal cycles. Second, the enclosed nature of the SCS basin allows only limited exchanges with the open ocean and brings about specific features in its basin-wise circulation and carbon cycling. Since both monsoonal climate and basin morphology have been evolved with geological time, the SCS must have had its own history in biogeochemical evolution and carbon reservoir changes.

However, as a relatively new aspect in the SCS studies, the available data are fragmentary and insufficient for making any systematical synthesis of the biogeo-chemical history of the SCS. Therefore, in the first part of the chapter we will briefly introduce the productivity and nutrient dynamics in the modern SCS and provide some examples of paleoproductivity reconstruction. The second part will be devoted to the carbon reservoir, starting from the carbon cycle in the modern

M. Zhao (✉)
State Key Laboratory of Marine Geology, Tongji University, Shanghai 200092, China; Key Laboratory of Marine Chemistry Theory and Technology of the Ministry of Education, College of Chemistry and Chemical Engineering, Ocean University of China, Qingdao 266100, China
e-mail: maxzhao@tongji.edu.cn; maxzhao@ouc.edu.cn

P. Wang
State Key Laboratory of Marine Geology, Tongji University, Shanghai 200092, China
e-mail: pxwang@tongji.edu.cn; pxwang@online.sh.cn

J. Tian
State Key Laboratory of Marine Geology, Tongji University, Shanghai 200092, China
e-mail: tianjun@tongji.edu.cn

J. Li
State Key Laboratory of Marine Geology, Tongji University, Shanghai 200092, China
e-mail: lijianru@tongji.edu.cn

P. Wang, Q. Li (eds.), *The South China Sea*, Developments in Paleoenvironmental Research 13, DOI 10.1007/978-1-4020-9745-4_7,
© Springer Science+Business Media B.V. 2009

SCS and then discussing the cycles of the carbon reservoir as revealed by carbon isotope records. The long eccentricity cycles in carbon reservoir found in the SCS are actually representative for the global ocean.

7.1 Productivity and Nutrient Dynamics in the Modern South China Sea (Zhao M.)

The SCS is an oligotrophic sea with both nitrate and phosphate in the euphotic layer usually below their detection limits (Gong et al. 1992; Ning et al. 2004; Wong et al. 2007). Since the average N:P ratio in the SCS is below the Redfield ratio of 16, the SCS productivity is ultimately limited by the supply of nitrate (Wu et al. 2003; Wong et al. 2007b). There are strong temporal and spatial variations of productivity in the SCS (Gong et al. 1992; Ning et al. 2004; Chen Y. 2005; Wong et al. 2007a), with estimates for annual productivity ranging from 100 to 1000 mgC $m^{-2}d^{-1}$.

Primary Productivity

Liu K. et al. (2002) presented the surface chlorophyll distribution for April, August, October and December as proxies for the seasonal variations of the SCS primary productivity (Fig. 7.1) and used a coupled physical-biogeochemical model to esti-mate the seasonal variations and annual averaged productivity of the SCS. The modeled annual averaged productivity of 280 mg $Cm^{-2}d^{-1}$ is smaller by 20% com-pared with the SeaWiFS data of 354 mg $Cm^{-2}d^{-1}$. Recently, it has been realized that photo-adaptation and benthic-pelagic coupling can significantly enhance marine phytoplankton production (Liu K. et al. 2007), an updated model incorporated these two effects predicted a mean annual production of 406 mg $Cm^{-2}d^{-1}$ for the SCS (Liu K. et al. 2007), a value that is higher but more comparable to the SeaWiFS data. Despite of the uncertainty in estimations, the seasonal variations in productiv-ity are always significant. The modeled productivity (Liu K. et al. 2002) effectively reflected the seasonal trends resembling those from the SeaWiFS data, the basin-wide SCS primary production is the highest in the winter with a secondary peak in the summer. Field measurements also confirm these seasonal trends. Thus, based on 8 cruises between 2000 and 2003 to the northern SCS north of 14 °N, Chen Y. (2005) found the winter was the most productive season of the year, and the primary pro-ductivity in the basin varied from of 550 mgCm^{-2} d^{-1} in winter, 260 mgCm^{-2} d^{-1} in spring, 190 mgCm^{-2} d^{-1} in summer, to 280 mgCm^{-2} d^{-1} in autumn.

The geographic variations of productivity are also remarkable. Based on com-prehensive cruise data of 61 sites covering the whole SCS basin, Ning et al. (2004) concluded that high production regions are associated with upwelling, such as those off the coast of central Vietnam, southeast of Hainan Island and the northern Sunda Shelf.

In general, the shelf is more productive than the slope and the basin. As Chen Y. (2005) noticed, the spring primary productivity in the shelf, slope and basin

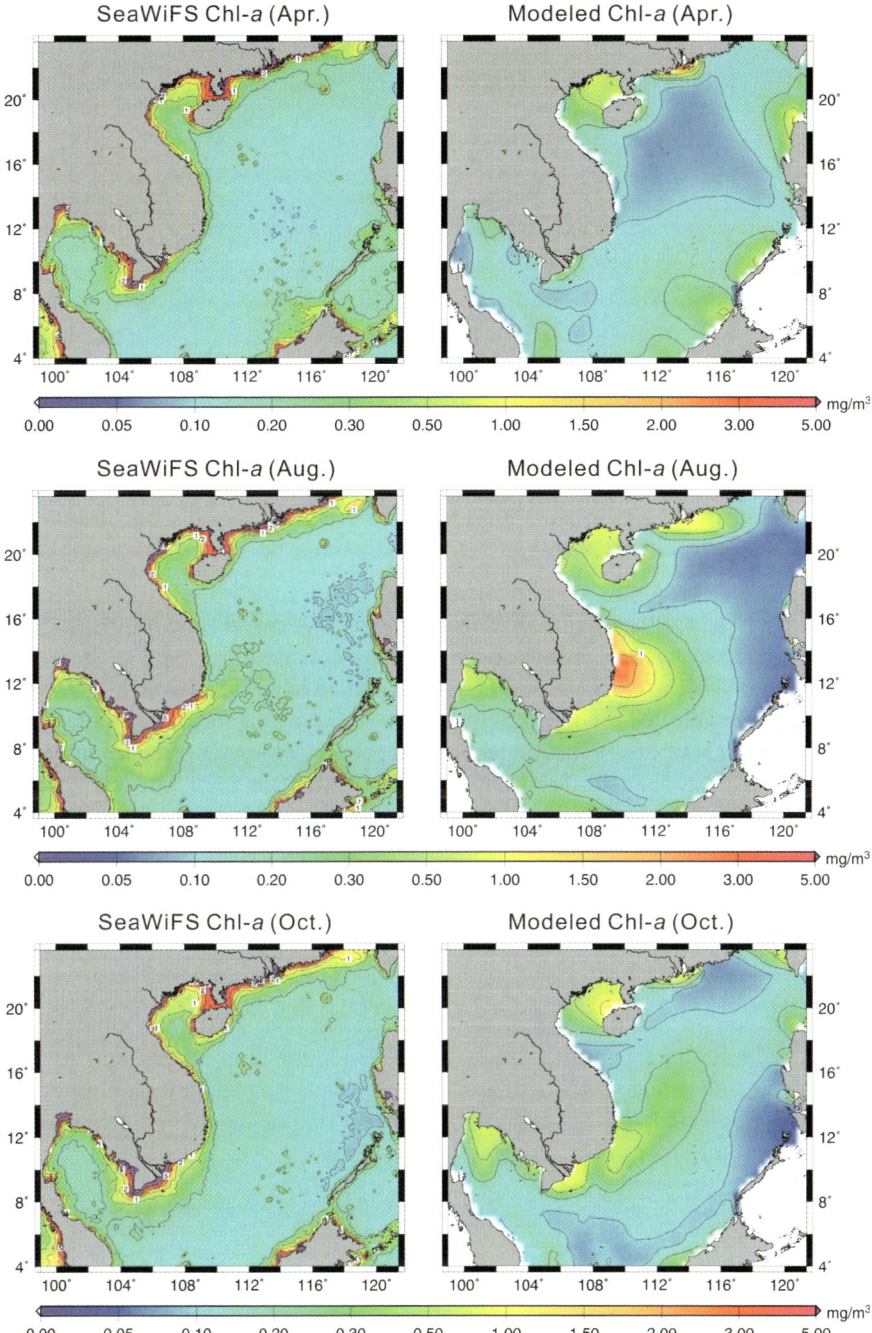

Fig. 7.1 (continued)

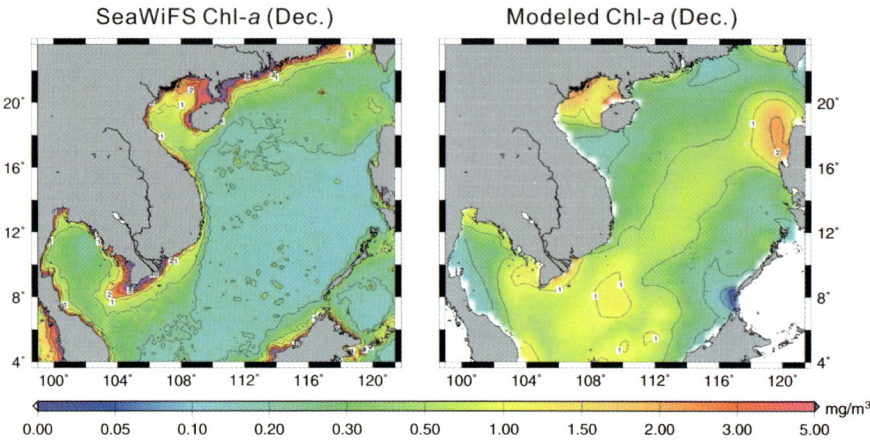

Fig. 7.1 Maps show the monthly average of SeaWiFS data and modeled sea surface chlorophyll distribution in the SCS for (previous page) April, August, October and (this page) December (Liu K. et al. 2002). Model output was generated by the improved biogeochemical model of Liu et al. (2007)

was 720, 340 and 490 mgCm^{-2} d^{-1}, respectively. This can also be illustrated with the higher phytoplankton abundance in shelf than in basin, as well as in winter than in summer (Table 7.1). Both the spatial and seasonal variations in SCS productivity are controlled by oceanographic and environmental processing supplying nutrients to euphotic zone. The modern oceanography of the SCS has been reviewed in Chapter 2. The main features that influence the SCS nutrient budgets, surface productivity, and carbon export are summarized here, with an emphasis on seasonal variabilities.

Nutrient Supplies

Like in other tropical seas, the SST is high in the SCS, the vertical stratification is persistently strong, and the photic zone is deep year round. Thus, the availability of light is not a limiting factor of primary production in the central SCS. Instead, it is oligotrophic and primary production is limited primarily by the availability of the nutrients (Wong et al. 2007a). In turn, the nutrient supply depends on the seasonal alternation of monsoon winds and water circulations.

Table 7.1 Abundance of phytoplankton assemblages in the SCS (unit: 10^3 cells^{-1}) is tabled with data from Chen Y. (2005)

	Shelf (Station 13)		Basin (Station 4)	
Phytoplankton	Summer (July 2001)	Winter (January 2003)	Summer (July 2000)	Winter (January 2003)
Diatoms	2.78	3.94	0.31	1.34
Dinoflagellates	1.44	1.72	0.83	0.54
Coccolithophorids	1.80	3.24	0.04	19.37

Water exchanges. Although the SCS exchanges water with the surrounding seas and the open Pacific Ocean through many sills, the main gateway for water exchange is through the Bashi Strait (Shaw 1989). For the surface water, there is a small net outflow from the SCS to the West Philippine Sea (WPS) in the wet season, but a net inflow in the dry season. For the intermediate water (350–1350 m), there is a net outflow into the WPS during the wet and dry seasons (see Table 2.3 and Fig. 2.15 in Chapter 2). Since the SCS water contains more nutrients than the tropical western Pacific water, these exchanges, especially the export of intermediate water from the SCS, play an important role in the SCS nutrient budgets and in partially keeping the SCS an oligotrophic region of the world oceans (Chen C. et al. 2001b). Overall, the SCS exports nutrients through the outflow of intermediate water and imports nutrients through the inflow of deep water from the West Pacific.

Upwelling. Three upwelling regions are recognized in the SCS: northwest Luzon, east of Vietnam, and north of Sunda Shelf. All three can be revealed by satellite and field measurements of lower sea surface temperatures and higher chlorophyll distributions (Fig. 7.1; also see Fig. 2.14). Since the upwelling of deep water brings nutrient-rich water to the surface, the seasonal and temporal variation of the monsoon-induced upwelling strength is one of the key parameters determining the SCS productivity. Modeled shipboard measurement and satellite (CZCS-SeaWiFS) data show that the strong seasonality of the SCS productivity is mostly influenced by the monsoon variability, which brings the nutrient to the surface layer (Fig. 7.1). Specifically, the east Vietnam upwelling is most intense in summer and the weakest during spring and autumn, while upwelling in both northwest Luzon and north of Sunda Shelf is strong in winter. Meso-scale eddy mixing (Su et al. 1999) is very important in bringing nutrients to surface waters, but eddy effect is often difficult to separate from those by upwelling in the northern SCS, as both processes are more intense during winter (Liu K. et al. 2002; Wu and Chiang 2007).

River discharge. As it is controlled by the monsoon climate, river discharge into the SCS is highest during summer. The nutrient brought in by rivers also enhances productivity, especially for the coastal and shelf regions with large rivers, such as those near the Pearl River and the Mekong River (see Table 4.1 in Chapter 4).

Dust input. A recent assessment of the aerosol input to the SCS on the basis of satellite data has shown multiple remote sources of fine aerosol particles there, including wind-blown dust from Asian deserts and biomass burning in Sumatra and Borneo. The results suggest that aerosols from continent are primarily transported by the winter monsoon and are confined to the northern SCS, whereas the dominant fine particles in the southern SCS are carried by the summer monsoon from the islands (Lin et al. 2007).

Many studies have demonstrated the effect of dust fertilization of the surface ocean (Boyd et al. 2007 and references herein). The SCS surface productivity is controlled by nitrogen, so nitrogen fixation by the cyanobacterium, *Trichodesmium*, could be an important source of bio-available nitrogen. In the tropical western Pacific, *Trichodesmium* fixation of nitrogen is limited by Fe-supply, thus dust inputs are expected to increase nitrogen fixation to increase surface productivity. However, this fertilizing effect is very limited in the SCS (Wu et al. 2003) despite the substantial supply of Fe to the SCS by winter monsoon winds (Duce, 1991). One study from the central SCS suggests that nitrogen fixation contributed less than 2% of the nitrate used in new production (Chen Y. 2005). Results of time series at the SEATS station of northern SCS revealed higher N:P ratios between September and April during the period of higher dust deposition (Wong et al. 2007b), suggesting iron fertilized nitrogen fixation producing nitrogen-rich organic matter. However, nitrogen fixation at the SEATS station could only enhance productivity by 3–13%, and the lack of iron-binding organic ligands in the SCS could be the factor limiting the bio-availability of iron and the growth of *Trichodesmium* (Wu et al. 2003).

Typhoon. Recent data has shown that tropical cyclones could enhance marine productivity significantly over a very short period of time (Lin et al. 2003). About 10–14 tropical cyclones traverse the SCS annually, and the passing of tropical cyclones can cause episodic injection of nutrients to the mixed layer and hence elevate primary production. As Lin I. et al. (2003) reported, the occurrence of the moderate tropical cyclone Kai Tak in summer 2000 led to a 30-fold increase in the concentration of surface chlorophyll a, and this one event alone could account for 2–4% of the annual new production in the SCS. However, it is difficult to estimate the long-term effect of typhoon, especially for events in the past.

The fertilization mechanisms discussed above result in both spatial and temporal variations of the SCS primary production. Many studies have been carried in the SCS to relate the surface productivity with the physical and chemical processes. Liu K. et al. (2002) presented the surface chlorophyll distribution for April, August, October and December as proxies for seasonal variations of the SCS primary productivity (Fig. 7.1). Although spatial variations exist, spring productivity is the lowest everywhere compared with the other three seasons, as no dominating fertilization mechanisms operate anywhere. In the summer, higher productivity is observed almost everywhere, but the increase is especially obvious off Vietnam and the Mekong River, and along the coast of southern China. The high productivity region off Vietnam is likely caused by both the summer monsoon-driven upwelling and the increased river discharge from the Mekong River. For the high productivity region along the southern China coast, increased river discharge likely remains as the main mechanism. Productivity for the fall season is similar to the spring. During the winter, productivity increased, especially for the two winter monsoon-driven upwelling regions: northwest Luzon and north of Sunda Shelf. However, productivity for areas northwest Luzon during winter could be enhanced by stronger monsoon-induced upwelling (Chen C. et al. 2006b) and by cold eddy effects which are also stronger during winter (Chen Y. et al. 2007b). Integrating satellite and field data, the basin-wide SCS primary production is highest in winter with a secondary peak in summer. These features underscore the importance of monsoons in controlling the SCS productivity. The broadly defined upwelling regions cover about 20% of the SCS, but they could account for about 30% of the total SCS productivity. Averaged annually, the summer and winter monsoon-dominated regions have similar productivities.

Community Structure, Export Productivity and Sedimentary Biogenic Content

Many studies have focused on the physical and chemical processes that control nutrient distribution and primary production of the SCS (Liu et al. 2002), but only a few detailed investigations have been carried on the abundance and distributions of the major phytoplankton groups (Hung and Tsai 1972; Huang 1991; Huang and Chen 1997; Ning et al. 2004; Chen Y. et al. 2007a,b). Knowledge on the community structure of the phytoplankton is very important in our understanding of how different phytoplankton groups related to different fertilization mechanisms in the SCS. This is also important concerning carbon export and burial since phytoplankters have different "biological pump" efficiencies (Buesseler et al. 2007). Firstly, large phytoplankters can transport carbon downward more efficiently (Boyd and Newton 1999; Buesseler et al. 2007). Secondly, growth and sedimentation of diatoms and dinoflagellates help to lower the CO_2 level of the ocean and atmosphere, while the growth of coccolithophorids can actually increase atmospheric CO_2 due to the release of CO_2 during the formation of calcium carbonate. Recently, it has been realized that picoplanktons are important producers in the ocean, especially in the tropical lower productivity regions (liu H. et al. 2007). However, their importance in carbon export and burial are still being debated (Buesseler et al. 2007; Richardson and Jackson 2007). Thus, a shift in community structure can alter the carbon sink efficiency, even if the total primary production remains the same.

In the SCS, earlier studies on the phytoplankton abundance and size-distribution in limited areas were carried out (Takahashi and Hori 1984; Hung and Tsai 1972; Huang 1991; Huang and Chen 1997). A more comprehensive survey of the biomass distribution covering almost the whole SCS basin for the summer and winter of 1998 has been reported (Ning et al. 2004). Simultaneous measurements of the physical, chemical and biological parameters offer the opportunity to assess how the phytoplankton biomass and production responded to seasonal physical and chemical changes. These results indicate that phytoplankton abundance was much higher in winter than in summer (Fig. 7.2). The winter increase was a result of increased surface nutrient flux caused by wind-driven upwelling and mixing. Among the major phytoplankton groups (diatoms, dinoflagellates and cyanophytes), diatoms were the most dominant during winter when total productivity was higher, they were also quite abundant in the northern and central SCS during summer when total productivity was lower. Another study for the central SCS (Chen Y. 2005) confirms that diatom abundance was higher during winter, but the phytoplankton assemblage was dominated by coccolithophores (Table 7.1). More important from a carbon sink perspective, if the percentage of diatoms increases as production increases during winter, then carbon export efficiency could also increase during the more productive seasons due as diatoms sink faster. Picoplanktons are the most important contributor to both the total biomass and production in the SCS (Ning et al. 2004; Chen Y. 2005), and their relative abundance increases as total production decreases in summer. It is very obvious that these small organisms are very important in the ocean

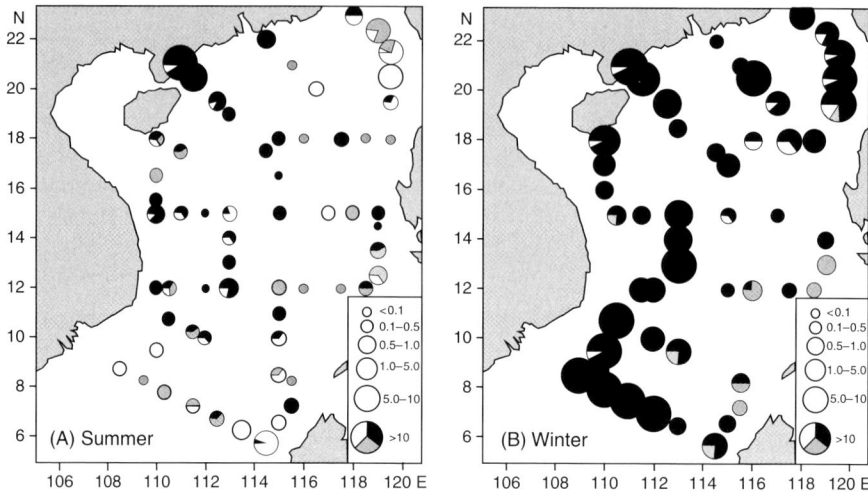

Fig. 7.2 Maps show the abundance changes in three major phytoplankton groups in the SCS for (**A**) summer and (**B**) winter of 1998 (from Ning et al. 2004). Diatoms are *shaded black*, dinoflagellates in *grey*, and cyanophytes in *white*

carbon cycle, the challenge is how to estimate their contribution to the downward carbon flux.

Sediment traps studies in the SCS were carried by several collaborative cruises between the University of Hamburg and the Second Institute of Oceanography of the State Oceanic Administration of China (Jennerjahn et al. 1992; Chen J. et al. 1998, 1999), and the detailed investigation of the downward fluxes of biogenic components was presented in a PhD thesis (Chen J. 2005). Modeling suggests that about 12% of the primary production carbon was exported from the photic zone (Liu K. et al. 2002), and sediment trap results confirm that only 1–2% the surface fixed organic carbon reaches the deep ocean, with POC degradation mostly occurring in the top 1000 m of the water column. Further remineralization at the deep water/sediment interface resulted in only 0.26% of the original TOC being buried in the sediment. Thus, the biological pump efficiency for the SCS is very low, only about 1/3 of that for the Arabian Sea (Haake et al. 1993), which is also controlled by monsoon climate. Part of the reason is that the SCS productivity is low, as it has been demonstrated that the percentage of organic carbon buried in sediment is positively correlated with surface primary productivity (Eppley and Peterson 1979). TOC degradation could be particularly high since deep water dissolved oxygen is high due to the high ventilation rate of the SCS deep water (see Chapter 2).

The biogenic carbonate/silica ratio in particulates from the SCS is relatively low compared with other oligotrophic regions (Nelson et al. 1995; Honda et al. 2002). This ratio has been used as a proxy for the community structure of the photic zone, with higher ratios indicating more contribution of coccolithophorids and lower ratios indicating more contributions from diatoms. Typically, lower ratios are observed in higher productivity regions where diatoms are more abundant. The higher carbonate/silica ratio in the SCS, especially in the central SCS, is in agreement with phytoplankton biomass study which shows coccolithophores were the main producers (Chen Y. 2005). However, this ratio has large annual variabilities, ranging from 3 to 0.6 during a three-year span from 1993 to 1996 (Wang R. et al. 2007). Since the water column loss of the biogenic carbonate and silica is also significant (72–95%), part of the reason for the variable carbonate/silica ratio is related to dissolution, not to productivity. Preferential dissolution of carbonate in the water column could also result in a lower carbonate/silica ratio in the SCS trap materials. Despite these uncertainties, sediment trap time series studies from central SCS indicate both TOC and biogenic silica are well correlated with primary productivity (Wang R. et al. 2007).

Since only a small percentage of the original surface production (often less than 1% for TOC) is preserved in the sediments, it is often debatable whether the sedimentary biogenic contents can be used as productivity proxies. Results from many studies confirm that, with the diverse environmental conditions bearing on surface water productivity, water column processes, sedimentation, and early diagenesis before and after burial, it is unlikely that the content or MAR of any biogenic proxy can be used as an absolute productivity indicator. However, some of these environmental effects can be eliminated by comparing the relative downcore variations of biogenic proxies, especially if the sedimentation rate is relatively constant. Thus, sedimentary biogenic content variations are best viewed as indicators of relative changes of surface productivity for given sites and regions.

One means to test the fidelity of these biogenic proxies is the correlation of their surface sediment content with measured surface productivity. For the SCS, sedimentary TOC content falls within 0.4–1.0%. Qualitatively, higher TOC values are beneath the higher productivity regions, with the most obvious areas beneath the three upwelling cells (Fig. 7.3). This correlation is in agreement with the global map, confirming that sedimentary TOC content is a time-smoothed proxy for surface productivity. However, sedimentary biogenic silica content did not reveal any significant correlation with surface productivity (Fig. 7.4). This is surprising since results from both biomass and particulates reveal increased diatom contributions for the more productive upwelling regions (see Figs. 7.1 and 7.2).

These results can be taken to indicate that diagenetic processes are more important in controlling the SCS sedimentary biogenic silica content. The environmental controls on opal production and dissolution in both the water column and sediments have been documented extensively (Nelson et al. 1995; Ragueneau et al. 2000). It is concluded that the efficiency of biogenic silica export fluxes, lateral advection of water masses, sediment redistribution, and spatial variations in the preservation efficiency of the biogenic silica all affect the use of opal as a productivity proxy. Lower opal export rate in the Atlantic Ocean compared with the Pacific Ocean could partially explain the lower opal content in the Atlantic Ocean sediment. On the other hand, the low ambient silicic acid concentration in waters off NW Africa could induce the growth of diatoms more susceptible to dissolution, which would further decrease their export rate. For the SCS, perhaps the diatoms growing under the less productive environments are also more susceptible to dissolution. Studies in the more productive Monterey Bay (Brzezinski et al. 2003) indicate diatom blooms are the main vectors of silica export in the sea. The majority of the silica produced

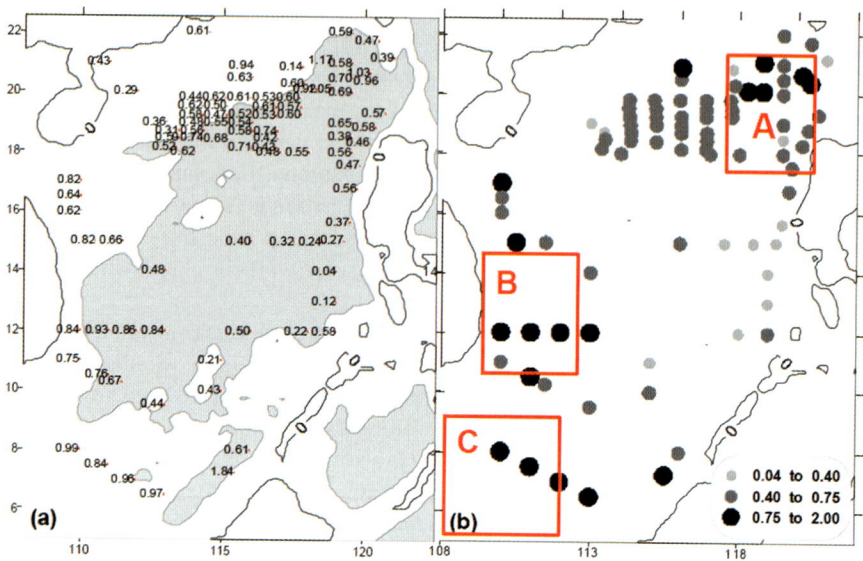

Fig. 7.3 Distribution of surface sedimentary TOC content (%) in the SCS is shown with (**a**) actual measuring results and (**b**) simplified ranges (Chen J. 2005). The three upwelling areas are also indicated

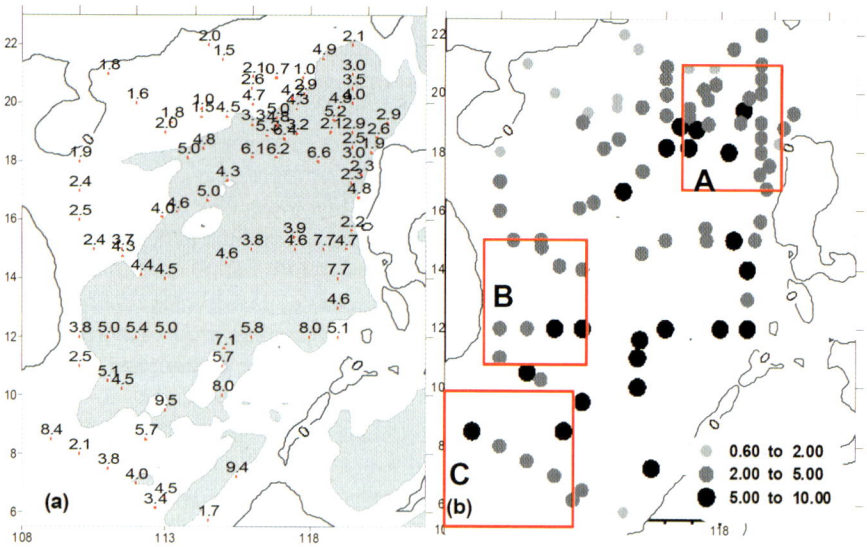

Fig. 7.4 Distribution of surface sediment biogenic silica content (%) in the SCS is shown with (**a**) actual measuring results and (**b**) simplified ranges (Chen J. 2005). The three upwelling areas are also indicated

during non-bloom periods is recycled in the photic zone. Perhaps the poor correlation between surface productivity and the sedimentary opal content in the SCS can be partially explained by the non-bloom nature of the SCS ecology.

Despite the limitations of these productivity proxies, they have been applied for paleoproductivity reconstruction. It is clear that, whenever possible, multiproxy studies using the same samples can offer better understanding of the processes controlling the sedimentary content of biogenic materials and their paleo-reconstruction application.

7.2 Paleoproductivity Reconstruction of the South China Sea (Zhao M.)

Reconstruction of productivity in the ocean is central to paleoceanography because of its important links to climate, carbon cycling, and biogeography. Nevertheless no proxy is completely free of secondary alternation and can be used as straightforward indicator of paleoproductivity. The use and misuse of proxies have been extensively discussed in the literature, including the merits and limitations of proxies for paleoproductivity (for reviews see Wefer et al. 1999; Hillaire-Marcel and de Vernal 2007). In the above section we have discussed some proxies for productivity in the SCS such as organic carbon and biogenic silica. The following part will review the productivity reconstruction based on various proxies.

Patterns of Productivity Changes During Glacial-Interglacial Oscillations

Records from Northern and Southern SCS

Many paleoproductivity proxies have been applied to the SCS reconstruction, and some general consensus on glacial/interglacial productivity changes for the northern SCS has been emerged based on these studies. Earlier studies used TOC content and mass accumulation rate (MAR) as productivity indicators and they revealed higher productivity during glacials (Huang C. et al. 1997a,b). Later studies confirm this pattern for the northern SCS, including using TOC and chlorins (for total productivity, Fig. 7.5), opal (for diatom productivity, Fig. 7.6), alkenones (for haptophyte productivity, Fig. 7.7) (Huang C. et al. 1997a,b; Lin et al. 1999a; Kienast et al. 2001a; Higginson et al. 2003; Wang R. et al. 2003, 2007). Carbon flux reconstruction (Jian et al. 2001b) and an organic carbon flux proxy based on benthic foraminiferal assemblages (Kuhnt et al. 1999) showed that although productivity was higher during the LGM for the northern SCS, the productivity peak was during Termination I.

For the southern SCS, earlier studies gave different patterns on glacial-interglacial productivity changes. Mass accumulation rates (MAR) of TOC of a transect of cores from the southeastern SCS revealed higher TOC contents during the LGM and it was concluded that productivity was two times higher (Thunell et al. 1992). This

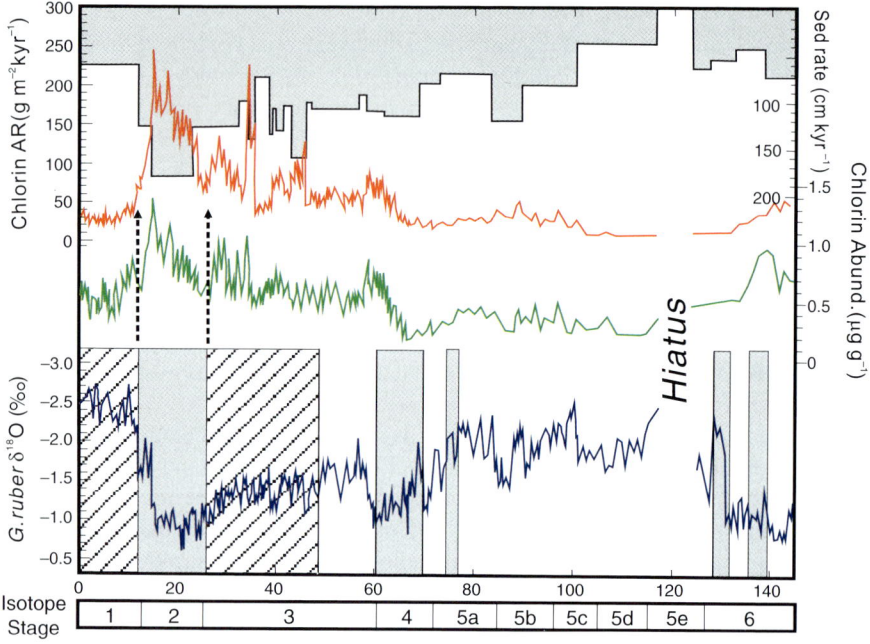

Fig. 7.5 Total chlorin abundance and accumulation rate (AR) are shown together with the planktonic $\delta^{18}O$ and sedimentation rate records at ODP Site 1144, the northern SCS. *Solid shading* indicates periods of sea level below −65 m; *hatched shading* highlights intervals of similar $\delta^{15}N$ records between the SCS and the eastern North Pacific (ODP Hole 889A) (modified from Higginson et al. 2003)

conclusion was supported by more records using TOC flux (Jian et al. 1999a; Fang et al. 2000), opal content (Jia et al. 2000) and benthic foraminiferal assemblage (Kuhnt et al. 1999), all showing higher productivity during the LGM for the southern SCS. However, TOC MAR was higher during interglacials compared with those for glacials in a core from the southwestern SCS, indicating higher productivity during interglacials (Jian et al. 2001b). The MARs of authigenic elements, such as Sr, Ba, P and Ca, also indicated higher productivity during interglacials for the southwestern SCS (Wei et al. 2003). Furthermore, a 1.6 Ma record of opal content record at ODP Site 1143 also revealed lower productivity for the southern SCS during glacials (Fig. 7.6; Wang and Li 2003). Alkenone content from the southern SCS revealed a decreasing productivity trend during the last glacial period extending beyond the LGM and into the Holocene, before productivity increased in the Holocene (Pelejero et al. 1999b; Kienast et al. 2001; Zhao M. et al. 2006). The high resolution record of MD972151 from the southwestern SCS showed no clear glacial-interglacial trends, even though the averaged productivity for haptophytes was slightly higher during the glacial (Fig. 7.7).

Some of these differences in the paleoproductivity records from the southern SCS could be related to the inherent problems connecting to such proxies as those caused by different preservation of the biogenic components, but they could also be related

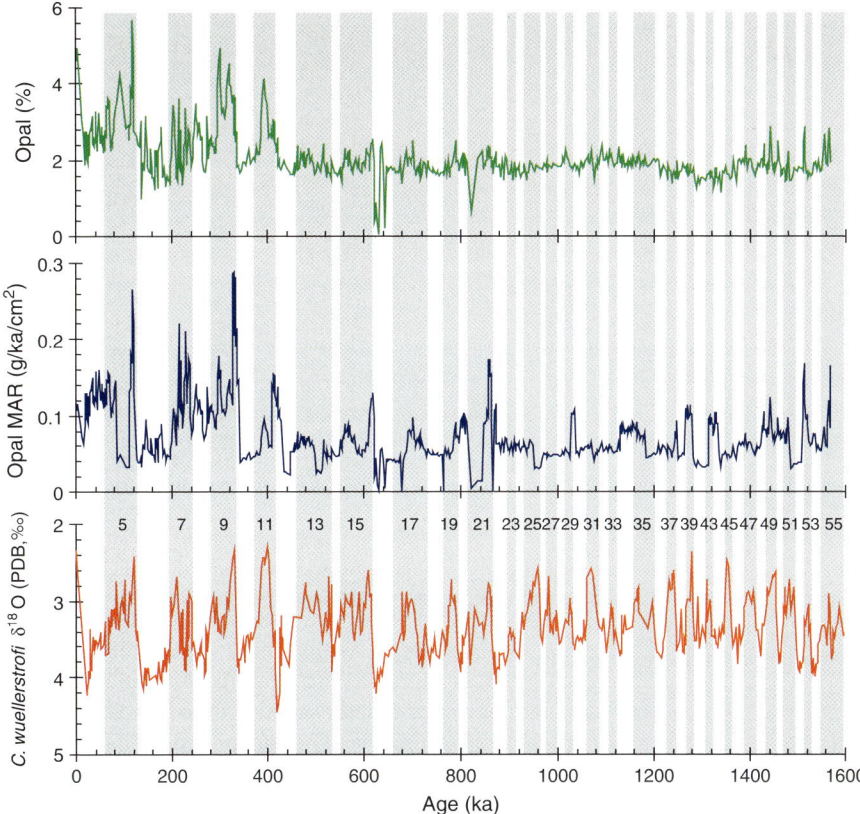

Fig. 7.6 Opal records are shown together with the benthic $\delta^{18}O$ curve for ODP Site 1143 from the southern SCS (Wang and Li 2003). Interglacial intervals (MIS 5 to MIS 55) are shaded

to the different meanings of these proxies. For example, while TOC and chlorin are best treated as proxies for total productivity, the content of opal and alkenone should be a proxy for siliceous plankton (such as diatoms) and calcareous plankton (such as haptophytes), respectively. Thus, the different productivity patterns for the SCS could be partially related to some change in the phytoplankton community structure. In the southeastern SCS, the TOC:carbonate rain ratio was proposed as a community structure proxy, and the higher ratio during glacials was suggested to reflect relatively more siliceous plankton production, as total productivity also increased. The opal record of ODP Site 1143 form the southern SCS (Wang and Li 2003) would indicate that diatom production was lower during glacials and higher during interglacials, a pattern in contrast to that from the southeastern SCS. However, the opal record from ODP Site 1143 does not rule out an increase of total productivity for the site during glacials, as diatoms are not the dominant primary producers in the oligotrophic SCS. Indeed, the alkenone record of core MD972151 (Zhao et al. 2006) from the southwestern SCS reveals generally higher haptophyte productivity during glacials. As calcareous planktons are the main primary producers in the SCS,

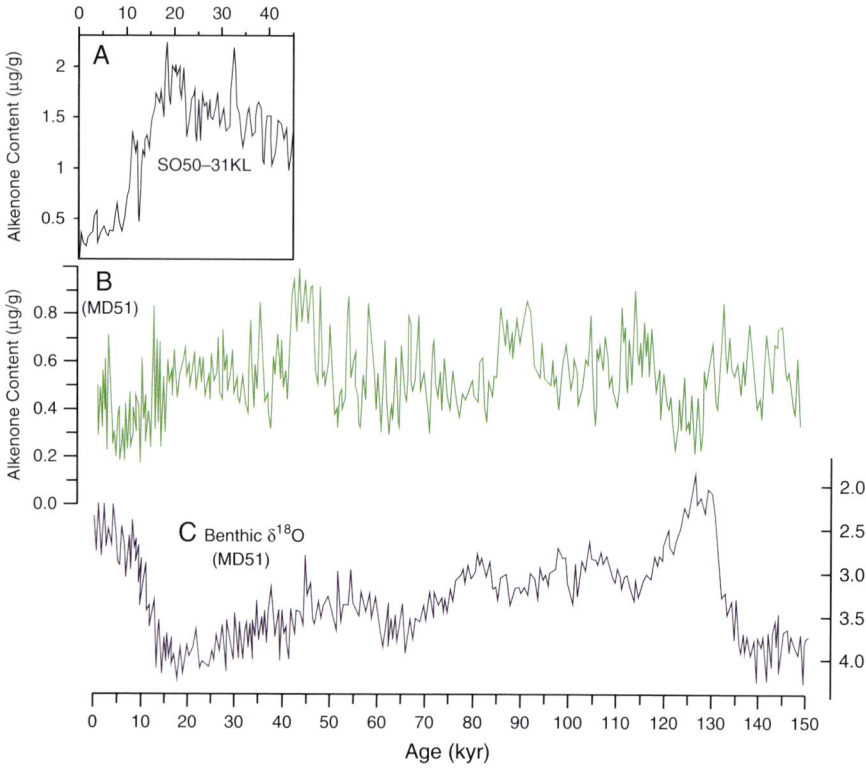

Fig. 7.7 (**A**) Alkenone content for core SO50-31KL from the northern SCS (Huang et al. 1997a). (**B**) Alkenone content and (**C**) benthic δ¹⁸O records for core MD972151 from the southwestern SCS (Zhao M. et al. 2006)

this record can be taken to imply higher total productivity during glacials. However, biogenic elemental records for core NS93-5, also from the southwestern SCS but north of core MD972151, reveals higher total productivity during interglacials (Wei et al. 2003).

These comparisons strengthen the argument that the southern SCS productivity has large spatial variability. However, productivity reconstruction using different proxies from different core locations also makes the comparison difficult and the inferences inconclusive. Recently, results of a multi-proxy reconstruction for core MD972142 from the southeastern SCS may help to resolve some of these inconsistencies (Shiau et al. 2008). The contents of TOC, carbonate, opal and alkenone were measured for the last 870 kyr (Fig. 7.8). A composite productivity index (CPI) was extracted from the principal component analysis of these biogenic components. The CPI record suggests higher total productivity during glacials, caused probably by strengthened winter monsoon, which brought more nutrients to the surface layer and caused mixing. The alkenone record generally parallels the CPI record, also suggesting higher haptophyte productivity during glacials. This parallelism, to a first order approximation, also suggests that calcareous planktons were the major producers

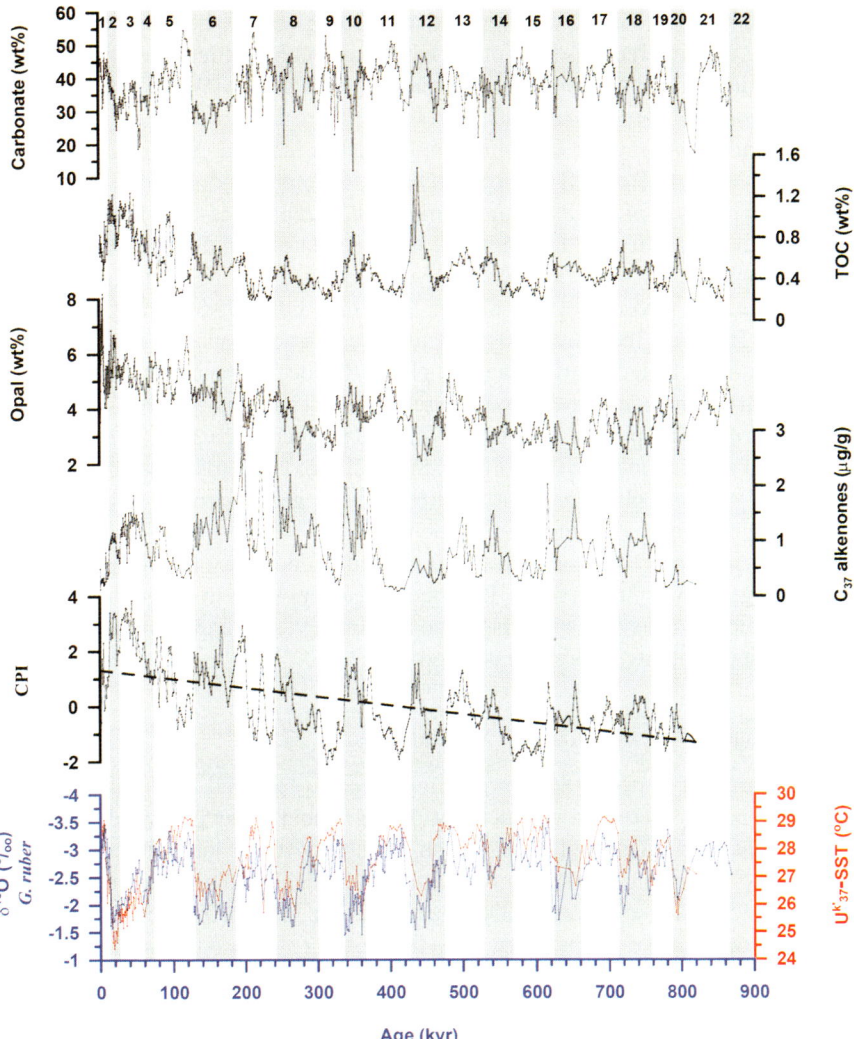

Fig. 7.8 Productivity proxy records are compared with the planktonic $\delta^{18}O$ curve for core MD972142 from the southeastern SCS (Shiau et al. 2008)

for this site. On the other hand, opal content from this site suggests higher diatom productivity during interglacials and lower productivity during glacials, in agreement with results from ODP Site 1143. The increase in diatom productivity during interglacials was linked with stronger summer monsoon and river inputs, however, it is not clear why diatoms did not respond to the winter monsoon fertilization during glacials when total productivity was higher. The face values of these records suggest major phytoplankton community structure changes for site MD972142 during glacial-interglacial cycles. Interestingly, the relative diatom contribution to the total productivity was lower during glacials when the total productivity was higher, and its relative contribution was higher during interglacials when total productivity was lower. This inference is in contrast with most published results, which suggest increased diatom contribution when productivity was higher (Werne et al. 2000).

Same core multi-proxy records from the northern SCS are few, and could not afford detailed assessment of community structure change. Recently, biomarker records have been generated for core MD052904 from the northern SCS (He et al. 2008). The contents of C_{37} alkenones, dinosterol, and brassicasterol have been used as productivity proxies for haptophytes, dinoflagellates and diatoms, respectively. These records reveal that individual phytoplankton group and total productivity all increased by several factors during the LGM compared with those for the Holocene. However, the community structure reconstruction using these biomarkers suggests minor changes during the last glacial-Holocene transition, in contrast to the result from the southern SCS discussed above. Thus, the SCS has at least two provinces regarding productivity and community structure changes during glacial-interglacial oscillations. For the northern SCS, all proxy records suggest higher total productivity during glacials, while limited data suggest relatively stable community structures. For the southern SCS (including the southeastern and southwestern SCS), some proxy records suggest higher total productivity during glacials, but others suggest lower productivity. But the limited data seem to suggest major community structure changes, with glacial conditions favoring haptophyte growth and interglacial conditions favoring diatom growth.

Implications on the Fertilization Mechanisms

The similarity of northern SCS productivity records revealed by different proxies suggests that the increased productivity during glacials was governed by similar mechanisms, mostly related to monsoon variability. Both marine and terrestrial records suggest stronger winter monsoon during glacials (Ding et al. 1998; Porter and An 1995; Wang and Wang 1990; Huang C. et al. 1997a; Zhao M. et al. 2006), which would increase productivity, as is the case for the modern ocean during the winter season (Liu K. et al. 2002; Ning et al. 2004; Wong et al. 2007b). Glacial increase in both winter monsoon strength and duration would increase upwelling and mixing to bring more nutrients to the surface.

Dust inputs into the SCS were expected to be higher due to the strengthened winter monsoon winds, coupled with drier (Sun and Li 1999; Zheng and Lei 1999) and expanded source region (Sun and Li 1999) on the continent. Dust inputs have been demonstrated to increase ocean productivity, mostly in high-nutrient low-chlorophyll (HNLC) regions (de Baar et al. 2005). For the SCS which is limited by nitrogen, increased dust inputs (especially Fe input) could increase nitrogen fixation and increase productivity. However, the proposed fertilization effect in the SCS has not been unambiguously proven. As the $\delta^{15}N$ of nitrogen fixed by cyanobacteria (ca. -1‰) is substantially lighter than the values for deep water nitrate (ca. 5‰), sedimentary bulk $\delta^{15}N$ values have been used to evaluate the contribution of different nitrogen sources for phytoplankton growth. An early analysis of several cores from the SCS reveals small $\delta^{15}N$ changes during the last climatic cycle, and these changes were not correlated with paleoceanographic and productivity changes

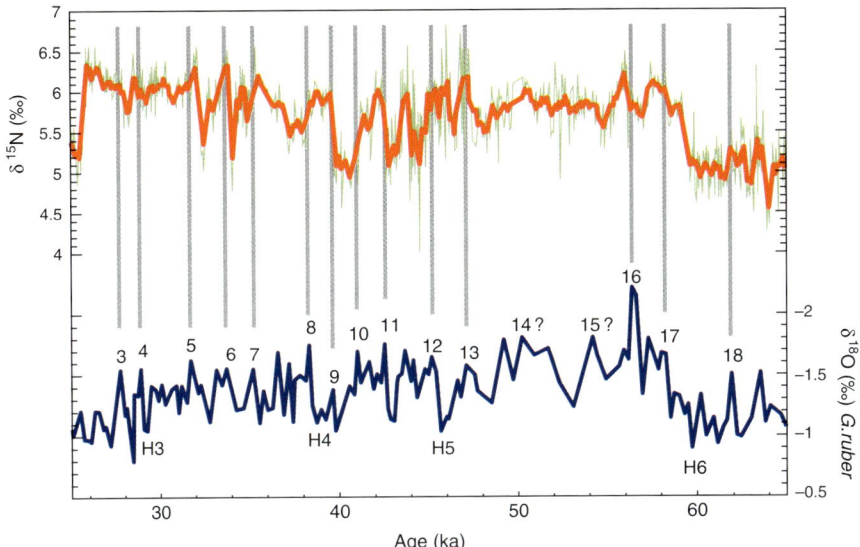

Fig. 7.9 Comparison of $\delta^{15}N$ and *G. ruber* $\delta^{18}O$ from ODP Site 1144 highlights millennial scale changes. Numbers in the lower panel indicate the timing of Daansgaard-Oeschger interstadial events, and H3, H4, H5 and H6 indicate the timing of Heinrich events (modified from Higginson et al. 2003)

(Kienast 2000). It was suggested that the $\delta^{15}N$ of nitrate in the SCS and in the western Pacific remained approximately constant, implying little contribution from cyanobacteria contribution. High resolution records of ODP Site 1144 from the northern SCS (Higginson et al. 2003) confirm that the $\delta^{15}N$ and chlorin (a total productivity proxy) were decoupled for much of the last 145 kyr. However, millennial scale $\delta^{15}N$ oscillations, especially during MIS 3, were correlated with D-O cycles (Fig. 7.9), suggesting climatic forcing of the $\delta^{15}N$ signal. It was proposed that regional (western Pacific) increases in dust enhanced N-fixation during cold periods were the cause of lighter $\delta^{15}N$ observed in the SCS. In addition, this study suggests glacial-interglacial changes in the marine nitrate inventory across the whole North Pacific (Higginson et al. 2003).

The role of riverine materials input to the northern SCS and their fertilization effect during glacials is also difficult to evaluate independently. The drier climate would decrease river discharge and reduce river-transported nutrient to the northern SCS, but the drop in sea-level increased the percentage of terrestrial materials to the continental slope region, where most of the paleo-records have been obtained so far.

Therefore, available data do not provide conclusive evidence for increased nutrient inventory during glacials for the SCS. For the northern SCS, increased productivity during glacials was mainly caused by intensified winter monsoon, which affected the mixed layer depth and increased the advection of nutrient from the deep water. Increased Fe deposition rate likely enhanced cyanobacteria fixation of nitrogen, but its contribution to the SCS productivity was limited.

The different glacial-interglacial productivity patterns for the southern SCS suggest that fertilization mechanisms are likely more complicated. Using present day pattern as an analogue and assuming that the same fertilization mechanisms operated during glacials, then productivity for summer monsoon-dominated south-western SCS region would decrease during glacials as summer monsoon weakened. However as discussed above, the reconstruction results were inconclusive. This suggests that other mechanisms should also be considered. Although the effect of river input is difficult to evaluate, the opal record for the southern SCS (ODP Site 1143) can shed some light on this. As this region is not dominated by monsoon-induced upwelling, it can be inferred that reduced river input had some effect on the decreases in diatom productivity during glacials. Since the intensified winter monsoon during glacials likely affected the whole SCS, its fertilization effect on the southern part of the SCS has to be considered. Records from the southeastern SCS (Thunell et al. 1992; Shiau et al. 2008) indicate total productivity increased during glacials, and this was most likely caused by the strengthened winter monsoon. Further evidence supporting this mechanism came from productivity records from the Sulu Sea (de Garidel-Thoron et al. 2001). Since there is no evidence of monsoon induced upwelling region in the Sulu Sea, its productivity change is mostly controlled by the winter-monsoon induced mixing. A 200 kyr record (Fig. 7.10) not only revealed major glacial increases in productivity, but several millennial scale productivity increase events were also correlated with lower $\delta^{18}O$ (colder temperature)

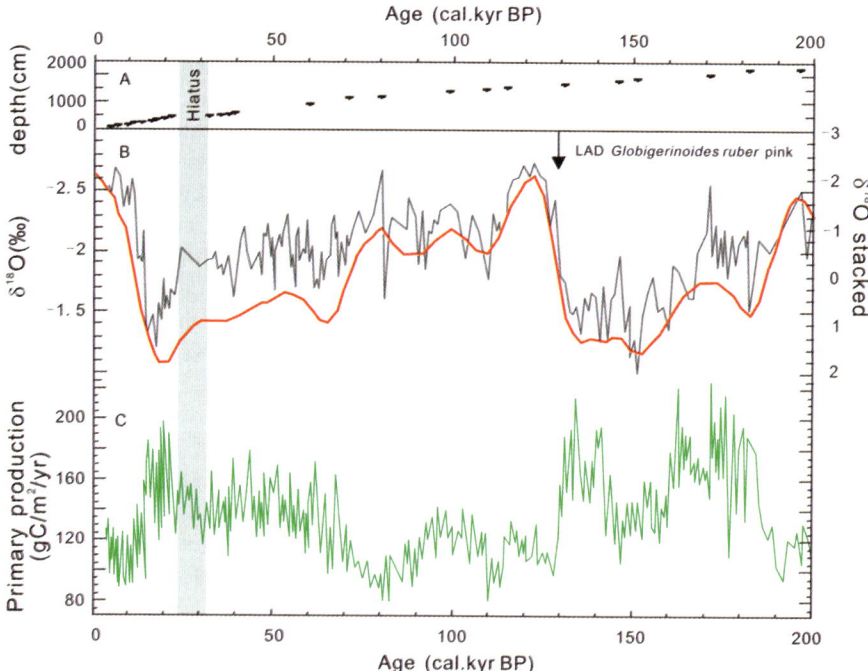

Fig. 7.10 Productivity estimate for core MD972141 from the Sulu Sea during the last 200 kyr is compared with the planktonic $\delta^{18}O$ from the same core and the globally stacked $\delta^{18}O$ record (red line) (de Garidel-Thoron et al. 2001)

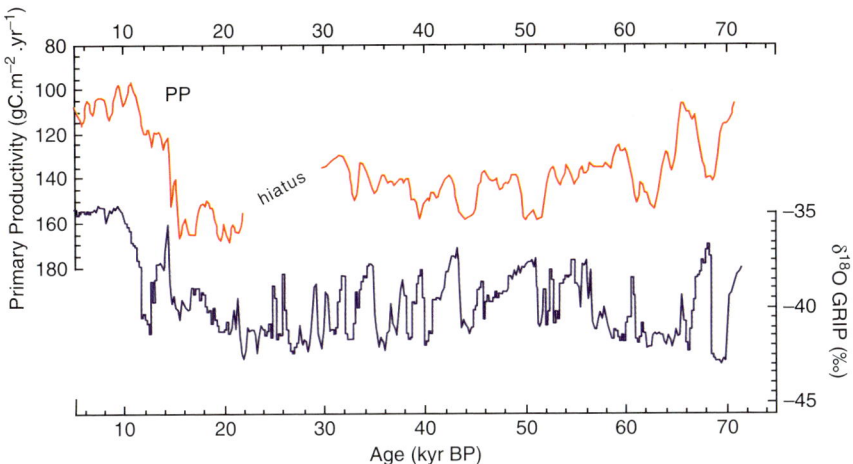

Fig. 7.11 Comparison of primary productivity estimate (PP) for core MD972141 from the Sulu Sea and d18O record of GRIP ice core during the last 75 kyr, highlighting millennial scale changes and correlations of higher PP in the Sulu with colder events in high-latitude (modified from de Garidel-Thoron et al. 2001)

events of the GRIP ice core record (Fig. 7.11). And based on age models, these higher productivity events (de Garidel-Thoron et al. 2001) were also correlated with increased winter monsoon events recorded in the Chinese Loess Plateau (Porter and An 1995).

In summary, almost all proxy records indicate that productivity was enhanced for the northern SCS during glacials due to increased upwelling/mixing in response to strengthened winter monsoon. For the southern SCS, productivity change was more local. For summer monsoon influenced regions, productivity was lower during glacials, mostly caused by reduced diatom contribution. Other regions reveal some evidence of increased total productivity, perhaps by more contribution from coccolithophorids. Multi-proxy reconstructions of the same cores offer opportunities to evaluate community structure changes, which can help to distinguish winter and summer monsoon finger-prints in the paleoproductivity records. Concerning carbon flux and its influence on pCO_2, productivity increases and especially the increases of diatom contribution for the northern SCS during glacials could have contributed to lower pCO_2 level. On the other hand, decreases in productivity and diatom contribution for the southern SCS during glacials suggest that the southern SCS could be a CO_2 source.

Pre-Pleistocene Paleoproductivity Changes

Along with geochemical proxies, micropaleontological sequences are widely used for paleoproductivity estimation in the SCS, especially for long-term changes. Not only the abundance of a certain group of phytoplankton such as diatom is sensitive to nutrient supply, but some other microfossil forms are also indicative of productivity. For example, the planktonic foraminifer *Neogloboquadrina dutertrei* is found to be a high-productivity species (Jian et al. 2001b, 2003), whereas higher percentage of *Florisphera profunda* in the nannoplankton assemblage implies lower productivity (Liu and Cheng 2001). Some genera of benthic foraminifera, such as *Bulimina, Buliminella, Globobulimina, Uvigerina*, belong to the "high productivity group", and their higher abundance indicates higher food supply from the surface ocean (Hess and Kuhnt 2005).

Results from ODP Leg 184 sites have revealed the pre-Pleistocene history of productivity changes in the SCS. Over the 33 myr, at least three time intervals are distinguished by high productivities: the Oligocene, the late Miocene, and the late-Pliocene to Pleistocene.

The Oligecene deposits at ODP Site 1148 is enriched in organic carbon (0.4–0.6%) as compared with those of the Miocene ($<0.2\%$) (Wang P. et al. 2003c), and the middle part of the sequence (ca. 29.5–27.5 Ma), contains unusually high abundances of siliceous microfossils (Wang R. et al. 2001; see Fig. 4.44 in Chapter 4). Two environmental factors might have been responsible for the Oligocene high productivity. Firstly, the narrow-gulf shape of the newly opened SCS basin was favorable for terrestrial nutrient supply. Secondly, the mid-Oligocene high productivity may have been associated with intensified water fractionation between the Pacific and Atlantic (see Fig. 4.45), since high values of biogenic opal about 30–28 Ma were also observed in the central equatorial Pacific.

The high productivity interval of the late Miocene is clearly seen from the deep-sea sequence over the last 12.5 myr at ODP 1143 in the southern SCS (Li J. et al. 2002a; Wang R. et al. 2004). Here the coeval increase of mass accumulation rate of TOC, biogenic opal and carbonate from $\sim$12.3 to 5.8 Ma, centred at 8.5–6 Ma, is broadly corresponding to the late Miocene–early Pliocene Biogenic Bloom Event (see Fig. 4.42). The event was originally discovered in the eastern equatorial Pacific, but it was later reported from practically all the low-latitude oceans. Although the origin of the event remains speculative, it is one of the most significant events in the Neogene tropical ocean (Diester-Haas et al. 2006).

The high-productivity interval of the last 3 myr has been noticed at all the sites of ODP Leg 184. At the northern Site 1148, increased percentages of the "high productivity group" in benthic foraminiferal fauna suggest enhanced carbon flux since $\sim$3 Ma (see Fig. 6.8 in Chapter 6). This is supported by the high content of TOC and biogenic opal in the southern SCS (see Fig. 4.42 in Chapter 4), and well in agreement with the remarkably raised proportion of *Neogloboquadrina dutertrei*, a high- productivity indicator, in the planktonic foraminiferal assemblage at ODP Sites 1146 and 1148 in the northern SCS (see Fig. 5.44 in Chapter 5). The enhanced productivity over the last 3 myr is not specific to the SCS, but a global phenomenon caused by the intensified nutrient supply in response to the boreal ice sheet development and/or intensified monsoon, especially strengthened winter monsoons.

7.3 Carbon Reservoir Changes (Wang P., Tian J. and Li J.)

Modern Carbon Cycling

There are three aspects of carbon cycling in the ocean: carbon exchange between the surface water and the atmosphere, within the water column, and between bottom water and sediments. The air-sea exchange of molecular CO_2 is crucial for the climate system because it controls the carbon concentration in the atmosphere and the pH of the ocean. It remains debatable, however, whether marginal seas, including the SCS, are global source or sink of atmospheric CO_2 (e.g., Thomas et al. 2004; Cai and Dai 2004). What we know for a certainty is the seasonal overturn of air-sea exchange in the SCS. Significant seasonal variations occur in total carbon dioxide (TCO_2) and total alkalinity (TA) in the upper layer, as shown by the four-year time-series at SEATS station (18 °N, 116 °S) in the northern SCS (Fig. 7.12) (Tseng et al. 2007). The air-sea exchanges of CO_2 overturn with seasonal alternation of monsoons: the open northern SCS serves as a sink of atmospheric CO_2 in winter, but as a weak source of CO_2 in the other three seasons. Overall, the SCS is considered as a very weak CO_2 source (Chen C. et al. 2006c), although there is little net annual exchange of CO_2 between the sea and its overlying atmosphere (Wong et al. 2007a).

The depth distribution of dissolved carbon in the SCS can be illustrated with measurements at the Station SEATS. As seen from Fig. 7.13, both total CO_2 (TCO_2) and IC/OC ratio increase rapidly in the upper part but remain fairly constant below 2000 m. According to a modeling estimation, approximately 28% of normalized

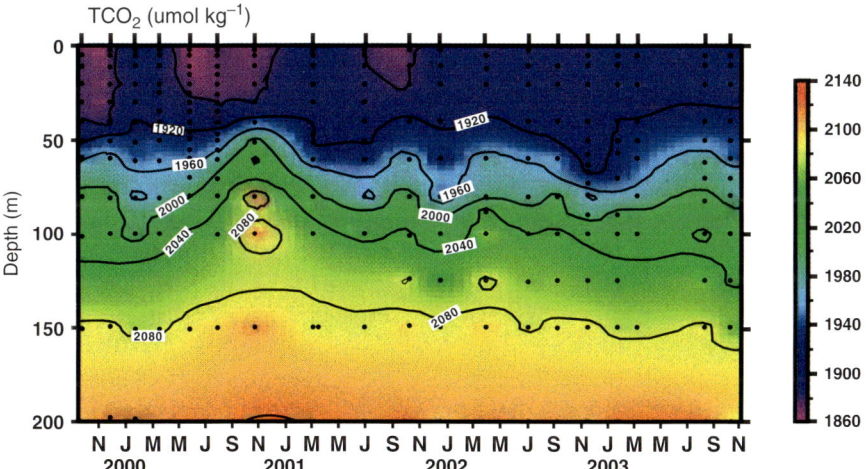

Fig. 7.12 Profile shows the variations in T CO_2 (total CO_2) in the upper 200 m at the SEATS station, northern SCS, between September 1999 and October 2003 (Tseng et al. 2007). Data points are indicated as solid dots below sampling times (short hash marks)

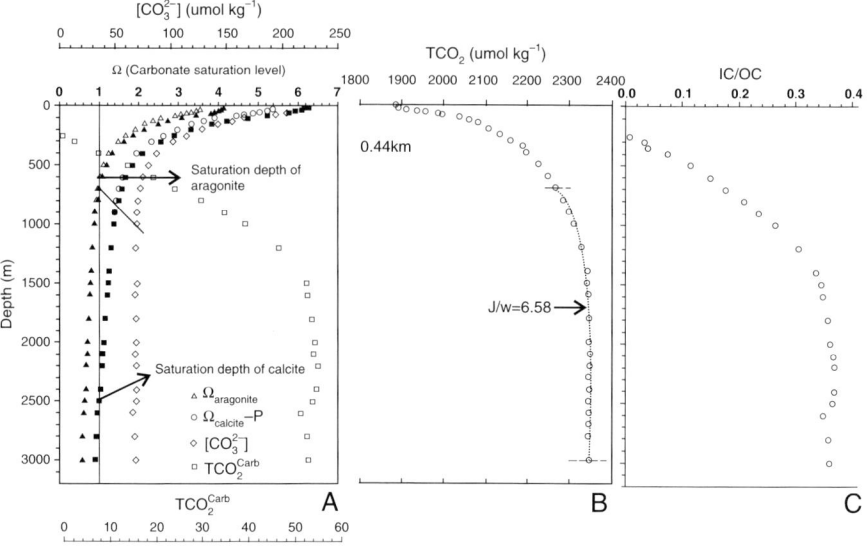

Fig. 7.13 Depth profiles of (**A**) aragonite ($\Omega_{\text{aragonite}}$), calcite ($\Omega_{\text{calcite}}$) saturation levels, concentration of carbonate ion ([CO_3^{2-}], open diamonds) and TCO_2^{carb} (TCO_2 produced from carbonate dissolution are compared with those of (**B**) TCO_2 (total CO_2) and (**C**) IC/OC ratio at the SEATS station, northern SCS (modified from Chou et al. 2007)

TCO_2($TCO_2 \times$ salinity/35) in the deep waters is from organic decomposition, and 72% from carbonate dissolution (Chou et al. 2007).

The carbonate chemistry of seawater is usually considered to be an important factor influencing calcium-carbonate precipitation only in deep waters. The recent studies, however, revealed that it could influence rate of calcification even for corals living in $CaCO_3$-supersaturated waters (Gattuso et al. 1998). At the SEATS station the degree of carbonate saturation with respect to aragonite and calcite, i.e. $\Omega_{\text{aragonite}}$ and Ω_{calcite} ($\Omega = [Ca^{2+}]\_[CO_3^{2-}]/Ksp$, where K_{sp} is the solubility-product constant) decreases rapidly with depth in the upper 1,000 m due to the decrease of [CO_3^{2-}], and the saturation depths of aragonite and calcite were placed at about 600 and 2,500 m, respectively, by Chou et al. (2007) (Fig. 7.13). The estimation for aragonite is similar to that of the previous authors (435 m by Han and Ma 1988, or 600 m by Chen C. et al. 2006), but the estimated depth of calcite saturation at 2500 m is much too deep as compared with the previous authors (1,630 m by Han and Ma 1988; or 1,600 m by Chen C. et al. 2006; see Chapter 4), although Chou et al. (2007) believe that the estimations are consistent with those reported in the same latitude of the Northwest Pacific by Feely et al. (2002).

In order to obtain a 3-dimensional view of the carbon distribution in the SCS, three cross-sections of normalized total CO_2 (NTCO$_2$) are presented in Fig. 7.14. All sections show a rapid increase of NTCO$_2$ in the upper part of the water column, from the surface to 1,000–1,500 m in the SCS, but much more uniform in the lower part. The west-east profile crossing the Bashi Strait (PR-20 line) reveals

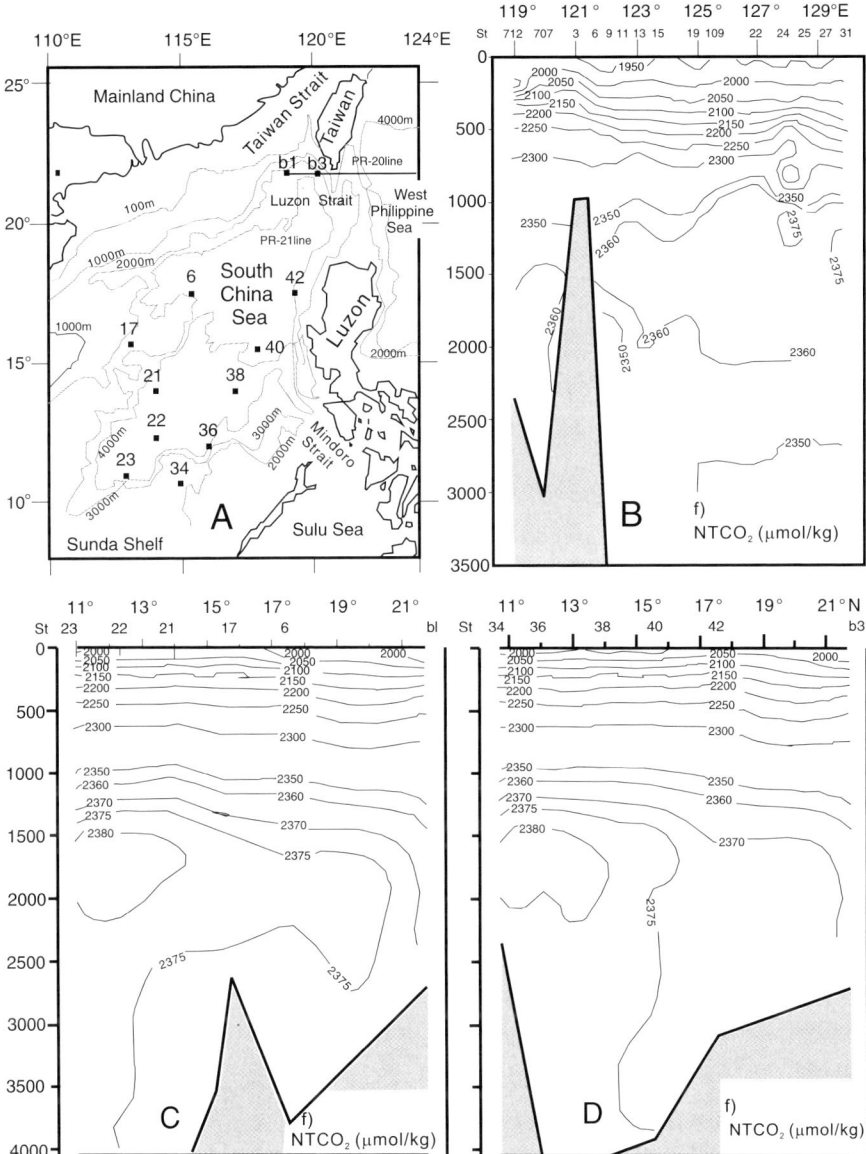

Fig. 7.14 Profiles (**B** to **D**) show normalized total CO_2 ($NTCO_2$) at various stations (**A**) along a W-E section about 21 °45′N or the PR-20 line (**B**), along a S-N section in the west from stations 23 to b1(**C**), and along a S-N section in the eastern SCS from stations 34 to b3 (**D**) (modified from Chen C. et al. 2006c)

the difference between the Western Pacific to the East and the SCS to the West (Fig. 7.14B). In the upper layers all the contours tend to shoal towards the west, because of intensive upwelling in the SCS basin (see Chapter 2). The $NTCO_2$ maximum commonly found in the open oceans at mid-depth prevails in the Western

Pacific around 1,200–1,400 m, but it is absent in the SCS again because of intensive upwelling and vertical mixing. In both of the W-E profiles the NTCO$_2$ concentrations at the same water depths are slightly higher in the southern part than in the northern, and a conspicuous feature is a NTCO$_2$ maximum at about 1800 m in the southern part (Fig. 7.14D), which is ascribed to the decomposition of organic matter (Chen C. et al. 2006a).

Turning to organic carbon, dissolved organic carbon (DOC) represents the major portion (up 80–95%) of total organic carbon (TOC) in the SCS, which is influenced significantly by the Kuroshio intrusion and continental fluxes. Except for the coastal zone, the concentration of DOC in the mixed layer of the northern SCS ranges from 70 to 85 mM, while that of particulate organic carbon (POC) ranges only from 1.6 to 4 mM. During summer with the highest river discharge, elevated concentrations of DOC and POC up to 132 and 13 mM, respectively, are found in the Zhujiang (Pearl River) plume. Distributions of DOC (4373 mM) and POC (1.1702 mM) are uniform in deep and bottom layers (>1,000 m) of the deep basin areas. In general, both DOC and POC are generally higher in shelf zones than in deep basins, and also higher in summer than in the other seasons particularly for coastal zones (Hung et al. 2007).

Surface distributions of DOC and POC in the SCS appear to be mainly controlled by physical (horizontal and vertical) mixing in addition to the influence of terrestrial inputs in the coastal zone. Surface production and vertical mixing primarily regulate DOC and POC distributions in the euphotic layer of deep basins (Hung et al. 2007). A model of organic carbon cycling was proposed by Wong et al. (2007) (Fig. 7.15). The available information suggests that the primary production (PP) in the SCS falls within the range of 120–170 g C m^{-2} yr^{-1}. Nitrate uptake studies suggest that new production is 19–37 g C m^{-2} yr^{-1} of the total productivity. Hence the recycled production is 83–151 g-C m^{-2} yr^{-1}, and the estimated export production ranges from 15 to 25 g C m^{-2} yr^{-1}. Sediment traps deployed at 1000 m in the northern SCS yielded a settling flux of organic carbon of 2 g C m^{-2} yr^{-1} (Jennerjahn et al. 1992) suggesting that about 90% of the organic matter produced in the euphotic

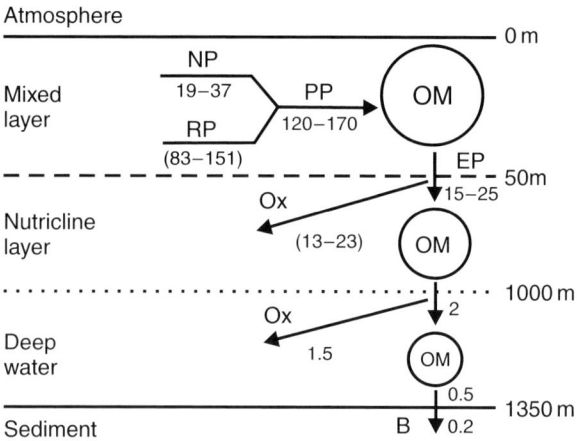

Fig. 7.15 A schematic diagram outlines the dynamics of particulate organic carbon in the SCS (Wong et al. 2007a). OM=organic matter; PP=net primary production; NP=new production; RP=recycled production; EP=export production; Ox=remineralization by the oxidation of organic matter; B=burial. Fluxes are in g C m^{-2} yr^{-1}

Table 7.2 Sedimentation rates and net burial of organic carbon (OC) and inorganic carbon (IC) are estimated for different areas in the SCS (modified from Chen C. et al. 2006a)

	Area (m^2)	Sedi. rate (cm/10^3 yr)	OC (%)	IC (%)	OC burial (10^{9}mol/yr)	IC burial (10^{9}mol/yr)
Shelves	18×10^{11}	62	0.71	28	346	2426
Slopes	8.3×10^{11}	11	0.55	27	20	174
Basin	8.3×10^{11}	3.3	0.50	7	6	15
Total	35×10^{11}				372	2615

zone is remineralized in the main nutricline between the bottom of the mixed layer (at about 50 m) and 1,000 m. After remineralition of organic carbon in the water column, only 0.5 g-C m^{-2} yr^{-1} survives oxidation and reaches the seabed at the average water depth of 1,350 m for the SCS, where the burial rate of organic carbon is estimated to be 0.2 g-C m^{-2} yr^{-1} (Wong et al. 2007). If the model holds, therefore, only 0.1–0.2% of photosynthetically fixed carbon in the mixed layer is likely to be preserved in the sediments.

Summarizing the available data, Chen C. et al. (2001b, 2006a) calculated the mass balance of carbon in the SCS. They found that the SCS receives $521,080 \times 10^9$ mol C and exports $534,181 \times 10^9$ mol C in the wet season (May–October, approximately the summer monsoon season), and receives $230,875 \times 10^9$ mol C and exports $229,015 \times 10^9$ mol C in the dry season (November-April, roughly the winter monsoon season). This means an excess of outflow in the wet season and of inflow in the dry season, clearly responding to monsoonal climate; and the unbalance in estimations is ascribed to the uncertainty in seawater fluxes (Chen C. et al. 2006a). Furthermore, they estimated the annual sedimentation rates and net burial of carbon in the SCS, as shown in Table 7.2, revealing the importance of shelves for carbon burial. Although the shelves account for only about 50% of the SCS area, they account for more than 90% of both OC and IC burials.

Late Quaternary $\delta^{13}C$ Cyclicity

The carbon reservoir of a marginal basin is biogeochemically controlled by external and internal factors. On geological time scales, the external factors include fluvial (or eolian) input and inter-basin exchanges of carbon and nutrients, the atmospheric carbon reservoir and the efficiency of sea-air exchanges, while internal factors include burial and remineralization of organic carbon, and deposition and dissolution of carbonates. Any change of one of the factors will cause variations in the carbon reservoir which can be deciphered from the basin sediment records. Carbonate content and carbon isotope ratios are two proxies most commonly used for recognition of carbon reservoir changes in the geological records. Since carbonate is discussed in Chapter 4, here we will focus on carbon isotope only.

Very little is known of the $\delta^{13}C$ distribution in the modern waters of the SCS. Lin et al. (1999b) analyzed $\delta^{13}C$ of ΣCO_2 in the water column at 9 stations in

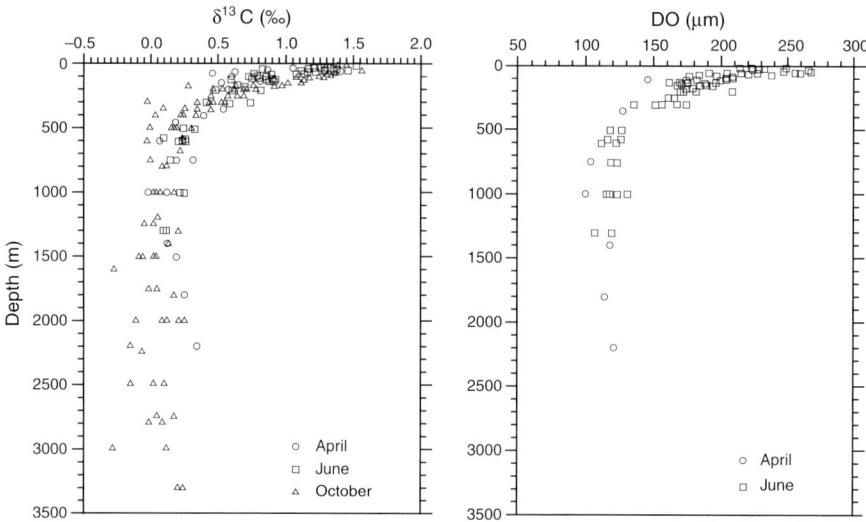

Fig. 7.16 Depth profiles of $\delta^{13}C$ of ΣCO_2 and dissolved oxygen (DO) show similar trends at 9 stations in the northeastern SCS (see text) (Lin et al. 1999)

the northeastern SCS (20–22 °N and 118–120 °E) in April, June, and October 1995 (Fig. 7.16). The results show $\delta^{13}C$ enrichment at the surface and depletion at depth, and the vertical gradient of $\delta^{13}C$ in the upper 500 m is significantly greater than it is below, as in the open ocean. The vertical profile of $\delta^{13}C$ is similar to that of oxygen concentrations (Fig. 7.16), suggesting biological productivity and organic matter degradation being the major controls on the amount and composition of the carbon reaching the deep sea (Lin et al. 1999).

Later, Lin et al. (2004) analyzed $\delta^{13}C$ variations in planktonic foraminifers in two-year sediment trap records (2000–2002) from the northern SCS, and the most depleted $\delta^{13}C$ values were found in winter months with enhanced nutrient supply by upwelling driven by the winter monsoon. When the $\delta^{13}C$ values of planktonic foraminifers in the surface sediments over the entire SCS are compared, the most depleted values are found in the NE part of the basin, which is again associated with the winter monsoon upwelling (Fig. 7.17; Cheng et al. 2005). The $\delta^{13}C$ values of the same species *Globigerinoides ruber* at the LGM maintain the same spatial pattern, implying the monsoon control both in glacial and postglacial SCS, although the glacial $\delta^{13}C$ values are generally heavier than those in the Holocene (Cheng et al. 2005). The more positive $\delta^{13}C$ values in the glaciation are well known to be related to the decrease in forest coverage and hence decreases in the terrestrial carbon reservoir.

Unlike the $\delta^{18}O$ records which are mostly used in stratigraphic correlations for their covariations through time, the $\delta^{13}C$ values are extensively used to reconstruct ocean circulations for their differences in space (Mackensen and Bickert 1999), and only little attention has been paid to the $\delta^{13}C$ variations of the global ocean. In the recent years, efforts have been made to single out the global signal in the SCS $\delta^{13}C$

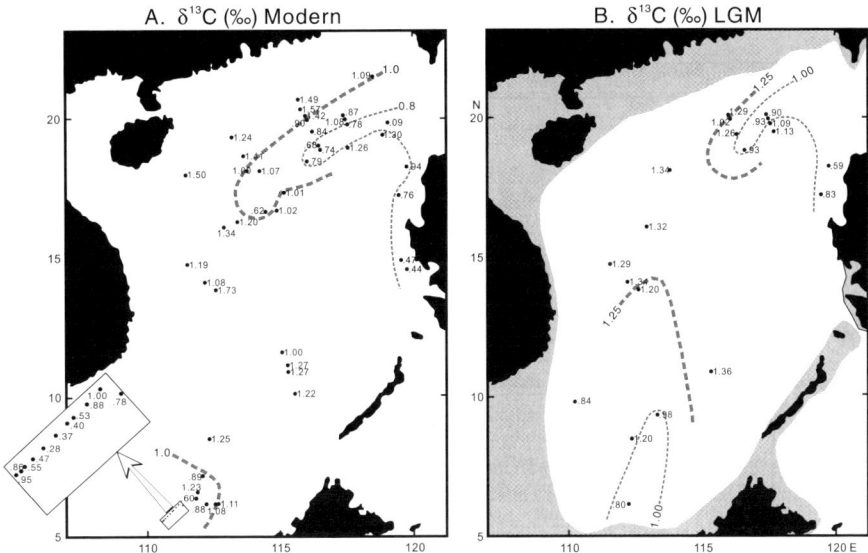

Fig. 7.17 The δ^{13}C distribution in planktonic foraminifer *Globigerinoides ruber* shows a "tongue" in the northeastern SCS in (**A**) modern surface sediments and (**B**) last glacial maximum (LGM) samples (Cheng et al. 2005). Gray area denotes exposed continental shelf at low sea-level stand

records, in particular for the late Quaternary which is covered by the most of δ^{13}C time series available. Figure 7.18 shows the δ^{13}C curves of planktonic foraminifers over the last 200 kyr from 8 sites in the SCS (for locations see Fig. 1.1 and Table 1.1, this volume).

All δ^{13}C values vary between 0 and 2‰, except for two outliers, and they exhibit a remarkably similar trend in downcore variations. A stacked 200 kyr carbon isotope record of the SCS (SCS PF δ^{13}C) based on the 8 series revealed clear cycles, with lighter values in glacial periods and heavier values in interglacials. Noticeable are three δ^{13}C minima, 0.6‰, 0.8‰ and 0.95‰ (Fig. 7.19), all occurring at the glacial terminations, MIS5/6, MIS3/4 and MIS 1/2. Furthermore, the SCS PF δ^{13}C displays a trend toward more positive values from MIS 6 to MIS 2 and a remarkable fit with precession cycles. When the SCS PF δ^{13}C is compared with δ^{13}C records from the east Pacific (ODP 677), the west Pacific (ODP 806), the north Atlantic (ODP 659), the South Ocean (TN57-6) and the Indian Ocean (ODP 758), the same trend toward more positive values and three δ^{13}C minima are observed in all the δ^{13}C curves (Fig. 7.19; Li and Wang 2006).

The occurrence of PF δ^{13}C minima at the glacial terminations is not specific to the SCS, and the same has been found in the late Quaternary δ^{13}C records from the sub-Antarctic to the Equator (Ninnemann and Charles 1997; Curry and Crowley 1987). The most probable cause is related to the Antarctic. According to the ice core record, the Antarctica became warm before the Arctic at the glacial termination, and the ocean δ^{13}C minima occurred at the beginning of the South Ocean warming. Presumably, the glacial stratification of the South Ocean broke down, and the upwelling of circum-Antarctic deep water resumed, carrying the ^{12}C stored in

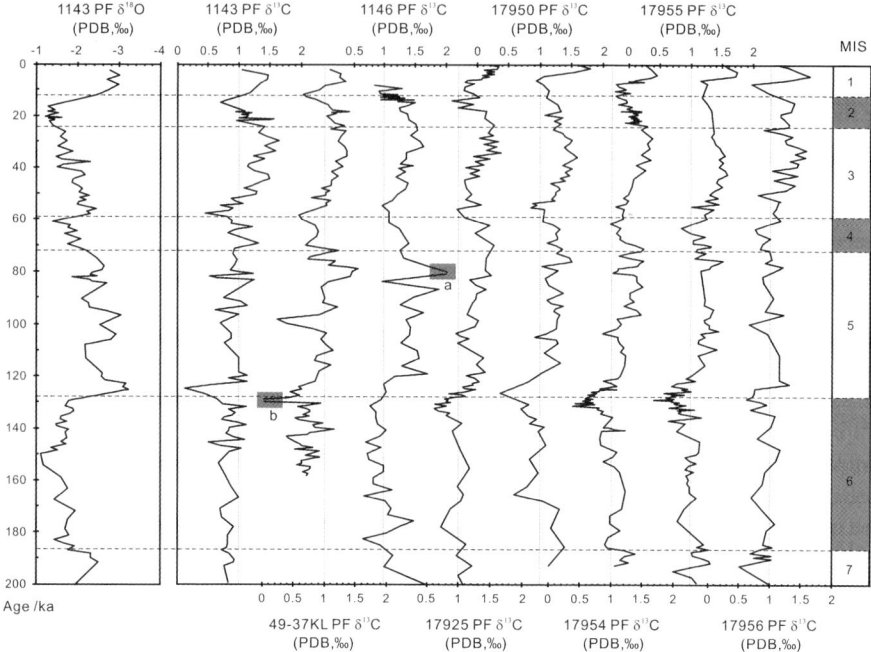

Fig. 7.18 The planktonic foraminiferal $\delta^{13}C$ records at 8 sites are compared together with the planktonic $\delta^{18}O$ curve of Site 1143 from the SCS, with a and b indicating $\delta^{13}C$ outliers (Li and Wang 2006). For site locations see Fig. 1.1 and Appendices 1.1 and 1.2

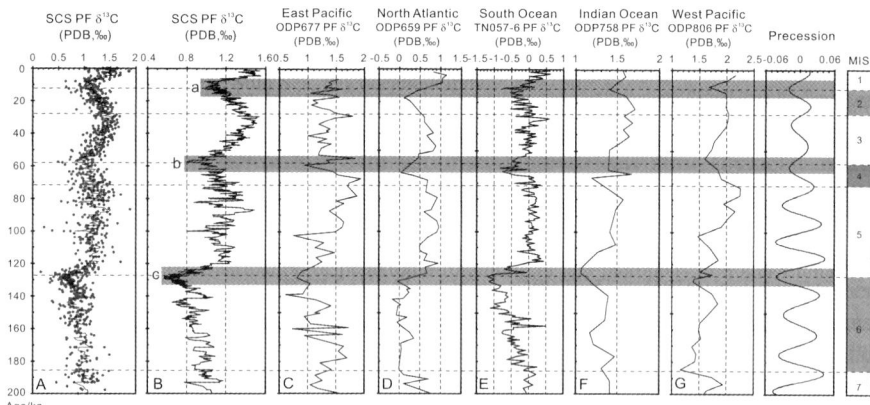

Fig. 7.19 Comparison of the SCS $\delta^{13}C$ curve with those from the global ocean over the last 200 kyr show three SCS PF $\delta^{13}C$ minima (horizontal bars a, b and c) (Li and Wang 2006). (**A** and **B**) SCS planktonic $\delta^{13}C$; (**C**) ODP 677 (Shackleton et al. 1990); (**D**) ODP659 (Tiedemann et al. 1994); (**E**) TN57-6 (Hodell et al. 2000); (**F**) ODP758 (Farrell and Janecek 1991); (**G**) ODP 806 (Schmidt et al. 1993)

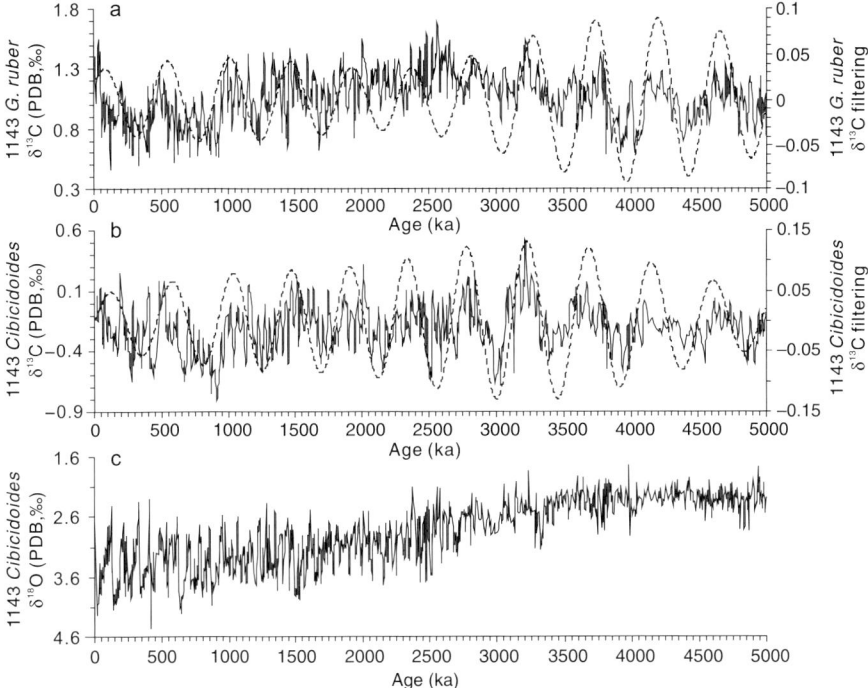

Fig. 7.20 ODP Site 1143 isotope records over the last 5 myr are superimposed by Gaussian filtering (*dash lines*) of the $\delta^{13}C$ records. (**a**) *G. ruber* $\delta^{13}C$ after 5-point Gaussian smoothing; (**b**) *Cibicidoides* $\delta^{13}C$ after 5-point Gaussian smoothing; (**c**) *Cibicidoides* $\delta^{18}O$. Gaussian filtering was performed with a central frequency of $0.0022\,kyr^{-1}$ and bandwidth of $0.00022\,kyr^{-1}$ (Wang P. et al. 2004)

the deep water upwards. The released ^{12}C spread and gave rise to the $\delta^{13}C$ minima in the Indo-Pacific and the south Atlantic (Spero and Lea 2002).

As to the trend toward more positive $\delta^{13}C$ values from MIS 6 to MIS 2, the explanation is related to the $\delta^{13}C$'s coherence with the precession cyclicity (Fig. 7.20). When this trend was first noticed twenty years ago in the north Atlantic and in the equatorial Pacific, Keigwin and Boyle (1985) ascribed it to a reduction in the amplitude of precessional cycles modulated by long eccentricity. Because the $\delta^{13}C$ change of the global ocean is largely controlled by tropical process (e.g., weathering) which depends heavily on precession forcing, therefore, the $\delta^{13}C$ long-term trend reflects the eccentricity control of the oceanic carbon reservoir, as will be discussed in the next section.

$\delta^{13}C$ Long-Eccentricity Cycles

One of the major contributions from the ODP Leg 184 to the SCS is the 5-myr record at Site 1143 in the Nansha coral reef area or "Spartly Islands". The upper section of the 191 m core yielded a 5-myr sequence of both planktonic and benthic

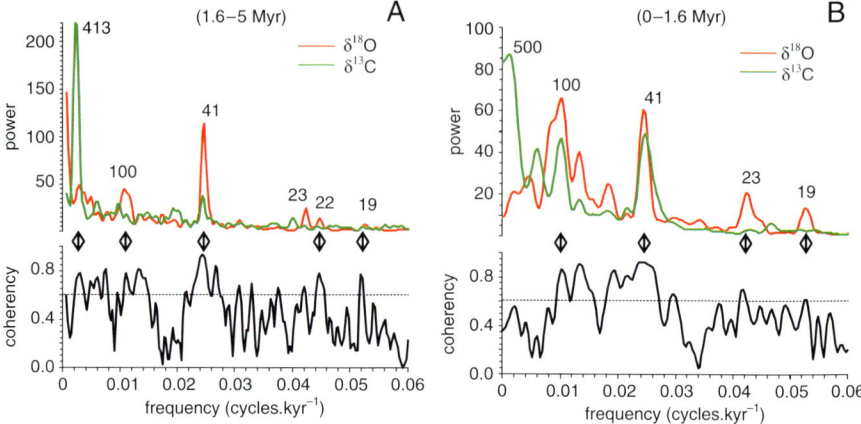

Fig. 7.21 Cross spectrums of $\delta^{18}O$ and $\delta^{13}C$ records at ODP 1143 are shown for two time intervals: (**A**) Pliocene, 1.6–5.0Ma, and (**B**) Quaternary, 0–1.6 Ma

isotope records with an average resolution of 2.5 kyr (Tian et al. 2002) (Fig. 7.20). This is the first high-resolution 5-myr long isotopic record in the Western Pacific, and it is unique in the global ocean to have both benthic and planktonic data. A distinctive feature of the $\delta^{13}C$ curves is the occurrence of long-period variations that are superimposed on the orbital periods at 10^5 time scales (Fig. 7.20) (Wang et al. 2003b; 2004). As seen from a visual inspection and from low-pass filtering in Fig. 7.20, both $\delta^{13}C$ time series display 400-kyr cycles in the Pliocene part, becoming somehow irregular only after 1.6 Ma, with two 500-kyr cycles during the last myr. This was confirmed by spectral analysis showing a significant peak with a period of $\sim$500 kyr in both the benthic and planktonic $\delta^{13}C$ records for the Quaternary, but with a period of $\sim$400 kyr in the Pliocene (Fig. 7.21).

To show the global nature of the above described long-term cycles of $\delta^{13}C$ maxima, available long-term $\delta^{13}C$ sequences from 5 widely distributed sites from different oceanic localities are compared (Fig. 7.22; Table 7.3). Noticeable is the close correspondence of the 400-kyr cycles with the long eccentricity over the heavier $\delta^{13}C$ values, labeled as $\delta^{13}Cmax$. These $\delta^{13}Cmax$ occurred at eccentricity minima, except for the last 1.6 myr when the cycles become longer (Fig. 7.22). A total of 12 $\delta^{13}Cmax$ events have been recognized within the last 5 myr, including four in the Pleistocene. In a descending order, the Pleistocene $\delta^{13}Cmax$ events include $\delta^{13}Cmax$-I at $\sim$30–70 kyr [MIS 3–4], $\delta^{13}Cmax$-II at 470–530 kyr [MIS 13], $\delta^{13}Cmax$-III at 950–1050 kyr [MIS 25–27] and $\delta^{13}Cmax$-IV at 1550–1650 kyr [MIS 53–57] (Wang P. et al. 2003c).

The 400-kyr long eccentricity cycles have been widely reported from deposits at least since the early Mesozoic (e.g., Olsen 1986). Based on continuous Paleogene isotopic records from ODP sites, the long eccentricity cycles in $\delta^{13}C$ were reported from the late Paleocene-early Eocene (Cramer et al. 2003), the Oligocene of the equatorial Pacific (Wade and Pälike 2004), the Oligocene/Miocene sections in Ceara

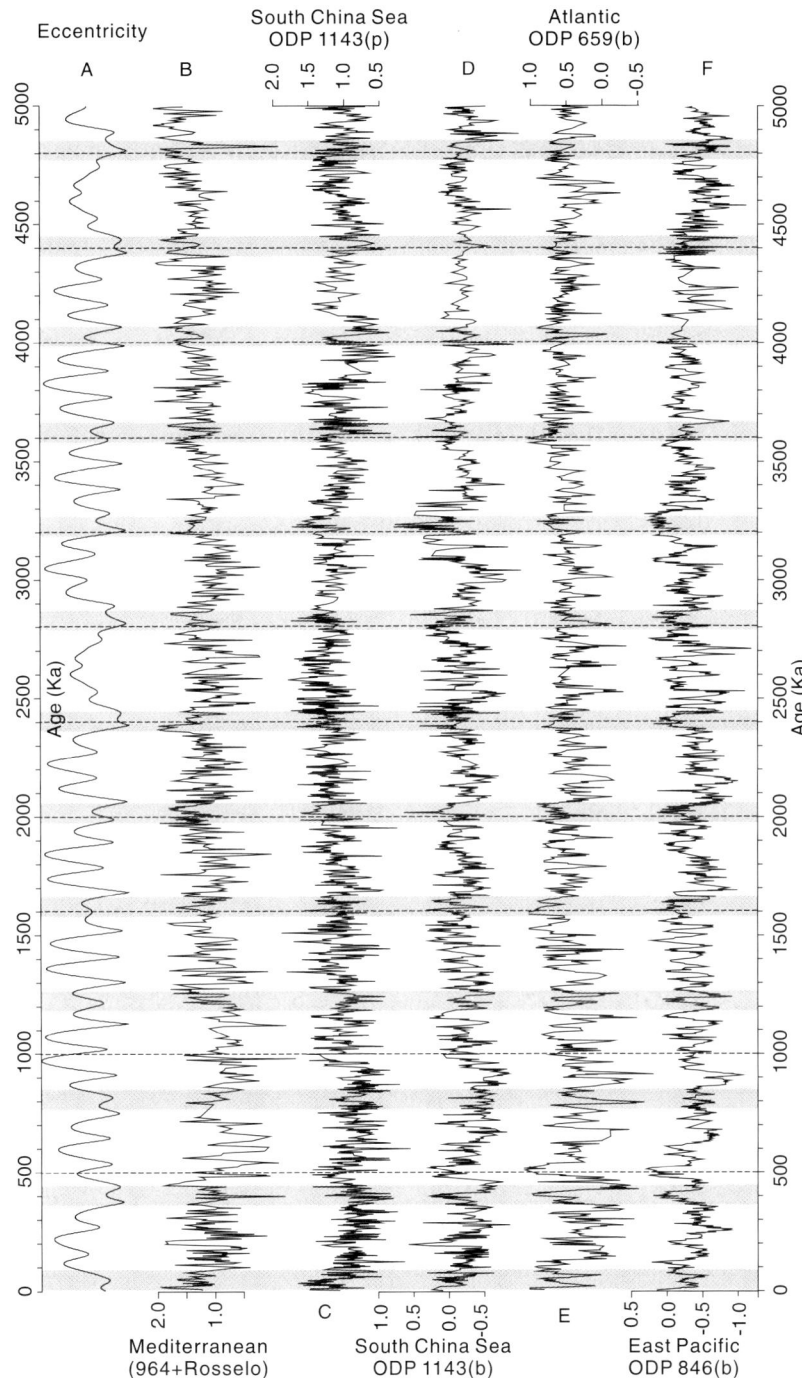

Fig. 7.22 Comparison of 5-myr time series of $\delta^{13}C$ records from various oceans reveals $\delta^{13}Cmax$ events (*vertical dash lines*) accompanying eccentricity minima (*grey bars*). (A) orbital eccentricity (Laskar 1990); (B) Mediterranean planktonic $\delta^{13}C$ record (Rossello composite and ODP 964: Lourens 1994; Howell et al. 1998); (C) SCS planktonic $\delta^{13}C$ record and (D) benthic $\delta^{13}C$ record (Tian et al. 2002; 2004d); (E) North Atlantic (ODP 659) benthic $\delta^{13}C$ record (Tiedemann et al. 1994); (F) Eastern Pacific (ODP 846) benthic $\delta^{13}C$ record (Shackleton et al. 1995)

Table 7.3 Site localities of the 5-myr δ^{13}C time series shown in Fig. 7.22 are listed together with sedimentation rate and sampling time resolution data

Ocean	Site	Location	Water depth (m)	Time span (Ma)	Sedi. rate (cm/kyr)	Foram type	Resolution (kyr)	Reference
SCS	1143	9°22′N 113°17′E	2772	0–5	3.9	P	2.6	Tian et al. 2002; Wang et al. 2004
E Pacific	846	3°06′S 90°49′W	3296	0–6	4.2	B	2.8 2.5	Shackleton et al. 1995
	849	0°11′N 110°31′W	3851	0–5	2.8	B	~4	Mix et al. 1995
N Atlantic	659	18°05′N 21°02′W	3070	0–5	2.9	B	~4	Tiedemann et al. 1994
Mediterranean	Rosello section 964	37°17′N 13°30′E 36°16′N 17°45′E	3657	1.2–5.3 0–3.9		P P		Lourens 1994; Lourens et al. 1996 Howell et al. 1998

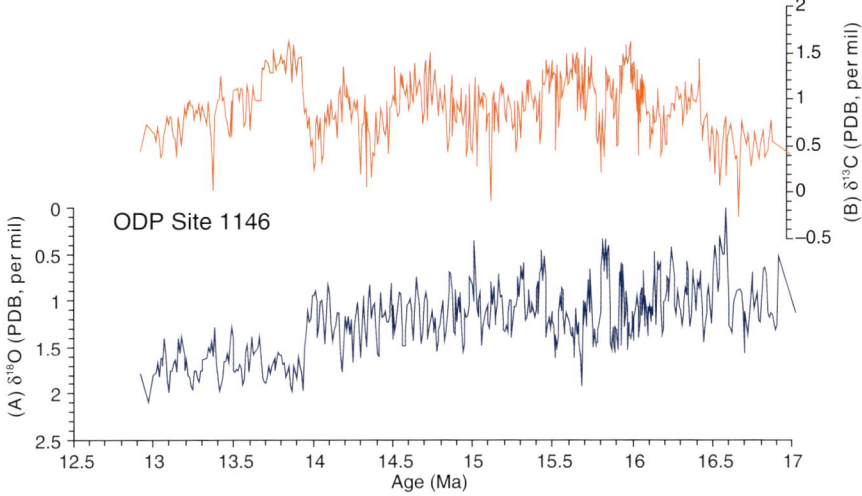

Fig. 7.23 Middle Miocene benthic foraminiferal $\delta^{18}O$ and $\delta^{13}C$ records from ODP Site 1146, northern SCS, co-vary over the long eccentricity periodicity (based on Holbourn et al. 2004)

Rise (Paul et al. 2000; Zachos et al. 2001b) and in the South Atlantic (Billups et al. 2004). In view of their particular strength, the 400-kyr cycles were likened to have been "heartbeat" of the Oligocene climate system (Pälike et al. 2006).

A remarkable feature of the long eccentricity cycles in Paleogene is the co-variations of $\delta^{18}O$ and $\delta^{13}C$ records over the 100-kyr and 400-kyr eccentricity cycles. This is observed also in the Miocene record at ODP Site 1146, northern SCS. A 4-myr high resolution benthic foraminiferal isotope sequence over the "Monterey excursion" period of the middle Miocene (17–13 Ma) shows clear 400-kyr eccentricity cycles both in $\delta^{13}C$ and $\delta^{18}O$ (Fig. 7.23) (Holbourn et al. 2004; 2005). Similar periodical variations were also found in middle Miocene records from other oceans (e.g., Woodruff and Savin 1991; Flower and Kennett 1993; Holbourn et al. 2005), implying a continuous coupled response of $\delta^{18}O$ and $\delta^{13}C$ to long eccentricity from the Paleogene to the Miocene at the least.

As for the ODP 1143 record from the SCS, the 400-kyr imprints on $\delta^{13}C$ and $\delta^{18}O$ records remain strong in the Pliocene, but disappeared around 1.6 Ma (Figs. 7.24, 7.25, and 7.26). After 1.6 Ma, the $\delta^{13}Cmax$ events no longer correspond to the eccentricity minima and $\delta^{18}O$ enrichment, but occurred at periods of high eccentricity and light $\delta^{18}O$, as $\delta^{13}Cmax$-II at 0.5 Ma and $\delta^{13}Cmax$-III at 1.0 Ma, and this has been found to be a feature common to the global ocean (Wang P. et al. 2003c; 2004). Furthermore, the pacing change in long-term cycles is not only restricted to isotopes, but it is also evident in a number of environmental records such as eolian detritus records in the Arabian Sea (DeMenocal 1995), dust flux in the Atlantic (Tiedemann et al. 1994), coccolithophore productivity (C_{37} alkenones concentration) in the equatorial Pacific (Lawrence et al. 2006), and opal mass accumulation rate in the northwest

Fig. 7.24 Comparison
between benthic (A and B)
and planktonic (C) isotope
records over the last 1.2 myr
from ODP Site 1143 shows
$\delta^{13}C$ maximum events (M,
grey bars) preceding the
Mid-Brunhes event (MBE)
and Mid-Pleistocene
Revolution (MPR) (Wang P.
et al. 2003c)

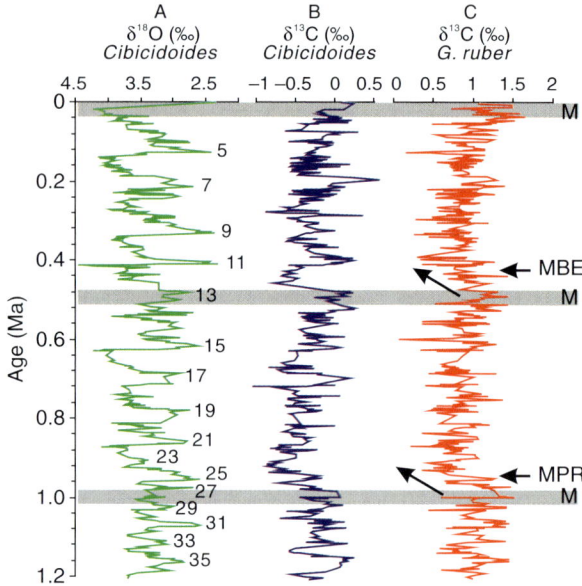

Pacific (Haug et al. 1999). All these proxies display a remarkable correlation with the long eccentricity and $\delta^{13}C$ records.

Comparison between the $\delta^{18}O$ and $\delta^{13}C$ records over the last 1 myr at Site 1143 (Fig. 7.24) revealed a link between the $\delta^{13}Cmax$ events and glacial cycles, as each $\delta^{13}Cmax$ was immediately followed by major changes in glacial cyclicity and the expansion of ice-sheet size expressed as positive $\delta^{18}O$ shift (Wang P. et al. 2003b,c). Thus, $\delta^{13}Cmax$-II at MIS 13 ca. 500 kyr ago was followed by a carbon shift that in turn led to a great expansion of ice-sheet at MIS 12/11 and the "Mid-Brunhes Event" (Jansen et al. 1986). Similarly, $\delta^{13}Cmax$-III about one myr ago and the subsequent carbon shift gave rise to the MIS 22 major glaciation and the "Mid-Pleistocene Revolution" (Berger et al. 1993) when the 40-kyr cyclicity was replaced by the 100-kyr dominance.

From a carbon perspective, therefore, the Quaternary period has passed through three major stages defined by four $\delta^{13}Cmax$ events, and each appears to represent a further step in ice-cap development (Fig. 7.25) (Wang P. et al. 2004). Interestingly, the triple division discussed here coincides with the three "states" in the South Atlantic Quaternary: the early "41 kyr state", the "interim state" and the late "100 kyr state" (Schmieder et al. 2000) (Fig. 7.25). The subdivision based on magnetic susceptibility logs of subtropical South Atlantic deep-water cores is inversely associated with carbonate content but is well correlated with the contemporary boundaries as in the SCS $\delta^{13}C$ record. The triple division is ubiquitous in all Quaternary sequences with sufficient length and resolution, although the mechanism behind it remains unclear (Wang P. et al. 2003b, 2004).

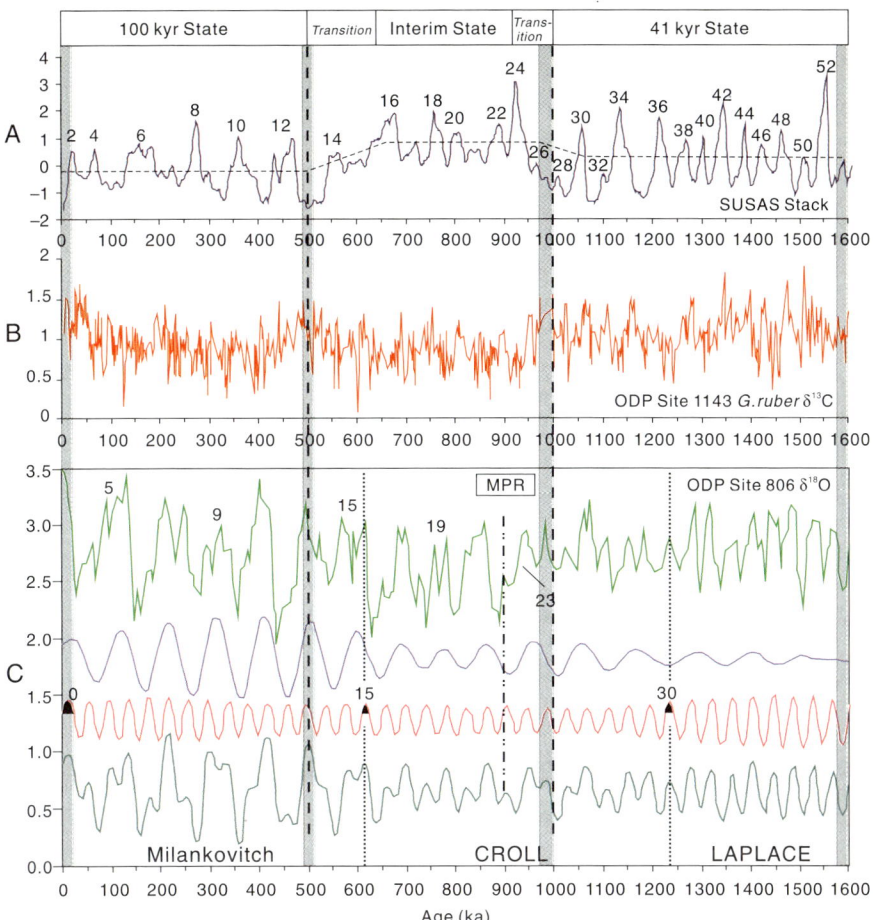

Fig. 7.25 The last 1.6 myr of the Quaternary can be subdivided into major climate stages based on the evolution of oceanic carbon system: (A) triple division into "100 kyr State", "Interim State" and "41 kyr State" based on subtropical South Atlantic susceptibility (SUSAS) stack (Schmieder et al. 2000); (B) triple division based on four δ^{13}Cmax events, represented by planktonic δ^{13}C of ODP Site 1143, SCS (Wang P. et al. 2003b, 2004); (C) triple division into "Milankovitch", "Croll", and "Laplace" chrons based on predominant climate cyclicity represented by benthic δ^{18}O of ODP 806, western tropical Pacific (Berger et al. 1993a)

Long-Term Trend of Carbon Isotopes

The longest sediment section cored during ODP Leg 184 to the SCS was at Site 1148 on the lowermost northern slope (water depth about 3,300 m). Isotopic analysis of the upper 457 m section has established a continuous high-resolution Neogene deep-sea sequence for the western Pacific, representing also the only late Cenozoic isotopic sequence in the global ocean from a single site. The isotope analyses of benthic foraminiferal and black carbon at ODP 1148 have generated a 23-myr

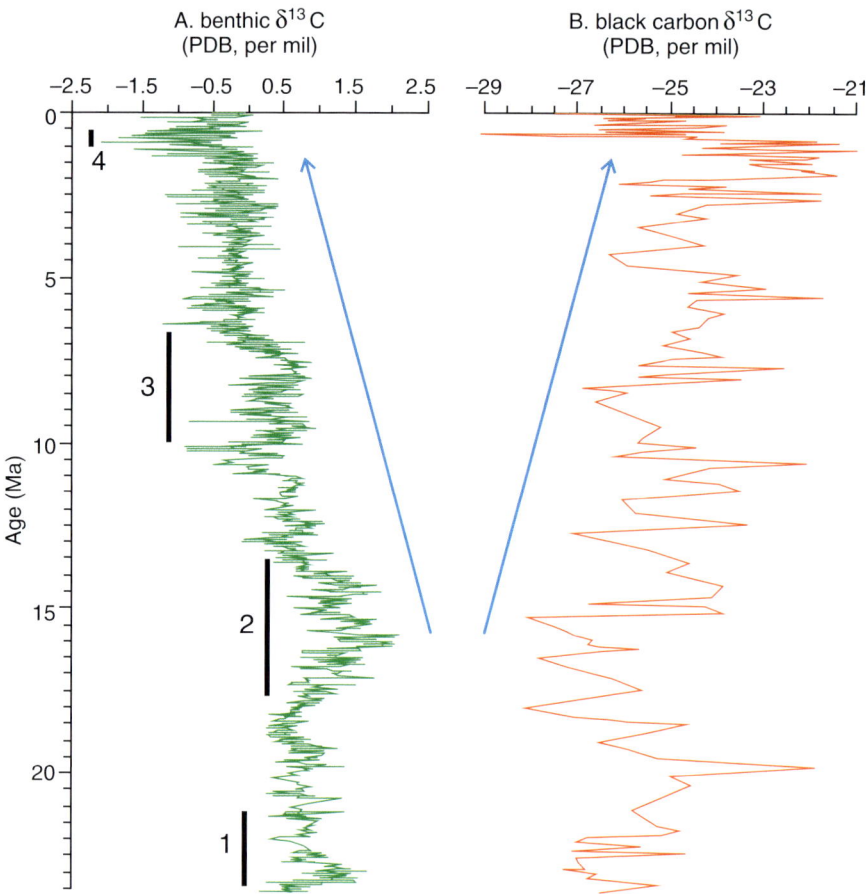

Fig. 7.26 23 myr carbon isotope sequences of (**A**) inorganic carbon and (**B**) organic carbon from ODP Site 1148 show opposite trends especially since ~16 Ma. The inorganic carbon record is represented by the benthic foraminiferal δ^{13}C (Zhao Q. et al. 2001b) and the organic carbon by the black carbon δ^{13}C (Jia et al. 2003). The marked intervals are: (1) the early Miocene δ^{13}C positive excursion, (2) the middle Miocene Monterey event, (3) the late Miocene "Biogenic Bloom" and (4) the "mid-Pleistocene climate transition"

sequence of inorganic and organic carbon isotopes for the Neogene (Fig. 7.26). An outstanding feature is the opposite trend between the two curves since the middle Miocene, with the benthic foraminiferal δ^{13}C depleting by >2‰ since 16 Ma (Zhao Q. et al. 2001b) and the black carbon δ^{13}C increasing by >4‰ over the same interval (Jia et al. 2003).

The declining trend of the ocean carbonate δ^{13}C since the middle Miocene was found more than 20 yr ago (Shackleton 1985) and confirmed by subsequent researches (Zachos et al. 2001). In contrast, the organic carbon δ^{13}C has become increasingly heavier over the last ~50 myr (Hayes et al. 1999) because of decreasing photosynthetic carbon isotope fractionation in the Cenozoic (Derry and France-Lanord 1996). As a consequence, the isotopic fractionation ε_{TOC} has fallen drastically since the middle Miocene (Hayes et al. 1999). This general trend in carbon reservoir change is perfectly recorded in the $\delta^{13}C$ sequences at Site 1148 (Fig. 7.26).

Compared to the compiled global benthic $\delta^{13}C$ curve (of Zachos et al. 2001), the SCS $\delta^{13}C$ values are more depleted in ^{13}C with larger amplitude variations. A number of $\delta^{13}C$ excursions observed in the Site 1148 $\delta^{13}C$ sequences are well correlated with those in other oceans, such as the early Miocene positive $\delta^{13}C$ excursion, the middle Miocene Monterey event, the late Miocene "Biogenic Bloom" and the "mid-Pleistocene climate transition" (Fig. 7.26).

The earliest Neogene $\delta^{13}C$ positive excursion is known as the carbon isotope maxima at the Oligocene/Miocene boundary (the CM-O/M event) lasting from the end of the Oligocene to $\sim$22.6 Ma (Billups et al. 2002), followed by a decline in $\delta^{13}C$. Despite the incomplete stratigraphic record of the latest Oligocene at Site 1148, the $\delta^{13}C$ decrease until 22.3 Ma is well recognizable. The Monterey carbon excursion (Vincent and Berger 1985) between the early and middle Miocene is outstandingly exhibited in the Site 1148 record and reflects the $\delta^{13}C$ fractionation between the Atlantic and Pacific oceans and the accumulation of diatom-rich deposits in the Pacific rim. As seen from Fig. 7.26B, the benthic $\delta^{13}C$ at Site 1148 increased since 17.8 Ma, reached a maximum at 16.0 Ma, then gradually decreased until 13.2 Ma. The same event is well observed also at the neighboring Site 1146, from which it was followed by a major $\delta^{18}O$ increase (the Mi3 event) after 13.8 Ma (Fig. 7.23) (Holbourn et al. 2004; 2005). A similar relationship between carbon cycling and ice sheet development existed for the earliest Miocene: the positive $\delta^{13}C$ excursion event was also followed by ice-sheet expansion as expressed by $\delta^{18}O$ increases at the Mi1a event (Zhao Q. et al. 2001b).

The $\delta^{13}C$ negative excursion between 10.2 and 9.6 Ma was recorded in the benthic $\delta^{13}C$ curve, 1.2 ‰ lighter than the Miocene average, but not in the planktonic $\delta^{13}C$ record. This $\delta^{13}C$ excursion has been related to a significant narrowing or temporary closure of the Central American passageway at 10–9.5 Ma and the development of the Antarctic Bottom Water, leading to "carbonate crush" in the equatorial Pacific (Roth et al. 2000). The subsequent increase in $\delta^{13}C$ corresponds to the "Biogenic Bloom" with high carbonate preservation (Fig. 7.26). The Messinian carbon shift at the end of Miocene, or the Chron 6 carbon shift (Vincent and Berger 1985), is a widely reported global event with the well known Mediterranean desiccation event as its follow-up. This is recorded at Site 1148 as a large amplitude negative excursion in benthic and planktonic $\delta^{13}C$ between 6.9 and 6.2 Ma (Fig. 7.26).

Along with the global ocean signals, there are also regional or local events in the SCS $\delta^{13}C$ record. For example, a drastic negative excursion of benthic $\delta^{13}C$ occurred at Site 1148 in the middle Pleistocene, between 1.0 and 0.4 Ma ("4" in Fig. 7.26), supposedly recording bottom water changes in the SCS due to the rise of the Bashi Strait, the sill separating the SCS from the Pacific (Wang P. et al. 2003a).

Summary

Given the fragmentary data available, it would be premature to suggest a history of productivity or biogeochemical evolution of the SCS. Nevertheless, the accumulated records have enabled us to find out some specifics in biogeochemistry of the SCS from the geological aspect. In general, the SCS records are largely common to the low-latitude Pacific region or to the global ocean. Long-term events such as the "Biogenic Bloom" from middle Miocene to early Pliocene and the enhanced productivity of the last 3 myr are clearly seen from the SCS records. The long-term carbon reservoir changes observed in the SCS are representative of the global ocean. Those include the opposite trend of organic and inorganic carbon isotope over the last 23 myr, and the long-eccentricity cycles of $\delta^{13}C$ and their Pleistocene obscuring.

The SCS is distinguished from other low-latitude oceans in biogeochemistry because of the prevailing monsoon climate and the enclosed basin. Seasonal variations are not typical to tropical oceans, but in the SCS the monsoon climate not only leads to remarkable seasonal variations in productivity, but also to its glacial/interglacial contrast because the enhanced winter monsoon in glacial times raises productivity. Since the winter monsoon prevails over the northern but not southern SCS, and the water exchanges with the Pacific takes place in the NE SCS, a clear north-south contrast has run through its history of productivity variations although the mechanism responsible for the difference is not yet clear. Much more paleo-records and observations of the modern processes are needed to elucidate the biogeochemistry and carbon reservoir history for the SCS.

Acknowledgments This work was supported by the Ministry of Science and Technology of China (NKBRSF Grant 2007CB815904; NKBRSF Grant 2007CB815902), the National Natural Science Foundation of China (Grants 40776029, 40776028, 40476027 and 40730844) and the Ministry of Education of China (FANEDD Grant No. 2005036 to Tian).

References

Billups K., Channell J. and Zachos J. 2002. Late Oligocene to early Miocene geochronology and paleoceanography from the subantarctic South Atlantic. Paleoceanography 17: 1, doi: 10.1029/2000PA000568.

Billups K., Pälike H., Channell J.E.T., Zachos J.C. and Shackleton N.J. 2004. Astronomic calibration of the late Oligocene through early Miocene geomagnetic polarity time scale. Earth Planet. Sci. Lett. 224: 33–44.

Berger A., Loutre M.F. and Laskar J. 1992. Stability of the astronomical frequencies over the Earth's history for paleoclimate studies. Science 255: 560–566.

Berger W.H., Bickert T., Jansen E., Wefer G. and Yasuda M. 1993a. The central mystery of the Quaternary Ice Age. Oceanus 36: 53–56.

Boyd P.W. and Newton P.P. 1999. Does planktonic community structure determine downward particulate organic carbon flux in different oceanic provinces? Deep-Sea Res. I 46: 63–91.

Boyd, P.W., Jickells, T., Law, C.S., Blain, S., Boyle, E.A., Buesseler, K.O., Coale, K.H., Cullen, J.J., de Baar, H.J.W., Follows, M., Harvey, M., Lancelot, C., Levasseur, M., Owens, N.P.J., Pollard, R., Rivkin, R.B., Sarmiento, J., Schoemann, V., Smetacek, V., Takeda, S., Tsuda, A., Turner, S., and Watson, A.J. 2007. Mesoscale iron enrichment experiments 1993–2005: Synthesis and future directions. Science 315: 612–617.

Brzezinski M.A., Jones J.L., Bidle K.D. and Azam F. 2003. The balance between silica production and silica dissolution in the sea: Insights from Monterey Bay, California, applied to the global data set. Limnol. Oceanogr. 48: 1846–1854.

Buesseler K.O., Lamborg C.H., Boyd P.W., Lam P.J., Trull T.W., Bidigare R.R., Bishop J.K.B., Casciotti K.L., Dehairs F., Elskens M., Honda M., Karl D.M., Siegel D.A., Silver M.W., Steinberg D.K., Valdes J., Van Mooy B. and Wilson S. 2007. Revisiting carbon flux through the ocean's twilight zone. Science 316: 567–570.

Cai, W.-J. and Dai, M. 2004. A Comment on "Enhanced open ocean storage of CO2 from shelf sea pumping." Science 306: 1477c.

Chen C.T., Wang C.H., Soong K.Y. and Wang B.J. 2001a. Water temperature records from corals near the nuclear power plant in southern Taiwan. Sci. China (D) 44(4): 356–362.

Chen C.T.A., Wang S.L., Wang B.J. and Pai S.C. 2001b. Nutrient budgets for the South China Sea basin. Mar. Chem. 75: 281–300.

Chen C.T.A., Hou W.P., Gamo T. and Wang S.L. 2006a. Carbonate-related parameters of subsurface waters in the West Philippine, South China and Sulu Seas. Mar. Chem. 99: 151–161.

Chen C.C., Shiah F.K., Chung S.W. and Liu K.K. 2006b. Winter phytoplankton blooms in the shallow mixed layer of the South China Sea enhanced by upwelling. J. Mar. Syst. 59: 97–110.

Chen C.T.A., Wang S.L., Chou W.C. and Sheu D.D. 2006c. Carbonate chemistry and projected future changes in pH and $CaCO_3$ saturation state of the South China Sea. Mar. Chem. 101: 277–305.

Chen J. 2005. Biogeochemistry of Settling Particles in the South China Sea and Its Significance on Paleo-environment Studies. PhD thesis, Tongji Univ., Shanghai, 136pp.

Chen J., Xu L., Zheng L., Wong H.K. and Jennerjahn T. 1998. Organic geochemical characteristics of time-series settling particulate matter in the northern South China Sea and their implications. Chinese J. Geochem. 17: 275–283.

Chen J., Wiesner M.G. and Wong H.K. 1999.Vertical changes of POC flux and indicators of early diagenesis in the South China Sea. Sci. China (D) 29(2):120–128.

Chen Y.L. 2005. Spatial and seasonal variations of nitrate-based new production and primary production in the South China Sea. Deep-Sea Res. I 52: 319–340.

Chen Y.L., Chen H.Y. and Chung C.W. 2007a. Seasonal variability of coccolithophore abundance and assemblage in the northern South China Sea. Deep-Sea Res. II 54: 1617–1633.

Chen Y.L., Chen H.Y., Lin I.I., Lee M.A. and Chang J. 2007b. Effects of cold eddy on phytoplankton production and assemblages in Luzon Strait bordering the South China Sea. J. Oceanogr. 63: 671–683.

Cheng X., Huang B., Jian Z., Zhao Q., Tian J. and Li J. 2005. Foraminiferal isotopic evidence for monsoonal activity in the South China Sea: a present-LGM comparison. Mar. Micropaleontol. 54: 125–139.

Chou W.C., Sheu D.D., Lee B.S., Tseng C.M., Chen C.T.A., Wang S.L. and Wong G.T.F. 2007. Depth distributions of alkalinity, TCO_2 and $\delta^{13}C$ TCO_2 at SEATS time-series site in the northern South China Sea. Deep-Sea Res. II 54: 1469–1485.

Cramer B.S., Wright J.D., Kent D.V. and Aubry M.P. 2003. Orbital climate forcing of $\delta^{13}C$ excursions in the late Paleocene-early Eocene (chrons C24n-C25n). Paleoceanography 18: doi: 10.1029/2003PA000909.

Curry W.B. and Crowley T.J. 1987. The $\delta^{13}C$ of equatorial Atlantic surface waters: Implications for ice age pCO_2 levels. Paleoceanography 2: 489–517.

de Baar H.J.W., Boyd P.W., Coale K.H., Landry M.R., Tsuda A., Assmy P., Bakker D.C.E., Bozec Y., Barber R.T., Brzezinski M.A., Buesseler K.O., Boye M., Croot P.L., Gervais F., Gorbunov M.Y., Harrison P.J., Hiscock W.T., Laan P., Lancelot C., Law C.S., Levasseur M., Marchetti A., Millero F.J., Nishioka J., Nojiri Y., van Oijen T., Riebesell U., Rijkenberg M.J.A., Saito H., Takeda S., Timmermans K.R., Veldhuis M.J.W., Waite A.M. and Wong C.S. 2005. Synthesis of iron fertilization experiments: From the iron age in the age of enlightenment. J. Geophysical Res.-Oceans 110: C09S16, doi:10.1029/2004JC002601.

de Garidel-Thoron T., Beaufort L., Linsley B.K. and Dannenmann S. 2001. Millennial-scale dynamics of the East-Asian winter monsoon during the last 200,000 years. Paleoceanography 16: 491–502.

DeMenocal P.B. 1995. Plio-Pleistocene African climate. Science 270: 53–59.

Derry L.A. and France-Lanord C. 1996. Neogene growth of the sedimentary organic carbon reservoir. Paleoceanography 11: 267–275.

Diester-Haass L., Billups K. and Emeis K.C. 2006. Late Miocene carbon isotope records and marine biological productivity: Was there a (dusty) link? Paleoceanography 21: PA4216, doi:10.1029/2006PA001267.

Ding Z.L., Sun J.M., Yang S.L. and Liu T.S. 1998a. Preliminary magnetostratigraphy of a thick eolian red clay-loess sequence at Lingtai, the Chinese Loess Plateau. Geophys. Res. Lett. 25: 1225–1228.

Ding Z.L., Rutter N.W., Liu T.S., Sun J.M., Ren J.Z., Rokosh D. and Xiong S.F. 1998b. Correlation of Dansgaard-Oeschger cycles between Greenland ice and Chinese loess. Paleoclimates 2: 281–291.

Duce, R.A. 1991. The atmospheric input of trace species to the world ocean. Global Biogeochemical Cycles 5: 193–259.

Eppley R. and Peterson B.J. 1979. Particulate organic matter flux and planktonic new production in the deep ocean. Nature 282: 677–680.

Fang D., Jian Z. and Wang P. 2000. The paleoproductivity recorded in the southern Nansha sea area for about 30 ka. Chinese Sci. Bull. 45: 227–230.

Farrell J.W. and Janecek T.R. 1991. Late Neogene paleoceanography and paleoclimatology of the northern Indian Ocean (Site 758). Proc. ODP Sci. Results 121: 297–355.

Feely R.A., Sabine C.L., Lee K., Millero F.L., Lamb M.F., Greeley D., Bullister J.L., Key R.M., Peng T.H., Kozyr A., Ono T. and Wong C.S. 2002. In situ calcium carbonate dissolution in the Pacific Ocean. Global Biogeochem. Cycles 16(4): 1144, doi:10.1029/2002GB001866.

Flower B.P. and Kennett J.P. 1993. Middle Miocene ocean-climate transition: high resolution oxygen and carbon isotopic records from DSDP Site 588A, southwest Pacific. Paleoceanography 8: 811–843.

Gattuso G.P., Frankignoulle M., Bourge I., Romaine S. and Buddemeier R.W. 1998. Effect of calcium carbonate saturation of seawater on coral calcification. Global Planet. Change 18: 37–46.

Gong G.C., Liu K.K., Liu C.T. and Pai S.C. 1992. The chemical hydrography of the South China Sea west of Luzon and a comparison with the West Philippine Sea. Terr. Atmos. Ocean. Sci. (TAO) Taipei 13: 587–602.

Haake B., Ittekkot V., Rixen T., Ramaswamy V., Nair R.R. and Curry W.B. 1993. Seasonality and interannual variability of particle fluxes to the deep Arabian Sea. Deep-Sea Res. I 40: 1323–1344.

Haug G.H., Sigman D.M., Tiedemann R., Pedersen T.F. and Sarnthein M. 1999. Onset of permanent stratification in the subarctic Pacific Ocean. Nature 401: 779–782.

Han W. and Ma K. 1988. Carbonate compensation depth, saturation horizon and lysocline in the northeast region of South China Sea. Tropical Oceanol. 7(3): 84–89 (in Chinese).

He J., Zhao M., Li L., Wang H. and Wang P. 2008. Biomarker evidence of relatively stable community structure in the northern South China Sea during the last glacial and Holocene. Terr. Atmos. Ocean. Sci. (TAO) Taipei 19(4): 377–387.

Hayes J.M., Strauss H. and Kaufman A.J. 1999. The abundance of ^{13}C in marine organic matter and isotopic fractionation in the global biogeochemical cycle of carbon during the past 800 Ma. Chem. Geol. 161: 103–125.

Hess S. and Kuhnt W. 2005. Neogene and Quaternary paleoceanographic changes in the southern South China Sea (Site 1143): the benthic foraminiferal record. Mar. Micropaleontol. 54: 63–87.

Higginson M., Maxwell J.R. and Altabet M.A. 2003. Nitrogen isotope and chlorin paleoproductivity records from the Northern South China Sea: remote vs. local forcing of millennial- and orbital-scale variability. Mar. Geol. 201: 223–250.

Hillaire-Marcel C. and de Vernal A. 2007. Methods in Late Cenozoic Paleoceanography: Introduction. In: Hillaire-Marcel C. and de Vernal A. (eds.), Proxies in Late Ceanozoic Paleoceanography. Elsevier, Amsterdam, pp. 1–15.

Hodell, A. D., Charles, D. C. and Ninnemann, S. U. 2000. Comparison of interglacial stages in the South Atlantic sector of the southern ocean for the past 450 kyr: implifications for Marine Isotope Stage (MIS) 11. Global and Planetary Change 24: 7–26.

Holbourn A., Kuhnt W. and Schulz M. 2004. Orbitally paced climate variability during the middle Miocene: high resolution benthic stable-isotope records from the tropical western Pacific. In: Clift P.D., Wang P., Hayes D. and Kuhnt W. (eds.), Continent-Ocean Interactions in the East Asian Marginal Seas. AGU Geophys. Monogr. 149, pp. 321–337.

Holbourn A., Kuhnt W., Schulz M. and Erlenkeuser H. 2005. Impacts of orbital forcing and atmospheric carbon dioxide on Miocene ice-sheet expansion. Nature 438: 483–487.

Honda M.C., Imai K., Nojiri Y., Hoshi F., Sugauara T. and Kusakabe M. 2002. The biological pump in the northwestern North Pacific based on fluxes and major components of particulate matter obtained by sediment-trap experiments (1997–2000). Deep-Sea Res. II 49: 5595–5625.

Howell M.W., Thunell R.C., Stefano E.D., Sprovieri R., Tappa E.J. and Sakamoto T. 1998. Stable isotope chronology and paleoceanographic history of Sites 963 and 964, Eastern Mediterranean Sea. Proc. ODP Sci. Results 160: 167–180.

Huang L. 1991. A preliminary study on the distribution of photosynthetic pigments and primary production in the sea area of Nansha Islands. In: Chen Q. (ed.), Collected Papers on Marine Biology Study of Nansha Islands and the Adjacent Sea Areas (II). China Ocean Press, Beijing, pp. 34–47 (in Chinese).

Huang L. and Chen C. 1997. Distribution of chlorophyll a and primary productivity of Nansha Islands sea area in winter. In: Chen Q. and Huang L. (eds.), Study on Ecological Processes of Nansha Islands Sea Area (I). China Sci. Press, Beijing, pp. 1–15 (in Chinese).

Huang C.Y., Liew P.M., Zhao M., Chang T.C., Kuo C.M., Chen M.T., Wang C.H. and Zhang L.F. 1997a. Deep sea and lake records of the Southeast Asian paleomonsoons for the last 25 thousands years. Earth Planet. Sci. Lett. 146: 59–72.

Huang C.Y., Wu S.F., Zhao M., Chen M.T., Wang C.H., Tu X. and Yuan P.B. 1997b. Surface ocean and monsoon climate variability in the South China Sea since the last glaciation. Mar. Micropaleotol. 32: 71–94.

Hung T. and Tsai C.C.H. 1972. Study on photosynthetic pigments and chemical nutrients in South China Sea. Acta Oceanogr. Taiwanica 2: 83–92.

Hung J.J., Wang S.M. and Chen Y.L. 2007. Biogeochemical controls on distributions and fluxes of dissolved and particulate organic carbon in the northern South China Sea. Deep-Sea Res. II, 54: 1486–1503

Jansen J.H.F., Kuijpers A. and Troelstra S.R. 1986. A Mid-Brunhes climatic event: Long-term changes in global atmosphere and ocean circulation. Science 232: 619–622.

Jennerjahn T.C., Liebezeit G. and Kempe S. 1992. Particle flux in the northern South China Sea. In: Jin X., Kudrass H.R. and Pautot G. (eds.), Marine Geology and Geophysics of South China Sea. China Ocean Press, Beijing, pp. 228–235.

Jia G., Jian Z., Peng P., Wang P. and Fu J. 2000. Biogenic silica records in core 17962 from southern South China Sea and their relation to paleoceanographical events. Geochimica 29(3): 293–296 (in Chinese).

Jia G., Peng P., Zhao Q. and Jian Z. 2003. Changes in terrestrial ecosystem since 30 Ma in East Asia: Stable isotope evidence from black carbon in the South China Sea. Geology 31: 1093–1096.

Jian Z., Wang L. and Kienast K. 1999a. Late Quaternary surface paleoproductivity and variations of the East Asian Monsoon in the South China Sea. Quat. Sci. 1: 32–40 (in Chinese).

Jian Z., Wang L., Kienast M., Sarnthein M., Kuhnt W., Lin H. and Wang P. 1999b. Benthic foraminiferal paleoceanography of the South China Sea over the last 40,000 years. Mar. Geol. 156: 159–186.

Jian Z., Cheng X., Zhao Q., Wang J. and Wang P. 2001a. Oxygen isotope stratigraphy and events in the northern South China Sea during the last 6 million years. Sci. China (D) 44(10): 952–960.

Jian Z., Huang B., Lin H. and Kuhnt W. 2001b. Late Quaternary upwelling intensity and East Asian monsoon forcing in the South China Sea. Quat. Res. 55: 363–370.

Jian Z., Zhao Q., Cheng X., Wang J., Wang P. and Su X. 2003. Pliocene-Pleistocene stable isotope and paleoceanographic changes in the northern South China Sea. Palaeogeogr. Palaeoclimatol. Palaeoecol. 193: 425–442.

Keigwin L.D. and Boyle A.E. 1985. Carbon isotopes in deep-sea benthic foraminifera: Precession and changes in low-latitude biomass. In: Sundquist E.T. and Broecker W. S. (eds.), The Carbon Cycle and Atmopspheric CO_2: Natural variations Archean to Present. AGU Geophys. Monogr. vol. 32, pp. 319–389.

Kienast M. 2000. Unchanged nitrogen isotopic composition of organic matter in the South China Sea during the last climatic cycle: Global implications. Paleoceanography 15: 244–253.

Kienast M., Calvert S.E., Pelejero C. and Grimalt J.O. 2001a. A critical review of marine sedimentary $^{13}C_{org}$-pCO_2 estimates: New palaeorecords form the South China Sea and a revisit of other low-latitude $^{13}C_{org}$-pCO_2 records. Global Biogeochem. Cycles 15: 113–127.

Kienast M., Steinke S., Stattegger K. and Calvert S.E. 2001b. Synchronous tropical South China Sea SST change and Greenland warming during deglaciation. Science 291: 2132–2134.

Kuhnt W., Hess S. and Jian Z. 1999. Quantitative composition of benthic foraminiferal assemblages as a proxy indicator for organic carbon flux rates in the South China Sea. Mar. Geol. 156: 123–157.

Laskar J. 1990. The chaotic motion of the solar system: A numerical estimate of the size of the chaotic zones. Icarus 88: 266–291.

Lawrence K.T., Liu Z. and Herbert T.D. 2006. Evolution of the eastern tropical Pacific through Plio-Pleistocene glaciation. Science 312: 79–83.

Li J., Wang R. and Li B. 2002a. Variations of opal accumulation rates and paleoproductivity over the past 12 Ma at ODP Site 1143, southern South China Sea. Chinese Sci. Bull. 47: 596–598.

Li J., Jin X. and Gao J. 2002b. Morpho-tectonic study on late-stage spreading of the Eastern Subbasin of South China Sea. Sci. China (D) 45: 978–989.

Li J. and Wang P. 2006. A 200-ka carbon isotope record from the South China Sea. Chinese Sci. Bull. 51(14): 1780–1784.

Lin A.T., Watts A.B. and Hesselbo S.P. 2003. Cenozoic stratigraphy and subsidence history of the South China Sea margin in the Chinese Taipei region. Basin Res. 15: 453–479.

Lin I.I., Liu W.T., Wu C.-C., Wong G.T.F., Hu C., Chen Z., Liang W.-D., Yang Y. and Liu K.K. 2003. New evidence for enhanced ocean primary production triggered by tropical cyclone. Geophys. Res. Lett. 30: 1718, doi:10.1029/2003GL017141.

Lin H., Lai C., Ting H., Wang L., Sarnthein M. and Huang J. 1999a. Late Pleistocene nutrients and sea surface productivity in the South China Sea: a record of teleconnections with northern hemisphere events. Mar. Geol. 156: 197–210.

Lin H.L., Wang L.W., Chung-Ho Wang C.H. and Gong G.C. 1999b. Vertical distribution of $\delta^{13}C$ of dissolved inorganic carbon in the northeastern South China Sea. Deep-Sea Res. I 46: 757–775.

Lin H.-L., Wang W.-C. and Hung G.-W. 2004. Seasonal variations of planktonic foraminiferal isotopic composition from sediment traps in the South China Sea. Mar. Micropaleontol. 53: 447–460.

Lin I.-I., Chen J.-P., Wong G.T.F., Huang C.-W. and Lien C.-C. 2007. Aerosol input to the South China Sea: results from the Moderate resolution imaging spectro-radiometer, the quick scatterometer, and the measurements of pollution in the troposphere sensor. Deep-Sea Res. II 54: 1589–1601.

Liu, C. and Cheng, X. 2001. Variations in upper ocean structure for the last 2 Ma of the Nansha area by means of calcareous nannofossils. Sci. China (D) 44: 905–911.

Liu K.-K., Chao S.-Y., Shaw P.-T., Gong G.-C., Chen C.-C. and Tang T.Y. 2002. Monsoon-forced chlorophyll distribution and primary production in the South China Sea: observations and a numerical study. Deep-Sea Res. I 49: 1387–1412.

Liu K.-K., Chen Y.-J., Tseng C.-M., Lin I.-I., Liu H.-B. and Snidvongs A. 2007. The significance of phytoplankton photo-adaptation and benthic-pelagic coupling to primary production
in the South China Sea: observations and numerical investigations. Deep-Sea Res. II 54:
1546–1574.

Lourens L.J. 1994. Astronomical forcing of Mediterranean climate during the last 5.3 million years.
PhD thesis, Utrecht University, Netherlands, 247pp.

Lourens, L.J., Antonarakou, A., Hilgen, F.J., Van Hoof, A., Vergnaud-Grazzini, C. and Zachariasse,
W.J. 1996. Evaluation of the Plio-Pleistocene astronomical timescale. Paleoceanography 11:
391–413.

Mackensen A. and Bickert T. 1999. Stable carbon isotopes in benthic foraminifera: Proxies for
deep and bottom water circulation and new production. In: Fischer G. and Wefer G. (eds.), Use
of Proxies in Paleoceanography: Examples from the South Atlantic. Springer, The Netherlands,
pp. 229–254.

Mix A.C., Le J. and Shackleton N.J. 1995. Benthic foraminiferal stable isotope stratigraphy of Site
846: 0–1.8 Ma. Proc. ODP Sci. Results 138: 839–854.

Nelson D.M., Tregure P., Brezinski M.A., Leynaert, A. and Quéguiner, B. 1995. Production
and dissolution of biogenic silica in the ocean: regional data and relationship to biogenic
sedimentation. Global Biogeochem. Cycles 9: 359–372.

Ning X., Chai F., Xue H., Chai Y., Liu C. and Shi J. 2004. Physical-biological oceanographic coupling influencing phytoplankton and primary production in the South China Sea. J. Geophys.
Res. 109: C10005, doi:10.1029/2004JC002365.

Ninnemann U.S. and Charles C.D. 1997. Regional differences in Quaternary Subantarctic
nutrient cycling: Link to intermediate and deep water ventilation. Paleoceanography 12:
560–567.

Olsen P.E. 1986. A 40 million-year lake record of early Mesozoic climatic forcing. Science 234:
842–848.

Paul H.A., Zachos J.C., Flower B.P. and Tripati A. 2000. Orbitally induced climate and geochemical variability across the Oligocene/Miocene boundary. Paleoceanography 15: 71–485.

Pälike H., Norris R.D., Herrle J.O., Wilson P.A., Coxall H.K., Lear C.H., Shackleton N.J., Tripati A.K. and Wade B.S. 2006. The heartbeat of the Oligocene climate system. Science 314:
1894–1898.

Pelejero C., Grimalt J.O., Heilig S., Kienast M. and Wang L. 1999a. High-resolution $U_{37}^{K'}$ temperature reconstructions in the South China Sea over the past 220 kyr. Paleoceanography 14:
224–231.

Pelejero C., Grimalt J.O., Sarnthein M., Wang L. and Flores J.A. 1999b. Molecular biomarker
record of sea surface temperature and climatic change in the South China Sea during the last
140,000 years. Mar. Geol. 156: 109–121.

Pelejero C., Kienast M., Wang L. and Grimalt J.O. 1999c. The flooding of Sundaland during the
last deglaciation: imprints in hemipelagic sediments from the southern South China Sea. Earth
Planet. Sci. Lett. 171: 661–671.

Porter S.C. and An Z. 1995. Correlation between climate events in the North Atlantic and China
during the last glaciation. Nature 375: 305–308.

Ragueneau O., Treguer P., Leynaert A., Anderson R.F., Brzezinski M.A., DeMaster D.J., Dugdale
R.C., Dymond J., Fischer G., François R., Heinze C., Maier-Reimer E., Martin-Jézéquel V.,
Nelson D. M. and Quéguiner B. 2000. A review of the Si cycle in the modern ocean: recent
progress and missing gaps in the application of biogenic opal as a paleoproductivity proxy.
Global Planet. Change 26: 317–365.

Richardson T.L. and Jackson G.A. 2007. Small phytoplankton and carbon export from the surface
ocean. Science 315: 838–840.

Roth J.M., Droxler A.W. and Kameo K. 2000. The Caribbean carbonate crash at the middle to late
Miocene transition: linkage to the establishment of the modern global ocean conveyor. Proc.
ODP Sci. Results 165: 249–273.

Schmidt H., Berger W.H., Bickert T. and Wefer G. 1993. Quaternary carbon isotope record of
pelagic foraminifers: Site 806, Ontong Java Plateau. Proc. ODP Sci. Res. 130: 397–409.

Schmieder F., von Dobeneck T. and Bleil U. 2000. The Mid-Pleistocene climate transition as documented in the deep South Atlantic Ocean: initiation, interimstate and terminal event. Earth Planet. Sci. Lett. 179: 539–549.

Shackleton N.J. 1985. Oceanic carbon isotope constraints on oxygen and carbon dioxide in the Cenozoic atmosphere. AGU Geophys. Monogr. 32: 412–418.

Shackleton, N.J., Berger, A. and Peltier, W.R. 1990. An alternative calibration of the lower Pleistocene timescale based on W. R. ODP site 677. Trans. R. Soc. Edinburgh Earth Sci. 81: 251–261.

Shackleton N.J., Hall M.A. and Pate D. 1995. Pliocene stable isotope stratigraphy of Site 846. Proc. ODP Sci. Results 138: 337–355.

Shaw P.T. 1989. The intrusion of water masses into the sea southwest of Taiwan. J. Geophys. Res. 94: 18213–18226.

Shiau L.J., Yu P.S., Wei K.Y., Yamamoto M., Lee T.Q., Yu E.F., Fang T.H. and Chen M.T. 2008. Sea surface temperature, productivity, and terrestrial flux variations of the southeastern South China Sea over the past 800000 years (IMAGES MD972142). Terr. Atmos. Ocean. Sci. (TAO) Taipei 19(4): 363–376.

Spero H.J. and Lea D.W. 2002. The cause of carbon isotope minimum events on glacial terminations. Science 296: 522–525.

Su J., Xu J., Cai S. and Wang O. 1999. Circulation and eddies of the South China Sea. In: Ding Y. and Li C. (eds.), The Eruption and Evolution of Monsoon and its Interaction with Oceans in the South China Sea. Meteorol. Press, Beijing, pp. 66–72.

Sun X. and Li X. 1999. A pollen record of the last 37 ka in deep sea core 17940 from the northern South China Sea. Mar. Geol. 156: 227–244.

Takahashi M. and Hori T. 1984. Abundance of picophytoplankton in the subsurface chlorophyll maximum layer in subtropical and tropical waters. Mar. Biol. 79: 177–186.

Tian J., Wang P., Cheng X. and Li Q. 2002. Astronomically tuned Plio-Pleistocene benthic $\delta^{18}O$ records from South China Sea and Atlantic-Pacific comparison. Earth Planet. Sci. Lett. 203: 1015–1029.

Tian J., Wang P. and Cheng X. 2004d. Development of the East Asian monsoon and Northern Hemisphere glaciation: Oxygen isotope records from the South China Sea. Quat. Sci. Rev. 23: 2007–2016.

Thomas, H., Bozec, Y., Elkalay, K., and de Baar, H. J. W. 2004. Enhanced Open Ocean Storage of CO2 from Shelf Sea Pumping. Science 304: 1005–1008.

Thunell R., Miao Q., Calvert S., Calvert S. and Pedersen T. 1992. Glacial-Holocene biogenic sedimentation patterns in the South China Sea: productivity variations and surface water pCO_2. Paleoceanography 7: 143–162.

Tiedemann R., Sarnthein M. and Shackleton N.J. 1994. Astronomic timescale for the Pliocene Atlantic $\delta^{18}O$ and dust flux records from Ocean Drilling Program Site 659. Paleoceanography 9: 619–638.

Tseng C.M., Wong G.T.F., Chou W.C., Lee B.S., Sheu D.D. and Liu K.K. 2007. Temporal variations in the carbonate system in the upper layer at the SEATS station. Deep-Sea Res. II 54: 1448–1468

Vincent E. and Berger W.H. 1985. Carbon dioxide and polar cooling in the Miocene: the Monterey hypothesis. Geophys. Monogr. 32: 455–468.

Wade B.S and Pälike H. 2004. Oligocene climate dynamics. Paleoceanography 19: PA4019, doi:10.1029/2004PA001042.

Wang L. and Wang P. 1990. Late Quaternary paleoceano-graphy of the South China Sea: glacial-interglacial contrasts in an enclosed basin. Paleoceanography 5: 77–90.

Wang P., Jian Z., Zhao Q., Li Q., Wang R., Liu Z., Wu G., Shao L., Wang J., Huang B., Fang D., Tian J., Li J., Li X., Wei G., Sun X., Luo Y., Su X., Mao S. and Chen M. 2003a. Evolution of the South China Sea and monsoon history revealed in deep-sea records. Chinese Sci. Bull. 48(23): 2549–2561.

Wang P., Tian J., Cheng X., Liu C. and Xu J. 2003b. Exploring cyclic changes of the ocean carbon reservoir. Chinese Sci. Bull. 48(23): 2536–2548.

Wang P., Tian J., Cheng X., Liu C. and Xu J. 2003c. Carbon reservoir change preceded major ice-sheets expansion at Mid-Brunhes Event. Geology 31: 239–242.

Wang P., Tian J., Cheng X., Liu C. and Xu J. 2004. Major Pleistocene stages in a carbon perspective: The South China Sea record and its global comparison. Paleoceanography 19: doi: 10.1029/2003PA000991.

Wang R. and Li J. 2003. Quaternary high resolution opal record and its paleoproductivity implication at ODP Site 1143, southern South China Sea. Chinese Sci. Bull. 48(4): 363–367.

Wang R., Fang D., Shao L., Chen M., Xia P. and Qi J. 2001. Oligocene biogenetic siliceous deposits on the slope of the northern South China Sea. Sci. China (D) 44(10): 912–918.

Wang R., Clemens S., Huang B. and Chen M. 2003. Late Quaternary paleoceanographic changes in the northern South China Sea (ODP Site 1146): radiolarian evidence. J. Quat. Sci. 18(8): 745–756.

Wang R., Li J. and Li B. 2004. Data report: Late Miocene–Quaternary biogenic opal accumulation at ODP Site 1143, southern South China Sea. In: Prell W.L., Wang P., Blum P., Rea D.K. and Clemens S.C. (eds.), Proc. ODP, Sci. Results 184 [Online].

Wang R., Jian Z., Xiao W., Tian J., Li J., Chen R., Zheng L. and Chen J. 2007. Quaternary biogenic opal records in the South China Sea: linkages to East Asian monsoon, global ice volume and orbital forcing. Sci. China (D) 50(5): 710–724.

Wefer G., Berger W.H., Bijma J. and Fischer G. 1999. Clues to ocean history: a brief overview of proxies. In: Fischer G. and Wefer G. (eds.), Use of Proxies in Paleoceanography: Examples from the South Atlantic. Springer-Verlag, Berlin Heidelberg, pp. 1–68.

Wei G., Liu Y., Li X., Shao L. and Liang X. 2003. Climatic impact on Al, K, Sc and Ti in marine sediments: Evidence from ODP Site 1144, South China Sea. Geochem. J. 37: 593–602.

Werne J.P., Hollander D.J., Lyons T.W. and Peterson L.C. 2000. Climate-induced variations in productivity and planktonic ecosystem structure from the Younger Dryas to Holocene in the Cariaco Basin, Venezuela. Paleoceanography 15: 19–29.

Wong G.T.F., Ku T.L., Mulholland M., Tseng C.M. and Wang D.P. 2007a. The South East Asian Time-series Study (SEATS) and the biogeochemistry of the South China Sea-An overview. Deep-Sea Res. II 54: 1434–1447.

Wong G.T.F., Tseng C.M., Wen L.S. and Chung S.W. 2007b. Nutrient dynamics and nitrate anomaly at the SEATS station. Deep-Sea Res. II 54: 1528–1545.

Woodruff F. and Savin S.M. 1991. Mid-Miocene isotope stratigraphy in the deep sea: high-resolution correlations, paleoclimatic cycles and sediment preservation. Paleoceanography 6: 755–806.

Wu, J.F., Chung, S.W., Wen, L.S., Liu, K.K., Chen, Y.L., Chen, H.Y. and Karl, D.M. 2003. Dissolved inorgonic phosphorus, dissolved iron, and Trichodesmium in the oligotrophic South China Sea. Global Biogeochem. Cycles 17: 1008, doi:10.1029/2002GB001924.

Wu C.R. and Chiang, T.L. 2007. Mesoscale eddies in the northern South China Sea. Deep-Sea Res. II 54: 1575–1588.

Zachos J., Pagani M., Sloan L., Thomas E. and Billups K. 2001a. Trends, rhythms, and aberrations in global climate 65 Ma to present. Science 292: 686–693.

Zachos J.C., Shackleton N.J., Revenaugh J.S., Pälike H., Flower B.P. 2001b. Climate response to orbital forcing across the Oligocene-Miocene boundary. Science 292: 274–278.

Zhao M., Huang C.Y., Wang C.C. and Wei G. 2006. A millennial-scale $U_{37}^{K'}$ sea-surface temperature record from the South China Sea (8 °N) over the last 150 kyr: Monsoon and sea-level influence. Paleogeogr. Paleoclimat. Paleoecol. 236: 39–55.

Zhao Q., Jian Z., Wang J., Cheng X., Huang B., Xu J., Zhou Z., Fang D. and Wang P. 2001a. Neogene oxygen isotopic stratigraphy, ODP Site 1148, northern South China Sea. Sci. China (D) 44(10): 934–942.

Zhao Q., Wang P., Cheng X., Wang J., Huang B., Xu J., Zhou Z. and Jian Z. 2001b. A record of Miocene carbon excursions in the South China Sea. Sci. China (D) 44: 943–951.

Zheng Z. and Lei Z.Q. 1999. A 400,000 year record of vegetational and climatic changes from a volcanic basin, Leizhou Peninsula, southern China. Palaeogeogr. Palaeoclimatol. Palaeoecol. 145: 339–362.

Chapter 8
History of the South China Sea – A Synthesis

Pinxian Wang and Qianyu Li

Introduction

An overwhelming part of our current knowledge on the geoclimatic history of the South China Sea (SCS) was generated over the last 20–30 years. The amount of paleoceanographic and paleoclimatic data available is tremendous and is still emerging rapidly. A comprehensive synthesis of these data is not easy because the proxy records need to be properly interpreted first. For example, to differentiate whether the observed changes in the SCS were caused by global, regional or local factors remains as a challenge.

For a long time, the prevalent history of the SCS basin has been reconstructed mainly based on geophysical data, and the history of the East Asian monsoon on terrestrial records, in particular from the Chinese Loess Plateau. Now, deep-sea sediment sequences recovered during ODP Leg 184 not only provide the first direct record of the formation and evolution of the SCS deep water basin but also the first long marine record of the East Asian monsoon history (Wang P. et al. 2003). The present chapter attempts to present an integrated history of the SCS basin and the East Asian Monsoon by synthesizing all available data presented in previous chapters of this volume. The perspective and importance of the SCS for recognition of continent-ocean interactions in the Asia–West Pacific in a historical aspect are also briefly discussed.

P. Wang (✉)
State Key Laboratory of Marine Geology, Tongji University, Shanghai 200092, China
e-mail: pxwang@online.sh.cn; pxwang@tongji.edu.cn

Q. Li
State Key Laboratory of Marine Geology, Tongji University, Shanghai 200092, China;
School of Earth and Environmental Sciences, The University of Adelaide,
South Australia 5005, Australia
e-mail: qli01@tongji.edu.cn; qianyu.li@adelaide.edu.au

P. Wang, Q. Li (eds.), *The South China Sea*, Developments in Paleoenvironmental Research 13, DOI 10.1007/978-1-4020-9745-4_8,
© Springer Science+Business Media B.V. 2009

8.1 Evolution of the South China Sea Basin

Broadly, the history of the SCS basin can be divided into three major stages: (1) a pre-spreading or rifting stage in the early Paleogene, (2) a seafloor spreading stage in the late Paleogene-early Miocene, and (3) a post-spreading or closing stage since the late Miocene (Chapter 2) (Table 8.1).

Table 8.1 Major stages of the SCS basin evolution are characterized by variations in basin morphology and sedimentation rates

Stage		Time (Ma)	Basin features	Sedimentation
Pre-spreading	Rifting	> 37	Rift basins only	Rapid
Spreading	Early spreading	37–28.5	A narrow deep water gulf	Rapid
	Ridge jump	28.5–23	Deepening and deformation	Slumping and hiatuses
	Late spreading	23–16	Increase in size, terminated by collision in the south	Slow; coral reef development
Post-spreading	Closing	16–6.5	Deformation in the south	Slow
		< 6.5	Closure on the eastern side	Slow
		< 3	Development of deltas and broad shelves	Increased terrigenous input

Pre-Spreading Stage in the Early Paleogene

The history of the SCS started from rifting in the Paleocene with the formation of a number of parallel rift basins along the South China continental margin. Filled with thick non-marine deposits of mainly Paleogene age, these rift basins now become the main hydrocarbon producers in the northern and western SCS. While most of the rift basins failed after the Paleogene, one or several of them in the northeast developed further into the modern South China Sea basin as a result of seafloor spreading. Earlier studies of magnetic anomalies in the SCS central basin indicated that seafloor spreading began at ~30 Ma, anomaly C11 (Chapter 2). However, the unexpected discovery of bathyal fauna in the lowermost core at ODP Site 1148, at the foot of the northern SCS slope, implies the onset of the SCS breakup at least 33 Ma ago, in the earliest Oligocene (Li Q. et al. 2006). The probable existence of oceanic crust in the northeastern corner with magnetic anomaly C17 (Fig. 2.25) suggests an inception of seafloor spreading as early as 37 Ma, in the late Eocene (Hsu et al. 2004). If confirmed, the history of the SCS should be extended back beyond the Oligocene to nearly the middle/late Eocene boundary. Probably, the earliest marine basin in the present northeast during the late Eocene indeed co-existed to the south with the Proto-South China Sea, a broad shallow sea basin between the Nansha terrain (Dangerous Grounds) and islands of Borneo and Palawan and further south (Fig. 2.21). As recorded in ODP 1148 cores, the late Paleogene marine basin in the place of the present northern SCS was a narrow gulf with rapid accumulation of bathyal deposits. Occasional transgressions during this stage might have left marine sediments with a broader spatial coverage, which explains the finding of an Eocene nannofossil zone in Well HJ15-11 from a neighboring rift center (Huang L. 1997).

Seafloor Spreading in the Oligocene-Early Miocene

Up to now, there is no unanimously accepted age model and mechanism for the opening of the SCS basin. Hypotheses range from crustal thermal variations to slab pull of Asia-Pacific plate collision to pulling effects of Indochina's southeast extrusion (Chapter 2). The difficulty lies partly in the absence of intense tectonic activities associated with the inception of the seafloor spreading likely due to the inheritance of the existing deep-water rift basins by the newly opening basin.

The early spreading stage in the early Oligocene was characterized by high sedimentation rates of > 60 m/myr (Chapter 3). This period lasted for about 5 myr or longer and terminated with an unconformity at ~25 Ma in the late Oligocene. All this was followed immediately by a very active tectonic regime at 25–23 Ma (anomaly C7 and younger), probably relating to the southeast extrusion of Indochina and a major ridge jump to the southwest (Chapter 2).

The late Oligocene experienced the strongest tectonic deformation in the entire process of the basin formation of the SCS. Along with widespread slumps, several unconformities have together erased a record of about 3 myr at ODP Site 1148. Geochemical analyses indicate a drastic shift of sedimentary source provenances from the southwest (Indochina-Sundaland) and possibly also the southeast to the north (i.e., mainland China) (Li X. et al. 2003). A similar sediment provenance change is also recorded in the Zhujiangkou Basin (Shao et al. 2007). Late Oligocene tectonic activities also caused diagenetic overprinting on fossil remnants. Below the slumped section at Site 1148, foraminiferal tests are re-crystallized and register strongly depleted oxygen isotope values, siliceous fossils have changed from opal-A toward the cristobalite dominated opal-CT, and fish teeth are brownish colored probably resulting from thermal alteration, all markedly different from those in the Miocene deposit overlying the slump (Wang P. et al. 2003). The late Oligocene unconformity is ubiquitous in many SCS shelf-slope basins, where it separates the Paleogene syn-rift, usually non-marine deposits from the Neogene post-rift marine deposits (Chapter 3) and represents a major tectonic event in East Asia.

Since 23 Ma, the SCS basin entered a stage of significant subsidence, with relatively low sedimentation rates of ~15 m/myr at Site 1148. Benthic foraminifera and ostracods from the site indicate a remarkable increase in water depth from an upper slope setting of ~2, 000 m in the Oligocene to a lower slope setting at ~3, 000 m in the early middle Miocene (Chapter 6) before spreading stopped at about 16 Ma, which was apparently responding to intensified Asia-Australia-Pacific collision. At the time, however, completely different depositional environments existed near-shore. In the Oligocene, near-shore sediments were characterized by thin- layered sandstones containing fresh-water phytoplankton (Wang P. et al. 2003). A reduction in terrigenous input coupled with increased topographic variations since the end of the Oligocene provided favorable conditions for reef development and, by the early to middle Miocene, carbonate reefs achieved the broadest distribution (Chapter 4). Seafloor spreading ended at ~16 Ma when the southward migrating Nansha terrains collided with the SCS southern margin including Borneo (Hutchison 2004). Marking this event in the south was a prominent middle Miocene unconformity, together with widespread subduction and/or subsidence along the southern margin. At Site 1148 in the northern SCS, the 16 Ma event caused abrupt changes in elemental ratios and in grain size of terrigenous sediments (Li X. et al. 2003; Shao et al. 2004).

Post-Spreading Stage Since the Late Miocene

Before and during spreading, the SCS basin was surrounded by China-Indochina landmass to its north and west, and by Sundaland to its southwest (Hall and Morley 2004), leaving only the eastern and southeastern sides open to the Pacific Ocean. After spreading, however, the sea basin has been successively blocked by island arcs including Borneo, Palawan, Philippines and Taiwan on its southeastern and eastern sides along the western margin of the Philippine Sea plate. The closure of the SCS was associated with Asia-Australia collision in the south and with the rotation of the Philippine Sea plate in the east (Hall 2002).

Australia-Asia collision began in Sulawesi at about 25 Ma, leading to the anticlockwise rotation of Borneo and perhaps also to the final stop of SCS spreading at 16 Ma and the subsequent final closure of the SCS in the south (Chapter 2). Comparison between benthic isotope curves from ODP Site 1148 and from other oceanic localities reveal lighter $\delta^{18}O$ values before 16 Ma and lighter $\delta^{13}C$ values after 16 Ma in the SCS records (Fig. 8.1). At ~16 Ma, benthic foraminifera changed from *Stilostomella* dominating to *Oridorsalis umbonatus* dominanting, and carbonate dissolution also enhanced with a dissolution D2 event (Chapter 6). These changes in isotopes, benthic fauna and dissolution imply a different deep water in the post-spreading SCS, possibly caused by a change in the deep water source after the closure from the south.

The anticlockwise rotation of the Philippine Sea plate resulted in a stepwise closing of the SCS basin on the eastern side (Hall 2002) (Chapter 2). Collision of the Luzon Arc with the Asian Plate started ~6.5 Ma ago and subsequently contributed to the emergence of Taiwan and formation of the present Bashi Strait (Huang C. et al. 1997) (Fig. 2.26). The closure isolated the sea basin almost completely from the open ocean, leaving the Bashi (Luzon) Strait as the only connection passage with a sill depth of ~2,400 m. The Bashi Strait and its intermediate sill depth play a critical role in the establishment of the modern SCS oceanographic features: a basin-wise gyre, upwelling along the eastern and western coasts, its sensitivity to monsoon-induced seasonality, and a short residence time (50 yr or less) of its uniform deep water (Chapter 2). Nevertheless, the closure in the south and in the east has mainly involved a stepwise process. Benthic foraminiferal turnover data, for example, imply that the modern sill depth in the Bashi Strait was formed over the mid-Pleistocene period, around 0.9 Ma (Li Q. et al. 2008).

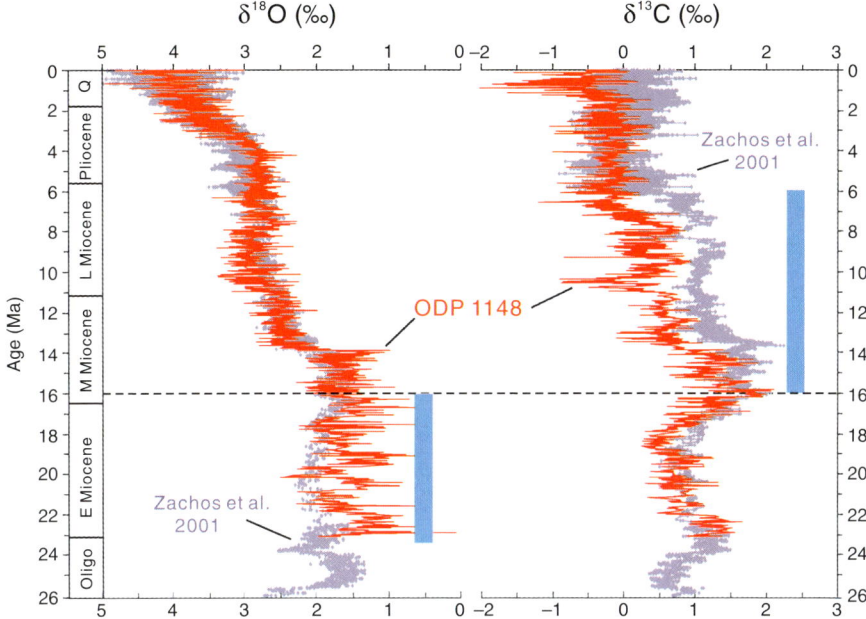

Fig. 8.1 Comparison of benthic foraminiferal oxygen and carbon isotopes between ODP Site 1148 and the global average from Zachos et al. (2001) reveals lighter SCS $\delta^{18}O$ before ~16 Ma and lighter SCS $\delta^{13}C$ after ~16 Ma (vertical blue bars), indicating an impact on the local record by different deep water regimes before and after the end of seafloor spreading at ~16 Ma

Another remarkable change in the post-spreading stage was the drastic increase of the sedimentation rates of terrigenous materials since the last 3 myr on the northern slope of the SCS (Fig. 8.2). This was related to continental uplift and to the formation of continental shelves and large deltas along the northern margin of the SCS (Wang P. et al. 2003).

8.2 Evolution of the East Asian Monsoon

Summer Monsoon and Chemical Weathering

Until recently, our knowledge of the monsoon history in East Asia relied largely on terrestrial records. Pollen and paleobotanical data from 120 on- and off-shore sites in China, together with lithological indicators, were used by Sun and Wang (2005) to reconstruct the distribution patterns of arid vs humid climates for five Cenozoic epochs. The results support an earlier notion that a broad arid zone was stretching across China in the Paleogene, but retreated to the northwest by the end of the Oligocene, indicating a transition from a planetary to monsoonal system in atmospheric circulation (Sun and Wang 2005). This is well in line with the discovery of the Miocene loess profile at Qin'an (Guo et al. 2002) and paleo-climate numerical modeling.

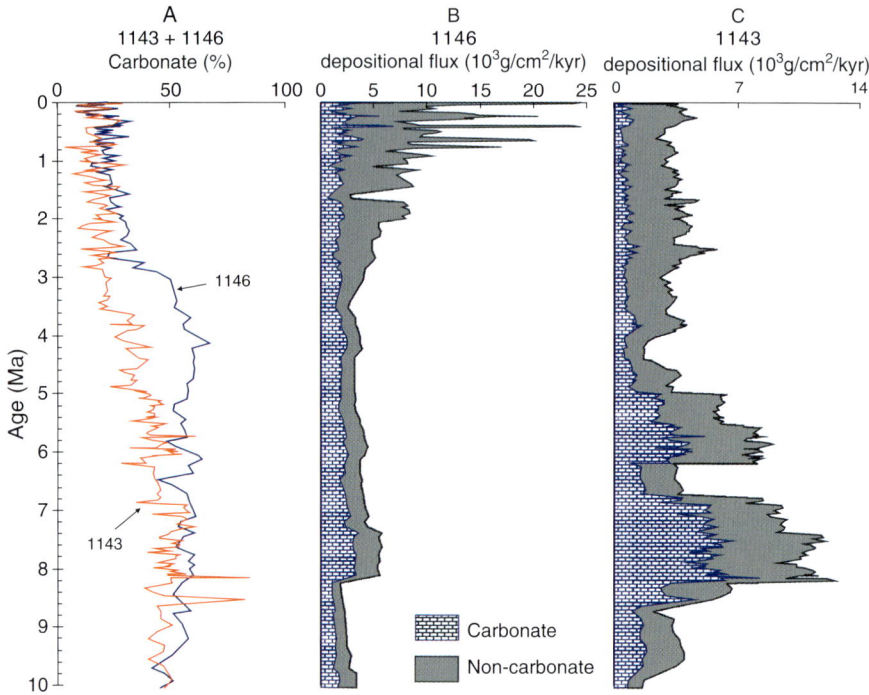

Fig. 8.2 Comparison between carbonate percentages at ODP Sites 1146 and 1143 (**A**) and mass accumulation rates of carbonate and non-carbonate at Site 1146 (**B**) and Site 1143 (**C**) indicates a shift of intensive depositional activities from the southwest (1143) at ∼8 Ma to the northern South China Sea (Site 1146) in the last 2 myr

Today, deep-sea sediment sequences from the SCS open up a new approach for exploring the monsoon history. These long sequences provide two kinds of information essential to reconstructing monsoon variability: records of terrestrial processes such as precipitation and weathering, and records of marine processes such as upwelling and productivity. At ODP Site 1148 in the northern SCS, for example, a series of element ratios (Al/Ti, Al/K, Rb/Sr and La/L) increased abruptly around 29.5 Ma, in the early stage of the seafloor expanding of the SCS (Li X. et al. 2003). As the element ratios are indicative of the intensity of chemical weathering, their increases at ∼29.5 Ma imply a period of enhanced humidity. Given the close tie between precipitation and summer monsoon in the region, further studies are needed to find out whether this was related to the inception of the East Asian Monsoon system.

The Neogene history of chemical weathering in South China was reconstructed on the basis of the chemical index of alteration (CIA) and elemental ratios (e.g., Ca/Ti, Na/Ti, Al/Ti, Al/Na, Al/K, and La/Sm) that are sensitive to chemical weathering from ODP Site 1148. The results indicate a warm and humid climate in South China during the early Miocene, but humidity decreased from the early Miocene to Present with several fluctuations centering respectively at about 15.7 Ma, 8.4 Ma,

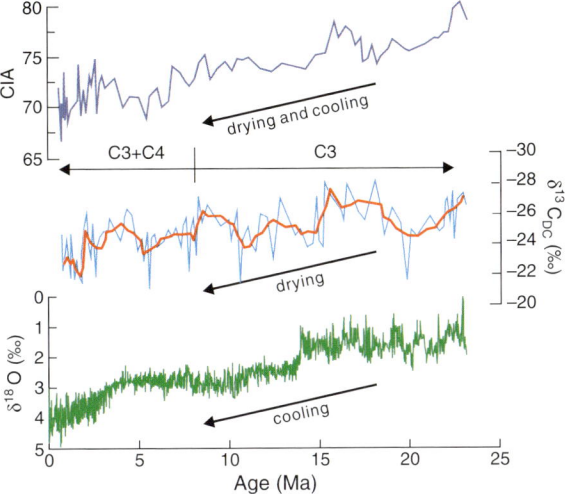

Fig. 8.3 Secular variations of the CIA (chemical index of alteration) values, black carbon $\delta^{13}C$, and benthic foraminifer $\delta^{18}O$ at ODP Site 1148 indicate weakening of East Asian summer monsoon on a continuous drying background over the last 20 myr (from Wei et al. 2006)

and 2.5 Ma, coincident with global cooling since the middle Miocene (Fig. 8.3) (Wei et al. 2006). A similar trend found in black carbon $\delta^{13}C$ records from the same site indicates an increase of C_4 plants in vegetation over the Neogene, which implies seasonal precipitation and intensification of seasonal contrast in humidity (Fig. 8.3) (Jia et al. 2003). Therefore, the influence of East Asian summer monsoon in this region has decreased continuously since the early Miocene on a general drying background (Wei et al. 2006).

Winter Monsoon and North-South Contrast

Oceanographically, the modern SCS can be divided into two parts by the cross-basin eastward jet at approximately 10 °N, 110 °E to 18 °N, 120 °E (Fig. 2.8). The prevailing climate control is the East Asian Monsoon in the northern part, whereas the southern part belongs in the Western Pacific Warm Pool (WPWP). As the influence of the winter monsoon is much more significant than that of summer monsoon in the northern SCS, the north-south contrast in the SCS climate has been mainly due to the development and intensity of the winter monsoon, as demonstrated by Plio-Pleistocene records at ODP Site 1146 from the north and ODP Site 1143 from the south.

The subsurface planktonic foraminifer *Neogloboquadrina dutertrei* is a useful proxy for winter monsoon in the SCS (Chapter 5). At Site 1146, its increase in percentage at 7.6 Ma and further increases from 3.2 to 2.0 Ma indicate several stages of enhanced winter monsoon and increased productivity. In contrast, the *Neogloboquadrina* group displays an opposite abundance trend at Site 1143 with lower percentages from ~4 to ~2 Ma (Fig. 5.17) (Li B. et al. 2004). The divergence between the two sites was likely caused by the late Pliocene development of the winter monsoon on one hand and that of the WPWP on the other.

The deep-dwelling planktonic foraminifer *Pulleniatina obliqueloculata* is characteristic of low-nutrient, warm water in the SCS. In Pleistocene interglacial periods, the relative abundance of this species increased in the northern but declined in the southern SCS (Fig. 8.4). However, its decline over interglacials at Site 1143 in the south did not start until ~0.85 Ma, during the mid-Pleistocene revolution (MPR), while its records before this time show no difference between the two sites (Fig. 8.4) (Xu et al. 2005). A similar post-MPR reversal also occurred in opal% records at Site 1143 (Wang R. et al. 2007), indicating a change in productivity (Fig. 8.4). Together, these observations imply that (1) a more saline subsurface water in the southern SCS during glacials after the MPR was probably caused by stronger stratification in the upper water column due to high precipitation, and (2) the East Asian winter monsoon has drastically increased its influence on the northern SCS climate after the MPR due to the enlarged boreal ice sheet. Therefore, coupled regional and global factors have driven further enhancement of the N-S contrast in the SCS region.

East and South Asian Monsoons

The afore-mentioned new findings are at odds with the previous notion that the East Asian Monsoon basically follows the path of the Indian Monsoon and has been strengthening since ~8 Ma and ~3 Ma. Indeed, climate seasonality in East Asia has been strengthened throughout the Neogene, but this is basically on account of the winter monsoon, not the summer monsoon. Marine productivity in the Indian Ocean is driven by the summer monsoon, but in the northern SCS it is mainly driven by the winter monsoon. For example, although an increase in the abundance of *N. dutertrei* group at 7.6 Ma represents a region-wide event in the SCS, its increases from 3.2 to 2.0 Ma appear to have begun first in the north, at Site 1146 (Fig. 5.27), indicating winter monsoon enhancement. These SCS records, therefore, demonstrate similar stages in the development of the East and South Asian monsoons during the late Neogene, with an enhanced winter monsoon over East Asia being the major difference (Wang P. et al. 2005).

As monsoon variability depends on seasonal and spatial distribution of insolation, orbital cyclicity is inherent in both winter and summer monsoon variations. While the winter monsoon in Asia responds to the development of boreal ice-sheet, the summer monsoon is closely associated with low-latitude processes and particularly sensitive to precession forcing and to eccentricity cycles which modulate the amplitude of climate precession (Ruddiman 2001). For example, fluctuations in the K/Si ratio from 3.2 Ma to 2.5 Ma at ODP Site 1145, northern SCS, show clearly

Fig. 8.4 North-South contrasts are recorded in the abundance variations of *P. obliqueloculata* and opal in their response to glacial cycles at northern Sites 1146 and 1144 and at southern Site 1143: (A) planktonic $\delta^{18}O$ at Site 1146, (B) *P. obliqueloculata*% at Site 1146, (C) benthic $\delta^{18}O$ at Site 1143, (D) *P. obliqueloculata*% at Site 1143, (E) planktonic $\delta^{18}O$ at Site 1144, (F) opal% at Site 1144, and (G.) opal% at Site 1143. *P. obliqueloculata*% data were from Xu et al. (2005), and opal% data from Wang R. et al. (2007)

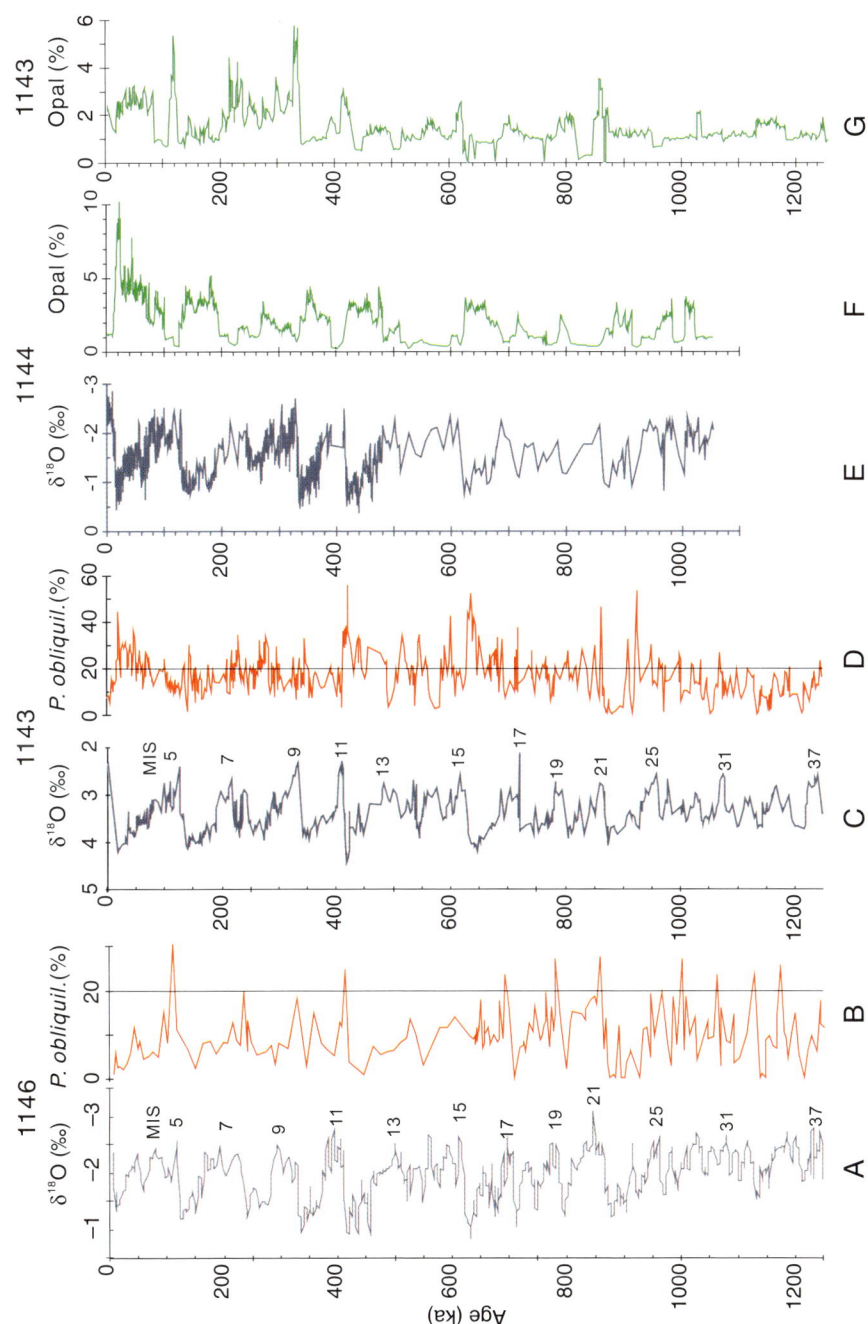

Fig. 8.4 (continued)

20 kyr precession and 400 kyr eccentricity cycles (Fig. 5.41) (Wehausen and Brumsack 2002). The 400 kyr long eccentricity cycle is best expressed in pre-Quaternary δ^{13}C records, such as at ODP 1143 (Wang P. et al. 2004). This long-eccentricity cyclicity in the oceanic carbon reservoir is believed to have originated from low-latitude chemical weathering, which in turn is affected by the summer monsoon (Chapter 7).

8.3 Evolution of Continent-Ocean Interactions

A series of marginal seas separates the largest Asia continent from the largest Pacific Ocean. All four major western Pacific marginal seas, the Sea of Okhotsk, the Sea of Japan, the East and South China Seas, are aligned in a NE-SW direction and connected to island arcs to the southeast. The marginal seas make up a distinct feature of the modern western Pacific and exert great influences on Asia-Pacific interactions (Wang P. et al. 2004).

Unlike the Sea of Japan or Sea of Okhotsk, the SCS produces neither deep nor intermediate water because of its low-latitude position and warm climate. Accordingly, deep water conditions in the SCS are favorable for preservation of carbonate and fossils which archive signals of Pacific influence. Ample sediment supply from a number of large rivers carries terrestrial information, and the resultant hemipelagic deposits are beneficial for high-resolution climate studies (Wang P. 1999). Meanwhile the basin-wide circulation gyre separates the central SCS from the river-dominated marginal parts, maintaining an oligotrophic, open-water condition in the basin. Therefore, the SCS is not only the largest marginal sea in the western Pacific but also the best witness and history narrator of Asia-Pacific interactions. Because its sediment sequence is better preserved than those of most western Pacific localities, the SCS has a critical role to play in providing paleoceanographic records of the West Pacific for many years to come.

The opening of the SCS, the Japan Sea and probably the Okhotsk Sea all dated back to the late Oligocene-middle Miocene (30–15 Ma), broadly corresponding to the onset of a period of major deformation in Tibet. A causal relationship has been proposed between the opening of marginal basins and India-Asia collision. Other theories include subduction-driven extensional forces due to collision between Eurasian and Pacific plates. However, the mechanism responsible for the opening of the marginal seas remains unclear partly because the role of Pacific oceanic tectonics is not well constrained. Nevertheless, the formation of the marginal basins has changed the trajectory of material flux and energy flow between the Asian continent and the Pacific Ocean. Even the formation of the WPWP and the Kuroshio Current depends on the presence of the island arcs as well as their associated marginal seas (Wang P. 2004). In the Paleogene, the area of the present SCS lied in the junction between the Tethys and Pacific oceans. A clearer picture of the Paleogene West Pacific without marginal seas is expected from in-depth studies of the Proto-SCS and the narrow gulf that subsequently evolved into the present northern SCS.

The hemipelagic deposits of the SCS contain rich information on the environmental evolution of the East Asian continent, ranging from reorganization of drainage systems (Chapter 4) to succession of vegetation covers (Chapter 5). The Red River, for example, has a small modern drainage area and exerts a very limited influence on the modern SCS oceanography, but the huge deltaic sediment package to over 14 km thick lying offshore the river mouth in the Yinggehai Basin (Chapter 3) demands the role of a more dynamic Paleo-Red River in the Cenozoic. The accelerated deposition rate of non-carbonate sediments off the modern Pearl River delta since 3 Ma (Fig. 8.2) indicates greater fluvial supply and enhanced transport of shelf deposits to the slope in responding to glacial cycles. Similarly, large amplitude changes in pollen assemblages in glacial cycles since MIS 6 suggest that the extensive northern SCS shelf formed only ca. 150 ka ago (Sun et al. 2003), which exemplifies the most recent development in sea-land interactions in the SCS.

References

Guo Z.T., Ruddiman W.F., Hao Q.Z., Wu H.B., Qiao Y.S., Zhu R.X., Peng S.Z., Wei J.J., Yuan B.Y. and Liu T.S. 2002. Onset of Asian desertification by 22 Myr ago inferred from loess deposits in China. Nature 416: 159–163.

Hall R. 2002. Cenozoic geological and plate tectonic evolution of SE Asia and the SW Pacific: computer-based reconstructions, model and animations. J. Asian Earth Sci. 20: 353–431.

Hall R. and Morley C.K. 2004. Sundaland basins. In: Clift P., Wang P., Kuhnt W. and Hayes D. (eds.), Continent-Ocean Interactions within East Asian Marginal Seas. AGU Geophys. Monogr. 149: 55–85.

Hsu S.K., Yeh Y.C., Doo W.B. and Tsai C.H. 2004. New bathymetry and magnetic lineations identifications in the northernmost South China Sea and their tectonic implications. Mar. Geophys. Res. 25: 29–44.

Huang C.Y., Wu W.Y., Chang C.P., Tsao S., Yuan P.B., Lin C.W. and Xia K.Y. 1997. Tectonic evolution of accretionary prism in the arc-continent collision terrane of Taiwan. Tectonophysics 281: 31–51.

Huang L. 1997. Calcareous nannofossil biostratigraphy in the Pearl River Mouth basin, South China Sea, and Neogene reticulofenestra coccolith size distribution pattern. Mar. Micropaleontol. 32: 3–29.

Hutchison C.S. 2004. Marginal basin evolution: the southern South China Sea. Mar. Petroleum Geol. 21: 1129–1148.

Jia G., Peng P., Zhao Q. and Jian Z. 2003. Changes in terrestrial ecosystem since 30 Ma in East Asia: Stable isotope evidence from black carbon in the South China Sea. Geology 31: 1093–1096.

Li B., Wang J., Huang B., Li Q., Jian Z. and Wang P. 2004. South China Sea surface water evolution over the last 12 Ma: A south-north comparison from ODP Sites 1143 and 1146. Paleoceanography 19: PA1009, doi:10.1029/2003PA000906.

Li Q., Wang P., Zhao Q., Shao L., Zhong G., Tian J., Cheng X., Jian Z. and Su X. 2006. A 33 Ma lithostratigraphic record of tectonic and paleoceanographic evolution of the South China Sea. Mar. Geol. 230: 217–235.

Li Q., Wang P., Zhao Q., Tian J., Cheng X., Jian Z. Zhong G. and Chen M. 2008. Paleoceanography of the mid-Pleistocene South China Sea. Quat. Sci. Rev. 27: 1217–1233.

Li X., Wei G., Shao L., Liu Y., Liang X., Jian Z., Sun M. and Wang P. 2003. Geochemical and Nd isotopic variations in sediments of the South China Sea: a response to Cenozoic tectonism in SE Asia. Earth Planet. Sci. Lett. 211: 207–220.

Ruddiman W.F. 2001. Earth's Climate. Past and Future. W.H. Freeman and Company, N.Y., 465pp.

Shao L., Li X., Geng J., Pang X., Lei Y., Qiao P., Wang L. and Wang H. 2007. Deep water bottom current deposition in the northern South China Sea. Sci. China (D) 50(7): 1060–1066.

Shao L., Li X., Wang P., Jian Z., Wei G., Pang X. and Liu Y. 2004. Sedimentary record of the tectonic evolution of the South China Sea since the Oligocene. Advances Earth Sci. 19: 539–544 (in Chinese).

Sun X. and Wang P. 2005. How old is the Asian monsoon system? – Palaeobotanical records from China. Palaeogeogr., Palaeoclimatol., Palaeoecol. 222: 181–222.

Sun X., Luo Y., Huang F., Tian J. and Wang P. 2003. Deep-sea pollen from the South China Sea: Pleistocene indicators of East Asian monsoon. Mar. Geol. 201: 97–118.

Wang P. 1999. Response of Western Pacific marginal seas to glacial cycles: Paleoceanographic and sedimentological features. Mar. Geol. 156: 5–39.

Wang P. 2004. Cenozoic deformation and the history of sea-land interactions in Asia. In: Clift P., Wang P., Kuhnt W. and Hayes D. (eds.), Continent-Ocean Interactions in the East Asian Marginal Seas. AGU Geophys. Monogr. 149: 1–22.

Wang P., Clemens S., Beaufort L., Braconnot P., Ganssen G., Jian Z., Kershaw P. and Sarnthein M. 2005. Evolution and variability of the Asian monsoon system: state of the art and outstanding issues. Quat. Sci. Rev. 24: 595–629.

Wang P., Tian J., Cheng X., Liu C. and Xu J. 2004. Major Pleistocene stages in a carbon perspective: The South China Sea record and its global comparison. Paleoceanography 19: doi: 10.1029/2003PA000991.

Wang P., Zhao Q., Jian Z., Cheng X., Huang W., Tian J., Wang J., Li Q., Li B. and Su X. 2003. Thirty million year deep-sea records in the South China Sea. Chinese Sci. Bull. 48(23): 2524–2535.

Wang R., Jian Z., Xiao W., Tian J., Li J., Chen R., Zheng L. and Chen J. 2007. Quaternary biogenic opal records in the South China Sea: linkages to East Asian monsoon, global ice volume and orbital forcing. Sci. China (D) 50(5): 710–724.

Wehausen R. and Brumsack H.J. 2002. Astronomical forcing of the East Asian monsoon mirrored by the composition of Pliocene South China Sea sediments. Earth Planet. Sci. Lett. 201: 621–636.

Wei G.J., Li X., Liu Y., Shao L. and Liang X. 2006. Geochemical record of chemical weathering and monsoon climate change since the early Miocene in the South China Sea. Paleoceanography 21: PA4214, doi: 10.1029/2006PA001300.

Xu J., Wang P., Huang B., Li Q. and Jian Z. 2005. Response of planktonic foraminifera to glacial cycles: Mid-Pleistocene change in the southern South China Sea. Mar. Micropaleontol. 54: 89–105.

Zachos J., Pagani M., Sloan L., Thomas E. and Billups K. 2001. Trends, rhythms, and aberrations in global climate 65 Ma to present. Science 292: 686–693.

Index